Universitext

Universitext

Universitext is a series of textbooks that presents material from a wide variety of mathematical disciplines at master's level and beyond. The books, often well class-tested by their author, may have an informal, personal, even experimental approach to their subject matter. Some of the most successful and established books in the series have evolved through several editions, always following the evolution of teaching curricula, into very polished texts.

Thus as research topics trickle down into graduate-level teaching, first textbooks written for new, cutting-edge courses may make their way into *Universitext*.

More information about this series at http://www.springer.com/series/223

Jean-Claude Hausmann

Mod Two Homology and Cohomology

Jean-Claude Hausmann
University of Geneva
Geneva
Switzerland

ISSN 0172-5939 ISSN 2191-6675 (electronic)
ISBN 978-3-319-09353-6 ISBN 978-3-319-09354-3 (eBook)
DOI 10.1007/978-3-319-09354-3

Library of Congress Control Number: 2014944717

JEL Classification Code: 55-01, 55N10, 55N91, 57R91, 57R20, 55U10, 55U25, 55S10, 57R19, 55R91

Springer Cham Heidelberg New York Dordrecht London

Printed on acid-free paper

Springer is part of Springer Science+Business Media (www.springer.com)

Contents

Chapter 1
Introduction

Mod 2 homology first occurred in 1908 in a paper of Tietze [196] (see also [40, pp. 41–42]). Several results were first established using this mod 2 approach, like the linking number for submanifolds in $\mathbb{R}^n$ (see Sect. 5.4.4), as well as Alexander duality [7]. One argument in favor of the choice of the mod 2 homology was its simplicity, as J.W. Alexander says in his introduction: "The theory of connectivity [homology] may be approached from two different angles depending on whether or not the notion of *sense* [orientation] is developed and taken into consideration. We have adopted the second and somewhat simpler point of view in this discussion in order to condense the necessary preliminaries as much as possible. A treatment involving the idea of sense would be somewhat more complicated but would follow along much the same lines."

Besides being simpler than its integral counterpart, mod 2 homology sometimes gives new theorems. The first historical main example is the generalization of Poincaré duality to all closed manifolds, whether orientable or not, a result obtained by Veblen and Alexander in 1913 [200]. As a consequence, the Euler characteristic of a closed odd-dimensional manifold vanishes.

The discoveries of Stiefel-Whitney classes in 1936–1938 and of Steenrod squares in 1947–1950 gave mod 2 cohomology the status of a major tool in algebraic topology, providing for instance the theory of spin structures and Thom's work on the cobordism ring.

These notes are an introduction, at graduate student's level, of mod 2 (co)homology (there will be essentially no other). They include classical applications (Brouwer fixed point theorem, Poincaré duality, Borsuk-Ulam theorem, Smith theory, etc) and less classical ones (face spaces, topological complexity, equivariant Morse theory, etc). The cohomology of flag manifolds is treated in details, including for Grassmannians the relationship between Stiefel-Whitney classes and Schubert calculus. Some original applications are given in Chap. 10.

Our approach is different than that of classical textbooks, in which mod 2 (co) homology is just a particular case of (co)homology with arbitrary coefficients. Also, most authors start with a full account of homology before approaching cohomology. In these notes, mod 2 (co)homology is treated as a subject by itself and we start

J.-C. Hausmann, *Mod Two Homology and Cohomology*, Universitext,
DOI 10.1007/978-3-319-09354-3_1

with cohomology and homology together from the beginning. The advantages of this approach are the following.

- The definition of a (co)chain is simple and intuitive: an (say, simplicial) *m-cochain* is a set of m-simplexes; an *m-chain* is a *finite* set of m-simplexes. The concept of cochain is simpler than that of chain (one less word in the definition...), more flexible and somehow more natural. We thus tend to consider cohomology as the main concept and homology as a (useful) tool for some arguments.
- Working with $\mathbb{Z}_2$ and its standard linear algebra is much simpler than working with $\mathbb{Z}$. For instance, the *Kronecker pairing* has an intuitive geometric interpretation occurring at the beginning which shows in an elementary way that cohomology is the dual of homology. Several computations, like the homology of surfaces, are quite easy and come early in the exposition. Also, the cohomology ring is *commutative*. The cup square $a \mapsto a \smile a$ is a linear map and may be also non-trivial in odd degrees, leading to important invariants.
- The absence of sign and orientation considerations is an enormous technical simplification (even of importance in computer algorithms computing homology). With much lighter computations and technicalities, the ideas of proofs are more apparent.

We hope that these notes will be, for students and teachers, a complement or companion to textbooks like those of Hatcher [82] or Munkres [155]. From our teaching experience, starting with mod 2 (co)homology and taking advantage of its above mentioned simplicity is a great help to grasp the ideas of the subject. The technical difficulties of signs and orientations for finer theories, like integral (co)homology, may then be introduced afterwards, as an adaptation of the more intuitive mod 2 (co)homology.

Not in this book The following tools are not used in these notes.

- Augmented (co)chain complexes. The reduced cohomology $\tilde{H}^*(X)$ is defined as coker $(H^*(pt) \to H^*(X))$ for the unique map $X \to pt$.
- Simplicial approximation.
- Spectral sequences (except in the proof of Proposition 7.2.17).

Also, we do not use advanced homotopy tools, like spectra, completions, etc. Because of this, some prominent problems using mod 2 cohomology are only briefly surveyed, like the work by Adams on the Hopf-invariant-one problem (p. 353), the Sullivan's conjecture (pp. 240 and 353) and the Kervaire invariant (Sect. 10.6).

Prerequisites The reader is assumed to have some familiarity with the following subjects:

- general point set topology (compactness, connectedness, etc).
- elementary language of categories and functors.
- simple techniques of exact sequences, like the five lemma.

- elementary facts about fundamental groups, coverings and higher homotopy groups (not much used).
- elementary techniques of smooth manifolds.

Acknowledgments A special thank is due to Volker Puppe who provided several valuable suggestions and simplifications. Michel Zisman, Pierre de la Harpe, Samuel Tinguely and Matthias Franz have carefully read several sections of these notes. The author is also grateful for useful comments to Jim Davis, Rebecca Goldin, André Haefliger, Tara Holm, Allen Knutson, Jérôme Scherer, Dirk Schütz, Andras Szenes, Vladimir Turaev, Paul Turner, Claude Weber and Saïd Zarati.

Chapter 2
Simplicial (Co)homology

Simplicial homology was invented by Poincaré in 1899 [162] and its mod 2 version, presented in this chapter, was introduced in 1908 by Tietze [196]. It is the simplest homology theory to understand and, for finite complexes, it may be computed algorithmically. The mod 2 version permits rapid computations on easy but non-trivial examples, like spheres and surfaces (see Sect. 2.4).

Simplicial (co)homology is defined for a simplicial complex, but is an invariant of the homotopy type of its geometric realization (this result will be obtained in different ways using singular homology: see Sect. 3.6). The first section of this chapter introduces classical techniques of (abstract) simplicial complexes. Since simplicial homology was the only existing (co)homology theory until the 1930s, simplicial complexes played a predominant role in algebraic topology during the first third of the 12th century (see the Introduction of Sect. 5.1). Later developments of (co)homology theories, defined directly for topological spaces, made this combinatorial approach less crucial. However, simplicial complexes remain an efficient way to construct topological spaces, also largely used in computer science.

2.1 Simplicial Complexes

In this section we fix notations and recall some classical facts about (abstract) simplicial complexes. For more details, see [179, Chap. 3].

A *simplicial complex* K consists of

- a set $V(K)$, the set of *vertices* of K.
- a set $\mathcal{S}(K)$ of finite non-empty subsets of $V(K)$ which is closed under inclusion: if $\sigma \in \mathcal{S}(K)$ and $\tau \subset \sigma$, then $\tau \in \mathcal{S}(K)$. We require that $\{v\} \in \mathcal{S}(K)$ for all $v \in V(K)$.

An element σ of $\mathcal{S}(K)$ is called a *simplex* of K ("simplexes" and "simplices" are admitted as plural of "simplex"; we shall use "simplexes", in analogy with "complexes"). If $\sharp(\sigma) = m+1$, we say that σ is of *dimension m* or that σ is an *m-simplex*. The set of m-simplexes of K is denoted by $\mathcal{S}_m(K)$. The set $\mathcal{S}_0(K)$ of 0-simplexes is

J.-C. Hausmann, *Mod Two Homology and Cohomology*, Universitext,
DOI 10.1007/978-3-319-09354-3_2

in bijection with $V(K)$, and we usually identify $v \in V(K)$ with $\{v\} \in \mathcal{S}_0(K)$. We say that K is of *dimension* $\leq n$ if $\mathcal{S}_m(K) = \emptyset$ for $m > n$, and that K is of *dimension n* (or *n-dimensional*) if it is of dimension $\leq n$ but not of dimension $\leq n-1$. A simplicial complex of dimension ≤ 1 is called a *simplicial graph*. A simplicial complex K is called *finite* if $V(K)$ is a finite set.

If $\sigma \in \mathcal{S}(K)$ and $\tau \subset \sigma$, we say that τ is a *face* of σ. As $\mathcal{S}(K)$ is closed under inclusion, it is determined by it subset $\mathcal{S}_{\max}(K)$ of *maximal* simplexes (if K is finite dimensional). A *subcomplex L* of K is a simplicial complex such that $V(L) \subset V(K)$ and $\mathcal{S}(L) \subset \mathcal{S}(K)$. If $S \subset \mathcal{S}(K)$ we denote by $\bar{S}$ the subcomplex generated by S, i.e. the smallest subcomplex of K such that $S \subset \mathcal{S}(\bar{S})$. The *m-skeleton* K^m of K is the subcomplex of K generated by the union of $\mathcal{S}_k(K)$ for $k \leq m$.

Let $\sigma \in \mathcal{S}(K)$. We denote by $\bar{\sigma}$ the subcomplex of K formed by σ and all its faces ($\overline{\{\sigma\}}$ in the above notation). The subcomplex $\dot{\sigma}$ of $\bar{\sigma}$ generated by the proper faces of σ is called the *boundary of* σ.

2.1.1 *Geometric realization.* The *geometric realization* $|K|$ of a simplicial complex K is, as a set, defined by

$$|K| := \{\mu : V(K) \to [0,1] \,\big|\, \textstyle\sum_{v \in V(K)} \mu(v) = 1 \text{ and } \mu^{-1}((0,1]) \in \mathcal{S}(K)\}\,.$$

We can thus see $|K|$ as the set of probability measures on $V(K)$ which are supported by the simplexes (this language is just used for comments and only in this section). There is a distance on $|K|$ defined by

$$d(\mu,\nu) = \sqrt{\sum_{v \in V(K)} [\mu(v) - \nu(v)]^2}$$

which defines the *metric topology* on $|K|$. The set $|K|$ with the metric topology is denoted by $|K|_d$. For instance, if $\sigma \in \mathcal{S}_m(K)$, then $|\bar{\sigma}|_d$ is isometric to the standard Euclidean simplex $\Delta^m = \{(x_0,\dots,x_m) \in \mathbb{R}^{m+1} \mid x_i \geq 0 \text{ and } \sum x_i = 1\}$.

However, a more used topology for $|K|$ is the *weak topology*, for which $A \subset |K|$ is closed if and only if $A \cap |\bar{\sigma}|_d$ is closed in $|\bar{\sigma}|_d$ for all $\sigma \in \mathcal{S}(K)$. The notation $|K|$ stands for the set $|K|$ endowed with the weak topology. A map f from $|K|$ to a topological space X is then continuous if and only if its restriction to $|\bar{\sigma}|_d$ is continuous for each $\sigma \in \mathcal{S}(K)$. In particular, the identity $|K| \to |K|_d$ is continuous, which implies that $|K|$ is Hausdorff. The weak and the metric topology coincide if and only if K is locally finite, that is each vertex is contained in a finite number of simplexes. When K is not locally finite, $|K|$ is not metrizable (see e.g. [179, Theorem 3.2.8]).

When a simplicial complex K is locally finite, has countably many vertices and is finite dimensional, it admits a *Euclidean realization*, i.e. an embedding of $|K|$ into some Euclidean space $\mathbb{R}^N$ which is piecewise affine. A map $f\colon |K| \to \mathbb{R}^N$ is *piecewise affine* if, for each $\sigma \in \mathcal{S}(K)$, the restriction of f to $|\bar{\sigma}|$ is an affine map. Thus, for each simplex σ, the image of $|\bar{\sigma}|$ is an affine simplex of $\mathbb{R}^N$. If $\dim K \leq n$, such a realization exists in $\mathbb{R}^{2n+1}$ (see e.g. [179, Theorem 3.3.9]).

If $\sigma \in \mathcal{S}(K)$ then $|\bar{\sigma}| \subset |K|$. We call $|\bar{\sigma}|$ the *geometric simplex* associated to σ. Its *boundary* is $|\dot{\sigma}|$. The space $|\sigma| - |\dot{\sigma}|$ is the *geometric open simplex* associated to σ. Observe that $|K|$ is the disjoint union of its geometric open simplexes.

There is a natural injection $i : V(K) \hookrightarrow |K|$ sending v to the Dirac measure with value 1 on v. We usually identify v with $i(v)$, seeing a simplex v as a point of $|K|$ (a geometric vertex). In this way, a point $\mu \in |K|$ may be expressed as a convex combination of (geometric) vertices:

$$\mu = \sum_{v \in V(K)} \mu(v)v \,. \tag{2.1.1}$$

2.1.2 Let K and L be simplicial complexes. Their *join* is the simplicial complex $K * L$ defined by

(1) $V(K * L) = V(K) \dot{\cup} V(L)$.
(2) $\mathcal{S}(K * L) = \mathcal{S}(K) \cup \mathcal{S}(L) \cup \{\sigma \cup \tau \mid \sigma \in \mathcal{S}(K) \text{ and } \tau \in \mathcal{S}(L)\}$.

Observe that, if $\sigma \in \mathcal{S}_r(K)$ and $\tau \in \mathcal{S}_s(L)$, then $\sigma \cup \tau \in \mathcal{S}_{r+s+1}(K * L)$. Also, $\overline{\sigma \cup \tau} = \bar{\sigma} * \bar{\tau}$ and $|K * L|$ the topological join of $|K|$ and $|L|$ (see p. 171).

2.1.3 *Stars, links, etc.* Let K be a simplicial complex and $\sigma \in \mathcal{S}(K)$. The *star* $\mathrm{St}(\sigma)$ *of* σ is the subcomplex of K generated by all the simplexes containing σ. The *link* $\mathrm{Lk}(\sigma)$ *of* σ is the subcomplex of K formed by the simplexes $\tau \in \mathcal{S}(K)$ such that $\tau \cap \sigma = \emptyset$ and $\tau \cup \sigma \in \mathcal{S}(K)$. Thus, $\mathrm{Lk}(\sigma)$ is a subcomplex of $\mathrm{St}(\sigma)$ and

$$\mathrm{St}(\sigma) = \bar{\sigma} * \mathrm{Lk}(\sigma) \,.$$

More generally, if L is a subcomplex of K, the *star* $\mathrm{St}(L)$ *of* L is the subcomplex of K generated by all the simplexes containing a simplex of L. The *link* $\mathrm{Lk}(L)$ *of* L is the subcomplex of K formed by the simplexes $\tau \in \mathcal{S}(\mathrm{St}(L)) - \mathcal{S}(L)$. One has $\mathrm{St}(L) = L * \mathrm{Lk}(L)$. The *open star* $\mathrm{Ost}(L)$ of L is the open neighbourhood of $|L|$ in $|K|$ defined by

$$\mathrm{Ost}(L) = \{\mu \in |K| \mid \mu(v) > 0 \text{ if } v \in V(L)\} \,.$$

This is the interior of $|\mathrm{St}(L)|$ in $|K|$.

2.1.4 *Simplicial maps.* Let K and L be two simplicial complexes. A *simplicial map* $f\colon K \to L$ is a map $f\colon V(K) \to V(L)$ such that $f(\sigma) \in \mathcal{S}(L)$ if $\sigma \in \mathcal{S}(K)$, i.e. the image of a simplex of K is a simplex of L. Simplicial complexes and simplicial maps form a category, the *simplicial category*, denoted by **Simp**.

A simplicial map $f\colon K \to L$ induces a continuous map $|f|\colon |K| \to |L|$ defined, for $w \in V(L)$, by

$$|f|(\mu)(w) = \sum_{v \in f^{-1}(w)} \mu(v) \,.$$

In other words, $|f|(\mu)$ is the pushforward of the probability measure μ on $|L|$. The geometric realization is thus a covariant functor from the simplicial category **Simp** to the topological category **Top** of topological spaces and continuous maps.

2.1.5 *Components.* Let K be a simplicial complex. We define an equivalence relation on $V(K)$ by saying that $v \sim v'$ if there exists $x_0, \dots, x_m \in V(K)$ with $x_0 = v$, $x_m = v'$ and $\{x_i, x_{i+1}\} \in \mathcal{S}(K)$. A maximal subcomplex L of K such that $V(L)$ is an equivalence class is called a *component* of K. The set of components of K is denoted by $\pi_0(K)$. As the vertices of a simplex are all equivalent, K is the disjoint union of its components and $\pi_0(K)$ is in bijection with $V(K)/\sim$. The relationship with $\pi_0(|K|)$, the set of (path)-components of the topological space $|K|$, is the following.

Lemma 2.1.6 *The natural injection $j\colon V(K) \to |K|$ descends to a bijection $\bar{j}$: $\pi_0(K) \xrightarrow{\approx} \pi_0(|K|)$.*

Proof The definition of the relation $\sim$ makes clear that j descends to a map $\bar{j}$: $\pi_0(K) \to \pi_0(|K)|$. Any point of $|K|$ is joinable by a continuous path to some vertex $j(v)$. Hence, $\bar{j}$ is surjective. To check the injectivity of $\bar{j}$, let $v, v' \in V(K)$ with $\bar{j}(v) = \bar{j}(v')$. There exists then a continuous path $c\colon [0, 1] \to |K|$ with $c(0) = j(v)$ and $c(1) = j(v')$. Consider the open cover $\{\mathrm{Ost}(w) \mid w \in V(K)\}$ of $|K|$. By compactness of $[0, 1]$, there exists $n \in \mathbb{N}$ and vertices $v_0, \dots, v_{n-1} \in V(K)$ such that $c([k/n, (k+1)/n]) \subset \mathrm{Ost}(v_k)$ for all $k = 0, \dots, n-1$. As $c(0) = j(v)$ and $c(1) = j(v')$, one deduces that $v_0 = v$ and $v_{n-1} = v'$. For $0 < k \le n-1$, one has $c(k/n) \in \mathrm{Ost}(v_{k-1}) \cap \mathrm{Ost}(v_k)$. This implies that $\{v_{k-1}, v_k\} \in \mathcal{S}(K)$ for all $k = 1, \dots, n-1$, proving that $v \sim v'$. □

A simplicial complex is called *connected* if it is either empty or has one component. Note that $|K|$ is locally path-connected for any simplicial complex K. Indeed, any point has a neighborhood of the form $|\mathrm{St}(v)|$ for some vertex v, and $|\mathrm{St}(v)|$ path-connected. Therefore, $|K|$ is path-connected if and only if $|K|$ is connected. Using Lemma 2.1.6, this proves the following lemma.

Lemma 2.1.7 *Let K be a simplicial complex. Then K is connected if and only if $|K|$ is a connected space.*

Finally, we note the functoriality of π_0. Let $f\colon K \to L$ be a simplicial map. If $v \sim v'$ for $v, v' \in V(K)$, then $f(v) \sim f(v')$, so f descends to a map $\pi_0 f$: $\pi_0(K) \to \pi_0(L)$. If $f\colon K \to L$ and $g\colon L \to M$ are two simplicial maps, then $\pi_0(g \circ f) = \pi_0 g \circ \pi_0 f$. Also, $\pi_0 \mathrm{id}_K = \mathrm{id}_{\pi_0(K)}$. Thus, π_0 is a covariant functor from the simplicial category **Simp** to the category **Set** of sets and maps.

2.1.8 *Simplicial order.* A *simplicial order* on a simplicial complex L is a partial order $\le$ on $V(L)$ such that each simplex is totally ordered. For example, a total order on $V(L)$, as in examples where vertices are labeled by integers, is a simplicial order. A simplicial order always exists, as a consequence of the well-ordering theorem.

2.1.9 *Triangulations.* A *triangulation* of a topological space X is a homeomorphism $h\colon |K| \to X$, where K is a simplicial complex. A topological space is *triangulable* if it admits a triangulation. It will be useful to have a good process to triangulate some subspaces of $\mathbb{R}^n$. A compact subspace A of $\mathbb{R}^n$ is a *convex cell* if it is the set of solutions of families of affine equations and inequalities

$$f_i(x) = 0,\ i = 1, \dots, r \quad \text{and} \quad g_j(x) \geq 0,\ j = 1, \dots, s\,.$$

A *face* B of A is a convex cell obtained by replacing some of the inequalities $g_j \geq 0$ by the equations $g_j = 0$. The *dimension* of B is the dimension of the smallest affine subspace of $\mathbb{R}^n$ containing B. A *vertex* of A is a cell of dimension 0. By induction on the dimension, one proves that a convex cell is the convex hull of its vertices (see e.g. [138, Theorem 5.2.2]).

A *convex-cell complex* P is a finite union of convex cells in $\mathbb{R}^n$ such that:

(i) if A is a cell of P, so are the faces of A;
(ii) the intersection of two cells of P is a common face of each of them.

The *dimension* of P is the maximal dimension of a cell of P. The r*-skeleton* P^r is the subcomplex formed by the cells of dimension $\leq r$. The 0-skeleton coincides with the set $V(P)$ of *vertices* of P.

A partial order $\leq$ on $V(P)$ is an *affine order for P* if any subset $R \in V(P)$ formed by affinely independent points is totally ordered. For instance, a total order on $V(P)$ is an affine order. The following lemma is a variant of [104, Lemma 1.4].

Lemma 2.1.10 *Let P be a convex-cell complex. An affine order $\leq$ for P determines a triangulation $h_{\leq}\colon |L_{\leq}| \xrightarrow{\approx} P$, where $L_{\leq}$ is a simplicial complex with $V(L_{\leq}) = V(P)$. The homeomorphism $h_{\leq}$ is piecewise affine and $\leq$ is a simplicial order on $L_{\leq}$.*

Proof The order $\leq$ being chosen, we drop it from the notations. For each subcomplex Q of P, we shall construct a simplicial complex $L(Q)$ and a piecewise affine homeomorphism $h_Q\colon |L(Q)| \to Q$ such that,

(i) $V(L(Q)) = V(Q)$;
(ii) if $Q' \subset Q$, then $L(Q') \subset L(Q)$ and $h_{Q'}$ is the restriction of h_Q to $|L(Q')|$.

The case $Q = P$ will prove the lemma. The construction is by induction on the dimension of Q, setting $L(Q) = Q$ and $h_Q = \mathrm{id}$ if $\dim Q = 0$.

Suppose that $L(Q)$ and h_Q have been constructed, satisfying (i) and (ii) above, for each subcomplex Q of P of dimension $\leq k-1$. Let A be a k-cell of K with minimal vertex a. Then A is the topological cone, with cone-vertex a, of the union B of faces of A not containing a. The triangulation $h_B\colon |L(B)| \to |B|$ being constructed by induction hypothesis, define $L(A)$ to be the join $L(B) * \{a\}$ and h_A to be the unique piecewise affine extension of h_B. Observe that, if C is a face of A, then h_C is the restriction to $L(C)$ of h_A. Therefore, this process may be used for each k-cell of P to construct $h_Q\colon |L(Q)| \to Q$ for each subcomplex Q of P with $\dim Q \leq k$. □

2.1.11 *Subdivisions.* Let Z be a set and $\mathcal{A}$ be a family of subsets of Z. A simplicial complex L such that

(a) $V(L) \subset Z$;
(b) for each $\sigma \in \mathcal{S}(L)$ there exists $A \in \mathcal{A}$ such that $\sigma \subset A$;

is called a $(Z, \mathcal{A})$*-simplicial complex*, or a Z-simplicial complex *supported by* $\mathcal{A}$.

Let K be a simplicial complex. Let N be a $(|K|, \mathcal{GS}(K))$-simplicial complex, where

$$\mathcal{GS}(K) = \{|\sigma| \mid \sigma \in \mathcal{S}(K)\}$$

is the family of geometric simplexes of K. A continuous map $j\colon |N| \to |K|$ is associated to N, defined by

$$j(\mu) = \sum_{w \in V(N)} \mu(w)w\,.$$

In other word, j is the piecewise affine map sending each vertex of N to to the corresponding point of $|K|$. A *subdivision* of a simplicial complex K is a $(|K|, \mathcal{GS}(K))$-simplicial complex N for which the associated map $j\colon |N| \to |K|$ is a homeomorphism (in other words, j is a triangulation of $|K|$).

Let N be a $(|K|, \mathcal{GS}(K))$-simplicial complex for a simplicial complex K. If L is a subcomplex of K, then

$$N_L = \{\sigma \in \mathcal{S}(N) | \sigma \subset |L|\}$$

is a $(|L|, \mathcal{GS}(L))$-simplicial complex. Its associated map $j_L\colon |N_L| \to |L|$ is the restriction of j to $|L|$. The following Lemma is useful to recognize a subdivision (compare [179, Chap. 3, Sect. 3, Theorem 4]).

Lemma 2.1.12 *Let N be a $(|K|, \mathcal{GS}(K))$-simplicial complex. Then N is a subdivision of K if and only if, for each $\tau \in \mathcal{S}(K)$, the simplicial complex $N_{\bar\tau}$ is finite and $j_{\bar\tau}\colon |N_{\bar\tau}| \to |\bar\tau|$ is bijective.*

Proof If N is a subdivision of K, then $j_{\bar\tau}$ is bijective since j is a homeomorphism. Also, $|N_{\bar\tau}| = j^{-1}(|\bar\tau|)$ is compact, so $N_{\bar\tau}$ is finite.

Conversely, The fact that $j_{\bar\tau}$ is bijective for each $\tau \in \mathcal{S}(K)$ implies that the continuous map j is bijective. If $N_{\bar\tau}$ is finite, then $j_{\bar\tau}$ is a continuous bijection between compact spaces, hence a homeomorphism. This implies that the map j^{-1}, restricted to each geometric simplex, is continuous. Therefore, j^{-1} is continuous since K is endowed with the weak topology. □

Seeing $V(K)$ as a subset of $|K|$, we get the following corollary.

Corollary 2.1.13 *Let N be a subdivision of K. Then $V(K) \subset V(N)$.*

A useful systematic subdivision process is the barycentric subdivision. Let $\sigma \in \mathcal{S}_m(K)$ be an m-simplex of a simplicial complex K. The *barycenter* $\hat{\sigma} \in |K|$ of σ is defined by

$$\hat{\sigma} = \frac{1}{m+1} \sum_{v \in \sigma} v\,.$$

The *barycentric subdivision* K' of K is the $(|K|, \mathcal{GS}(K))$-simplicial complex where

- $V(K') = \{\hat{\sigma} \in |K| \mid \sigma \in \mathcal{S}(K)\}$;
- $\{\hat{\sigma}_0, \dots, \hat{\sigma}_m\} \in \mathcal{S}_m(K')$ whenever $\sigma_0 \subset \cdots \subset \sigma_m$ ($\sigma_i \neq \sigma_j$ if $i \neq j$).

Using Lemma 2.1.12, the reader can check that K' is a subdivision of K. Observe that the partial order "$\leq$" defined by

$$\hat{\sigma} \leq \hat{\tau} \iff \sigma \subset \tau \tag{2.1.2}$$

is a simplicial order on K'.

2.2 Definitions of Simplicial (Co)homology

Let K be a simplicial complex. In this section, we give the definitions of the homology $H_*(K)$ and cohomology $H^*(K)$ of K under the various and peculiar forms available when the coefficients are in the field $\mathbb{Z}_2 = \{0, 1\}$.

Definition 2.2.1 (*subset definitions*)

(a) An *m-cochain* is a subset of $\mathcal{S}_m(K)$.
(b) An *m-chain* is a finite subset of $\mathcal{S}_m(K)$.

The set of m-cochains of K is denoted by $C^m(K)$ and that of m-chains by $C_m(K)$. By identifying $\sigma \in \mathcal{S}_m(K)$ with the singleton $\{\sigma\}$, we see $\mathcal{S}_m(K)$ as a subset of both $C_m(K)$ and $C^m(K)$. Each subset A of $\mathcal{S}_m(K)$ is determined by its characteristic function $\chi_A\colon \mathcal{S}_m(K) \to \mathbb{Z}_2$, defined by

$$\chi_A(\sigma) = \begin{cases} 1 & \text{if } \sigma \in A \\ 0 & \text{otherwise.} \end{cases}$$

This gives a bijection between subsets of $\mathcal{S}_m(K)$ and functions from $\mathcal{S}_m(K)$ to $\mathbb{Z}_2$. We see such a function as a colouring (0 = white and 1 = black). The following "colouring definition" is equivalent to the subset definition:

Definition 2.2.2 (*colouring definitions*)

(a) An *m-cochain* is a function $a\colon \mathcal{S}_m(K) \to \mathbb{Z}_2$.
(b) An *m-chain* is a function $\alpha\colon \mathcal{S}_m(K) \to \mathbb{Z}_2$ with finite support.

The colouring definition is used in low-dimensional graphical examples to draw (co)chains in black (bold lines for 1-(co)chains).

Definition 2.2.2 endow $C^m(K)$ and $C_m(K)$ with a structure of a $\mathbb{Z}_2$-vector space. The singletons provide a basis of $C_m(K)$, in bijection with $\mathcal{S}_m(K)$. Thus, Definition 2.2.2b is equivalent to

Definition 2.2.3 $C_m(K)$ is the $\mathbb{Z}_2$-vector space with basis $\mathcal{S}_m(K)$:

$$C_m(K) = \bigoplus_{\sigma \in \mathcal{S}_m(K)} \mathbb{Z}_2\, \sigma\,.$$

We shall pass from one of Definitions 2.2.1, 2.2.2 or 2.2.3 to another without notice; the context usually prevents ambiguity. We consider $C_*(K) = \oplus_{m\in\mathbb{N}} C_m(K)$ and $C^*(K) = \oplus_{m\in\mathbb{N}} C^m(K)$ as graded $\mathbb{Z}_2$-vector spaces. The convention $C_{-1}(K) = C^{-1}(K) = 0$ is useful.

We now define the *Kronecker pairing* on (co)chains

$$C^m(K) \times C_m(K) \xrightarrow{\langle\, ,\, \rangle} \mathbb{Z}_2$$

by the equivalent formulae

$$\begin{aligned} \langle a, \alpha\rangle &= \sharp(a \cap \alpha) \pmod 2 && \text{using Definition 2.2.1a and b} \\ &= \textstyle\sum_{\sigma\in\alpha} a(\sigma) && \text{using Definitions 2.2.1a and 2.2.2b} \\ &= \textstyle\sum_{\sigma\in\mathcal{S}_m(K)} a(\sigma)\alpha(\sigma) && \text{using Definitions 2.2.2a and b.} \end{aligned} \tag{2.2.1}$$

Lemma 2.2.4 *The Kronecker pairing is bilinear and the map $a \mapsto \langle a, \rangle$ is an isomorphism between $C^m(K)$ and $C_m(K)^\sharp = hom(C_m(K), \mathbb{Z}_2)$.*

Proof The bilinearity is obvious from the third line of Eq. (2.2.1). Let $0 \neq a \in C^m(K)$. This means that, as a subset of $\mathcal{S}_m(K)$, a is not empty. If $\sigma \in a$, then $\langle a, \sigma\rangle \neq 0$, which proves the injectivity of $a \mapsto \langle a, \rangle$. As for its surjectivity, let $h \in \hom(C_m(K), \mathbb{Z}_2)$. Using the inclusion $\mathcal{S}_m(K) \hookrightarrow C_m(K)$ given by $\tau \mapsto \{\tau\}$, define

$$a = \{\tau \in \mathcal{S}_m(K) \mid h(\tau) = 1\}\,.$$

For each $\sigma \in \mathcal{S}_m(K)$ the equation $h(\sigma) = \langle a, \sigma\rangle$ holds true. As $\mathcal{S}_m(K)$ is a basis of $C_m(K)$, this implies that $h = \langle a, \rangle$. □

We now define the boundary and coboundary operators. The *boundary operator* $\partial\colon C_m(K) \to C_{m-1}(K)$ is the $\mathbb{Z}_2$-linear map defined by

$$\partial(\sigma) = \{(m-1)\text{-faces of } \sigma\} = \mathcal{S}_{m-1}(\bar{\sigma}), \quad \sigma \in \mathcal{S}_m(K). \tag{2.2.2}$$

Formula (2.2.2) is written in the language of Definition 2.2.1b. Using Definition 2.2.3, we get

$$\partial(\sigma) = \sum_{\tau \in \mathcal{S}_{m-1}(\bar{\sigma})} \tau \,. \tag{2.2.3}$$

The *coboundary operator* $\delta : C^m(K) \to C^{m+1}(K)$ is defined by the equation

$$\langle \delta a, \alpha \rangle = \langle a, \partial \alpha \rangle \,. \tag{2.2.4}$$

The last equation indeed defines δ by Lemma 2.2.4 and δ may be seen as the Kronecker adjoint of ∂. In particular, if $\sigma \in \mathcal{S}_m(K)$ and $\tau \in \mathcal{S}_{m-1}(K)$ then

$$\tau \in \partial(\sigma) \;\Leftrightarrow\; \tau \subset \sigma \;\Leftrightarrow\; \sigma \in \delta(\tau) \,. \tag{2.2.5}$$

The first equivalence determines the operator ∂ since $\mathcal{S}_m(K)$ is a basis for $C_m(K)$. The second equivalence determines δ if $\mathcal{S}_{m-1}(K)$ is finite. Note that the definition of δ may also be given as follows: if $a \in C^m(K)$, then

$$\delta(a) = \{\tau \in \mathcal{S}_{m+1}(K) \mid \sharp\,(a \cap \partial(\tau)) \text{ is odd}\} \,.$$

Let $\sigma \in \mathcal{S}_m(K)$. Each $\tau \in \mathcal{S}_{m-2}(K)$ with $\tau \subset \sigma$ belongs to the boundary of exactly two $(m-1)$-simplexes of σ. Using Eq. (2.2.3), this implies that $\partial \circ \partial = 0$. By Eq. (2.2.4) and Lemma 2.2.4, we get $\delta \circ \delta = 0$. We define the $\mathbb{Z}_2$-vector spaces

- $Z_m(K) = \ker(\partial : C_m(K) \to C_{m-1}(K))$, the *m-cycles* of K.
- $B_m(K) = \text{image}\,(\partial : C_{m+1}(K) \to C_m(K))$, the *m-boundaries* of K.
- $Z^m(K) = \ker(\delta : C^m(K) \to C^{m+1}(K))$, the *m-cocycles* of K.
- $B^m(K) = \text{image}\,(\delta : C^{m-1}(K) \to C^m(K))$, the *m-coboundaries* of K.

For example, Fig. 2.1 shows a triangulation K of the plane, with $V(K) = \mathbb{Z} \times \mathbb{Z}$. The bold line is a cocycle a which is a coboundary: $a = \delta B$, with $B = \{\{(m,n)\} \mid (m,n) \in V(K) \text{ and } m \leq 0\}$, drawn in bold dots.

Since $\partial \circ \partial = 0$ and $\delta \circ \delta = 0$, one has $B_m(K) \subset Z_m(K)$ and $B^m(K) \subset Z^m(K)$. We form the quotient vector spaces

- $H_m(K) = Z_m(K)/B_m(K)$, the mth *-homology vector space of* K.
- $H^m(K) = Z^m(K)/B^m(K)$, the mth *-cohomology vector space of* K.

As for the (co)chains, the notations $H_*(K) = \oplus_{m \in \mathbb{N}} H_m(K)$ and $H^*(K) = \oplus_{m \in \mathbb{N}} H^m(K)$ stand for the (co)homology seen as graded $\mathbb{Z}_2$-vector spaces. By

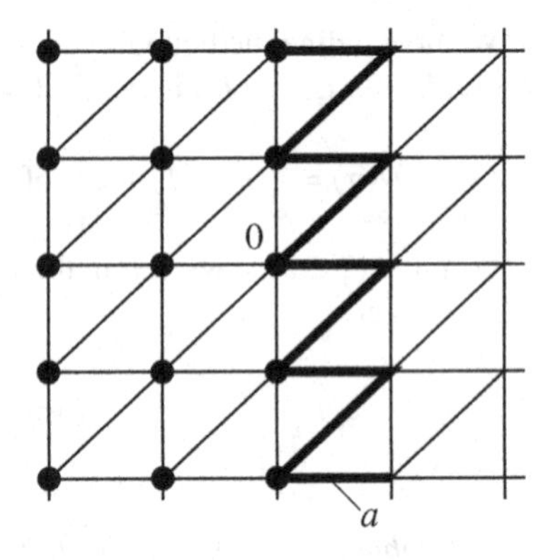

Fig. 2.1 A triangulation K of the plan, with $V(K) = \mathbb{Z} \times \mathbb{Z}$

convention, $H_{-1}(K) = H^{-1}(K) = 0$. Also, the homology and the cohomology are in duality via the Kronecker pairing:

Proposition 2.2.5 (Kronecker duality) *The Kronecker pairing on (co)chains induces a bilinear map*

$$H^m(K) \times H_m(K) \xrightarrow{\langle\,,\,\rangle} \mathbb{Z}_2 .$$

Moreover, the correspondence $a \mapsto \langle a, \ \rangle$ is an isomorphism

$$H^m(K) \underset{\approx}{\xrightarrow{\mathbf{k}}} hom(H_m(K), \mathbb{Z}_2) .$$

Proof Instead of giving a direct proof, which the reader may do as an exercise, we will take advantage of the more general setting of *Kronecker pairs*, developed in the next section. In this way, Proposition 2.2.5 follows from Proposition 2.3.5. □

2.3 Kronecker Pairs

All the vector spaces in this section are over an arbitrary fixed field $\mathbb{F}$. The dual of a vector space V is denoted by $V^\sharp$.

A *chain complex* is a pair (C_*, ∂), where

- C_* is a graded vector space $C_* = \bigoplus_{m\in\mathbb{N}} C_m$. We add the convention that $C_{-1} = 0$.
- $\partial : C_* \to C_*$ is a linear map of degree -1, i.e. $\partial(C_m) \subset C_{m-1}$, satisfying $\partial\circ\partial = 0$. The operator ∂ is called the *boundary* of the chain complex.

A *cochain complex* is a pair (C^*, δ), where

- C^* is a graded vector space $C^* = \bigoplus_{m\in\mathbb{N}} C^m$. We add the convention that $C^{-1} = 0$.
- $\delta : C^* \to C^*$ is a linear map of degree $+1$, i.e. $\partial(C^m) \subset C^{m+1}$, satisfying $\delta\circ\delta = 0$. The operator δ is called the *coboundary* of the cochain complex.

A *Kronecker pair* consists of three items:

(a) a chain complex (C_*, ∂).
(b) a cochain complex (C^*, δ).
(c) a bilinear map

$$C^m \times C_m \xrightarrow{\langle , \rangle} \mathbb{F}$$

satisfying the equation

$$\langle \delta a, \alpha \rangle = \langle a, \partial \alpha \rangle \,. \tag{2.3.1}$$

for all $a \in C^m$ and $\alpha \in C_{m+1}$ and all $m \in \mathbb{N}$. Moreover, we require that the map $\mathbf{k}$: $C^m \to C_m^\sharp$, given by $\mathbf{k}(a) = \langle a, \ \rangle$, is an isomorphism.

Example 2.3.1 Let K be a simplicial complex. Its simplicial (co)chain complexes $(C^*(K), \delta)$, $(C_*(K), \partial)$, together with the pairing $\langle \, , \rangle$ of Sect. 2.2 is a Kronecker pair, with $\mathbb{F} = \mathbb{Z}_2$, as seen in Lemma 2.2.4 and Eq. (2.2.4).

Example 2.3.2 Let (C_*, ∂) be a chain complex. One can define a cochain complex (C^*, δ) by $C^m = C_m^\sharp$ and $\delta = \partial^\sharp$ and then get a bilinear map (pairing) $\langle , \rangle$ by the evaluation: $\langle a, \alpha \rangle = a(\alpha)$. These constitute a Kronecker pair. Actually, via the map $\mathbf{k}$, any Kronecker pair is isomorphic to this one. The reader may use this fact to produce alternative proofs of the results of this section.

We first observe that, as the Kronecker pairing is non-degenerate, chains and cochains mutually determine each other:

Lemma 2.3.3 *Let* $\big((C^*, \delta), (C_*, \partial), \langle \, , \rangle\big)$ *be a Kronecker pair.*

(a) *Let* $a, a' \in C^m$*. Suppose that* $\langle a, \alpha \rangle = \langle a', \alpha \rangle$ *for all* $\alpha \in C_m$*. Then* $a = a'$.
(b) *Let* $\alpha, \alpha' \in C_m$*. Suppose that* $\langle a, \alpha \rangle = \langle a, \alpha' \rangle$ *for all* $a \in C^m$*. Then* $\alpha = \alpha'$.
(c) *Let* $\mathcal{S}_m$ *be a basis for* C_m *and let* $f\colon \mathcal{S}_m \to \mathbb{F}$ *be a map. Then, there is a unique* $a \in C^m$ *such that* $\langle a, \sigma \rangle = f(\sigma)$ *for all* $\sigma \in \mathcal{S}_m$.

Proof In Point (a), the hypotheses imply that $\mathbf{k}(a) = \mathbf{k}(a')$. As $\mathbf{k}$ is injective, this shows that $a = a'$.

In Point (b), suppose that $\alpha \neq \alpha'$. Let $A \in (C_m)^\sharp$ such that $A(\alpha - \alpha') \neq 0$. Then, $\langle a, \alpha \rangle \neq \langle a, \alpha' \rangle$ for $a = \mathbf{k}^{-1}(A) \in C^m$.

Finally, the condition $\tilde{a}(\sigma) = f(\sigma)$ for all $\sigma \in \mathcal{S}_m$ defines a unique $\tilde{a} \in C_m^\sharp$ and $a = \mathbf{k}^{-1}(\tilde{a})$. □

As is Sect. 2.2, we consider the $\mathbb{Z}_2$-vector spaces

- $Z_m = \ker(\partial : C_m \to C_{m-1})$, the *m-cycles* (of C_*).
- $B_m = $ image $(\partial : C_{m+1} \to C_m)$, the *m-boundaries*.
- $Z^m = \ker(\delta : C^m \to C^{m+1})$, the *m-cocycles*.
- $B^m = $ image $(\delta : C^{m-1} \to C^m)$, the *m-coboundaries*.

Since $\partial \circ \partial = 0$ and $\delta \circ \delta = 0$, one has $B_m \subset Z_m$ and $B^m \subset Z^m$. We form the quotient vector spaces

- $H_m = Z_m/B_m$, the *m*th-*homology group* (*or vector space*).
- $H^m = Z^m/B^m$, the *m*th-*cohomology group* (*or vector space*).

We consider the (co)homology as graded vector spaces: $H_* = \oplus_{m \in \mathbb{N}} H_m$ and $H^* = \oplus_{m \in \mathbb{N}} H^m$.

The cocycles and coboundaries may be detected by the pairing:

Lemma 2.3.4 *Let* $a \in C^m$. *Then*

(i) $a \in Z^m$ *if and only if* $\langle a, B_m \rangle = 0$.
(ii) $a \in B^m$ *if and only if* $\langle a, Z_m \rangle = 0$.

Proof Point (i) directly follows from Eq. (2.3.1) and the fact that **k** is injective. Also, if $a \in B^m$, Eq. (2.3.1) implies that $\langle a, Z_m \rangle = 0$. It remains to prove the converse (this is the only place in this lemma where we need vector spaces over a field instead just module over a ring). We consider the exact sequence

$$0 \to Z_m \to C_m \to B_{m-1} \xrightarrow{\partial} 0\,. \tag{2.3.2}$$

Let $a \in C^m$ such that $\langle a, Z_m \rangle = 0$. By (2.3.2), there exists $a_1 \in B^\sharp_{m-1}$ such that $\langle a, \ \rangle = a_1 \circ \partial$. As we are dealing with vector spaces, B_{m-1} is a direct summand of C_{m-1}. We can thus extend a_1 to $a_2 \in C^\sharp_{m-1}$. As **k** is surjective, there exists $a_3 \in C^{m-1}$ such that $\langle a_3, \ \rangle = a_2$. For all $\alpha \in C_m$, one then has

$$\langle \delta a_3, \alpha \rangle = \langle a_3, \partial \alpha \rangle = a_2(\partial \alpha) = a_1(\partial \alpha) = \langle a, \alpha \rangle\,.$$

As **k** is injective this implies that $a = \delta a_3 \in B^m$. □

Let us restrict the pairing $\langle\, ,\rangle$ to $Z^m \times Z_m$. Formula (2.3.1) implies that

$$\langle Z^m, B_m \rangle = \langle B^m, Z_m \rangle = 0\,.$$

Hence, the pairing descends to a bilinear map $H^m \times H_m \xrightarrow{\langle\, ,\rangle} \mathbb{F}$, giving rise to a linear map $\mathbf{k}\colon H^m \to H^\sharp_m$, called the *Kronecker pairing on* (*co*)*homology*. We see H_* and H^* as (co)chain complexes by setting $\partial = 0$ and $\delta = 0$.

Proposition 2.3.5 $(H_*, H^*, \langle\, ,\rangle)$ *is a Kronecker pair.*

Proof Equation (2.3.1) holds trivially since ∂ and δ both vanish. It remains to show that $\mathbf{k}\colon H^m \to H^\sharp_m$ is bijective.

Let $a_0 \in H^\sharp_m$. Pre-composing a_0 with the projection $Z_m \twoheadrightarrow H_m$ produces $a_1 \in Z^\sharp_m$. As Z_m is a direct summand in C_m, one can extend a_1 to $a_2 \in C^\sharp_m$. Since $(C_*, C^*, \langle\, ,\rangle)$ is a Kronecker pair, there exists $a \in C^m$ such that $\langle a, \ \rangle = a_2$. The cochain a satisfies $\langle a, B_m \rangle = a_2(B_m) = 0$ which, by Lemma 2.3.4, implies that $a \in Z^m$. The cohomology class $[a] \in H^m$ of a then satisfies $\langle [a], \ \rangle = a_0$. Thus, $\mathbf{k}$ is surjective.

For the injectivity of $\mathbf{k}$, let $b \in H^m$ with $\langle b, H_m \rangle = 0$. Represent b by $\tilde{b} \in Z^m$, which then satisfies $\langle \tilde{b}, Z_m \rangle = 0$. By Lemma 2.3.4, $\tilde{b} \in B^m$ and thus $b = 0$. □

Let (C_*, ∂) and $(\bar{C}_*, \bar{\partial})$ be two chain complexes. A map $\varphi \colon C_* \to \bar{C}_*$ is a *morphism of chain complexes* or a *chain map* if it is linear map of degree 0 (i.e. $\varphi(C_m) \subset \bar{C}_m$) such that $\varphi \circ \partial = \bar{\partial} \circ \varphi$. This implies that $\varphi(Z_m) \subset \bar{Z}_m$ and $\varphi(B_m) \subset \bar{B}_m$. Hence, φ induces a linear map $H_*\varphi : H_m \to \bar{H}_m$ for all m.

In the same way, let (C^*, δ) and $(\bar{C}^*, \bar{\delta})$ be two cochain complexes. A linear map $\phi \colon \bar{C}^* \to C^*$ of degree 0 is a *morphism of cochain complexes* or a *cochain map* if $\phi \circ \bar{\delta} = \delta \circ \phi$. Hence, ϕ induces a linear map $H^*\phi : \bar{H}^m \to H^m$ for all m.

Let $\mathcal{P} = (C_*, \partial, C^*, \delta, \langle \, , \rangle)$ and $\bar{\mathcal{P}} = (\bar{C}_*, \bar{\partial}, \bar{C}^*, \bar{\delta}, \langle \, , \rangle^-)$ be two Kronecker pairs. A *morphism of Kronecker pairs*, from $\mathcal{P}$ to $\bar{\mathcal{P}}$, consists of a pair (φ, ϕ) where $\varphi \colon C_* \to \bar{C}_*$ is a morphism of chain complexes and $\phi \colon \bar{C}^* \to C^*$ is a morphism of cochain complexes such that

$$\langle a, \varphi(\alpha) \rangle^- = \langle \phi(a), \alpha \rangle \, . \tag{2.3.3}$$

Using the isomorphisms $\mathbf{k}$ and $\bar{\mathbf{k}}$, Eq. (2.3.3) is equivalent to the commutativity of the diagram

$$\begin{array}{ccc} \bar{C}^* & \xrightarrow{\phi} & C^* \\ \approx \big\downarrow \bar{\mathbf{k}} & & \approx \big\downarrow \mathbf{k} \\ \bar{C}_*^\sharp & \xrightarrow{\varphi^\sharp} & C_*^\sharp \end{array} \, . \tag{2.3.4}$$

Lemma 2.3.6 *Let $\mathcal{P}$ and $\bar{\mathcal{P}}$ be Kronecker pairs as above. Let $\varphi \colon C_* \to \bar{C}_*$ be a morphism of chain complex. Define $\phi \colon \bar{C}^* \to C^*$ by Eq. (2.3.3) (or Diagram (2.3.4)). Then the pair (φ, ϕ) is a morphism of Kronecker pairs.*

Proof Obviously, ϕ is a linear map of degree 0 and Eq. (2.3.3) is satisfied. It remains to show that ϕ is a morphism of cochain-complexes. But, if $b \in C_m(\bar{K})$ and $\alpha \in C_{m+1}(K)$, one has

$$\begin{aligned} \langle \delta\phi(b), \alpha \rangle &= \langle \phi(b), \partial\alpha \rangle = \langle b, \varphi(\partial\alpha) \rangle^- = \langle b, \bar{\partial}\varphi(\alpha) \rangle^- \\ &= \langle \bar{\delta} b, \varphi(\alpha) \rangle^- = \langle \phi(\bar{\delta} b), \alpha \rangle \, , \end{aligned}$$

which proves that $\delta\phi(b) = \phi(\bar{\delta} b)$. □

A morphism (φ, ϕ) of Kronecker pairs determines a morphism of Kronecker pairs $(H_*\varphi, H^*\phi)$ from $(H_*, H^*, \langle\,,\rangle)$ to $(\bar{H}_*, \bar{H}^*, \langle\,,\rangle^-)$. This process is functorial:

Lemma 2.3.7 *Let (φ_1, ϕ_1) be a morphism of Kronecker pairs from $\mathcal{P}$ to $\bar{\mathcal{P}}$ and let (φ_2, ϕ_2) be a morphism of Kronecker pairs from $\bar{\mathcal{P}}$ to $\dot{\mathcal{P}}$. Then*

$$(H_*\varphi_2 \circ H_*\varphi_1, H^*\phi_1 \circ H^*\phi_2) = (H_*(\varphi_2 \circ \varphi_1), H^*(\phi_2 \circ \phi_1))$$

Proof That $H_*\varphi_2 \circ H_*\varphi_1 = H_*(\varphi_2 \circ \varphi_1)$ is a tautology. For the cohomology equality, we use that

$$\begin{aligned}\langle H^*\phi_1 \circ H^*\phi_2(a), \alpha\rangle &= \langle H^*\phi_2(a), H_*\varphi_1(\alpha)\rangle = \langle a, H_*\varphi_2 \circ H_*\varphi_1(\alpha)\rangle \\ &= \langle a, H_*(\varphi_2 \circ \varphi_1)(\alpha)\rangle = \langle H^*(\phi_2 \circ \phi_1))(a), \alpha\rangle\end{aligned}$$

holds for all $a \in \bar{H}^*$ and all $\alpha \in H_*$. □

We finish this section with some technical results which will be used later.

Lemma 2.3.8 *Let $f: U \to V$ and $g: V \to W$ be two linear maps between vector spaces. Then, the sequence*

$$U \xrightarrow{f} V \xrightarrow{g} W \tag{2.3.5}$$

is exact at V if and only if the sequence

$$U^\sharp \xleftarrow{f^\sharp} V^\sharp \xleftarrow{g^\sharp} W^\sharp \tag{2.3.6}$$

is exact at $V^\sharp$.

Proof As $f^\sharp \circ g^\sharp = (g \circ f)^\sharp$, then $f^\sharp \circ g^\sharp = 0$ if and only if $g \circ f = 0$.

On the other hand, suppose that $\ker g \subset \operatorname{image} f$. We shall prove that $\ker f^\sharp \subset \operatorname{image} g^\sharp$. Indeed, let $a \in \ker f^\sharp$. Then, $a(\operatorname{image} f) = 0$ and, using the inclusion $\ker g \subset \operatorname{image} f$, we deduce that $a(\ker g) = 0$. Therefore, a descends to a linear map $\bar{a}: V/\ker g \to \mathbb{F}$. The quotient space $V/\ker g$ injects into W, so there exists $b \in W^\sharp$ such that $a = b \circ g = g^\sharp(b)$, proving that $a \in \operatorname{image} g^\sharp$.

Finally, suppose that $\ker g \not\subset \operatorname{image} f$. Then there exists $a \in V^\sharp$ such that $a(\operatorname{image} f) = 0$, i.e., $a \in \ker f^\sharp$, and $a(\ker g) \neq 0$, i.e. $a \notin \operatorname{image} g^\sharp$. This proves that $\ker f^\sharp \not\subset \operatorname{image} g^\sharp$. □

Lemma 2.3.9 *Let (φ, ϕ) be a morphism of Kronecker pairs from $\mathcal{P} = (C_*, \partial, C^*, \delta, \langle\,,\rangle)$ to $\bar{\mathcal{P}} = (\bar{C}_*, \bar{\partial}, \bar{C}^*, \bar{\delta}, \langle\,,\rangle^-)$. Then the pairings $\langle\,,\rangle$ and $\langle\,,\rangle^-$ induce bilinear maps*

$$\operatorname{coker} \phi \times \ker \varphi \xrightarrow{\langle,\rangle} \mathbb{F} \;\; \textit{and} \;\; \ker \phi \times \operatorname{coker} \varphi \xrightarrow{\langle,\rangle^-} \mathbb{F}$$

such that the induced linear maps

$$\operatorname{coker}\phi \xrightarrow{\mathbf{k}} (\ker\varphi)^\sharp \quad and \quad \ker\phi \xrightarrow{\bar{\mathbf{k}}} (\operatorname{coker}\varphi)^\sharp$$

are isomorphisms.

Proof Equation (2.3.3) implies that $\langle \phi(C^*), \ker\varphi\rangle = 0$ and $\langle \ker\phi, \varphi(C_*)\rangle^- = 0$, whence the induced pairings. Consider the exact sequence

$$0 \longrightarrow \ker\varphi \longrightarrow C_* \xrightarrow{\varphi} \bar{C}_* \longrightarrow \operatorname{coker}\varphi \longrightarrow 0 .$$

By Lemma 2.3.8, passing to the dual preserves exactness. Using Diagram (2.3.4), one gets a commutative diagram

$$\begin{array}{ccccccccc}
0 & \longleftarrow & (\ker\varphi)^\sharp & \longleftarrow & C_*^\sharp & \xleftarrow{\varphi^\sharp} & \bar{C}_*^\sharp & \longleftarrow & (\operatorname{coker}\varphi)^\sharp & \longleftarrow & 0 \\
& & \uparrow \mathbf{k} & & \approx\uparrow \mathbf{k} & & \approx\uparrow \bar{\mathbf{k}} & & \uparrow \bar{\mathbf{k}} & & \\
0 & \longleftarrow & \operatorname{coker}\phi & \longleftarrow & H^* & \xleftarrow{\phi} & \bar{C}^* & \longleftarrow & \ker\phi & \longleftarrow & 0
\end{array} \quad . \tag{2.3.7}$$

By diagram-chasing, the two extreme up-arrows are bijective (one can also invoke the famous *five-lemma*: see e.g. [179, Chap. 4, Sect. 5, Lemma 11]). □

Corollary 2.3.10 *Let (φ, ϕ) be a morphism of Kronecker pairs from $(C_*, \partial, C^*, \delta, \langle\,,\rangle)$ to $(\bar{C}_*, \bar{\partial}, \bar{C}^*, \bar{\delta}, \langle\,,\rangle^-)$. Then the pairings $\langle\,,\rangle$ and $\langle\,,\rangle^-$ on (co)homology induce bilinear maps*

$$\operatorname{coker} H^*\phi \times \ker H_*\varphi \xrightarrow{\langle,\rangle} \mathbb{F} \quad and \quad \ker H^*\phi \times \operatorname{coker} H_*\varphi \xrightarrow{\langle,\rangle^-} \mathbb{F}$$

such that the induced linear maps

$$\operatorname{coker} H^*\phi \xrightarrow{\mathbf{k}} (\ker H_*\varphi)^\sharp \quad and \quad \ker H^*\phi \xrightarrow{\bar{\mathbf{k}}} (\operatorname{coker} H_*\varphi)^\sharp$$

are isomorphisms.

Proof The morphism (ϕ, φ) induces a morphism of Kronecker pairs $(H^*\phi, H_*\varphi)$ from $(H^*, H_*, \langle\,,\rangle)$ to $(\bar{H}^*, \bar{H}_*, \langle\,,\rangle^-)$. Corollary 2.3.10 follows then from Lemma 2.3.9 applied to $(H^*\phi, H_*\varphi)$. □

Corollary 2.3.10 implies the following

Corollary 2.3.11 *Let (φ, ϕ) be a morphism of Kronecker pairs from $(C_*, \partial, C^*, \delta, \langle\,,\rangle)$ to $(\bar{C}_*, \bar{\partial}, \bar{C}^*, \bar{\delta}, \langle\,,\rangle^-)$. Then*

(a) *$H^*\phi$ is surjective if and only if $H_*\varphi$ is injective.*
(b) *$H^*\phi$ is injective if and only if $H_*\varphi$ is surjective.*
(c) *$H^*\phi$ is bijective if and only if $H_*\varphi$ is bijective.*

2.4 First Computations

2.4.1 Reduction to Components

Let K be a simplicial complex. We have seen in 2.1.5 that K is the disjoint union of its components, whose set is denoted by $\pi_0(K)$. Therefore, $\mathcal{S}_m(K) = \coprod_{L\in\pi_0(K)} \mathcal{S}_m(L)$ which, by Definition 2.2.3, gives a canonical isomorphism

$$\bigoplus_{L\in\pi_0(K)} C_m(L) \xrightarrow{\approx} C_m(K)\,.$$

This direct sum decomposition commutes with the boundary operators, giving a canonical isomorphism

$$\bigoplus_{L\in\pi_0(K)} H_*(L) \xrightarrow{\approx} H_*(K)\,. \tag{2.4.1}$$

As for the cohomology, seeing an m-cochain as a map $\alpha\colon \mathcal{S}_m(K) \to \mathbb{Z}_2$ (Definition 2.2.2) the restrictions of α to $\mathcal{S}_m(L)$ for all $L \in \pi_0(K)$ gives an isomorphism

$$C^m(K) \xrightarrow{\approx} \prod_{L\in\pi_0(K)} C^m(L)$$

commuting with the coboundary operators. This gives an isomorphism

$$H^*(K) \xrightarrow{\approx} \prod_{L\in\pi_0(K)} H^*(L)\,. \tag{2.4.2}$$

The isomorphisms of (2.4.1) and (2.4.2) permit us to reduce (co)homology computations to connected simplicial complexes. They are of course compatible with the Kronecker duality (Proposition 2.2.5). A formulation of these isomorphisms using simplicial maps is given in Proposition 2.5.3.

2.4.2 0-Dimensional (Co)homology

Let K be a simplicial complex. The *unit cochain* $\mathbf{1} \in C^0(K)$ is defined by $\mathbf{1} = \mathcal{S}_0(K)$, using the subset definition. In the language of colouring, one has $\mathbf{1}(v) = 1$ for all

$v \in V(K) = \mathcal{S}_0(K)$, that is all vertices are black. If $\beta = \{v, w\} \in \mathcal{S}_1(K)$, then

$$\langle \delta\mathbf{1}, \beta \rangle = \langle \mathbf{1}, \partial\beta \rangle = \mathbf{1}(v) + \mathbf{1}(w) = 0\,,$$

which proves that $\delta(\mathbf{1}) = 0$ by Lemma 2.2.4. Hence, $\mathbf{1}$ is a cocycle, whose cohomology class is again denoted by $\mathbf{1} \in H^0(K)$.

Proposition 2.4.1 *Let K be a non-empty connected simplicial complex. Then,*

(i) $H^0(K) = \mathbb{Z}_2$, *generated by* $\mathbf{1}$ *which is the only non-vanishing* 0*-cocycle.*
(ii) $H_0(K) = \mathbb{Z}_2$. *Any* 0*-chain α is a cycle, which represents the non-zero element of $H_0(K)$ if and only if $\sharp\alpha$ is odd.*

Proof If K is non-empty the unit cochain does not vanish. As $C^{-1}(K) = 0$, this implies that $\mathbf{1} \neq 0$ in $H^0(K)$.

Let $a \in C^0(K)$ with $a \neq 0, \mathbf{1}$. Then there exists $v, v' \in V(K)$ with $a(v) \neq a(v')$. Since K is connected, there exists $x_0, \dots, x_m \in V(K)$ with $x_0 = v$, $x_m = v'$ and $\{x_i, x_{i+1}\} \in \mathcal{S}(K)$. Therefore, there exists $0 \le k < m$ with $a(x_k) \neq a(x_{k+1})$. This implies that $\{x_k, x_{k+1}\} \in \delta a$, proving that $\delta a \neq 0$. We have thus proved (i).

Now, $H_0(K) = \mathbb{Z}_2$ since $H^0(K) \approx H_0(K)^\sharp$. Any $\alpha \in C_0(K)$ is a cycle since $C_{-1}(K) = 0$. It represents the non-zero homology class if and only if $\langle \mathbf{1}, \alpha \rangle = 1$, that is if and only if $\sharp\alpha$ is odd. □

Corollary 2.4.2 *Let K be a simplicial complex. Then $H^0(K) \approx \mathbb{Z}_2^{\pi_0(K)}$.*

Here, $\mathbb{Z}_2^{\pi_0(K)}$ denotes the set of maps from $\pi_0(K)$ to $\mathbb{Z}_2$. The isomorphism of Corollary 2.4.2 is natural for simplicial maps (see Corollary 2.5.6).

Proof By Proposition 2.4.1 and its proof, $H^0(K) = Z^0(K)$ is the set of maps from $V(K)$ to $\mathbb{Z}_2$ which are constant on each component. Such a map is determined by a map from $\pi_0(K)$ to $\mathbb{Z}_2$ and conversely. □

2.4.3 Pseudomanifolds

An *n-dimensional pseudomanifold* is a simplicial complex M such that

(a) every simplex of M is contained in an n-simplex of M.
(b) every $(n-1)$-simplex of M is a face of exactly two n-simplexes of M.
(c) for any $\sigma, \sigma' \in \mathcal{S}_n(M)$, there exists a sequence $\sigma = \sigma_0, \dots, \sigma_m = \sigma'$ of n-simplexes such that σ_i and σ_{i+1} have an $(n-1)$-face in common for $i \le 1 < m$.

Example 2.4.3 (1) Let m be an integer with $m \geq 3$. The *polygon* $\mathcal{P}_m$ is the 1-dimensional pseudomanifold for which $V(\mathcal{P}_m) = \{0, 1, \dots, m-1\} = \mathbb{Z}/m\mathbb{Z}$ and $\mathcal{S}_1(\mathcal{P}_m) = \{\{k, k+1\} \mid k \in V(\mathcal{P}_m)\}$. It can be visualized in the complex plane as the equilateral m-gon whose vertices are the mth roots of the unity.

(2) Consider the triangulation of S^2 given by an icosahedron. Choose one pair of antipodal vertices and identify them in a single point. This gives a quotient simplicial complex K which is a 2-dimensional pseudomanifold. Observe that $|K|$ is not a topological manifold.

Pseudomanifolds have been introduced in 1911 by Brouwer [22, p. 477], for his work on the degree and on the invariance of the dimension. They are also called *n-circuits* in the literature. Proposition 2.4.4 below and its proof, together with Proposition 2.4.1, shows that n-dimensional pseudomanifolds satisfy Poincaré duality in dimensions 0 and n.

Let M be a finite n-dimensional pseudomanifold. The n-chain $[M] = \mathcal{S}_n(M) \in C_n(M)$ is called the *fundamental cycle* of M (it is a cycle by Point (b) of the above definition). Its homology class, also denoted by $[M] \in H_n(M)$ is called the *fundamental class* of M.

Proposition 2.4.4 *Let M be a finite non-empty n-dimensional pseudomanifold. Then,*

(i) *$H_n(M) = \mathbb{Z}_2$, generated by $[M]$ which is the only non-vanishing n-cycle.*
(ii) *$H^n(M) = \mathbb{Z}_2$. Any n-cochain a is a cocycle, and $[a] \neq 0$ in $H^n(M)$ if and only if $\sharp a$ is odd.*

Proof We define a simplicial graph L with $V(L) = \mathcal{S}_n(M)$ by setting $\{\sigma, \sigma'\} \in \mathcal{S}_1(L)$ if and only if σ and σ' have an $(n-1)$-face in common. The identification $\mathcal{S}_n(M) = V(L)$ produces isomorphisms

$$\tilde{F}_n : C_n(M) \xrightarrow{\approx} C^0(L) \text{ and } \tilde{F}^n : C^n(M) \xrightarrow{\approx} C_0(L) . \tag{2.4.3}$$

(As M is finite, so is L and $C_*(L)$ is equal to $C^*(L)$, using Definition 2.2.2) On the other hand, by Point (b) of the definition of a pseudomanifold, one gets a bijection $\tilde{F}\colon \mathcal{S}_{n-1}(M) \xrightarrow{\approx} \mathcal{S}_1(L)$. It gives rise to isomorphisms

$$\tilde{F}_{n-1} : C_{n-1}(M) \xrightarrow{\approx} C^1(L) \text{ and } \tilde{F}^{n-1} : C^{n-1}(M) \xrightarrow{\approx} C_1(L) . \tag{2.4.4}$$

The isomorphisms of (2.4.3) and (2.4.4) satisfy

$$\tilde{F}_{n-1} \circ \partial = \delta \circ \tilde{F}_n \text{ and } \partial \circ \tilde{F}^{n-1} = F^n \circ \delta .$$

Since $C_{n+1}(M) = 0$ by Point (a) of the definition of a pseudomanifold, the above isomorphisms give rise to isomorphisms

$$F_* : H_n(M) \xrightarrow{\approx} H^0(L) \text{ and } F^* : H^n(M) \xrightarrow{\approx} H_0(L)$$

with $F_*([M]) = \mathbf{1}$. By Point (c) of the definition of a pseudomanifold, the graph L is connected. Therefore, Proposition 2.4.4 follows from Proposition 2.4.1. □

The proof of Proposition 2.4.4 actually gives the following result.

Proposition 2.4.5 *Let M be a finite non-empty simplicial complex satisfying Conditions (a) and (b) of the definition of an n-dimensional pseudomanifold. Then, M is a pseudomanifold if and only if $H_n(M) = \mathbb{Z}_2$.*

2.4.4 Poincaré Series and Polynomials

A graded $\mathbb{Z}_2$-vector space $A_* = \bigoplus_{i\in\mathbb{N}} A_i$ is of *finite type* if A_i is finite dimensional for all $i \in \mathbb{N}$. In this case, the *Poincaré series* of A_* is the formal power series defined by

$$\mathfrak{P}_t(A_*) = \sum_{i\in\mathbb{N}} \dim A_i t^i \in \mathbb{N}[[t]].$$

When $\dim A_* < \infty$, the series $\mathfrak{P}_t(A_*)$ is a polynomial, also called the *Poincaré polynomial* of A_*.

A simplicial complex K is of *finite (co)homology type* if $H_*(K)$ (or, equivalently, $H^*(K)$) is of finite type. In this case, the *Poincaré series* of K is that of $H_*(K)$. The (co)homology of a simplicial complex of finite (co)homology type is, up to isomorphism, determined by its Poincaré series, which is often the shortest way to describe it. The number $\dim H_m(K)$ is called the m-th *Betti number of K*. The vector space $C_*(K)$ is endowed with the basis $\mathcal{S}(K)$ for which the matrix of the boundary operator is given explicitly. Thus, the Betti numbers may be effectively computed by standard algorithms of linear algebra.

2.4.5 (Co)homology of a Cone

The simplest non-empty simplicial complex is a point whose (co)homology is obviously

$$H^m(pt) \approx H_m(pt) \approx \begin{cases} 0 & \text{if } m > 0 \\ \mathbb{Z}_2 & \text{if } m = 0 . \end{cases} \tag{2.4.5}$$

In terms of Poincaré polynomial: $\mathfrak{P}_t(pt) = 1$. Let L be a simplicial complex. The *cone on L* is the simplicial complex CL defined by $V(CL) = V(L) \cup \{\infty\}$ and

$$\mathcal{S}_m(CL) = \mathcal{S}_m(L) \cup \{\sigma \cup \{\infty\} \mid \sigma \in \mathcal{S}_{m-1}(L)\} .$$

Note that CL is the join $CL \approx L * \{\infty\}$.

Proposition 2.4.6 *The cone CL on a simplicial complex L has its (co)homology isomorphic to that of a point. In other words, $\mathfrak{P}_t(CL) = 1$.*

Proof By Kronecker duality, it is enough to prove the result on homology. The cone CL is obviously connected and non-empty (it contains ∞), so $H_0(CL) = \mathbb{Z}_2$.

Define a linear map $D\colon C_m(CL) \to C_{m+1}(CL)$ by setting, for $\sigma \in \mathcal{S}_m(CL)$:

$$D(\sigma) = \begin{cases} \sigma \cup \{\infty\} & \text{if } \infty \notin \sigma \\ 0 & \text{if } \infty \in \sigma \,. \end{cases}$$

Hence, $D \circ D = 0$. If $\infty \notin \sigma$, the formula

$$\partial D(\sigma) = D(\partial\sigma) + \sigma \tag{2.4.6}$$

holds true in $C_m(CL)$ (and has a clear geometrical interpretation). Suppose that $\infty \in \sigma$ and $\dim\sigma \geq 1$. Then $\sigma = D(\tau)$ with $\tau = \sigma - \{\infty\}$. Using Formula (2.4.6) and that $D \circ D = 0$, one has

$$D(\partial\sigma) + \sigma = D(\partial D(\tau)) + \sigma = D(D(\partial\tau) + \tau) + D(\tau) = 0\,.$$

Therefore, Formula (2.4.6) holds also true if $\infty \in \sigma$, provided $\dim\sigma \geq 1$. This proves that

$$\partial D(\alpha) = D(\partial\alpha) + \alpha \quad \textit{for all } \alpha \in C_m(CL) \textit{ with } m \geq 1\,. \tag{2.4.7}$$

Now, if $\alpha \in C_m(CL)$ satisfies $\partial\alpha = 0$, Formula (2.4.7) implies that $\alpha = \partial D(\alpha)$, which proves that $H_m(CL) = 0$ if $m \geq 1$. □

As an application of Proposition 2.4.6, let A be a set. The *full complex* $\mathcal{F}A$ on A is the simplicial complex for which $V(\mathcal{F}A) = A$ and $\mathcal{S}(\mathcal{F}A)$ is the family of *all* finite non-empty subsets of A. If A is finite and non-empty, then $\mathcal{F}A$ is isomorphic to a simplex of dimension $\sharp A - 1$. Denote by $\dot{\mathcal{F}}A$ the subcomplex of $\mathcal{F}A$ generated by the proper (i.e. $\neq A$) subsets of A. For instance, $\dot{\mathcal{F}}A = \mathcal{F}A$ if A is infinite.

Corollary 2.4.7 *Let A be a non-empty set. Then*

(i) *$\mathcal{F}A$ has its (co)homology isomorphic to that of a point, i.e. $\mathfrak{P}_t(\mathcal{F}A) = 1$.*
(ii) *If $3 \leq \sharp A \leq \infty$, then $\mathfrak{P}_t(\dot{\mathcal{F}}A) = 1 + t^{\sharp A - 1}$.*
(iii) *If $\sharp A = 2$, then $\mathfrak{P}_t(\dot{\mathcal{F}}A) = 2$.*

Proof As A is not empty, $\mathcal{F}A$ is isomorphic to the cone over $\mathcal{F}A$ deprived of one of its elements. Point (i) then follows from Proposition 2.4.6. Let $n = \sharp A - 1$. The chain complex of $\mathcal{F}A$ looks like a sequence

$$0 \to C_n(\mathcal{F}A) \xrightarrow{\partial_n} C_{n-1}(\mathcal{F}A) \xrightarrow{\partial_{n-1}} \cdots \to C_0(\mathcal{F}A) \to 0\,,$$

which, by (i), is exact except at $C_0(\mathcal{F}A)$. One has $C_n(\mathcal{F}A) = \mathbb{Z}_2$, generated by the $A \in \mathcal{S}_n(\mathcal{F}A)$. Hence, $\ker \partial_{n-1} \approx \mathbb{Z}_2$. As the chain complex $C_*(\dot{\mathcal{F}}A)$ is the same as that of $\mathcal{F}A$ with C_n replaced by 0, this proves (ii). If $\sharp A = 2$, then $\dot{\mathcal{F}}A$ consists of two 0-simplexes and Point (iii) follows from (2.4.5) to (2.4.1). □

2.4.6 The Euler Characteristic

Let K be a finite simplicial complex. Its *Euler characteristic* $\chi(K)$ is defined as

$$\chi(K) = \sum_{m\in\mathbb{N}} (-1)^m \, \sharp \mathcal{S}_m(K) \in \mathbb{Z}\,.$$

Proposition 2.4.8 *Let K be a finite simplicial complex. Then*

$$\chi(K) = \sum_{m\in\mathbb{N}} (-1)^m \dim H_m(K) = \sum_{m} (-1)^m \dim H^m(K)\,.$$

As in the definition of the Poincaré polynomial, the number $\dim H_m(K)$ is the dimension of $H_m(K)$ as a $\mathbb{Z}_2$-vector space. In other words, $\dim H_m(K)$ is the m-th Betti number of K. Proposition 2.4.8 holds true for the (co)homology with coefficients in any field $\mathbb{F}$, though the Betti numbers depend individually on $\mathbb{F}$.

Proof By Kronecker duality, only the first equality requires a proof. Let c_m, z_m, b_m and h_m be the dimensions of $C_m(K)$, $Z_m(K)$, $B_m(K)$ and $H_m(K)$. Elementary linear algebra gives the equalities

$$\begin{cases} c_m = z_m + b_{m-1} \\ z_m = b_m + h_m\,. \end{cases}$$

We deduce that

$$\chi(K) = \sum_{m\in\mathbb{N}} (-1)^m c_m = \sum_{m\in\mathbb{N}} (-1)^m h_m + \sum_{m\in\mathbb{N}} (-1)^m b_m + \sum_{m\in\mathbb{N}} (-1)^m b_{m-1}\,.$$

As $b_{-1} = 0$, the last two sums cancels each other, proving Proposition 2.4.8. □

Corollary 2.4.9 *Let K be a finite simplicial complex. Then*

$$\chi(K) = \mathfrak{P}_t(K)_{t=-1}\,.$$

The following additive formula for the Euler characteristic is useful.

Lemma 2.4.10 *Let K be a simplicial complex. Let K_1 and K_2 be two subcomplexes of K such that $K = K_1 \cup K_2$. Then,*

$$\chi(K) = \chi(K_1) + \chi(K_2) - \chi(K_1 \cap K_2)\,.$$

Proof The formula follows directly from the equations $\mathcal{S}_m(K) = \mathcal{S}_m(K_1) \cup \mathcal{S}_m(K_2)$ and $\mathcal{S}_m(K_1 \cap K_2) = \mathcal{S}_m(K_1) \cap \mathcal{S}_m(K_2)$. □

2.4.7 Surfaces

A *surface* is a manifold of dimension 2. In this section, we give examples of triangulations of surfaces and compute their (co)homology. Strictly speaking, the results would hold only for the given triangulations, but we allow us to formulate them in more general terms. For this, we somehow admit that

- a connected surface is a pseudomanifold of dimension 2. This will be established rigorously in Corollary 5.2.7 but the reader may find a proof as an exercise and this is easy to check for the particular triangulations given below.
- up to isomorphism, the (co)homology of a simplicial complex K depends only of the homotopy type of $|K|$. This will be proved in Sect. 3.6. In particular, the Euler characteristic of two triangulations of a surface coincide.

The 2-Sphere

The 2-sphere S^2 being homeomorphic to the boundary of a 3-simplex, it follows from Corollary 2.4.7 that:

$$\mathfrak{P}_t(S^2) = 1 + t^2 .$$

The Projective Plane

The *projective plane* $\mathbb{R}P^2$ is the quotient of S^2 by the antipodal map. The triangulation of S^2 as a regular icosahedron being invariant under the antipodal map, it gives a triangulation of $\mathbb{R}P^2$ given in Fig. 2.2. Note that the border edges appear twice,

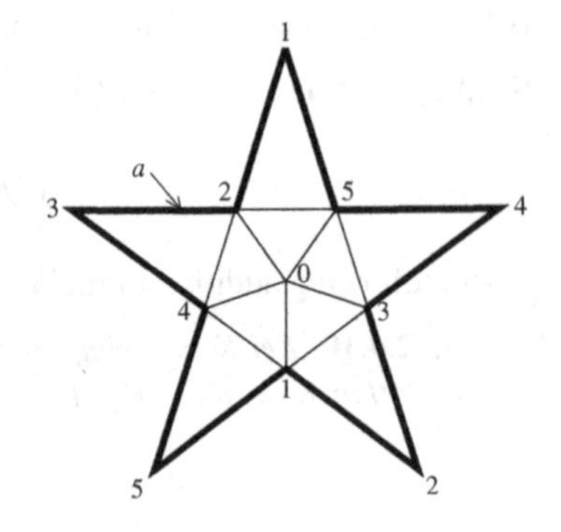

Fig. 2.2 A triangulation of $\mathbb{R}P^2$

showing as expected that $\mathbb{R}P^2$ is the quotient of a 2-disk modulo the antipodal involution on its boundary.

Being a quotient of an icosahedron, the triangulation of Fig. 2.2 has 6 vertices, 15 edges and 10 facets, thus $\chi(\mathbb{R}P^2) = 1$. Using that $\mathbb{R}P^2$ is a connected 2-dimensional pseudomanifold, we deduce that

$$\mathfrak{P}_t(\mathbb{R}P^2) = 1 + t + t^2 . \tag{2.4.8}$$

To identify the generators of $H^1(\mathbb{R}P^2) \approx \mathbb{Z}_2$ and $H_1(\mathbb{R}P^2)$, we define

$$a = \alpha = \big\{\{1, 2\}, \{2, 3\}, \{3, 4\}, \{4, 5\}, \{5, 1\}\big\} \subset \mathcal{S}_1(\mathbb{R}P^2) . \tag{2.4.9}$$

We see $a \in C^1(\mathbb{R}P^2)$ and $\alpha \in C_1(\mathbb{R}P^2)$. The cochain a is drawn in bold on Fig. 2.2, where it looks as the set of border edges, since each of its edges appears twice on the figure. It is easy to check that $\delta(a) = 0$ and $\partial(\alpha) = 0$. As $\sharp\alpha = 5$ is odd, one has $\langle a, \alpha\rangle = 1$, showing that a is the generator of $H^1(\mathbb{R}P^2) = \mathbb{Z}_2$ and α is the generator of $H_1(\mathbb{R}P^2) = \mathbb{Z}_2$.

The 2-Torus

The 2-torus $T^2 = S^1 \times S^1$ is the quotient of a square whose opposite sides are identified. A triangulation of T^2 is described (in two copies) in Fig. 2.3. This triangulation has 9 vertices, 27 edges and 18 facets, which implies that $\chi(T^2) = 0$. Since T^2 is a connected 2-dimensional pseudomanifold, we deduce that

$$\mathfrak{P}_t(T^2) = (1 + t)^2 .$$

In Fig. 2.3 are drawn two chains $\alpha, \beta \in C_1(T^2)$ given by

$$\alpha = \big\{\{3, 8\}, \{8, 9\}, \{9, 3\}\big\} \text{ and } \beta = \big\{\{5, 7\}, \{7, 9\}, \{9, 5\}\big\} .$$

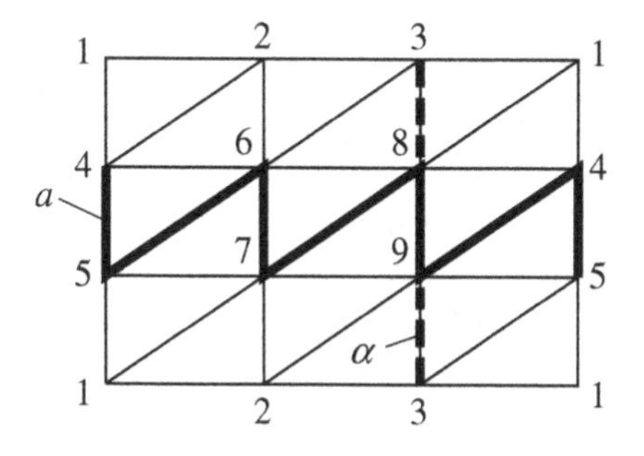

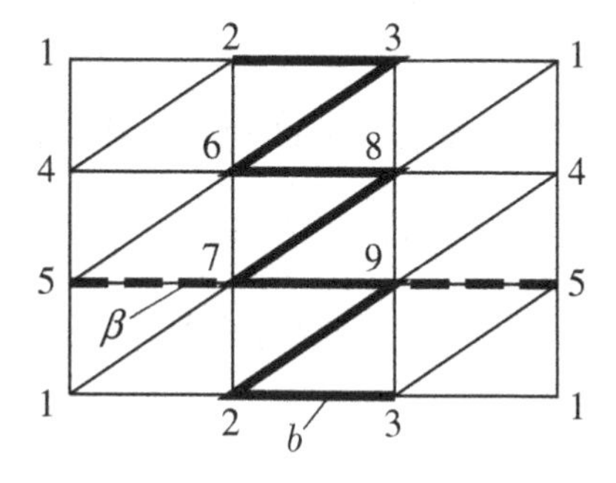

Fig. 2.3 Two copies of a triangulation of the 2-torus T^2, showing generators of $H^1(T^2)$ and $H_1(T^2)$

We also drew two cochains $a, b \in C^1(T^2)$ defined as

$$a = \{\{4,5\},\{5,6\},\{6,7\},\{7,8\},\{8,9\},\{9,4\}\}$$

and

$$b = \{\{2,3\},\{3,6\},\{6,8\},\{8,7\},\{7,9\},\{9,2\}\}.$$

One checks that $\partial\alpha = \partial\beta = 0$ and that $\delta a = \delta b = 0$. Therefore, they represent classes $a, b \in H^1(T^2)$ and $\alpha, \beta \in H_1(T^2)$. The equalities

$$\langle a, \alpha \rangle = 1 \ , \ \langle a, \beta \rangle = 0 \ , \ \langle b, \alpha \rangle = 0 \ , \ \langle b, \beta \rangle = 1$$

imply that a, b is a basis of $H^1(T^2)$ and α, β is a basis of $H_1(T^2)$.

If we consider a and b as 1-chains (call them $\tilde{a}$ and $\tilde{b}$), we also have $\partial\tilde{a} = \partial\tilde{b} = 0$. Note that

$$\langle a, \tilde{b} \rangle = 1 \ , \ \langle a, \tilde{a} \rangle = 0 \ , \ \langle b, \tilde{b} \rangle = 0 \ , \ \langle b, \tilde{a} \rangle = 1$$

This proves that $\tilde{a} = \beta$ and $\tilde{b} = \alpha$ in $H_1(T^2)$.

The Klein Bottle

A triangulation of the Klein bottle K is pictured in Fig. 2.4. As the 2-torus, the Klein bottle is the quotient of a square with opposite side identified, one of these identifications "reversing the orientation". One checks that $\chi(K) = 0$. Since K is a connected 2-dimensional pseudomanifold, the (co)homology of K is abstractly isomorphic to that of T^2:

$$\mathfrak{P}_t(K) = (1+t)^2$$

(In Chap. 3, $H^*(T^2)$ and $H^*(K)$ will be distinguished by their cup product: see p. 138). In Fig. 2.4 the dotted lines show two 1-chains $\alpha, \beta \in C_1(K)$ given by

$$\alpha = \{\{3,8\},\{8,9\},\{9,3\}\} \text{ and } \beta = \{\{5,7\},\{7,9\},\{9,5\}\}. \tag{2.4.10}$$

The bold lines describe two 1-cochains $a, b \in C^1(K)$ defined as

$$a = \{\{4,5\},\{5,6\},\{6,7\},\{7,8\},\{8,9\},\{9,5\}\} \tag{2.4.11}$$

and

$$b = \{\{2,3\},\{3,6\},\{6,8\},\{8,7\},\{7,9\},\{9,2\}\}. \tag{2.4.12}$$

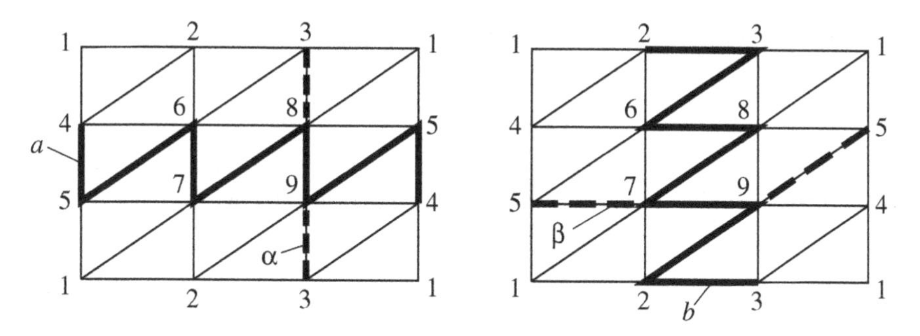

Fig. 2.4 Two copies of a triangulation of the Klein bottle K, showing generators of $H^1(K)$ and $H_1(K)$

One checks that $\partial\alpha = \partial\beta = 0$ and that $\delta a = \delta b = 0$. Therefore, they represent classes $a, b \in H^1(K)$ and $\alpha, \beta \in H_1(K)$. The equalities

$$\langle a, \alpha\rangle = 1\ ,\ \langle a, \beta\rangle = 1\ ,\ \langle b, \alpha\rangle = 0\ ,\ \langle b, \beta\rangle = 1$$

imply that a, b is a basis of $H^1(K)$ and α, β is a basis of $H_1(K)$.

As in the case of T^2, we may regard a and b as 1-chains (call them $\tilde{a}$ and $\tilde{b}$). Here $\partial\tilde{b} = 0$ but $\partial\tilde{a} = \{4\} + \{5\} \neq 0$.

Other Surfaces

Let K_1 and K_2 be two simplicial complexes such that $|K_1|$ and $|K_2|$ are surfaces. A simplicial complex L with $|L|$ homeomorphic to the connected sum $|K_1|\sharp|K_2|$ may be obtained in the following way: choose 2-simplexes $\sigma_1 \in K_1$ and $\sigma_2 \in K_2$. Let $L_i = K_i - \sigma_i$ and let L be obtained by taking the disjoint union of L_1 and L_2 and identifying $\dot{\sigma}_1$ with $\dot{\sigma}_2$. Thus, $L = L_1 \cup L_2$ and $L_0 = L_1 \cap L_2$ is isomorphic to the boundary of a 2-simplex.

By Lemma 2.4.10, one has

$$\begin{aligned} \chi(L) &= \chi(L_1) + \chi(L_2) - \chi(L_0) \\ &= \chi(K_1) - 1 + \chi(K_2) - 1 - 0 \\ &= \chi(K_1) + \chi(K_2) - 2\,. \end{aligned} \tag{2.4.13}$$

The *orientable surface* Σ_g *of genus* g is defined as the connected sum of g copies of the torus T^2. By Formula (2.4.13), one has

$$\chi(\Sigma_g) = 2 - 2g\,. \tag{2.4.14}$$

As Σ_g is a 2-dimensional connected pseudomanifold, one has

$$\mathfrak{P}_t(\Sigma_g) = 1 + 2gt + t^2 .$$

The *nonorientable surface* $\bar{\Sigma}_g$ *of genus* g is defined as the connected sum of g copies of $\mathbb{R}P^2$. For instance, $\bar{\Sigma}_1 = \mathbb{R}P^2$ and $\bar{\Sigma}_2$ is the Klein bottle. Formula (2.4.13) implies

$$\chi(\bar{\Sigma}_g) = 2 - g . \tag{2.4.15}$$

As $\bar{\Sigma}_g$ is a 2-dimensional connected pseudomanifold, one has

$$\mathfrak{P}_t(\bar{\Sigma}_g) = 1 + gt + t^2 .$$

2.5 The Homomorphism Induced by a Simplicial Map

Let $f: K \to L$ be a simplicial map between the simplicial complexes K and L. Recall that f is given by a map $f: V(K) \to V(L)$ such that $f(\sigma) \in \mathcal{S}(L)$ if $\sigma \in \mathcal{S}(K)$, i.e. the image of an m-simplex of K is an n-simplex of L with $n \leq m$. We define C_*f: $C_*(K) \to C_*(L)$ as the degree 0 linear map such that, for all $\sigma \in \mathcal{S}_m(K)$, one has

$$C_*f(\sigma) = \begin{cases} f(\sigma) & \text{if } f(\sigma) \in \mathcal{S}_m(L) \quad \text{(i.e. if } f_{|\sigma} \text{ is injective)} \\ 0 & \text{otherwise.} \end{cases} \tag{2.5.1}$$

We also define $C^*f: C^*(L) \to C^*(K)$ by setting, for $a \in C^m(L)$,

$$C^*f(a) = \big\{\sigma \in \mathcal{S}_m(K) \mid f(\sigma) \in a\big\} . \tag{2.5.2}$$

In the following lemma, we use the same notation for the (co)boundary operators ∂ and δ and the Kronecker product $\langle\ ,\ \rangle$, both for K of for L.

Lemma 2.5.1 *Let* $f: K \to L$ *be a simplicial map. Then*

(a) $C_*f \circ \partial = \partial \circ C_*f$.
(b) $\delta \circ C^*f = C^*f \circ \delta$.
(c) $\langle C^*f(b), \alpha\rangle = \langle b, C_*f(\alpha)\rangle$ *for all* $b \in C^*(L)$ *and all* $\alpha \in C_*(K)$.

In other words, the couple (C_*f, C^*f) *is a morphism of Kronecker pairs.*

Proof To prove (a), let $\sigma \in \mathcal{S}_m(K)$. If f restricted to σ is injective, it is straightforward that $C_*f \circ \partial(\sigma) = \partial \circ C_*f(\sigma)$. Otherwise, we have to show that $C_*f \circ \partial(\sigma) = 0$. Let us label the vertices $v_0, v_1, \ldots, v_m$ of σ in such a way that $f(v_0) = f(v_1)$. Then, $C_*f \circ \partial(\sigma)$ is a sum of two terms: $C_*f \circ \partial(\sigma) = C_*f(\tau_0) + C_*f(\tau_1)$, where $\tau_0 = \{v_1, v_2, \ldots, v_m\}$ and $\tau_1 = \{v_0, v_2, \ldots, v_m\}$. As $C_*f(\tau_0) = C_*f(\tau_1)$, one has

$C_* f \circ \partial(\sigma) = 0$. Thus, Point (a) is established. Point (c) can be easily deduced from Definitions (2.5.1) and (2.5.2), taking for α a simplex of K. Point (b) then follows from Points (a) and (c), using Lemma 2.3.6 and its proof. □

By Lemma 2.5.1 and Proposition 2.3.5, the couple $(C_* f, C^* f)$ determines linear maps of degree zero

$$H_* f : H_*(K) \to H_*(L) \text{ and } H^* f : H^*(L) \to H^*(K)$$

such that

$$\langle H^* f(a), \alpha \rangle = \langle a, H_* f(\alpha) \rangle \text{ for all } a \in H^*(L) \text{ and } \alpha \in H_*(K). \qquad (2.5.3)$$

Lemma 2.5.2 (Functoriality) *Let $f\colon ZK \to L$ and $g\colon L \to M$ be simplicial maps. Then $H_*(g \circ f) = H_* g \circ H_* f$ and $H^*(g \circ f) = H^* f \circ H^* g$. Also $H_* \mathrm{id}_K = \mathrm{id}_{H_*(K)}$ and $H^* \mathrm{id}_K = \mathrm{id}_{H^*(K)}$*

In other words, H^* and H_* are functors from the simplicial category **Simp** to the category **GrV** of graded vector spaces and degree 0 linear maps. The cohomology is contravariant and the homology is covariant.

Proof For $\sigma \in \mathcal{S}(K)$, the formula $C_*(g \circ f)(\sigma) = C_* g \circ C_* f(\sigma)$ follows directly from Definition (2.5.1). Therefore $C_*(g \circ f) = C_* g \circ C_* f$ and then $H_*(g \circ f) = H_* g \circ H_* f$. The corresponding formulae for cochains and cohomology follow from Point (c) of Lemma 2.5.1. The formulae for id_K is obvious. □

Simplicial maps and components. Let K be a simplicial complex. For each component $L \in \pi_0(K)$ of K, the inclusion $i_L\colon L \to K$ is a simplicial map. The results of Sect. 2.4.1 may be strengthened as follows.

Proposition 2.5.3 *Let K be a simplicial complex. The family of simplicial maps $i_L\colon L \to K$ for $L \in \pi_0(K)$ gives rise to isomorphisms*

$$H^*(K) \xrightarrow[\approx]{(H^* i_L)} \prod_{L \in \pi_0(K)} H^*(L)$$

and

$$\bigoplus_{L \in \pi_0(K)} H_*(L) \xrightarrow[\approx]{\sum H_* i_L} H_*(K).$$

The homomorphisms $H^0 f$ and $H_0 f$. We use the same notation $\mathbf{1} \in H^0(K)$ and $\mathbf{1} \in H^0(L)$ for the classes given by the unit cochains.

Lemma 2.5.4 *Let $f\colon K \to L$ be a simplicial map. Then $H^0 f(\mathbf{1}) = \mathbf{1}$.*

Proof The formula $C^0 f(\mathbf{1}) = \mathbf{1}$ in $C^0(K)$ follows directly from Definition (2.5.2). □

Corollary 2.5.5 *Let $f: K \to L$ be a simplicial map with K and L connected. Then*

$$H^0 f : \mathbb{Z}_2 = H^0(L) \to H^0(K) = \mathbb{Z}_2$$

and

$$H_0 f : \mathbb{Z}_2 = H_0(K) \to H_0(L) = \mathbb{Z}_2$$

are the identity isomorphism.

Proof By Proposition 2.4.1, the generator of $H^0(L)$ (or $H^0(K)$) is the unit cocycle $\mathbf{1}$. By Lemma 2.5.4, this proves the cohomology statement. The homology statement also follows from Proposition 2.4.1, since $H_0(K)$ and $H_0(L)$ are generated by a cycle consisting of a single vertex. □

More generally, one has $H^0(L) \approx \mathbb{Z}_2^{\pi_0(L)}$ and $H^0(K) \approx \mathbb{Z}_2^{\pi_0(K)}$ by Corollary 2.4.2. Using this and Lemma 2.5.4, one gets the following corollary.

Corollary 2.5.6 *Let $f: K \to L$ be a simplicial map. Then $H^0 f: \mathbb{Z}_2^{\pi_0(L)} \to \mathbb{Z}_2^{\pi_0(K)}$ is given by $H^0 f(\lambda) = \lambda \circ \pi_0 f$.*

The degree of a map. Let $f: K \to L$ be a simplicial map between two finite connected n-dimensional pseudomanifolds. Define the *degree* $\deg(f) \in \mathbb{Z}_2$ by

$$\deg(f) = \begin{cases} 0 & \text{if } H^n f = 0 \\ 1 & \text{otherwise.} \end{cases} \tag{2.5.4}$$

By Proposition 2.4.4, $H^n(K) \approx H^n(L) \approx \mathbb{Z}_2$. Thus, $\deg(f) = 1$ if and only if $H^n f$ is the (only possible) isomorphism between $H^n(K)$ and $H^n(L)$. By Kronecker duality, the homomorphism $H_n f$ may be used instead of $H^n f$ in the definition of $\deg(f)$. Our degree is sometimes called the mod 2 *degree*, since, for oriented pseudomanifolds, it is the mod 2 reduction of a degree defined in $\mathbb{Z}$ (see, e.g. [179, Exercises of Chap. 4]).

Let $f: K \to L$ be a simplicial map between two finite n-dimensional pseudomanifolds. For $\sigma \in \mathcal{S}_n(L)$, define

$$d(f, \sigma) = \sharp\{\tau \in \mathcal{S}_n(K) \mid f(\tau) = \sigma\} \in \mathbb{N}. \tag{2.5.5}$$

As an example, let $K = L = \mathcal{P}_4$, the polygon of Example 2.4.3 with 4 edges. Let f be defined by $f(0) = 0$, $f(1) = 1$, $f(2) = 2$, $f(3) = 1$. Then, $d(f, \{0, 1\}) = d(f, \{1, 2\}) = 2$, $d(f, \{2, 3\}) = d(f, \{3, 0\}) = 0$ and $\deg(f) = 0$. This example illustrates the following proposition.

Proposition 2.5.7 *Let $f: K \to L$ be a simplicial map between two finite n-dimensional pseudomanifolds which are connected. For any $\sigma \in \mathcal{S}_n(L)$, one has*

$$\deg(f) = d(f, \sigma) \bmod 2\,.$$

Proof By Proposition 2.4.4, $H^n(L) = \mathbb{Z}_2$ is generated by the cocycle formed by the singleton σ and $C^n f(\sigma)$ represents the non-zero element of $H^n(K)$ if and only if $\sharp C^n f(\sigma) = d(f, \sigma)$ is odd. □

The interest of Proposition 2.5.7 is 2-fold: first, it tells us that $\deg(f)$ may be computed using any $\sigma \in \mathcal{S}_m(L)$ and, second, it asserts that $d(f, \sigma)$ is independent of σ. Proposition 2.5.7 is the mod 2 context of the identity between the degree introduced by Brouwer in 1910, [22, p.419], and its homological interpretation due to Hopf in 1930, [98, Sect. 2]. For a history of the notion of the degree of a map, see [40, pp.169–175].

Example 2.5.8 Let $f: T^2 \to K$ be the two-fold cover of the Klein bottle K by the 2-torus T^2, given in Fig. 2.5. In formulae: $f(i) = i = f(\bar{i})$ for $i = 1, \ldots, 9$.

The 1-dimensional (co)homology vector spaces of T^2 and K admit the bases:

(i) $\tilde{\mathcal{V}} = \{[\tilde{a}], [\tilde{b}]\} \subset H^1(T^2)$, where $\tilde{a}$ is drawn in Fig. 2.5 and

$$\tilde{b} = \big\{\{2,3\}, \{3,6\}, \{6,8\}, \{8,7\}, \{7,9\}, \{9,2\}\big\}.$$

(ii) $\tilde{\mathcal{W}} = \{[\tilde{\alpha}], [\tilde{\beta}]\} \subset H_1(T^2)$, where $\tilde{\alpha}$ is drawn in Fig. 2.5 and

$$\tilde{\beta} = \big\{\{5,7\}, \{7,9\}, \{9,\bar{4}\}, \{\bar{4},\bar{7}\}, \{\bar{7},\bar{9}\}, \{\bar{9},5\}\big\}.$$

(iii) $\mathcal{V} = \{[a], [b]\} \subset H^1(K)$, where a and b are defined in Eqs. (2.4.11) and (2.4.12) (a drawn in Fig. 2.5).

(iv) $\mathcal{W} = \{[\alpha], [\beta]\} \subset H_1(K)$, where α and β are defined in Eq. (2.4.10) (α drawn in Fig. 2.5).

The matrices for $C^* f$ and $C_* f$ in these bases are

$$C^* f = \begin{pmatrix} 1 & 0 \\ 0 & 0 \end{pmatrix} \quad \text{and} \quad C_* f = \begin{pmatrix} 1 & 0 \\ 0 & 0 \end{pmatrix}.$$

Note that, under the isomorphism $\mathbf{k}: H^1(-) \xrightarrow{\approx} H_1(-)^\sharp$, the bases $\tilde{\mathcal{V}}$ and $\mathcal{V}$ are dual of $\tilde{\mathcal{W}}$ and $\mathcal{W}$; therefore, the matrix of $C^* f$ is the transposed of that of $C_* f$.

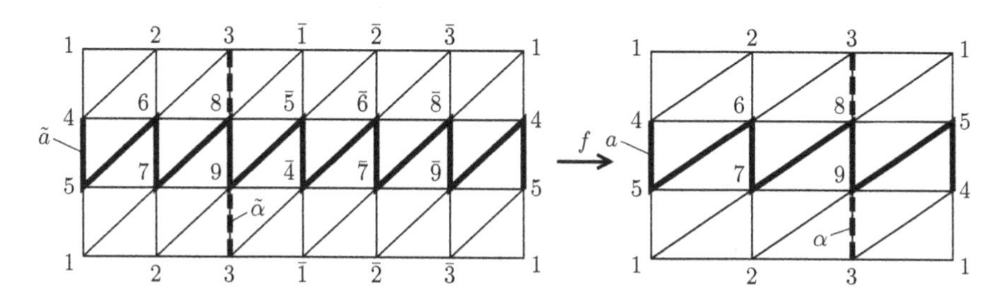

Fig. 2.5 Two-fold cover $f: T^2 \to K$ over the triangulation K of the Klein bottle given in Fig. 2.4

Now, T^2 and K are 2-dimensional pseudomanifolds and $d(f, \sigma) = 2$ for each $\sigma \in \mathcal{S}_2(K)$. By Proposition 2.5.7, $\deg(f) = 0$ and both $H^*f : H^2(K) \to H^2(T^2)$ and $H_*f : H_2(T^2) \to H_2(K)$ vanish.

Contiguous maps. Two simplicial maps $f, g\colon K \to L$ are called *contiguous* if $f(\sigma) \cup g(\sigma) \in \mathcal{S}(L)$ for all $\sigma \in \mathcal{S}(K)$. We denote by $\tau(\sigma)$ the subcomplex of L generated by the simplex $f(\sigma)\cup g(\sigma) \in \mathcal{S}(L)$. For example, the inclusion $K \hookrightarrow CK$ of a simplicial complex K into its cone and the constant map of K onto the cone vertex of CK are contiguous.

Proposition 2.5.9 *Let $f, g\colon K \to L$ be two simplicial maps which are contiguous. Then $H_*f = H_*g$ and $H^*f = H^*g$.*

Proof By Kronecker Duality, using Diagram (2.3.4), it is enough to prove that $H_*f = H_*g$. By induction on m, we shall prove the following property:

Property $\mathcal{H}(m)$: there exists a linear map $D\colon C_m(K) \to C_{m+1}(L)$ such that:

(i) $\partial D(\alpha) + D(\partial\alpha) = C_*f(\alpha) + C_*g(\alpha)$ for each $\alpha \in C_m(K)$.
(ii) for each $\sigma \in \mathcal{S}_m(K)$, $D(\sigma) \in C_{m+1}(\tau(\sigma)) \subset C_{m+1}(L)$.

We first prove that Property $\mathcal{H}(m)$ for all m implies that $H_*f = H_*g$. Indeed, we would then have a linear map $D\colon C_*(K) \to C_{*+1}(L)$ satisfying

$$C_*f + C_*g = \partial \circ D + D \circ \partial\,. \tag{2.5.6}$$

Such a map D is called a *chain homotopy* from C_*f to C_*g. Let $\beta \in Z_*(K)$. By Eq. (2.5.6), one has $C_*f(\beta) + C_*g(\beta) = \partial D(\beta)$ which implies that $H_*f([\beta]) + H_*g([\beta])$ in $H_*(L)$.

We now prove that $\mathcal{H}(0)$ holds true. We define $D\colon C_0(K) \to C_1(L)$ as the unique linear map such that, for $v \in V(K)$:

$$D(\{v\}) = \begin{cases} \{f(v), g(v)\} = \tau(\{v\}) & \text{if } f(v) \neq g(v) \\ 0 & \text{otherwise.} \end{cases}$$

Formula (i) being true for any $\{v\} \in \mathcal{S}_0(K)$, it is true for any $\alpha \in C_0(K)$. Formula (ii) is obvious.

Suppose that $\mathcal{H}(m-1)$ holds true for $m \geq 1$. We want to prove that $\mathcal{H}(m)$ also holds true. Let $\sigma \in \mathcal{S}_m(K)$. Observe that $D(\partial\sigma)$ exists by $\mathcal{H}(m-1)$. Consider the chain $\zeta \in C_m(L)$ defined by

$$\zeta = C_*f(\sigma) + C_*g(\sigma) + D(\partial\sigma)$$

Using $\mathcal{H}(m-1)$, one has

$$\begin{aligned} \partial\zeta &= \partial C_*f(\sigma) + \partial C_*g(\sigma) + \partial D(\partial\sigma) \\ &= C_*f(\partial\sigma) + C_*g(\partial\sigma) + D(\partial\partial\sigma) + C_*f(\partial\sigma) + C_*g(\partial\sigma) \\ &= 0\,. \end{aligned}$$

On the other hand, $\zeta \in C_m(\tau(\sigma))$. As $m \geq 1$, $H_m(\tau(\sigma)) = 0$ by Corollary 2.4.7. There exists then $\eta \in C_{m+1}(\tau(\sigma))$ such that $\zeta = \partial\eta$. Choose such an η and set $D(\sigma) = \eta$. This defines $D\colon C_m(K) \to C_{m+1}(L)$ which satisfies (i) and (ii), proving $\mathcal{H}(m)$. □

Remark 2.5.10 The chain homotopy D in the proof of Proposition 2.5.9 is not explicitly defined. This is because several of these exist and there is no canonical way to choose one (see [155, p. 68]). The proof of Proposition 2.5.9 is an example of the technique of acyclic carriers which will be developed in Sect. 2.9.

Remark 2.5.11 Let $f, g\colon K \to L$ be two simplicial maps which are contiguous. Then $|f|, |g|\colon |K| \to |L|$ are homotopic. Indeed, the formula $F(\mu, t) = (1-t)|f|(\mu) + t|g|(\mu)$ ($t \in [0,1]$) makes sense and defines a homotopy from $|f|$ to $|g|$.

2.6 Exact Sequences

In this section, we develop techniques to obtain long (co)homology exact sequences from short exact sequences of (co)chain complexes. The results are used in several forthcoming sections. All vector spaces in this section are over a fixed arbitrary field $\mathbb{F}$.

Let (C_1^*, δ_1), (C_2^*, δ_2) and (C^*, δ) be cochain complexes of vector spaces, giving rise to cohomology graded vector spaces H_1^*, H_2^* and H^*. We consider morphisms of cochain complexes $J\colon C_1^* \to C^*$ and $I\colon C^* \to C_2^*$ so that

$$0 \to C_1^* \xrightarrow{J} C^* \xrightarrow{I} C_2^* \to 0 \tag{2.6.1}$$

is an exact sequence. We call (2.6.1) a *short exact sequence of cochain complexes*. Choose a **GrV**-morphism $S\colon C_2^* \hookrightarrow C^*$ which is a section of I. The section S cannot be assumed in general to be a morphism of cochain complexes. The linear map $\delta \circ S\colon C_2^m \to C^{m+1}$ satisfies

$$I \circ \delta \circ S(a) = \delta_2 \circ I \circ S(a) = \delta_2(a)\,,$$

thus $\delta \circ S(Z_2^m) \subset J(C_1^{m+1})$. We can then define a linear map $\tilde{\delta}^*\colon Z_2^m \to C_1^{m+1}$ by the equation

$$J \circ \tilde{\delta}^* = \delta \circ S\,. \tag{2.6.2}$$

If $a \in Z_2^m$, then $J \circ \delta_1(\tilde{\delta}^*(a)) = \delta \circ \delta(S(a)) = 0$. Therefore, $\tilde{\delta}^*(Z_2^m) \subset Z_1^{m+1}$. Moreover, if $b \in C_2^{m-1}$ and $a = \delta_2(b)$, then

$$I \circ \delta \circ S(b) = \delta_2 \circ I \circ S(b) = a\,,$$

whence $\delta \circ S(b) = S(a) + J(c)$ for some $c \in C_1^m$. Therefore $\tilde{\delta}^*(a) = \delta_1(c)$, which shows that $\tilde{\delta}^*(B_2^*) \subset B_1^*$. Hence, $\tilde{\delta}^*$ induces a linear map

$$\delta^* : H_2^* \to H_1^{*+1}$$

which is called the *cohomology connecting homomorphism* for the short exact sequence (2.6.1).

Lemma 2.6.1 *The connecting homomorphism $\delta^* : H_2^* \to H_1^{*+1}$ does not depend on the linear section S.*

Proof Let S': $C_2^m \to C^m$ be another section of I, giving rise to $\tilde{\delta}'^*$: $Z_2^m \to Z_1^{m+1}$, via the equation $J \circ \tilde{\delta}'^* = \delta \circ S'$. Let $a \in Z_2^m$. Then

$$S'(a) = S(a) + J(u)$$

for some $u \in C_1^m$. Therefore, the equations

$$J \circ \tilde{\delta}'^*(a) = \delta(S(a)) + \delta(J(u)) = \delta(S(a)) + J(\delta_1(u))$$

hold in C^{m+1}. This implies that $\tilde{\delta}'^*(a) = \tilde{\delta}^*(a) + \delta_1(u)$ in Z_1^{m+1}, and then $\delta'^*(a) = \delta^*(a)$ in H_1^{m+1}. □

Proposition 2.6.2 *The long sequence*

$$\cdots \to H_1^m \xrightarrow{H^*J} H^m \xrightarrow{H^*I} H_2^m \xrightarrow{\delta^*} H_1^{m+1} \xrightarrow{H^*J} \cdots$$

is exact.

The exact sequence of Proposition 2.6.2 is called the *cohomology exact sequence*, associated to the short exact of cochain complexes (2.6.1).

Proof The proof involves 6 steps.

1. $H^*I \circ H^*J = 0$ As $H^*I \circ H^*J = H^*(I \circ J)$, this comes from that $I \circ J = 0$.
2. $\delta^* \circ H^*I = 0$ Let $b \in Z^m$. Then $I(b + S(I(b))) = 0$. Hence, $b + S(I(b)) = J(c)$ for some $c \in C_1^m$. Therefore,

$$J \circ \tilde{\delta}^* \circ I(b) = \delta(S(I(b)) = \delta(b + J(c)) = \delta(b) + J \circ \delta_1(c) = J \circ \delta_1(c)\,,$$

   which proves that $\tilde{\delta}^* \circ I(b) = \delta_1(c)$, and then $\delta^* \circ H^*I = 0$ in H_1^*.
3. $H^*J \circ \delta^* = 0$ Let $a \in Z_2^m$. Then, $J \circ \tilde{\delta}^*(a) = \delta(S(a)) \subset B^{m+1}$, so $H^*J \circ \delta^*([a]) = 0$ in $H^{m+1}(K)$.
4. $\ker H^*J \subset \text{Image}\,\delta^*$ Let $a \in Z_1^{m+1}$ representing $[a] \in \ker H^*J$. This means that $J(a) = \delta(b)$ for some $b \in C^m$. Then, $I(b) \in Z_2^m$ and $S(I(b)) = b + J(c)$

for some $c \in C_1^m$. Therefore,

$$\delta \circ S \circ I(b) = \delta(b) + \delta(J(c)) = J(a) + J(\delta_1(c)) .$$

As J is injective, this implies that $\tilde{\delta}^*(I(c)) = a + \delta_1(c)$, proving that $\delta^*([I(c)]) = [a]$.

5. ker $H^*I \subset$ Image H^*J Let $a \in Z^m$ representing $[a] \in \ker H^*I$. This means that $I(a) = \delta_2(b)$ for some $b \in C_2^{m-1}$. Let $c = \delta(S(b)) \in C^m$. One has $I(a+c) = 0$, so $a+c = J(e)$ for some $e \in C_1^m$. As $\delta(a+c) = 0$ and J is injective, the cochain e is in Z_1^m. As $c \in B^m$, $H^*J([e]) = [a]$ in H^m.
6. ker $\delta^* \subset$ Image H^*I Let $a \in Z_2^m$ representing $[a] \in \ker \delta^*$. This means that $\tilde{\delta}^*(a) = \delta_1(b)$ for some $b \in C_1^m$. In other words,

$$\delta(S(a)) = J(\delta_1(b)) = \delta(J(b)) .$$

Hence, $c = J(b) + S(a) \in Z^m$ and $H^*I([c]) = [a]$. □

We now prove the naturality of the connecting homomorphism in cohomology. We are helped by the following intuitive interpretation of δ^*: first, we consider C_1^* a cochain subcomplex of C^* via the injection J. Second, a cocycle $a \in Z_2^m$ may be represented by a cochain in $\tilde{a} \in C^m$ such that $\delta(\tilde{a}) \in C_1^*$. Then, $\delta^*([a]) = [\delta(\tilde{a})]$. More precisely:

Lemma 2.6.3 *Let*

$$0 \to C_1^* \xrightarrow{J} C^* \xrightarrow{I} C_2^* \to 0$$

be a short exact sequence of cochain complexes. Then

(a) $I^{-1}(Z_2^m) = \{b \in C^m \mid \delta(b) \in J(C_1^{m+1})\}$.
(b) *Let* $a \in Z_2^m$ *representing* $[a] \in H_2^m$. *Let* $b \in C^m$ *with* $I(b) = a$. *Then* $\delta^*([a]) = [J^{-1}(\delta(b))]$ *in* H_1^{m+1}.

Proof Point (a) follows from the fact that I is surjective and from the equality $\delta_2 \circ I = I \circ \delta$. For Point (b), choose a section $S\colon C_2^m \to C_m$ of I. By Lemma 2.6.1, $\delta^*([a]) = [J^{-1}(\delta(S(a))]$. The equality $I(b) = a$ implies that $b = S(a) + J(c)$ for some $c \in C_1^m$. Therefore,

$$[J^{-1}(\delta(b))] = [J^{-1}(\delta \circ S(a))] + [\delta_1(c)] = \delta^*([a]) .$$ □

Let us consider a commutative diagram

$$\begin{array}{ccccccccc} 0 & \longrightarrow & \bar{C}_1^* & \xrightarrow{\bar{J}} & \bar{C}^* & \xrightarrow{\bar{I}} & \bar{C}_2^* & \longrightarrow & 0 \\ & & \downarrow{\scriptstyle F_1} & & \downarrow{\scriptstyle F} & & \downarrow{\scriptstyle F_2} & & \\ 0 & \longrightarrow & C_1^* & \xrightarrow{J} & C^* & \xrightarrow{I} & C_2^* & \longrightarrow & 0 \end{array} \tag{2.6.3}$$

of morphisms of cochain complexes, where the horizontal sequences are exact. This gives rise to two connecting homomorphisms $\bar{\delta}^*\colon \bar{H}_2^* \to \bar{H}_1^{*+1}$ and $\delta^*\colon H_2^* \to H_1^{*+1}$.

Lemma 2.6.4 (Naturality of the cohomology exact sequence) *The diagram*

$$\begin{array}{ccccccccccc}
\cdots \longrightarrow & \bar{H}_1^m & \xrightarrow{H^*\bar{J}} & \bar{H}^m & \xrightarrow{H^*\bar{I}} & \bar{H}_2^m & \xrightarrow{\bar{\delta}^*} & \bar{H}_1^{m+1} & \xrightarrow{H^*\bar{J}} & \cdots \\
& \downarrow{\scriptstyle H^*F_1} & & \downarrow{\scriptstyle H^*F} & & \downarrow{\scriptstyle H^*F_2} & & \downarrow{\scriptstyle H^*F_1} & & \\
\cdots \longrightarrow & H_1^m & \xrightarrow{H^*J} & H^m & \xrightarrow{H^*I} & H_2^m & \xrightarrow{\delta^*} & H_1^{m+1} & \xrightarrow{H^*J} & \cdots
\end{array}$$

is commutative.

Proof The commutativity of two of the square diagrams follows from the functoriality of the cohomology: $H^*F \circ H^*\bar{J} = H^*J \circ H^*F_1$ since $F \circ \bar{J} = J \circ F_1$ and $H^*F_2 \circ H^*\bar{I} = H^*I \circ H^*F$ since $F_2 \circ \bar{I} = I \circ F$. It remains to prove that $H^*F_1 \circ \bar{\delta}^* = H^*\delta^* \circ F_2$.

Let $a \in \bar{Z}_2^m$ representing $[a] \in \bar{H}_2^m$. Let $b \in \bar{C}^m$ with $\bar{I}(b) = a$. Then, $I \circ F(b) = F_2(a)$. Using Lemma 2.6.3, one has

$$\begin{aligned}
\delta^* \circ H^*F_2([a]) &= [J^{-1} \circ \delta \circ F(b)] \\
&= [J^{-1} \circ F \circ \bar{\delta}(b)] \\
&= [F_1 \circ \bar{J}^{-1} \circ \bar{\delta}(b)] \\
&= H^*F_1 \circ \bar{\delta}^*([a]) .
\end{aligned}$$

□

We are now interested in the case where the cochain complexes (C_i^*, δ_i) and (C^*, δ) are parts of Kronecker pairs

$$\mathcal{P}_1 = \big((C_1^*, \delta_1), (C_{*,1}, \partial_1), \langle\,,\rangle_1\big) \quad , \quad \mathcal{P}_2 = \big((C_2^*, \delta_2), (C_{*,2}, \partial_2)\langle\,,\rangle_2\big)$$

and

$$\mathcal{P} = \big((C^*, \delta), (C_*, \partial), \langle\,,\rangle\big) .$$

Let us consider two morphism of Kronecker pairs, (J, j) from $\mathcal{P}$ to $\mathcal{P}_1$ and (I, i) from $\mathcal{P}_2$ to $\mathcal{P}$. We suppose that the two sequences

$$0 \to C_1^* \xrightarrow{J} C^* \xrightarrow{I} C_2^* \to 0 \tag{2.6.4}$$

and

$$0 \to C_{*,2} \xrightarrow{i} C_* \xrightarrow{j} C_{*,1} \to 0 \tag{2.6.5}$$

are exact sequences of (co)chain complexes. Note that, by Lemma 2.3.8, (2.6.4) is exact if and only if (2.6.5) is exact. Exact sequence (2.6.4) gives rise to the cohomology connecting homomorphism $\delta^*\colon H_2^* \to H_1^{*+1}$. We construct a homology connecting homomorphism in the same way. Choose a linear section $s\colon C_{*,1} \to C_*$ of j, not required to be a morphism of chain complexes. As in the cohomology setting, one can defines $\tilde{\partial}_*\colon Z_{m+1,1} \to Z_{m,2}$ by the equation

$$i \circ \tilde{\partial}_* = \partial \circ s\,. \tag{2.6.6}$$

We check that $\tilde{\partial}_*(B_{m+1,1}) \subset B_{m,2}$. Hence $\tilde{\partial}_*$ induces a linear map

$$\partial_* : H_{*+1,1} \to H_{*,2}$$

called the *homology connecting homomorphism* for the short exact sequence (2.6.5).

Lemma 2.6.5 *The connecting homomorphism $\partial_*\colon H_{*+1,1} \to H_{*,2}$ does not depend on the linear section s.*

Proof The proof is analogous to that of Lemma 2.6.1 and is left as an exercise to the reader. □

Lemma 2.6.6 *The connecting homomorphisms $\delta^*\colon H_2^m \to H_1^{m+1}$ and $\partial_*\colon H_{m+1,1} \to H_{m,1}$ satisfy the equation*

$$\langle \delta^*(a), \alpha\rangle_1 = \langle a, \partial_*(\alpha)\rangle_2$$

for all $a \in H_2^m$, $\alpha \in H_{m+1,1}$ and all $m \in \mathbb{N}$. In other words, (δ^*, ∂_*) is a morphism of Kronecker pairs from $(H_1^*, H_{*,1}, \langle\,,\rangle_1)$ to $(H_2^*, H_{*,2}, \langle\,,\rangle_2)$.

Proof Let $\tilde{a} \in Z_2^m$ represent a and $\tilde{\alpha} \in Z_{m+1,1}$ represent α. Choose linear sections S and s of I and j. Using Formulae (2.6.2) and (2.6.6), one has

$$\begin{aligned}
\langle \delta^*(a), \alpha\rangle_1 &= \langle \tilde{\delta}^*(\tilde{a}), \tilde{\alpha}\rangle_1 \\
&= \langle \tilde{\delta}^*(\tilde{a}), j \circ s(\tilde{\alpha})\rangle_1 \\
&= \langle J \circ \tilde{\delta}^*(\tilde{a}), s(\tilde{\alpha})\rangle \\
&= \langle S(\tilde{a}), \partial \circ s(\tilde{\alpha})\rangle \\
&= \langle S(\tilde{a}), i \circ \tilde{\partial}_*(\tilde{\alpha})\rangle \\
&= \langle I \circ S(\tilde{a}), \tilde{\partial}_*(\tilde{\alpha})\rangle_2 \\
&= \langle \tilde{a}, \tilde{\partial}_*(\tilde{\alpha})\rangle_2 = \langle a, \partial_*(\alpha)\rangle_2\,.
\end{aligned}$$
□

Proposition 2.6.7 *The long sequence*

$$\cdots \to H_{m,2} \xrightarrow{H_*i} H_m \xrightarrow{H_*j} H_{m,1} \xrightarrow{\partial_*} H_{m-1,2} \xrightarrow{H_*i} \cdots$$

is exact.

The exact sequence of Proposition 2.6.7 is called the *homology exact sequence* associated to the short exact of chain complexes (2.6.5). It can be established directly, in an analogous way to that of Proposition 2.6.2. To make a change, we shall deduce Proposition 2.6.7 from Proposition 2.6.2 by Kronecker duality.

Proof By our hypotheses couples (I, i) and (J, j) are morphisms of Kronecker pairs, and so is (δ^*, ∂_*) by Lemma 2.6.6. Using Diagram (2.3.4), we get a commutative diagram

$$\begin{array}{ccccccccc}
\cdots \longleftarrow & (H_{m,2})^\sharp & \xleftarrow{(H_*i)^\sharp} & (H_m)^\sharp & \xleftarrow{(H_*j)^\sharp} & (H_{m,1})^\sharp & \xleftarrow{\partial_*^\sharp} & H^\sharp_{m-1,2} & \longleftarrow \cdots \\
& \approx\uparrow \mathbf{k} & & \approx\uparrow \mathbf{k} & & \approx\uparrow \mathbf{k} & & \approx\uparrow \mathbf{k} & \\
\cdots \longleftarrow & H_1^m & \xleftarrow{H^*I} & H^m & \xleftarrow{H^*J} & H_1^m & \xleftarrow{\delta^*} & H_2^{m-1} & \longleftarrow \cdots
\end{array}.$$

By Proposition 2.6.2, the bottom sequence of the above diagram is exact. Thus, the top sequence is exact. By Lemma 2.3.8, the sequence of Proposition 2.6.7 is exact. □

Let us consider commutative diagrams

$$\begin{array}{ccccccccc}
0 \longrightarrow & \bar{C}_1^* & \xrightarrow{\bar{J}} & \bar{C}^* & \xrightarrow{\bar{I}} & \bar{C}_2^* & \longrightarrow 0 \\
& \downarrow F_1 & & \downarrow F & & \downarrow F_2 & \\
0 \longrightarrow & C_1^* & \xrightarrow{J} & C^* & \xrightarrow{I} & C_2^* & \longrightarrow 0
\end{array} \qquad (2.6.7)$$

and

$$\begin{array}{ccccccccc}
0 \longleftarrow & \bar{C}_{*,1} & \xleftarrow{\bar{j}} & \bar{C}_* & \xleftarrow{\bar{i}} & \bar{C}_{*,2} & \longleftarrow 0 \\
& \uparrow f_1 & & \uparrow f & & \uparrow f_2 & \\
0 \longleftarrow & C_{*,1} & \xleftarrow{j} & C_* & \xleftarrow{i} & C_{*,2} & \longleftarrow 0
\end{array} \qquad (2.6.8)$$

such that the horizontal sequences are exact, F_i and F are morphisms of cochain complexes and f_i and f are morphisms of cochain complexes.

Lemma 2.6.8 (Naturality of the homology exact sequence) *Suppose that (F_i, f_i) and (F, f) are morphisms of Kronecker pairs. Then, the diagram*

$$\begin{array}{ccccccccc}
\cdots \longrightarrow & H_{m,2} & \xrightarrow{H_*i} & H_m & \xrightarrow{H_*j} & H_{m,1} & \xrightarrow{\partial_*} & H_{m-1,2} & \xrightarrow{H_*i} \cdots \\
& \downarrow H_*f_2 & & \downarrow H_*f & & \downarrow H_*f_1 & & \downarrow H_*f_2 & \\
\cdots \longrightarrow & \bar{H}_{m,2} & \xrightarrow{H_*\bar{i}} & \bar{H}_m & \xrightarrow{H_*\bar{j}} & \bar{H}_{m,1} & \xrightarrow{\partial_*} & \bar{H}_{m-1,2} & \xrightarrow{H_*\bar{i}} \cdots
\end{array}$$

is commutative.

Proof By functoriality of the homology, the square diagrams not involving ∂_* commute. It remains to show that $\bar{\partial}_* \circ H_* f_1 = H_* f_2 \circ \partial_*$. As $H^* F_1 \circ \bar{\delta}^* = \delta^* \circ H^* F_2$ by Lemma 2.6.4, one has

$$\langle a, \bar{\partial}_* \circ H_* f_1(\alpha)\rangle_2 = \langle H^* F_1 \circ \bar{\delta}^*(a), \alpha\rangle_1 = \langle \delta^* \circ H^* F_2(a), \alpha\rangle_1 = \langle a, H_* f_1 \circ \partial_*(\alpha)\rangle_2$$

for all $a \in \bar{H}_2^{m-1}$ and $\alpha \in H_{m,1}$. By Lemma 2.3.3, this implies that $\bar{\partial}_* \circ H_* f_1 = H_* f_2 \circ \partial_*$. □

2.7 Relative (Co)homology

A *simplicial pair* is a couple (K, L) where K is a simplicial complex and L is a subcomplex of K. The inclusion $i: L \hookrightarrow K$ is a simplicial map. Let $a \in C^m(K)$. If, using Definition 2.2.1a of Sect. 2.2, we consider a as a subset of $\mathcal{S}_m(K)$, then $C^* i(a) = a \cap \mathcal{S}_m(L)$. If we see a as a map $a: \mathcal{S}_m(K) \to \mathbb{Z}_2$, then $C^* i(a)$ is the restriction of a to $\mathcal{S}_m(L)$. We see that $C^* i: C^*(K) \to C^*(L)$ is surjective. Define

$$C^m(K, L) = \ker \left(C^m(K) \xrightarrow{C^* i} C^m(L)\right)$$

and $C^*(K, L) = \oplus_{m \in \mathbb{N}} C^m(K, L)$. This definition implies that

- $C^m(K, L)$ is the set of subsets of $\mathcal{S}_m(K) - \mathcal{S}_m(L)$;
- if K is a finite simplicial complex, $C^m(K, L)$ is the vector space with basis $\mathcal{S}_m(K) - \mathcal{S}_m(L)$.

As $C^* i$ is a morphism of cochain complexes, the coboundary $\delta: C^*(K) \to C^*(K)$ preserves $C^*(K, L)$ and gives rise to a coboundary $\delta: C^*(K, L) \to C^*(K, L)$ so that $(C^*(K, L), \delta)$ is a cochain complex. The cocycles $Z^*(K, L)$ and the coboundaries $B^*(K, L)$ are defined as usual, giving rise to the definition

$$H^m(K, L) = Z^m(K, L)/B^m(K, L).$$

The graded $\mathbb{Z}_2$-vector space $H^*(K, L) = \oplus_{m \in \mathbb{N}} H^m(K, L)$ is the *simplicial relative cohomology* of the simplicial pair (K, L).

When useful, the notations δ_K, δ_L and $\delta_{K,L}$ are used for the coboundaries of the cochain complexes $C^*(K)$, $C^*(L)$ and $C^*(K, L)$. We denote by j^* the inclusion j^*: $C^*(K, L) \hookrightarrow C^*(K)$, which is a morphism of cochain complexes, and use the same notation j^* for the induced linear map $j^*: H^*(K, L) \to H^*(K)$ on cohomology. We also use the notation i^* for both $C^* i$ and $H^* i$. We get thus a short exact sequence of cochain complexes

$$0 \to C^*(K, L) \xrightarrow{j^*} C^*(K) \xrightarrow{i^*} C^*(L) \to 0. \qquad (2.7.1)$$

If $a \in C^m(L)$, any cochain $\bar{a} \in C^m(K)$ with $i^*(\bar{a}) = a$ is called a *extension of* a as a cochain in K. For instance, the 0*-extension* of a is defined by $\bar{a} = a \in \mathcal{S}_m(L) \subset \mathcal{S}_m(K)$. Using Sect. 2.6, Exact sequence (2.7.1) gives rise to a (*simplicial cohomology*) *connecting homomorphism*

$$\delta^* : H^*(L) \to H^{*+1}(K, L)\,.$$

It is induced by a linear map $\tilde{\delta}^*\colon Z^m(L) \to Z^{m+1}(K, L)$ characterized by the equation $j^* \circ \tilde{\delta}^* = \delta_K \circ S$ for some (or any) linear section $S\colon C^m(L) \to C^m(K)$ of i^*, not required to be a morphism of cochain complex. For instance, one can take $S(a)$ to be the 0-extension of a. Using that $C^*(K, L)$ is a chain subcomplex of $C^*(K)$, the following statement makes sense and constitutes a useful recipe for computing the connecting homomorphism δ^*.

Lemma 2.7.1 *Let $a \in Z^m(L)$ and let $\bar{a} \in C^m(K)$ be any extension of a as an m-cochain of K. Then, $\delta_K(\bar{a})$ is an $(m+1)$-cocycle of (K, L) representing $\delta^*(a)$.*

Proof Choose a linear section $S\colon C^m(L) \to C^m(K)$ such that $S(a) = \bar{a}$. The equation $j^* \circ \tilde{\delta}^* = \delta_K \circ S$ proves the lemma. □

We can now use Proposition 2.6.2 and get the following result.

Proposition 2.7.2 *The long sequence*

$$\cdots \to H^m(K, L) \xrightarrow{j^*} H^m(K) \xrightarrow{i^*} H^m(L) \xrightarrow{\delta^*} H^{m+1}(K, L) \xrightarrow{j^*} \cdots$$

is exact.

The exact sequence of Proposition 2.7.2 is called the *simplicial cohomology exact sequence*, or just the *simplicial cohomology sequence*, of the simplicial pair (K, L).

We now turn our interest to homology. The inclusion $L \hookrightarrow K$ induces an inclusion $i_*\colon C_*(L) \hookrightarrow C_*(K)$ of chain complexes. We define $C_m(K, L)$ as the quotient vector space

$$C_m(K, L) = \operatorname{coker}\big(i_* : C_m(L) \hookrightarrow C_m(K)\big)\,.$$

As i_* is a morphism of chain complexes, $C_*(K, L) = \oplus_{m\in\mathbb{N}} C_m(K, L)$ inherits a boundary operator $\partial = \partial_{K,L}\colon C_*(K, L) \to C_{*-1}(K, L)$. The projection $j_*\colon C_*(K) \to C_*(K, L)$ is a morphism of chain complexes and one gets a short exact sequence of chain complexes

$$0 \to C_*(L) \xrightarrow{i_*} C_*(K) \xrightarrow{j_*} C_*(K, L) \to 0\,. \tag{2.7.2}$$

The cycles and boundaries $Z_*(K, L)$ and $B_*(K, L)$ are defined as usual, giving rise to the definition

$$H_m(K,L) = Z_m(K,L)/B_m(K,L)\,.$$

The graded $\mathbb{Z}_2$-vector space $H_*(K,L) = \oplus_{m\in\mathbb{N}} H_m(K,L)$ is the *relative homology* of the simplicial pair (K,L). As before, the notations ∂_K and ∂_L may be used for the boundary operators in $C_*(K)$ and $C_*(L)$ and i_* and j_* are also used for the induced maps in homology.

Since the linear map $i_*\colon C_*(L) \hookrightarrow C_*(K)$ is induced by the inclusion of bases $\mathcal{S}(L) \hookrightarrow \mathcal{S}(K)$, the quotient vector space $C_*(K,L)$ may be considered as the vector space with basis $\mathcal{S}(K) - \mathcal{S}(L)$. This point of view provides a tautological linear map $s\colon C_*(K,L) \to C_*(K)$, which is a section of j_* but not a morphism of chain complexes.

The Kronecker pairings for K and L are denoted by $\langle\,,\rangle_K$ and $\langle\,,\rangle_L$, both at the levels of (co)chains and of (co)homology. As $\langle j^*(K,L), i_*(L)\rangle_K = 0$, we get a bilinear map

$$C^m(K,L) \times C_m(K,L) \xrightarrow{\langle,\rangle_{K,L}} \mathbb{Z}_2\,.$$

The formula

$$\langle a,\alpha\rangle_{K,L} = \langle j^*(a), s(\alpha)\rangle_K \tag{2.7.3}$$

holds for all $a \in C^m(K,L)$, $\alpha \in C_m(K,L)$ and all $m \in \mathbb{N}$. Observe also that the formula

$$\langle S(b), i_*(\beta)\rangle_K = \langle b,\beta\rangle_L \tag{2.7.4}$$

holds for all $b \in C^m(L)$, $\beta \in C_m(L)$ and all $m \in \mathbb{N}$.

Lemma 2.7.3 $\big(C^*(K,L), \delta_{K,L}, C_*(K,L), \partial_{K,L}, \langle\,,\rangle_{K,L}\big)$ *is a Kronecker pair.*

Proof We first prove that $\langle \delta_{K,L}(a),\alpha\rangle_{K,L} = \langle a, \partial_{K,L}(\alpha)\rangle_{K,L}$ for all $a \in C^m(K,L)$ and all $\alpha \in C_{m+1}(K,L)$ and all $m \in \mathbb{N}$. Indeed, one has

$$\begin{aligned}\langle \delta_{K,L}(a),\alpha\rangle_{K,L} &= \langle j^*\circ\delta_{K,L}(a), s(\alpha)\rangle_K \\ &= \langle \delta_K\circ j^*(a), s(\alpha)\rangle_K \\ &= \langle j^*(a), \partial_K\circ s(\alpha)\rangle_K\end{aligned} \tag{2.7.5}$$

Observe that $j_*\circ\partial_K\circ s(\alpha) = \partial_{K,L}(\alpha)$ and therefore $\partial_K\circ s(\alpha) = s\circ\partial_{K,L}(\alpha) + i_*(c)$ for some $c \in C_m(L)$. Hence, the chain of equalities in (2.7.5) may be continued

$$\begin{aligned}\langle \delta_{K,L}(a),\alpha\rangle_{K,L} &= \langle j^*(a), \partial_K\circ s(\alpha)\rangle_K \\ &= \langle j^*(a), s\circ\partial_{K,L}(\alpha) + i_*(c)\rangle_K \\ &= \langle j^*(a), s\circ\partial_{K,L}(\alpha)\rangle_K + \underbrace{\langle j^*(a), i_*(c)\rangle_K}_{0} \\ &= \langle a, \partial_{K,L}(\alpha)\rangle_{K,L}\,.\end{aligned} \tag{2.7.6}$$

It remains to prove that the linear map $\mathbf{k}$: $C^*(K,L) \to C_*(K,L)^\sharp$ given by $\mathbf{k}(a) = \langle a, \; \rangle$ is an isomorphism. As the inclusion $i: L \hookrightarrow K$ is a simplicial map, the couple (C^*i, C_*i) is a morphism of Kronecker pairs by Lemma 2.5.1 and the result follows from Lemma 2.3.9. □

Passing to homology then produces three Kronecker pairs with vanishing (co)boundary operators:

$$\mathcal{P}_L = (H^*(L), H_*(L), \langle , \rangle_L), \quad \mathcal{P}_K = (H^*(K), H_*(K), \langle , \rangle_K)$$

and

$$\mathcal{P}_{K,L} = (H^*(K,L), H_*(K,L), \langle , \rangle_{K,L}) \,.$$

Using Sect. 2.6, short exact sequence (2.7.2) gives rise to the (*simplicial homology*) *connecting homomorphism*

$$\partial_* : H_*(K,L) \to H_{*-1}(L) \,.$$

It is induced by a linear map $\tilde{\partial}$: $Z_m(K,L) \to Z_{m-1}(L)$ characterized by the equation

$$j^* \circ \tilde{\partial}_* = \partial_K \circ s \,,$$

using the section s of j_* defined above (or any other one).

Lemma 2.7.4 *The following couples are morphisms of Kronecker pairs:*

(a) (i^*, i_*), *from* $\mathcal{P}_L$ *to* $\mathcal{P}_K$.
(b) (j^*, j_*), *from* $\mathcal{P}_K$ *to* $\mathcal{P}_{K,L}$.
(c) (δ^*, ∂_*), *from* $\mathcal{P}_{K,L}$ *to* $\mathcal{P}_L$.

Proof As the inclusion $L \hookrightarrow K$ is a simplicial map, Point (a) follows from Lemma 2.5.1. Point (c) is implied by Lemma 2.6.6. To prove Point (b), let $a \in C^m(K,L)$ and $\alpha \in C_m(K)$. Observe that $s(j_*(\alpha)) = \alpha + i_*(\beta)$ for some $\beta \in C_m(L)$ and that $\langle j^*(a), i_*(\beta)\rangle_K = 0$. Therefore:

$$\langle a, j_*(\alpha)\rangle_{K,L} = \langle j^*(a), s \circ j_*(\alpha)\rangle_K = \langle j^*(a), \alpha\rangle_K$$ □

Proposition 2.6.7 now gives the following result.

Proposition 2.7.5 *The long sequence*

$$\cdots \to H_m(L) \xrightarrow{i_*} H_m(K) \xrightarrow{j_*} H^m(K,L) \xrightarrow{\partial_*} H_{m-1}(L) \xrightarrow{i_*} \cdots$$

is exact.

The exact sequence of Proposition 2.7.5 is called the *(simplicial) homology exact sequence*, or just the *(simplicial) cohomology sequence*, of the simplicial pair (K, L).

We now study the naturality of the (co)homology sequences. Let (K, L) and (K', L') be simplicial pairs. A *simplicial map f of simplicial pairs* from (K, L) to (K', L') is a simplicial map $f_K\colon K \to K'$ such that the restriction of f to L is a simplicial map $f_L\colon L \to L'$. The morphism $C^* f_K\colon C^*(K') \to C^*(K)$ then restricts to a morphism of cochain complexes $C^* f\colon C^*(K', L') \to C^*(K, L)$ and the morphism $C_* f_K\colon C_*(K) \to C_*(K')$ descends to a morphism of chain complexes $C_* f\colon C_*(K, L) \to C_*(K', L')$. The couples $(C^* f_K, C_* f_K)$ and $(C^* f_L, C_* f_L)$ are morphisms of Kronecker pairs by Lemma 2.5.1. We claim that $(C_* f, C_* f)$ is a morphism of Kronecker pair from $(C^*(K, L), \dots)$ to $(C^*(K', L'), \dots)$. Indeed, let $a \in C^m(K', L')$ and $\alpha \in C_m(K, L)$. One has

$$\begin{aligned} \langle C^* f(a), \alpha\rangle_{K,L} &= \langle j^* \circ C^* f(a), s(\alpha)\rangle_K \\ &= \langle C^* f_K \circ j'^*(a), s(\alpha)\rangle_K \\ &= \langle j'^*(a), C_* f_K \circ s(\alpha)\rangle_{K'} \\ &= \langle j'^*(a), C_* f_K \circ s(\alpha)\rangle_{K'} \end{aligned} \tag{2.7.7}$$

and

$$\langle a, C_* f(\alpha)\rangle_{K',L'} = \langle j'^*(a), s' \circ C_* f(\alpha)\rangle_{K'} \tag{2.7.8}$$

The equation $j'_* \circ s' \circ C_* f(\alpha) = j'_* \circ C_* f_K \circ s(\alpha) = *f(\alpha)$ implies that $s' \circ C_* f(\alpha) = C_* f_K \circ s(\alpha) + i'_*(\beta)$ for some $\beta \in C_m(L')$. As $\langle j'^*(a), i'_*(\beta)\rangle_{K'} = 0$, Equations (2.7.7) and (2.7.8) imply that $\langle C^* f(a), \alpha\rangle_{K,L} = \langle a, C_* f(\alpha)\rangle_{K',L'}$.

Lemmas 2.6.4 and 2.6.8 then imply the following

Proposition 2.7.6 *The cohomology and homology sequences are natural with respect to simplicial maps of simplicial pairs. In other words, given a simplicial map of simplicial pairs $f\colon (K, L) \to (K', L')$, the following diagrams*

$$\begin{array}{ccccccccccc} \cdots \longrightarrow & H^m(K', L') & \xrightarrow{j'^*} & H^m(K') & \xrightarrow{i'^*} & H^m(L') & \xrightarrow{\delta'^*} & H^{m+1}(K', L') & \xrightarrow{j'^*} & \cdots \\ & \downarrow H^* f & & \downarrow H^* f_K & & \downarrow H^* f_L & & \downarrow H^* f & & \\ \cdots \longrightarrow & H^m(K, L) & \xrightarrow{j^*} & H^m(K) & \xrightarrow{i^*} & H^m(L) & \xrightarrow{\delta^*} & H^{m+1}(K, L) & \xrightarrow{j^*} & \cdots \end{array}$$

and

$$\begin{array}{ccccccccccc} \cdots \longrightarrow & H_m(L) & \xrightarrow{i^*} & H_m(K) & \xrightarrow{j^*} & H_m(K, L) & \xrightarrow{\partial_*} & H_{m-1}(L) & \xrightarrow{i^*} & \cdots \\ & \downarrow H_* f_L & & \downarrow H_* f_K & & \downarrow H_* f & & \downarrow H_* f_L & & \\ \cdots \longrightarrow & H_m(L') & \xrightarrow{i'_*} & H^m(K') & \xrightarrow{j'_*} & H_m(K', L') & \xrightarrow{\partial'_*} & H_{m-1}(L') & \xrightarrow{i'_*} & \cdots \end{array}$$

are commutative.

We finish this section by the exact sequences for a triple. A *simplicial triple* is a triplet (K, L, M) where K is a simplicial complex, L is a subcomplex of K and M is a subcomplex of L. A *simplicial map f of simplicial triples*, from (K, L, M) to (K', L', M') is a simplicial map $f_K\colon K \to K'$ such that the restrictions of f_K to L and M are simplicial maps $f_L\colon L \to L'$ and $f_M\colon M \to M'$.

A simplicial triple $T = (K, L, M)$ gives rise to pair inclusions

$$(L, M) \xrightarrow{i} (K, M) \xrightarrow{j} (K, L)$$

and to a commutative diagram

$$\begin{array}{ccccccccc} 0 & \longrightarrow & C^*(K,L) & \xrightarrow{j^*_{K,L}} & C^*(K) & \xrightarrow{i^*_{K,L}} & C^*(L) & \longrightarrow & 0 \\ & & \big\downarrow{\scriptstyle C^*j} & & \big\updownarrow{=}\,{\scriptstyle \mathrm{id}} & & \big\downarrow{\scriptstyle i^*_{L,M}} & & \\ 0 & \longrightarrow & C^*(K,M) & \xrightarrow{j^*_{K,M}} & C^*(K) & \xrightarrow{i^*_{K,M}} & C^*(M) & \longrightarrow & 0 \end{array} \qquad (2.7.9)$$

where the horizontal lines are exact sequences of cochain complexes. A diagram-chase shows that the morphism $i^*_{K,L} \circ j^*_{K,M}$, which sends $C^*(K, M)$ to $C^*(L)$, has image $C^*(L, M)$ and kernel the image of C^*j. This morphism coincides with C^*i. We thus get a short exact sequence of cochain complexes

$$0 \to C^*(K, L) \xrightarrow{C^*j} C^*(K, M) \xrightarrow{C^*i} C^*(L, M) \to 0. \qquad (2.7.10)$$

The same arguments with the chain complexes gives a short exact sequence

$$0 \to C_*(L, M) \xrightarrow{C_*i} C_*(K, M) \xrightarrow{C_*j} C_*(K, L) \to 0. \qquad (2.7.11)$$

As above in this section, short exact sequences (2.7.10) and (2.7.11) produces connecting homomorphisms $\delta_T\colon H^*(L, M) \to H^{*+1}(K, L)$ and $\partial_T : H_*(K, L) \to C_{*-1}(L, M)$. They satisfy $\langle \delta_T(a), \alpha\rangle = \langle a, \partial_T(\alpha)\rangle$ as well as following proposition.

Proposition 2.7.7 ((Co)homology sequences of a simplicial triple) *Let $T = (K, L, M)$ be a simplicial triple. Then,*

(a) *the sequences*

$$\cdots \to H^m(K, L) \xrightarrow{H^*j} H^m(K, M) \xrightarrow{H^*i} H^m(L, M) \xrightarrow{\delta_T} H^{m+1}(K, L) \xrightarrow{H^*j} \cdots$$

and

$$\cdots \to H_m(L, M) \xrightarrow{H_*i} H_m(K, M) \xrightarrow{H_*j} H_m(K, L) \xrightarrow{\partial_T} H_{m-1}(L, M) \xrightarrow{H_*i} \cdots$$

are exact.

(b) *the exact sequences of Point (a) are natural for simplicial maps of simplicial triples.*

Remark 2.7.8 As $H^*(\emptyset) = 0$, we get a canonical **GrV**-isomorphisms $H^*(K, \emptyset) \xrightarrow{\approx} H^*(K)$, etc. Thus, the (co)homology sequences for the triple $(K, L, \emptyset)$ give back those of the pair (K, L)

$$\cdots \to H^m(K, L) \xrightarrow{H^*j} H^m(K) \xrightarrow{H^*i} H^m(L) \xrightarrow{\delta^*} H^{m+1}(K, L) \xrightarrow{H^*j} \cdots \tag{2.7.12}$$

and

$$\cdots \to H_m(L) \xrightarrow{H_*i} H_m(K) \xrightarrow{H_*j} H^m(K, L) \xrightarrow{\partial_*} H_{m-1}(L) \xrightarrow{H_*i} \cdots \tag{2.7.13}$$

where $i \colon L \to K$ and $j \colon (K, \emptyset) \to (K, L)$ denote the inclusions. This gives a more precise description of the morphisms j^* and j_* of Propositions 2.7.2 and 2.7.5.

2.7.9 *Historical note.* The relative homology was introduced by S. Lefschetz in 1927 in order to work out the Poincaré duality for manifolds with boundary (see, e.g. [40, p. 58], [51, p. 47]). The use of exact sequences occurred in several parts of algebraic topology after 1941 (see, e.g. [40, p. 86], [51, p. 47]). The (co)homology exact sequences play an essential role in the axiomatic approach of Eilenberg-Steenrod, [51].

2.8 Mayer-Vietoris Sequences

Let K be a simplicial complex with two subcomplexes K_1 and K_2. We suppose that $K = K_1 \cup K_2$ (i.e. $\mathcal{S}(K) = \mathcal{S}(K_1) \cup \mathcal{S}(K_2)$). We call (K, K_1, K_2) a *simplicial triad*. Then, $K_0 = K_1 \cap K_2$ is a subcomplex of K_1, K_2 and K, with $\mathcal{S}(K_0) = \mathcal{S}(K_1) \cap \mathcal{S}(K_2)$. The Mayer-Vietoris sequences relate the (co)homology of X to that of X_i, generalizing Lemma 2.4.10. The various inclusions are denoted as follows

$$\begin{array}{ccc} K_0 & \xrightarrow{i_1} & K_1 \\ {\scriptstyle i_2}\downarrow & & \downarrow{\scriptstyle j_1} \\ K_2 & \xrightarrow{j_2} & K\,. \end{array} \tag{2.8.1}$$

The notations i_1^*, j_1^*, ..., stand for both C^*i_1, C^*j_1, etc, and H^*i_1, H^*j_1, etc. The same holds for chains and homology: i_{1*} for both C_*i_1 and H_*i_1, etc. Diagram (2.8.1) induces two diagrams

$$\begin{array}{ccc} C^*(K) & \xrightarrow{j_1^*} & C^*(K_1) \\ {\scriptstyle j_2^*}\downarrow & & \downarrow{\scriptstyle i_1^*} \\ C^*(K_2) & \xrightarrow{i_2^*} & C^*(K_0) \end{array} \quad \text{and} \quad \begin{array}{ccc} C_*(K_0) & \xrightarrow{i_{1*}} & C_*(K_1) \\ {\scriptstyle i_{2*}}\downarrow & & \downarrow{\scriptstyle j_{1*}} \\ C_*(K_2) & \xrightarrow{j_{2*}} & C_*(K)\,. \end{array}$$

The cohomology diagram is Cartesian (pullback) and the homology diagram is co-Cartesian (pushout). Therefore, the sequence

$$0 \to C^*(K) \xrightarrow{(j_1^*, j_2^*)} C^*(K_1) \oplus C^*(K_2) \xrightarrow{i_1^* + i_2^*} C^*(K_0) \to 0 \tag{2.8.2}$$

is an exact sequence of cochain complexes and the sequence

$$0 \to C_*(K_0) \xrightarrow{(i_{1*}, i_{2*})} C_*(K_1) \oplus C_*(K_2) \xrightarrow{j_{1*} + j_{2*}} C_*(K) \to 0 \tag{2.8.3}$$

is an exact sequence of chain complexes.

Consider the Kronecker pairs $(C^*(K_i), C_*(K_i), \langle\,,\,\rangle_i)$ for $i = 0, 1, 2$, and the Kronecker pair $(C^*(K), C_*(K), \langle\,,\,\rangle)$. A bilinear map

$$\langle\,,\,\rangle_\oplus : \left[C^*(K_1) \oplus C^*(K_2)\right] \times \left[C_*(K_1) \oplus C_*(K_2)\right] \to \mathbb{Z}_2$$

is defined by

$$\langle (a_1, a_2), (\alpha_1, \alpha_2) \rangle_\oplus = \langle a_1, \alpha_1 \rangle_1 + \langle a_2, \alpha_2 \rangle_2 \,.$$

We check that $(C^*(K_1) \oplus C^*(K_2), C_*(K_1) \oplus C_*(K_2), \langle\,,\,\rangle_\oplus)$ is a Kronecker pair and that the couples $((j_1^*, j_2^*), j_1^* + j_2^*)$ and $(i_1^* + i_2^*, (i_1^*, i_2^*))$ are morphisms of Kronecker pairs. By Sect. 2.6, there exist linear maps $\delta_{MV}\colon H^*(K_0) \to H^{*+1}(K)$ and $\partial_{MV}\colon H_*(K) \to H_{*-1}(K_0)$ which, by Propositions 2.6.2 and 2.6.7, give the following proposition.

Proposition 2.8.1 (Mayer-Vietoris sequences) *The long sequences*

$$\cdots \to H^m(K) \xrightarrow{(j_1^*, j_2^*)} H^m(K_1) \oplus H^m(K_2) \xrightarrow{i_1^* + i_2^*} H^m(K_0) \xrightarrow{\delta_{MV}} H^{m+1}(K) \to \cdots$$

and

$$\cdots \to H_m(K_0) \xrightarrow{(i_{1*}, i_{2*})} H_m(K_1) \oplus H_m(K_2) \xrightarrow{j_{1*} + j_{2*}} H_m(K) \xrightarrow{\partial_{MV}} H_{m-1}(K_0) \to \cdots$$

are exact.

The homomorphisms δ_{MV} and ∂_{MV} are called the *Mayer-Vietoris connecting homomorphisms* in (co)homology. By Lemma 2.6.6, they satisfy $\langle \delta_{MV}(a), \alpha \rangle =$

$\langle a, \partial_{MV}(\alpha)\rangle_0$ for all $a \in H^m(K_0)$, all $\alpha \in H_{m+1}(k)$ and all $m \in \mathbb{N}$. To define the connecting homomorphisms, one must choose a linear section S of $i_1^* + i_2^*$ and s of $j_{1*} + j_{2*}$. One can choose $S(a) = (S_1(a), 0)$, where $S_1\colon C^*(K) \to C^*(K_1)$ is the tautological section of i_1^* given by the inclusion $\mathcal{S}(K_0) \hookrightarrow \mathcal{S}(K_1)$ (see Sect. 2.7). A choice of s is given, for $\sigma \in \mathcal{S}(K)$, by

$$s(\sigma) = \begin{cases} (\sigma, 0) & \text{if } \sigma \in \mathcal{S}(K_1) \\ (0, 0) & \text{if } \sigma \notin \mathcal{S}(K_1)\,. \end{cases}$$

These choices produce linear maps $\tilde{\delta}_{MV}\colon Z^*(K_0) \to Z^{*+1}(K)$ and $\tilde{\partial}_{MV}\colon Z_*(K) \to Z_{*-1}(K_0)$, representing δ_{MV} and ∂_{MV} and defined by the equations

$$(j_1^*, j_2^*)\circ\tilde{\delta}_{MV} = (\delta_1, \delta_2)\circ S \quad\text{and}\quad (i_{1*}, i_{2*})\circ\tilde{\partial}_{MV} = (\partial_1, \partial_2)\circ s\,.$$

(The apparent asymmetry of the choices has no effect by Lemma 2.6.1 and its homology counterpart: exchanging 1 and 2 produces other sections, giving rise to the same connecting homomorphisms.)

Finally, the Mayer-Vietoris sequences are natural for maps of simplicial triads. If $\mathcal{T} = (K, K_1, K_2)$ and $\mathcal{T}' = (K', K_1', K_2')$ are simplicial triads and if $f\colon K \to K'$ is a simplicial map such that $f(K_i) \subset K_i'$, then the Mayer Vietoris sequences of $\mathcal{T}$ and $\mathcal{T}'$ are related by commutative diagrams, as in Proposition 2.7.6. This is a direct consequence of Lemmas 2.6.4 and 2.6.8.

2.9 Appendix A: An Acyclic Carrier Result

The powerful technique of acyclic carriers was introduced by Eilenberg and MacLane in 1953 [50], after earlier work by Lefschetz. Proposition 2.9.1 below is a very particular example of this technique, adapted to our needs. For a full development of acyclic carriers, see, e.g., [155, Chap. 1,Sect. 13].

Let (C_*, ∂) and $(\bar{C}_*, \bar{\partial})$ be two chain complexes and let $\varphi\colon C_* \to \bar{C}_*$ be a morphism of chain complexes. We suppose that C_m is equipped with a basis $\mathcal{S}_m$ for each m and denote by $\mathcal{S}$ the union of all $\mathcal{S}_m$. An *acyclic carrier* A_* for φ with respect to the basis $\mathcal{S}$ is a correspondence which associates to each $s \in \mathcal{S}$ a subchain complex $A_*(s)$ of $\bar{C}_*$ such that

(a) $\varphi(s) \in A_*(s)$.
(b) $H_0(A_*(s)) = \mathbb{Z}_2$ and $H_m(A_*(s)) = 0$ for $m > 0$.
(c) let $s \in \mathcal{S}_m$ and $t \in \mathcal{S}_{m-1}$ such that t occurs in the expression of ∂s in the basis $\mathcal{S}_{m-1}$. Then $A_*(t)$ is a subchain complex of $A_*(s)$ and the inclusion $A_*(t) \subset A_*(s)$ induces an isomorphism on H_0.
(d) if $s \in \mathcal{S}_0 \subset C_0 = Z_0$, then $H_0\varphi(s) \neq 0$ in $H_0(A_*(s))$.

Proposition 2.9.1 *Let φ and φ' be two morphisms of chain complexes from (C_*, ∂) to $(\bar{C}_*, \bar{\partial})$. Suppose that φ and φ' admit the same acyclic carrier A_* with respect to some basis $\mathcal{S}$ of C_*. Then $H_*\varphi = H_*\varphi'$.*

Proof The proof is similar to that of Proposition 2.5.9. By induction on m, we shall prove the following property:

Property $\mathcal{H}(m)$: there exists a linear map $D\colon C_m \to \bar{C}_{m+1}$ such that:

(i) $\bar{\partial}D(\alpha) + D(\partial\alpha) = \varphi(\alpha) + \varphi'(\alpha)$ for all $\alpha \in C_m$.
(ii) for each $s \in \mathcal{S}_m$, $D(s) \in A_{m+1}(s)$.

Property $\mathcal{H}(m)$ for all m implies that $H_*\varphi = H_*\varphi'$. Indeed, we then have a linear map $D\colon C_* \to \bar{C}_{*+1}$ satisfying

$$\varphi + \varphi' = \bar{\partial}\circ D + D\circ\partial\,. \tag{2.9.1}$$

Let $\beta \in Z_*$. By Eq. (2.9.1), one has $\varphi(\beta) + \varphi(\beta) = \bar{\partial}D(\beta)$ which implies that $H_*\varphi([\beta]) + H_*\varphi'([\beta])$ in $\bar{H}_*$.

Let us prove $\mathcal{H}(0)$. Let $s \in \mathcal{S}_0$. In $H_0(A_*(s)) = \mathbb{Z}_2$, one has $H_0\varphi(s) \neq 0$ and $H_0\varphi'(s) \neq 0$. Therefore $H_*\varphi(s) = H_*\varphi'(s)$ in $H_0(A_*(s))$. This implies that $\varphi(s) + \varphi'(s) = \bar{\partial}(\eta_s)$ for some $\eta_s \in A_1(s)$. We set $D(s) = \eta_s$. This procedure, for each $s \in \mathcal{S}_0$, provides a linear map $D\colon C_0 \to \bar{C}_1$, which, as $\partial C_0 = 0$, satisfies $\varphi(s) + \varphi'(s) = \bar{\partial}D(\alpha) + D(\partial(\alpha))$.

We now prove that $\mathcal{H}(m-1)$ implies $\mathcal{H}(m)$ for $m \geq 1$. Let $s \in \mathcal{S}_m$. The chain $D(\partial s)$ exists in $A_m(s)$ by $\mathcal{H}(m-1)$. Let $\zeta \in A_m(s)$ defined by

$$\zeta = \varphi(s) + \varphi'(s) + D(\partial s)$$

Using $\mathcal{H}(m-1)$, one checks that $\partial\zeta = 0$. Since $H_m(A_*(s)) = 0$, there exists $\nu \in A_{m+1}(s)$ such that $\zeta = \partial\nu$. Choose such an element ν and set $D(\sigma) = \nu$. This defines $D\colon C_m \to \bar{C}_{m+1}$ which satisfies (i) and (ii), proving $\mathcal{H}(m)$. □

2.10 Appendix B: Ordered Simplicial (Co)homology

This technical section may be skipped in a first reading. It shows that simplicial (co)homology may be defined using larger sets of (co)chains, based on ordered simplexes. This will be used for comparisons between simplicial and singular (co)homology (see § 17) and to define the cup and cap products in Chap. 4.

Let K be a simplicial complex. Define

$$\hat{\mathcal{S}}_m(K) = \{(v_0, \dots, v_m) \in V(K)^{m+1} \mid \{v_0, \dots, v_m\} \in \mathcal{S}(K)\}\,.$$

Observe that $\dim\{v_0, \dots, v_m\} \leq m$ and may be strictly smaller if there are repetitions amongst the v_i's. An element of $\hat{\mathcal{S}}_m(K)$ is an *ordered m-simplex* of K.

The definitions of ordered (co)chains and (co)homology are the same those for the simplicial case (see Sect. 2.2), replacing the simplexes by the ordered simplexes. We thus set

Definition 2.10.1 (*subset definitions*)

(a) An *ordered m-cochain* is a subset of $\hat{\mathcal{S}}_m(K)$.
(b) An *ordered m-chain* is a finite subset of $\hat{\mathcal{S}}_m(K)$.

The set of ordered m-cochains of K is denoted by $\hat{C}^m(K)$ and that of ordered m-chains by $\hat{C}_m(K)$. As in Sect. 2.2, Definition 2.10.1 are equivalent to

Definition 2.10.2 (*colouring definitions*)

(a) An *ordered m-cochain* is a function $a\colon \hat{\mathcal{S}}_m(K) \to \mathbb{Z}_2$.
(b) An *ordered m-chain* is a function $\alpha\colon \hat{\mathcal{S}}_m(K) \to \mathbb{Z}_2$ with finite support.

Definition 2.10.2 endow $\hat{C}^m(K)$ and $\hat{C}_m(K)$ with a structure of a $\mathbb{Z}_2$-vector space. The singletons provide a basis of $\hat{C}_m(K)$, in bijection with $\hat{\mathcal{S}}_m(K)$. Thus, Definition 2.10.2.b is equivalent to

Definition 2.10.3 $\hat{C}_m(K)$ is the $\mathbb{Z}_2$-vector space with basis $\hat{\mathcal{S}}_m(K)$:

$$\hat{C}_m(X) = \bigoplus_{\sigma \in \hat{\mathcal{S}}_m(X)} \mathbb{Z}_2\, \sigma\,.$$

We consider the graded $\mathbb{Z}_2$-vector spaces $\hat{C}_*(K) = \oplus_{m\in\mathbb{N}} \hat{C}_m(K)$ and $\hat{C}^*(K) = \oplus_{m\in\mathbb{N}} \hat{C}^m(K)$. The *Kronecker pairing* on ordered (co)chains

$$\hat{C}^m(K) \times \hat{C}_m(K) \xrightarrow{\langle\,,\,\rangle} \mathbb{Z}_2$$

is defined, using the various above definitions, by the equivalent formulae

$$\begin{aligned} \langle a, \alpha\rangle &= \sharp(a \cap \alpha) \pmod 2 && \text{using Definition 2.10.1a and b} \\ &= \textstyle\sum_{\sigma\in\alpha} a(\sigma) && \text{using Definitions 2.10.1a and 2.10.2b} \\ &= \textstyle\sum_{\sigma\in S_m(K)} a(\sigma)\alpha(\sigma) && \text{using Definitions 2.10.2a and b.} \end{aligned} \tag{2.10.1}$$

As in Lemma 2.2.4, we check that the map $\mathbf{k}\colon \hat{C}^m(K) \to \hat{C}_m(K)^\sharp$, given by $\mathbf{k}(a) = \langle a,\ \rangle$, is an isomorphism.

The *boundary operator* $\hat{\partial}\colon \hat{C}_m(K) \to \hat{C}_{m-1}(K)$ is the $\mathbb{Z}_2$-linear map defined, for $(v_0, \dots, v_m) \in \hat{\mathcal{S}}_m(K)$ by

$$\hat{\partial}(v_0, \dots, v_m) = \sum_{i=0}^{m} (v_0, \dots, \hat{v}_i, \dots, v_m)\,, \tag{2.10.2}$$

where $(v_0, \dots, \hat{v}_i, \dots, v_m) \in \hat{\mathcal{S}}_{m-1}$ is the m-tuple obtained by removing v_i. The *coboundary operator* $\hat{\delta} : C^m(K) \to C^{m+1}(K)$ is defined by the equation

$$\langle \hat{\delta} a, \alpha \rangle = \langle a, \hat{\partial} \alpha \rangle . \tag{2.10.3}$$

With these definition, $(\hat{C}_*(K), \hat{\partial}, \hat{C}^*(K), \hat{\delta}, \langle\, ,\rangle)$ is a Kronecker pair. We define the vector spaces of *ordered cycles* $\hat{Z}_*(K)$, *ordered boundaries* $\hat{B}_*(K)$, *ordered cocycles* $\hat{Z}^*(K)$, *ordered coboundaries* $\hat{B}^*(K)$, *ordered homology* $\hat{H}_*(K)$ and *ordered cohomology* $\hat{H}^*(K)$ as in Sect. 2.3. By Proposition 2.3.5, the pairing on (co)chain descends to a pairing

$$H^m(K) \times H_m(K) \xrightarrow{\langle\, ,\rangle} \mathbb{Z}_2$$

so that the map $\mathbf{k}$: $\hat{H}^m \to \hat{H}_m^\sharp$, given by $\mathbf{k}(a) = \langle a,\ \rangle$, is an isomorphism (*ordered Kronecker duality*).

Example 2.10.4 Let $K = pt$ be a point. Then, $\hat{\mathcal{S}}_m(pt)$ contains one element for each integer m, namely the $(m+1)$-tuple $(pt, \dots, pt)$. Then, $\hat{C}_m(pt) = \mathbb{Z}_2$ for all $m \in \mathbb{N}$ and the chain complex looks like

$$\cdots \xrightarrow{\approx} \hat{C}_{2k+1}(pt) \xrightarrow{0} \hat{C}_{2k}(pt) \xrightarrow{\approx} \hat{C}_{2k-1}(pt) \xrightarrow{0} \cdots \xrightarrow{\approx} \hat{C}_1(pt) \xrightarrow{0} \hat{C}_0(pt) \to 0 .$$

Therefore,

$$\hat{H}^*(pt) \approx \hat{H}_*(pt) \approx \begin{cases} 0 & \text{if } * > 0 \\ \mathbb{Z}_2 & \text{if } * = 0 . \end{cases}$$

One sees that, for a simplicial complex reduced to a point, the ordered (co)homology and the simplicial (co)homology are isomorphic.

Example 2.10.5 The *unit cochain* $\mathbf{1} \in \hat{C}^0(K)$ is defined as $\mathbf{1} = \hat{\mathcal{S}}_0(K)$. It is a cocycle and defines a class $\mathbf{1} = \hat{H}^0(K)$. If K is non-empty and connected, then $\hat{H}^0(K) \approx \mathbb{Z}_2$ generated by $\mathbf{1}$. Then $H_0(K) \approx \mathbb{Z}_2$ by Kronecker duality; one has $\hat{Z}_0(K) = \hat{C}_0(K)$ and $\alpha \in \hat{Z}_0(K)$ represents the non-zero element of $H_0(K)$ if and only if $\sharp\alpha$ is odd. The proofs are the same as for Proposition 2.4.1.

Example 2.10.6 Let L be a simplicial complex and CL be the cone on L. Then

$$\hat{H}^*(CL) \approx \hat{H}_*(CL) \approx \begin{cases} 0 & \text{if } * > 0 \\ \mathbb{Z}_2 & \text{if } * = 0 . \end{cases}$$

The proof is the same as for Proposition 2.4.6, even simpler, since D: $\hat{C}_m(CL) \to \hat{C}_{m+1}(CL)$ is defined, for $(v_0, \dots, v_m) \in \hat{\mathcal{S}}_m(CL)$ by the single line formula $D(v_0, \dots, v_m) = (\infty, v_0, \dots, v_m)$.

Let f: $L \to K$ be a simplicial map. We define $\hat{C}_* f$: $\hat{C}_*(L) \to \hat{C}_*(K)$ as the degree 0 linear map such that

$$\hat{C}_* f(v_0, \ldots, v_m) = (f(v_0), \ldots, f(v_m))$$

for all $(v_0, \ldots, v_m) \in \hat{\mathcal{S}}(L)$. The degree 0 linear map $\hat{C}^* f \colon \hat{C}^*(K) \to \hat{C}^*(L)$ is defined by

$$\langle \hat{C}^* f(a), \alpha \rangle = \langle a, \hat{C}_* f(\alpha) \rangle \, .$$

By Lemma 2.3.6, $(\hat{C}^* f, \hat{C}_* f)$ is a morphism of Kronecker pairs.

We now construct a functorial isomorphism between the ordered and non-ordered (co)homologies, its existence being suggested by the previous examples. Define $\psi_* \colon \hat{C}_*(K) \to C_*(K)$ by

$$\psi_*((v_0, \ldots, v_m)) = \begin{cases} \{v_0, \ldots, v_m\} & \text{if } v_i \neq v_j \text{ for all } i \neq j \\ 0 & \text{otherwise.} \end{cases}$$

We check that ψ is a morphism of chain complexes. We define $\psi^* : C^*(K) \to \hat{C}^*(K)$ by requiring that the equation $\langle \psi^*(a), \alpha \rangle = \langle a, \psi_*(\alpha) \rangle$ holds for all $a \in C^*(K)$ and all $\alpha \in \hat{C}_*(K)$. By Lemma 2.3.6, ψ^* is a morphism of cochain complexes and (ψ_*, ψ^*) is a morphism of Kronecker pairs between $(\hat{C}_*(K), \hat{C}^*(K))$ and $(C_*(K), C^*(K))$. It thus defines a morphism of Kronecker pairs $(H_*\psi, H^*\psi)$ between $(\hat{H}_*(K), \hat{H}^*(K))$ and $(H_*(K), H^*(K))$.

To define a morphism of Kronecker pairs in the other direction, choose a simplicial order $\leq$ on K (see 2.1.8). Define $\phi_{\leq *} \colon C_*(K) \to \hat{C}_*(K)$ as the unique linear map such that

$$\phi_{\leq *}(\{v_0, \ldots, v_m\}) = (v_0, \ldots, v_m) \, ,$$

where $v_0 \leq v_1 \leq \cdots \leq v_m$. We check that $\phi_{\leq *}$ is a morphism of chain complexes and define $\phi_{\leq}{}^* \colon \hat{C}^*(K) \to C^*(K)$ by requiring that the equation $\langle \phi_{\leq}{}^*(a), \alpha \rangle = \langle a, \phi_{\leq *}(\alpha) \rangle$ holds for all $a \in \hat{C}^*(K)$ and all $\alpha \in C_*(K)$. By Lemma 2.3.6, $(\phi_{\leq *}, \phi_{\leq}{}^*)$ is a morphism of Kronecker pairs between $(C_*(K), C^*(K))$ and $(\hat{C}_*(K), \hat{C}^*(K))$. It then defines a morphism of Kronecker pairs $(H_*\phi_{\leq}, H^*\phi_{\leq})$ between $(H_*(K), H^*(K))$ and $(\hat{H}_*(K), \hat{H}^*(K))$.

Proposition 2.10.7 *$H_*\psi \circ H_*\phi_{\leq} = \mathrm{id}_{H_*(K)}$ and $H_*\phi_{\leq} \circ H_*\psi = \mathrm{id}_{\hat{H}_*(K)}$.*

Proof As $\psi_* \circ \phi_{\leq *} = \mathrm{id}_{C_*(K)}$, the first equality follows from Lemma 2.3.7. For the second one, let $(v_0, \ldots, v_m) \in \hat{\mathcal{S}}_m(K)$. Let $\sigma = \{v_0, \ldots, v_m\} \in \mathcal{S}_k(K)$ with $k \leq m$. Clearly, $\phi_{\leq *} \circ \psi_*(v_0, \ldots, v_m) \in \hat{C}_*(\bar{\sigma})$. By what was seen in Examples 2.10.5 and 2.10.6, the correspondence $(v_0, \ldots, v_m) \mapsto \hat{C}_*(\overline{\{v_0, \ldots, v_m\}})$ is an acyclic carrier A_*, with respect to the basis $\hat{\mathcal{S}}_*(K)$, for both $\mathrm{id}_{\hat{C}(K)}$ and $\phi_{\leq *} \circ \psi_*$. Therefore, the equality $H_*\phi_{\leq} \circ H_*\psi = \mathrm{id}_{\hat{H}_*(K)}$ follows by Lemma 2.3.7 and Proposition 2.9.1.

Applying Kronecker duality to Proposition 2.10.7 gives the following

Corollary 2.10.8 *$H^*\psi \circ H^*\phi_\leq = \mathrm{id}_{\hat{H}^*(K)}$ and $H^*\phi_\leq \circ H^*\psi = \mathrm{id}_{H^*(K)}$.*

Corollary 2.10.9 *$H_*\psi$ and $H^*\psi$ are isomorphisms.*

Corollary 2.10.10 *$H_*\phi_\leq$ and $H^*\phi_\leq$ are isomorphisms which do not depend on the simplicial order $\leq$.*

Proof This follows from Proposition 2.10.7 and Corollary 2.10.8, since $H_*\psi$ and $H^*\psi$ do not depend on $\leq$. □

We shall see in Sect. 4.1 that $H^*\psi$ and $H^*\phi_\leq$ are isomorphisms of graded $\mathbb{Z}_2$-algebras. We now prove that they are also natural with respect to simplicial maps. Let $f: L \to K$ be a simplicial map. Let $\hat{C}_* f: \hat{C}_*(L) \to \hat{C}_*(K)$ be the unique linear map such that

$$\hat{C}_* f((v_0, \dots, v_m)) = (f(v_0), \dots, f(v_m))$$

for each $(v_0, \dots, v_m) \in \hat{\mathcal{S}}_m(K)$. Doing this for each $m \in \mathbb{N}$ produces a **GrV**-morphism $\hat{C}_* f: \hat{C}_*(L) \to \hat{C}_*(K)$. The formula $\hat{\partial} \circ \hat{C}_* f = \hat{C}_* f \circ \hat{\partial}$ is straightforward (much easier than that for non-ordered chains). Hence, we get a **GrV**-morphism $\hat{H}_* f: \hat{H}_*(L) \to \hat{H}_*(K)$. A **GrV**-morphism $\hat{C}^* f: \hat{C}^*(K) \to \hat{C}^*(L)$ is defined by the equation $\langle \hat{C}^* f(a), \alpha \rangle = \langle a, \hat{C}_* f(\alpha) \rangle$ required to hold for all $a \in \hat{C}^m(L)$, $\alpha \in \hat{C}_m(K)$ and all $m \in \mathbb{N}$. It is a cochain map and induces a **GrV**-morphism $\hat{H}^* f: \hat{H}^*(K) \to \hat{H}^*(L)$, Kronecker dual to $H_* f$.

Proposition 2.10.11 *Let $f: L \to K$ be a simplicial map. Let $\leq$ be a simplicial order on K and $\leq'$ be a simplicial order on L. Then the diagrams*

$$\begin{array}{ccc} \hat{H}_*(L) & \xrightarrow{\hat{H}_* f} & \hat{H}_*(K) \\ {\scriptstyle H_*\phi_{\leq'}} \uparrow \downarrow {\scriptstyle H_*\psi} & & {\scriptstyle H_*\phi_{\leq}} \uparrow \downarrow {\scriptstyle H_*\psi} \\ H_*(L) & \xrightarrow{H_* f} & H_*(K) \end{array} \quad \text{and} \quad \begin{array}{ccc} \hat{H}^*(K) & \xrightarrow{\hat{H}^* f} & \hat{H}_*(L) \\ {\scriptstyle H^*\psi} \uparrow \downarrow {\scriptstyle H^*\phi^{\leq}} & & {\scriptstyle H^*\psi} \uparrow \downarrow {\scriptstyle H^*\phi^{\leq'}} \\ H_*(K) & \xrightarrow{H^* f} & H_*(L) \end{array}$$

are commutative.

Proof By Kronecker duality, only the homology statement requires a proof. It is enough to prove that $H_* f \circ H_*\psi = H_*\psi \circ \hat{H}_* f$ since the formula $\hat{H}_* f \circ H_*\phi_{\leq'} = H_*\phi_\leq \circ H_* f$ will follow by Corollary 2.10.8. Finally, the formula $\hat{C}_* f \circ C_*\phi_{\leq'} = C_*\phi_\leq \circ C_* f$ is straightforward. □

The above isomorphism results also work in relative ordered (co)homology. Let (K, L) be a simplicial pair. Denote by $i: L \hookrightarrow K$ the simplicial inclusion. We define the $\mathbb{Z}_2$-vector space of *relative ordered (co)chain* by

$$\hat{C}^m(K, L) = \ker \left(\hat{C}^m(K) \xrightarrow{\hat{C}^* i} \hat{C}^m(L) \right)$$

and

$$\hat{C}_m(K,L) = \text{coker}\left(i_* : \hat{C}_m(L) \hookrightarrow \hat{C}_m(K)\right).$$

These inherit (co)boundaries $\hat{\delta} : \hat{C}^*(K,L) \to \hat{C}^*(K,L)$ and $\hat{\partial} = \hat{C}_*(K,L) \to \hat{C}_{*-1}(K,L)$ which give rise to the definition of *relative ordered (co)homology* $\hat{H}^*(k,L)$ and $\hat{H}_*(K,L)$. Connecting homomorphisms $\hat{\delta}_*\colon \hat{H}^*(L) \to \hat{H}^{*+1}(K,L)$ and $\hat{\partial}_*\colon \hat{H}_*(K,L) \to \hat{H}_{*-1}(L)$ are defined as in Sect. 2.7, giving rise to long exact sequences. Our homomorphisms $\psi_*\colon \hat{C}_*(K) \to C_*(K)$ and $\phi_{\leq *}\colon C_*(K) \to \hat{C}_*(K)$ satisfy $\psi_*(\hat{C}_*(L)) \subset C_*(L)$ and $\phi_{\leq *}(C_*(L) \subset \hat{C}_*(L)$, giving rise to homomorphisms on relative (co)chains and relative (co)homology $H_*\psi\colon \hat{H}_*(K,L) \to H - *(K,L)$, *etc*. Proposition 2.10.7 and Corollary 2.10.8 and their proofs hold in relative (co)homology. Hence, as for Corollaries 2.10.9 and 2.10.10, we get

Corollary 2.10.12 *$H_*\psi : \hat{H}_*(K,L) \to H_*(K,L)$ and $H^*\psi : H^*(K,L) \to \hat{H}^*(K,L)$ are isomorphisms.*

Corollary 2.10.13 *$H_*\phi_\leq : H_*(K,L) \to \hat{H}_*(K,L)$ and $H^*\phi_\leq : \hat{H}^*(K,L) \to \hat{H}^*(K,L)$ are isomorphisms which do not depend on the simplicial order $\leq$.*

2.11 Exercises for Chapter 2

2.1. Let $\mathcal{F}_n$ be the full complex on the set $\{0, 1, \dots, n\}$ (see p. 24). What are the 2-simplexes of the barycentric subdivision $\mathcal{F}_2'$ of $\mathcal{F}_2$? How many n-simplexes does $\mathcal{F}_n'$ contain?

2.2. Compute the Euler characteristic and the Poincaré polynomial of the k-skeleton $\mathcal{F}_n^k$ of $\mathcal{F}_n$.

2.3. Let X be a metric space and let $\varepsilon > 0$. The *Vietoris-Rips complex* X_ε of X is the simplicial complex whose simplexes are the finite non-empty subset of X whose diameter is $< \varepsilon$ (the diameter of $A \subset X$ is the least upper bound of $d(x,y)$ for $x, y \in A$). In particular, $V(X_\varepsilon) = X$.

(a) Describe $|X_\varepsilon|$ for various ε when X is the set of vertices of a cube of edge 1 in $\mathbb{R}^3$. In particular, if $\sqrt{2} < \varepsilon \leq \sqrt{3}$, show that $|X_\varepsilon|$ is homeomorphic to S^3.

(b) Let X be the space n-th roots of unity, with the distance $d(x,y)$ being the minimal length of an arc of the unit circle joining x to y. Suppose that $4\pi/n < \varepsilon \leq 6\pi/n$.

  (i) If n = 6, show that $|X_\varepsilon|$ is homeomorphic to S^2.
  (ii) If $n \geq 7$ is odd, show that $|X_\varepsilon|$ is homeomorphic to a Möbius band.
  (iii) If $n \geq 7$ is even, show that $|X_\varepsilon|$ is homeomorphic to $S^1 \times [0,1]$.

Note: the complex X_ε was introduced by Vietoris in 1927 [201]. After its re-introduction by E. Rips for studying hyperbolic groups, it has been popularized under the name of *Rips complex*. For some developments and applications, see [84, 129] and Wikipedia's page "Vietoris-Rips complex".

2.4. Let $\ell = (\ell_1, \dots, \ell_n) \in \mathbb{R}^n_{>0}$. A subset J of $\{1, \dots, n\}$ is called ℓ-*short* (or just *short*) if $\sum_{i \in J} \ell_i < \sum_{i \notin J} \ell_i$. Show that short subsets are the simplexes of a simplicial complex $\mathrm{Sh}(\ell)$ with $V(\mathrm{Sh}(\ell)) \subset J$ (used in Sect. 10.3). Describe $\mathrm{Sh}(1,1,1,1,3)$, $\mathrm{Sh}(1,1,3,3,3)$ and $\mathrm{Sh}(1,1,1,1,1)$. Compute their Euler characteristics and their Poincaré polynomials.

2.5. Let K be the simplicial complex with $V(K) = \mathbb{Z}$ and $\mathcal{S}_1(K) = \{\{r, r+1\} \mid r \in \mathbb{Z}\}$ ($|K| \approx \mathbb{R}$). Then $\mathcal{S}_1(K)$ is a 1-cocycle. Find all the cochains $a \in C^0(K)$ such that $\mathcal{S}_1(K) = \delta(a)$.

2.6. Find a simplicial pair (K, L) such that $|K|$ is homeomorphic to $S^1 \times I$ and $|L| = \mathrm{Bd}\,|K|$. In the spirit of Sect. 2.4.7, compute the simplicial cohomology of K and of (K, L) and find (co)cycles generating $H_*(K)$, $H_*(K, L)$, $H^*(K)$ and $H^*(K, L)$. Write completely the (co)homology sequence of (K, L).

2.7. Same exercise as before with $|K|$ the Möbius band and $|L| = \mathrm{Bd}\,|K|$.

2.8. Let $f\colon K \to L$ be a simplicial map between simplicial complexes. Suppose that L is connected and K is non-empty. Show that $H_0 f$ is surjective.

2.9. Let m, n, q be positive integers. If $m = nq$, the quotient map $\mathbb{Z} \to \mathbb{Z}/n\mathbb{Z}$ descends to a map $\mathbb{Z}/m\mathbb{Z} \to \mathbb{Z}/n\mathbb{Z}$, giving rise to a simplicial map $f\colon \mathcal{P}_m \to \mathcal{P}_n$ between the simplicial polygons $\mathcal{P}_m$ and $\mathcal{P}_n$ (see Example 2.4.3). Compute $H^* f$.

2.10. Let M be an n-dimensional pseudomanifold. Let σ and σ' be two distinct n-simplexes of M. Find $a \in C^{n-1}(M)$ such that $\delta(a) = \{\sigma, \sigma'\}$.

2.11. Let M be a finite non-empty n-dimensional pseudomanifold. Let $\gamma \in Z_{n-1}(M)$ which is a boundary. Prove that γ is the boundary of exactly two n chains.

2.12. Let $f\colon M \to N$ be a simplicial map between finite n-dimensional pseudomanifolds. Show that the following two conditions are equivalent.

   (a) $H_n f \neq 0$.
   (b) There exists $\sigma \in \mathcal{S}(N)$ such that $\sharp f^{-1}(\{\sigma\})$ is odd.

2.13. Let $\{\pm 1\}$ be the 0-dimensional simplicial complex with vertices -1 and 1. Let K be a simplicial complex. The *simplicial suspension* ΣK is the join $K * \{\pm 1\}$.

   (a) Let $\mathcal{P}_4$ be the polygon complex with 4-edges (see Example 2.4.3). Show that $\mathcal{P}_4 * K$ is isomorphic to the double suspension $\Sigma(\Sigma K)$. [Hint: show that the join operation is associative: $(K * L) * M \approx K * (L * M)$.]
   (b) Prove that the suspension of a pseudomanifold is a pseudomanifold.
   (c) Prove that the correspondence $K \mapsto \Sigma K$ gives a functor from **Simp** to itself.

2.14. Let A be a finite set. Show that $\dot{\mathcal{F}} A$ is a pseudomanifold.

2.15. Let M be an n-dimensional pseudomanifold which is infinite. What is $H_n(M)$?

2.16. Let (K, K_1, K_2) be a simplicial triad. Suppose that K_1 and K_2 are connected and that $K_1 \cap K_2$ is not empty. Show that K is connected.

2.17. Let (K, K_1, K_2) be a simplicial triad and let $K_0 = K_1 \cap K_2$.

(a) Prove that the homomorphism $H_*(K_1, K_0) \to H_*(K, K_2)$ induced by the inclusion is an isomorphism (*simplicial excision*).
(b) Write the commutative diagram involving the homology sequences of (K_1, K_0) and (K, K_2). Using (a), construct out of this diagram the Mayer-Vietoris sequence for the triad (K, K_1, K_2).

2.18. Deduce the additivity formula for the Euler characteristic of Lemma 2.4.10 from the Mayer-Vietoris sequence.

2.19. Let M_1 and M_2 be two finite n-dimensional pseudomanifolds. Let $\sigma_i \in \mathcal{S}(M_i)$ and let $h: \sigma_1 \to \sigma_2$ be a bijection. The *simplicial connected sum* $M = M_1 \sharp M_2$ (using h) is the simplicial complex defined by

$$V(M) = V(M_1) \dot{\cup} V(M_2) / \{v \sim h(v) \text{ for } v \in \sigma_1\}$$

and

$$\mathcal{S}(M) = \big(\mathcal{S}(M_1) - \{\sigma_1\}\big) \dot{\cup} \big(\mathcal{S}(M_2) - \{\sigma_2\}\big) .$$

Prove that M is a pseudomanifold. Compute $H_*(M)$ in terms of $H_*(M_1)$ and $H_*(M_2)$.

Chapter 3
Singular and Cellular (Co)homologies

3.1 Singular (Co)homology

Singular (co)homology provides a functor associating to a topological space X a graded $\mathbb{Z}_2$-vector space, whose isomorphism class depends only on the homotopy type of X. Such functors, from **Top** to categories of algebraic objects, constitute the main subject of algebraic topology.

Invented by Eilenberg in 1944 [49] after earlier attempts by Lefschetz, singular homology is formally akin to simplicial homology. However, in order to make computations for non-trivial examples, we need to establish some properties, such as homotopy and excision, which require some work. When K is a simplicial complex, the simplicial homology of K and the singular cohomology of $|K|$ are isomorphic in several ways, some of them being functorial (see Sect. 3.6). Singular (co)homology is especially powerful and relevant for spaces having the homotopy type of a *CW-complex*, a notion introduced in Sect. 3.4. For such spaces, singular (co)homology is isomorphic to other (co)homology theories (see Sect. 3.7) and the cohomology functor H^n is representable by the Eilenberg-MacLane space $K(\mathbb{Z}_2, n)$ (see Sect. 3.8).

3.1.1 Definitions

The *standard Euclidean m-simplex* Δ^m is defined by

$$\Delta^m = \{(x_0, \ldots, x_m) \in \mathbb{R}^{m+1} \mid x_i \geq 0 \text{ and } \sum x_i = 1\},$$

endowed with the induced topology from that of $\mathbb{R}^{n+1}$. In particular, $\Delta^m = \emptyset$ if $m < 0$. Let X be a topological space. A *singular m-simplex of X* is a continuous map $\sigma : \Delta^m \to X$. The set of singular m-simplexes of X is denoted by $\mathcal{S}_m(X)$.

J.-C. Hausmann, *Mod Two Homology and Cohomology*, Universitext,
DOI 10.1007/978-3-319-09354-3_3

The definitions of singular (co)chains and (co)homology are copied from those for the simplicial case (see Sect. 2.2), replacing simplicial simplexes by singular ones. We thus set

Definition 3.1.1 (*subset definitions*)

(a) A *singular m-cochain* of X is a subset of $\mathcal{S}_m(X)$.
(b) A *singular m-chain* of X is a finite subset of $\mathcal{S}_m(X)$.

The set of singular m-cochains of X is denoted by $C^m(X)$ and that of singular m-chains by $C_m(X)$. As in Sect. 2.2, Definition 3.1.1 are equivalent to

Definition 3.1.2 (*colouring definitions*)

(a) A *singular m-cochain* is a function $a : \mathcal{S}_m(X) \to \mathbb{Z}_2$.
(b) A *singular m-chain* is a function $\alpha : \mathcal{S}_m(X) \to \mathbb{Z}_2$ with finite support.

Definition 3.1.2 endow $C^m(X)$ and $C_m(X)$ with a structure of a $\mathbb{Z}_2$-vector space. The singletons provide a basis of $C_m(X)$, in bijection with $\mathcal{S}_m(X)$. Thus, Definition 3.1.2b is equivalent to

Definition 3.1.3 $C_m(X)$ is the $\mathbb{Z}_2$-vector space with basis $\mathcal{S}_m(X)$:

$$C_m(X) = \bigoplus_{\sigma \in \mathcal{S}_m(X)} \mathbb{Z}_2\, \sigma .$$

We consider the graded vector spaces $C_*(X) = \oplus_{m\in\mathbb{N}} C_m(X)$ and $C^*(X) = \oplus_{m\in\mathbb{N}} C^m(X)$. By convention, $C^m(X) = C_m(X) = 0$ if $m < 0$ (so the index m could be taken in $\mathbb{Z}$ in the previous formulae).

The *Kronecker pairing* on singular (co)chains

$$C^m(X) \times C_m(X) \xrightarrow{\langle\, ,\, \rangle} \mathbb{Z}_2$$

is defined, using the various above definitions, by the equivalent formulae

$$\begin{aligned} \langle a, \alpha \rangle &= \sharp(a \cap \alpha)(\text{mod } 2) && \text{using Definition 3.1.1a and b} \\ &= \textstyle\sum_{\sigma \in \alpha} a(\sigma) && \text{using Definition 3.1.1a and 3.1.2b} \\ &= \textstyle\sum_{\sigma \in \mathcal{S}_m(X)} a(\sigma)\alpha(\sigma) && \text{using Definition 3.1.2a and b.} \end{aligned} \tag{3.1.1}$$

As in Lemma 2.2.4, we check that the map $\mathbf{k} : C^m(X) \to C_m(X)^\sharp$, given by $\mathbf{k}(a) = \langle a, \ \rangle$, is an isomorphism.

Let $m, i \in \mathbb{N}$ with $0 \le i \le m$. Define the *i-th face inclusion* $\epsilon_i : \Delta^{m-1} \to \Delta^m$ by

$$\epsilon_i(x_0, \ldots, x_{m-1}) = (x_0, \ldots, x_{i-1}, 0, x_{i+1}, \ldots, x_{m-1}).$$

The *boundary operator* $\partial : C_m(X) \to C_{m-1}(X)$ is the $\mathbb{Z}_2$-linear map defined, for $\sigma \in \mathcal{S}_m(X)$ by

$$\partial(\sigma) = \sum_{i=0}^{m} \sigma \circ \epsilon_i. \tag{3.1.2}$$

Lemma 3.1.4 $\partial \circ \partial = 0$.

Proof By linearity, it suffices to prove that $\partial \circ \partial(\sigma) = 0$ for $\sigma \in \mathcal{S}_m(X)$. One has

$$\partial \circ \partial(\sigma) = \partial(\sum_{i=0}^{m} \sigma \circ \epsilon_i) = \sum_{(i,j)\in A} \sigma \circ \epsilon_i \circ \epsilon_j, \tag{3.1.3}$$

where $A = \{0, \dots, m\} \times \{0, \dots, m-1\}$. The set $B = \{(i, j) \in A \mid i \leq j\}$ is in bijection with $A - B$, via the map $(i, j) \mapsto (j + 1, i)$. But if $(i, j) \in B$, then $\epsilon_i \circ \epsilon_j = \epsilon_{j+1} \circ \epsilon_i$, which implies that $\partial \circ \partial = 0$. □

The *coboundary operator* $\delta : C^m(X) \to C^{m+1}(X)$ is defined by the equation

$$\langle \delta a, \alpha \rangle = \langle a, \partial \alpha \rangle. \tag{3.1.4}$$

With these definition, $((C_*(X), \partial), (C^*(X), \delta), \langle\,,\,\rangle)$ is a Kronecker pair. We define the vector spaces of *singular cycles*$Z_*(X)$, *singular boundaries* $B_*(X)$, *singular cocycles*$Z^*(X)$, *singular coboundaries* $B^*(X)$, *singular homology*$H_*(X)$ and *singular cohomology*$H^*(X)$ as in Sect. 2.3. By Proposition 2.3.5, the pairing on (co)chain descends to a pairing

$$H^m(X) \times H_m(X) \xrightarrow{\langle\,,\,\rangle} \mathbb{Z}_2$$

so that the map $\mathbf{k} : H^m \to H_m^\sharp$, given by $\mathbf{k}(a) = \langle a,\ \rangle$, is an isomorphism (*Kronecker duality* in singular (co)homology). The Kronecker pairing extends to a bilinear map $H^*(X) \times H_*(X) \xrightarrow{\langle\,,\,\rangle} \mathbb{Z}_2$ by setting $\langle a, \alpha \rangle = 0$ if $a \in H^p(X)$ and $\alpha \in H_q(X)$ with $p \neq q$.

Example 3.1.5 If X is the empty space, then $\mathcal{S}_m(X) = \emptyset$ for all m and thus $H^*(\emptyset) = H_*(\emptyset) = 0$. Let $X = pt$ be a point. Then, $\mathcal{S}_m(pt)$ contains one element for each $m \in \mathbb{N}$, namely the constant singular simplex $\Delta^m \to pt$. Then, $C_m(pt) = \mathbb{Z}_2$ for all $m \in \mathbb{N}$ and the chain complex looks like

$$\cdots \xrightarrow{\approx} C_{2k+1}(pt) \xrightarrow{0} C_{2k}(pt) \xrightarrow{\approx} C_{2k-1}(pt) \xrightarrow{0} \cdots \xrightarrow{\approx} C_1(pt) \xrightarrow{0} C_0(pt) \to 0.$$

Therefore,

$$H^*(pt) \approx H_*(pt) \approx \begin{cases} 0 & \text{if } * > 0 \\ \mathbb{Z}_2 & \text{if } * = 0. \end{cases} \tag{3.1.5}$$

Example 3.1.6 Let K be a simplicial complex. Choose a simplicial order "$\leq$" for K. To an m-simplex $\sigma = \{v_0, \dots, v_m\} \in \mathcal{S}_m(K)$, with $v_0 \leq \cdots \leq v_m$, we associate the singular m-simplex $R_\leq(\sigma) : \Delta^m \to |K|$ defined by

$$R_\leq(\sigma)(t_0, \dots, t_m) = \sum_{i=0}^{m} t_i v_i. \tag{3.1.6}$$

The linear combination in (3.1.6) makes sense since $\{v_0, \dots, v_m\}$ is a simplex of K. This defines a map $R_\leq : \mathcal{S}_m(K) \to \mathcal{S}_m(|K|)$ which extends to a linear map

$$R_{\leq,*} : C_*(K) \to C_*(|K|).$$

This map will be used several times in this chapter. The formula $\partial \circ R_{\leq,*} = R_{\leq,*} \circ \partial$ is obvious, so $R_{\leq,*}$ is a chain map from $(C_*(K), \partial)$ to $(C_*(|K|), \partial)$. We shall prove, in Theorem 3.6.3, that $R_{\leq,*}$ induces an isomorphism between the simplicial (co)homology) of K and the singular (co)homology) of $|K|$.

Example 3.1.7 As the affine simplex Δ^0 is a point, one can identify a singular 0-simplex of X with its image, a point of X. This gives a bijection $\mathcal{S}_0(X) \approx X$ and a bijection between subsets of X and singular 0-cochains. For $B \subset X$ and $x \in X$, one has $\langle B, x\rangle = \chi_B(x)$, where χ_B stands for the characteristic function for B. The 1-cochain δB is the *connecting cochain* for B: if $\beta \in \mathcal{S}_1(X)$, then

$$\langle \delta(B), \beta\rangle = \langle B, \partial\beta\rangle = \langle B, \beta(1,0)\rangle + \langle B, \beta(0,1)\rangle. \tag{3.1.7}$$

In other words $\langle \delta(B), \beta\rangle = 1$ if and only if the (non-oriented) path β connects a point in B to a point in $X - B$. Observe that $\delta(B) = \delta(X - B)$.

Following Example 3.1.7, the *unit cochain* $\mathbf{1} \in C^0(X)$ is defined by $\mathbf{1} = \mathcal{S}_0(X) \approx X$. By Eq. (3.1.7) $\langle \delta\mathbf{1}, \beta\rangle = 0$ for all $\beta \in \mathcal{S}_1(X)$. This proves that $\delta(\mathbf{1}) = 0$ by Lemma 2.2.4. Hence, $\mathbf{1}$ is a cocycle, whose cohomology class is again denoted by $\mathbf{1} \in H^0(X)$.

Proposition 3.1.8 *Let X be a non-empty path-connected space. Then,*

(i) $H^0(X) = \mathbb{Z}_2$, *generated by* $\mathbf{1}$ *which is the only non-vanishing singular 0-cocycle.*
(ii) $H_0(X) = \mathbb{Z}_2$. *Any 0-chain α is a cycle, which represents the non-zero element of $H_0(X)$ if and only if $\sharp\alpha$ is odd.*

Proof The proof is analogous to that of Proposition 2.4.1. If X is non-empty the unit cochain does not vanish and, as $C^{-1}(X) = 0$, $\mathbf{1} \neq 0$ in $H^0(X)$.

Let $a \in C^0(X)$ with $a \neq 0, \mathbf{1}$. Then there exists $x, y \in X = \mathcal{S}(X)$ with $a(x) \neq a(y)$. Since X is path-connected, there exists $\sigma \in \mathcal{S}_1(X)$ with $\sigma(1,0) = x$ and $\sigma(0,1) = y$. As in Eq. (3.1.7), this proves that $\langle \delta(a), \sigma \rangle \neq 0$ so a is not a cocycle. This proves (i).

Now, $H_0(X) = \mathbb{Z}_2$ since $H^0(X) \approx H_0(X)^\sharp$. Any $\alpha \in C_0(X)$ is a cycle since $C_{-1}(X) = 0$. It represents the non-zero homology class if and only if $\langle \mathbf{1}, \alpha \rangle = 1$, that is if and only if $\sharp\alpha$ is odd. □

The *reduced (singular) cohomology* $\tilde{H}^*(X)$ and *homology* $\tilde{H}_*(X)$ of a topological space X are the graded $\mathbb{Z}_2$-vector spaces defined by

$$\begin{aligned} \tilde{H}^*(X) &= \operatorname{coker}\left(H^*p : H^*(pt) \to H^*(X)\right) \\ \tilde{H}_*(X) &= \ker\left(H_*p : H_*(X) \to H_*(pt)\right) \end{aligned} \tag{3.1.8}$$

where $p : X \to pt$ denotes the constant map to a point. In particular, $\tilde{H}^*(pt) = 0 = \tilde{H}_*(pt)$. One checks that the Kronecker pairing induces a bilinear map $\langle \, , \rangle : \tilde{H}^m(X) \times \tilde{H}_m(X) \to \mathbb{Z}_2$ such that the correspondence $a \mapsto \langle a, \ \rangle$ gives an isomorphism $\mathbf{k} : \tilde{H}^m(X) \xrightarrow{\approx} \tilde{H}_m(X)^\sharp$.

The full strength of Definition (3.1.8) appears in other (co)homology theories, such as equivariant cohomology (see p. 266). For the singular cohomology, as $H^*(pt) = \mathbb{Z}_2\mathbf{1}$, one gets

$$\tilde{H}^m(X) = \begin{cases} H^0(X)/\mathbb{Z}_2\mathbf{1} & \text{if } m = 0 \\ H^m(X) & \text{if } m \neq 0 \end{cases}$$

and

$$\tilde{H}_m(X) = \begin{cases} \ker\left(H_0(X) \xrightarrow{\langle \mathbf{1}, \rangle} \mathbb{Z}_2\right) & \text{if } m = 0 \\ H_m(X) & \text{if } m \neq 0. \end{cases}$$

Thus, by Proposition 3.1.8, $\tilde{H}^0(X) = 0 = \tilde{H}_0(X)$ if X is path-connected (see also Corollary 3.1.12).

Let $f : Y \to X$ be a continuous map between topological spaces. It induces a map $\mathcal{S}f : \mathcal{S}(Y) \to \mathcal{S}(X)$ defined by $\mathcal{S}f(\sigma) = f \circ \sigma$. The linear map $C^*f : C^*(X) \to C^*(Y)$ is, using Definition 3.1.2, defined by $C^*f(a) = a \circ \mathcal{S}(f)$. As for $C_*f : C_*(Y) \to C_*(X)$, it is the linear map extending $\mathcal{S}f$, using Definition 3.1.3. One checks that the couple (C^*f, C_*f) is a morphism of Kronecker pair. It thus defines linear maps of degree zero $H^*f : H^*(X) \to H^*(Y)$ and $H_*f : H_*(Y) \to H_*(X)$. The functorial properties are easy to prove: H^* and H_* are functors from the category **Top** of topological spaces to the category **GrV** of graded vector spaces (see Proposition 3.1.22 for a more general statement). Also, for any map $f : Y \to X$, the diagram

$$\begin{array}{ccc} Y & \xrightarrow{f} & X \\ & {\scriptstyle p}\searrow \quad \swarrow{\scriptstyle p} & \\ & pt & \end{array} \tag{3.1.9}$$

is obviously commutative. This implies that the reduced cohomology $\tilde{H}^*$ and homology $\tilde{H}_*$ are also functors from **Top** to **GrV**. The notations $H^* f$, $H_* f$, $\tilde{H}^* f$ and $\tilde{H}_* f$ are sometimes shortened in f^* and f_*.

As in Lemma 2.5.4, we prove the following

Lemma 3.1.9 *Let $f : Y \to X$ be a continuous map. Then $H^0 f(\mathbf{1}) = \mathbf{1}$.*

Lemma 3.1.9 implies the following result.

Lemma 3.1.10 *Let (X, Y) be a topological pair with X path-connected. Denote by $i : Y \to X$ the inclusion. Then there are exact sequences*

$$0 \to H^0(X) \xrightarrow{H^* i} H^0(Y) \to \tilde{H}^0(Y) \to 0$$

and

$$0 \to \tilde{H}_0(Y) \to H_0(Y) \xrightarrow{H_* i} H_0(X) \to 0.$$

We now prove some general results useful to compute the (co)homology of a space. Let X be a topological space which is a disjoint union:

$$X = \dot{\bigcup}_{j \in \mathcal{J}} X_j.$$

By this we mean that the above equality holds as sets and that each X_j is open (and therefore closed) in X. Denote the inclusion by $i_j : X_j \to X$. The equality $\mathcal{S}_m(X) = \dot{\bigcup}_{j \in \mathcal{J}} i_j(\mathcal{S}_m(X_j))$ implies the following proposition.

Proposition 3.1.11 *The family of inclusions $i_j : X_j \to X$ for $j \in \mathcal{J}$ gives rise to isomorphisms*

$$H^*(X) \xrightarrow[\approx]{(H^* i_j)} \prod_{j \in \mathcal{J}} H^*(X_j)$$

and

$$\bigoplus_{j \in \mathcal{J}} H_*(X_j) \xrightarrow[\approx]{\sum H_* i_j} H_*(X).$$

Corollary 3.1.12 *Let X be a topological space which is locally path-connected. Then, the family of inclusions $i_Y : Y \to X$ for $Y \in \pi_0(X)$ gives rise to isomorphisms*

$$H^*(X) \xrightarrow[\approx]{(H^*i_Y)} \prod_{Y\in\pi_0(X)} H^*(Y)$$

and

$$\bigoplus_{Y\in\pi_0(K)} H_*(Y) \xrightarrow[\approx]{\sum H_* i_Y} H_*(X)\,.$$

Proof As X is locally path-connected, each $Y \in \pi_0(X)$ is open in X and X is topologically the disjoint union of its path-connected components. Corollary 3.1.12 then follows from Proposition 3.1.11. □

Corollary 3.1.13 *Let X be a topological space which is locally path-connected. Then,*

$$\tilde{H}^0(X) = 0 \Leftrightarrow \tilde{H}_0(X) = 0 \Leftrightarrow X \text{ is path-connected.}$$

Also,

$$H^0(X) = \tilde{H}^0(X) \oplus \mathbb{Z}_2 \text{ and } H_0(X) = \tilde{H}_0(X) \oplus \mathbb{Z}_2$$

if X is not empty.

In the same spirit of reducing the computations of $H^*(X)$ to those of smaller subspaces, another consequence of the definition of the singular (co)homology is the following proposition.

Proposition 3.1.14 *Let X be a topological space. Let $\mathcal{K}$ be the set of compact subspaces of X, partially ordered by inclusion. Then, the natural homomorphisms*

$$J_* : \varinjlim_{K\in\mathcal{K}} H_*(K) \to H_*(X)$$

and

$$J^* : H^*(X) \to \varprojlim_{K\in\mathcal{K}} H^*(K)$$

are isomorphisms.

Here, $\varinjlim$ denotes the *direct limit* (also called *inductive limit* or *colimit*) and $\varprojlim$ denotes the *inverse limit* (also called *projective limit* or just *limit*) in **GrV**.

Proof Let $A \in H_r(X)$, represented by $\alpha \in Z_r(X)$. Then, α is a finite set of r-simplexes of X and $K = \bigcup_{\sigma\in\alpha} \sigma(\Delta^r)$ is a compact subspace of X. One can see $\alpha \in Z_r(K)$, so J_* is onto. Now, let K be a compact subspace of X and $A \in H_r(K)$ mapped to 0 under $H_r(K) \to H_r(X)$. Represent A by $\alpha \in Z_r(K)$ and let $\beta \in$

$C_{r+1}(X)$ with $\alpha = \partial(\beta)$. As before, there exists a compact subset L of X containing K with $\beta \in C_{r+1}(L)$, so A is mapped to 0 under $H_r(K) \to H_r(L)$. This proves that J_* is injective. Finally, the bijectivity of J^* is deduced from that of J_* by Kronecker duality. □

Remark 3.1.15 In Proposition 3.1.14 the morphism J_* is an isomorphism for the homology with any coefficients. The morphism J^* is always surjective but, in general not injective (except for coefficients in a field, like $\mathbb{Z}_2$). Its kernel is expressible using the derived functor $\varprojlim^1$ (see e.g. [82, Theorem 3F.8]). The same considerations hold true for the following corollary.

Corollary 3.1.16 *Let X be a topological space and let $\mathcal{A}$ be a family of subspaces of X, partially ordered by the inclusion. Suppose that each compact subspace of X is contained in some $A \in \mathcal{A}$. Then, the homomorphisms*

$$j_* : \varinjlim_{A \in \mathcal{A}} H_*(A) \to H_*(X)$$

and

$$j^* : H^*(X) \to \varprojlim_{A \in \mathcal{A}} H^*(A)$$

are isomorphisms.

Proof The hypothesis that each compact $K \subset X$ is contained in some $A \in \mathcal{A}$ implies a factorization of the homomorphism J_* of Proposition 3.1.14:

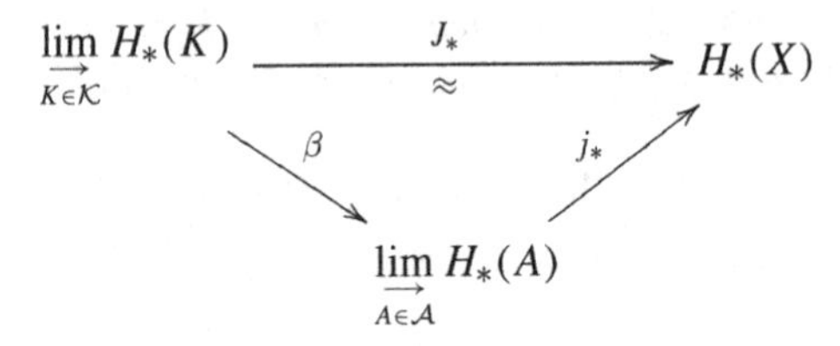

The same hypothesis implies that β is onto, whence j_* is an isomorphism. The assertion for j^* comes from Kronecker duality. □

3.1.2 Relative Singular (Co)homology

A (*topological*) *pair* is a couple (X, Y) where X is a topological space and Y is a subspace of X. The inclusion $i : Y \hookrightarrow X$ is a continuous map. Let $a \in C^m(X)$. If, using Definition 3.1.1a, we consider a as a subset of $\mathcal{S}_m(X)$, then $C^*i(a) = a \cap \mathcal{S}_m(Y)$. If we see a as a map $a : \mathcal{S}_m(X) \to \mathbb{Z}_2$, then $C^*i(a)$ is the restriction of a to $\mathcal{S}_m(Y)$. We see that $C^*i : C^*(X) \to C^*(Y)$ is surjective. Define

$$C^m(X,Y) = \ker\left(C^m(X) \xrightarrow{C^*i} C^m(Y)\right)$$

and $C^*(X,Y) = \oplus_{m\in\mathbb{N}} C^m(X,Y)$. As C^*i is a morphism of cochain complexes, the coboundary $\delta : C^*(X) \to C^*(X)$ preserves $C^*(X,Y)$ and gives rise to a coboundary $\delta : C^*(X,Y) \to C^*(X,Y)$ so that $(C^*(X,Y),\delta)$ is a cochain complex. The cocycles $Z^*(X,Y)$ and the coboundaries $B^*(X,Y)$ are defined as usual, giving rise to the definition

$$H^m(X,Y) = Z^m(X,Y)/B^m(X,Y).$$

The graded $\mathbb{Z}_2$-vector space $H^*(X,Y) = \oplus_{m\in\mathbb{N}} H^m(X,Y)$ is the *relative (singular) cohomology* of the pair (X,Y). Observe that $H^*(X,\emptyset) = H^*(X)$. We denote by j^* the inclusion $j^* : C^*(X,Y) \hookrightarrow C^*(X)$, which is a morphism of cochain complexes, and use the same notation j^* for the induced linear map $j^* : H^*(X,Y) \to H^*(X)$ on cohomology. We also use the notation i^* for both C^*i and H^*i. We get thus a short exact sequence of cochain complexes

$$0 \to C^*(X,Y) \xrightarrow{j^*} C^*(X) \xrightarrow{i^*} C^*(Y) \to 0. \tag{3.1.10}$$

If $a \in C^m(Y)$, any cochain $\bar{a} \in C^m(X)$ with $i^*(\bar{a}) = a$ is called a *extension of* a as a singular cochain in X. For instance, the 0*-extension* of a is defined by $\bar{a} = a \in \mathcal{S}_m(Y) \subset \mathcal{S}_m(X)$.

With chains, the inclusion $Y \hookrightarrow X$ induces an inclusion $i_* : C_*(Y) \hookrightarrow C_*(X)$ of chain complexes. We define $C_m(X,Y)$ as the quotient vector space

$$C_m(X,Y) = \operatorname{coker}\left(i_* : C_m(Y) \hookrightarrow C_m(X)\right).$$

As i_* is a morphism of chain complexes, $C_*(X,Y) = \oplus_{m\in\mathbb{N}} C_m(X,Y)$ inherits a boundary operator $\partial = \partial_{X,Y} : C_*(X,Y) \to C_{*-1}(X,Y)$. The projection $j_* : C_*(X) \to\to C_*(X,Y)$ is a morphism of chain complexes and one obtains a short exact sequence of chain complexes

$$0 \to C_*(Y) \xrightarrow{i_*} C_*(X) \xrightarrow{j_*} C_*(X,Y) \to 0. \tag{3.1.11}$$

The cycles and boundaries $Z_*(X,Y)$ and $B_*(X,Y)$ are defined as usual, giving rise to the definition

$$H_m(X,Y) = Z_m(X,Y)/B_m(X,Y).$$

The graded $\mathbb{Z}_2$-vector space $H_*(X,Y) = \oplus_{m\in\mathbb{N}} H_m(X,Y)$ is the *relative (singular) homology* of the pair (X,Y). Observe that $H_*(X,\emptyset) = H_*(X)$. The notations i_* and j_* are also used for the induced maps in homology.

As in Sects. 2.6 and 2.7 of Chap. 2, one gets a pairing $\langle\, ,\rangle : H^m(X,Y) \times H_m(X,Y) \to \mathbb{Z}_2$ which makes $(H^m(X,Y), H_m(X,Y), \langle\, ,\rangle)$ a Kronecker pair. Also, the *singular (co)homology connecting homomorphisms*

$$\delta^* : H^*(Y) \to H^{*+1}(X,Y). \quad \text{and} \quad \partial_* : H_*(X,Y) \to H_{*-1}(Y)$$

are defined and satisfy $\langle \delta^*(a), \alpha\rangle = \langle a, \partial_*(\alpha)\rangle$. The proof of the following lemma is the same as that of Lemma 2.7.1.

Lemma 3.1.17 *Let $a \in Z^m(Y)$ and let $\bar{a} \in C^m(X)$ be any extension of a as a singular m-cochain of X. Then, $\delta_X(\bar{a})$ is a singular $(m+1)$-cocycle of (X,Y) representing $\delta^*(a)$.*

Remark 3.1.18 A class in $A \in H_m(X,Y)$ is represented by a *relative singular cycle*, i.e. a singular chain $\alpha \in C_m(X)$ such that $\partial(\alpha)$ is a singular chain (cycle) of Y. The homology class of $\partial(\alpha)$ in $H_{n-1}(Y)$ is $\partial_*(A)$. This is the Kronecker dual statement of Lemma 3.1.17.

As for the simplicial (co)homology (see Sect. 2.7), the results of Sect. 2.6 give the following *(singular) (co)homology exact sequences*, or just the *(co)homology sequence*, of the pair (X,Y).

Proposition 3.1.19 ((Co)homology exact sequences of a pair) *Let (X,Y) be a topological pair. Then, the sequences*

$$\cdots \to H^m(X,Y) \xrightarrow{j^*} H^m(X) \xrightarrow{i^*} H^m(Y) \xrightarrow{\delta^*} H^{m+1}(X,Y) \xrightarrow{j^*} \cdots$$

and

$$\cdots \to H_m(Y) \xrightarrow{i_*} H_m(X) \xrightarrow{j_*} H_m(X,Y) \xrightarrow{\partial_*} H_{m-1}(Y) \xrightarrow{i_*} \cdots$$

are exact.

These exact sequences are also available for reduced (co)homology. For this, the reduced (co)homology of a pair is defined as follows: when $Y \neq \emptyset$, then $\tilde{H}^*(X,Y) = H^*(X,Y)$ and $\tilde{H}_*(X,Y) = H_*(X,Y)$; otherwise $\tilde{H}^*(X,\emptyset) = \tilde{H}^*(X)$ and $\tilde{H}_*(X,\emptyset) = \tilde{H}_*(X)$.

Proposition 3.1.20 (Reduced (co)homology sequences of a pair) *The exact sequences of Proposition* 3.1.19 hold with reduced (co)homology.

Proof An argument is only required around $m = 0$. For the homology exact sequence, consider the commutative diagram:

$$\begin{array}{ccccccccc}
\cdots \longrightarrow & \tilde{H}_1(X,Y) & \xrightarrow{\tilde{\partial}_*} & \tilde{H}_0(Y) & \xrightarrow{\tilde{i}_*} & \tilde{H}_0(X) & \xrightarrow{\tilde{j}_*} & \tilde{H}_0(X,Y) & \longrightarrow 0 \\
& \downarrow = & & \downarrow & & \downarrow & & \downarrow = & \\
\cdots \longrightarrow & H_1(X,Y) & \xrightarrow{\partial_*} & H_0(Y) & \xrightarrow{i_*} & H_0(X) & \xrightarrow{j_*} & H_0(X,Y) & \longrightarrow 0 \\
& & & \downarrow \langle \mathbf{1}, \rangle & & \downarrow \langle \mathbf{1}, \rangle & & & \\
& & & \mathbb{Z}_2 & \xrightarrow{=} & \mathbb{Z}_2 & & &
\end{array}$$

The commutativity of the bottom square is due to Lemma 3.1.9. As $i_* \circ \partial_* = 0$, $\langle \mathbf{1}, \partial(\alpha)\rangle = 0$ for all $\alpha \in H_1(X,Y)$ and therefore $\tilde{\partial} : \tilde{H}_1(X,Y) \to \tilde{H}_1(Y)$ exists. Since the sequence of the second line is exact, an easy diagram-chase shows that the sequence of the first line is exact as well.

The reduced cohomology sequence can be established in an analogous way or deduced from the homology one by Kronecker duality, using Lemma 2.3.8. □

Remark 3.1.21 Let (X, Y) be a topological pair with Y path-connected and non-empty. By Proposition 3.1.20 and its proof, we get the isomorphisms

$$j_* : \tilde{H}_0(X) \xrightarrow{\approx} H_0(X,Y) \quad \text{and} \quad j^* : H^0(X,Y) \xrightarrow{\approx} \tilde{H}^0(X). \tag{3.1.12}$$

Also, if $Y = \{x\}$, we get the isomorphisms

$$j_* : \tilde{H}_*(X) \xrightarrow{\approx} H_*(X,x) \quad \text{and} \quad j^* : H^*(X,x) \xrightarrow{\approx} \tilde{H}^*(X). \tag{3.1.13}$$

A direct proof of (3.1.13), say for cohomology, is given by the diagram.

$$\begin{array}{ccccc}
 & & H^*(pt) & & \\
 & & \downarrow p^* & \searrow \approx & \\
H^*(X,x) & \xrightarrow{j^*} & H^*(X) & \xrightarrow{i^*} & H^*(x) \\
 & \searrow \approx & \downarrow & & \\
 & & \tilde{H}^*(X) & &
\end{array} \tag{3.1.14}$$

where the line and the column are exact and $p : X \to pt$ is the constant map onto a point. We see that the choice of $x \in X$ produces a supplementary vector subspace to $p^*(H^*(pt))$ in $H^*(X)$.

We now study the naturality of the relative (co)homology and of the exact sequences. Let (X, Y) and (X', Y') be topological pairs. A *map f of (topological) pairs* from (X, Y) to (X', Y') is a continuous map $f : X \to X'$ such $f(Y) \subset Y'$. With these maps, topological pairs constitute a category $\mathbf{Top}_2$. The correspondence $X \mapsto (X, \emptyset)$ makes $\mathbf{Top}$ a full subcategory of $\mathbf{Top}_2$.

Let $f : (X, Y) \to (X', Y')$ be a map of topological pairs. The morphism $C^*f : C^*(X') \to C^*(X)$ then restricts to a morphism of cochain complexes $C^*f : C^*(X', Y') \to C^*(X, Y)$ and the morphism $C_*f : C_*(X) \to C_*(X')$ descends to a morphism of chain complexes $C_*f : C_*(X, Y) \to C_*(X', Y')$. As in Sect. 2.7, we prove that (C_*f, C_*f) is a morphism of Kronecker pair. One then gets degree zero linear maps $H^*f : H^*(X', Y') \to H^*(X, Y)$ and $H_*f : H^*(X, Y) \to H_*(X', Y')$ satisfying $\langle H^*a, \alpha\rangle = \langle a, H_*\alpha\rangle$ for all $a \in H^m(X', Y')$, $\alpha \in H_m(X, Y)$ and all $m \in \mathbb{N}$. Functorial properties are easy, so we get the following

Proposition 3.1.22 *The relative singular cohomology $H^*(,)$ is a contravariant functor from the category* **Top**$_2$ *to the category* **GrV** *of graded $\mathbb{Z}_2$-vector spaces. The relative singular homology $H_*(,)$ is a covariant functor between these categories. The same holds true for the reduced singular (co)homology.*

As for Proposition 2.7.6, we can prove the following

Proposition 3.1.23 *The (co)homology sequences are natural with respect to maps of topological pairs.*

Here is a special form of the cohomology sequence of a pair.

Proposition 3.1.24 *Let A and B be topological spaces. Then the cohomology sequence of the pair $(A\dot{\cup}B, A)$ cuts into short exact sequences and there is a commutative diagram*

$$\begin{array}{ccccccccc} 0 & \longrightarrow & H^*(B) & \longrightarrow & H^*(A\dot{\cup}B) & \longrightarrow & H^*(A) & \longrightarrow & 0 \\ & & \downarrow \approx & & \downarrow \mathrm{id} & & \downarrow \mathrm{id} & & \\ 0 & \longrightarrow & H^*(A\dot{\cup}B, A) & \longrightarrow & H^*(A\dot{\cup}B) & \longrightarrow & H^*(A) & \longrightarrow & 0 \end{array} \quad . \tag{3.1.15}$$

Proof If $i_A : A \to A\dot{\cup}B$ and $i_B : B \to A\dot{\cup}B$ denote the inclusions, Proposition 3.1.11 provides a commutative diagram

$$\begin{array}{ccc} H^*(A\dot{\cup}B) & \xrightarrow[\approx]{(i_A^*, i_B^*)} & H^*(A) \times H^*(B) \\ & {\scriptstyle i_A^*}\searrow \quad \swarrow{\scriptstyle \mathrm{proj}_1} & \\ & H^*(A) & \end{array} .$$

This proves that i_a^* is surjective, which cuts the cohomology sequence of $(A\dot{\cup}B, A)$, giving the bottom line of (3.1.15). Also, $\ker i_A^*$ is the image of $H^*(B)$ under the monomorphism $j : H^*(B) \to H^*A\dot{\cup}B)$ given by $j(u) = (i_A^*, i_B^*)^{-1}(0, u)$, which we placed in the top line of (3.1.15). □

As in simplicial (co)homology, the exact sequences of a pair generalize to that of a triple. A (*topological*) *triple* is a triplet (X, Y, Z) where X is a topological spaces and Y, Z are subspaces of X with $Z \subset Y$. A *map f of triples*, from (X, Y, Z) to (X', Y', Z') is a continuous map $f : X \to X'$ such that $f(Y) \subset Y'$ and $f(Z) \subset Z'$.

A triple $T = (X, Y, Z)$ gives rise to pair inclusions

$$(Y, Z) \xrightarrow{i} (X, Z) \xrightarrow{j} (X, Y)$$

and to a commutative diagram

$$\begin{array}{ccccccccc} 0 & \longrightarrow & C^*(X, Y) & \xrightarrow{j^*_{X,Y}} & C^*(X) & \xrightarrow{i^*_{X,Y}} & C^*(Y) & \longrightarrow & 0 \\ & & \downarrow{\scriptstyle C^*j} & & = \downarrow{\scriptstyle \mathrm{id}} & & \downarrow{\scriptstyle i^*_{Y,Z}} & & \\ 0 & \longrightarrow & C^*(X, Z) & \xrightarrow{j^*_{X,Z}} & C^*(X) & \xrightarrow{i^*_{X,Z}} & C^*(Z) & \longrightarrow & 0 \end{array} \tag{3.1.16}$$

where the horizontal lines are exact sequences of cochain complexes as in (2.7.9), we get a short exact sequence of cochain complexes

$$0 \to C^*(X, Y) \xrightarrow{C^*j} C^*(X, Z) \xrightarrow{C^*i} C^*(Y, Z) \to 0. \tag{3.1.17}$$

The same arguments with the chain complexes gives a short exact sequence

$$0 \to C_*(Y, Z) \xrightarrow{C_*i} C_*(X, Z) \xrightarrow{C_*j} C_*(X, Y) \to 0. \tag{3.1.18}$$

As in Sect. 2.7 of Chap. 2, short exact sequences (3.1.17) and (3.1.18) produces connecting homomorphisms $\delta_T : H^*(Y, Z) \to H^{*+1}(X, Y)$ and $\partial_T : H_*(X, Y) \to C_{*-1}(Y, Z)$. They satisfy $\langle \delta_T(a), \alpha \rangle = \langle a, \partial_T(\alpha) \rangle$ as well as following proposition.

Proposition 3.1.25 ((Co)homology sequences of a triple) *Let $T = (X, Y, Z)$ be a triple. Then,*

(a) *the sequences*

$$\cdots \to H^m(X, Y) \xrightarrow{H^*j} H^m(X, Z) \xrightarrow{H^*i} H^m(Y, Z) \xrightarrow{\delta_T} H^{m+1}(X, Y) \xrightarrow{H^*j} \cdots$$

and

$$\cdots \to H_m(Y, Z) \xrightarrow{H_*i} H_m(X, Z) \xrightarrow{H_*j} H_m(X, Y) \xrightarrow{\partial_T} H_{m-1}(Y, Z) \xrightarrow{H_*i} \cdots$$

are exact.

(b) *the exact sequences of Point* (a) *are natural for maps of triples.*

Remark 3.1.26 As $H^*(\emptyset) = 0$, we get a canonical **GrV**-isomorphisms $H^*(X, \emptyset) \xrightarrow{\approx} H^*(X)$, etc. Thus, the (co)homology sequences for the triple $(X, Y, \emptyset)$ give back those of the pair (X, Y)

$$\cdots \to H^m(X,Y) \xrightarrow{H^*j} H^m(X) \xrightarrow{H^*i} H^m(Y) \xrightarrow{\delta^*} H^{m+1}(X,Y) \xrightarrow{H^*j} \cdots \tag{3.1.19}$$

and

$$\cdots \to H_m(Y) \xrightarrow{H_*i} H_m(X) \xrightarrow{H_*j} H^m(X,Y) \xrightarrow{\partial_*} H_{m-1}(Y) \xrightarrow{H_*i} \cdots \tag{3.1.20}$$

where $i : Y \to X$ and $j : (X, \emptyset) \to (X, Y)$ denote the inclusions. This gives a more precise description of the morphisms j^* and j_* of Proposition 3.1.19.

We now draw a few consequences of Proposition 3.1.19. A topological pair (X, Y) is of *finite (co)homology type* if its singular homology (or, equivalently, cohomology) is of finite type. In this case, the *Poincaré series* of (X, Y) is that of $H_*(X, Y)$:

$$\mathfrak{P}_t(X,Y) = \sum_{i\in\mathbb{N}} \dim H_i(X,Y)\, t^i = \sum_{i\in\mathbb{N}} \dim H^i(X,Y)\, t^i \in \mathbb{N}[[t]].$$

Corollary 3.1.27 *Let (X, Y, Z) be a topological triple. Suppose that two of the pairs (X, Y), (Y, Z) and (X, Z) are of finite cohomology type. Then, the third pair is of finite cohomology type and there is $Q_t \in \mathbb{N}[[t]]$ such that the equality*

$$\mathfrak{P}_t(X,Y) + \mathfrak{P}_t(Y,Z) = \mathfrak{P}_t(X,Z) + (1+t)\, Q_t, \tag{3.1.21}$$

holds in $\mathbb{N}[[t]]$.

Proof This follows from the cohomology sequence of $T = (X, Y, Z)$ and elementary linear algebra. If $\delta_T^k : H^k(Y, Z) \to H^{k+1}(X, Y)$ denotes the connecting homomorphism, one checks that (3.1.21) holds true for

$$Q_t = \sum t^k \operatorname{codim} \delta_T^k. \qquad \square$$

Corollary 3.1.27 implies straightforwardly the following result.

Corollary 3.1.28 *Let (X, Y, Z) be a topological triple. Suppose that* $\dim H^*(Y, Z) < \infty$ *and that* $\dim H^*(X, Y) < \infty$. *Then* $\dim H^*(X, Z) < \infty$ *and*

$$\dim H^*(X,Z) \le \dim H^*(X,Y) + \dim H^*(Y,Z).$$

Corollary 3.1.16 has the following generalization with relative (co)homology.

Proposition 3.1.29 *Let (X, Y) be a topological pair. Let $\mathcal{A}$ be family of subspaces of X, partially ordered by inclusion. Suppose that each compact subspace of X is contained in some $A \in \mathcal{A}$. Then, the natural homomorphisms*

$$J_* : \varinjlim_{A\in\mathcal{A}} H_*(A, A\cap Y) \xrightarrow{\approx} H_*(X,Y)$$

and

$$J^* : H^*(X,Y) \xrightarrow{\approx} \varprojlim_{A\in\mathcal{A}} H^*(A, A\cap Y)$$

are isomorphisms.

Proof By Kronecker duality, only the bijectivity of J_* must be proven. Let $\mathcal{H}_r(Y) = \varinjlim_{A\in\mathcal{A}} H_r(A\cap Y), \mathcal{H}_r(X) = \varinjlim_{A\in\mathcal{A}} H_r(A)$, and $\mathcal{H}_r(X,Y) = \varinjlim_{A\in\mathcal{A}} H_r(A, A\cap Y)$. For each $A \in \mathcal{A}$, one has the homology sequence of the pair $(A, A\cap Y)$. By naturality of these exact sequences under inclusions, one gets the diagram:

$$\begin{array}{ccccccccc}
\mathcal{H}_r(Y) & \longrightarrow & \mathcal{H}_r(X) & \longrightarrow & \mathcal{H}_r(X,Y) & \xrightarrow{\partial} & \mathcal{H}_{r-1}(Y) & \longrightarrow & \mathcal{H}_{r-1}(X) \\
\approx\downarrow & & \approx\downarrow & & \downarrow J_* & & \approx\downarrow & & \approx\downarrow \\
H_r(Y) & \longrightarrow & H_r(X) & \longrightarrow & H_r(X,Y) & \xrightarrow{\partial} & H_{r-1}(Y) & \longrightarrow & H_{r-1}(X)
\end{array}$$

The top horizontal line is exact because the direct limit of exact sequences is exact. The bijectivity of the vertical arrows comes from Corollary 3.1.16. By the five-lemma, one deduces that J_* is an isomorphism. □

3.1.3 The Homotopy Property

Let $f, g : (X,Y) \to (X',Y')$ be two maps between topological pairs. Let $I = [0,1]$. A *homotopy* between f and g is a map of pairs $F : (X \times I, Y \times I) \to (X', Y')$ such that $F(x,0) = f(x)$ and $F(x,1) = g(x)$. If such a homotopy exists, we say that f and g are *homotopic*.

Proposition 3.1.30 (Homotopy property) *Let $f, g : (X,Y) \to (X',Y')$ be two maps between topological pairs which are homotopic. Then $H_* f = H_* g$ and $H^* f = H^* g$.*

Proof Note that $H_* f = H_* g$ implies $H^* f = H^* g$ by Kronecker duality, using Diagram (2.3.4). We shall construct a $\mathbb{Z}_2$-linear map $D : C_*(X) \to C_{*+1}(X')$ such that

$$C_* f + C_* g = \partial \circ D + D \circ \partial, \tag{3.1.22}$$

i.e. D is a chain homotopy from $C^* f$ to $C^* g$. The map D will satisfy $D(C_*(Y)) \subset C_{*+1}(Y')$ and will so induce a linear map $D : C_*(X,Y) \to C_{*+1}(X',Y')$ satisfying (3.1.22). As in the proof of Proposition 2.5.9, this will prove that $H_* f = H_* g$. That $H^* f = H^* g$ is then deduced by Kronecker duality, using Diagram (2.3.4). Let $F : (X \times I, Y \times I) \to (X', Y')$ be a homotopy from f to g.

By linearity, it is enough to define D on singular simplexes. Let $\sigma : \Delta^m \to X$ be a singular m-simplex of X. Consider the convex-cell complex $P = \Delta^m \times I$. One has

$V(P) = V(\Delta^m) \times \{0, 1\}$. Using the natural total order on $V(\Delta^m)$, we can define an affine order on P by deciding that the elements of $V(\Delta^m) \times \{1\}$ are greater than those of $V(\Delta^m) \times \{0\}$. Lemma 2.1.10 thus provides a triangulation $h_{\leq} : |L_{\leq}(P)| \xrightarrow{\approx} P$, with $V(L_{\leq}(P)) = V(P)$. Set $L = L_{\leq}(P)$ and $h = h_{\leq}$. The order $\leq$ becomes a simplicial order on L, giving rise to a chain map $R_{\leq,*} : C_*(L) \to C_*(P)$ from the simplicial chains of L to the singular chains of P (see Example 3.1.6). Consider $\mathcal{S}_{m+1}(L)$ as an $(m+1)$-simplicial cochain of L and define $D(\sigma)$ to be the image of $\mathcal{S}_{m+1}(L)$ under the composed map

$$\mathcal{S}_{m+1}(L) \xrightarrow{R_{\leq,*}} \mathcal{S}_{m+1}(|L|) \xrightarrow{C_*h} \mathcal{S}_{m+1}(P) \xrightarrow{C_*(\sigma\times\mathrm{id})} \mathcal{S}_{m+1}(X \times I) \xrightarrow{C_*F} \mathcal{S}_{m+1}(X'). \tag{3.1.23}$$

Observe that, if $\tau \in \mathcal{S}_m(L)$ such that $h(|\bar{\tau}|)$ hits the interior of P, then τ is the face of exactly two $(m+1)$-simplexes of L. Therefore, $\partial(\mathcal{S}_{m+1}(L)) = \mathcal{S}_m(L(\mathrm{Bd}\, P))$. But

$$\mathrm{Bd}\, P = \Delta^m \times \{0\} \cup \Delta^m \times \{1\} \cup \mathrm{Bd}\, \Delta^m \times I.$$

As all the maps in (3.1.23) are chain maps, this permits us to prove that

$$\partial \circ D(\sigma) = C_* f(\sigma) + C_* g(\sigma) + D \circ \partial(\sigma) \tag{3.1.24}$$

As (3.1.24) holds true for all $\sigma \in \mathcal{S}(X)$, it implies (3.1.22). □

Remark 3.1.31 In the proof of Proposition 3.1.30, the chain homotopy D is not unique. Some authors (e.g. [43, 155, 179]) just give an existence proof, based on an easy case of the acyclic carrier's technique (like in our proof of Proposition 2.5.9). We used above an explicit triangulation of $\Delta^m \times I$. The same triangulation occurs the proof of [82, p. 112], presented differently for the sake of sign's control. The idea of such triangulations of $\Delta^m \times I$ will be used again in the proof of the small simplex theorem 3.1.34.

A map of pairs $f : (X, Y) \to (X', Y')$ is a *homotopy equivalence* if there exists a map of pairs $g : (X', Y') \to (X, Y)$ such that $g \circ f$ is homotopic to $\mathrm{id}_{(X,Y)}$ and $f \circ g$ is homotopic to $\mathrm{id}_{(X',Y')}$. The pairs (X, Y) and (X', Y') are then called *homotopy equivalent*. Two spaces X and X' are *homotopy equivalent* if the pairs $(X, \emptyset)$ and $(X', \emptyset)$ are homotopy equivalent. Two homotopy equivalent spaces (or pairs) are also said to have the same *homotopy type*.

By functoriality (Proposition 3.1.22), Proposition 3.1.30 implies that (co)homology is an invariant of homotopy type:

Corollary 3.1.32 (Homotopy invariance of (co)homology) *Let $f : (X, Y) \to (X', Y')$ be a homotopy equivalence. Then $H_* f : H_*(X, Y) \to H_*(X', Y')$ and $H^* f : H^*(X', Y') \to H^*(X, Y)$ are isomorphisms.*

A (non-empty) topological space X is *contractible* if there exists a homotopy from id_X to a constant map. For instance, the *cone CX over a space X*

$$CX = \big(X \times I\big)/\big(X \times \{1\}\big), \tag{3.1.25}$$

with the quotient topology, is contractible. A homotopy from id_{CX} to a constant map is given by $F((x,\tau),t) = [x, t + (1-t)\tau]$.

Corollary 3.1.33 *The (co)homology of a contractible space is isomorphic to that of a point:*

$$H^*(X) \approx H_*(X) \approx \begin{cases} 0 & \text{if} * > 0 \\ \mathbb{Z}_2 & \text{if} * = 0. \end{cases}$$

Proof Let $x_0 \in X$ such that there exists a homotopy from id_X to the constant map onto x_0. Then, the inclusion $\{x_0\} \to X$ is a homotopy equivalence and Corollary 3.1.33 follows from Corollary 3.1.32. □

For a direct proof of Corollary 3.1.33, see Exercise 3.2.

3.1.4 Excision

Let X be a topological space. Let $\mathcal{B}$ be a family of subspaces of X. A map $f : L \to X$ is called $\mathcal{B}$-*small* if $f(L)$ is contained in an element of $\mathcal{B}$. Let $\mathcal{S}_m^{\mathcal{B}}(X)$ be the set of singular m-simplexes of X which are $\mathcal{B}$-small. The vector spaces of (co)chains $C_{\mathcal{B}}^m(X)$ and $C_m^{\mathcal{B}}(X)$ are defined as in Sect. 3.1, using $\mathcal{B}$-small m-simplexes. We get, in the same way, a pairing $\langle\,,\rangle : C_{\mathcal{B}}^m(X) \times C_m^{\mathcal{B}}(X) \to \mathbb{Z}_2$ identifying $C_{\mathcal{B}}^m(X)$ to $C_m^{\mathcal{B}}(X)^\sharp$. The boundary of a $\mathcal{B}$-small simplex is a $\mathcal{B}$-small chain, so $(C_*^{\mathcal{B}}(X), \partial)$ is a subcomplex of chains of $(C_*(X), \partial)$, the inclusion being denoted by $i_*^{\mathcal{B}} : C_{\mathcal{B}}^m(X) \to C^m(X)$. Define $\delta : C_{\mathcal{B}}^m(X) \to C_{\mathcal{B}}^{m+1}(X)$ by $\langle \delta(a), \alpha\rangle = \langle a, \partial(\alpha)\rangle$ and $i_{\mathcal{B}}^* C^m(X) \to C_{\mathcal{B}}^m(X)$ by $\langle i_{\mathcal{B}}^*(a), \alpha\rangle = \langle a, i_*^{\mathcal{B}}(\alpha)\rangle$. Then, $((C_{\mathcal{B}}^*(X), \delta), (C_*^{\mathcal{B}}(X), \partial), \langle\,,\rangle)$ is a Kronecker pair.

The (co)homologies obtained by these definitions are denoted by $H_{\mathcal{B}}^*(X)$ and $H_*^{\mathcal{B}}(X)$. One uses the notations $i_*^{\mathcal{B}} : H_m^{\mathcal{B}}(X) \to H_m(X)$ and $i_{\mathcal{B}}^* : H^m(X) \to H_{\mathcal{B}}^m(X)$ for the induced linear maps. The following result is very useful.

Proposition 3.1.34 (Small simplexes theorem) *Let X be a topological space with a family $\mathcal{B}$ of subspaces of X, whose interiors cover X. Then $i_*^{\mathcal{B}} : H_*^{\mathcal{B}}(X) \to H_*(X)$ and $i_{\mathcal{B}}^* : H^*(X) \to H_{\mathcal{B}}^*(X)$ are isomorphisms.*

The proof of Proposition 3.1.34 uses iterations of the subdivision operator, a chain map $\mathrm{sd}_* : C_*(X,Y) \to C_*(X,Y)$ which replaces chains by chains with "smaller" simplexes. Intuitively, sd_* replaces a singular simplex $\sigma : \Delta^m \to X$ by the sum of σ restricted to the barycentric subdivision of Δ^m.

More precisely, consider the standard simplex Δ^m as the geometric realization of the full complex $\mathcal{F}_m$ over the set $\{0, 1, \ldots, m\}$. The barycentric subdivision $\mathcal{F}_m'$

is endowed with its natural simplicial order $\leq$ of (2.1.2), p. 11. As explained in Example 3.1.6, we get a chain map

$$R_* = R_{\leq,*} : C_*(\mathcal{F}'_m) \to C_*(|\mathcal{F}'_m|) = C_*(\Delta^m).$$

Let $\sigma \in \mathcal{S}_m(X)$. As a continuous map from Δ^m to X, σ induces $C_*\sigma : C_*(\Delta^m) \to C_*(X)$. Define

$$\operatorname{sd}_*(\sigma) = C_*\sigma(\mathcal{S}_m(\mathcal{F}'_m)).$$

This formula determines a unique linear map $\operatorname{sd}_* : C_*(X) \to C_*(X)$ which is clearly a chain map. If Y is a subspace of X, then $\operatorname{sd}_*(C_*(Y)) \subset C_*(X)$, so we get a chain map $\operatorname{sd}_* : C_*(X, Y) \to C_*(X, Y)$, giving rise to a **GrV**-morphism

$$\operatorname{sd}_* : H_*(X, Y) \to H_*(X, Y).$$

By Kronecker duality, we get a cochain map $\operatorname{sd}^* : C^*(X, Y) \to C^*(X, Y)$ and a **GrV**-morphism $\operatorname{sd}^* : H^*(X, Y) \to H^*(X, Y)$ satisfying $\langle \operatorname{sd}^*(a), \alpha \rangle = \langle a, \operatorname{sd}_*(\alpha) \rangle$ for all $a \in C^*(X, Y)$ and $\alpha \in C_*(X, Y)$.

Observe that sd sends $C_*^{\mathcal{B}}(X, Y)$ into $C_*^{\mathcal{B}}(X, Y)$ and thus $H_*^{\mathcal{B}}(X, Y)$ into $H_*^{\mathcal{B}}(X, Y)$ and $H^*_{\mathcal{B}}(X, Y)$ into $H^*_{\mathcal{B}}(X, Y)$

Lemma 3.1.35 *The subdivision operators* $\operatorname{sd}_* : H_*(X, Y) \to H_*(X, Y)$ *and* $\operatorname{sd}^* : H^*(X, Y) \to H^*(X, Y)$ *are equal to the identity. The same holds true for* $\operatorname{sd}_* : H_*^{\mathcal{B}}(X, Y) \to H_*^{\mathcal{B}}(X, Y)$ *and* $\operatorname{sd}^* : H^*_{\mathcal{B}}(X, Y) \to H^*_{\mathcal{B}}(X, Y)$.

Proof We shall construct a $\mathbb{Z}_2$-linear map $D : C_*(X) \to C_{*+1}(X)$ such that

$$\mathrm{id} + \operatorname{sd}_* = \partial \circ D + D \circ \partial. \tag{3.1.26}$$

In other words, D is a chain homotopy from id to sd_* (see p. 34). The map D will satisfy $D(C_*(Y)) \subset C_{*+1}(Y)$ and will so induce a linear map $D : C_*(X, Y) \to C_{*+1}(X, Y)$ satisfying (3.1.26). As in the proof of Proposition 2.5.9, this will prove that $\operatorname{sd}_* = \mathrm{id}$. That $\operatorname{sd}^* = \mathrm{id}$ is then implied by Kronecker duality, using Diagram (2.3.4). Also, the map D will satisfy $D(C_*^{\mathcal{B}}(X)) \subset C_{*+1}^{\mathcal{B}}(X)$.

By linearity, it is enough to define D on singular simplexes. The proof is similar to that of Proposition 3.1.30 (an idea of V. Puppe). Let $\sigma : \Delta^m \to X$ be a singular m-simplex of X. Consider the convex-cell complex $P = \Delta^m \times I$, where the upper face $\Delta^m \times \{1\}$ is replaced by its barycentric subdivision $|\mathcal{F}'_m|$. One has $V(P) = V(\Delta^m) \times \{0\} \dot\cup V(\mathcal{F}'_m) \times \{1\}$. We use the natural total order on $V(\Delta^m) \times \{0\}$ and the natural simplicial order on $V(\mathcal{F}'_m) \times \{1\}$ (see (2.1.2)). Deciding in addition that the elements of $V(\mathcal{F}'_m) \times \{1\}$ are greater than those of $V(\Delta^m) \times \{0\}$ provides an affine order $\leq$ on P. Lemma 2.1.10 thus constructs a triangulation $h_{\leq} : |L_{\leq}(P)| \xrightarrow{\approx} P$, with $V(L_{\leq}(P)) = V(P)$. Set $L = L_{\leq}(P)$ and $h = h_{\leq}$. Seeing $\leq$ as a simplicial order on L gives rise to a chain map $R_{\leq,*} : C_*(L) \to C_*(P)$ from the simplicial chains of L to the singular chains of P (see Example 3.1.6). Consider $\mathcal{S}_{m+1}(L)$ as an $(m+1)$-simplicial cochain of L and define $D(\sigma)$ to be the image of $\mathcal{S}_{m+1}(L)$ under the composed map

$$\mathcal{S}_{m+1}(L) \xrightarrow{R_{\leq,*}} \mathcal{S}_{m+1}(|L|) \xrightarrow{C_*h} \mathcal{S}_{m+1}(P) \xrightarrow{C_*(\hat{\sigma})} \mathcal{S}_{m+1}(X), \tag{3.1.27}$$

where $\hat{\sigma} : P \to X$ is the map $\hat{\sigma}(x,t) = \sigma(x)$. Observe that the inclusion $|\mathcal{F}'_m| \subset \Delta^m \times \{1\}$ is already the piecewise affine triangulation of $\Delta^m \times \{1\}$ determined by the simplicial order on $\mathcal{F}'_m$. Therefore, the construction of the proof of Lemma 2.1.10 leaves $\Delta^m \times \{1\}$ unchanged. Formula (3.1.26) is then deduced as in the proof of Proposition 3.1.30. Finally, if σ is $\mathcal{B}$-small, so is the map $\hat{\sigma}$. Hence $D(\sigma) \in C^{\mathcal{B}}_{m+1}(X)$, which proves the lemma for the $\mathcal{B}$-small (co)homology. □

Proof of Proposition 3.1.34 By Kronecker duality, using Corollary 2.3.11, only the homology statement must be proved. Let $\mathrm{sd}^k = \mathrm{sd} \circ \cdots \circ \mathrm{sd}$ (k times). We shall need the following statement.

Claim let $\alpha \in C_*(X)$. Then, there exists $k(\alpha) \in \mathbb{N}$ such that $\mathrm{sd}^k(\alpha) \in C^{\mathcal{B}}_*(X)$ for all $k \geq k(\alpha)$.

Let us show that the claim implies Proposition 3.1.34. Let $\bar{\alpha} \in H_m(X,Y)$ represented by $\alpha \in C_m(X)$ with $\partial(\alpha) \in C_{m-1}(Y)$. The claim implies that, for k big enough, $\mathrm{sd}^k(\alpha) \in C^{\mathcal{B}}_m(X)$, and thus $\partial(\alpha) \in C_{m-1}(Y)$. This implies that $\mathrm{sd}^k(\bar{\alpha})$ is in the image of $i^{\mathcal{B}}_*$. By Lemma 3.1.35, $\mathrm{sd}^k(\bar{\alpha}) = \bar{\alpha}$, so $\bar{\alpha}$ is in the image of $i^{\mathcal{B}}_*$, which proves that $i^{\mathcal{B}}_*$ is surjective. For the injectivity, let $\bar{\beta} \in H^{\mathcal{B}}_m(X,Y)$ with $i^{\mathcal{B}}_*(\beta) = 0$. Represent $\bar{\beta}$ by $\beta \in C^{\mathcal{B}}_m(X)$ with $\partial(\beta) \in C_{m-1}(Y)$. The hypothesis $i^{\mathcal{B}}_*(\beta) = 0$ says that $\beta = \partial(\gamma) + \omega$ with $\gamma \in C_{m+1}(X)$ and $\omega \in C_m(Y)$. The claim tells us that, for k big enough, $\mathrm{sd}^k(\gamma) \in C^{\mathcal{B}}_{m+1}(X)$ (and, so, $\mathrm{sd}^k(\omega) \in C^{\mathcal{B}}_m(Y)$). This implies that $\mathrm{sd}^k(\bar{\beta}) = 0$ in $H^{\mathcal{B}}_m(X,Y)$. But $\mathrm{sd}^k(\bar{\beta}) \in C^{\mathcal{B}}_m(X)$ and Lemma 3.1.35 tells us that sd^k coincides with the identity of $H^{\mathcal{B}}_m(X,Y)$. Thus, $\bar{\beta} = 0$ for all $\bar{\beta} \in \ker i^{\mathcal{B}}_*$.

It remains to prove the claim. Let $\rho(m,k)$ be the maximal distance between two points of a simplex of the k-th barycentric subdivision of Δ^m. An elementary argument of Euclidean geometry shows that

$$\rho(m,k) \leq \rho(m,0) \left(\frac{m}{m+1}\right)^k \tag{3.1.28}$$

(of course, $\rho(m,0) = \sqrt{2}$). For details, see e.g., [155, Proof of Theorem 15.4] or [82, p. 120]. By hypothesis, the family $\dot{\mathcal{B}} = \{\mathrm{int}\, B \mid B \in \mathcal{B}\}$ is an open covering of X. Consider the induced open covering $\sigma^{-1}\dot{\mathcal{B}}$ of Δ^m. By (3.1.28), $\rho(m,k) \to 0$ when $k \to \infty$. Using a Lebesgue number for the open covering $\sigma^{-1}\dot{\mathcal{B}}$, this proves the claim.

The main application of the small simplexes theorem is the invariance under excision (see also Sect. 3.1.6).

Proposition 3.1.36 (Excision property) *Let (X,Y) be a topological pair. Let U be a subspace of X with $\bar{U} \subset \mathrm{int}\, Y$. Then, the linear maps induced by inclusions*

$$i^* : H^*(X, Y) \xrightarrow{\approx} H^*(X - U, Y - U) \;\; \textit{and} \;\; i_* : H_*(X - U, Y - U) \xrightarrow{\approx} H_*(X, Y)$$

are isomorphisms.

Proof By Corollary 2.3.11, i_* is an isomorphism if and only if i^* is an isomorphism. We shall prove that i_* is an isomorphism.

Let $\mathcal{B} = \{Y, X - U\}$. One has a commutative diagram

$$\begin{array}{ccccccccc} 0 & \longrightarrow & C_*(Y) & \longrightarrow & C_*^{\mathcal{B}}(X) & \longrightarrow & C_*^{\mathcal{B}}(X)/C_*(Y) & \longrightarrow & 0 \\ & & = \downarrow \text{id} & & \downarrow i_*^{\mathcal{B}} & & \downarrow I_*^{\mathcal{B}} & & \\ 0 & \longrightarrow & C_*(Y) & \longrightarrow & C_*(X) & \longrightarrow & C_*(X)/C_*(Y) & \longrightarrow & 0 \end{array}$$

where all arrows are induced by inclusions and the horizontal lines are short exact sequences of chain complexes. As in Sect. 2.6, this gives a commutative diagram between the corresponding long homology sequences

$$\begin{array}{ccccccccccc} \cdots & \longrightarrow & H_m(Y) & \longrightarrow & H_m^{\mathcal{B}}(X) & \longrightarrow & H_m(C_*^{\mathcal{B}}(X)/C_*(Y)) & \longrightarrow & H_{m-1}(Y) & \longrightarrow & \cdots \\ & & = \downarrow \text{id} & & \downarrow i_*^{\mathcal{B}} & & \downarrow I_*^{\mathcal{B}} & & = \downarrow \text{id} & & \\ \cdots & \longrightarrow & H_m(Y) & \longrightarrow & H_m(X) & \longrightarrow & H_m(X, Y) & \longrightarrow & H_{m-1}(Y) & \longrightarrow & \cdots \end{array}$$

As $\bar{U} \subset \operatorname{int} Y$, the family $\mathcal{B} = \{Y, X - U\}$ satisfy the hypotheses of Proposition 3.1.34 and $i_*^{\mathcal{B}}$ is an isomorphism. By the five-lemma, $I_*^{\mathcal{B}}$ is an isomorphism. Therefore, it suffices to show that $H_*(X - U, Y - U) \to H_*(C_*^{\mathcal{B}}(X)/C_*(Y))$ is an isomorphism. But it is easy to see that this is already the case at the chain level:

$$C_*(X - U, Y - U) = C_*(X - U)/C_*(Y - U) \xrightarrow{\approx} C_*^{\mathcal{B}}(X)/C_*(Y). \qquad \square$$

3.1.5 Well Cofibrant Pairs

Let (Z, Y) be a topological pair and denote by $i : Y \to Z$ the inclusion. A (continuous) map $r : Z \to Y$ is called a *retraction* if $r \circ i = \mathrm{id}_Y$. It is a *retraction by deformation* if $i \circ r$ is homotopic to the identity of Z. A retraction by deformation is thus a homotopy equivalence.

Note that Z retracts by deformation on Y if and only if there is a homotopy $h : Z \times I \to Z$ which, for all $(z, t) \in Z \times I$, satisfies $h(z, 0) = z$, $h(z, 1) \in Y$ and $h(y, t) = y$ when $y \in Y$. A topological pair (X, A) is called *good* if A is closed in X and if there is a neighbourhood V of A which retracts by deformation onto A. For instance, $(X, \emptyset)$ is a good pair ($V = \emptyset$ and $h(x, t) = x$).

Good pairs were introduced in [82] (with the additional condition that A is non-empty). Earlier books rather rely on the notion of cofibration, developed in the 1960s essentially by Puppe and Steenrod (see [185] for references). Both are useful in different circumstances, so we introduce below the mixed notion of a *well cofibrant pair*, especially useful in equivariant homotopy theory (see e.g. Chap. 7). We begin by cofibrant pairs, starting with the following lemma.

Lemma 3.1.37 *For a topological pair (X, A), the following conditions are equivalent.*

(1) *There is a retraction from $X \times I$ onto $X \times \{0\} \cup A \times I$.*
(2) *Let $f : X \to Z$ and $F_A : A \times I \to Z$ be continuous maps such that $F_A(a, 0) = f(a)$. Then, F_A extends to a continuous map $F : X \times I \to Z$ such that $F(x, 0) = f(x)$ for all $x \in X$.*

Proof We give below the easier proof available when A is closed in X (for a proof without this hypothesis: see [39, (1.19)]). Let $r : X \times I \to X \times \{0\} \cup A \times I$ be a retraction. Given f and F_A as in (2), define the map $g : X \times \{0\} \cup A \times I \to Z$ by $g(x, 0) = f(x)$ and $g(a, t) = F_A(a, t)$. If A is closed, then g is continuous and the map $F = g \circ r$ satisfies the required condition. Hence, (1) implies (2). Conversely, if f and F_A are the inclusions of X and $A \times I$ into $Z = X \times \{0\} \cup A \times I$, the extension F given by (2) is a continuous retraction from $X \times I$ onto Z. □

A pair (X, A) with A closed in X which satisfies (1) or (2) of Lemma 3.1.37 is called *cofibrant*. According to the literature, the inclusion $A \hookrightarrow X$ is a *cofibration*, or satisfies the *absolute homotopy extension property* (*AHEP*) (see e.g. [44, 73]). See e.g. [38, Chap. 5] for other characterizations and properties of cofibrant pairs.

As a motivation of our concept of *well cofibrant pair*, we first give an example.

Example 3.1.38 Mapping cylinder neighbourhoods. Let (X, A) be a topological pair. A neighbourhood V of A is called a *mapping cylinder neighbourhood* if there is a continuous map $\varphi : \dot{V} \to A$ (where $\dot{V}$ is the frontier of V) and a homeomorphism $\psi : M_\varphi \to V$ where

$$M_\varphi = [(\dot{V} \times I) \dot{\cup} A] / \{(x, 0) \sim \varphi(x) \mid x \in \dot{V}\}$$

is the mapping cylinder of φ. The homeomorphism ψ is required to satisfy $\psi(x, 1) = x$ and $\psi(x, 0) = \varphi(x)$ for all $x \in \dot{V}$. Here are examples of mapping cylinder neighbourhoods

- if X is a smooth manifold and A is a smooth submanifold of codimension ≥ 1, then a closed tubular neighbourhood of A [95, Sect. 4.6] is a mapping cylinder neighbourhood.
- if A is the boundary of a smooth manifold X, then a collar neighbourhood of A [95, Sect. 4.6] is a mapping cylinder neighbourhood.
- a subcomplex of a CW-complex admits a mapping cylinder neighbourhood. The proof of this will be given in Lemma 3.4.2.

Given a mapping cylinder neighbourhood as above, a continuous retraction $F : X \times I \to X \times \{0\} \cup A \times I$ is defined by

$$F(x,t) = \begin{cases} (\varphi(v), t(1-2\tau)) & \text{if } x = \psi(v,\tau) \text{ with } \tau \le 1/2. \\ (\psi(v, 2\tau - 1), 0) & \text{if } x = \psi(v,\tau) \text{ with } \tau \ge 1/2. \\ (x, 0) & \text{if } x \in X - \operatorname{int} V. \end{cases}$$

Let $u : X \to I$ defined by

$$u(x) = \begin{cases} 2\tau & \text{if } x = \psi(v,\tau) \text{ with } \tau \le 1/2. \\ 1 & \text{otherwise.} \end{cases}$$

Let $h : X \times I \to X$ defined by $h = p_X \circ F$, where $p_X : X \times I \to X$ is the projection. Then, $u(h(x)) \le u(x)$ which implies that, for all $T < 1$, h restricts to a strong deformation retraction form $u^{-1}([0, T])$ onto $A = u^{-1}(0)$. Hence, (X, A) is a good and cofibrant pair.

A topological pair (X, A) is *well cofibrant* if there exists continuous maps $u : X \to I$ and $h : X \times I \to X$ such that

(1) $A = u^{-1}(0)$ (in particular, A is closed in X).
(2) $h(x, 0) = x$ for all $x \in X$.
(3) $h(a, t) = a$ for all $(a, t) \in A \times I$.
(4) $h(x, 1) \in A$ for all $x \in X$ such that $u(x) < 1$.
(5) $u(h(x, t)) \le u(x)$ for all $(x, t) \in X \times I$.

We say that (u, h) is a *presentation of* (X, A) as a well cofibrant pair. Conditions (1–4) define a *NDR-pair* (neighbourhood deformation retract pair) in the sense of [38, 140, 185] (see also Remarks 3.1.42 (b)).

The pairs $(X, \emptyset)$ and (X, X) are well cofibrant. One takes $u(x) = 1$ for $(X, \emptyset)$, $u(x) = 0$ for (X, X) and $h(x, t) = x$ for both pairs. Another basic example of well cofibrant pairs is given by the following lemma.

Lemma 3.1.39 *Suppose that $A \subset X$ admits a mapping cylinder neighbourhood in X. Then, (X, A) is well cofibrant.*

Proof The pair (u, h) in Example 3.1.38 is a presentation of (X, A) as a well cofibrant pair. □

Lemma 3.1.40 *Let (X, A) and (Y, B) be two well cofibrant pairs. Then, the "product pair" $(X \times Y, A \times Y \cup X \times B)$ is well cofibrant.*

The following proof, coming from that of [185, Theorem 6.3], will be convenient for the equivariant setting (see Lemma 7.2.12).

Proof Let (u, h) and (v, j) be presentations of (X, A) and (Y, B) as well cofibrant pairs. Define $w : X \times Y \to I$ by $w(x, y) = u(x)v(y)$. Define $q : X \times Y \times I \to X \times Y$ by

$$q(x, y, t) = \begin{cases} (x, y) & \text{if } (x, y) \in A \times B. \\ \big(h(x, t), j(y, \frac{u(x)}{v(y)}t)\big) & \text{if } v(y) \geq u(x) \text{ and } v(y) > 0. \\ \big(h(x, \frac{v(y)}{u(x)}t), j(y, t)\big) & \text{if } v(y) \leq u(x) \text{ and } u(x) > 0. \end{cases}$$

One checks that (w, q) is a presentation of $(X \times Y, A \times Y \cup X \times B)$ as a well cofibrant pair. Details for (1)–(4) are given in [185, p. 144] and (5) is obvious. □

Lemma 3.1.41 *Let (X, A) be a well cofibrant pair. Then, (X, A) is good and cofibrant.*

Proof Let (u, h) be a presentation of (X, A) as a well cofibrant pair. As noticed in Example 3.1.38, the condition $u(h(x, t)) \leq u(x)$ implies that, for all $T < 1$, h restricts to a strong deformation retraction form $u^{-1}([0, T])$ onto A. Since $A = u^{-1}(0)$, it is closed. Hence, (X, A) is good. To see that (X, A) is cofibrant, let $(Y, B) = (I, \{0\})$ presented as well cofibrant pair by (v, j) where $v(y) = y/2$ and $j(y, t) = (1 - t)y$. Let (w, q) be the presentation of

$$(X \times Y, A \times Y \cup X \times B) = (X \times I, X \times \{0\} \cup A \times I)$$

as a well cofibrant pair given in the proof of Lemma 3.1.40. As

$$w(x, y) = u(x)y/2 < 1,$$

the formula $r(x, y) = q(x, y, 1)$ defines a retraction

$$X \times I \xrightarrow{r} X \times \{0\} \cup A \times I. \tag{3.1.29}$$

By Lemma 3.1.37, (X, A) is cofibrant. □

Remarks 3.1.42 (a) The fact that the retraction r of (3.1.29) is a strong deformation retraction should not be a surprise. If $r = (r_1, r_2) : X \times I \to X \times \{0\} \cup A \times I \subset X \times I$ is any retraction, then the map $R : X \times I \to X \times I$ defined by

$$R(x, t, s) = \big(r_1(x, (1 - s)t), st + (1 - s)r_2(x, t)\big)$$

is a homotopy from $\mathrm{id}_{X \times I}$ to r [73, Lemma 16.28].
(b) The proof of Lemma 3.1.41 shows that a NDR-pair is cofibrant. The converse is also true (see [140, Sect. 6.4]).

If (X, A) is a topological pair, we denote by X/A the quotient space where all points of A are identified in a single class. The projection $\pi : (X, A) \to (X/A, A/A)$ is a map of pairs.

Lemma 3.1.43 *Let (X, A) be a well cofibrant pair and let $B \subset A$. Then $(X/B, A/B)$ is well cofibrant. In particular, the pair $(X/A, A/A)$ is well cofibrant.*

Proof Let (u, h) be a presentation of (X, A) as a well cofibrant pair. By (1) and (3), u and h descend to continuous maps $\bar{u} : X/B \to I$ and $\bar{h} : (X/B) \times I \to X/B$, giving a presentation $(\bar{u}, \bar{h})$ of $(X/B, A/B)$. □

Lemma 3.1.44 *Let (X, A) be a cofibrant pair such that A is contractible. Then the quotient map $X \to X/A$ is a homotopy equivalence.*

Proof If A is contractible, there is a continuous map $F_A : A \times I \to A \subset X$ such that $F_A(a, 0) = a$ and $F(A \times 1) = \{a_0\}$. As (X, A) is cofibrant, there is a continuous map $F : X \times I \to X$ such that $F(x, 0) = x$ and $F(a, t) = F_A(a, t)$ for $a \in A$. $F_{|X\times 1}$ admits a factorization

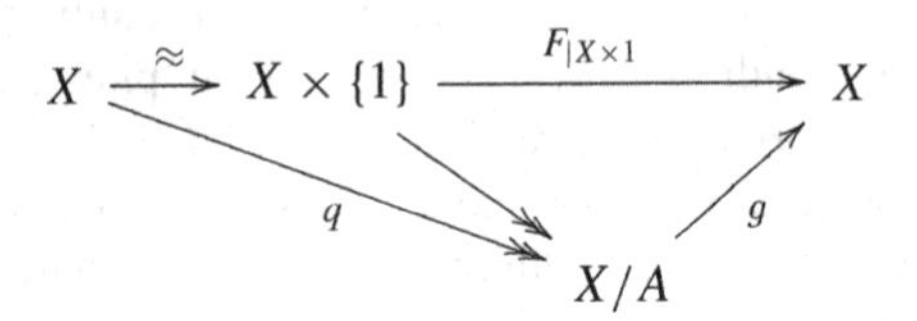

Using F, $g \circ q$ is homotopic to id_X. On the other hand, as $F(A \times I) \subset A$, the map $q \circ F$ descends to a continuous map $\bar{F} : X/A \times I \to X/A$ which is a homotopy from $\mathrm{id}_{X/A}$ to $q \circ g$. □

Proposition 3.1.45 *Let (X, A) be a well cofibrant pair. Then, the homomorphisms*

$$\pi^* : H^*(X/A, A/A) \to H^*(X, A) \;\; \textit{and} \;\; \pi_* : H_*(X, A) \to H_*(X/A, A/A)$$

are isomorphisms.

Proof By Corollary 2.3.11, π_* is an isomorphism if and only if π^* is an isomorphism. We shall prove that π^* is an isomorphism (the proof is the same for both). There is nothing to prove if $A = \emptyset$, so we assume that A is not empty.

By Lemma 3.1.41, (X, A) is cofibrant. Let r be a retraction from $X \times I$ to $X \times \{0\} \cup A \times I$. Let $CA = \big(A \times I\big)/\big(A \times \{1\}\big)$ be the cone over A and let $\hat{X} = X \cup_A CA$. As A is closed, r extends (by the identity on $CA \times I$) to a continuous retraction from $\hat{X} \times I$ onto $\hat{X} \times \{0\} \cup CA \times I$. Hence, the pair $(\hat{X}, CA)$ is cofibrant. But

$$H^*(\hat{X}, CA) \xrightarrow[\text{excision}]{\approx} H^*(\hat{X} - [A \times \{1\}], CA - [A \times \{1\}]) \xrightarrow[\text{homotopy}]{\approx} H^*(X, A).$$

On the other hand, $X/A = \hat{X}/CA$. Set $\bar{X} = \hat{X}/CA$ and $\bar{C} = CA/CA \approx \{pt\}$. The quotient map $q : (\hat{X}, CA) \to (\bar{X}, \bar{C})$ provides a morphism of exact sequences

$$\begin{array}{ccccccccc} H^{k-1}(\bar{X}) & \longrightarrow & H^{k-1}(\bar{C}) & \longrightarrow & H^k(\bar{X}, \bar{C}) & \longrightarrow & H^k(\bar{X}) & \longrightarrow & H^k(\bar{C}) \\ \approx \downarrow q^* & & \approx \downarrow q^* & & \downarrow q^* & & \approx \downarrow q^* & & \approx \downarrow q^* \\ H^{k-1}(\hat{X}) & \longrightarrow & H^{k-1}(CA) & \longrightarrow & H^k(\hat{X}, CA) & \longrightarrow & H^k(\hat{X}) & \longrightarrow & H^k(CA) \end{array} .$$

As $\bar{C}$ and CA are contractible, $q^* : H^*(\bar{C}) \to H^*(CA)$ is an isomorphism and so is $q^* : H^*(\bar{X}) \to H^*(\hat{X})$ by Lemma 3.1.44. By the five lemma, $q^* : H^k(\bar{X}, \bar{C}) \to H^k(\hat{X}, CA)$ is an isomorphism, which proves Proposition 3.1.45. □

Remark 3.1.46 The proof of Proposition 3.1.45 uses only that the pair (X, A) is cofibrant. Another proof exists using that (X, A) is a good pair (see [82, Proposition 2.22] or Proposition 7.2.15). It is interesting to note that these relatively short proofs both use almost all the axioms of a cohomology theory (see Sect. 3.9): functoriality, homotopy, excision and functorial exactness.

Corollary 3.1.47 *Let (X, A) be a well cofibrant pair with A non-empty. Then,*

$$\pi^* : \tilde{H}^*(X/A) \xrightarrow{\approx} H^*(X, A) \quad \text{and} \quad \pi_* : H_*(X, A) \xrightarrow{\approx} \tilde{H}_*(X/A)$$

are isomorphisms.

Proof If $A \neq \emptyset$, then A/A is a point. Therefore, by (3.1.13), $\tilde{H}^*(X/A) \xrightarrow{\approx} H^*(X/A, A/A)$ and $H_*(X/A, A/A) \xrightarrow{\approx} \tilde{H}_*(X/A)$. The results then follows form Proposition 3.1.45. □

Corollary 3.1.48 *Let (X, A) be a well cofibrant pair. Denote by $i : A \to X$ the inclusion and by $j : X \to X/A$ the quotient map. Then, there is a functorial exact sequence in reduced cohomology*

$$\cdots \to \tilde{H}^{k-1}(X) \xrightarrow{\tilde{H}^* i} \tilde{H}^{k-1}(A) \xrightarrow{\tilde{\delta}^*} \tilde{H}^k(X/A) \xrightarrow{\tilde{H}^* j} \tilde{H}^k(X) \xrightarrow{\tilde{H}^* i} \tilde{H}^k(A) \to \cdots$$

The corresponding sequence exists in reduced homology.

Proof The result is obvious if A is empty. Otherwise, this comes from the exact sequence of Proposition 3.1.20 together with the isomorphism $\tilde{H}^*(X/A) \xrightarrow{\approx} H^*(X, A) \xrightarrow{\approx} \tilde{H}^*(X, A)$ of Corollary 3.1.47. □

One application of well cofibrant pairs is the suspension isomorphism. Let X be a topological space. The *suspension* ΣX *of* X is the quotient space

$$\Sigma X = CX/\big(X \times \{0\}\big)$$

where CX is the cone over X (see (3.1.25)). The pairs $(CX, X \times 0)$ and $(\Sigma X, X \times \frac{1}{2})$ are well cofibrant by Lemma 3.1.39, since the subspaces admits mapping cylinder neighbourhoods. The *(cohomology) suspension homomorphism* is the degree-1 linear map $\Sigma^* : \tilde{H}^m(X) \to \tilde{H}^{m+1}(\Sigma X)$ given by the composition

$$\Sigma^* : \tilde{H}^m(X) \xrightarrow{\delta^*} \tilde{H}^{m+1}(CX, X \times 0) \approx \tilde{H}^{m+1}(\Sigma X, [X \times 0]) \xrightarrow{j^*} \tilde{H}^{m+1}(\Sigma X), \tag{3.1.30}$$

where the middle isomorphism comes from Proposition 3.1.45.

Proposition 3.1.49 *For any topological space X, the suspension homomorphism*

$$\Sigma^* : \tilde{H}^m(X) \to \tilde{H}^{m+1}(\Sigma X)$$

is an isomorphism for all m.

Because of Proposition 3.1.49, the homomorphism Σ^* is called the *suspension isomorphism* (in cohomology). Observe that Formula (3.1.30) can also be used to define $\Sigma^* : H^m(X) \to H^{m+1}(\Sigma X)$. By Proposition 3.1.49, this unreduced suspension homomorphism is an isomorphism if $m \geq 1$.

Proof If X is empty, so is ΣX and Proposition 3.1.49 is trivial. Suppose then that $X \neq \emptyset$. As $[X \times 0]$ is a point, its reduced cohomology vanish and j^* is an isomorphism in Formula (3.1.30), by the reduced cohomology sequence of the pair $(\Sigma X, [X \times 0])$. As CX is contractible, its reduced cohomology also vanish and δ^* is an isomorphism in Formula (3.1.30), by the reduced cohomology sequence of the pair $(CX, X \times 0)$. Finally, note that ΣX is path-connected, so $\tilde{H}^m(\Sigma X) = 0$ for $m \leq 0$, so Proposition 3.1.49 is also true for $m \leq 0$. □

Analogously, we define $\Sigma_* : \tilde{H}_{m+1}(\Sigma X) \to \tilde{H}_m(\Sigma X)$ by the composition

$$\Sigma_* : \tilde{H}_{m+1}(\Sigma X) \xrightarrow{j_*} \tilde{H}_{m+1}(\Sigma X, [X \times 0]) \approx \tilde{H}_{m+1}(CX, X \times 0) \xrightarrow{\partial_*} \tilde{H}_m(X) \tag{3.1.31}$$

which satisfies $\langle \Sigma^*(a), \alpha \rangle = \langle a, \Sigma_*(\alpha) \rangle$ for all $a \in \tilde{H}^m(X)$ and all $\alpha \in \tilde{H}_{m+1}(X)$. By Proposition 3.1.49 (or directly), we deduce that Σ_* is an isomorphism, called the *suspension isomorphism* (in homology).

Let (X_j, x_j) $(j \in \mathcal{J})$ be a family of *pointed spaces*, i.e. $x_i \in X_i$. Their *bouquet* (or *wedge*) $X = \bigvee_{j \in \mathcal{J}} X_j$ is defined as the quotient space

$$X = \bigvee_{j \in \mathcal{J}} X_j = \dot{\bigcup}_{j \in \mathcal{J}} X_j \Big/ \dot{\bigcup}_{j \in \mathcal{J}} \{x_j\}.$$

By naming $x \in X$ the equivalence class $x = \dot{\bigcup}_{j \in \mathcal{J}} \{x_j\}$, the couple (X, x) is a pointed space. For each $j \in \mathcal{J}$, one has a pointed inclusion $i_j : (X_j, x_j) \to (X, x)$. The bouquet plays the role of a sum in the category of pointed spaces and pointed maps: if $f^j : (X_j, x_j) \to (Y, y)$ are continuous pointed maps, then there is a unique continuous pointed map $f : (X, x) \to (Y, y)$ such that $f \circ i_j = f_j$.

A *well pointed space* is a pointed space (X, x) such that $(X, \{x\})$ is a well cofibrant pair. Observe that this definition is stronger than that in other textbooks.

Lemma 3.1.50 *If (X_j, x_j) $(j \in \mathcal{J})$ are well pointed spaces, then their wedge (X, x) is a well pointed space.*

Proof Let (u^j, h^j) be a presentation of (X_j, x_j) as a well cofibrant pair. Then $(\dot{\bigcup}_{j\in\mathcal{J}} u^j, \dot{\bigcup}_{j\in\mathcal{J}} h^j)$ is a presentation of $(\dot{\bigcup}_{j\in\mathcal{J}} X^j, \dot{\bigcup}_{j\in\mathcal{J}} \{x^j\})$ as a well cofibrant pair. By Lemma 3.1.43, the quotient pair $(X, \{x\})$ is well cofibrant, so (X, x) is a well pointed space. □

Proposition 3.1.51 *Let (X_j, x_j), with $j \in \mathcal{J}$, be a family of well pointed spaces. Then, the family of inclusions $i_j : X_j \to X = \bigvee_{j\in\mathcal{J}} X_j$, for $j \in \mathcal{J}$, gives rise to isomorphisms on reduced (co)homology*

$$\tilde{H}^*(X) \xrightarrow[\approx]{(\tilde{H}^* i_j)} \prod_{j\in\mathcal{J}} \tilde{H}^*(X_j)$$

and

$$\bigoplus_{j\in\mathcal{J}} \tilde{H}_*(X_j) \xrightarrow[\approx]{\sum \tilde{H}_* i_j} \tilde{H}_*(X).$$

Proof It is enough to establish that $\sum \tilde{H}_* i_j$ is an isomorphism. The cohomology statement can be proved analogously or by Kronecker duality, using Diagram (2.3.4).

Write, as above, $x = \dot{\bigcup}_{j\in\mathcal{J}} \{x_i\} \in X$. The map of pairs $(X_j, \{x_i\}) \to (X, x)$ give rise, for each $m \in \mathbb{N}$, to a commutative diagram between exact sequences

$$\begin{array}{ccccccc}
\bigoplus_{j\in\mathcal{J}} H_{m+1}(\{x_j\}) & \longrightarrow & \bigoplus_{j\in\mathcal{J}} H_{m+1}(X_j) & \longrightarrow & \bigoplus_{j\in\mathcal{J}} H_{m+1}(X_j, \{x_j\}) & \xrightarrow{\partial_*} \\
\downarrow \approx & & \downarrow \approx & & \downarrow \bigoplus H_* i_j & \\
H_{m+1}(\dot{\bigcup}_{j\in\mathcal{J}} \{x_j\}) & \longrightarrow & H^m(\dot{\bigcup}_{j\in\mathcal{J}} X_j) & \longrightarrow & H_{m+1}(\dot{\bigcup}_{j\in\mathcal{J}} X_j, \dot{\bigcup}_{j\in\mathcal{J}} \{x_j\}) & \xrightarrow{\partial_*}
\end{array}$$

$$\begin{array}{ccccc}
\xrightarrow{\partial_*} & \bigoplus_{j\in\mathcal{J}} H_m(\{x_j\}) & \longrightarrow & \bigoplus_{j\in\mathcal{J}} H_m(X_j) \\
 & \downarrow \approx & & \downarrow \approx \\
\xrightarrow{\partial_*} & H_m(\dot{\bigcup}_{j\in\mathcal{J}} \{x_j\}) & \longrightarrow & H_m(\dot{\bigcup}_{j\in\mathcal{J}} X_j)
\end{array}$$

The isomorphisms for the vertical arrows are due to Proposition 3.1.11. By the five-lemma, $\bigoplus H_* i_j$ is an isomorphism. As (X_j, x_j) is well pointed, the pair $(\dot{\bigcup}_{j\in\mathcal{J}} X_j, \dot{\bigcup}_{j\in\mathcal{J}} \{x_j\})$ is well cofibrant. By Proposition 3.1.45, the quotient map $q : (\dot{\bigcup}_{j\in\mathcal{J}} X_j,$

$\dot{\bigcup}_{j\in\mathcal{J}}\{x_j\}) \to (X, x)$ induces an isomorphism on homology. One has the commutative diagram

$$\begin{array}{ccc} \bigoplus_{j\in\mathcal{J}} \tilde{H}_*(X_j) & \xrightarrow{\sum \tilde{H}_* i_j} & \tilde{H}_*(X) \\ \approx\downarrow & & \approx\downarrow \\ \bigoplus_{j\in\mathcal{J}} H_*(X_j, \{x_j\}) \xrightarrow[\approx]{\bigoplus H_* i_j} & H_*(\dot{\bigcup}_{j\in\mathcal{J}} X_j, \dot{\bigcup}_{j\in\mathcal{J}}\{x_j\}) & \xrightarrow[\approx]{H_* q} H_*(X, \{x\}) , \end{array}$$

where the vertical arrows are isomorphisms by Remark 3.1.21. Therefore, $\sum \tilde{H}_* i_j$ is an isomorphism. □

The (co)homology of $X = \bigvee_{j\in\mathcal{J}} X_j$ may somehow be also controlled using the projection $\pi_j : X \to X_j$ defined by

$$\pi_j(z) = \begin{cases} z & \text{if } z \in X_j \\ x & \text{otherwise,} \end{cases} \tag{3.1.32}$$

where $x = \dot{\bigcup}_{j\in\mathcal{J}}\{x_i\} \in X$. As $\pi_j \circ i_j = \mathrm{id}_{X_j}$, Proposition 3.1.51 implies the following

Proposition 3.1.52 *Let (X_j, x_j), with $j \in \mathcal{J}$, be a family of well pointed spaces. Then, the composition*

$$\bigoplus_{j\in\mathcal{J}} \tilde{H}^*(X_j) \xrightarrow{\sum \tilde{H}^* \pi_j} \tilde{H}^*(X) \xrightarrow[\approx]{(\tilde{H}^* i_j)} \prod_{j\in\mathcal{J}} \tilde{H}^*(X_j)$$

is the inclusion of the direct sum into the product. Also, the composition

$$\bigoplus_{j\in\mathcal{J}} \tilde{H}_*(X_j) \xrightarrow[\approx]{\sum \tilde{H}_* i_j} \tilde{H}_*(X) \xrightarrow{(\tilde{H}_* \pi_j)} \prod_{j\in\mathcal{J}} \tilde{H}_*(X_j)$$

is the inclusion of the direct sum into the product.

In particular, if $\mathcal{J}$ is finite, then

$$\bigoplus \tilde{H}^*(X_j) \xrightarrow[\approx]{\sum \tilde{H}^* \pi_j} \tilde{H}^*(X) \quad \text{and}$$

$$\tilde{H}_*(X) \xrightarrow[\approx]{(\tilde{H}_* \pi_j)} \prod_{j\in\mathcal{J}} \tilde{H}_*(X_j)$$

are isomorphisms.

3.1.6 Mayer-Vietoris Sequences

Let X be a topological space. Let $\mathcal{B} = \{X_1, X_2\}$ be a collection of two subspaces of X. Write $X_0 = X_1 \cap X_2$ and

$$\begin{array}{ccc} X_0 & \xrightarrow{i_1} & X_1 \\ \downarrow{\scriptstyle i_2} & & \downarrow{\scriptstyle j_1} \\ X_2 & \xrightarrow{j_2} & X \end{array}$$

for the inclusions. We call (X, X_1, X_2, X_0) a *Mayer-Vietoris data*. The sequence of cochain complexes

$$0 \to C^*(X) \xrightarrow{(C^*j_1, C^*j_2)} C^*(X_1) \oplus C^*(X_2) \xrightarrow{C^*i_1 + C^*i_2} C^{\mathcal{B}}_*(X_0) \to 0$$

is then exact, as well as the sequence of chain complexes

$$0 \to C_*(X_0) \xrightarrow{(C_*i_1, C_*i_2)} C_*(X_1) \oplus C_*(X_2) \xrightarrow{C_*j_1 + C_*j_2} C^{\mathcal{B}}_*(X) \to 0$$

By Sect. 2.6, these short exact sequences give rise to connecting homomorphisms

$$\delta_{MV} : H^*(X_0) \to H^{*+1}_{\mathcal{B}}(X) \text{ and } \partial_{MV} : H^{\mathcal{B}}_*(X) \to H_{*-1}(X_0)$$

involved in long (co)homology exact sequences. If the interiors of X_1 and X_2 cover X, the theorem of small simplexes (3.1.34) implies that $H^*_{\mathcal{B}}(X) \approx H^*(X)$ and $H^{\mathcal{B}}_*(X) \approx H_*(X)$. Therefore, we obtain the following proposition.

Proposition 3.1.53 (Mayer-Vietoris sequences I) *Let (X, X_1, X_2, X_0) be a Mayer-Vietoris data. Suppose that $X = \operatorname{int} X_1 \cup \operatorname{int} X_2$. Then, the long sequences*

$$\to H^m(X) \xrightarrow{(H^*j_1, H^*j_2)} H^m(X_1) \oplus H^m(X_2) \xrightarrow{H^*i_1 + H^*i_2} H^m(X_0) \xrightarrow{\delta_{MV}} H^{m+1}(X) \to$$

and

$$\to H_m(X_0) \xrightarrow{(H_*i_1, H_*i_2)} H_m(X_1) \oplus H_m(X_2) \xrightarrow{H_*j_1 + H_*j_2} H^m(X) \xrightarrow{\partial_{MV}} H_{m-1}(X_0) \to$$

are exact.

These Mayer-Vietoris sequences are natural for maps $f : X \to X'$ such that $f(X_i) \subset X'_i$.

The hypotheses of Proposition 3.1.53 may not by directly satisfied in usual situations. Here is a variant which is more useful in practice.

Proposition 3.1.54 (Mayer-Vietoris sequences II). *Let (X, X_1, X_2, X_0) be a Mayer-Vietoris data, with X_i closed in X. Suppose that $X = X_1 \cup X_2$ and that (X_i, X_0) is a good pair for $i = 1, 2$. Then, the long sequences*

$$\to H^m(X) \xrightarrow{(H^*j_1, H^*j_2)} H^m(X_1) \oplus H^m(X_2) \xrightarrow{H^*i_1 + H^*i_2} H^m(X_0) \xrightarrow{\delta_{MV}} H^{m+1}(X) \to$$

and

$$\to H_m(X_0) \xrightarrow{(H_*i_1, H_*i_2)} H_m(X_1) \oplus H_m(X_2) \xrightarrow{H_*j_1 + H_*j_2} H^m(X) \xrightarrow{\partial_{MV}} H_{m-1}(X_0) \to$$

are exact.

Proof Choose a neighbourhood U_i of X_0 in X_i admitting a retraction by deformation onto X_0, called $\rho^i_t : U_i \to U_i$, $t \in I$. Let $X'_1 = X_1 \cup U_2$, $X'_2 = X_2 \cup U_1$ and $X'_0 = X'_1 \cap X'_2 = X_0 \cup U_1 \cup U_2$. We claim that $X = \operatorname{int} X'_1 \cup \operatorname{int} X'_2$. Indeed, as U_2 is a neighbourhood of X_0 in X_2, there exists an open set V_2 of X such that $X_0 \subset V_2 \cap X_2 \subset U_2$. As X_2 is closed, $X_1 - X_2 = X - X_2$ is open in X. Therefore

$$X_1 \subset (X_1 - X_2) \cup (V_2 \cap X_2) = (X_1 - X_2) \cup V_2 \subset \operatorname{int} X'_1.$$

In the same way, $X_2 \subset \operatorname{int} X'_2$. Hence, $X = \operatorname{int} X'_1 \cup \operatorname{int} X'_2$.

As $X = \operatorname{int} X'_1 \cup \operatorname{int} X'_2$, the Mayer-Vietoris sequences of Proposition 3.1.53 hold true with (X'_1, X'_2, X'_0). But X'_1 retracts by deformation onto X_1, using the retraction $\hat\rho^1_t : X'_1 \to X'_1$ given by

$$\hat\rho^1_t = \begin{cases} \rho^2_t(x) & \text{if } x \in U_2 \\ x & \text{if } x \in X_1. \end{cases}$$

In the same way, X'_2 retracts by deformation onto X_2 and X'_0 retracts by deformation onto X_0. This proves Proposition 3.1.54. □

For Mayer-Vietoris sequences with other hypotheses, see Exercise 3.11, from which Proposition 3.1.54 may also be deduced.

3.2 Spheres, Disks, Degree

So far, we have not encountered any space whose (co)homology is not zero in positive dimensions. The unit sphere S^n in $\mathbb{R}^{n+1}$ will be the first example. The shortest way to describe the (co)homology of such simple spaces is by giving their Poincaré polynomials. The definitions are the same as for simplicial complexes. A topological pair (X, Y) is of *finite (co)homology type* if its singular homology (or, equivalently, cohomology) is of finite type. In this case, the *Poincaré series* of (X, Y) (or of X if Y is empty) is that of $H_*(X, Y)$:

$$\mathfrak{P}_t(X,Y) = \sum_{i\in\mathbb{N}} \dim H_i(X,Y)\, t^i = \sum_{i\in\mathbb{N}} \dim H^i(X,Y)\, t^i \in \mathbb{N}[[t]].$$

When the series is a polynomial, we speak of the *Poincaré polynomial* of (X, Y).

Proposition 3.2.1 *The Poincaré polynomial of the sphere S^n is*

$$\mathfrak{P}_t(S^n) = 1 + t^n.$$

Proof The sphere S^0 consists of two points, so the result for $n = 0$ follows from (3.1.5) and Corollary 3.1.12. We can then propagate the result by the suspension isomorphism $\Sigma^* : \tilde{H}^*(S^n) \xrightarrow{\approx} \tilde{H}^*(S^{n+1})$ (see Proposition 3.1.49), since S^{n+1} is homeomorphic to ΣS^n. The homology statement uses the homology suspension isomorphism $\Sigma_* : \tilde{H}_*(S^{n+1}) \xrightarrow{\approx} \tilde{H}^*(S^n)$. □

As a consequence of Proposition 3.2.1, the sphere S^n is not contractible, though it is path-connected if $n > 0$. Also S^n and S^p are not homotopy equivalent if $n \neq p$. A useful corollary of Proposition 3.2.1 is the following

Corollary 3.2.2 $\mathfrak{P}_t(D^n, S^{n-1}) = t^n$.

Proof This follows from Proposition 3.2.1 and the (co)homology exact sequence of the pair (D^n, S^{n-1}). □

It will be useful to have explicit cycles for the generators of $H_n(D^n, S^{n-1})$ and $H_n(S^n)$. Let $\dot{\Delta}^n = \Delta^n - \text{int}\,\Delta^n$ be the topological boundary of the standard simplex Δ^n. The identity map $i_n : \Delta^n \to \Delta^n$ is a relative cycle of $(\Delta^n, \dot{\Delta}^n)$, representing a class $[i_n] \in H_n(\Delta^n, \dot{\Delta}^n)$. The boundary $\partial(i_n)$ belongs to $Z_{m-1}(\dot{\Delta}^n)$ and represents $\partial_*([i_n])$ in $H_{n-1}(\dot{\Delta}^n)$.

Proposition 3.2.3 *For all $n \in \mathbb{N}$, the following two statements hold true*:

A_n: *$[i_n]$ is the non-zero element of $H_n(\Delta^n, \dot{\Delta}^n) = \mathbb{Z}_2$.*
B_n: *$[\partial(i_{n+1})]$ is the non-zero element of $\tilde{H}_n(\dot{\Delta}^{n+1}) = \mathbb{Z}_2$.*

Proof Statements A_n and B_n are proven together, by induction on n, as follows:

(a) A_0 and B_0 are true.
(b) A_n implies B_n.
(c) B_n implies A_{n+1}.

As the affine simplex Δ^0 is a point and $\dot{\Delta}^0$ is empty, Statement $A(0)$ follows from the discussion in Example 3.1.5. To prove B_0, observe that $\dot{\Delta}^1$ consists of two points p and q. Identifying a singular 0-simplex with a point, one has $\partial(i_1) = p+q$, which represents a non-vanishing element of $H_0(\dot{\Delta}^1)$. But $\langle \mathbf{1}, p+q\rangle = 0$, which shows that $[\partial(i_1)] \neq 0$ in $\tilde{H}_0(\dot{\Delta}^1)$.

Let us prove (b). Consider the inclusion $\epsilon : \Delta^n \hookrightarrow \dot{\Delta}^{n+1}$ given by $\epsilon(t_0, \ldots, t_n) = (t_0, \ldots, t_n, 0)$. Let $\Lambda_n = \text{adh}\,(\dot{\Delta}^{n+1} - \epsilon(\Delta^n))$. Consider the homomorphisms:

$$\tilde{H}_n(\dot{\Delta}^{n+1}) \xrightarrow[\approx]{j*} H_n(\dot{\Delta}^{n+1}, \Lambda^n) \xleftarrow[\approx]{H_*\epsilon} H_n(\Delta^n, \dot{\Delta}^n).$$

The arrow j_* is bijective, as in (3.1.13), since Λ_m is contractible; the arrow $H_*\epsilon$ is bijective by excision and homotopy. As in $H_*(\dot{\Delta}^{n+1}, \Lambda^n)$ we neglect the singular chains in Λ_n, one has

$$j_*([\partial(i_{n+1})]) = H_*\epsilon\,([i_n])$$

which proves (b).

To prove (c), we use that $\partial_* : \tilde{H}_{n+1}(\Delta^{n+1}, \dot{\Delta}^{n+1}) \xrightarrow{\approx} \tilde{H}_n(\dot{\Delta}^n)$ is an isomorphism, since Δ^{n+1} is contractible, and that $\partial_*([i_{n+1}]) = [\partial(i_{n+1})]$. □

For the sphere S^1, Proposition 3.2.3 has the following corollary.

Corollary 3.2.4 *Let $\sigma : \Delta^1 \to S^1$ given by $\sigma(t, 1-t) = e^{2i\pi t}$. Then $\sigma \in C_1(S^1)$ is a singular 1-cycle of S^1 and its homology class is the non-zero element of $H_1(S^1) = \mathbb{Z}_2$.*

Proof Since $\sigma(1, 0) = \sigma(0, 1)$, the 1-cochain σ is a cycle. The map σ factors as

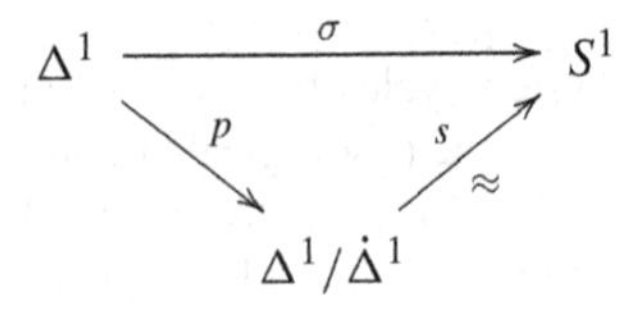

where s is a homeomorphism. Under the composed homomorphism

$$H_1(\Delta^1, \dot{\Delta}^1) \xrightarrow[\approx]{H_*p} H_1(\Delta^1/\dot{\Delta}^1, [\dot{\Delta}^1]) \xrightarrow{\approx} H_1(\Delta^1) \xrightarrow[\approx]{H_*s} H_1(S^1)\,,$$

the class $[i_1]$ goes to $[\sigma]$. By Proposition 3.2.3, $[i_1]$ is a generator of $H_1(S^1)$, which proves Corollary 3.2.4. □

Let $f : S^n \to S^n$ be a continuous map. The linear map $\tilde{H}_n f : \tilde{H}_n(S^n) \to \tilde{H}_n(S^n)$ is a map between $\mathbb{Z}_2$ and itself. The *degree* $\deg(f) \in \mathbb{Z}_2$ *of* f by

$$\deg(f) = \begin{cases} 0 & \text{if } \tilde{H}_n f = 0 \\ 1 & \text{otherwise.} \end{cases}$$

One can define the same degree using $\tilde{H}^n f$. For instance, the degree of a homeomorphism is 1 and the degree of a constant map is 0. Let $f, g : S^n \to S^n$. By

Proposition 3.1.30, $\deg(f) = \deg(g)$ if $f, g : S^n \to S^n$ are homotopic. Also, using that $H_n(g \circ f) = H_n g \circ H_n f$ one gets

$$\deg(g \circ f) = \deg(g) \cdot \deg(f) \tag{3.2.1}$$

These simple remarks have the following surprisingly strong consequences. (For a refinement of Proposition 3.2.5 below using the integral degree (see [155, Theorems 21.4 and 21.5]).

Proposition 3.2.5 *Let $f : S^n \to S^n$ be a continuous map with* $\deg f = 0$. *Then,*

(a) *f admits a fixed point.*
(b) *there exists $x \in S^n$ with $f(x) = -x$.*

Proof Suppose that there is no fixed point. Then f is homotopic to the antipodal map $a(x) = -x$: a homotopy is obtained by following the arc of great circle from $f(x)$ to $-x$ not containing x. Therefore $\deg f = \deg a = 1$ since a is a homeomorphism. If $f(x) \neq -x$ for all x, then $\deg f = 1$ because f is homotopic to the identity (following the arc of great circle from $f(x)$ to x not containing $-x$). □

We now give three recipes to compute the degree of a map from S^n to itself. A point $u \in S^n$ is a *topological regular value* for $f : S^n \to S^n$ if there is a neighbourhood U of u such that U is "*evenly covered*" by f. By this, we mean that $f^{-1}(U)$ is a disjoint union of U_j, indexed by a set $\mathcal{J}$, such that, for each $j \in \mathcal{J}$, the restriction of f to U_j is a homeomorphism from U_j to U. In particular, $f^{-1}(u)$ is a discrete closed subset of S^n indexed by $\mathcal{J}$, so $\mathcal{J}$ is finite since S^n is compact. For instance, a point u which is not in the range of f is a topological regular value of f (with $\mathcal{J}$ empty). For a topological regular value u of f, we define the *local degree* $d(f, u) \in \mathbb{N}$ *of f at u* by

$$d(f, u) = \sharp f^{-1}(u).$$

Proposition 3.2.6 *Let $f : S^n \to S^n$ be a continuous map. For any topological regular value u of f, one has*

$$\deg(f) = d(f, u) \mod 2.$$

Example 3.2.7 The map $S^1 \to S^1$ given by $z \mapsto z^k$ has degree the residue class of k mod 2.

Proof of Proposition 3.2.6 When $n = 0$, each of the two points of S^0 is a regular value of f and the equality of Proposition 3.2.6 is easy to check by examination of the various cases. We then suppose that $n > 0$.

If u is a topological regular value, there is a neighbourhood B of u which is evenly covered by f and which is homeomorphic to a closed disk D^n. Its preimage $\tilde{B} = f^{-1}(B)$ is a finite disjoint union of n-disks B_j, indexed by $j \in \mathcal{J}$. Define $\bar{\mathcal{J}} = \mathcal{J} \dot{\cup} \{0\}$ and set $B_0 = B$.

For $j \in \bar{\mathcal{J}}$, define $V_j = \overline{S^n - B_j}$ and set $\tilde{V} = \overline{S^n - \tilde{B}}$. Consider the quotient spaces $S^n_j = S^n/V_j$ ($j \in \bar{\mathcal{J}}$), which are homeomorphic to S^n. Thus, $S^n/\tilde{V} \approx \bigvee_{j\in\mathcal{J}} S^n_j$ is homeomorphic to a bouquet of $\sharp\mathcal{J}$ copies of S^n. Denote the quotient maps by $\varpi_j : S^n \to S^n_j$ and $\varpi : S^n \to S^n/\tilde{V} \approx \bigvee_{j\in\mathcal{J}} S^n_j$.

If $u_j \in U_j$ is the point such that $f(u_j) = u$ ($u_0 = u$), then $S^n - \{u_j\}$ is a neighbourhood of V_j which retracts by deformation onto V_j. Therefore, (S^n, V_j) is a good pair and, as $S^n - \{u_j\}$ is homeomorphic to $\mathbb{R}^n$, the space V_j is contractible. Also, V_j admits a mapping cylinder neighbourhood in S^n, so the pair (S^n, V_j) is well cofibrant by Lemma 3.1.39. By the reduced homology sequence of the pairs (S^n, V_j) and $(S^n_j, [V_j])$ and Proposition 3.1.45, we get three isomorphisms in the commutative diagram

$$\begin{array}{ccc} \tilde{H}_n(S^n) & \xrightarrow{H_n\varpi_j} & \tilde{H}_n(S^n_j) \\ \downarrow \approx & & \downarrow \approx \\ \tilde{H}_n(S^n, V_j) & \xrightarrow[\approx]{} & \tilde{H}_n(S^n_j, [V_j]), \end{array}$$

which shows that $H_n\varpi_j$ is an isomorphism.

Let $0 \neq \alpha \in H_n(S^n)$, and, for $j \in \bar{\mathcal{J}}$, let $0 \neq \alpha_j \in H_n(S^n_j)$. The map f descends to a continuous map $\bar{f} : S^n/\tilde{V} \to S^n_0$. Let us consider the commutative diagram:

$$\begin{array}{ccc} \tilde{H}_n(S^n) & \xrightarrow{H_n\varpi} & \bigoplus_{j\in\mathcal{J}} \tilde{H}_n(S^n_j) \\ \downarrow \tilde{H}_n f & & \downarrow H_n \bar{f} \\ \tilde{H}_n(S^n) & \xrightarrow[\approx]{\tilde{H}_n\varpi_0} & \tilde{H}_n(S^n_0) \end{array} \tag{3.2.2}$$

The restriction of $\bar{f}$ to S^n_j is a homeomorphism, so $H_n\bar{f}(\alpha_j) = \alpha_0$. Let $\pi_k : \bigvee_{j\in\mathcal{J}} S^n_j \to S^n_k$ be the projection onto the kth component (see Eq. (3.1.32)). Then $\varpi_j = \pi_j \circ \varpi$. By Proposition 3.1.52, this implies that $\tilde{H}_n\varpi(\alpha) = (\alpha_j)$. Then, $\tilde{H}_n\bar{f} \circ \tilde{H}_n\varpi(\alpha) = d(f, u)\,\alpha_0$. On the other hand, $\tilde{H}_n\varpi_0 \circ \tilde{H}_n f(\alpha) = \deg(f)\,\alpha_0$. As Diagram (3.2.2) is commutative, this proves Proposition 3.2.6.

The second recipe is the following lemma.

Lemma 3.2.8 *Let $f : S^n \to S^n$ be a continuous map, with $n > 0$. Let $B_1, \dots, B_k$ be disjoint embedded closed n-disks of S^n with boundary $\dot{B}_i$. Let V be the closure of $S^n - \bigcup B_i$. Suppose that f sends V onto a single point $v \in S^n$ and thus induces continuous maps $f_i : S^n \approx B_i/\dot{B}_i \to S^n$. Then*

$$\deg f = \sum_{i=1}^{k} \deg f_i .$$

Proof Let $S_i^n = B_i/\dot{B}_i$, homeomorphic to S^n. The map f factors in the following way:

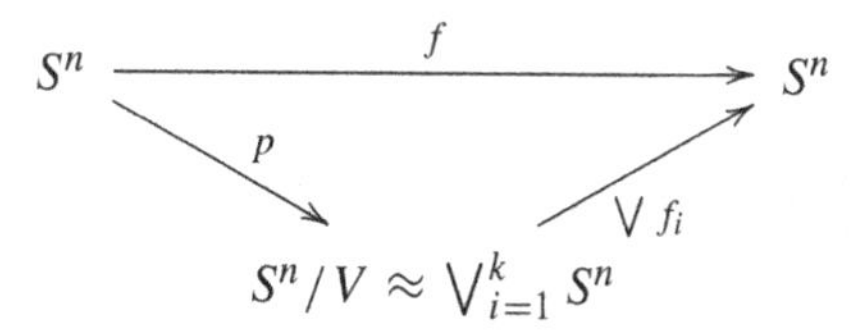

Obviously, $H_* p([S^n]) = \sum_{i=1}^k [S_i^n]$. Hence,

$$\deg f\,[S^n] = \sum_{i=1}^{k} H_* f_i([S_i^n]) = \big(\sum_{i=1}^{k} \deg f_i\big)\,[S^n].$$

□

The third recipe concerns the self-maps of S^1. Recall the elementary way to prove that $[S^1, S^1] \approx \mathbb{Z}$. Let $f : S^1 \to S^1$ be a (continuous) map. As $t \mapsto \exp(2i\pi t)$ is a local homeomorphism $\mathbb{R} \to S^1$, there exists a map $g : I \to \mathbb{R}$ such that $f(\exp(2i\pi t)) = \exp(2i\pi g(t))$. The integer

$$\mathrm{DEG}\,(f) = g(1) - g(0) \in \mathbb{Z}$$

depends only on the homotopy class of f. This defines a bijection

$$\mathrm{DEG} : [S^1, S^1] \xrightarrow{\approx} \mathbb{Z}. \tag{3.2.3}$$

For instance, for the map $f(z) = z^n$, one can choose $g(t) = nt$. Thus, $\mathrm{DEG}\,(f) = n$ if and only if f is homotopic to $z \mapsto z^n$.

Proposition 3.2.9 *For a map* $f : S^1 \to S^1$,

$$\deg(f) = \mathrm{DEG}\,(f) \bmod 2.$$

Proof If $\mathrm{DEG}\,(f) = n$, then f is homotopic to $z \mapsto z^n$. This map satisfies $\deg(f) = n \bmod 2$ by Proposition 3.2.6. □

Remarks 3.2.10 (a) Our degree is the reduction mod 2 of the integral degree obtained using integral homology (see e.g. [82, Sect. 2.2]). Proposition 3.2.6 would also hold for the integral degree, provided one takes into account the orientations in the definition of the local degree.
(b) A continuous map $f : S^n \to S^n$ may not have any topological regular value. For example, S. Ferry constructed a map $f : S^3 \to S^3$ with (integral) degree 2 so that the preimage of every point is connected [62].
(c) Suppose that $f = |g|$, where $g : K \to L$ is a simplicial map, with $|K|$ and $|L|$ homeomorphic to S^n. Let $\tau \in \mathcal{S}_n(L)$ and u be a point in the interior of $|\bar{\tau}|$. Then u is a regular value and $d(f, u) = d(g, \tau)$ (see Eq. (2.5.5)).

(d) Let $f : S^n \to S^n$ be a smooth map. Any smooth regular value is a topological regular value. Then, $\deg(f)$ coincides with the degree mod 2 of f as presented in e.g. [152, Sect. 4].

3.3 Classical Applications of the mod 2 (Co)homology

At our stage of development of homology, textbooks usually present a couple of classical applications in topology. Several of them only require $\mathbb{Z}_2$-homology, other need the integral homology. We discuss this matter in this section.

Retractions and Brouwer's fixed point Theorem. Let X be a topological space with a subspace Y. A *continuous retraction* of X onto Y is a continuous map $r : X \to Y$ extending the identity of Y. In other words, $r \circ i = \mathrm{id}_Y$, where $i : Y \to X$ denotes the inclusion. Therefore, $H_*r \circ H_*i = \mathrm{id}$ and $H^*i \circ H^*r = \mathrm{id}$, which implies the following lemma.

Lemma 3.3.1 *If there exists a continuous retraction from X onto Y, then $H_*i : H_*(Y) \to H_*(X)$ is injective and $H_*i : H^*(X) \to H^*(Y)$ is surjective. The same holds true for the reduced (co)homology.*

As $\tilde{H}_*(D^n) = 0$ while $\tilde{H}_{n-1}(S^{n-1}) = \mathbb{Z}_2$, Lemma 3.3.1 has the following corollary.

Proposition 3.3.2 *There is no continuous retraction of the n-disk D^n onto its boundary S^{n-1}.*

The most well known corollary of Proposition 3.3.2 is the *fixed point theorem* proved by Luitzen Egbertus Jan Brouwer around 1911 (see [40, Chap. 3]).

Corollary 3.3.3 *A continuous map from the disk D^n to itself has at least one fixed point.*

Proof Suppose that $f(x) \neq x$ for all $x \in D^n$. Then a retraction $r : D^n \to S^{n-1}$ is constructed using the following picture, contradicting Proposition 3.3.2.

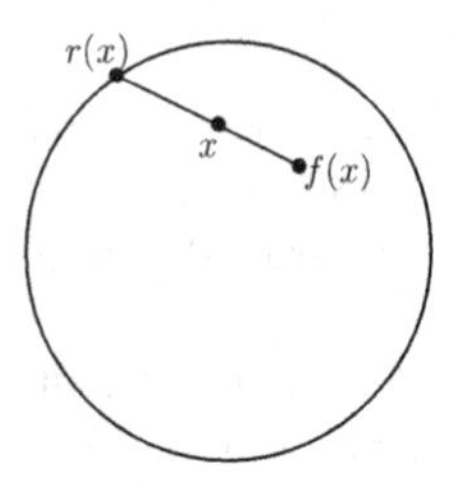

□

Brouwer's theorem says that, for a map $f : D^n \to D^n$, the equation $f(x) = x$ admits a solution under the only hypothesis that f is continuous. Given the possible wildness of a continuous map, this is a very deep theorem. It is impressive that such a result is due to the fact that $H_n(D^n) = 0$ and $H_n(S^{n-1}) = \mathbb{Z}_2$.

Invariance of dimension. An n-dimensional *topological manifold* is a topological space such that each point has an open neighbourhood homeomorphic to $\mathbb{R}^n$. The following result is known as the *topological invariance of the dimension* and goes back to the work of Brouwer in 1911 (see [40, Chap. II]).

Theorem 3.3.4 *Suppose that a non-empty m-dimensional topological manifold is homeomorphic to an n-dimensional topological manifold. Then $m = n$.*

Proof Let M be a non-empty m-dimensional topological manifold and N be an n-dimensional topological manifold. Let $h : M \to N$ be a homeomorphism. Let $x \in M$. Then h restricts to a homeomorphism from $M - \{x\}$ onto $N - \{h(x)\}$. Hence, $H_*h : H_*(M, M - \{x\}) \to H_*(N, N - \{h(x)\})$ is an isomorphism. But this contradicts the fact that

$$H_k(M, M - \{x\}) = \begin{cases} \mathbb{Z}_2 & \text{if } k = 0, m \\ 0 & \text{otherwise} \end{cases} \quad \text{and} \quad H_k(N, N - \{h(x)\}) = \begin{cases} \mathbb{Z}_2 & \text{if } k = 0, n \\ 0 & \text{otherwise.} \end{cases} \tag{3.3.1}$$

Indeed, it enough to prove (3.3.1) in the case of M. As x has a neighbourhood in M homeomorphic to $\mathbb{R}^m$, it has a neighbourhood B homeomorphic to closed m-ball. By the excision of $M - B$ and homotopy, one has

$$H_*(M, M - \{x\}) \approx H_*(B, B - \{x\}) \approx H_*(B, \mathrm{Bd}\, B)$$

and (3.3.1) follows from Corollary 3.2.2. □

Balls and spheres in spheres. The following results concerns the complements of k-balls or k-spheres in S^n.

Proposition 3.3.5 *Let $h : D^k \to S^n$ be an embedding. Then $\tilde{H}_*(S^n - h(D^k)) = 0$.*

Proof We follow the classical proof (see e.g. [82, Proposition 2b.1]), which goes by induction on k. For $k = 0$, D^0 is a point and $S^n - h(D^0)$ is then contractible. For the induction step, suppose that $\tilde{H}_i(S^n - h(D^k))$ contains a non-zero element α_0. We use the homeomorphism $D^k \approx D^{k-1} \times I_0$ with $I_0 = [0, 1]$. Then $S^n - h(D^k) = A \cup B$ with $A = S^n - (D^{k-1} \times [0, 1/2])$ and $B = S^n - (D^{k-1} \times [1/2, 1])$. Since, by induction hypothesis, $\tilde{H}_*(A \cap B) = \tilde{H}_*(S^n - h(D^{k-1} \times \{1/2\}) = 0$, the Mayer-Vietoris sequence implies that for $I_1 = [0, 1/2]$ or $I_1 = [1/2, 1]$, the homomorphism $H_i(S^n - h(D^k \times I_0)) \to H_i(S^n - h(D^k \times I_1))$ sends α_0 to $0 \neq \alpha_1 \in H_i(S^n - (D^{k-1} \times I_1))$. Iterating this process produces a nested sequence I_j of closed intervals converging to a point $p \in I$ and a non-zero element $\{\alpha_j\} \in \lim_{\to} \tilde{H}_*(X_j)$ where $X_j = S^n - (D^{k-1} \times I_j)$. Set $X = S^n - (D^{k-1} \times \{p\})$. As each compact subspace of X is contained in some X_j, Corollary 3.1.16 implies that $\lim_{\to} \tilde{H}_*(X_j)$ is isomorphic to $\tilde{H}_*(X)$, contradicting the induction hypothesis. □

Proposition 3.3.6 *Let $h : S^k \to S^n$ be an embedding with $k < n$. Then $\tilde{H}_*(S^n - h(S^k)) \approx \tilde{H}_*(S^{n-k-1})$.*

Proof The proof is by induction on k. The sphere S^0 consisting of two points, $S^n - h(S^0)$ is homotopy equivalent to S^{n-1}. For the induction step, write S^k as the union of two hemispheres $D_\pm$. The Mayer-Vietoris sequence for $S^n - h(D_\pm)$ together with Proposition 3.3.5 gives the isomorphism $\tilde{H}_*(S^n - h(S^k)) \approx \tilde{H}_{*-1}(S^n - h(S^{k-1}))$. □

The case $k = n - 1$ in Proposition 3.3.6 gives the following corollary.

Corollary 3.3.7 (Generalized Jordan Theorem) *Let $h : S^{n-1} \to S^n$ be an embedding. Then $S^n - h(S^{n-1})$ has two path-connected components.*

Remarks 3.3.8 (a) Topological arguments show that, in Corollary 3.3.7, $h(S^{n-1})$ is the common frontier of each of the components of its complement (see e.g. [155, Theorem 6.3]). For a discussion about the possible homotopy types of these components, see e.g. [155, Sect. 36] or [82, Sect. 2B].
(b) A well known consequence of the generalized Jordan theorem is the *invariance of domain* : *if U is an open set in $\mathbb{R}^n$, then its image $h(U)$ under an embedding $h : U \to \mathbb{R}^n$ is an open set in $\mathbb{R}^n$*. This can be deduced from Corollary 3.3.7 by a purely topological argument (see, e.g. [155, Theorem 36.5] or [82, Theorem 2B.3]).

Unavailable applications. Some applications cannot be obtained using $\mathbb{Z}_2$-homology. The most well known are the following.

(a) *The antipodal map in S^{2n} is not homotopic to the identity* and its consequence, the *non-existence of non-zero vector fields on even-dimensional spheres* (see, e.g. [82, Theorem 2.28] or [155, Corollary 21.6]). This requires the (co)homology with coefficients in $\mathbb{Z}$ or in a field of characteristic $\neq 2$.
(b) The determination of $[S^1, S^1]$ and the fundamental theorem of algebra (using $H^1(S^1; \mathbb{Z})$) (see, e.g., [155, Exercise 2,Sect. 21] or [82, Theorem 1.8]).

3.4 CW-Complexes

CW-complexes were introduced and developed by J.H.C. Whitehead in the years 1940–1950 [40, p.221]. The spaces having the homotopy type of a CW-complex (*CW-space*) are closed under several natural construction (see [147]). They are the spaces for which many functors of algebraic topology, like singular (co)homology, are reasonably efficient.

Let Y be a topological space and let (Z, A) be a topological pair. Let $\varphi : A \to Y$ be a continuous map. Consider the space

$$Z \cup_\varphi Y = Z \dot{\cup} Y / \{z = \varphi(z) \mid z \in A\},$$

endowed with the quotient topology. The space Y is naturally embedded into $Z \cup_\varphi Y$. We say that $Z \cup_\varphi Y$ is *obtained from* Y *by attachment* (*or adjunction*) *of* Z, using the *attaching map* φ. When (Z, A) is homeomorphic to $(\Lambda \times D^n, \Lambda \times S^{n-1})$, where Λ is a set (considered as a discrete space), we say that Z is obtained from Y by *attachment of n-cells*, indexed by Λ. For $\lambda \in \Lambda$, the image of $\{\lambda\} \times \operatorname{int} D^n$ in X is the *open cell* indexed by λ.

A *CW-structure* on the space X is a filtration

$$\emptyset = X^{-1} \subset X^0 \subset X^1 \subset \cdots \subset X = \bigcup_{n \in \mathbb{N}} X^n , \tag{3.4.1}$$

such that, for each n, the space X^n is homeomorphic to a space obtained from X^{n-1} by attachment of n-cells, indexed by a set $\Lambda_n = \Lambda_n(X)$. A space endowed with a CW-structure is a *CW-complex*. We see Λ_n as the *set of n-cells of X*. The space X^n is called the *n-skeleton* of X. The topology of X is supposed to be the *weak topology*: a subspace $A \subset X$ is open (or closed) if and only if $A \cap X^n$ is open (or closed) for all $k \in \mathbb{N}$.

If X is a CW-complex, a subspace $Y \subset X$ is a *subcomplex* of X if $Y^n = Y \cap X^n$ is obtained from $Y^{n-1} = Y \cap X^{n-1}$ by attaching n-cells, indexed by $\Lambda_n(Y) \subset \Lambda_n(X)$ and using the same attaching maps. For instance, the skeleta of X are subcomplexes of X. A topological pair (X, Y) formed by a CW-complex X and a subcomplex Y is called a *CW-pair*.

Let X be a CW-complex. With the above definition, the following properties hold true:

(1) X is a Hausdorff space.
(2) for each n and each $\lambda \in \Lambda_k$, there exists a continuous map $\varphi_\lambda : (D^n, S^{n-1}) \to (X^n, X^{n-1}) \subset (X, X^{n-1})$ such that its restriction to $\operatorname{int} D^n$ is an embedding from $\operatorname{int} D^n$ into X. Indeed, such a map, called a *characteristic map* for the n-cell λ, may be obtained by choosing a homeomorphism between (X^n, X^{n-1}) and $([\Lambda \times D^n] \cup_\varphi X^{n-1}, X^{n-1})$.
(3) a map $f : X \to Z$ to the topological space Z is continuous if and only if its restriction to each skeleton is continuous. Also, f is continuous if and only if $f \circ \varphi_\lambda$ is continuous for any characteristic map φ_λ and any cell λ.
(4) each subcomplex of X is a closed subset of X.
(5) X^0 is a discrete space.
(6) A compact subset of a CW-complex meets only finitely many cells. In consequence, a CW-complex is compact if and only if it is finite, i.e. it contains a finite number of cells.

These properties are easy to prove (see, e.g. [82, pp. 519–523]).

Proposition 3.4.1 *A CW-pair* (X, A) *is well cofibrant.*

The literature contains many proofs that a CW-pair is good (see e.g. [82, Proposition A.5] or [64, Proposition 1.3.1], or cofibrant (see e.g. [73, Proposition

14.13] or [38, Proposition 8.3.9]). The proof of Proposition 3.4.1 uses the following lemma.

Lemma 3.4.2 *Let Z be a space obtained from a space Y by attaching a collection of n-cells. Then Y admits in Z a mapping cylinder neighbourhood (see Example* 3.1.38).

Proof Let $\varphi : \Lambda \times D^n \to Y$ be the attaching map. Let $C^n = \{x \in D^n \mid |x| \geq 1/2\}$. Then, Z contains $V = (\Lambda \times C^n) \cup_\varphi Y$ as a closed neighbourhood of Y. The reader will check that V is homeomorphic to the mapping cylinder of φ. □

Proof of Proposition 3.4.1 Let $\bar{X}^n = X^n \cup A$. By Lemmas 3.4.2 and 3.1.39, the pair $(\bar{X}^n, \bar{X}^{n-1})$ is well cofibrant for all n. Let (v^n, g^n) be a presentation of $(\bar{X}^n, \bar{X}^{n-1})$ as a well cofibrant pair. As $\bar{X}^0$ is the disjoint union of A with a discrete set, we may assume that $v^0(x) = 1$ if $x \notin A$. Let W^n be the closure of $(v^n)^{-1}([0, 1))$. For $n \geq 1$, by replacing v^n by $\min\{2v^n, 1\}$ if necessary, we may assume that g^n restricts to a strong deformation retraction of W^n onto $\bar{X}^{n-1}$.

We now define a map $u : X \to I$ by constructing, inductively on $n \in \mathbb{N}$, its restriction to $\bar{X}^n$, denoted by u^n. We set $u^0 = v^0$ and

$$u^n(x) = \begin{cases} \min\{1, v^n(x) + u^{n-1}(g^n(x, 1))\} & \text{if } x \in W^n \\ 1 & \text{if } x \in X^n - \operatorname{int} W^n. \end{cases}$$

We check that u^n is continuous. If $x \in X^n$, then $u^k(x) = u^n(x)$ for $k \geq n$, therefore u is well defined and continuous. The space $V = u^{-1}([0, 1)) = \bigcup_n V^n$ is a closed neighbourhood of A in X, where $V^n = V \cap \bar{X}^n \subset W^n$.

Define $h^n : V^n \times I \to V^n$ by

$$h^n(x, t) = \begin{cases} x & \text{if } t \leq 1/2^{n+1} \\ g^n(x, 2^{n+1}t - 1) & \text{if } 1/2^{n+1} \leq t \leq 1/2^n \\ g^n(x, 1) & \text{if } t \geq 1/2^n. \end{cases}$$

Define $h_t^n : V^n \to V^n$ by $h_t^n(x) = h(x, t)$. If $x \in V^n$, define $h(x, t) = h_t^1 \circ \cdots \circ h_t^n(x)$. Note that $h_t^k(x) = x$ for $k > n$ so, if $x \in V^n \subset V^m$, then $h_t^1 \circ \cdots \circ h_t^n(x) = h_t^1 \circ \cdots \circ h_t^m(x)$. Therefore, $h : V \times I \to V$ is well defined and continuous.

The pair (u, h) satisfies all the conditions for being a presentation of (X, A) as a well cofibrant pair, except that h is only defined on $V \times I$ instead of $X \times I$. To fix that, choose a continuous map $\alpha : I \to I$ such that $\alpha([0, 1/2]) = \{1\}$ and α vanish on a neighbourhood of 1. Let $\bar{h} : X \times I \to X$ and $\bar{u} : X \to I$ defined by $\bar{h}(x, t) = h\big(x, \alpha(u(x))t\big)$ and $\bar{u}(x) = u(x)$. One checks that $(\bar{u}, \bar{h})$ is a presentation of (X, A) as a well cofibrant pair.

Here are classical examples of CW-complexes.

Example 3.4.3 The sphere S^n has an obvious CW-structure with one 0-cell and one n-cell (attached trivially).

Example 3.4.4 Observe that the sphere

$$S^n = \{x = (x_0, x_1, \ldots, x_n) \in \mathbb{R}^{n+1} \mid |x|^2 = 1\}$$

is obtained from S^{n-1} by adjunction of two $(n+1)$-cells $D^{n+1}_{\pm}$, attached by the identity map of S^{n-1}. Indeed, the embeddings $D^n_{\pm} \to S^n$ given by $y = (y_1, \ldots, y_n) \mapsto (\pm\sqrt{1-|y|^2}, y_1, \ldots, y_n)$ extend the inclusion $S^{n-1} \hookrightarrow S^n$ and provide a homeomorphism between $S^{n-1} \cup D^n_{\pm}$ and S^n. Starting from $S^0 = \{\pm 1\}$, we thus get a CW-structure on S^n with two cells in each dimension and whose k-skeleton is S^k. Taking the inductive limit S^∞ of those S^n gives a CW-complex known as the *infinite dimensional sphere*. This is a contractible space (see e.g. [82, Example 1.B.3p. 88]).

Example 3.4.5 The CW-structure on S^n of Example 3.4.4 is invariant under the antipodal map. It then descends to a CW-structure on the projective space $\mathbb{R}P^n = S^n/\{x \sim -x\}$, having one cell in each dimension. Its k-th skeleton is $\mathbb{R}P^k$ and the $(k+1)$-cell is attached to $\mathbb{R}P^k$ by the projection map $S^k \to \mathbb{R}P^k$. This is called the *standard CW-structure* on $\mathbb{R}P^n$. Taking the inductive limit $\mathbb{R}P^\infty$ of these CW-complexes gives a CW-complex known as the *infinite dimensional (real) projective space*. Analogous CW-decompositions for complex and quaternionic projective spaces are given in Sect. 6.1.

Example 3.4.6 If X and Y are CW-complexes, a CW-structure on $X \times Y$ may be defined, with $(X \times Y)^n = \bigcup_{p+q=n} X^p \times X^q$ and $\Lambda_n(X \times Y) = \dot{\bigcup}_{p+q=n} \Lambda_p(X) \times \Lambda_q(Y)$ (see [64, Theorem 2.2.2]). The weak topology may have more open sets than the product topology so the identity $i : (X \times Y)_{\text{CW}} \to (X \times Y)_{\text{prod}}$ is only a continuous bijection. If X or Y is finite, or if both are countable, then i is a homeomorphism (see [64, p. 60]). These consideration are not important for us since the two topologies have the same compact sets. Therefore, they have the same singular simplexes, whence i induces an isomorphism on singular (co)homology.

We now establish a few lemmas useful for the cellular (co)homology. Let X be a CW-complex. Fix an integer n and choose, for each $\lambda \in \Lambda_n$, a characteristic maps $\varphi_\lambda : (D^n, S^{n-1}) \to (X^n, X^{n-1})$. These maps produce a *global characteristic map*

$$\varphi^n : (\Lambda_n \times D^n, \Lambda_n \times S^{n-1}) \to (X^n, X^{n-1}).$$

Lemma 3.4.7 *Let X be a CW-complex and let $n \in \mathbb{N}$. Let φ^n be a global characteristic map for the n-cells. Then*

(i) $H_*\varphi^n : H_*(\Lambda_n \times D^n, \Lambda_n \times S^{n-1}) \xrightarrow{\approx} H_k(X^n, X^{n-1})$ *is an isomorphism.*
(ii) $H^*\varphi^n : H^*(X^n, X^{n-1}) \xrightarrow{\approx} H^k(\Lambda_n \times D^n, \Lambda_n \times S^{n-1})$ *is an isomorphism.*

Proof By Kronecker duality, using Corollary 2.3.11, only statement (i) must be proved. The proof for $n = 0$ is easy and left to the reader, so we assume that $n \geq 1$.

By Lemma 3.4.2, (X^n, X^{n-1}) is a well cofibrant pair. As $n \geq 1$, the space X^{n-1} is not empty (unless $X = \emptyset$, a trivial case). The continuous map

$$\hat{\varphi} : \Lambda_n \times D^n / \Lambda_n \times S^{n-1} \to X^n / X^{n-1}$$

induced by φ^n is a homeomorphism, both spaces being homeomorphic to a bouquet of copies of S^n indexed by Λ_n. In the commutative diagram

$$\begin{array}{ccc} H_*(\Lambda_n \times D^n, \Lambda_n \times S^{n-1}) & \xrightarrow{\approx} & \tilde{H}_*(\Lambda_n \times D^n / \Lambda_n \times S^{n-1}) \\ \downarrow{\scriptstyle H_*\varphi^n} & & \approx \downarrow{\scriptstyle H_*\hat{\varphi}} \\ H_*(X^n, X^{n-1}) & \xrightarrow{\approx} & \tilde{H}_*(X^n / X^{n-1}) \end{array},$$

the horizontal maps are isomorphisms by Corollary 3.1.47. Therefore, $H_*\varphi^n$ is an isomorphism. □

Corollary 3.4.8 *Let X be a CW-complex and let $n \in \mathbb{N}$. Then*

(i)

$$H_k(X^n, X^{n-1}) \approx \begin{cases} \bigoplus_{\Lambda_n} \mathbb{Z}_2 & \text{if } k = n. \\ 0 & \text{if } k \neq n. \end{cases}$$

(ii)

$$H^k(X^n, X^{n-1}) \approx \begin{cases} \prod_{\Lambda_n} \mathbb{Z}_2 & \text{if } k = n. \\ 0 & \text{if } k \neq n. \end{cases}$$

Proof Again, the easy case $n = 0$ is left to the reader. If $n > 0$, we use that, as noticed in the proof of Lemma 3.4.7, the map

$$X^n / X^{n-1} \to \bigvee_{\Lambda_n} S^n$$

is a homeomorphism. Corollary 3.4.8 then follows from Proposition 3.1.51. □

Lemma 3.4.9 *Let X be a CW-complex and let $n \in \mathbb{N}$. Then*

(i) *the homomorphism $H_k(X^n) \to H_k(X)$ induced by the inclusion is an isomorphism for $k < n$ and is surjective for $k = n$.*
(ii) *the homomorphism $H^k(X) \to H^k(X^n)$ induced by the inclusion is an isomorphism for $k < n$ and is injective for $k = n$.*

Proof By Kronecker duality, using Corollary 2.3.11, only statement (i) must be proved. The homomorphisms induced by inclusions form a sequence

$$H_k(X^k) \twoheadrightarrow H_k(X^{k+1}) \xrightarrow{\approx} H_k(X^{k+2}) \xrightarrow{\approx} \cdots \to H_k(X). \tag{3.4.2}$$

The bijectivity or surjectivity of $H_k(X^r) \to H_k(X^{r+1})$ is deduced from the homology sequence of the pair (X^{r+1}, X^r) and Corollary 3.4.8. By Proposition 3.1.29, $H_*(X)$ is the direct limit of $H_*(K)$, for all compact sets K of X. Using that each compact set of X is contained in some skeleton, one checks that $H_*(X)$ is the direct limit of $H_*(X^k)$. By (3.4.2), this proves (i). □

Lemma 3.4.10 *Let X be a CW-complex and let $n \in \mathbb{N}$. Then $H^k(X^n) = H_k(X^n) = 0$ if $k > n$.*

Proof The proof is by induction on n. The lemma is true for $n = 0$ since X^0 is a discrete set. The induction step uses the exact sequence of the pair (X^n, X^{n-1}) together with Corollary 3.4.8. □

Let (X, Y) be a CW-pair. Let $\mathcal{M} = \{(r, s) \in \mathbb{N} \times \mathbb{N} \mid r \geq s\}$ endowed with the lexicographic order. The pairs (X^r, Y^s) $((r, s) \in \mathcal{M})$, together with the inclusion $(X^r, Y^s) \hookrightarrow (X^{r'}, Y^{s'})$ when $(r, s) \leq (r', s')$, forms a direct system. The inclusions $j_{r,s} : (X^r, Y^s) \hookrightarrow (X, Y)$ induce a **GrV**-morphism

$$J_* : \varinjlim_{(r,s)\in\mathcal{M}} H_*(X^r, Y^s) \to H_*(X, Y)$$

and a **GrA**-morphism

$$J^* : H^*(X, Y) \to \varprojlim_{K\in\mathcal{K}} H^*(X^r, Y^s).$$

To get a more general result, which will be useful, we can take the product with an arbitrary topological space Z.

Proposition 3.4.11 *Let (X, Y) be a CW-pair and $\mathcal{M}$ be as above. Let Z be a topological space. Then, the* **GrV***-morphism*

$$J_* : \varinjlim_{(r,s)\in\mathcal{M}} H_*(X^r \times Z, Y^s \times Z) \xrightarrow{\approx} H_*(X \times Z, Y \times Z)$$

and the **GrA***-morphism*

$$J^* : H^*(X \times Z, Y \times Z) \xrightarrow{\approx} \varprojlim_{K\in\mathcal{K}} H^*(X^r \times Z, Y^s \times Z).$$

are isomorphisms.

Proof By Kronecker duality, only the homology statement needs a proof. Let K be a compact subspace of $X \times Z$. By Property (6) of p. 97, K is contained in $X^r \times Z$ for some integer r. Hence, if Y is empty, Proposition 3.4.11 follows from Corollary 3.1.16. When $Y \neq \emptyset$, we use the long exact sequences in homology and the five lemma, as in the proof of Proposition 3.1.29. □

3.5 Cellular (Co)homology

Let X be a CW-complex. For $m \in \mathbb{N}$, the *m-cellular (co)chain vector spaces* $\dot{C}_m(X)$ and $\dot{C}^m(X)$ are defined as

$$\dot{C}_m(X) = H_m(X^m, X^{m-1}) \text{ and } \dot{C}^m(X) = H^m(X^m, X^{m-1}).$$

The *cellular boundary operator* $\dot{\partial} : C_m(X) \to \dot{C}_{m-1}(X)$ is defined by the composed homomorphism

$$\dot{\partial} : H_m(X^m, X^{m-1}) \xrightarrow{\partial} H_{m-1}(X^{m-1}) \to H_{m-1}(X^{m-1}, X^{m-2}).$$

The expression for $\dot{\partial} \circ \dot{\partial}$ contains the sequence $H_{m-1}(X^{m-1}) \to H_{m-1}(X^{m-1}, X^{m-2}) \xrightarrow{\partial} H_{m-2}(X^{m-2})$ and then $\dot{\partial} \circ \dot{\partial} = 0$.

The *cellular co-boundary operator* $\dot{\delta} : C^m(X) \to \dot{C}^{m+1}(X)$ is defined by the composed homomorphism

$$\dot{\delta} : H^m(X^m, X^{m-1}) \to H^m(X^m) \xrightarrow{\delta} H^{m+1}(X^{m+1}, X^m).$$

with again $\dot{\delta} \circ \dot{\delta} = 0$. (Co)cycles $\dot{Z}_m$, $\dot{Z}^m$ and (co)boundaries $\dot{B}_m$, $\dot{B}^m$ are defined as usual, which leads to the definition

$$\dot{H}_m(X) = \dot{Z}_m(X) / \dot{B}_m(X) \text{ and } \dot{H}^m(X) = \dot{Z}^m(X) / \dot{B}^m(X).$$

The graded $\mathbb{Z}_2$-vector space $\dot{H}_*(X)$ is the *cellular homology* of the CW-complex X and the graded $\mathbb{Z}_2$-vector space $\dot{H}^*(X)$ is its *cellular cohomology*. The Kronecker pairing

$$H^m(X^m, X^{m-1}) \times H_m(X^m, X^{m-1}) \xrightarrow{\langle\,,\rangle} \mathbb{Z}_2$$

gives a pairing

$$\dot{C}^m(X) \times \dot{C}_m(X) \xrightarrow{\langle\,,\rangle} \mathbb{Z}_2$$

which makes $((\dot{C}^*(X), \dot{\delta}), (\dot{C}_*(X), \dot{\partial}), \langle\,,\rangle)$ a Kronecker pair.

In the language of former sections, the cellular (co)chains admit the usual equivalent definitions:

Definition 3.5.1 (*subset definitions*)

(a) A *cellular m-cochain* is a subset of Λ_m.
(b) A *cellular m-chain* is a finite subset of Λ_m.

Definition 3.5.2 (*colouring definitions*)

(a) A *cellular m-cochain* is a function $a : \Lambda_m \to \mathbb{Z}_2$.
(b) A *cellular m-chain* is a function $\alpha : \Lambda_m \to \mathbb{Z}_2$ with finite support.

Definition 3.5.2b is equivalent to

Definition 3.5.3 $\dot{C}_m(X)$ is the $\mathbb{Z}_2$-vector space with basis Λ_m:

$$\dot{C}_m(X) = \bigoplus_{\lambda \in \Lambda_m(X)} \mathbb{Z}_2 \, \lambda.$$

The *Kronecker pairing* on (co)chains admits the usual equivalent formula

$$\begin{aligned} \langle a, \alpha \rangle &= \sharp(a \cap \alpha) \pmod 2 \\ &= \textstyle\sum_{\sigma \in \alpha} a(\sigma). \end{aligned} \tag{3.5.1}$$

We now give a formula for the cellular boundary operator $\dot{\partial} : \dot{C}_m(X) \to \dot{C}_{m-1}(X)$. By Definition 3.1.3, it is enough to define $\dot{\partial}(\lambda)$ for $\lambda \in \Lambda_m$. Choose an attaching map $\varphi_\lambda : S^{m-1} \to X^{m-1}$ for the m-cell λ. When $m = 1$, the formula for $\dot{\partial}(\lambda)$ is easy:

$$\dot{\partial}(\lambda) = \begin{cases} 0 & \text{if } \sharp\varphi_\lambda(S^0) = 1 \\ \varphi_\lambda(S^0) & \text{otherwise (using the subset definition).} \end{cases} \tag{3.5.2}$$

Let us now suppose that $m > 1$. For $\mu \in \Lambda_{m-1}$, define $\varphi_{\lambda,\mu} : S^{m-1} \to S^{m-1}$ as the composed map:

$$S^{m-1} \xrightarrow{\varphi_\lambda} X^{m-1} \twoheadrightarrow X^{m-1}/X^{m-2} \approx \bigvee_{\Lambda_{m-1}} S^{m-1} \xrightarrow{\pi_\mu} S^{m-1},$$

where π_μ is the projection onto the μ-th component. Using the colouring definition of cellular chains, we must give, for each $\mu \in \Lambda_{m-1}$, the value $\dot{\partial}(\lambda)(\mu) \in \mathbb{Z}_2$.

Lemma 3.5.4 *For $m > 1$, the cellular boundary operator $\dot{\partial} : \dot{C}_m(X) \to \dot{C}_{m-1}(X)$ is the unique linear map satisfying*

$$\dot{\partial}(\lambda)(\mu) = \deg(\varphi_{\lambda,\mu}). \tag{3.5.3}$$

for each $\lambda \in \Lambda_m$.

Proof The attaching map $\varphi_\lambda : S^{m-1} \to X^{m-1}$ extends to a characteristic map $\hat{\varphi}_\lambda : D^m \to X^m$. Consider the commutative diagram:

$$\begin{array}{ccccc} H_m(D^m, S^{m-1}) & \xrightarrow{H_*\hat{\varphi}_\lambda} & H_m(X^m, X^{m-1}) & & \\ \approx \downarrow \partial & & \downarrow \partial & \searrow \dot{\partial} & \\ H_{m-1}(S^{m-1}) & \xrightarrow{H_*\varphi_\lambda} & H_{m-1}(X^{m-1}) & \longrightarrow & H_{m-1}(X^{m-1}/X^{m-2}) \\ & & & & \downarrow H_*\pi_\mu \\ & & & & H_{m-1}(S^{m-1}) \end{array}$$

Let α be the generator of $H_m(D^m, S^{m-1}) = \mathbb{Z}_2$ and let β be that of $H_{m-1}(S^{m-1})$. Using Lemma 3.4.7 and its proof, one sees that

(a) $\lambda \in \dot{C}_m(X)$ corresponds to $H_*\hat{\varphi}_\lambda(\alpha) \in H_m(X^m, X^{m-1})$.
(b) if $\gamma \in H_{m-1}(X^{m-1}/X^{m-2})$, then $H_*\pi_\mu(\gamma) = \gamma(\mu) \cdot \beta$ (we use the colouring definition and see γ as a function from Λ_{m-1} to $\mathbb{Z}_2$).

As $\partial(\alpha) = \beta$, one has

$$\begin{aligned} \dot{\partial}(\lambda)(\mu) \cdot \beta &= H_*\pi_\mu \circ \dot{\partial} \circ H_*\hat{\varphi}_\lambda(\alpha) \\ &= H_*\pi_\mu \circ H_*\varphi_\lambda(\beta) \\ &= \deg(\varphi_{\lambda,\mu}) \cdot \beta, \end{aligned}$$

which proves the lemma. □

Formulae (3.5.2) and (3.5.3) for the cellular boundary operator take a special form when X is a *regular CW-complex*, i.e. when each cell λ admits a characteristic map φ_λ which is an embedding onto a subcomplex of X. A cell of this subcomplex is called a *face* of λ.

Lemma 3.5.5 *Let X be a regular CW-complex. Let $\lambda \in \Lambda_m$ and $\mu \in \Lambda_{m-1}$. Then*

$$\partial(\lambda)(\mu) = \begin{cases} 1 & \text{if } \mu \text{ is a face of } \lambda \\ 0 & \text{otherwise.} \end{cases}$$

Proof When $m = 1$, this follows from (3.5.2), where the case $\sharp\varphi_\lambda(S^0) = 1$ does not happen since X is regular. When $m > 1$, we use Lemma 3.5.4 and compute the degree of $\varphi_{\lambda,\mu}$ using Proposition 3.2.6: since φ_λ is an embedding, any topological regular value of $\varphi_{\lambda,\mu}$ has exactly one element in its preimage. □

We now prove the main result of this section.

Theorem 3.5.6 *Let X be a CW-complex. Then, the cellular and the singular (co)homology of X are isomorphic:*

$$\dot{H}_*(X) \approx H_*(X) \;\; \textit{and} \;\; \dot{H}^*(X) \approx H^*(X).$$

Proof We consider the commutative diagram:

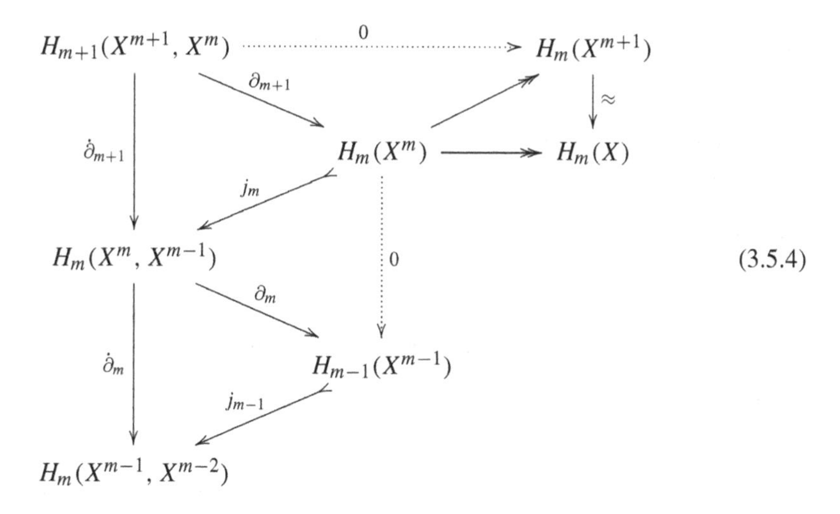

(3.5.4)

The properties of arrows (surjective, injective, bijective) come from Lemmas 3.4.9, 3.4.10 and Corollary 3.4.8. From Diagram (3.5.4), we get

$$H_m(X) \overset{\approx}{\leftarrow} H_m(X^{m+1}) \approx H_m(X^m)/\mathrm{Im}\,\partial_{m+1} \underset{\approx}{\overset{j_m}{\longrightarrow}} \ker \dot{\partial}_m/\mathrm{Im}\,\dot{\partial}_{m+1} = \dot{H}_m(X)$$

As the isomorphism $H_*(X) \approx \dot{H}_*(X)$ does not come from a morphism of chain complex, we cannot invoke Kronecker duality to deduce the isomorphism in cohomology. Instead, we consider the Kronecker dual of Diagram (3.5.4)

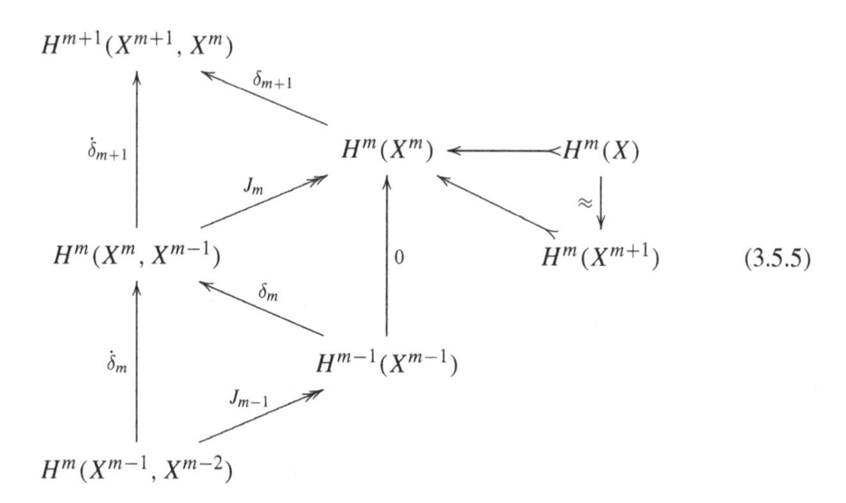

(3.5.5)

which gives

$$H^m(X) \xrightarrow{\approx} H^m(X^{m+1}) \xrightarrow{\approx} \ker \delta_{m+1} \approx J_m^{-1}(\ker \delta_{m+1})/\mathrm{Im}\, \delta_m$$
$$= \ker \dot{\delta}_{m+1}/\mathrm{Im}\, \dot{\delta}_m = \dot{H}^m(X). \quad \square$$

Here are some applications of the isomorphism between cellular and singular (co)homology.

Corollary 3.5.7 *Let X be a CW-complex with no m-dimensional cell. Then* $H_m(X) = H^m(X) = 0$.

Proof If $\Lambda_m(X) = \emptyset$, then $\dot{C}_m(X) = \dot{C}^m(X) = 0$, which implies $\dot{H}_m(X) = \dot{H}^m(X) = 0$, and then $H_m(X) = H^m(X) = 0$ by Theorem 3.5.6. □

Corollary 3.5.8 *Let X be a CW-complex with k cells of dimension m. Then* $\dim H_m(X) = \dim H^m(X) \leq k$.

Proof One has $\dim \dot{H}_m(X) \leq \dim \dot{Z}_m(X) \leq \dim \dot{C}_m(X) = k$. Therefore, $\dim H_m(X) \leq k$ by Theorem 3.5.6. The result on cohomology is proven in the same way or deduced by Kronecker duality. □

A CW-complex is *finite* if it has a finite number of cells.

Corollary 3.5.9 *Let X be a compact CW-complex. Then*

$$\dim H_*(X) = \dim H^*(X) < \infty.$$

Proof By the weak topology, a compact CW-complex is finite (see Remark (6) p. 97). Hence, $\dot{C}_*(X)$ is a finite dimensional vector space, and so is $\dot{Z}_*(X)$ and $\dot{H}_*(X)$. Corollary 3.5.9 then follows from Theorem 3.5.6 and Kronecker duality. □

Let X be a finite CW-complex. Its *Euler characteristic* $\chi(X)$ is defined as

$$\chi(X) = \sum_{m\in\mathbb{N}} (-1)^m \sharp\Lambda_m(X) \in \mathbb{Z}.$$

Proposition 3.5.10 *Let X be a finite CW-complex. Then*

$$\chi(X) = \sum_{m\in\mathbb{N}} (-1)^m \dim H_m(X) = \sum_m (-1)^m \dim H^m(X).$$

Proof If we use the cellular (co)homology, the proof of Proposition 3.5.10 is the same as that of Proposition 2.4.8. The result then follows from Theorem 3.5.6. □

A CW-complex X (or a CW-structure on X) is called *perfect* if the cellular boundary vanishes. For instance, if X does not have cells in consecutive dimensions, then it is perfect. Also, the standard CW-structure on $\mathbb{R}P^n$ ($n \leq \infty$) is perfect (see e.g.

Proposition 6.1.1). If X is a perfect CW-complex, $\dot{C}_*(X) = \dot{H}_*(X)$ and the identification between the singular and cellular homologies, out of Diagram (3.5.4), is particularly simple:

$$H_m(X) \stackrel{\approx}{\leftarrow} H_m(X^m) \stackrel{\approx}{\rightarrow} \dot{H}_m(X^m, X^{m-1}) \stackrel{\approx}{\rightarrow} \dot{H}_m(X) \stackrel{\approx}{\leftarrow} \dot{C}_m(X). \tag{3.5.6}$$

The natural functoriality of cellular (co)homology is for cellular maps. If X and Y are CW-complexes, a continuous map $f : Y \to X$ is *cellular* if $f(Y^m) \subset X^m$ for all $m \in \mathbb{N}$. We thus get **GrV**-morphisms $\dot{C}_* f$ and $\dot{C}^* f$ making the following diagrams commute

$$\begin{array}{ccc} \dot{C}_m(Y) & \xrightarrow{\dot{C}_* f} & \dot{C}_m(X) \\ \downarrow = & & \downarrow = \\ H_m(Y^m, Y^{m-1}) & \xrightarrow{H_* f} & H_m(X^m, X^{m-1}) \end{array} \qquad \begin{array}{ccc} \dot{C}^m(Y) & \xleftarrow{\dot{C}^* f} & \dot{C}^m(X) \\ \uparrow = & & \uparrow = \\ H^m(Y^m, Y^{m-1}) & \xleftarrow{\dot{C}^* f} & H^m(X^m, X^{m-1}) \end{array}$$

They satisfy $\langle \dot{C}^* f(a), \alpha \rangle = \langle a, \dot{C}_* f(\alpha) \rangle$ for all $a \in \dot{C}^*(X)$ and $\alpha \in \dot{C}^*(X)$. It is useful to have a formula for $C_* f$, using that

$$\dot{C}_m(Y) = \bigoplus_{\lambda \in \Lambda_m(Y)} \mathbb{Z}_2 \, \lambda \text{ and } \dot{C}_m(X) = \bigoplus_{\mu \in \Lambda_m(X)} \mathbb{Z}_2 \, \mu.$$

For $\lambda \in \Lambda_m(Y)$ and $\mu \in \Lambda_m(X)$, consider the map $f_{\lambda,\mu} : S^m \to S^m$ defined by the composition

$$f_{\lambda,\mu} : S^m \xrightarrow{j_\lambda} \bigvee_{\lambda \in \Lambda_m(Y)} S^m \approx Y^m / Y^{m-1} \xrightarrow{f} X^m / X^{m-1} \approx \bigvee_{\mu \in \Lambda_m(X)} S^m \xrightarrow{\pi_\mu} S^m,$$

where j_λ is the inclusion of the λ-component and π_μ the projection onto the μ-component.

Lemma 3.5.11 *For $m \geq 1$, $\dot{C}_* f : C_m(Y) \to C_m(X)$ is the unique linear map such that*

$$\dot{C}_* f(\lambda) = \sum_{\mu \in \Lambda_m(X)} \deg(f_{\lambda,\mu}) \, \mu.$$

Proof The map $f : (Y^m, Y^{m-1}) \to (X^m, X^{m-1})$ induces a map $\bar{f} : \bigvee_{\lambda \in \Lambda_m(Y)} S^m \to \bigvee_{\mu \in \Lambda_m(X)} S^m$ making the following diagram commute

$$\begin{array}{ccccc} \dot{C}_m(Y) & \xrightarrow{\approx} & H_m(Y^m, Y^{m-1}) & \xrightarrow{\approx} & H_m(\bigvee_{\lambda \in \Lambda_m(Y)} S^m) \\ \downarrow \dot{C}_* f & & \downarrow H_* f & & \downarrow H_* \bar{f} \\ \dot{C}_m(X) & \xrightarrow{\approx} & H_m(Y^m, Y^{m-1}) & \xrightarrow{\approx} & H_m(\bigvee_{\mu \in \Lambda_m(X)} S^m) \end{array}$$

As in the proof of Lemma 3.5.4, one checks that, under the top horizontal isomorphisms, $\lambda \in \dot{C}_m(Y)$ corresponds to $H_* j_\lambda([S^{m-1}])$. Also, if $\gamma \in H_{m-1}(X^{m-1}/X^{m-2})$, then $H_*\pi_\mu(\gamma) = \gamma(\mu) \cdot [S^{m-1}]$ (seeing γ as a function from $\Lambda_m(X)$ to $\mathbb{Z}_2$ by the colouring definition). Hence,

$$\begin{aligned}\dot{C}_* f(\lambda)(\mu) = H_*\pi_\mu \circ H_* \bar{f} \circ H_* j_\lambda([S^{m-1}]) &= H_* f_{\lambda,\mu}([S^{m-1}]) \\ &= \deg(f_{\lambda,\mu})\,[S^{m-1}],\end{aligned}$$

which proves the lemma. □

3.5.12 *Homology-cell complexes.* The results of this section and the previous one are also valid for complexes where cells are replaced by homology cells. A well cofibrant pair $(B, \dot{B})$ is a *homology n-cell* if $\tilde{H}_*(B) = 0$ and $H_*(\dot{B}) \approx H_*(S^{n-1})$. This **GrV**-isomorphism is "abstract", i.e. not assumed to be given by any continuous map. It follows that $H_*(B, \dot{B}) \approx H_*(D^n, S^{n-1})$. We also say that B is a *homology n-cell* with *boundary* $\dot{B}$. If $\lambda \in \Lambda$ is indexing a family of homology n-cell $\{(B(\lambda), \dot{B}(\lambda))\}$ and if $\varphi_\lambda : \dot{B}(\lambda) \to Y$ is a family of continuous maps, we say that the quotient space

$$X = Y \cup_\varphi \left(\dot{\bigcup}_{\lambda\in\Lambda} B(\lambda)\right)$$

is obtained from Y by attaching homology n-cells (they may be different for various λ's). We identify Λ with the set of homology n-cells.

A *homology-cell complex* is defined as in p. 97 with attachments of n-cells replaced by attachments of a set $\Lambda_n(X)$ of homology n-cells. The cellular (co)homology $\dot{H}_*(X)$ and $\dot{H}^*(X)$ are defined accordingly and Theorem 3.5.6 holds true, with the same proof. Homology-cell structures are used in the proof of Poincaré duality (see Sect. 5.2).

3.6 Isomorphisms Between Simplicial and Singular (Co)homology

Let K be a simplicial complex. In this section, we prove three theorems showing that the simplicial (co)homology of K and the singular (co)homology of $|K|$ are isomorphic.

Theorem 3.6.1 *Let K be a simplicial complex. Then*

$$H_*(K) \approx H_*(|K|) \text{ and } H^*(K) \approx H^*(|K|)$$

Proof The geometric realization $|K|$ of K is naturally endowed with a structure of a regular CW-complex, with $|K|^m = |K^m|$, $\Lambda_m(|K|) = \mathcal{S}_m(K)$, with a canonical characteristic map for the m-cell $\sigma \in \mathcal{S}_m(K)$ given by the inclusion of $|\bar{\sigma}|$ into $|K|$. Thus, $\dot{C}_m(|K|) = C_m(K)$ and, using Lemma 3.5.5, the diagram

$$\begin{array}{ccc} \dot{C}_m(|K|) & \xrightarrow{=} & C_m(K) \\ \dot{\partial}\downarrow & & \downarrow\partial \\ \dot{C}_{m-1}(|K|) & \xrightarrow{=} & C_{m-1}(K) \end{array}$$

is commutative. Therefore, $\dot{H}_*(|K|) = H_*(K)$ and, by Theorem 3.5.6, the singular homology $H_*(|K|)$ and the simplicial homology $H_*(K)$ are isomorphic. The equality $\dot{H}^*(|K|) = H^*(K)$ is deduced from $\dot{H}_*(|K|) = H_*(K)$ by Kronecker duality and, using by Theorem 3.5.6 again, the singular cohomology $H^*(|K|)$ and the simplicial cohomology $H^*(K)$ are also isomorphic. □

We now go to the second isomorphism theorem, which uses the ordered simplicial (co)homology of Sect. 2.10. To an ordered m-simplex $(v_0, \dots, v_m) \in \hat{S}_m(K)$, we associate the singular m-simplex $R(v_0, \dots, v_m) : \Delta^m \to |K|$ defined by

$$R(v_0, \dots, v_m)(t_0, \dots, t_m) = \sum_{i=0}^{m} t_i v_i \,. \tag{3.6.1}$$

The linear combination in (3.6.1) makes sense since $\{v_0, \dots, v_m\}$ is a simplex of K. This defines a map $R : \hat{S}_m(K) \to S_m(|K|)$ which extends to a linear map

$$R_* : \hat{C}_*(K) \to C_*(|K|) \,.$$

The formula $\partial \circ R = R \circ \hat{\partial}$ is obvious, so R is a morphism of chain complexes from $(\hat{C}_*(K), \hat{\partial})$ to $(C_*(|K|), \partial)$. Define the linear map $R^* : C^*(|K|) \to \hat{C}^*(K)$ by $\langle R^*(a), \alpha\rangle = \langle a, R_*(\alpha)\rangle$. By Lemma 2.3.6, (R^*, R_*) is a morphism of Kronecker pair. We also denote by R^* and R_* the induced linear maps on (co)homology:

$$R_* : \hat{H}_*(K) \to H_*(|K|) \text{ and } R^* : H^*(|K|) \to \hat{H}^*(K) \,.$$

If $f : L \to K$ be a simplicial map, the formulae

$$R_* \circ \hat{C}_* f = C_*|f| \circ R_* \text{ and } \hat{C}^* f \circ R^* = R^* \circ C^*|f|$$

are easy to check. They induce the formulae

$$R_* \circ \hat{H}_* f = H_*|f| \circ R_* \text{ and } \hat{H}^* f \circ R^* = H^* \circ C^*|f| \tag{3.6.2}$$

on (co)homology. In particular, if f is the inclusion of a subcomplex L of K, the above considerations permit us to construct degree zero linear maps

$$R_* : \hat{H}_*(K, L) \to H_*(|K|, |L|) \text{ and } R^* : H^*(|K|, |L|) \to \hat{H}^*(K, L)$$

so that (R^*, R_*) is a morphism of Kronecker pair. Finally, if $f : (K, L) \to (K', L')$ is a simplicial map of simplicial pairs, then Formulae (3.6.2) hold true in relative (co)homology.

Theorem 3.6.2 *Let (K, L) be a simplicial pair. Then the linear maps*

$$R_* : \hat{H}_*(K, L) \xrightarrow{\approx} H_*(|K|, |L|) \text{ and } R^* : H^*(|K|, |L|) \xrightarrow{\approx} \hat{H}^*(K, L)$$

are isomorphisms. They are functorial for simplicial maps of simplicial pairs

Proof The functoriality has already been established. By Kronecker duality, it is enough to prove that R_* is an isomorphism. The proof goes through a couple of particular cases.

Case 1: $(K, L) = (\mathcal{F}A, \dot{\mathcal{F}}A)$, where $\mathcal{F}A$ is the full complex on the finite set of $m+1$ elements $A = \{v_0, \dots, v_m\}$ (see p. 24), which is isomorphic to an m-simplex. Then $(|\mathcal{F}A|, |\dot{\mathcal{F}}A|) \approx (D^m, S^{m-1})$. By Corollaries 2.4.7 and 3.2.2,

$$\hat{H}_k(\mathcal{F}A, \dot{\mathcal{F}}A) = H_k(\mathcal{F}A, \dot{\mathcal{F}}A) = H_k(|\mathcal{F}A|, |\dot{\mathcal{F}}A|) = 0$$

if $k \neq m$ and

$$\hat{H}_m(\mathcal{F}A, \dot{\mathcal{F}}A) \approx H_m(\mathcal{F}A, \dot{\mathcal{F}}A) \approx H_m(|\mathcal{F}A|, |\dot{\mathcal{F}}A|) \approx \mathbb{Z}_2.$$

Thus, it is enough to prove that $R_* : \hat{H}_m(\mathcal{F}A, \dot{\mathcal{F}}A) \to H_m(|\mathcal{F}A|, |\dot{\mathcal{F}}A|)$ is not trivial. The vector space $\hat{H}_m(\mathcal{F}A, \dot{\mathcal{F}}A)$ is generated by the ordered simplex $\hat{\sigma} = (v_0, \dots, v_m)$. Let $r = R(\hat{\sigma}) : (\Delta^m, \dot{\Delta}^m) \to |\mathcal{F}A|, |\dot{\mathcal{F}}A|)$. One has $[r] = H_*r([i_m])$ where i_m is the identity map of $(\Delta^m, \dot{\Delta}^m)$. But $[i_m] \neq 0$ in $H_m(\Delta^m, \dot{\Delta}^m)$ by Proposition 3.2.3 and r is a homeomorphism of pairs. Thus $R_*(\hat{\sigma}) \neq 0$ in $H_m(|\mathcal{F}A|, |\dot{\mathcal{F}}A|)$.

Case 2: $(K, L) = (K^m, K^{m-1})$ with $m \geq 1$. The non-vanishing homology groups are

$$\hat{H}_m(K^m, K^{m-1}) \approx H_m(K^m, K^{m-1}) \approx H_m(|K^m|, |K^{m-1}|) \approx \bigoplus_{\mathcal{S}_m(K)} \mathbb{Z}_2$$

For each $\sigma \in \mathcal{S}_m(K)$ choose an ordered simplex $\hat{\sigma} = (v_0, \dots, v_m)$ with $\{v_0, \dots, v_m\} = \sigma$. Then $\hat{H}_m(K^m, K^{m-1}) \approx H_m(K^m, K^{m-1})$ is the $\mathbb{Z}_2$-vector space with basis $\{\hat{\sigma} \mid \sigma \in \mathcal{S}_m(K)\}$. Denote by $\bar{\sigma}$ the subcomplexes of K generated by σ and by $\dot{\bar{\sigma}}$ the subcomplex of the proper faces of $\bar{\sigma}$. The map $r_\sigma : (|\bar{\sigma}|, |\dot{\bar{\sigma}}|) \to (|K^m|, |K^{m-1}|)$ is a characteristic map for the m-cells of $|K|$ corresponding to σ. The union r^m of the r_σ is then a global characteristic map for the m-cells of $|K|$. Let us consider the commutative diagram

$$\begin{array}{ccc}
\bigoplus_{\sigma\in\mathcal{S}_m(K)} \hat{H}_m(\bar{\sigma},\dot{\bar{\sigma}}) & \xrightarrow[\approx]{R_*} & \bigoplus_{\sigma\in\mathcal{S}_m(K)} H_m(|\bar{\sigma}|,|\dot{\bar{\sigma}}|) \\
\downarrow\approx & & \approx\downarrow H_*r^m \\
\hat{H}_m(K^m,K^{m-1}) & \xrightarrow{R_*} & H_m(|K^m|,|K^{m-1}|)
\end{array}$$

The bijectivity of the left vertical arrow was seen above. That of the right vertical arrow is Lemma 3.4.7. The bijectivity of the top horizontal arrow comes from Case 1. Hence, $R_* : \hat{H}_m(K^m, K^{m-1}) \to H_m(|K^m|, |K^{m-1}|)$ is an isomorphism.

Case 3: $(K, L) = (K^m, \emptyset)$. This is proven by induction on m, the case $m = 0$ being obvious. By the naturality of R_*, one has the commutative diagram of exact sequences:

$$\begin{array}{ccccccccc}
\hat{H}_{*+1}(K^m,K^{m-1}) & \xrightarrow{\partial_*} & \hat{H}_*(K^{m-1}) & \longrightarrow & \hat{H}_*(K^m) & \longrightarrow & \hat{H}_*(K^m,K^{m-1}) & \xrightarrow{\partial_*} & \hat{H}_{*-1}(K^{m-1}) \\
R_*\downarrow\approx & & R_*\downarrow\approx & & R_*\downarrow & & R_*\downarrow\approx & & R_*\downarrow\approx \\
H_{*+1}(K^m,K^{m-1}) & \xrightarrow{\partial_*} & H_*(K^{m-1}) & \longrightarrow & H_*(K^m) & \longrightarrow & H_*(K^m,K^{m-1}) & \xrightarrow{\partial_*} & H_{*-1}(K^{m-1})
\end{array}$$

(one has to check that the diagrams with ∂_* are commutative). The bijectivity of the vertical arrows come by induction hypothesis and by case 2. By the five-lemma, $R_* : \hat{H}_*(K_m) \to H_*(|K^m|)$ is an isomorphism.

General case. We first prove that $R_* : \hat{H}_m(K) \to H_m(|K|)$ is an isomorphism for all m. By the naturality of R_*, one has the commutative diagram:

$$\begin{array}{ccc}
\hat{H}_m(K^{m+1}) & \xrightarrow[\approx]{R_*} & H_m(|K^{m+1}|) \\
\downarrow\approx & & \downarrow\approx \\
\hat{H}_m(K) & \xrightarrow{R_*} & H_m(|K|)\,.
\end{array}$$

The bijectivity of the left vertical arrow is obvious. That of the right vertical arrow is Lemma 3.4.9. The bijectivity of the top horizontal arrow was established in Case 3. Therefore, the bottom horizontal arrow is bijective. Finally, the general case (K, L) is deduced from the absolute cases using, as in Case 3, the homology sequences of the pair (K, L) and the five-lemma. □

Four our third isomorphism theorem, choose a simplicial order $\leq$ on K. Define a map $R_\leq : \mathcal{S}(K) \to \mathcal{S}(|K|)$ by $R_\leq(\sigma) = R(\hat{\sigma})$ where, if $\sigma = \{v_0, \dots, v_m\}$, then $\hat{\sigma} = (v_0, \dots, v_m)$ with $v_0 \leq \dots \leq v_m$. As above, we check that $R_\leq$ induces linear maps $R_{\leq,*} : H_*(K, L) \to H_*(|K|, |L|)$ and $R^*_\leq : H^*(|K|, |L|) \to H^*(K, L)$ of degree zero.

Theorem 3.6.3 *Let (K, L) be a simplicial pair. For any simplicial order $\leq$ on K, the linear maps*

$$R_{\leq,*} : H_*(K, L) \xrightarrow{\approx} H_*(|K|, |L|) \text{ and } R^*_{\leq} : H^*(|K|, |L|) \xrightarrow{\approx} H^*(K, L)$$

are isomorphisms. Moreover, these isomorphisms do not depend on the simplicial order $\leq$.

Proof By Kronecker duality, only the homology statement requires a proof. By our definitions, one has the commutative diagram

$$\begin{array}{ccc} H_*(K, L) & \xrightarrow{R_{\leq,*}} & H_*(|K|, |L|) \\ & {\scriptstyle H_*\phi_{\leq}} \searrow_{\approx} \quad \nearrow^{R_*}_{\approx} & \\ & \hat{H}_*(K, L) & \end{array}$$

The bijectivity of the arrows come from Corollary 2.10.13 and Theorem 3.6.3. Therefore, $R_{\leq,*}$ is an isomorphism. As $H_*\phi_{\leq}$ is independent of $\leq$ by Corollary 2.10.13, so is $R_{\leq,*}$. □

3.7 CW-Approximations

It is sometimes useful to know that any space has the (co)homology of a CW-complex or of a simplicial complex (see e.g. p. 132, 161 and 326 in this book). We give below classical functorial results about that. Relationships with similar constructions in the literature are discussed in Remarks 3.7.5 at the end of the section.

We shall need a standard notion of category theory: natural transformations. Let **a** and **b** be two (covariant) functors from a category **C** to a category **C**′. A *natural transformation* associates to each object X in **C** a morphism $\phi_X : \mathbf{a}(X) \to \mathbf{b}(X)$ in **C**′ such that the diagram

$$\begin{array}{ccc} \mathbf{a}(X) & \xrightarrow{\mathbf{a}(f)} & \mathbf{a}(Y) \\ \downarrow{\scriptstyle \Phi_X} & & \downarrow{\scriptstyle \Phi_Y} \\ \mathbf{b}(X) & \xrightarrow{\mathbf{b}(f)} & \mathbf{b}(Y) \end{array} \tag{3.7.1}$$

is commutative for every morphism $f : X \to Y$ in **C**.

We first consider the category of CW-spaces and cellular maps. It is denoted by **CW** and, as usual, by $\mathbf{CW}_2$ for pairs of CW-complexes. We denote by **j** be the inclusion morphism from $\mathbf{CW}_2$ to $\mathbf{Top}_2$.

Theorem 3.7.1 *There is a covariant functor* $\mathbf{cw} : (X, Y) \to (X^{\mathrm{CW}}, Y^{\mathrm{CW}})$ *from* $\mathbf{Top}_2 \to \mathbf{CW}_2$ *and a natural transformation* $\phi = \phi_{(X,Y)} : (X^{\mathrm{CW}}, Y^{\mathrm{CW}}) \to (X, Y)$ *from* $\mathbf{j} \circ \mathbf{cw}$ *to the identity functor of* $\mathbf{Top}_2$, *such that* $H_*\phi$ *and* $H^*\phi$ *are isomorphisms.*

The construction in the proof below is sometimes called in the literature the *thick geometric realization of the singular complex of* X.

Proof We start with some preliminaries. If Δ^m is the standard m-simplex and $I \subset \{0, 1, \dots, m\}$, we set

$$\Delta^m_I = \{(t_0, \dots, t_m) \in \Delta^m \mid t_i = 0 \text{ if } i \notin I\}$$

which is a simplex of dimension $\sharp I - 1$. This gives rise to an obvious inclusion map $\epsilon_I : \Delta^{\sharp I - 1} \to \Delta^m$.

Let X be a topological space. The space X^{CW} is defined as the quotient space

$$X^{\mathrm{CW}} = \dot{\bigcup}_{m \geq 0} (\mathcal{S}_m(X) \times \Delta^m) / \sim \tag{3.7.2}$$

where $\sim$ is the equivalence relation $(\sigma, \epsilon_I(u)) \sim (\sigma \circ \epsilon_I, u)$ for all $\sigma \in \mathcal{S}_m(X)$, $I \subset \{0, \dots, m\}$ and $u \in \Delta^{\sharp I - 1}$. Then X^{CW} is a CW-complex whose k-skeleton is

$$(X^{\mathrm{CW}})^k = \dot{\bigcup}_{0 \leq m \leq k} (\mathcal{S}_m(X) \times \Delta^m) / \sim .$$

In particular, $(X^{\mathrm{CW}})^0$ is just the space X endowed with the discrete topology. The k-cells are indexed by $\mathcal{S}_k(X)$. The characteristic map for the k-cell corresponding to $\sigma \in \mathcal{S}_k(X)$ is the restriction to $\sigma \times \Delta^k$ of the quotient map from the disjoint union in (3.7.2) onto $(X^{\mathrm{CW}})^k$.

A continuous $f : X_1 \to X_2$ determines a cellular map $f^{\mathrm{CW}} : X_1^{\mathrm{CW}} \to X_2^{\mathrm{CW}}$ induced by $f^{\mathrm{CW}}(\sigma, u) = (f \circ \sigma, u)$. Note that, if Y is a subspace of X, then Y^{CW} is a subcomplex of X^{CW}. We thus check that **cw** is a covariant functor $(X, Y) \to (X^{CW}, Y^{CW})$ from $\mathbf{Top}_2$ to $\mathbf{CW}_2$.

For $\sigma \in \mathcal{S}_m(X)$, one has a continuous map $\phi_\sigma : \{\sigma\} \times \Delta^m \to X$ defined by $\phi_\sigma(\sigma, u) = \sigma(u)$. The disjoint union of those ϕ_σ descends to a continuous map $\phi : X^{\mathrm{CW}} \to X$, or $\phi : (X^{\mathrm{CW}}, Y^{\mathrm{CW}}) \to (X, Y)$. One has

$$\phi \circ f^{\mathrm{CW}}(\sigma, u) = \phi(f \circ \sigma, u) = f \circ \sigma(u) = f \circ \phi(\sigma, u)$$

which amounts, using (3.7.1), to ϕ being a natural transformation from $\mathbf{j} \circ \mathbf{cw}$ to the identity functor of $\mathbf{Top}_2$.

It remains to prove that $H_*\phi$ is a **GrV**-isomorphism (that $H^*\phi$ is a **GrA**-isomorphism will follow by Kronecker duality). We start with the absolute case $Y = \emptyset$. We shall construct a diagram

$$\begin{array}{ccc} H_*(X^{\mathrm{CW}}) & \xrightarrow{H_*\phi} & H_*(X) \\ & \nwarrow^{\beta} \quad \swarrow_{\alpha} & \\ & \dot{H}_*(X^{\mathrm{CW}}) & \end{array} \tag{3.7.3}$$

such that $H_*\phi \circ \beta \circ \alpha = \mathrm{id}$ and α and β are isomorphisms. The bijection $\mathcal{S}(X) \xrightarrow{=}$ {cells of X^{CW}} extends to a linear map $\alpha : C_*(X) \to \dot{C}_*(X^{\mathrm{CW}})$ which satisfies $\alpha \circ \delta = \dot{\delta} \circ \alpha$ and thus induces the isomorphism $\alpha : H_*(X) \xrightarrow{\approx} \dot{H}_*(X^{\mathrm{CW}})$. For β, one associates to the k-cell of X^{CW} indexed by σ the map

$$\Delta^k \xrightarrow{\approx} \{\sigma\} \times \Delta^k \xrightarrow{\text{char.map}} X^{\mathrm{CW}}$$

which is an element of $\mathcal{S}_k(X^{\mathrm{CW}})$. This extends to a linear map $\beta : \dot{C}_k(X^{\mathrm{CW}}) \to C_k(X^{\mathrm{CW}})$. Again, we check that $\beta \circ \dot{\delta} = \delta \circ \beta$. We thus get the linear map $\beta : \dot{H}_*(X^{\mathrm{CW}}) \to H_*(X^{\mathrm{CW}})$. The equation $H_*\phi \circ \beta \circ \alpha = \mathrm{id}$ is straightforward.

It remains to prove that β is an isomorphism. To simplify the notation, write $\hat{X} = X^{\mathrm{CW}}$. Note that β induces linear maps $\beta_k : \dot{H}_*(\hat{X}^k) \to H_*(\hat{X}^k)$ and $\beta_{k+1,k} : \dot{H}_*(\hat{X}^{k+1}, \hat{X}^k) \to H_*(\hat{X}^{k+1}, \hat{X}^k)$. Obviously, $\dot{H}_*(\hat{X}) = \lim_k \dot{H}_*(\hat{X}^k)$. By Corollary 3.1.16, one also has that $H_*(\hat{X}) = \lim_k H_*(\hat{X}^k)$. Therefore, it suffices to show that β_k is an isomorphism for all k. This is done by induction on k. It is obviously true for $k = 0$, since $\hat{X}^0$ is a discrete space. For the induction step, suppose that $\beta_k : \dot{H}_i(\hat{X}^k) \to H_i(\hat{X}^k)$ is an isomorphism for all $i \in \mathbb{N}$. Then, $\beta_{k+1} : \dot{H}_i(\hat{X}^{k+1}) \to H_i(\hat{X}^{k+1})$ is an isomorphism for all i, except perhaps for $i = k, k+1$ where we must consider the commutative diagram

$$\begin{array}{ccccccccc}
0 \to & \dot{H}_{k+1}(\hat{X}^{k+1}) & \to & \dot{H}_{k+1}(\hat{X}^{k+1}, \hat{X}^k) & \to & \dot{H}_k(\hat{X}^k) & \to & \dot{H}_k(\hat{X}^{k+1}) & \to 0 \\
& \downarrow \beta_{k+1} & & \downarrow \beta_{k+1,k} & & \approx \downarrow \beta_k & & \downarrow \beta_{k+1} & \\
0 \to & H_{k+1}(\hat{X}^{k+1}) & \to & H_{k+1}(\hat{X}^{k+1}, \hat{X}^k) & \to & H_k(\hat{X}^k) & \to & H_k(\hat{X}^{k+1}) & \to 0
\end{array} \tag{3.7.4}$$

where the horizontal lines are the cellular and singular homology exact sequences of the pair $(\hat{X}^{k+1}, \hat{X}^k)$. By the five lemma, it thus suffices to prove that $\beta_{k+1,k}$ is an isomorphism. One has the commutative diagram

$$\begin{array}{ccc}
\dot{H}_{k+1}(\hat{X}^{k+1}, \hat{X}^k) & \xleftarrow{\approx} & \bigoplus_{\sigma \in \mathcal{S}_{k+1}(X)} \dot{H}_{k+1}(\{\sigma\} \times (\Delta^{k+1}, \mathrm{Bd}\Delta^{k+1})) \\
\downarrow \beta_{k+1,k} & & \downarrow \oplus\beta_\sigma \\
H_{k+1}(\hat{X}^{k+1}, \hat{X}^k) & \xleftarrow{\approx} & \bigoplus_{\sigma \in \mathcal{S}_{k+1}(X)} H_{k+1}(\{\sigma\} \times (\Delta^{k+1}, \mathrm{Bd}\Delta^{k+1}))
\end{array} \tag{3.7.5}$$

where β_σ sends the $(k+1)$-cell $\{\sigma\} \times (\Delta^{k+1}$ (generator of $\dot{H}_{k+1}(\{\sigma\} \times (\Delta^{k+1}, \mathrm{Bd}\Delta^{k+1})) = \mathbb{Z}_2$) to the tautological singular simplex $\Delta^{k+1} \to \{\sigma\} \times \Delta^{k+1}$. The latter is the generator of $H_{k+1}(\{\sigma\}\times(\Delta^{k+1}, \mathrm{Bd}\Delta^{k+1})) = \mathbb{Z}_2$ (see Proposition 3.2.3). Hence, $\beta_{k+1,k}$ is an isomorphism.

We have proven that $H_*\phi : H_*(X^{\mathrm{CW}}) \to H_*(X)$ is a **GrV**-isomorphism for all topological space X. Using the homology sequences and the five lemma, this implies that $H_*\phi : H_*(X^{\mathrm{CW}}, Y^{\mathrm{CW}}) \to H_*(X, Y)$ is a **GrV**-isomorphism for all topological pairs (X, Y). □

A slightly more sophisticated construction for the functor of Theorem 3.7.2 gives the following result. Let **RCW** be the category of regular CW-complexes and cellular maps and let **j** be the inclusion morphism from $\mathbf{RCW}_2$ to $\mathbf{Top}_2$.

Theorem 3.7.2 *There is a covariant functor* **rcw** : $(X, Y) \to (X^{\mathrm{RCW}}, Y^{\mathrm{RCW}})$ *from* $\mathbf{Top}_2 \to \mathbf{RCW}_2$ *and a natural transformation* $\phi = \phi_{(X,Y)} : (X^{\mathrm{RCW}}, Y^{\mathrm{RCW}}) \to (X, Y)$ *from* $\mathbf{j} \circ \mathbf{rcw}$ *to the identity functor of* $\mathbf{Top}_2$*, such that* $H_*\phi$ *and* $H^*\phi$ *are isomorphisms.*

The proof of Theorem 3.7.2 requires some preliminaries.

Let $\mathcal{F}\mathbb{N}$ be the full simplicial complex with vertex set the integers $\mathbb{N}$. If X is a topological space, the set of $\mathbb{N}$-*singular simplexes* of X is defined by

$$\mathbb{N}\mathcal{S}(X) = \{(s, \tau) \mid s \in \mathcal{S}(\mathcal{F}\mathbb{N}) \text{ and } \tau : |\bar{s}| \to X \text{ is a continuous map}\}.$$

where $\bar{s}$ is the simplicial complex formed by s and all its faces (see p. 6). Let $\mathbb{N}\mathcal{S}_n(X)$ be the subset of $\mathbb{N}\mathcal{S}(X)$ formed by those pairs (s, τ), where s is of dimension n and let $\mathbb{N}C_n(X)$ be the $\mathbb{Z}_2$-vector space with basis $\mathbb{N}\mathcal{S}_n(X)$. Using the facets of $\bar{s}$, we define a boundary operator $\partial : \mathbb{N}C_n(X) \to \mathbb{N}C_{n-1}(X)$ making $\mathbb{N}C_*(X)$ a chain complex. The homology of this chain complex is the $\mathbb{N}$-*singular homology of* X, denoted by $\mathbb{N}H_*(X)$. The relative homology $\mathbb{N}H_*(X, Y)$ is defined as in Sect. 3.1.2.

The order on $\mathbb{N}$ provides a simplicial order on $\mathcal{F}\mathbb{N}$. Thus, if $s \in \mathcal{S}_n(\mathcal{F}\mathbb{N})$, there is a canonical homeomorphism $h_s : |\bar{s}| \xrightarrow{\approx} \Delta^n$ (see (3.1.6)). We define maps $\mu : \mathcal{S}_n(X) \to \mathbb{N}\mathcal{S}_n(X)$ and $\nu : \mathbb{N}\mathcal{S}_n(X) \to \mathcal{S}_n(X)$ by:

- $\mu(\sigma) = (s_0, \sigma \circ h_{s_0(n)})$, where $s_0(n) = \{0, 1, \dots, n\}$ and
- $\nu(s, \tau) = \tau \circ h_s^{-1}$.

The linear extensions $C_*\mu : C_n(X) \to \mathbb{N}C_n(X)$ and $C_*\nu : \mathbb{N}C_n(X) \to C_n(X)$ commute with the boundary operators and are thus morphisms of chain complexes. The constructions extend to pairs and we get **GrV**-morphisms $H_*\mu : H_n(X, Y) \to \mathbb{N}H_n(X, Y)$ and $H_*\nu : \mathbb{N}H_n(X, Y) \to H_n(X, Y)$.

Lemma 3.7.3 $H_*\mu : H_n(X, Y) \to \mathbb{N}H_n(X, Y)$ *and* $H_*\nu : \mathbb{N}H_n(X, Y) \to H_n(X, Y)$ *are isomorphisms, inverse of each other.*

Proof Clearly, $\nu \circ \mu = \mathrm{id}$ on $\mathcal{S}(X)$, thus $H_*\nu \circ H_*\mu = \mathrm{id}$ on $H_*(X, Y)$. To see that $H_*\mu \circ H_*\nu = \mathrm{id}$ on $\mathbb{N}H_*(X, Y)$, we first restrict ourselves to the absolute case $Y = \emptyset$. We shall prove that $C_*\mu \circ C_*\nu$ and the identity of $\mathbb{N}C_*(X)$ admit a common acyclic carrier A_* with respect to the basis $\mathbb{N}\mathcal{S}(X)$. The condition that $H_*\mu \circ H_*\nu = \mathrm{id}$ on $\mathbb{N}H_*(X)$ then follows from Proposition 2.9.1.

For $(s, \tau) \in \mathbb{N}\mathcal{S}_n(X)$ and $k \in \mathbb{N}$, define $B_k(s, \tau)$ by

$$B_k(s, \tau) = \{(t, \tau \circ |p|) \mid t \in \mathcal{S}_k(\mathcal{F}\mathbb{N}) \text{ and } p : \bar{t} \to \bar{s} \text{ is a simplicial map }\}.$$

Let $A_k(s, \tau)$ be the $\mathbb{Z}_2$-vector space with basis $B_k(s, \tau)$. Using the restriction of p to the facets of t, one defines a boundary operator $\partial : A_k(s, \tau) \to A_{k-1}(s, \tau)$ making $A_*(s, \tau)$ a subchain complex of $\mathbb{N}C_*(X)$.

For $(s, \tau) \in \mathbb{N}\mathcal{S}(X)$, one has $(s\tau) \in A_*(s, \tau)$ ($p = \mathrm{id}_{\bar{s}}$ and

$$\mu \circ \nu(s, \tau) = (s_0, \tau \circ h_s^{-1} \circ h_{s_0(n)}) \in A_*(s, \tau),$$

since $h_s^{-1} \circ h_{s_0(n)} = |p|$ for $p : \bar{s} \to \bar{s}_0$ the unique simplicial isomorphism preserving the order. The conditions for the correspondence $(s, \tau) \mapsto A_*(s, \tau)$ being an acyclic carrier (see Sect. 2.9) are easy to check once we know that $H_0(A_*(s)) = \mathbb{Z}_2$ and $H_m(A_*(s)) = 0$ for $m > 0$ which we prove below.

Consider the simplicial complex $K(s)$ with vertex set $V(K(s)) = \mathbb{N} \times V(s)$ and whose k-simplexes are the sets $\{(n_0, s_0), \dots, (n_k, s_k)\}$ with $n_0 < \cdots < n_k$. We check that the correspondence sending $(t, \tau \circ |p|)$ to the graph of $p : V(t) \to V(s)$ induces an isomorphism of chain complex between $A_*(s, \tau)$ and the simplicial chain of $K(s)$. We have thus to prove that $H_*(K(s)) \approx H_*(pt)$. But $K(s)$ is the union of $K_n(s)$, where $K_n(s)$ is the union of all simplexes of $K(s)$ with vertices in $\{0, \dots, n\} \times V(S)$. The inclusion $V(K_n(s)) \hookrightarrow V(K_{n+1}(s))$ together with the map $k \mapsto (n+1, k)$ provides a bijection $V(K_n(s)) \dot{\cup} V(s) \xrightarrow{\approx} V(K_{n+1}(s))$ and a simplicial isomorphism

$$K_n(s) * \bar{s}^0 \xrightarrow{\approx} K_{n+1}(s),$$

where $\bar{s}^0$ is the 0-skeleton of $\bar{s}$. Hence, the inclusion $K_n(s) \hookrightarrow K_{n+1}(s)$ factors through the cones on $K_n(s)$ contained in the join $K_n(s) * \bar{s}^0$. Therefore, $H_*(K(s)) \approx \varinjlim H_*(K_n(s)) \approx H_*(pt)$.

We have thus proved Lemma 3.7.3 in the case $Y = \emptyset$. Using the homology sequences, this proves that $H_*\mu : H_n(X, Y) \to \mathbb{N}H_n(X, Y)$ and $H_*\nu : \mathbb{N}H_n(X, Y) \to H_n(X, Y)$ are both isomorphisms. But we have already noted that $H_*\nu \circ H_*\mu = \mathrm{id}$ on $H_*(X, Y)$. Therefore, $H_*\mu \circ H_*\nu = \mathrm{id}$ on $\mathbb{N}H_*(X, Y)$. □

We are now ready for the proof of Theorem 3.7.2.

Proof of Theorem 3.7.2 If $t \subset s$ are simplexes of $\mathcal{F}\mathbb{N}$, we denote by $i_{t,s} : \bar{t} \to \bar{s}$ the simplicial map given by the inclusion. The space X^{RCW} is defined as the quotient space

$$X^{\mathrm{RCW}} = \dot{\bigcup}_{(s,\tau) \in \mathbb{N}\mathcal{S}(X)} \big(\{(s, \tau))\} \times |\bar{s}|\big) \Big/ \sim \tag{3.7.6}$$

where $\sim$ is the equivalence relation $((s, \tau), |i_{t,s}|(u)) \sim (t, \tau \circ |i_{t,s}|), u)$ for all $(s, \tau) \in \mathbb{N}\mathcal{S}(X)$, all subsimplex t of s and all $u \in |\bar{t}|$. As in the proof of Theorem 3.7.1, X^{RCW} is a naturally a CW-complex. The characteristic map for the k-cell corresponding to

$(s, \tau) \in \mathbb{N}\mathcal{S}_k(X)$ is the restriction to $\{(s, \tau))\} \times |\bar{s}|$ of the quotient map in (3.7.6). In particular, $(X^{\text{RCW}})^0$ is the set $\mathbb{N} \times X$ endowed with the discrete topology. Because of the role of $\mathbb{N}$ in the indexing of the cells, one checks that X^{RCW} is a regular CW-complex. For $(s, \tau) \in \mathbb{N}\mathcal{S}_m(X)$, one has a continuous map $\phi_{(s,\tau)} : \{(s, \tau)\} \times |\bar{s}| \to X$ defined by $\phi_{(s,\tau)}((s, \tau), u) = \tau(u)$. The disjoint union of those evaluation maps descends to a continuous map $\phi : X^{\text{RCW}} \to X$, if Y is a subspace of X, then Y^{RCW} is a subcomplex of X^{RCW}. The functoriality of the correspondence $(X, Y) \mapsto (X^{\text{RCW}}, Y^{\text{RCW}})$, as well as that ϕ is a natural transformation from $\mathbf{j} \circ \mathbf{rcw}$ to the identity functor of $\mathbf{Top_2}$, are established as in the proof of Theorem 3.7.1.

We now prove that $\mathbb{N}H_*\phi : \mathbb{N}H_*(X^{\text{RCW}}) \to \mathbb{N}H_*(X)$ is a **GrV**-isomorphism, following the pattern of the proof of Theorem 3.7.1. Similarly to (3.7.3), we construct the diagram

$$\begin{array}{ccc} \mathbb{N}H_*(X^{\text{RCW}}) & \xrightarrow{\mathbb{N}H_*\phi} & \mathbb{N}H_*(X) \\ & \nwarrow^{\mathbb{N}\beta} \quad \swarrow_{\mathbb{N}\alpha} & \\ & \dot{H}_*(X^{\text{RCW}}) & \end{array} \tag{3.7.7}$$

such that $\mathbb{N}H_*\phi \circ \mathbb{N}\beta \circ \mathbb{N}\alpha = \text{id}$ and $\mathbb{N}\alpha$ and $\mathbb{N}\beta$ are isomorphisms. As in (3.7.3), the bijection $\mathbb{N}\mathcal{S}(X) \xrightarrow{=} \{\text{cells of } X^{\text{RCW}}\}$ gives the isomorphism $\mathbb{N}\alpha$. The grv-morphism β comes from associating to the k-cell of X^{RCW} indexed by (s, τ) the map

$$|\bar{s}| \xrightarrow{\approx} \{(s, \tau)\} \times |\bar{s}| \xrightarrow{\text{char.map}} X^{\text{RCW}}$$

which is an element of $\mathbb{N}\mathcal{S}_k(X^{\text{RCW}})$. The equation $\mathbb{N}H_*\phi \circ \mathbb{N}\beta \circ \mathbb{N}\alpha = \text{id}$ is straightforward.

The proof that $\mathbb{N}\beta$ is an isomorphism is quite similar to that (for β) in the proof of Theorem 3.7.1. Indeed, using Lemma 3.7.3, $\mathbb{N}H_*()$ is a homology theory and thus diagrams like in (3.7.4) and (3.7.5) do exist; the isomorphism $H_*\mu$: $\mathbb{N}H_*(\Delta^{k-1}, \text{Bd}\Delta^{k+1}) fl{\approx} H_*(\Delta^{k-1}, \text{Bd}\Delta^{k+1})$ is also explicit enough and permits to proceed as in the proof of Theorem 3.7.1. Details are left to the reader.

The **GrV**-isomorphism $H_*\mu$ of Lemma 3.7.3 is natural; one has thus a commutative diagram

$$\begin{array}{ccc} \mathbb{N}H_*(X^{RCW}) & \xrightarrow[\approx]{\mathbb{N}H_*\phi} & \mathbb{N}H_*(X) \\ \approx\big\downarrow \mu & & \approx\big\downarrow \mu \\ H_*(X^{RCW}) & \xrightarrow{H_*\phi} & H_*(X) \end{array}$$

which shows that $H_*\phi$ is an isomorphism. The relative case is obtained as at the end of the proof of Theorem 3.7.1 and that $H^*\phi$ is a **GrA**-isomorphism comes from Kronecker duality. □

Theorem 3.7.4 *There is a covariant functor* symp : $(X, Y) \to (K_X, K_Y)$ *from* $\mathbf{Top}_2 \to \mathbf{Simp}_2$ *and a natural transformation* $\phi = \phi_{(X,Y)} : (|K_X|, |K_Y|) \to (X, Y)$ *from* $|\,\mathbf{symp}|$ *to the identity functor of* $\mathbf{Top}_2$ *such that* $H_*\phi$ *and* $H^*\phi$ *are isomorphisms.*

Proof Let X be a topological space. By the proof of Theorem 3.7.2, the regular CW-complex X^{RCW} comes equipped with characteristic embeddings $\varphi_{(s,\tau)} : \{(s, \tau))\} \times |\bar{s}| \to X^{RCW}$ $((s, \tau)) \in \mathbb{N}\mathcal{S}(X))$, satisfying the following condition: if t is a face of s with simplicial inclusion $i_{t,s} : \bar{t} \to \bar{s}$, then $\varphi_{(t,\tau\circ|i_{t,s}|)} = \varphi_{(s,\tau)} \circ |i_{t,s}|$. The only missing thing to make X^{RCW} a simplicial complex is that several simplexes may have the same boundary. But this can be avoided by taking the barycentric subdivision of each cell, with the characteristic embedding $\varphi'_{(s,\tau)} : \{(s, \tau))\} \times |(\bar{s})'| \to X^{RCW}$. We thus get a functorial triangulation of X^{RCW}. □

Remarks 3.7.5 The following facts about the above constructions should be noted.

(a) The proof of Theorem 3.7.1 goes back to Giever [67]; for a more recent treatment, see [73, p. 146]. Theorem 3.7.2 may be obtained from Theorem 3.7.1 by subdivision techniques in semi-simplicial complexes (see [73, Theorem 16.41]). Our proof of Theorem 3.7.2 is different.
(b) The construction X^{CW} in the proof of Theorem 3.7.1 is sometimes called in the literature the *thick geometric realization of the singular complex of* X. A quotient $\overline{X^{\mathrm{CW}}}$ of X^{CW} was introduced by Milnor [146], in which the degenerate simplexes are collapsed. Thus, $\bar{X}^{\mathrm{CW}}$ has one k-cell for each *non-degenerate* singular k-simplex of X. Under mild conditions, the Milnor functor behaves well with products (see [146, Sect. 2]).
(c) The maps ϕ of Theorems 3.7.1, 3.7.2 and 3.7.4 are actually weak homotopy equivalences (see e.g. [73, Corollary 16.43]). Such maps are called *CW-approximations* [82] or *resolutions* [73]. In particular, if X is itself a CW-complex, these maps are homotopy equivalences by the Whitehead theorem [82, Theorem 4.5] (but they are not homeomorphisms). Somehow simpler (but not functorial) proofs that a spaces has the weak homotopy type of a CW-complex may be found in e.g. [82, Proposition 4.13] or [73, Proposition 16.4].
(d) By its construction in the proof of Theorem 3.7.2, X^{RCW} is a regular Δ-set is the sense of [82, pp. 533–34]. In this appendix of [82], the reader may find enlightening considerations related to our constructions in this section.

3.8 Eilenberg-MacLane Spaces

The Eilenberg-MacLane spaces are used to make the cohomology $H^*(-)$ a representable functor. With $\mathbb{Z}_2$ as coefficients, they admit an *ad hoc* presentation given below, which only uses the material developed in this book. The equivalence with the usual definition using the homotopy groups is proven at the end of the section.

A CW-complex $\mathcal{K}$ is an *Eilenberg-MacLane space in degree m* if

(i) $H^m(\mathcal{K}) = \mathbb{Z}_2$; we denote by ι the generator of $H^m(\mathcal{K})$.
(ii) for any CW complex X, the correspondence $f \mapsto H^*f(\iota)$ gives a bijection

$$\phi : [X, \mathcal{K}] \xrightarrow{\approx} H^m(X),$$

where $[X, \mathcal{K}]$ denotes the set of homotopy classes of continuous maps from X to $\mathcal{K}$.

If $f : X \to \mathcal{K}$ is a map, the class $H^*f(\iota)$ is said to be *represented by* f. Property (ii) says that the functor $H^*(-)$ would be *representable* by $\mathcal{K}$ in the sense of category theory [134].

The notation $K(\mathbb{Z}_2, m)$ is usual for a CW-complex which is an Eilenberg-MacLane space in degree m. We shall also use the notation $\mathcal{K}_m$. The unambiguity of these notations is guaranteed by the following existence and uniqueness result.

Proposition 3.8.1 (a) *For any integer m, there exists an Eilenberg-MacLane space in degree m.*
(b) *Let $\mathcal{K}_m$ and $\mathcal{K}'_m$ be two Eilenberg-MacLane spaces in degree m. Then, there exists a homotopy equivalence $g : \mathcal{K}'_m \to \mathcal{K}_m$ whose homotopy class is unique.*

Example 3.8.2 By Corollary 3.1.12, we see that the point is an Eilenberg-MacLane space in degree 0.

Proof We start by the uniqueness statement (b). Let $\mathcal{K}$ and $\mathcal{K}'$ be two Eilenberg-MacLane spaces in degree m. Then, there is a bijection

$$\mathbb{Z}_2 = H^m(\mathcal{K}') \approx [\mathcal{K}', \mathcal{K}]$$

under which the constant maps corresponds to 0. Let $g : \mathcal{K}' \to \mathcal{K}$ be a continuous map representing the non-vanishing class (unique up to homotopy). In the same way, let $h : \mathcal{K} \to \mathcal{K}'$ represent the non-vanishing class of $\mathbb{Z}_2 = H^m(\mathcal{K}) \approx [\mathcal{K}, \mathcal{K}']$. Then, $g \circ h$ represent the non-vanishing class of $\mathbb{Z}_2 = H^m(\mathcal{K}) \approx [\mathcal{K}, \mathcal{K}]$ and $h \circ g$ represent the non-vanishing class of $\mathbb{Z}_2 = H^m(\mathcal{K}') \approx [\mathcal{K}', \mathcal{K}']$. As $\mathrm{id}_{\mathcal{K}}$ and $\mathrm{id}_{\mathcal{K}'}$ do the same, we deduce that $g \circ h$ is homotopic to $\mathrm{id}_{\mathcal{K}}$ and $h \circ g$ is homotopic to $\mathrm{id}_{\mathcal{K}'}$. Therefore, h and g are homotopy equivalences.

We now construct an Eilenberg-MacLane space $\mathcal{K}$ in degree $m \geq 1$ ($\mathcal{K}_0 = pt$, as noticed in Example 3.8.2). Its m-skeleton $\mathcal{K}^m$ is the sphere S^m, with one 0-cell $\{v\}$ and one m-cell called ε. Then, for each map $\varphi : S^m \to \mathcal{K}^m$ of degree 0, an $(m+1)$-cell is attached to to $\mathcal{K}^m$ via φ, thus getting $\mathcal{K}^{m+1}$. Finally, for $k \geq m+2$, $\mathcal{K}^k$ is constructed by induction by attaching to $\mathcal{K}^{k-1}$ a k-cell for *each* continuous map $f : S^{k-1} \to \mathcal{K}^{k-1}$.

As the $(m+1)$-cells of $\mathcal{K}$ are attached to $\mathcal{K}^m$ by maps of degree 0, the cellular boundary $\dot{\partial} : \dot{C}^{n+1}(\mathcal{K}) \to \dot{C}^n(\mathcal{K})$ vanishes by Lemma 3.5.4. Therefore, $H_m(\mathcal{K}) = \mathbb{Z}_2$ by Theorem 3.5.6 and $H^m(\mathcal{K}) = \mathbb{Z}_2$ by Kronecker duality. The singleton $\{\varepsilon\}$, seen as a cellular m-cycle of $\mathcal{K}$, is called $\underline{\iota} \in Z_m(\mathcal{K})$. Seen as an m-cocycle, we denote it by $\bar{\iota} \in \dot{Z}^m(\mathcal{K})$ (it represents $\iota \in \dot{H}^m(\mathcal{K})$).

Let us prove the surjectivity of $\phi : [X, \mathcal{K}] \to H^m(X)$. Let $\bar{a} \in H^m(X)$, represented by a cellular cocycle $a \in \dot{C}^m(X) \subset \dot{C}^m(X^m)$. We shall construct a map $f : X \to \mathcal{K}$ such that $H^* f(\iota) = \bar{a}$. Let $j : D^m \to \mathcal{K}$ be a characteristic map for the unique m-cell of $\mathcal{K}$. The map f sends X^{m-1} to the point $\{v\} = \mathcal{K}^0$. Its restriction to an m-cell e of X is equal to j if $e \in a$ and the constant map onto v otherwise. This gives a map $f : X^m \to \mathcal{K}^m$ which, by construction and Lemma 3.5.11, satisfies

$$\dot{C}_* f(\alpha) = \langle a, \alpha \rangle \, \underline{\iota} \tag{3.8.1}$$

for all $\alpha \in \dot{C}_m(X)$. Hence

$$\langle \dot{C}^* f(\bar{\iota}), \alpha \rangle = \langle \bar{\iota}, \dot{C}_* f(\alpha) \rangle = \langle \bar{\iota}, \langle a, \alpha \rangle \underline{\iota} \rangle = \langle a, \alpha \rangle$$

for all $\alpha \in \dot{C}_m(X)$. By Lemma 2.3.3, we deduce that $\dot{C}^* f(\bar{\iota}) = a$.

To extend f to X^{m+1}, let $\lambda \in \Lambda_{m+1}(X)$ with attaching map $\varphi_\lambda : S^m \to X^m$. As a is a cocycle, one has $\langle a, \partial\lambda \rangle = \langle \delta(a), \lambda \rangle = 0$. Using Lemmas 3.5.4 and 3.5.11 together with Eq. (3.8.1), we get that $f_\lambda = f \circ \varphi_\lambda : S^m \to \mathcal{K}^m \approx S^m$ is a map of degree 0. By construction of $\mathcal{K}$, an $(m+1)$-cell e is attached to $\mathcal{K}^m$ via f_λ, so f may be extended to λ, using a characteristic map for e extending f_λ. This produces a cellular map $f^{m+1} : X^{m+1} \to \mathcal{K}^{m+1}$. Finally, suppose, by induction on $k \geq m+1$, that f^{m+1} extends to $f^k : X^k \to \mathcal{K}^k$. Let $\lambda \in \Lambda_{k+1}(X)$ with attaching map $\varphi_\lambda : S^k \to X^k$. Set $g_\lambda = f^k \circ \varphi_\lambda$. By construction of $\mathcal{K}$, there exists $e_{g_\lambda} \in \Lambda_{k+1}(\mathcal{K})$ with attaching map g_λ. Thus f^k may be extended to the cell λ, using a characteristic map for e_λ extending g_λ. Doing this for each $\lambda \in \Lambda_{k+1}(X)$ produces the desired extension $f^{k+1} : X^{k+1} \to \mathcal{K}^{k+1}$. The surjectivity of $\phi : [X, \mathcal{K}] \to H^m(X)$ is thus established.

For the injectivity of ϕ, let $f_0, f_1 : X \to \mathcal{K}$ such that $H^* f_0(\iota) = H^* f_1(\iota)$. Since any map between CW-complexes is homotopic to a cellular map (see e.g. [64, Theorem 2.4.11]), we may assume that f_0 and f_1 are cellular. We must construct a homotopy $F : X \times I \to \mathcal{K}$ between f_0 and f_1, which will be done cell by cell. As $f_0(X^{m-1}) = f_1(X^{m-1}) = \{v\}$, the maps f_0 and f_1 descend to cellular maps from X/X^{m-1} to $\mathcal{K}$. Hence, we can assume that $X^{m-1} = X^0$ is a single point w, with $f_0(w) = f_1(w) = v$. The homotopy F is defined to be constant on w: $F(w,t) = v$.

As X^{m-1} is a point, the homology class $\dot{H}_* f_0(\iota) = \dot{H}_* f_1(\iota)$ is represented by a single cellular cocycle $a \in \dot{C}^m(X)$ ($\dot{B}^m(X) = 0$). Let $\lambda \in \Lambda_m(X)$ with characteristic map $\hat{\varphi}_\lambda : S^m \to X$. By Lemma 3.5.11, one has, for $j = 0, 1$:

$$\langle a, \lambda \rangle = \langle C^* f_j(\bar{\iota}), \lambda \rangle = \langle \bar{\iota}, C_* f_j(\lambda) \rangle = \langle \bar{\iota}, \deg(f_j \circ \hat{\varphi}_\lambda) \underline{\iota} \rangle = \deg(f_j \circ \hat{\varphi}_\lambda). \tag{3.8.2}$$

Let Σ^m be the boundary of $D^m \times I$, homeomorphic to S^m. A map $\dot{F}_\lambda : \Sigma^m \to \mathcal{K}^m$ is defined by

$$F_\lambda(x,t) = \begin{cases} f_0(\hat{\varphi}_\lambda(x)) & \text{if } t = 0 \\ f_1(\hat{\varphi}_\lambda(x)) & \text{if } t = 1 \\ v & \text{if } x \in S^{m-1}. \end{cases} \quad (3.8.3)$$

Using (3.8.2) together with Lemma 3.2.8 (with $B_1 = D^m \times \{0\}$ and $B_2 = D^m \times \{1\}$), we deduce that $\deg F_\lambda = 0$. Then, there is an $(m+1)$-cell of $\mathcal{K}$ is attached to $\mathcal{K}^m$ with F_λ. This implies that F extends to $\hat{F}_\lambda : D^m \times I \to \mathcal{K}^{m+1}$ which is a homotopy from f_0 to f_1 over X^m union the cell λ. Doing this for each $\lambda \in \Lambda_m(X)$ produces a homotopy $F^m : X^m \times I \to \mathcal{K}^{m+1}$ between f_0 and f_1. We can thus assume, by induction on $k \geq m$, that a homotopy $F^k : X^k \times I \to \mathcal{K}^{k+1}$ between f_0 and f_1 has been constructed. We must extend it to $F^{k+1} : X^{k+1} \times I \to \mathcal{K}^{k+2}$, which can be done individually over each cell $\lambda \in \Lambda_{k+1}(X)$. We define $F_\lambda : \Sigma^{k+1} \to \mathcal{K}^{k+1}$ as in (3.8.3). As $k+1 > m$, a $(k+2)$-cell of $\mathcal{K}$ is attached to $\mathcal{K}^{k+1}$ with F_λ, which permits us, as above, to extend the homotopy F^k over the cell λ.

The proof of Proposition 3.8.1 is now complete. □

The above construction of an Eilenberg-MacLane space uses a lot of cells so we may expect that the (co)homology of $\mathcal{K}_n$ is complicated. It was computed by Serre [175, Sect. 2], whose result will be given in Theorem 8.5.5. In degree 1 however, we have the following simple example of an Eilenberg-MacLane space.

Proposition 3.8.3 *The projective space $\mathbb{R}P^\infty$ is an Eilenberg-MacLane space in degree* 1 ($\mathbb{R}P^\infty \approx K(\mathbb{Z}_2, 1)$).

Proof We use the standard CW-structure on $\mathcal{K} = \mathbb{R}P^\infty$ of Example 3.4.5, with one cell in each dimension and so that $\mathcal{K}^k = \mathbb{R}P^k$. Let $p_k : S^1 \to S^1$ given by $p_k(z) = z^k$. The following properties hold true:

(i) the 2-cell of $\mathcal{K}$ is attached to $\mathcal{K}^1 \approx S^1$ by the map p_2 which, by Proposition 3.2.6 is of degree 0.
(ii) each map $g : S^1 \to \mathcal{K}^1$ of degree 0 is null-homotopic (i.e. homotopic to a constant map) in $\mathcal{K}^2$. Indeed, it is classical that any map from S^1 to $S^1 \approx S^1$ is homotopic to p_k for some integer k (see e.g. [136, Theorem 5.1] or [82, Theorem 1.7]). By Proposition 3.2.6, $\deg p_k = 0$ if and only if $k = 2r$. Point (i) implies that $g = p_2$ is null homotopic and so is $p_{2r} = p_2 \circ p_r$.
(iii) for $k \geq 2$, each map $g : S^k \to \mathcal{K}^k$ is null-homotopic into $\mathcal{K}^{k+1}$. Indeed, the lifting property of covering spaces tells us that g admits a lifting

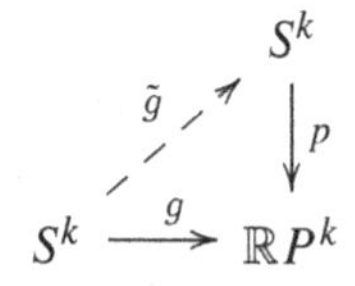

and the $(k+1)$-cell of $\mathcal{K}$ is attached via the covering map p.

By Point (i), $H^1(\mathcal{K}) = \mathbb{Z}_2$. Points (ii) and (iii) imply that the argument of the proof of Proposition 3.8.1 may be used to prove that $\phi : [X, \mathbb{R}P^1] \to H^1(X)$ is a bijection. Hence, $\mathcal{K} = \mathbb{R}P^\infty$ is an Eilenberg-MacLane space in degree 1. □

Corollary 3.8.4 *Let $f : \mathbb{R}P^n \to \mathbb{R}P^k$ be a continuous map, with $n < k \leq \infty$. Then f is either homotopic to a constant map or to the inclusion $\mathbb{R}P^n \hookrightarrow \mathbb{R}P^k$.*

Proof The lemma is true for $k = \infty$ by Proposition 3.8.3. Therefore, there is a homotopy from the composition of f with the inclusion $\mathbb{R}P^k \hookrightarrow \mathbb{R}P^\infty$ to either a constant map or the inclusion. Making this homotopy cellular (see [207, Corollary 4.7,p. 78]) produce a homotopy whose range is in $\mathbb{R}P^{n+1}$. □

We finish this section with the relationship between our definition of Eilenberg-MacLane spaces and the usual one involving the homotopy groups. Recall that the i-th homotopy group $\pi_i(X, x)$ of a pointed space (X, x) is defined by $\pi_i(X, x) = [S^i, X]^\bullet$, for some fixed base point in S^n. Below, the base points are omitted from the notation.

Proposition 3.8.5 *A CW-complex X is an Eilenberg-MacLane space $\mathcal{K}_m$ if and only if $\pi_i(X) = 0$ if $i \neq m$ and $\pi_m(X) = \mathbb{Z}_2$.*

Proof We first prove that $\mathcal{K}_m$ satisfies the conditions. By Propositions 3.8.1 and 3.8.3, the space $\mathcal{K}_1$ is homotopy equivalent to $\mathbb{R}P^1$. The statement then follows using the 2-fold covering $S^\infty \to \mathbb{R}P^\infty$ and the fact that S^∞ is contractible (see [82, Example 1.B.3p. 88]). By Proposition 3.8.1 and its proof, the space $\mathcal{K}_m$ admits a CW-structure whose $(m-1)$-skeleton is a point. Thus, when $m > 1$, $\mathcal{K}_m$ is simply connected and $[S^i, \mathcal{K}_m]^\bullet \approx [S^i, \mathcal{K}_m]$ (see [82, proposition 4A.2]). The cohomology of S^i, computed in Proposition 3.8.1, implies that the set $[S^i, \mathcal{K}_m] \approx H^m(S^i)$ contains one element if $i \neq m$ and two elements if $i = m$.

Conversely, if X is a CW-complex satisfying $\pi_i(X) = 0$ if $i \neq m$ and $\pi_m(X) = \mathbb{Z}_2$, we must prove that X is homotopy equivalent to $\mathcal{K}_m$. This requires techniques not developed in this book. When $m = 1$, there exists a map $f : X \to ro\infty \approx \mathcal{K}_1$ inducing an isomorphism on the fundamental group (see (4.3.1)). The map f then induces an isomorphism on all the homotopy groups, what is called a *weak homotopy equivalence*. By the Whitehead theorem [82, Theorem 4.5], a weak homotopy equivalence between connected CW complexes is a homotopy equivalence. When $m > 1$, let $\alpha : S^m \to X$ representing the non-zero element of $\pi_m(X)$. By the Hurewicz theorem [82, Theorem 4.32], the *integral* homology $H_m(X; \mathbb{Z}) = \mathbb{Z}_2$ and, from the universal coefficient theorem [82, Theorem 3B.5], it follows that $H_m(X) = \mathbb{Z}_2$ and $H_*\alpha : H_m(S^m) \to H_m(X)$ is an isomorphism. By Kronecker duality, $H^m(X) = \mathbb{Z}_2$ and $H^*\alpha : H^m(X) \to H^m(S^m)$ is an isomorphism. Let $g : X \to \mathcal{K}_m$ representing the non-zero element of $H^m(X)$. As $H^*\alpha : H^m(X) \to H^m(S^m)$ is an isomorphism, the map g induces an isomorphism from $\pi_m(X)$ to $\pi_m(\mathcal{K}_m)$ (we have proved above that $\pi_m(\mathcal{K}_m) = \mathbb{Z}_2$). Hence, g is a weak homotopy equivalence and therefore a homotopy equivalence by the Whitehead theorem. □

3.9 Generalized Cohomology Theories

The axiomatic viewpoint for (co)homology was initiated by Eilenberg and Steenrod in the late 1940s [51, 52] and had a great impact on the general understanding of the theory. We give below a version in the spirit of [82, Sect. 2.3 and Chap. 3]. Our application will be the Künneth theorem 4.6.7.

A *cohomology theory* is a contravariant functor h^* from the category $\mathbf{Top}_2$ of topological pairs to the category **GrV** of graded $\mathbb{Z}_2$-vector spaces, together with a natural *connecting homomorphism* $\delta^* : h^*(A) \to h^{*+1}(X, A)$ (the notation $h^*(A)$ stands for $h^*(A, \emptyset)$). In addition, the following axioms must be satisfied.

(1) *Homotopy axiom*: if $f, g : (X, A) \to (X', A')$ are homotopic, then $h^* f = h^* g$.
(2) *Exactness axiom*: for each topological pair (X, A) there is a long exact sequence

$$\cdots \xrightarrow{\delta^*} h^m(X, A) \to h^m(X) \to h^m(A) \xrightarrow{\delta^*} h^{m+1}(X, A) \to \cdots$$

where the unlabeled arrows are induced by inclusions. This exact sequence is functorial, i.e. if $f : (X', A') \to (X, A)$ is a map of pair, there is a commutative diagram

$$\begin{array}{ccccccccc}
\cdots \longrightarrow & h^*(X) & \longrightarrow & h^*(A) & \xrightarrow{\delta_*} & h^{*+1}(X, A) & \longrightarrow & h^{*+1}(X) & \longrightarrow \cdots \\
& \downarrow{\scriptstyle h^* f} & & \downarrow{\scriptstyle h^* f} & & \downarrow{\scriptstyle h^* f} & & \downarrow{\scriptstyle h^* f} & \\
\cdots \longrightarrow & h^*(X') & \longrightarrow & h^*(A') & \xrightarrow{\delta_*} & h^{*+1}(X', A') & \longrightarrow & h^{*+1}(X') & \longrightarrow \cdots
\end{array}$$

(3) *Excision axiom*: let (X, A) be a topological pair, with U be a subspace of X satisfying $\bar{U} \subset \operatorname{int} A$. Then, the **GrV**-morphism induced by inclusions $i^* : h^*(X, A) \to h^*(X - U, A - U)$ is an isomorphism.
(4) *Disjoint union axiom*: for a disjoint union $(X, A) = \bigcup_{j \in \mathcal{J}} (X_j, A_j)$ the homomorphism

$$h^*(X, A) \to \prod_{j \in \mathcal{J}} h^*(X_j, A_j)$$

induced by the family of inclusions $(X_j, A_j) \hookrightarrow (X, A)$ is an isomorphism.

Example 3.9.1 The singular cohomology H^* is a generalized cohomology theory. Axioms (1)–(3) are fulfilled, as seen in Sects. 3.1.2–3.1.4. The disjoint union axiom corresponds to Proposition 3.1.11 for a pair $(X, \emptyset)$; it may be extended to arbitrary topological pairs, using the exactness axiom and the five lemma.

K-theory and cobordism are other examples of generalized cohomology theories.

Let h^* and k^* be two cohomology theories. A *natural transformation* μ from h^* to k^* is a natural transformation of functors commuting with the connecting homomorphisms. In particular, for each topological pair (X, A), one has a commutative diagram of exact sequences:

$$\begin{array}{ccccccccccc}
\cdots & \longrightarrow & h^*(X) & \longrightarrow & h^*(A) & \xrightarrow{\delta_*} & h^{*+1}(X,A) & \longrightarrow & h^{*+1}(X) & \longrightarrow & \cdots \\
& & \downarrow{\scriptstyle\mu} & & \downarrow{\scriptstyle\mu} & & \downarrow{\scriptstyle\mu} & & \downarrow{\scriptstyle\mu} & & \\
\cdots & \longrightarrow & k^*(X) & \longrightarrow & k^*(A) & \xrightarrow{\delta_*} & k^{*+1}(X,A) & \longrightarrow & k^{*+1}(X) & \longrightarrow & \cdots
\end{array} \tag{3.9.1}$$

The aim of this section is to prove the following theorem.

Proposition 3.9.2 *Let h^* and k^* be two cohomology theories and let μ be a natural transformation from h^* to k^*. Suppose that $\mu : h^*(pt) \xrightarrow{\approx} k^*(pt)$ is an isomorphism. Then $\mu : h^*(X, A) \xrightarrow{\approx} k^*(X, A)$ is an isomorphism for all CW-pairs (X, A) where X is finite dimensional.*

The hypothesis that X is finite dimensional is not necessary in Proposition 3.9.2 (see [82, Proposition 3.19]), but it simplifies the proof considerably. Proposition 3.9.2 is enough for the applications in this book (see Sect. 4.6).

Proof We essentially recopy the proof of [82, Proposition 3.19]. By Diagram (3.9.1) and the five-lemma, it suffices to show that μ is an isomorphism when $A = \emptyset$. The proof goes by induction on the dimension of X. When X is 0-dimensional, the result holds by hypothesis and by the axiom for disjoint unions. Diagram (3.9.1) for $(X, A) = (X^m, X^{m-1})$ and the five-lemma reduce the induction step to showing that μ is an isomorphism for the pair (X^m, X^{m-1}). Let $\varphi^m : \Lambda_m \times (D^m, S^{m-1}) \to (X, X^{m-1})$ be a global characteristic maps for all the m cells of X. Like in the proof of Lemma 3.4.7, the axioms (essentially excision) imply that $h^*\varphi^m$ and $k^*\varphi^m$ are isomorphisms so, by naturality, it suffices to show that μ is an isomorphism for $\Lambda_m \times (D^m, S^{m-1})$. The axiom for disjoint unions gives a further reduction to the case of the pair (D^m, S^{m-1}). Finally, this case follows by applying the five-lemma to Diagram (3.9.1), since D^m is contractible and hence is covered by the 0-dimensional case, and S^{n-1} is $(n-1)$-dimensional. □

3.10 Exercises for Chapter 3

3.1. Give the list of the (maximal) simplexes of the triangulation of $\Delta^m \times I$ used in the proof of Proposition 3.1.30. Draw them for $m = 1, 2$. Same question for the triangulation used in the proof of Lemma 3.1.35.

3.2. Let X be a topological space.

(a) Show that X is contractible if and only if there is a correspondence $f \mapsto \hat{f}$ associating to a continuous map $f : A \to X$ a continuous extension $\hat{f} : CA \to X$, where CA is the cone over A. This correspondence is natural in the following sense: if $g : B \to A$ is a continuous map, then $\widehat{f \circ g} = \hat{f} \circ Cg$.

(b) Show that if X is contractible then, for any CW-pair (A, B), any continuous map $f : B \to X$ admits a continuous extension $g : A \to X$.

(c) Using (a), find a direct proof of that the (co)homology of a contractible space is isomorphic to that of a point (Corollary 3.1.33). [Hint: use that $\Delta^{n+1} \approx C\Delta^n$.]

3.3. Let X be a 2-sphere or a 2-torus. Let A be a non-empty subset of X containing n points. Compute $H_*(X - A)$ and $H_*(X, A)$.

3.4. Find topological pairs (X, Y) and (X', Y') such that $H_*(X, Y) \not\approx H_*(X', Y')$ while X is homeomorphic to X' and Y is homeomorphic to Y'.

3.5. Let X be a topological space. Let A be a subspace of X which is open and closed. Show that (X, A) is well cofibrant.

3.6. Show that there is no continuous retraction from the Möbius band onto its boundary.

3.7. Show that the Klein bottle K is made out of two copies of the Möbius band glued along their common boundaries. Compute $H_*(K)$, using the Mayer-Vietoris exact sequence for this decomposition.

3.8. Let $f : S^n \to S^n$ be a continuous map such that no antipodal pair of points goes to an antipodal pair of points. Show that the degree of f is 0.

3.9. Let (X, Y, Z) be a topological triple. Draw a commutative diagram linking the cohomology sequences of the pairs (X, Y), (X, Z), (Y, Z) and that of the triple (X, Y, Z).

3.10. Let (X, X_1, X_2, X_0) be a Mayer-Vietoris data with $X = X_1 \cup X_2$. Suppose that X is a CW-complex and that X_i are subcomplexes. Find a short proof of the existence of the Mayer-Vietoris for the cellular (co)homology. [Hint: analogous to the simplicial case.]

3.11. Let (X, X_1, X_2, X_0) be a Mayer-Vietoris data with $X = X_1 \cup X_2$. Suppose that the homomorphism $H_*(X_1, X_0) \to H_*(X, X_2)$ induced by the inclusion is an isomorphism. Deduce the Mayer-Vietoris (co)homology sequences for (X, X_1, X_2, X_0).

3.12. Using a tubular neighbourhood and the Mayer-Vietoris sequence, compute the homology of the complement of a (smooth) knot in S^3.

3.13. Let X be a countable CW-complex. Show that $H_*(X)$ is countable. Is it true for $H^*(X)$?

3.14. For $n \in \mathbb{N}_{\geq 1}$, consider the circle $C_n := \{z \in \mathbb{C} \mid |z - 1/n| = 1/n\}$. The *Hawaiian earring* is the subspace B of $\mathbb{C}$ consisting of the union of C_n for $n \geq 1$.

(a) Show that $H_1(B)$ surjects onto $\prod_{n=1}^{\infty} \mathbb{Z}_2$.
(b) Show that $[B, \mathbb{R}P^\infty]$ is countable.
(c) Deduce from (a) and (b) that B does not have the homotopy type of a CW-complex.

3.15. Let R_q, $q \in \mathbb{N}_{\geq 1}$ be a sequence of $\mathbb{Z}_2$-vector spaces. Find a path-connected space X such that $H_q(X) \approx R_q$.

3.16. Let X be a 2-dimensional CW-complex with a single 0-cell, m 1-cells and n 2-cells. Show that $m = n$ if and only if $b_1(X) = b_2(X)$.

3.17. Find perfect CW-decompositions for the 2-torus and the Klein bottle.

3.18. Let $P = \langle A|R\rangle$ be a presentation of a group G with a set A of generators and a set B of relators. The *presentation complex* X_P is the 2-dimensional complex obtained from a bouquet C_A of circles indexed by A by attaching, for each relator $r \in R$, a 2-cell according to $r \in \pi_1(C_A)$. Hence, $\pi_1(X_P) \approx G$. Compute $H_*(X_P)$ in the following cases.

- $P_1 = \langle a, b, c \,|\, abc^{-1}b^{-1}, bca^{-1}c^{-1}\rangle$ and $P_2 = \langle x, y \,|\, xyxy^{-1}x^{-1}y^{-1}\rangle$ (two presentations of the trefoil knot group).
- $P_3 = \langle a, b, c \,|\, a^5, b^3, (ab)^2\rangle$ (a presentation of the alternate group A_5).

3.19. Let $\mathcal{K} = K(\mathbb{Z}_2, n)$ be an Eilenberg-MacLane space. Let $f : \mathcal{K} \to \mathcal{K}$ be a continuous map. Show that f is either homotopic to the identity or to a constant map.

3.20. Let $\mathcal{K}_n = K(\mathbb{Z}_2, n)$ be an Eilenberg-MacLane space. Let X be a non-contractible CW-complex. Suppose that there are continuous maps $X \xrightarrow{j} \mathcal{K}_n \xrightarrow{r} X$ such that $r \circ j$ is homotopic to the identity. Show that X and $\mathcal{K}_n$ have the same homotopy type.

Chapter 4
Products

So far, the reader may not have been impressed by the essential differences between homology and cohomology: the latter is dual to the former via the Kronecker pairing, so they are even isomorphic for spaces of finite homology type. However, cohomology is a definitely more powerful invariant than homology, thanks to its *cup product*, making $H^*(-)$ a graded $\mathbb{Z}_2$-algebra. Thus, the homotopy types of two spaces with isomorphic homology may sometimes be distinguished by the algebra-structure of their cohomology. Simple examples are provided by $\mathbb{R}P^2$ versus $S^1 \vee S^2$, or by the 2-torus versus the Klein bottle.

In this chapter, we present the cup product for simplicial and singular cohomology, out of which the cap and cross products are derived, with already many applications (more will come in other chapters).

Cohomology and its cup product occurred in 1935 (40 years after homology) in the independent works of Kolmogoroff and Alexander, soon revisited and improved by Čhech and by Whitney [29, 209]. These people were all present in the international topology conference held in Moscow, September 1935. Vivid recollections of this memorable meeting were later written by Hopf and by Whitney [102, 211]. For surveys of the interesting history of cohomology and products, see [40, Chap. IV] and [137].

4.1 The Cup Product

4.1.1 The Cup Product in Simplicial Cohomology

Let K be a simplicial complex. Choose a simplicial order $\leq$ on K. Let $a \in C^p(K)$ and $b \in C^q(K)$. Using Point (c) of Lemma 2.3.3, we define a cochain $a \smile_{\leq} b \in C^{p+q}(K)$ by the formula

$$\langle a \smile_{\leq} b, \sigma\rangle = \langle a, \{v_0, \dots, v_p\}\rangle \, \langle b, \{v_p, \dots, v_{p+q}\}\rangle ,$$

J.-C. Hausmann, *Mod Two Homology and Cohomology*, Universitext,
DOI 10.1007/978-3-319-09354-3_4

required to be valid for all $\sigma = \{v_0, \ldots, v_{p+q}\} \in \mathcal{S}_{p+q}(K)$, with $v_0 < v_1 < \cdots < v_{p+q}$. This defines a map

$$C^p(K) \times C^q(K) \xrightarrow{\smile_\leq} C^{p+q}(K)\,.$$

We can see $\smile_\leq$ as a composition law on $C^*(K)$:

$$C^*(K) \times C^*(K) \xrightarrow{\smile_\leq} C^*(K)\,.$$

Lemma 4.1.1 *$(C^*(K), +, \smile_\leq)$ is a (non-commutative) graded $\mathbb{Z}_2$-algebra.*

Proof The associativity and distributivity properties are obvious. The neutral element for $\smile_\leq$ is the unit cochain $\mathbf{1} \in C^0(K)$. □

Lemma 4.1.2 *$\delta(a \smile_\leq b) = \delta a \smile_\leq b + a \smile_\leq \delta b$.*

Proof Set $a \in C^p(K)$, $b \in C^q(K)$ and $\sigma = \{v_0, \ldots, v_{p+q+1}\} \in \mathcal{S}_{p+q+1}(K)$ with $v_0 < v_1 < \cdots < v_{p+q+1}$. One has

$$\begin{aligned}\langle \delta a \smile_\leq b, \sigma\rangle &= \langle \delta a, \{v_0, \ldots, v_{p+1}\}\rangle\, \langle b, \{v_{p+1}, \ldots, v_{p+q+1}\}\rangle \\ &= \langle a, \partial\{v_0, \ldots, v_{p+1}\}\rangle\, \langle b, \{v_{p+1}, \ldots, v_{p+q+1}\}\rangle \\ &= \sum_{i=0}^{p+1} \langle a, \{v_0, \ldots, \hat{v}_i, \ldots, v_{p+1}\}\rangle\, \langle b, \{v_{p+1}, \ldots, v_{p+q+1}\}\rangle\,. \end{aligned} \tag{4.1.1}$$

In the same way,

$$\langle a \smile_\leq \delta b, \sigma\rangle = \sum_{i=p}^{p+q+1} \langle a, \{v_0, \ldots, v_p\}\rangle\, \langle b, \{v_p, \ldots, \hat{v}_i, \ldots, v_{p+q+1}\}\rangle\,. \tag{4.1.2}$$

The last term in the sum of (4.1.1) is equal to the first term in the sum of (4.1.2). Hence, these terms cancel when adding up the two sums and the remaining terms are those of $\langle a \smile_\leq b, \partial(\sigma)\rangle = \langle \delta(a \smile_\leq b), \sigma\rangle$. □

Lemma 4.1.2 implies that $Z^*(K) \smile_\leq Z^*(K) \subset Z^*(K)$, $B^*(K) \smile_\leq Z^*(K) \subset B^*(K)$ and $Z^*(K) \smile_\leq B^*(K) \subset B^*(K)$. Therefore, $\smile_\leq$ induces a map $H^p(K) \times H^q(K) \xrightarrow{\smile} H^{p+q}(K)$, seen as a composition law on $H^*(K)$:

$$H^*(K) \times H^*(K) \xrightarrow{\smile} H^*(K)$$

called the *cup product* on simplicial cohomology. The notation $\smile$ and the name *cup* product (the latter due to the former) were first used by Whitney [209]. It follows from Lemma 4.1.1 that $(H^*(K), +, \smile)$ is a graded $\mathbb{Z}_2$-algebra. Dropping the index "$\leq$" is justified by the following proposition.

Proposition 4.1.3 *The cup product on $H^*(K)$ does not depend on the simplicial order "$\leq$".*

Proof The procedure to define the cup product may be done with the ordered cochains. For $a \in \hat{C}^p(K)$ and $b \in \hat{C}^q(K)$, we define $a \smile b \in \hat{C}^{p+q}(K)$ by the formula

$$\langle a \smile b, \sigma \rangle = \langle a, (v_0, \ldots, v_p) \rangle \, \langle b, (v_p, \ldots, v_{p+q}) \rangle \, ,$$

required to be valid for all $(v_0, \ldots, v_{p+q}) \in \hat{S}_{p+q}(K)$. This defines a graded $\mathbb{Z}_2$-algebra structure on $\hat{C}^*(K)$. The formula $\delta(a \smile b) = \delta a \smile b + a \smile \delta b$ is proven as for Lemma 4.1.2, whence a graded algebra structure on $\hat{H}^*(K)$. These definitions imply that the isomorphism

$$H^*\phi_{\leq} : (\hat{H}^*(K), +, \smile) \xrightarrow{\approx} (H^*(K), +, \smile_{\leq})$$

of Sect. 2.10 is an isomorphism of graded algebras. As, by Corollary 2.10.10, $H^*\phi_{\leq}$ is independent of the simplicial order "$\leq$", so is the cup product on $H^*(K)$. □

Corollary 4.1.4 (Commutativity of the cup product) *The cup product in simplicial cohomology is commutative, i.e. $a \smile b = b \smile a$ for all $a, b \in H^*(K)$.*

Proof Let $\tilde{a}, \tilde{b} \in Z^*(K)$ representing a, b. Let "$\leq$" be a simplicial order on K. In $Z^*(K)$, one has

$$\tilde{a} \smile_{\leq} \tilde{b} = \tilde{b} \smile_{\geq} \tilde{a} \, ,$$

where "$\geq$" is the opposite order of "$\leq$". By Proposition 4.1.3, this proves Corollary 4.1.4. □

The commutativity of the cup product is an important feature of the mod2 cohomology. In other coefficients, holds true only up to signs.

Let **GrA** be the category whose objects are commutative graded $\mathbb{Z}_2$-algebras and whose morphisms are algebra maps. Corollary 4.1.4 says that $H^*(K)$ is an object of **GrA**. There is an obvious forgetful functor from **GrA** to **GrV**.

Proposition 4.1.5 (Functoriality of the cup product) *Let $f : L \to K$ be a simplicial map. Then $H^*f : H^*(K) \to H^*(L)$ is multiplicative: $H^*f(a \smile b) = H^*f(a) \smile H^*f(b)$ for all $a, b \in H^*(K)$.*

Proof The proof of Proposition 4.1.3 shows that $\hat{H}^*(K)$ is an object of **GrA**. Using Corollary 2.10.8, it also shows that the isomorphism $H^*\psi : H^*(K) \to \hat{H}^*(K)$ is a **GrA**-isomorphism. Let $a \in \hat{C}^p(K)$ and $b \in \hat{C}^q(K)$. Then, for all $\sigma = (v_0, \ldots, v_{p+q}) \in \hat{S}_{p+q}(K)$, one has

$$\begin{aligned} \langle \hat{C}^* f(a \smile b), \sigma \rangle &= \langle a \smile b, (f(v_0), \ldots, f(v_{p+q})) \rangle \\ &= \langle a, (f(v_0), \ldots, f(v_p)) \rangle \, \langle b, (f(v_p), \ldots, f(v_{p+q})) \rangle \end{aligned}$$

$$= \langle \hat{C}^* f(a), (v_0, \dots, v_p) \rangle \, \langle \hat{C}^* f(b), (v_p, \dots, v_{p+q}) \rangle$$
$$= \langle \hat{C}^* f(a) \smile \hat{C}^* f(b), \sigma \rangle .$$

By Lemma 2.3.3, this implies that $\hat{C}^* f(a \smile b) = \hat{C}^* f(a) \smile \hat{C}^* f(b)$. We deduce that $\hat{H}^* f : \hat{H}^*(K) \to \hat{H}^*(L)$ is multiplicative. Using Proposition 2.10.11, this implies that $H^* f$ is multiplicative. □

Corollary 4.1.6 *The simplicial cohomology is a contravariant functor from* **Simp** *to* **GrA**.

The cup product may also be defined in relative simplicial cohomology. Let L_1 and L_2 be two subcomplexes of K. For any simplicial order "$\leq$" on K, one has

$$C^*(K, L_1) \smile_{\leq} C^*(K, L_2) \subset C^*(K, L_1 \cup L_2) .$$

Hence, we get a map

$$H^*(K, L_1) \times H^*(K, L_2) \xrightarrow{\smile} H^*(K, L_1 \cup L_2)$$

which is bilinear and commutative. In particular, we get relative cup products

$$H^*(K, L) \times H^*(K) \xrightarrow{\smile} H^*(K, L) \text{ and } H^*(K) \times H^*(K, L) \xrightarrow{\smile} H^*(K, L)$$

which are related as described by the following two lemmas.

Lemma 4.1.7 *Let (K, L) be a simplicial pair. Denote by $j : (K, \emptyset) \to (K, L)$ the inclusion. Let $a \in H^p(K, L)$ and $b \in H^q(K, L)$. Then, the equality*

$$H^* j(a) \smile b = a \smile b = a \smile H^* j(b)$$

holds in $H^{p+q}(K, L)$.

Proof Denote also by $a \in Z^p(K, L)$ and $b \in Z^q(K, L)$ cocycles representing the cohomology classes a and b. Choose a simplicial order on K and let $\sigma = \{v_0, \dots, v_{p+q}\} \in \mathcal{S}_{p+q}(K) - \mathcal{S}_{p+q}(L)$ with $v_0 < \cdots < v_{p+q}$. Let $\sigma_1 = \{v_0, \dots, v_p\}$ and $\sigma_2 = \{v_p, \dots, v_{p+q}\}$. One has

$$\langle C^* j(a) \smile b, \sigma \rangle = \langle C^* j(a), \sigma_1 \rangle \langle b, \sigma_2 \rangle = \langle a, C_* j(\sigma_1) \rangle \langle b, \sigma_2 \rangle \tag{4.1.3}$$

and

$$\langle a \smile b, \sigma \rangle = \langle a, \sigma_1 \rangle \langle b, \sigma_2 \rangle . \tag{4.1.4}$$

If $\sigma_1 \in \mathcal{S}_p(L)$, then $C_* j(\sigma_1) = 0$ and the right hand sides of (4.1.3) and (4.1.4) both vanish. If $\sigma_1 \notin \mathcal{S}_p(L)$, then $C_* j(\sigma_1) = \sigma_1$ and the right hand sides of (4.1.3) and (4.1.4) are equal. As $C_{p+q}(K, L)$ is the vector space with basis

$\mathcal{S}_{p+q}(K) - \mathcal{S}_{p+q}(L)$, this proves that $H^*j(a) \smile b = a \smile b$. The other equation is proven similarly. □

The proof of the following lemma, quite similar to that of Lemma 4.1.7, is left to the reader (Exercise 4.1).

Lemma 4.1.8 *Let (K, L) be a simplicial pair. Denote by $j : (K, \emptyset) \to (K, L)$ the inclusion. Let $a \in H^p(K)$ and $b \in H^q(K, L)$. Then, the equality*

$$H^*j(a \smile b) = a \smile H^*j(b)$$

holds in $H^{p+q}(K)$.

There is also a relationship between the relative cup product and the connecting homomorphism δ^* of a simplicial pair.

Lemma 4.1.9 *Let (K, L) be a simplicial pair. Denote by $i : L \to K$ the inclusion. Let $a \in H^p(K)$ and $b \in H^q(L)$. Then, the equality*

$$\delta^*(b \smile H^*i(a)) = \delta^*b \smile a$$

holds true in $H^{p+q+1}(K, L)$.

Proof Denote also by $a \in Z^p(L)$ and $b \in Z^q(L)$ the cocycles representing the cohomology classes a and b. Let $\bar{b} \in C^q(K)$ be an extension of the cochain b. The cochain $\bar{b} \smile a \in C^{p+q}(K)$ is then an extension of $b \smile C^*i(a)$. By Lemma 2.7.1, $\delta_K(\bar{b} \smile a) \in Z^{p+q+1}(K, L)$ represents $\delta^*(b \smile H^*i(a))$, where $\delta_K : C^*(K) \to C^{*+1}(K)$ is the coboundary homomorphism for K. As a is a cocycle, one has $\delta_K(\bar{b} \smile a) = \delta_K(\bar{b}) \smile a$. By Lemma 2.7.1 again, $\delta_K(\bar{b}) \smile a$ represents the cohomology class $\delta^*b \smile a$. This proves the lemma. □

4.1.2 The Cup Product in Singular Cohomology

Let X be a topological space and let $\sigma : \Delta^m \to X$ be an element of $\mathcal{S}_m(X)$. For $0 \le p, q \le m$, we define ${}^p\sigma \in \mathcal{S}_p(X)$ and $\sigma^q \in \mathcal{S}_q(X)$ by

$$\begin{aligned} {}^p\sigma(t_0, \dots, t_p) &= \sigma(t_0, \dots, t_p, 0 \dots, 0) \text{ and} \\ \sigma^q(t_0, \dots, t_q) &= \sigma(0, \dots, 0, t_0, \dots, t_q) \,. \end{aligned}$$

The singular simplexes ${}^p\sigma$ and σ^q are called the *front* and *back faces* of σ. Let $a \in C^p(X)$ and $b \in C^q(X)$. Using Point (c) of Lemma 2.3.3, we define a cochain $a \smile b \in C^{p+q}(X)$ by the formula

$$\langle a \smile b, \sigma \rangle = \langle a, {}^p\sigma \rangle \, \langle b, \sigma^q \rangle \,, \tag{4.1.5}$$

required to be valid for all $\sigma \in \mathcal{S}_{p+q}(X)$. This defines a bilinear map

$$C^p(X) \times C^q(X) \xrightarrow{\smile} C^{p+q}(X). \tag{4.1.6}$$

The formula of Lemma 4.1.2 holds true, with the same proof. Hence, we get a *cup product in singular cohomology*: $H^p(X) \times H^q(X) \xrightarrow{\smile} H^{p+q}(X)$, giving rise to a composition law

$$H^*(X) \times H^*(X) \xrightarrow{\smile} H^*(X).$$

Proposition 4.1.10 *$(H^*(X), +, \smile)$ is a commutative graded $\mathbb{Z}_2$-algebras.*

Proof The associativity and distributivities are easily deduced from the definitions, like for the cup product in simplicial cohomology. If X is empty, then $H^*(X) = 0$ and there is nothing to prove. Otherwise, the neutral element for $\smile$ is the class of the unit cochain $\mathbf{1} \in H^0(X)$. Proving the commutativity directly is rather difficult. We shall use that the singular cohomology of X is that of a simplicial complex (see Theorem 3.7.4), together with Proposition 4.1.11, whose proof is straightforward. □

Proposition 4.1.11 *Let K be a simplicial complex. For any simplicial order $\leq$ on K, the isomorphism*

$$R^*_{\leq} : H^*(|K|) \xrightarrow{\approx} H^*(K)$$

of Theorem 3.6.3 *is an isomorphism of graded algebras.*

Proposition 4.1.12 *The singular cohomology is a contravariant functor from* **TOP** *to* **GrA**.

Proof By Proposition 4.1.10, we already know that $H^*(X)$ is an object of **GrA**. We also know, by Proposition 3.1.19, that $H^*()$ is a contravariant functor from **TOP** to **GrV**. It remains to prove the multiplicativity of $H^*f : H^*(X) \to H^*(Y)$ for a continuous map $f : Y \to X$. If $\sigma \in \mathcal{S}_{p+q}(X)$, then $f \circ {}^p\sigma = {}^p(f \circ \sigma)$ and $f \circ \sigma^q = (f \circ \sigma)^q$. Thus, the proof that $C^*f(a \smile b) = C^*f(a) \smile C^*f(b)$ is the same as for Proposition 4.1.5. □

To get relative cup products as in simplicial cohomology, some hypothesis related to the techniques of small simplexes (Sect. 3.1.4) is required. Let Y_1 and Y_2 be subspaces of a topological space X. Let $Y = Y_1 \cup Y_2$ and $\mathcal{B} = \{Y_1, Y_2\}$. We say that (Y_1, Y_2) is an *excisive couple* if $H^*(Y) \to H^*_{\mathcal{B}}(Y)$ is an isomorphism.

Lemma 4.1.13 *A couple (Y_1, Y_2) of subspaces of X is excisive if and only if the inclusion $(Y_1, Y_1 \cap Y_2) \hookrightarrow (Y_1 \cup Y_2, Y_2)$ induces an isomorphism in (co)homology.*

Proof Let $Y = Y_1 \cup Y_2$ and $\mathcal{B} = \{Y_1, Y_2\}$ as above. There is a morphism

$$\begin{array}{ccccccccc} 0 & \longrightarrow & C^*(Y, Y_2) & \longrightarrow & C^*(Y) & \longrightarrow & C^*(Y_2) & \longrightarrow & 0 \\ & & \downarrow & & \downarrow & & \downarrow & & \\ 0 & \longrightarrow & C^*(Y_1, Y_1 \cap Y_2) & \longrightarrow & C^*_{\mathcal{B}}(Y) & \longrightarrow & C^*(Y_2) & \longrightarrow & 0 \end{array}$$

of short exact sequences of singular cochain complex. It induces a morphism of the associated long exact sequences on cohomology which, by the five-lemma, implies the result. □

Lemma 4.1.14 *Let* (Y_1, Y_2) *be an excisive couple of a topological space* X. *Then,* (4.1.6) *defines a relative cup product*

$$H^*(X, Y_1) \times H^*(X, Y_2) \xrightarrow{\smile} H^*(X, Y_1 \cup Y_2) \tag{4.1.7}$$

which is bilinear. The analogues of Lemmas 4.1.7–4.1.9 *hold true.*

Proof Let $Y = Y_1 \cup Y_2$ and $\mathcal{B} = \{Y_1, Y_2\}$. Equation (4.1.5) gives a bilinear map

$$C^*(X, Y_1) \times C^*(X, Y_2) \xrightarrow{\smile} C^*(X, Y^{\mathcal{B}})$$

where $C^*(X, Y^{\mathcal{B}}) = \ker(C^*(X) \to C^*_{\mathcal{B}}(Y))$. There is a commutative diagram

$$\begin{array}{ccccccccc} H^k(X) & \longrightarrow & H^k(Y) & \longrightarrow & H^{k+1}(X, Y) & \longrightarrow & H^{k+1}(X) & \longrightarrow & H^{k+1}(Y) \\ \downarrow{\approx} & & \downarrow & & \downarrow & & \downarrow{\approx} & & \downarrow \\ H^k(X) & \longrightarrow & H^k_{\mathcal{B}}(Y) & \longrightarrow & H^{k+1}(X, Y^{\mathcal{B}}) & \longrightarrow & H^{k+1}(X) & \longrightarrow & H^{k+1}_{\mathcal{B}}(Y) \end{array}$$

where the lines are exact. By the five-lemma, if $H^*(Y) \to H^*_{\mathcal{B}}(Y)$ is an isomorphism, so is $H^*(X, Y) \to H^*(X, Y^{\mathcal{B}})$, which gives (4.1.7). The properties of the relative cup product listed at the end of Lemma 4.1.14 are proved as in the simplicial case. □

Remark 4.1.15 The couple (Y_1, Y_2) is excisive in X if and only if it is excisive in $Y_1 \cup Y_2$. Thus, by Proposition 3.1.34, (Y_1, Y_2) is excisive when Y_1 and Y_2 are both open. Also, (Y_1, Y_2) is excisive when one of the subspaces Y_i is contained in the other, for instance if one is empty or if $Y_1 = Y_2$. In some situations, the hypothesis can be fulfilled by enlarging Y_i to Y_i' without changing the homotopy type, and then (4.1.7) makes sense. As in Proposition 3.1.54 and its proof, this is the case if X is a CW-complex and Y_i are subcomplexes. Note that, if (Y_1, Y_2) is excisive, then the Mayer-Vietoris sequence for $(Y_1 \cup Y_2, Y_1, Y_2, Y_1 \cap Y_2)$ holds true, by Lemma 4.1.13 and Exercise 3.11.

4.2 Examples

4.2.1 Disjoint Unions

Let X be a topological space which is a disjoint union:

$$X = \dot{\bigcup}_{j\in\mathcal{J}} X_j \, .$$

By Proposition 3.1.11, the family of inclusions $i_j : X_j \to X$ induce an isomorphism in **GrV**

$$H^*(X) \xrightarrow[\approx]{(H^*i_j)} \prod_{j\in\mathcal{J}} H^*(X_j) \, .$$

By Proposition 4.1.12, H^*i_j is a homomorphism of algebras for each $j \in \mathcal{J}$. Hence, the above map (H^*i_j) is an isomorphism of graded algebras.

4.2.2 Bouquets

Let (X_j, x_j), with $j \in \mathcal{J}$, be a family of well pointed spaces which are path-connected. By Proposition 3.1.54, the family of inclusions $i_j : X_j \to X = \bigvee_{j\in\mathcal{J}} X_j$, for $j \in \mathcal{J}$, gives rise to isomorphisms on reduced cohomology

$$\tilde{H}^*(X) \xrightarrow[\approx]{(\tilde{H}^*i_j)} \prod_{j\in\mathcal{J}} \tilde{H}^*(X_j) \, .$$

The reduced and unreduced cohomologies share the same *positive parts*: $H^{>0}() = \tilde{H}^{>0}()$. As each space X_j is path-connected, so is the their bouquet X. Thus $H^{>0}(X) = \tilde{H}^*(X)$ and we get a **GrV**-morphism

$$H^{>0}(X) \xrightarrow[\approx]{(H^*i_j)} \prod_{j\in\mathcal{J}} \tilde{H}^{>0}(X_j) \, .$$

Being induced by continuous maps, (H^*i_j) is multiplicative. As X is path-connected, this produces the **GrA**-isomorphism

$$H^*(X) \xrightarrow{\approx} \mathbb{Z}_2 \mathbf{1} \oplus \prod_{j\in|} \tilde{H}^{>0}(X_j) \, . \tag{4.2.1}$$

When $\mathcal{J}$ is finite, one can also use the projections $\pi_j : X \to X_j$ defined in (3.1.32). By Proposition 3.1.52, they produce a **GrV**-isomorphism

$$\bigoplus_{j\in\mathcal{J}} H^{>0}(X_j) \xrightarrow[\approx]{\sum H^*\pi_j} H^{>0}(X) \qquad (\mathcal{J} \text{ finite}).$$

Being induced by continuous maps, $\sum H^*\pi_j$ is multiplicative. As X is path-connected, this produces the **GrA**-isomorphism

$$\mathbb{Z}_2\mathbf{1} \oplus \bigoplus_{j\in\mathcal{J}} H^{>0}(X_j) \xrightarrow{\approx} H^*(X) \qquad (\mathcal{J} \text{ finite}). \tag{4.2.2}$$

4.2.3 Connected Sum(s) of Closed Topological Manifolds

A closed n-dimensional *topological manifold* is a compact space such that each point has an open neighbourhood homeomorphic to $\mathbb{R}^n$.

Let M_1 and M_2 be two closed n-dimensional topological manifolds. We suppose that M_1 and M_2 are connected. Let $B_j \subset M_j$ be two embedded compact n-balls with boundary S_j. We suppose that each ball B_j is nicely embedded in a bigger ball; this implies that (M_j, B_j) and (M_j, S_j) are good pairs. Given a homeomorphism $h : B_1 \xrightarrow{\approx} B_2$, form the closed topological manifold

$$M = M_1 \sharp_h M_2 = (M_1 - \operatorname{int} B_1) \cup_h (M_2 - \operatorname{int} B_2).$$

The manifold M is called a *connected sum* of M_1 and M_2. Though connected topological manifolds are homogeneous for nicely embedded balls (see e.g. [95, Theorem 6.7]), the homeomorphism type of M may depend on h: for example, if $\bar{h}$ is obtained from h by precomposition with a homeomorphism of B_1 which reverses the orientation, then $M_1 \sharp_h M_2$ does not have, in general, the same homotopy type as $M_1 \sharp_{\bar{h}} M_2$ (for instance, if $M_1 = M_2 = \mathbb{C}P^2$, the two cases are distinguished by their integral intersection form). In most applications in the literature, the connected sum is defined for oriented manifolds and one requires that h reverses the orientation; this makes the oriented homeomorphism type of $M_1 \sharp M_2$ well defined. However, by Proposition 4.2.1 below, the mod 2-cohomology algebra of $M_1 \sharp_h M_2$ does not depend on h, up to algebra isomorphism.

If each M_j admit a triangulation $|K_j| \approx M_j$, then K_j is a connected n-dimensional pseudomanifold (see Corollary 5.2.7). The connected sum may be done in the world of pseudomanifolds, using n-simplexes for the balls B_j. By Proposition 2.4.4, $H_n(M_j) = \mathbb{Z}_2$, generated by the fundamental class $[M_i]$. The statement $H_n(M_j) = \mathbb{Z}_2$ also holds for closed connected topological manifolds (see e.g. [82, Theorem3.26]). We denote by $[M_j]^\sharp$ the generator of $H^n(M_i) = \mathbb{Z}_2$.

Proposition 4.2.1 *Under the above hypotheses, the cohomology ring $H^*(M_1 \sharp_h M_2)$ is isomorphic to the quotient of $\mathbb{Z}_2\,\mathbf{1} \oplus H^{>0}(M_1) \oplus H^{>0}(M_2)$ by the ideal generated by $[M_1]^\sharp + [M_2]^\sharp$:*

$$H^*(M_1 \sharp_h M_2) \approx \mathbb{Z}_2\,\mathbf{1} \oplus H^{>0}(M_1) \oplus H^{>0}(M_2) \Big/ ([M_1]^\sharp + [M_2]^\sharp)\,.$$

In particular, under this isomorphism, the classes $[M_1]^\sharp$ and $[M_2]^\sharp$ both correspond to the fundamental class $[M]^\sharp$ of M.

Proof Form the space $\hat{M} = M_1 \cup_h M_2$ and let $B \subset \hat{M}$ be the common image of B_1 and B_2 in $\hat{M}$, with boundary S. As (M_j, B_j) are good pairs, Proposition 3.1.54 provides a Mayer-Vietoris sequence for $(\hat{M}, M_1, M_2, B)$. As B has the cohomology of a point, one gets a multiplicative **GrV**-isomorphism:

$$\alpha : H^{>0}(\hat{M}) \xrightarrow{\approx} H^{>0}(M_1) \oplus H^{>0}(M_2)\,.$$

As M_j is connected, so is $\hat{M}$ and α extend to a **GrA**-isomorphism

$$\hat{\alpha} : H^*(\hat{M}) \xrightarrow{\approx} \mathbb{Z}_2\,\mathbf{1} \oplus H^{>0}(M_1) \oplus H^{>0}(M_2)\,.$$

Let $M = M_1 \sharp_h M_2 \subset \hat{M}$. The pair $(\hat{M}, M)$ is obviously a good pair, whence, by excision and homotopy, the isomorphism $H^*(\hat{M}, M) \approx H^*(B, S)$. The non-zero part of $H^*(B, S)$ is $H^n(B, S) = \mathbb{Z}_2$. Therefore, the homomorphism $\beta^* : H^k(\hat{M}) \to H^k(M)$ induced by the inclusion is an isomorphism, except possibly for $k = n - 1$ or n. In these degrees, the cohomology sequence of $(\hat{M}, M)$ looks like

$$\begin{array}{ccccccccc}
H^{n-1}(\hat{M}) & \xrightarrow{\beta^*} & H^{n-1}(M) & \xrightarrow{\delta^*} & H^n(\hat{M}, M) & \longrightarrow & H^n(\hat{M}) & \xrightarrow{\beta^*} & H^n(M) & \longrightarrow 0 \\
 & & & & \downarrow\approx & & \downarrow\approx & & \downarrow\approx & \\
 & & & & \mathbb{Z}_2 & & \mathbb{Z}_2 \oplus \mathbb{Z}_2 & & \mathbb{Z}_2 &
\end{array}$$

Therefore, $\beta^* : H^k(\hat{M}) \to H^k(M)$ is an isomorphism for $k \le n - 1$ and the **GrA** homomorphism $\beta^* : H^*(\hat{M}) \to H^*(M)$ is onto. The kernel of $\beta^* : H^n(\hat{M}) \to H^n(M)$ is of dimension 1 and, by symmetry ($M_1 \cup_h M_2 = M_2 \cup_{h^{-1}} M_1$), it must be generated by $[M_1]^\sharp + [M_2]^\sharp$. □

Remark 4.2.2 If we work simplicially with pseudomanifolds, the fact that $\ker(\beta^* : H^n(\hat{M}) \to H^n(M))$ contains $[M_1]^\sharp + [M_2]^\sharp$ may be seen directly. Indeed the n-cocycle consisting of the n-simplex B_j represents $[M_j]^\sharp$ by Proposition 2.4.4. Hence, the n-cocycle $\{B\}$ represents $[M_1]^\sharp + [M_2]^\sharp$ in $H^n(\hat{M})$ and is in $\ker \beta^*$.

4.2.4 Cohomology Algebras of Surfaces

We start with the triangulation M of $\mathbb{R}P^2$ drawn in Fig. 2.2, p. 26. We use the simplicial order given by the numeration $0, \dots, 5$ of the vertices. The computation of $H^*(M)$ is given in (2.4.8) and the generator of $H^1(M) = \mathbb{Z}_2$ is given by the cocycle a given in (2.4.9):

$$a = \alpha = \{\{1,2\},\{2,3\},\{3,4\},\{4,5\},\{5,1\}\} \subset \mathcal{S}_1(\mathbb{R}P^2)\,.$$

We see that, in $C^2(M)$,

$$a \smile a = \{\{1,2,3\},\{2,3,4\},\{3,4,5\}\}\,.$$

As $a \smile a$ contains an odd number of 2-simplexes, Proposition 2.4.4 implies that $a \smile a$ is the generator $[M]^\sharp$ of $H^2(M)$. Therefore, one gets a **GrA**-isomorphism

$$H^*(\mathbb{R}P^2) \approx \mathbb{Z}_2[a]/(a^3)$$

from $H^*(\mathbb{R}P^2)$ to the quotient of the polynomial ring $\mathbb{Z}_2[a]$ by the ideal generated by a^3. Using (4.2.1), this shows that $\mathbb{R}P^2$ and $S^1 \vee S^2$ do not have the same homotopy type though they have the same Betti numbers.

Our next example is the torus T^2. We use the triangulation given in Fig. 2.3 on p. 27 which shows two 1-cocycles $a, b \in C^1(T^2)$ whose cohomology classes, again denoted by a and b, form a basis of $H^1(T^2) \approx \mathbb{Z}_2 \oplus \mathbb{Z}_2$. One checks that the following equations hold in $C^2(T^2)$:

$$\begin{aligned} a \smile a &= \{\{4,5,6\},\{5,6,7\},\{6,7,8\},\{7,8,9\}\} \\ b \smile b &= \{\{2,3,6\},\{3,6,8\}\} \\ a \smile b &= \{\{6,7,8\}\} \\ b \smile a &= \{\{7,8,9\}\}\,. \end{aligned}$$

In $H^2(T^2) = \mathbb{Z}_2$, generated by $[T^2]^\sharp$, Proposition 2.4.4 implies that

$$a \smile a = b \smile b = 0 \quad \text{and} \quad a \smile b = b \smile a = [T^2]^\sharp\,.$$

Observe that $a \smile b \neq b \smile a$ in $C^2(T^2)$, the equality only holding true in cohomology. We get a **GrA**-isomorphism

$$H^*(T^2) \approx \mathbb{Z}_2[a,b]/(a^2, b^2)\,.$$

Our third example is the Klein bottle K, using the triangulation given in Fig. 2.4 on p. 29: analogously to the case of the torus, Fig. 2.4 shows two 1-cocycles

$a, b \in C^1(K)$ whose cohomology classes, again denoted by a and b, form a basis of $H^1(K) \approx \mathbb{Z}_2 \oplus \mathbb{Z}_2$. The following equations hold in $C^2(K)$:

$$\begin{aligned} a \smile a &= \big\{\{4,5,6\}, \{5,6,7\}, \{6,7,8\}, \{7,8,9\}, \{4,5,9\}\big\} \\ b \smile b &= \big\{\{2,3,6\}, \{3,6,8\}\big\} \\ a \smile b &= \big\{\{6,7,8\}\big\} \\ b \smile a &= \big\{\{7,8,9\}\big\}. \end{aligned}$$

In $H^2(K) = \mathbb{Z}_2$, generated by $[K]^\sharp$, Proposition 2.4.4 implies

$$a \smile a = [K]^\sharp \,, \; b \smile b = 0 \quad \text{and} \quad a \smile b = b \smile a = [K]^\sharp \,.$$

Though $H^*(T^2)$ and $H^*(K)$ are **GrV**-isomorphic, we see that they are not **GrA**-isomorphic. Indeed, for a space X, consider the *cup-square map*

$$H^*(X) \xrightarrow{\smile^2} H^*(X)$$

given by $\smile^2(x) = x \smile x$. Note that this map is linear, since the ground field is $\mathbb{Z}_2$. Our above computations show that $\smile^2 = 0$ for $X = T^2$ but not for $X = K$. It does not vanish either for $X = \mathbb{R}P^2$, as seen above. Now, it is classical that a connected closed surface X is a connected sum of tori if X is orientable and a connected sum of projective spaces otherwise. Hence, Proposition 4.2.1 implies that the orientability of a connected surface may be seen on its cohomology algebra:

Proposition 4.2.3 *Let M be a closed connected surface. Then M is orientable if and only if its cup-square map $H^1(M) \to H^2(M)$ vanishes.*

Remark 4.2.4 As a consequence of Wu's formula, we shall see in Corollary 9.8.5 that Proposition 4.2.3 generalizes in the following way: a closed connected n-dimensional manifold M is orientable if and only if the linear map $\mathrm{sq}^1 : H^{n-1}(M) \to H^n(M)$ vanishes.

Finally, we see that closed surfaces are distinguished by their cohomology algebra.

Proposition 4.2.5 *Two closed surfaces are diffeomorphic if and only if their cohomology algebra are* **GrA***-isomorphic.*

Proof By Proposition 4.2.3, the cohomology algebra determines whether a closed surface M is orientable or not. If M is orientable, then M is a connected sum of m tori and, by Proposition 4.2.1 $H^1(M) \approx \mathbb{Z}_2^{2m}$. If M is not orientable, then M is a connected sum of m projective planes and, by Proposition 4.2.1 $H^1(M) \approx \mathbb{Z}_2^m$. □

4.3 Two-Fold Coverings

4.3.1 H^1, Fundamental Group and 2-Fold Coverings

Let (Y, y) and (Y', y') be two pointed spaces. Let $[Y, Y']^\bullet$ be the set of homotopy classes of pointed maps from Y to Y' (the homotopies also preserving the base point). Let $F : [Y, Y']^\bullet \to [Y, Y']$ be the obvious forgetful map.

Let (X, x) be a pointed topological space. We first define a map $e : H^1(X) \to \text{map}(\pi_1(X, x), \mathbb{Z}_2)$. Let $a \in H^1(X)$. If $c : S^1 \to X$ is a pointed map representing $[c] \in [S^1, X]^\bullet = \pi_1(X, x)$, we set $e(a)([c]) = H^*c(a) \in H^1(S^1) = \mathbb{Z}_2$. As $H^*c = H^*c'$ if c is homotopic to c', the map is well defined. Observe that $\text{map}(\pi_1(X, x), \mathbb{Z}_2)$ is naturally a $\mathbb{Z}_2$-vector space, containing $\hom(\pi_1(X, x), \mathbb{Z}_2)$ as a linear subspace.

Lemma 4.3.1 *Let X be a connected CW-complex, pointed by $x \in X^0$. Then the map e is an isomorphism*

$$e : H^1(X) \xrightarrow{\approx} \hom(\pi_1(X, x), \mathbb{Z}_2)\,.$$

Proof We first prove that the image of e lies in $\hom(\pi_1(X, x), \mathbb{Z}_2)$. The multiplication in $\pi_1(X, x) = [S^1, X]^\bullet$ may be expressed using the comultiplication $\mu : S^1 \twoheadrightarrow S^1/S^0 \approx S^1 \vee S^1$. Then $[c][c'] = [(c \vee c')\circ\mu]$. Using that $H^1(S^1 \vee S^1) \approx H^1(S^1) \times H^1(S^1)$ (see Proposition 3.1.51), one has

$$e(a)([c][c']) = H^*\mu\big(e(a)([c]), e(a)([c'])\big) = e(a)([c]) + e(a)([c'])$$

for all $a \in H^1(X)$. This proves that $e([c][c']) = e([c]) + e([c'])$. The equality $e(a + b)([c]) = e(a)([c]) + e(b)([c])$ is obvious, so e is a homomorphism.

Let us consider $\mathbb{R}P^\infty$ with its standard CW-structure of Example 3.4.5, with one cell in each dimension, pointed by its 0-cell a. Van Kampen's Theorem implies that $\pi_1(\mathbb{R}P^\infty, a) = \mathbb{Z}_2$. The fundamental group functor gives rise to a map

$$[X, \mathbb{R}P^\infty]^\bullet \xrightarrow{\approx} \hom(\pi_1(X, x), \mathbb{Z}_2) \tag{4.3.1}$$

which is a bijection. Indeed, the bijectivity is established in the same way as, in Proposition 3.8.3, the fact that $\phi{:}[X, \mathbb{R}P^\infty] \to H^1(X)$ is a bijection. The forgetful map $F : [X, \mathbb{R}P^\infty]^\bullet \to [X, \mathbb{R}P^\infty]$ and the homomorphism e fit in the commutative diagram

$$\begin{array}{ccc}
[X, \mathbb{R}P^\infty]^\bullet & \xrightarrow{F} & [X, \mathbb{R}P^\infty] \\
\downarrow{\approx} & & \downarrow{\approx} \\
\hom(\pi_1(X, x), \mathbb{Z}_2) & \xleftarrow{e} & H^1(X)
\end{array} \tag{4.3.2}$$

The map F is surjective: if $f : X \to \mathbb{R}P^\infty$, any path γ from $f(x)$ to a extends to homotopy from f to a pointed map. This follows from the fact that $(X, \{a\})$ is cofibrant (see Proposition 3.4.1). Hence, the commutativity of Diagram (4.3.2) implies that e (and F) are bijective. □

We now turn our attention to 2-fold coverings. The reader is assumed some familiarity with the theory of covering spaces, as presented in many textbooks (see e.g. [179, Chap. 2] or [82, Sect. 1.3]).

Let X be a connected CW-complex, pointed by $x \in X^0$. Two covering projections $p_i : X_i \to X$ are *equivalent* if there exists a homeomorphism $h : X_1 \to X_2$ such that $p_2 \circ h = p_1$. Denote by $\mathrm{Cov}_2(X)$ the set of equivalence classes of 2-fold coverings of X.

Let $p : \tilde{X} \to X$ be a 2-fold covering. Choose $\tilde{x} \in p^{-1}(\{x\})$. Then $p_*(\pi_1(\tilde{X}, \tilde{x}))$ is a subgroup of index ≤ 2 of $\pi_1(X, x)$. Let $\mathrm{Grp}_2(\pi_1(X, x))$ be the set of such subgroups. A subgroup of index ≤ 2 being normal, the subgroup $p_*(\pi_1(\tilde{X}, \tilde{x}))$ does not depend on the choice of $\tilde{x} \in p^{-1}(\{x\})$. We thus get a map

$$\mathrm{Cov}_2(X) \xrightarrow{\approx} \mathrm{Grp}_2(\pi_1(X, x))$$

which is a bijection (see, e.g., [82, Theorem 1.38]). For example, the trivial 2-fold covering $\{\pm 1\} \times X \to X$ corresponds to the whole group $\pi_1(X, x)$ which is of index $1 \leq 2$. An element $H \in \mathrm{Grp}_2(\pi_1(X, x))$ is the kernel of a unique homomorphism $\pi_1(X, x) \twoheadrightarrow \pi_1(X, x)/H \hookrightarrow \mathbb{Z}_2$. This gives a bijection

$$\mathrm{Grp}_2(\pi_1(X, x)) \xrightarrow{\approx} \hom(\pi_1(X, x), \mathbb{Z}_2)\,.$$

If $f : X \to \mathbb{R}P^\infty$ is a continuous map, one can form the pullback diagram

$$\begin{array}{ccc} \hat{X} & \xrightarrow{\hat{f}} & S^\infty \\ \hat{p}\downarrow & & \downarrow p_\infty \\ X & \xrightarrow{f} & \mathbb{R}P^\infty\,. \end{array} \tag{4.3.3}$$

In detail, $\hat{X} = \{(u, z) \in X \times S^\infty \mid f(u) = p_\infty(z)\}$, with $\hat{p}(u, z) = u$ and $\hat{f}(u, z) = z$. We say that the covering projection $\hat{p}$ is *induced* from p_∞ by the map f and write $\hat{p} = f^* p_\infty$. Observe that $\hat{p}$ correspond to the subgroup $\ker \pi_1 f$. Thus, homotopic maps induce equivalent coverings and we get a map $\mathrm{ind} : [X, \mathbb{R}P^\infty] \to \mathrm{Cov}_2(X)$. These various maps, together with those of (4.3.2) sit in the commutative diagram

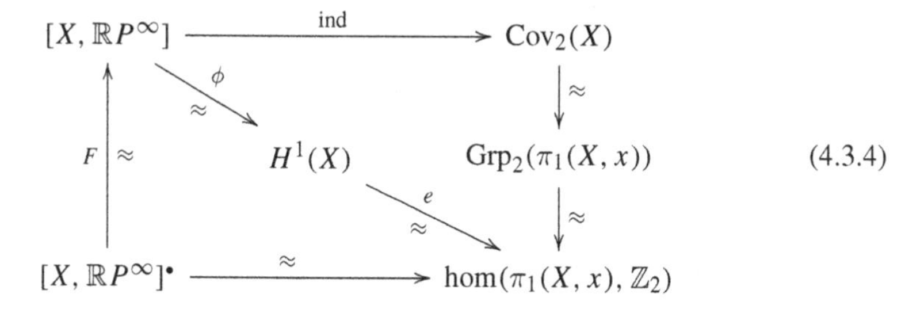

(4.3.4)

The commutativity of Diagram (4.3.4) implies the following proposition.

Proposition 4.3.2 ind : $[X, \mathbb{R}P^\infty] \to \mathrm{Cov}_2(X)$ *is a bijection.*

Let $p : \tilde{X} \to X$ be a 2-fold covering. A continuous map $f : X \to \mathbb{R}P^\infty$ such that p is equivalent to $f^* p_\infty$ is called a *characteristic map* for the covering p. Proposition 4.3.2 implies the following corollary.

Corollary 4.3.3 *Let X be a connected CW-complex. Then, any 2-fold covering admits a characteristic map. Two such characteristic maps are homotopic.*

Let $p : \tilde{X} \to X$ be a 2-fold covering. The correspondence which, over each $x \in X$, exchanges the two points of $p^{-1}(x)$ defines a homeomorphism $\tau : \tilde{X} \to \tilde{X}$, which is an involution (i.e. $\tau \circ \tau = \mathrm{id}$) without fixed point. Also, τ is a *deck transformation*, i.e. $p \circ \tau = p$. We call τ the *deck involution* of p. For the covering $p_\infty : S^\infty \to \mathbb{R}P^\infty$, one has $\tau(z) = -z$.

Lemma 4.3.4 *A continuous map $f : X \to \mathbb{R}P^\infty$ is a characteristic map for the 2-fold covering $p : \tilde{X} \to X$ if and only if there exists a commutative diagram*

$$\begin{array}{ccc} \tilde{X} & \xrightarrow{\tilde{f}} & S^\infty \\ {\scriptstyle p}\downarrow & & \downarrow{\scriptstyle p_\infty} \\ X & \xrightarrow{f} & \mathbb{R}P^\infty , \end{array}$$

where $\tilde{f}$ is a continuous map such that $\tilde{f} \circ \tau(v) = -\tilde{f}(v)$.

Proof Let $\hat{X} \to X$ be the covering induced by f (see (4.3.3)). If f is a characteristic map for p, there is a homeomorphism $g : \tilde{X} \xrightarrow{\approx} \hat{X}$ such that $\hat{p} \circ g = p$. Therefore, g satisfies $g \circ \tau = \hat{\tau} \circ g$. As $\hat{f} \circ \hat{\tau}(v) = -\hat{f}(v)$, the map $\tilde{f} = \hat{f} \circ g$ satisfies the requirements of Lemma 4.3.4. Conversely, given $\tilde{f}$, let $g : \tilde{X} \to \hat{X}$ given by $g(v) = (p(v), \tilde{f}(v))$. The map g satisfies $\hat{p} \circ g = p$ and $g \circ \tau = \hat{\tau} \circ g$. Hence, g is surjective and is a covering projection. Since both p and $\hat{p}$ are 2-folds coverings, g is a homeomorphism and p is equivalent to $\hat{p}$. □

Example 4.3.5 The inclusion $i : \mathbb{R}P^n \to \mathbb{R}P^\infty$ is covered by the τ-equivariant map $\tilde{i} : S^n \to S^\infty$. By Lemma 4.3.4, the map i is characteristic map for the covering $S^n \to \mathbb{R}P^n$. In particular, the identity of $\mathbb{R}P^\infty$ is a characteristic map for the covering $S^\infty \to \mathbb{R}P^\infty$.

4.3.2 The Characteristic Class

Diagram (4.3.4) together with Proposition 4.3.2 provides a bijection

$$w : \mathrm{Cov}_2(X) \xrightarrow{\approx} H^1(X)\,. \tag{4.3.5}$$

This associates to a 2-fold covering $p : \tilde{X} \to X$ its *characteristic class* $w(p) \in H^1(X)$. For instance, the characteristic class $w(p_\infty)$ for the covering $p_\infty : S^\infty \to \mathbb{R}P^\infty$ is the non-zero element $\iota \in H^1(\mathbb{R}P^\infty) = \mathbb{Z}_2$. Indeed, as $H^1(\mathbb{R}P^\infty) = \mathbb{Z}_2$, the set $\mathrm{Cov}_2(\mathbb{R}P^\infty)$ has two elements and the trivial covering corresponds to 0. As S^∞ is connected, p_∞ is not the trivial covering, hence $w(p_\infty) = \iota$. The following lemma is obvious.

Lemma 4.3.6 *Let $p : \tilde{X} \to X$ be a 2-fold covering over a CW-complex. Then*

(1) *if $f : X \to \mathbb{R}P^\infty$ is a characteristic map for the covering p, then $w(p) = H^*f(w(p_\infty)) = H^*f(\iota)$.*
(2) *if $g : Y \to X$ is a continuous map, then $w(g^*p) = H^*g(w(p))$.*
(3) *p is the trivial covering if and only if $w(p) = 0$.*

Let us give geometric descriptions of the characteristic class $w(p)$. Choose a set-theoretic section $b : X \to \tilde{X}$ of p and let $B = b(X) \subset \tilde{X}$. We consider B as a singular 0-cochain of $\tilde{X}$. Using the coboundary $\tilde{\delta} : C^0(\tilde{X}) \to C^1(\tilde{X})$, we get $\tilde{\delta}B \in C^1(\tilde{X})$ which is the connecting 1-cochain for B: for $\tilde{\sigma} \in \mathcal{S}_1(\tilde{X})$, $\langle \tilde{\delta}(B), \tilde{\sigma} \rangle = 1$ if and only if the (non-oriented) path $\tilde{\sigma}$ connects a point in B to a point in $X - B$ (see Example 3.1.4). Observe that $\tilde{\delta}(B) = \tilde{\delta}(X - B) = \tilde{\delta}(\tau(B))$, where τ is the deck involution of p. Hence

$$\begin{aligned}\langle \tilde{\delta}(B), \tau\circ\tilde{\sigma} \rangle &= \langle \tilde{\delta}(B), C_*\tau(\tilde{\sigma}) \rangle = \langle C^*\tau\circ\tilde{\delta}(B), \tilde{\sigma} \rangle \\ &= \langle \tilde{\delta}\circ C^*\tau(B), \tilde{\sigma} \rangle = \langle \tilde{\delta}(\tau(B)), \tilde{\sigma} \rangle \\ &= \langle \tilde{\delta}(B), \tilde{\sigma} \rangle\,.\end{aligned}$$

Thus, $\langle \tilde{\delta}(B), \tilde{\sigma} \rangle$ depends only on $p\circ\sigma \in \mathcal{S}_1(X)$. This permits us to define a singular 1-cochain $w_b(p) \in C^1(X)$ by the formula

$$\langle w_b(p), \sigma \rangle = \langle \tilde{\delta}(B), \tilde{\sigma} \rangle$$

where $\tilde{\sigma} \in \mathcal{S}_1(\tilde{X})$ is any lifting of $\sigma \in \mathcal{S}_1(X)$.

Proposition 4.3.7 *$w_b(p)$ is a 1-cocycle representing $w(p) \in H^1(X)$.*

Proof Let $w_b = w_b(p)$. Let $\sigma_2 \in \mathcal{S}_2(X)$. If $\tilde{\sigma}_2 \in \mathcal{S}_2(\tilde{X})$ is a lifting of σ_2, then the 1-simplexes in $\tilde{\partial}(\tilde{\sigma}_2)$ are liftings of those in $\partial(\sigma_2)$. Therefore

$$\langle \delta(w_b), \sigma_2 \rangle = \langle w_b, \partial(\sigma_2) \rangle = \langle \tilde{\delta}(B), \tilde{\partial}(\tilde{\sigma}_2) \rangle = 0 , \tag{4.3.6}$$

which proves that w_b is a cocycle.

We next prove that the cohomology class $w_b \in H^1(X)$ of w_b does not depend on the set-theoretic section b. Let $b' : X \to X'$ another such section, giving $B' = b'(X) \in C^0(\tilde{X})$. Define $r \in C^0(X)$ by

$$\langle r, \{x\} \rangle = \langle B, \tilde{x} \rangle + \langle B', \tilde{x} \rangle ,$$

where $\tilde{x}$ is a chosen element in $p^{-1}(\{x\})$. If $\tilde{\tilde{x}}$ is another choice, one has $\langle B, \tilde{\tilde{x}} \rangle = \langle B, \tilde{x} \rangle + 1$ and $\langle B', \tilde{\tilde{x}} \rangle = \langle B', \tilde{x} \rangle + 1$ in $\mathbb{Z}_2$, so r is well defined. Let $\sigma \in \mathcal{S}_1(X)$ with end points u and v. Let $\tilde{\sigma} \in \mathcal{S}_1(\tilde{X})$ be a lifting of σ with end points $\tilde{u}$ and $\tilde{v}$. Then

$$\begin{aligned} \langle w_b + w_{b'}, \sigma \rangle &= \langle \tilde{\delta}(B) + \tilde{\delta}(B'), \tilde{\sigma} \rangle \\ &= \langle B, \tilde{u} \rangle + \langle B', \tilde{u} \rangle + \langle B, \tilde{v} \rangle + \langle B', \tilde{v} \rangle \\ &= \langle \delta(r), \sigma \rangle . \end{aligned}$$

This proves that $w_{b'} = w_b + \delta(r)$ and thus $[w_{b'}] = [w_b]$. Denote by $\bar{w}(p) \in H^1(X)$ the cohomology class $[w_b]$.

We can now prove that $\bar{w}(p) = \bar{w}(p')$ if $p' : \tilde{X}' \to X$ is a 2-fold covering equivalent to p. Indeed, if $h : \tilde{X} \xrightarrow{\approx} \tilde{X}'$ is a homeomorphism such that $p' \circ h = p$, then, $w_b(p) = w_{h \circ b}(p')$, which implies that $\bar{w}(p) = \bar{w}(p')$.

Choosing a characteristic map $f : X \to \mathbb{R}P^\infty$ for p, we now have $\bar{w}(p) = \bar{w}(\hat{p})$, where $\hat{p} : \hat{X} \to X$ is the induced covering of Diagram (4.3.3). Choose a set-theoretic section $b_0 : \mathbb{R}P^\infty \to S^\infty$ of p_∞ and set $B_0 = p_\infty(\mathbb{R}P^\infty)$ and $\bar{w}_0 = [w_{b_0}(p_\infty)] \in H^1(\mathbb{R}P^\infty)$. This gives rise to a set-theoretic section $\hat{b}$ of $\hat{p}$ by the formula $\hat{b}(x) = (x, b_0 \circ f(x))$. It satisfies $\hat{B} = \hat{b}(X) = \hat{f}^{-1}(B_0)$, where $\hat{f} : \hat{X} \to S^\infty$ is the map covering f, as in (4.3.3). Let $\sigma \in \mathcal{S}_1(X)$ with a lifting $\hat{\sigma} \in \mathcal{S}_1(\hat{X})$. Then, $\hat{f} \circ \hat{\sigma}$ is a lifting of $f \circ \sigma$ in $\mathcal{S}_1(S^\infty)$ and we have

$$\begin{aligned} \langle C * f(w_{b_0}), \sigma \rangle &= \langle w_{b_0}, f \circ \sigma \rangle = \langle \delta_0(B_0), \hat{f} \circ \hat{\sigma} \rangle \\ &= \langle \delta_0(B_0), C_* \hat{f}(\hat{\sigma}) \rangle = \langle C^* \hat{f} \circ \delta_0(B_0), \hat{\sigma} \rangle \\ &= \langle \hat{\delta} \circ C^* \hat{f}(B_0), \hat{\sigma} \rangle = \langle \hat{\delta}(\hat{f}^{-1}(B_0)), \hat{\sigma} \rangle = \langle \hat{\delta}(\hat{B}), \hat{\sigma} \rangle \\ &= \langle w_{\hat{b}}(\hat{p}), \sigma \rangle . \end{aligned}$$

Hence,

$$\bar{w}(\hat{p}) = C^* f(\bar{w}_0) . \tag{4.3.7}$$

Together with Lemma 4.3.6, Eq. (4.3.7) reduces the proof of Proposition 4.3.7, to showing that $\bar{w}_0 = w(p_\infty)$. As $0 \neq w(p_\infty) \in H^1(\mathbb{R}P^\infty) = \mathbb{Z}_2$, it is enough to prove that $\bar{w}_0 \neq 0$. By Eq. (4.3.7) again, it is enough to find a covering $q : \tilde{Y} \to Y$ over some CW-complex Y for which $\bar{w}(q) \neq 0$.

We take for q the double covering $q : S^1 \to S^1$. Let $\sigma : \Delta^1 \to S^1$ given by $\sigma(t, 1-t) = e^{2i\pi t}$. As $\tilde{\sigma}(0,1) \neq \tilde{\sigma}(1,0)$, one has $\langle \bar{w}(q), \sigma \rangle \neq 0$. But σ is a 1-cocycle representing the generator of $H^1(S^1) = \mathbb{Z}_2$ (see Corollary 3.2.4). Hence, $\bar{w}(q) \neq 0$. □

Remark 4.3.8 Let $p : \tilde{X} \to X$ be a two-fold covering over a CW-complex X. To describe the characteristic class $w(p)$ in the cellular cohomology of X, choose a section $b : X^0 \to \tilde{X}$ and see $B = b(X^0) \subset \tilde{X}^0$ as a cellular 0-cochain of $\tilde{X}$ (for the cellular decomposition induced from that of X). Let $\varphi : \Lambda_1(X) \times I \to X$ be a global characteristic maps for the 1-cells of X. Let $\tilde{\varphi} : \Lambda_1(X) \times I \to \tilde{X}$ be the lifting of φ for which $\varphi(\lambda, 0) \in B$. Consider the cellular 1-cochain $w_B \in \dot{C}^1(X)$ defined, for $\lambda \in \Lambda_1(X)$, by

$$\langle w_B, \lambda \rangle \begin{cases} 1 & \text{if } \varphi(\lambda, 1) \notin B \\ 0 & \text{otherwise.} \end{cases}$$

Note that $\langle w_B, \lambda \rangle = \langle \tilde{\delta}(B), \tilde{\lambda} \rangle$ where $\tilde{\lambda}$ is any one-cell of $\tilde{X}$ above λ, which, as in (4.3.6), proves that w_B is a cellular cocycle. We claim that $[w_B] \in \dot{H}^1(X)$ corresponds to $w(p) \in H^1(X)$, under the identification of $\dot{H}^1(X)$ and $H^1(X)$ as the same subgroup of $H^1(X^1)$ (see (3.5.5)). We can thus suppose that $X = X^1$. We can also suppose that X is connected. If T is a maximal tree of X^1, then the quotient map $X^1 \to X^1/T$ is a homotopy equivalence by Proposition 3.4.1 and Lemma 3.1.44. The covering p is then induced from one over X^1/T, so we can assume that X^1 is a bouquet of circles indexed by Λ_1. For each one cell λ, a characteristic map $\varphi_\lambda : D^1 \to X^1$ gives a singular 1-simplex of X^1 (identifying D^1 with Δ^1). If $\tilde{\varphi}_\lambda$ is a lifting of φ_λ, one has

$$\langle w_B, \lambda \rangle = \langle w_{\hat{b}}, \varphi_\lambda \rangle = \begin{cases} 1 & \text{if } \tilde{\varphi}_\lambda \text{ is a loop} \\ 0 & \text{otherwise,} \end{cases} \tag{4.3.8}$$

where $\hat{b} : X \to \tilde{X}$ is a set theoretic section of p extending b. As $\{[\varphi_\lambda] \mid \lambda \in \Lambda_1\}$ is a basis for $H^1(X^1)$, Eq. (4.3.8) implies that $[w_B] \in \dot{H}^1(X^1)$ corresponds to $w(p) \in H^1(X)$.

4.3.3 The Transfer Exact Sequence of a 2-Fold Covering

Let $p:\tilde{X} \to X$ be a 2-fold covering projection with deck involution τ. To each singular simplex $\sigma : \Delta^m \to X$, one can associate the set of the two liftings of

σ into $\tilde{X}$. This defines a map from $\mathcal{S}_m(X)$ to $C_m(\tilde{X})$, extending to a linear map $\mathrm{tr}_* : C_m(X) \to C_m(\tilde{X})$. The map tr_* is clearly a chain map. By Sect. 2.3, this gives risen to two **GrV**-morphisms

$$\mathrm{tr}_* : H_m(X) \to H_m(\tilde{X}) \text{ and } \mathrm{tr}^* : H^m(\tilde{X}) \to H^m(X)$$

satisfying $\langle \mathrm{tr}^*(a), \alpha\rangle = \langle a, \mathrm{tr}_*(\alpha)\rangle$. The linear maps tr^* and tr_* are called the *transfer* homomorphisms for the covering p.

The transfer homomorphism in cohomology and the characteristic class $w(p) \in H^1(X)$ are related by the following exact sequence.

Proposition 4.3.9 (Transfer exact sequence) *The sequence*

$$\cdots \to H^m(X) \xrightarrow{H^*p} H^m(\tilde{X}) \xrightarrow{\mathrm{tr}^*} H^m(X) \xrightarrow{w(p)\smile -} H^{m+1}(X) \xrightarrow{H^*p} \cdots$$

is exact. It is functorial with respect to induced coverings.

Proof The sequence

$$0 \to C_*(X) \xrightarrow{\mathrm{tr}_*} C_*(\tilde{X}) \xrightarrow{C_*p} C_*(X) \to 0\,.$$

is clearly an exact sequence of chain complexes and it is functorial with respect to induced coverings. By Kronecker duality, it gives a short exact sequence of cochain complexes

$$0 \to C^*(X) \xrightarrow{C^*p} C^*(\tilde{X}) \xrightarrow{\mathrm{tr}^*} C^*(X) \to 0\,. \tag{4.3.9}$$

By Proposition 2.6.2, this gives rise to a connecting homomorphism $d^* : H^*(X) \to H^{*+1}(X)$ and a functorial long exact sequence

$$\cdots \to H^m(X) \xrightarrow{H^*p} H^m(\tilde{X}) \xrightarrow{\mathrm{tr}^*} H^m(X) \xrightarrow{d^*} H^{m+1}(X) \xrightarrow{H^*p} \cdots\,.$$

It just remains to identify d^* with $w(p) \smile -$.

To construct the connecting homomorphism d^* we need a **GrV**-section of tr^* in Sequence (4.3.9). Choose a set-theoretic section $b : X \to \tilde{X}$ of p. If $\sigma : \Delta^m \to X$ is a singular 1-simplex of X, define $b^\times(\sigma) : \Delta^m \to \tilde{X}$ to be the unique lifting of σ with $b^\times(\sigma)(1, 0, \ldots, 0) \in b(X)$. This defines a map $b^\times : \mathcal{S}(X) \to \mathcal{S}(\tilde{X})$. If $a \in C^m(X)$, we consider a as a subset of $\mathcal{S}_m(X)$ and so its direct image $b^\times(a) \subset \mathcal{S}_m(\tilde{X})$ is an m-cochain of $\tilde{X}$. This determines a **GrV**-morphism $b^\times : C^*(X) \to C^*(\tilde{X})$ which is a section of tr^*. By Eq. (2.6.2), the connecting homomorphism d^* is determined by the equation

$$\langle C^*p \circ d^*(a), \beta\rangle = \langle \tilde{\delta} \circ b^\times(a), \beta\rangle\,, \tag{4.3.10}$$

for all $a \in C^m(X)$ and $\beta \in \mathcal{S}_m(\tilde{X})$, where $\tilde{\delta} : C^*(\tilde{X}) \to C^{*+1}(\tilde{X})$ is the coboundary. The equality $\langle C^*p \circ d^*(a), \beta \rangle = \langle d^*(a), C_*p(\beta) \rangle$ together with (4.3.10) shows that $\langle \tilde{\delta} \circ b^\times(a), \beta \rangle$ depends only on $C_*p(\beta) = p \circ \beta = \bar{\beta}$. Therefore, by taking $\tau \circ \beta$ instead of β if necessary, we may assume that $\beta \notin b^\times(\mathcal{S}_m(X))$. Then, the faces $\beta \circ \epsilon_i$ of β are not in $b^\times(\mathcal{S}_m(X))$, except possibly for $i = 0$ and

$$\langle \tilde{\delta} \circ b^\times(a), \beta \rangle = \langle b^\times(a), \tilde{\partial}(\beta) \rangle = \langle b^\times(a), \beta \circ \epsilon_0 \rangle \,. \qquad (4.3.11)$$

The number $\langle b^\times(a), \beta \circ \epsilon_0 \rangle$ equals 1 if and only if $\beta(0, 1, 0, \dots, 0) \in b(X)$ and $\bar{\beta} \circ \epsilon_0 \in a$. In other words, if and only if the front and back faces of $\bar{\beta}$ satisfy ${}^1\bar{\beta} \in w_b(p)$ (see Sect. 4.3.2) and $\bar{\beta}^{m-1} \in a$. Hence, Eqs. (4.3.10) and (4.3.11) imply that

$$\langle d^*(a), \bar{\beta} \rangle = \langle b^\times(a), \beta \circ \epsilon_0 \rangle = \langle w_b(p) \smile a, \bar{\beta} \rangle$$

for all $a \in C^m(X)$ and $\bar{\beta} \in \mathcal{S}_m(X)$. This proves that $d^*(-) = w_b(p) \smile -$ in $C^*(X)$. By Proposition 4.3.7, this implies that $d^*(-) = w(p) \smile -$ in $H^*(X)$. □

An important application of the transfer exact sequence is the determination of the cohomology ring of $\mathbb{R}P^n$.

4.3.4 *The Cohomology Ring of* $\mathbb{R}P^n$

Let $\mathbb{Z}_2[a]$ be the polynomial ring over a formal variable a in degree 1. This is an object of **GrA**, as well as its truncation $\mathbb{Z}_2[a]/(a^{n+1})$, the quotient of $\mathbb{Z}_2[a]$ by the ideal generated by a^{n+1}. By Proposition 3.8.3, $H^1(\mathbb{R}P^\infty) = \mathbb{Z}_2$, generated by the class ι. Therefore, there is a **GrA**-morphism $\mathbb{Z}_2[a] \to H^*(\mathbb{R}P^\infty)$ sending a^k to ι^k, where the latter denotes the cup product of k copies of ι. The composition $\mathbb{Z}_2[a] \to H^*(\mathbb{R}P^\infty) \to H^*(\mathbb{R}P^n)$ factors by a **GrA**-morphism $\mathbb{Z}_2[a]/(a^{n+1}) \to H^*(\mathbb{R}P^n)$.

Proposition 4.3.10 *The above* **GrA**-*morphisms*

$$\mathbb{Z}_2[a] \to H^*(\mathbb{R}P^\infty) \text{ and } \mathbb{Z}_2[a]/(a^{n+1}) \to H^*(\mathbb{R}P^n)$$

are **GrA**-*isomorphisms. In particular, the* **GrA**-*morphism* $H^*(\mathbb{R}P^\infty) \to H(\mathbb{R}P^n)$, *induced by the inclusion, is surjective.*

Proof As S^∞ is contractible [82, Example 1.B.3 p. 88], the transfer exact sequence of the covering $p_\infty : S^\infty \to \mathbb{R}P^\infty$ shows that the cup product with $w(p_\infty) \in H^1(\mathbb{R}P^\infty)$ gives an isomorphism $H^*(\mathbb{R}P^\infty) \xrightarrow{\approx} H^{*+1}(\mathbb{R}P^\infty)$. In particular, $w(p_\infty)$ is the generator of $H^1(\mathbb{R}P^\infty)$. This proves the statement for $\mathbb{R}P^\infty$.

For $\mathbb{R}P^n$ we use the covering $p : S^n \to \mathbb{R}P^n$. The transfer exact sequence proves at once that the cup product with $w(p) \in H^1(\mathbb{R}P^n)$ gives an isomorphism $H^m(\mathbb{R}P^n) \xrightarrow{\approx} H^{m+1}(\mathbb{R}P^n)$ for $0 \le m < n - 1$. As $\mathbb{R}P^n$ has a CW-structure with one k-cell for $0 \leq k \leq n$, the end of the transfer exact sequence of the covering $p : S^n \to \mathbb{R}P^n$ involves the $\mathbb{Z}_2$-vector spaces

$$0 \to \underbrace{H^{n-1}(\mathbb{R}P^n)}_{\dim=1} \xrightarrow{w(p)\smile-} \underbrace{H^n(\mathbb{R}P^n)}_{\dim\leq 1} \xrightarrow{H^*p} \underbrace{H^n(S^n)}_{\dim=1} \xrightarrow{\text{tr}^*} \underbrace{H^n(\mathbb{R}P^n)}_{\dim\leq 1} \to 0$$

Thus, the cup product with $w(p)$ is also an isomorphism $H^{n-1}(\mathbb{R}P^n) \xrightarrow{\approx} H^n(\mathbb{R}P^n)$. This proves the proposition for $\mathbb{R}P^n$. □

4.4 Nilpotency, Lusternik-Schnirelmann Categories and Topological Complexity

Let X be a topological space. A subspace U of X is *categorical* if the inclusion $U \hookrightarrow X$ is homotopic to a constant map. The *Lusternik-Schnirelmann category* **cat** (X) is the minimal cardinality of an open covering of X with categorical subspaces. Some authors (see, e.g. [35]) adopt a different normalization for the Lusternik-Schnirelmann category, equal to one less than the definition above.

For a survey paper about the Lusternik-Schnirelmann category, see [108]. Amongst its properties, **cat** (X) is an invariant of the homotopy type of X. For example, **cat** $(X) = 1$ if and only if X is contractible and **cat** $(S^n) = 2$. More generally, one has the following result (see [108, Proposition 1.2] for a more general statement and a different proof).

Proposition 4.4.1 *Let X be a connected CW-complex of dimension n. Then* **cat** $(X) \leq n + 1$.

Proof By induction on the dimension of X, the statement being obvious if $\dim X = 0$. By Proposition 3.4.1, we can write X as the union of two open sets $X = C \cup Z$, where C is the disjoint union of the open n-cells of X and Z retracting by deformation on $X^{(n-1)}$. By induction hypothesis, Z admits an open covering with $\leq n$ categorical subspaces. If $n \geq 1$, C is a categorical open set of X, which proves that **cat** $(X) \leq n + 1$. □

Let X be a topological space and B be a vector subspace of $H^*(X)$. The *nilpotency class* **nil** B of B is the minimal integer m such that

$$\underbrace{B \smile \cdots \smile B}_{m} = 0\,.$$

If no such integer exists, we set **nil** $B = \infty$.

Proposition 4.4.2 *Let X be a topological space. Then* **nil** $H^{>0}(X) \leq$ **cat** (X).

Proof Let $U_1, \ldots, U_m \subset X$ be open subspaces of X which are categorical. By the homotopy property, the homomorphism $H^{>0}(X) \to H^{>0}(U_i)$ induced by the inclusion vanishes. Hence, the exact sequence of the pair (X, U_i) implies that the restriction homomorphism $H^{>0}(X, U_i) \to H^{>0}(X)$ is surjective. Then, in the diagram

$$\begin{array}{ccc} \prod_{i=1}^{m} H^{>0}(X, U_i) & \twoheadrightarrow & \prod_{i=1}^{m} H^{>0}(X) \\ \downarrow \smile & & \downarrow \smile \\ H^{>0}(X, U_1 \cup \cdots \cup U_m) & \longrightarrow & H^*(X) \end{array},$$

which is commutative by the functoriality of the cup product, the upper horizontal arrow is surjective. But, if $X = U_1 \cup \cdots \cup U_m$, the lower left vector space vanishes. This proves that $\mathbf{nil}\, H^{>0}(X) \le m$. □

Corollary 4.4.3 *The Lusternik-Schnirelmann category* $\mathbf{cat}\,(\mathbb{R}P^n)$ *of* $\mathbb{R}P^n$ *is equal to* $n+1$.

Proof As seen in Example 3.4.5, the projective space $\mathbb{R}P^n$ is a CW-complex of dimension n. Therefore, $\mathbf{cat}\,(\mathbb{R}P^n) \le n+1$ by Proposition 4.4.1. On the other hand, by Proposition 4.4.2, $\mathbf{cat}\,(\mathbb{R}P^n) \ge \mathbf{nil}\, H^{>0}(\mathbb{R}P^n)$ and $\mathbf{nil}\, H^{>0}(\mathbb{R}P^n) = n+1$ by Proposition 4.3.10. □

Another classical consequence of Proposition 4.4.2 is the vanishing of the cup products in a suspension.

Corollary 4.4.4 *Let Y be a topological space. Then, all cup products in* $H^{>0}(\Sigma Y)$ *vanish.*

Proof As ΣY is the union of two cones, $\mathbf{cat}\,(\Sigma Y) \le 2$, which proves the corollary. □

The Lusternik-Schnirelmann category admits several generalizations, for instance the category of a map (see, e.g. [108, Sect. 7]). Here, we introduce the category $\mathbf{cat}\,(X, A)$ of a topological pair (X, A). A subspace U of a topological space X is *A-categorical* if the inclusion $U \hookrightarrow X$ is homotopic to a map with value in A. Then $\mathbf{cat}\,(X, A)$ is defined to be the minimal cardinality of an open covering of X with A-categorical subspaces. For instance, X is path-connected, then

$$\mathbf{cat}\,(X, A) \le \mathbf{cat}\,(X) = \mathbf{cat}\,(X, pt) \tag{4.4.1}$$

Lemma 4.4.5 $\mathbf{cat}\,(X, A)$ *is an invariant of the homotopy type of the pair* (X, A).

Proof Let $f : (X, A) \to (X', A')$ be a homotopy equivalence of pair. It suffices to prove that, if U' is a subset of X' which is A'-categorical, then $U = f^{-1}(U')$

is A-categorical in X. This will imply that $\mathbf{cat}\,(X, A) \leq \mathbf{cat}\,(X', A')$. Homotopy equivalence being an equivalence relation, we also get $\mathbf{cat}\,(X', A') \leq \mathbf{cat}\,(X, A)$.

Let $g : (X', A') \to (X, A)$ be a homotopy inverse of f. Let $\beta_t : U' \to X'$ be a homotopy with $\beta_0(v) = v$ and $\beta_1(U') \subset A'$. Then, the map $\alpha_t : U \to X$ defined by $\alpha_t(u) = g \circ \beta_t \circ f(u)$ is a homotopy satisfying $\alpha_0(u) = g \circ f(u)$ and $\alpha_1(U) \subset A$. As $g \circ f$ is homotopic to $\mathrm{id}_{(X,A)}$, this proves that U is A-categorical. □

Proposition 4.4.2 generalizes in the following statement.

Proposition 4.4.6 *Let (X, A) be a topological pair with X path-connected. Then*

$$\mathbf{nil}\, B \leq \mathbf{cat}\,(X, A)\,,$$

where $B = \ker\{H^*(X) \to H^*(A)\}$.

Proof As X is path-connected, $B \subset H^{>0}(X)$. Let $U_1, \ldots, U_m \subset X$ be open subspaces of X which are A-categorical. Then, the homomorphism $H^*(X) \to H^*(U_i)$ factors:

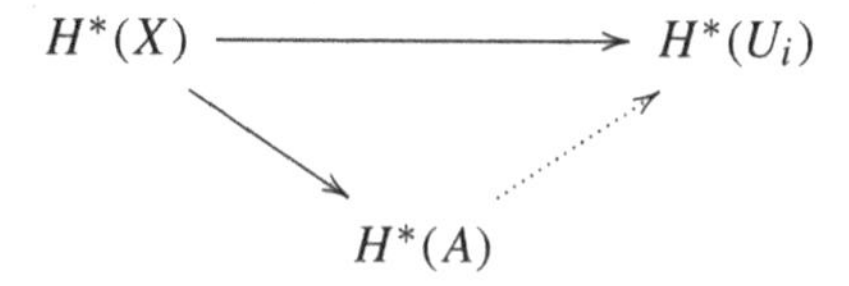

Therefore, if $a \in B$, then a is in the image of $H^{>0}(X, U_i)$. The proof of Proposition 4.4.6 is then the same as that of Proposition 4.4.2. □

This category for pairs is related to the *topological complexity*, a notion of mathematical robotics introduced by Farber [54, 55]. Let Y be a topological space and PY be the space of continuous paths $\gamma : I \to Y$, endowed with the compact-open topology. Let $\pi : PY \to Y \times Y$ be the origin-end map: $\pi(\gamma) = (\gamma(0), \gamma(1))$. A *motion planning algorithm* is a section of π. It is not possible to find a *continuous* motion planning algorithm unless Y is contractible [54, Theorem 1]. The *topological complexity* $\mathbf{TC}\,(Y)$ is the minimal cardinality of an open covering $\mathcal{U}$ of $Y \times Y$ such that $\pi : PY \to Y \times Y$ admits a continuous section over each $U \in \mathcal{U}$. Let ΔY be the diagonal subset of $Y \times Y$. The following proposition is the contents of [55, Corollary 18.2].

Proposition 4.4.7 $\mathbf{TC}\,(Y) = \mathbf{cat}\,(Y \times Y, \Delta Y)$.

In consequence, $\mathbf{TC}\,(Y)$ is an invariant of the homotopy type of Y.

Proof Let $U \subset Y \times Y$. Suppose that a continuous section $s : U \to PY$ of π exists. Then $\sigma(y, y', t) = (s(y, y')(t), y')$ satisfies $\sigma(y, y', 0) = (y, y')$ and $\sigma(y, y', 1) = (y', y') \in \Delta Y$, showing that U is ΔY-categorical. Conversely, if $c(t) = (c_1(t), c_2(t)) \in Y \times Y$ is a path from (y, y') to $(u, u) \in \Delta Y$, then the path $c_1\, c_2^{-1}$ joins y to y'. This process being continuous in (y, y'), it provides a sections of π over ΔY-categorical subsets of $Y \times Y$. □

Proposition 4.4.7 together with (4.4.1) implies that $\mathbf{TC}(Y) \leq \mathbf{cat}(Y \times Y)$. The inequality $\mathbf{cat}(Y) \leq \mathbf{TC}(Y)$ also holds true [55, Lemma 9.2], but is not a consequence of Proposition 4.4.7.

If Y is path-connected, Propositions 4.4.7 and 4.4.6 give the inequality

$$\mathbf{TC}(Y) \geq \mathbf{nil} \ker H^* j \,, \tag{4.4.2}$$

where $j : \Delta Y \to Y \times Y$ denotes the inclusion. We shall see with the Künneth theorem that there is a commutative diagram (see Remark 4.6.1):

$$\begin{array}{ccc} H^*(Y \times Y) & \xrightarrow{H^* j} & H^*(\Delta Y) \\ \approx \downarrow \times & & \downarrow \approx \\ H^*(Y) \otimes H^*(Y) & \xrightarrow{\smile} & H^*(Y) \end{array} . \tag{4.4.3}$$

(to use the Künneth theorem, we need that Y is of finite cohomology type). Under the cross product, the image of $\ker H^* j$ in the ring $H^*(Y) \otimes H^*(Y)$ is the ideal of the *divisors of zero for the cup product*. The inequality (4.4.2) thus corresponds to [54, Theorem 7].

For results concerning the topological complexity of the projective space, see the end of Sect. 6.2.2.

4.5 The Cap Product

Let K be a simplicial complex. Choose a simplicial order $\leq$ on K. We define the *cap product*

$$C^p(K) \times C_n(K) \xrightarrow{\frown_\leq} C_{n-p}(K)$$

to be the unique bilinear map such that

$$a \frown_\leq \{v_0, \dots, v_n\} = \langle a, \{v_0, \dots, v_p\} \rangle \{v_p, \dots, v_n\} \tag{4.5.1}$$

for all $a \in C^p(K)$ and all $\{v_0, \dots, v_n\} \in \mathcal{S}_n(K)$, with $v_0 < v_1 < \cdots < v_n$ (this makes sense if $n \geq p$; otherwise, the cap product just vanishes). If $a \in C^p(K)$, $b \in C^{n-p}(K)$ and $\gamma \in C_n(K)$ the following formula follows directly from the definitions

$$\langle a \smile_\leq b, \gamma \rangle = \langle b, a \frown_\leq \gamma \rangle \,. \tag{4.5.2}$$

Lemma 4.5.1 *If $a \in C^p(K)$ and $\gamma \in C_n(K)$, then*

$$\partial(a \frown_\leq \gamma) = \delta(a) \frown_\leq \gamma + a \frown_\leq \partial(\gamma) \,.$$

Proof Let $q = n - p$ and $b \in C^q(K)$. Denote $\smile_\leq$ and $\frown_\leq$ by just $\smile$ and $\frown$. Using (4.5.2), one has

$$\langle \delta(a \smile b), \gamma\rangle = \langle (a \smile b, \partial(\gamma)\rangle = \langle b, a \frown \partial(\gamma)\rangle . \tag{4.5.3}$$

In the other hand

$$\begin{aligned}\langle \delta(a \smile b), \gamma\rangle &= \langle \delta(a) \smile b, \gamma\rangle + \langle a \smile \delta(b), \gamma\rangle \\ &= \langle b, \delta(a) \frown \gamma\rangle + \langle \delta(b), a \frown \gamma\rangle \\ &= \langle b, \delta(a) \frown \gamma\rangle + \langle b, \partial(a \frown \gamma)\rangle .\end{aligned} \tag{4.5.4}$$

Equations (4.5.3) and (4.5.4) imply that

$$\langle b, \partial(a \frown \gamma)\rangle = \langle b, \delta(a) \frown \gamma + a \frown \partial(\gamma)\rangle .$$

for all $b \in C^q(K)$. By Lemma 2.3.3, this implies Lemma 4.5.1. □

Lemma 4.5.1 implies that $Z^*(K) \frown_\leq Z_*(K) \subset Z_*(K)$, $B^*(K) \frown_\leq Z_*(K) \subset B_*(K)$ and $Z^*(K) \frown_\leq B_*(K) \subset B_*(K)$. Therefore, $\frown_\leq$ induces a map $H^p(K) \times H_n(K) \xrightarrow{\frown} H_{n-p}(K)$, or

$$H^*(K) \times H_*(K) \xrightarrow{\frown} H_*(K)$$

called the *cap product* (on simplicial cohomology). As in the case of the cup product we drop the index "$\leq$" from the notation because of the following proposition.

Proposition 4.5.2 *The cap product on $H^*(K) \times H_*(K) \xrightarrow{\frown} H_*(K)$ does not depend on the simplicial order "$\leq$".*

Proof Let $\leq$ and $\leq'$ be two simplicial orders on K. Let $a \in H^p(K)$ and $\gamma \in H_n(K)$. For any $b \in H^{n-p}(K)$, Formula (4.5.2) and Proposition 4.1.3 imply that

$$\begin{aligned}\langle b, a \frown_\leq \gamma\rangle &= \langle a \smile_\leq b, \gamma\rangle \\ &= \langle a \smile_{\leq'} b, \gamma\rangle \\ &= \langle b, a \frown_{\leq'} \gamma\rangle .\end{aligned}$$

By Part (b) of Lemma 2.3.3, this implies that $a \frown_\leq \gamma = a \frown_{\leq'} \gamma$ in $H_{n-p}(K)$. □

Proposition 4.5.3 *The cap product $H^*(K) \times H_*(K) \xrightarrow{\frown} H_*(K)$ endows $H_*(K)$ with a structure of $H^*(K)$-module.*

Proof By definition, $\frown$ is bilinear and the equality $\mathbf{1} \frown \gamma = \gamma$ is obvious. It remains to prove that

$$(a \smile b) \frown \gamma = a \frown (b \frown \gamma) \tag{4.5.5}$$

for all $a \in H^p(K)$, $b \in H^q(K)$ and $\gamma \in H_n(K)$. As the cup product is associative and commutative (Corollary 4.1.4), one has, for any $c \in H^{n-p-q}(K)$,

$$\begin{aligned}\langle c, (a \smile b) \frown \gamma\rangle &= \langle c \smile (a \smile b), \gamma\rangle \\ &= \langle (c \smile a) \smile b, \gamma\rangle \\ &= \langle c \smile a, b \frown \gamma\rangle \\ &= \langle c, a \frown (b \frown \gamma)\rangle .\end{aligned}$$

By Lemma 2.3.3, this proves Eq. (4.5.5). □

Proposition 4.5.4 (Functoriality of the cap product) *Let $f : L \to K$ be a simplicial map. Then, the formula*

$$a \frown H_*f(\gamma) = H_*f(H^*f(a) \frown \gamma)$$

holds in $H_*(K)$ for all $a \in H^*(K)$ and all $\gamma \in H_*(L)$.

Proof Suppose that $a \in H^p(K)$ and $\gamma \in H_n(L)$. Using the functoriality of the cup product established in Proposition 4.1.5, one has, for any $b \in H^{n-p}(K)$,

$$\begin{aligned}\langle b, a \frown H_*f(\gamma)\rangle &= \langle a \smile b, H_*f(\gamma)\rangle \\ &= \langle H^*f(a \smile b), \gamma\rangle \\ &= \langle H^*f(a) \smile H^*f(b), \gamma\rangle \\ &= \langle H^*f(b), H^*f(a) \frown \gamma\rangle \\ &= \langle b, H_*f(H^*f(a) \frown \gamma)\rangle .\end{aligned}$$

By Part (b) of Lemma 2.3.3, this proves Proposition 4.5.4. □

There are several version of the cap product in relative simplicial (co)homology. Let (K, L) be a simplicial pair. Choose a simplicial order $\leq$ on K. We note first that $C^*(K) \frown_{\leq} C_*(L) \subset C_*(L)$, whence a cap product

$$H^p(K) \times H_n(K, L) \xrightarrow{\frown} H_{n-p}(K, L) . \tag{4.5.6}$$

One may also compose

$$H^p(K, L) \times H_n(K) \xrightarrow{j^* \times id} H^p(K) \times H_n(K) \xrightarrow{\frown} H_{n-p}(K)$$

to obtain a cap product

$$H^p(K, L) \times H_n(K) \xrightarrow{\frown} H_{n-p}(K) . \tag{4.5.7}$$

The latter cap product may be post-composed with $H_{n-p}(K) \to H_{n-p}(K, L)$ and get a cap product

$$H^p(K, L) \times H_n(K) \xrightarrow{\frown} H_{n-p}(K, L) . \tag{4.5.8}$$

As the restriction of $C^p(K) \times C_n(K) \xrightarrow{\frown} C_{n-p}(K)$ to $C^p(K, L) \times C_n(L)$ vanishes, we obtain a cap product

$$H^p(K, L) \times H_n(K, L) \xrightarrow{\frown} H_{n-p}(K) . \tag{4.5.9}$$

As for Formula (4.5.5), the equation

$$(a \smile b) \frown \gamma = a \frown (b \frown \gamma) \tag{4.5.10}$$

holds true in $H_{n-p-q}(K)$ for all $a \in H^p(K, L)$, $b \in H^q(K, L)$ and $\gamma \in H_n(K, L)$. The cap products (4.5.7) and (4.5.9) are used in (4.5.10).

More generally, suppose that L is the union of two subcomplexes $L = L_1 \cup L_2$. Then, the restriction of $C^p(K) \times C_n(K) \xrightarrow{\frown} C_{n-p}(K)$ to $C^p(K, L_1) \times C_n(L)$ has image contained in $C_{n-p}(L_2)$. This gives a cap product

$$H^p(K, L_1) \times H_n(K, L) \xrightarrow{\frown} H_{n-p}(K, L_2) . \tag{4.5.11}$$

The functoriality holds for a simplicial map $f : (K', L') \to (K, L)$ satisfying $f(L'_i) \subset L_i$ for $i = 1, 2$: the formula

$$a \frown H_* f(\gamma) = H_* f(H^* f(a) \frown \gamma) \tag{4.5.12}$$

holds in $H_*(K, L_2)$ for all $a \in H^*(K, L_1)$ and all $\gamma \in H_*(K', L')$. The proof is the same as for Proposition 4.5.4.

The next two lemmas express the compatibility between these relative cap products, the absolute one and the connecting homomorphisms for a simplicial pair (K, L).

Lemma 4.5.5 *Let (K, L) be a simplicial pair. Denote by $i : L \to K$ and $j : (K, \emptyset) \to (K, L)$ the inclusions. Let $x \in H_n(K, L)$. Then, for all integer p, the diagram*

$$\begin{array}{ccccccc}
H^p(K, L) & \xrightarrow{H^* j} & H^p(K) & \xrightarrow{H^* i} & H^p(L) & \xrightarrow{\delta^*} & H^{p+1}(K, L) \\
\downarrow{\frown x} & & \downarrow{\frown x} & & \downarrow{\frown \partial_* x} & & \downarrow{\frown x} \\
H_{n-p}(K) & \xrightarrow{H_* j} & H_{n-p}(K, L) & \xrightarrow{\partial_*} & H_{n-p-1}(L) & \xrightarrow{H_* i} & H_{n-p-1}(K)
\end{array}$$

is commutative.

Proof For the left hand square diagram, let $a \in H^p(K, L)$ and $b \in H^{n-p}(K, L)$. One has

$$\langle b, H^*j(a) \frown x\rangle = \langle b \smile H^*j(a), x\rangle$$

and

$$\langle b, H_*j(a \frown x)\rangle = \langle H^*j(b), a \frown x\rangle = \langle H^*j(b) \smile a, x\rangle .$$

Hence, the left hand square diagram commutes if and only if $b \smile H^*j(a) = H^*j(b) \smile a$, which was established in Lemma 4.1.7.

For the middle square diagram, let $a \in H^p(K)$ and $b \in H^{n-p-1}(L)$. One has

$$\langle b, H^*i(a) \frown \partial_* x\rangle = \langle b \smile H^*i(a), \partial_* x\rangle = \langle \delta^*(b \smile H^*i(a)), x\rangle .$$

On the other hand:

$$\langle b, \partial_*(a \frown x)\rangle = \langle \delta^*(b), a \frown x\rangle = \langle \delta^*(b) \smile a, x\rangle .$$

The commutativity of the middle square diagram is thus equivalent to the formula $\delta^*(b \smile H^*i(a)) = \delta^*(b) \smile a$ holding true in $H^{n-p}(K, L)$ for all $a \in H^p(K)$ and $b \in H^{n-p-1}(L)$. This formula was proven in Lemma 4.1.9. In the same way, we see that the commutativity of the right hand square diagram is a consequence of Lemma 4.1.9 (intertwining the role of a and b). □

Lemma 4.5.6 *Let (K, L) be a simplicial pair. Denote by $j : (K, \emptyset) \to (K, L)$ the pair inclusion. Then, the equation*

$$H_*j(a \frown \alpha) = a \frown H_*j(\alpha) .$$

holds true in $H_{n-p}(K, L)$ for all $a \in H^p(K)$ and all $\alpha \in H_n(K)$.

Proof It is then enough to prove that $\langle b, H_*j(a \frown \alpha)\rangle = \langle b, a \frown H_*j(\alpha)\rangle$ for all $b \in H^{n-p}(K, L)$. But,

$$\begin{aligned}
\langle b, H_*j(a \frown \alpha)\rangle &= \langle H^*j(b), a \frown \alpha\rangle \\
&= \langle H^*j(b) \smile a, \alpha\rangle \\
&= \langle H^*j(b \smile a), \alpha\rangle \quad \text{by Lemma 4.1.8} \\
&= \langle b \smile a, H_*j(\alpha)\rangle \\
&= \langle b, a \frown H_*j(\alpha)\rangle .
\end{aligned}$$

□

The cap product is also defined in the singular (co)homology of a space X. On the (co)chain level, it is the unique bilinear map

$$C^p(X) \times C_n(X) \xrightarrow{\frown} C_{n-p}(X)$$

such that

$$a \frown \sigma = \langle a, {}^p\sigma \rangle \, \sigma^q$$

for all $a \in C^p(X)$ and all $\sigma \in \mathcal{S}_n(X)$, where the back and front faces ${}^p\sigma$ and σ^q are defined as in p. 131. If $a \in C^p(X)$, $b \in C^{n-p}(X)$ and $\gamma \in C_n(X)$ the following formula follows directly from the definition

$$\langle a \smile b, \gamma \rangle = \langle b, a \frown \gamma \rangle \, . \tag{4.5.13}$$

Therefore, as for the simplicial cap product, properties follows from those of the cup product. The formula $\partial(a \frown \gamma) = \delta(a) \frown \gamma + a \frown \partial(\gamma)$ is proved as for Lemma 4.5.1 and we get an induced bilinear map $H^p(X) \times H_n(X) \xrightarrow{\frown} H_{n-p}(X)$, or

$$H^*(X) \times H_*(X) \xrightarrow{\frown} H_*(X)$$

called the *cap product in singular (co)homology*. This cap product endows $H_*(X)$ with a structure of $H^*(X)$-module, as in Proposition 4.5.3 and is functorial for continuous maps $f : Y \to X$, as for Proposition 4.5.4.

For a topological pair (X, Y), the three relative versions of the cap products:

$$H^p(X) \times H_n(X, Y) \xrightarrow{\frown} H_{n-p}(X, Y) \, , \tag{4.5.14}$$

$$H^p(X, Y) \times H_n(X) \xrightarrow{\frown} H_{n-p}(X, Y) \tag{4.5.15}$$

and

$$H^p(X, Y) \times H_n(X, Y) \xrightarrow{\frown} H_{n-p}(X) \tag{4.5.16}$$

hold true, as for (4.5.6)–(4.5.9). When $Y = Y_1 \cup Y_2$, a relative cap product analogous to (4.5.11)

$$H^p(X, Y_1) \times H_n(X, Y) \xrightarrow{\frown} H_{n-p}(X, Y_2) \, . \tag{4.5.17}$$

is available under some conditions, for instance if (Y, Y_i) is a good pair for $i = 1, 2$, so one can use the small simplexes technique, as for the Mayer-Vietoris sequence in Proposition 3.1.54. The functoriality formula (4.5.12) as well as the analogues of Lemmas 4.5.5 and 4.5.6 hold true.

Finally, the simplicial and singular cap products are intertwined by the isomorphisms

$$R_{\leq,*} : H_*(K) \xrightarrow{\approx} H_*(|K|) \text{ and } R^*_{\leq} : H^*(|K|) \xrightarrow{\approx} H^*(K)$$

of Theorem 3.6.3. For any simplicial order $\leq$, the equation

$$a \frown R_{\leq,*}(\gamma) = R_{\leq,*}(R^*_{\leq}(a) \frown \gamma) \tag{4.5.18}$$

holds in $H_*(|K|)$ for all $a \in H_*(|K|)$ and all $\gamma \in H_*(K)$: The proof of (4.5.18) is straightforward for γ a simplex of K.

4.6 The Cross Product and the Künneth Theorem

Let X and Y be topological spaces. Results computing $H^*(X \times Y)$ in terms of $H^*(X)$ and $H^*(Y)$ are known as *Künneth theorems* (or *Künneth formulas*). This generic name comes from the thesis of Hermann Künneth in 1923 (see [40, pp. 55–56]). To give an example, when X and Y are discrete spaces, the cohomology rings are concentrated in dimension 0 and

$$H^0(X) = \mathbb{Z}_2^X \,,\ H^0(Y) = \mathbb{Z}_2^Y \,,\ H^0(X \times Y) = \mathbb{Z}_2{}^{X\times Y} \,. \tag{4.6.1}$$

The *cross product* of maps

$$\mathbb{Z}_2^X \times \mathbb{Z}_2^Y \xrightarrow{\times} \mathbb{Z}_2{}^{X\times Y}$$

defined by $(f \times g)(x, y) = f(x)g(y)$ is bilinear. The associated linear map

$$\mathbb{Z}_2^X \otimes \mathbb{Z}_2^Y \xrightarrow{\times} \mathbb{Z}_2{}^{X\times Y} \tag{4.6.2}$$

is also called the *cross product*. The map (4.6.2) is clearly injective. It is not surjective if both X and Y are infinite; for instance, if $X = Y$ is infinite, it is easy to see that the characteristic function of the diagonal in $X \times X$ is not in the image of $\times$. On the other hand, suppose that X or Y is finite (say Y). Let $F : X \times Y \to \mathbb{Z}_2$. For $y \in Y$, define $F_y : X \to Z_2$ by $F_y(x) = F(x, y)$ and let χ_y be the characteristic function of $\{y\}$. Then

$$F = \sum_{y\in Y} F_y \times \chi_y \,.$$

Thus, if Y is finite, the cross product of (4.6.2) is an isomorphism.

Such finiteness conditions will occur in the statements of this section, under the form that Y should be of finite cohomology type (see Definition p. 88).

Observe that, under the identification of (4.6.1), the cross product $H^0(X) \times H^0(Y) \to H^0(X \times Y)$ satisfies the formula

$$f \times g = \pi_X^* f \smile \pi_Y^* g \,,$$

where π_X and π_Y are the projections of $X \times Y$ onto X and Y.

More generally, let X and Y be two topological spaces. Using the usual tensor product $\otimes$ of vector spaces over $\mathbb{Z}_2$, we define the *tensor product* of the $\mathbb{Z}_2$-algebras $(H^*(X), +, \smile)$ and $(H^*(Y), +, \smile)$ as the $\mathbb{Z}_2$-algebra $(H^*(X) \otimes H^*(Y), +, \bullet)$ defined by

$$[H^*(X) \otimes H^*(Y)]^m = \bigoplus_{i+j=m} H^i(X) \otimes H^j(Y),$$

with the product

$$(a_1 \otimes b_1) \bullet (a_2 \otimes b_2) = (a_1 \smile a_2) \otimes (b_1 \smile b_2). \tag{4.6.3}$$

The projections $\pi_X : X \times Y \to X$ et $\pi_Y : X \times Y \to Y$ give **GrA**-morphisms $\pi_X^* : H^*(X) \to H^*(X \times Y)$ et $\pi_Y^* : H^*(Y) \to H^*(X \times Y)$. This permits us to define a bilinear map

$$H^*(X) \times H^*(Y) \xrightarrow{\times} H^*(X \times Y)$$

by

$$a \times b = \times(a, b) = \pi_X^*(a) \smile \pi_Y^*(b) \tag{4.6.4}$$

called the *cross product*. By the universal property of the tensor product (analogous to that for vector spaces), this gives a **GrV**-morphism

$$H^*(X) \otimes H^*(Y) \xrightarrow{\times} H^*(X \times Y),$$

also called the *cross product*.

Remark 4.6.1 Let $\Delta : X \to X \times X$ be the diagonal map $\Delta(x) = (x, x)$. The composition

$$H^*(X) \times H^*(X) \xrightarrow{\times} H^*(X \times X) \xrightarrow{\Delta^*} H^*(X) \tag{4.6.5}$$

is equal to the cup product (see also Diagram (4.4.3)) This relation, due to Lefschetz (see [183, pp. 38–41] for historical considerations), was quite influential: in some books (e.g. [136, 179]), the cross product is introduced first using homological algebra (the Eilenberg-Zilber theorem) and the cup product is defined via Formula (4.6.5). Our opposite approach follows the viewpoint of [74, 82].

Under some hypotheses, the cross product may be defined in relative cohomology. Let (X, A) and (Y, B) be topological pairs. The projections π_X and π_Y give homomorphisms $\pi_X^* : H^*(X, A) \to H^*(X \times Y, A \times Y)$ and $\pi_Y^* : H^*(Y, B) \to H^*(X \times Y, B \times X)$. Suppose that A or B is empty, or one of the pairs (X, A) or

(Y, B) is a good pair. Then formula (4.6.4) defines a relative cross product

$$H^*(X, A) \otimes H^*(Y, B) \xrightarrow{\times} H^*(X \times Y, A \times Y \cup X \times B). \tag{4.6.6}$$

Indeed, we must just check that the relative cup product

$$H^*(X \times Y, A \times Y) \otimes H^*(X \times Y, B \times X) \xrightarrow{\smile} H^*(X \times Y, A \times Y \cup X \times B).$$

is defined. By Lemma 4.1.14, it is enough to show that $(A \times Y, X \times B)$ is excisive in $X \times Y$. This is obvious if A or B is empty. Otherwise, suppose that one of the pair, say (Y, B), is a good pair. Let V be a neighbourhood of B in Y which retracts by deformation onto B. Let $Z = A \times Y \cup X \times B$. Then, $\overline{A \times (Y - V)} \subset \mathrm{int}_Z(A \times Y)$. By excision of $A \times (Y - V)$ and homotopy, we get isomorphisms

$$H^*(Z, A \times Y) \xrightarrow{\approx} H^*(A \times V \cup X \times B, A \times V) \xrightarrow{\approx} H^*(X \times B, A \times B).$$

By Lemma 4.1.13, this implies that $(A \times Y, X \times B)$ is excisive.

We first establish the functoriality of the cross product. In Lemmas 4.6.2 and 4.6.3 below, we assume that the conditions for the relative cross product to be defined are satisfied.

Lemma 4.6.2 *Let $f : (X', A') \to (X, A)$ and $g : (Y', B') \to (Y, B)$ be maps of pairs. Then, for all $a \in H^*(X, A)$ and $b \in H^*(Y, B)$ the following formula holds:*

$$H^*(f \times g)(a \times b) = H^*f(a) \times H^*g(b).$$

Proof As $\pi_X \circ (f \times g) = f \circ \pi_{X'}$ and $\pi_Y \circ (f \times g) = g \circ \pi_{Y'}$, one has

$$\begin{aligned} H^*(f \times g)(a \times b) &= H^*(f \times g)\big(H^*\pi_X(a) \smile H^*\pi_Y(b)\big) \\ &= H^*(f \times g)(H^*\pi_X(a)) \smile H^*(f \times g)(H^*\pi_Y(b))) \\ &= H^*\pi_{X'} \circ H^*f(a) \smile H^*\pi_{Y'} \circ H^*g(b) \\ &= H^*f(a) \times H^*g(b). \end{aligned}$$

□

Formula (4.6.3) provides a product "•" on $H^*(X, A) \otimes H^*(Y, B)$.

Lemma 4.6.3 *The cross product $H^*(X, A) \otimes H^*(Y, B) \xrightarrow{\times} H^*(X \times Y, A \times Y \cup X \times B)$ is multiplicative. In particular, the cross product $H^*(X) \otimes H^*(Y) \xrightarrow{\times} H^*(X \times Y)$ is a* **GrA***-morphism.*

Proof

$$\begin{aligned} \times\big((a_1 \otimes b_1) \bullet (a_2 \otimes b_2)\big) &= (a_1 \smile a_2) \times (b_1 \smile b_2) \\ &= \pi_X^*(a_1 \smile a_2) \smile \pi_Y^*(b_1 \smile b_2) \\ &= \pi_X^*(a_1) \smile \pi_X^*(a_2) \smile \pi_Y^*(b_1) \smile \pi_Y^*(b_2) \end{aligned}$$

$$= \pi_X^*(a_1) \smile \pi_Y^*(b_1) \smile \pi_X^*(a_2) \smile \pi_Y^*(b_2)$$
$$= (a_1 \times b_1) \smile (a_2 \times b_2)$$
$$= \times(a_1 \otimes b_1) \smile \times(a_2 \otimes b_2) \qquad \square$$

Remark 4.6.4 In the proof of Lemma 4.6.3, we have established that

$$(a_1 \smile a_2) \times (b_1 \smile b_2) = (a_1 \times b_1) \smile (a_2 \times b_2)$$

for all $a_i \in H^*(X, A)$ and $b_j \in H^*(Y, B)$.

Observe that the Kronecker pairing

$$[H^*(X) \otimes H^*(Y)] \times [H_*(X) \otimes H_*(Y)] \xrightarrow{\langle\,,\,\rangle} \mathbb{Z}_2$$

given by

$$\langle a \otimes b, \alpha \otimes \beta \rangle = \langle a, \alpha \rangle \langle b, \beta \rangle \tag{4.6.7}$$

is a bilinear map (By convention, $\langle a, \alpha \rangle = 0$ if $a \in H^p(-)$ and $\alpha \in H_q(-)$ with $p \neq q$.)

Lemma 4.6.5 *Let X and Y be topological spaces with Y of finite cohomology type. Then, for all $n \in \mathbb{N}$, the linear map*

$$\bigoplus_{p+q=n} H^p(X) \otimes H^q(Y) \xrightarrow{\mathbf{k}} [\bigoplus_{p+q=n} H_p(X) \otimes H_q(Y)]^\sharp$$

given by $\mathbf{k}(a \otimes b) = \langle a \otimes b, - \rangle$ is an isomorphism.

Proof It suffices to prove that $\mathbf{k} : H^p(X) \otimes H^q(Y) \to [H_p(X) \otimes H_q(Y)]^\sharp$ is an isomorphism for all integers p, q. As $H^r(-) \approx H_r(-)^\sharp$ via the Kronecker pairing, this amounts to prove that, for vector spaces V and W, the homomorphism

$$k : V^\sharp \otimes W^\sharp \to [V \otimes W]^\sharp ,$$

given by $k(r \otimes s)(v \otimes w) = r(v)s(w)$, is an isomorphism when W is finite dimensional. This classical fact (true over any base field) is easily proven by induction on dim W (see, e.g., [43, Chap. VI, Proposition 10.18] for a proof in a more general setting). $\square$

The following lemma permits us to define a Kronecker dual $\underline{\times}$ to the cross product, called the *homology cross product*.

Lemma 4.6.6 *Let X and Y be topological spaces, with Y of finite cohomology type. Then, there exists a unique* **GrV***-homomorphism*

$$\underline{\times} : H_*(X \times Y) \to H_*(X) \otimes H_*(Y)$$

such that the equation

$$\langle a \times b, \gamma \rangle = \langle a \otimes b, \underline{\times}(\gamma) \rangle \tag{4.6.8}$$

holds true for all $a \in H^*(X)$, $b \in H^*(Y)$ and $\gamma \in H_*(X \times Y)$.

Proof Let $\gamma \in H_*(X \times Y)$. The uniqueness of $\underline{\times}(\gamma)$ is guaranteed by Lemma 4.6.5, using Lemma 2.3.3. For its existence, let $\mathcal{M}$ be a basis for $H_*(Y)$ and let $\mathcal{M}^* = \{m^* \in H^*(Y) \mid m \in \mathcal{M}\}$ be the dual basis for the Kronecker pairing. We define

$$\underline{\times}(\gamma) = \sum_{m \in \mathcal{M}} H_*\pi_X\big(H^*\pi_Y(m^*) \frown \gamma\big) \otimes m\,. \tag{4.6.9}$$

We must check that (4.6.8) holds true for all $a \in H^*(X)$, $b \in H^*(Y)$. It is enough to check it for $b = n^*$ with $n \in \mathcal{M}$. As $\langle n^*, m \rangle = 1$ if $m = n$ and 0 otherwise, one has

$$\begin{aligned}
\langle a \otimes n^*, \tilde{\underline{\times}}(\gamma) \rangle &= \langle a \otimes n^*, \textstyle\sum_{m \in \mathcal{M}} H_*\pi_X\big(H^*\pi_Y(m^*) \frown \gamma\big) \otimes m \rangle \\
&= \textstyle\sum_{m \in \mathcal{M}} \langle a, H_*\pi_X\big(H^*\pi_Y(m^*) \frown \gamma\big) \rangle \langle n^*, m \rangle \\
&= \langle H^*\pi_X(a), H^*\pi_Y(n^*) \frown \gamma \rangle \\
&= \langle H^*\pi_X(a) \smile H^*\pi_Y(n^*), \gamma \rangle \\
&= \langle a \times n^*, \gamma \rangle\,.
\end{aligned}$$

(Remark: the uniqueness of $\underline{\times}(\gamma)$ shows that the right member of (4.6.9) does not depend on the choice of the basis $\mathcal{M}$.) □

Theorem 4.6.7 (Künneth Theorem) *Let X and Y be topological spaces. Suppose that Y is of finite cohomology type. Then, the cross product*

$$\times : (H^*(X) \otimes H^*(Y), +, \bullet) \xrightarrow{\approx} (H^*(X \times Y), +, \smile)$$

is a **GrA***-isomorphism and the homology cross product*

$$\underline{\times} : H_*(X \times Y) \to H_*(X) \otimes H_*(Y)$$

is a **GrV***-isomorphism.*

The finiteness condition on one of the space (here Y) is necessary in the cohomology statement, as seen in the beginning of the section. It is used in the proof through the following lemma.

Lemma 4.6.8 *Let $\mathcal{V}$ be a family of vector spaces over a field $\mathbb{F}$. Let W be a finite dimensional $\mathbb{F}$-vector space. Then the linear map*

$$\Phi : \Big(\prod_{V\in\mathcal{V}} V\Big)\otimes W \to \prod_{V\in\mathcal{V}} (V\otimes W)$$

given by

$$\Phi((v)\otimes w) = (v\otimes w)$$

is an isomorphism.

Proof The proof is by induction on $n = \dim W$. The case $n = 1$ follows from the canonical isomorphism $T\otimes\mathbb{F}\approx T$ for any vector space T. The induction step uses that, in the category of $\mathbb{F}$-vector spaces, tensor and Cartesian products commute with direct sums. □

Proof (Proof of the Künneth theorem) By Lemma 4.6.3, we know that the cross product is a **GrA**-morphism. It is then enough to prove that it is a **GrV**-isomorphism. Assuming that Y is of finite cohomology type, the proof goes as follows.

(1) We prove that the cross product is a **GrV**-isomorphism when X is a finite dimensional CW-complex.
(2) By Kronecker duality, Point (1) implies that the homology cross product is a **GrV**-isomorphism when X is a finite dimensional CW-complex. Any compact subspace of $X\times Y$ is contained in $X^n\times Y$ for some $n\in\mathbb{N}$. Therefore, $H_*(X\times Y)$ is the direct limit of $H_*(X^n\times Y)$ by Corollary 3.1.16. Also, $H_*(X)$ is the direct limit of $H_*(X^n)$. The homology cross product being natural by Lemma 4.6.2 and Kronecker duality, we deduce that $\underline{\times}$ is a **GrV**-isomorphism when X is any CW-complex. By Kronecker duality, the cross product is a **GrV**-isomorphism for any CW-complex X.
(3) If X is any space, there is a map $f_X : \hat{X}\to X$, where $\hat{X}$ is a CW-complex and f is a weak homotopy equivalence, i.e. the induced map on the homotopy groups $\pi_* f : \pi_*(\hat{X}, u) \xrightarrow{\approx} \pi_*(X, f(u))$ is an isomorphism for all $u\in\hat{X}$ (see [82, p. 352] or Remark 3.7.5). As $\pi_*(A\times B, (a, b)) \xrightarrow{\approx} \pi_*(A, a)\times\pi_*(B, b)$, the map $f_X\times\mathrm{id} : \hat{X}\times Y\to X\times Y$ is also a weak homotopy equivalence. But, weak homotopy equivalences induce isomorphisms on singular (co)homology (see [82, Proposition 4.21]). The diagram

$$\begin{array}{ccc} H^*(X)\otimes H^*(Y) & \xrightarrow{\times} & H^*(X\times Y) \\ {\scriptstyle H^*f_X\otimes\mathrm{id}}\downarrow\approx & & \approx\downarrow{\scriptstyle H^*(f_X\times\mathrm{id})} \\ H^*(\hat{X})\otimes H^*(Y) & \xrightarrow[\approx]{\times} & H^*(\hat{X}\times Y) \end{array}$$

is commutative by Lemma 4.6.2. This proves that the cross product in singular cohomology is a **GrA**-isomorphism for any space X. The corresponding diagram for the homology cross product, or Kronecker duality, proves that the homology cross product is a **GrV**-isomorphism.

It thus remains to prove Point (1). We follow the idea of [82, p. 218]. Let us fix the topological space Y. To a topological pair (X, A), we associate two graded $\mathbb{Z}_2$-vector spaces:

$$h^*(X, A) = H^*(X, A) \otimes H^*(Y) \text{ and } k^*(X, A) = H^*(X \times Y, A \times Y) .$$

Using Proposition 3.9.2, Point (1) follows from the following lemma. □

Lemma 4.6.9 *Let Y be a topological space of finite cohomology type. Then*

(a) *k^* and h^* are two generalized cohomology theories in the sense of Sect. 3.9, with $h^*(pt) \approx k^*(pt) \approx H^*(Y)$.*
(b) *The cross product provides a natural transformation from h^* to k^*, restricting to an isomorphism $h^*(pt) \xrightarrow{\approx} k^*(pt)$.*

Proof If $f : (X, A) \to (X', A')$ is a continuous map of pairs, we define $h^*f = H^*f \otimes \mathrm{id}_{H*(Y)}$ and $k^*f = H^*(f \times \mathrm{id}_Y)$. This makes h^* and k^* functors from $\mathbf{TOP_2}$ to $\mathbf{GrV}$. The connecting homomorphism $\delta_h^* : h^*(A) \to h^{*+1}(X, A)$ and $\delta_k^* : k^*(A) \to k^{*+1}(X, A)$ are defined by

$$\delta_h^* = \delta^* \otimes \mathrm{id}_{H*(Y)} \text{ and } \delta_k^* = \delta^* : H^*(A \times Y) \to H^{*+1}(X \times Y, A \times Y) ,$$

using the homomorphism δ^* of singular cohomology; δ_h^* and δ_k^* are then functorial for continuous maps.

We now check that Axioms (1)–(3) of p. 123 hold both for h^* and k^*. The homotopy and excision axioms are clear. For a topological pair (X, A), the long exact sequence for k^* is that in singular cohomology for the pair $(X \times Y, A \times Y)$. The exact sequence for h^* is obtained by tensoring with $H^*(Y)$ the exact sequence of (X, A) for H^*. We use that a direct sum of exact sequences is exact and that, over a field, tensoring with a vector space preserves exactness. The disjoint union axiom holds trivially for k^*. For h^*, we use that $H^m(Y)$ if of finite dimension for all m and Lemma 4.6.8. Thus, both h^* and k^* are generalized cohomology theories.

We now check Point (b). Let $f : (X', A') \to (X, A)$ be a continuous map of pairs. We must prove that the following diagram

$$\begin{array}{ccc} h^*(X, A) & \xrightarrow{\times} & k^*(X, A) \\ \downarrow{\scriptstyle h^*f} & & \downarrow{\scriptstyle k^*f} \\ h^*(X', A') & \xrightarrow{\times} & k^*(X', A') \end{array} \tag{4.6.10}$$

is commutative. This amounts to show that

$$H^*f(a) \times y = H^*(f \times \mathrm{id}_Y)(a \times y) \tag{4.6.11}$$

for all $a \in H^*(X, A)$ and $y \in H^*(Y)$. This follows from Lemma 4.6.2.

For the second part of Point (b), we must show the commutativity of the diagram

$$\begin{array}{ccc} h^*(A) & \xrightarrow{\delta_h^*} & h^{*+1}(X, A) \\ \downarrow{\times} & & \downarrow{\times} \\ k^*(A) & \xrightarrow{\delta_k^*} & k^{*+1}(X, A)\,. \end{array} \tag{4.6.12}$$

This is equivalent to the commutativity of the diagram

$$\begin{array}{ccc} H^p(A) \times H^q(Y) & \xrightarrow{\delta^* \times \mathrm{id}} & H^{p+1}(X, A) \times H^q(Y) \\ \downarrow{\times} & & \downarrow{\times} \\ H^{p+q}(A \times Y) & \xrightarrow{\delta_\times^*} & H^{p+q+1}(X \times Y, A \times Y)\,, \end{array} \tag{4.6.13}$$

for all $p, q \in \mathbb{N}$. Here, we have introduced more precise notations, distinguishing the connecting homomorphisms in singular cohomology $\delta^* : H^*(A) \to H^{*+1}(X, A)$ and $\delta_\times^* : H^*(A \times Y) \to H^{*+1}(X \times Y, A \times Y)$. We shall also distinguish the homomorphisms $\pi_X^* : H^*(X) \to H^*(X \times Y)$ and $\bar{\pi}_X^* : H^*(A) \to H^*(A \times Y)$ induced by the projections onto A and X, as well as the homomorphisms $\pi_Y^* : H^*(Y) \to H^*(X \times Y)$ and $\bar{\pi}_Y^* : H^*(Y) \to H^*(A \times Y)$ induced by the projections onto Y. Analogous notations are used for cochains. The commutativity of Diagram (4.6.13) is thus equivalent to the formula

$$\pi_X^* \circ \delta^*(a) \smile \pi_Y^*(y) = \delta_\times^*\left(\bar{\pi}_X^*(a) \smile \bar{\pi}_Y^*(y)\right) \quad \text{for all } a \in H^p(A)\,, y \in H^q(Y)\,. \tag{4.6.14}$$

Let $\tilde{a} \in Z^p(A)$ and $\tilde{y} \in Z^q(Y)$ represent a and y. Let $\bar{a} \in C^p(X)$ be an extension of $\tilde{a}$ as a p-cochain of X. By the recipe of Lemma 3.1.17, $\delta(\bar{a})$ is a cocycle of $C^{p+1}(X, A)$ representing $\delta^*(a)$. Thus, the left hand member of (4.6.14) is represented by the cocycle

$$\pi_X^* \circ \delta(\bar{a}) \smile \pi_Y^*(\tilde{y})\,. \tag{4.6.15}$$

To compute the right hand member, we need an extension of $\bar{\pi}_X^*(\tilde{a}) \smile \bar{\pi}_Y^*(\tilde{y})$ as a cochain of $X \times Y$. But, as cochains of $X \times Y$, $\pi_X^*(\bar{a})$ is an extension of $\bar{\pi}_X^*(\tilde{a})$ and the cocycle $\pi_Y(\tilde{y})$ is an extension of $\bar{\pi}_Y^*(\tilde{y})$. Therefore, $\pi_X^*(\bar{a}) \smile \pi_Y(\tilde{y})$ is an extension of $\bar{\pi}_X^*(\tilde{a}) \smile \bar{\pi}_Y^*(\tilde{y})$. By Lemma 3.1.17, the right hand member of (4.6.14) is then represented by the cocycle $\delta_\times\big(\pi_X^*(\bar{a}) \smile \pi_Y(\tilde{y})\big)$. As $\delta_\times(\pi_Y(\tilde{y})) = 0$, one has

$$\delta_\times\big(\pi_X^*(\bar{a}) \smile \pi_Y(\tilde{y})\big) = \delta_\times \circ \pi_X^*(\bar{a}) \smile \pi_Y(\tilde{y}) = \pi_X^* \circ \delta(\bar{a}) \smile \pi_Y^*(\tilde{y}) \qquad (4.6.16)$$

Comparing (4.6.15) and (4.6.16) proves Formula (4.6.14) and then the commutativity of Diagram (4.6.12). □

Under some hypotheses, there are relative versions of the Künneth theorem, generalizing Theorem 4.6.7.

Theorem 4.6.10 (Relative Künneth theorem) *Let (X, A) be a topological pair. Let (Y, B) be a good pair such that Y and B are of finite cohomology type. Then, the cross product*

$$\times : H^*(X, A) \otimes H^*(Y, B) \xrightarrow{\approx} H^*(X \times Y, A \times Y \cup X \times B) \qquad (4.6.17)$$

is a **GrA**-*isomorphism.*

The classical proof of the Künneth theorem (see e.g. [179]) gives the more general statement that (4.6.17) is an isomorphism if $(X \times B, A \times Y)$ is excisive in $X \times Y$ and (Y, B) is of finite cohomology type. If (X, A) and (Y, B) are CW-pairs, the condition that (Y, B) is of finite cohomology type is also sufficient (see [82, Theorem 3.21]; see also Corollary 4.7.25 below).

Proof As (Y, B) is a good pair, the relative cross product (4.6.6) is defined. Let $Z = A \times Y \cup X \times B$ and let $p \in \mathbb{N}$. Let us consider the commutative diagram.

$$\begin{array}{ccccc}
H^p(X,A) \otimes H^{q-1}(Y) & & \xrightarrow[\approx]{\times} & & H^{p+q-1}(X \times Y, A \times Y) \\
\downarrow & & & & \downarrow \\
H^p(X,A) \otimes H^{q-1}(B) & \xrightarrow[\approx]{\times} & H^{p+q-1}(X \times B, A \times B) & \xleftarrow[\approx]{J^*} & H^{p+q-1}(Z, A \times Y) \\
\downarrow & & & & \downarrow \\
H^p(X,A) \otimes H^q(Y,B) & & \xrightarrow{\times} & & H^{p+q}(X \times Y, Z) \\
\downarrow & & & & \downarrow \\
H^p(X,A) \otimes H^q(Y) & & \xrightarrow[\approx]{\times} & & H^{p+q}(X \times Y, A \times Y) \\
\downarrow & & & & \downarrow \\
H^p(X,A) \otimes H^q(B) & \xrightarrow[\approx]{\times} & H^{p+q}(X \times B, A \times B) & \xleftarrow[\approx]{J^*} & H^{p+q}(Z, A \times Y)
\end{array}$$

The left column is the cohomology sequence for (Y, B) tensored by $H^p(X, A)$. It is still exact since we work in the category of $\mathbb{Z}_2$-vector spaces. The right column is the cohomology sequence for the triple $(X \times Y, Z, A \times Y)$. The homomorphism J^*, induced by inclusion, is an isomorphism: if V is a neighbourhood of B in Y which retracts by deformation onto B, J^* is the composition

$$H^*(A\times Y \cup X\times B, A\times Y) \xrightarrow{\approx} H^*(A\times V \cup X\times B, A\times V) \xrightarrow{\approx} H^*(X\times B, A\times B).$$

The left arrow is an isomorphism by excision of $A \times (Y - V)$ and the right one by the homotopy property. As Y and B are of finite cohomology type, the cross products involving the absolute cohomology $H^*(Y)$ or $H^*(B)$ are isomorphisms, as established during the proof of Theorem 4.6.7. By the five-lemma, this proves that the middle cross product is an isomorphism. □

4.7 Some Applications of the Künneth Theorem

4.7.1 Poincaré Series and Euler Characteristic of a Product

One application of the Künneth theorem is the multiplicativity of Poincaré series and Euler characteristic.

Proposition 4.7.1 *Let X and Y be spaces of finite cohomology type. Then, $X \times Y$ is of finite cohomology type and*

$$\mathfrak{P}_t(X \times Y) = \mathfrak{P}_t(X) \cdot \mathfrak{P}_t(Y). \tag{4.7.1}$$

If X and Y are finite complexes, then

$$\chi(X \times Y) = \chi(X) \cdot \chi(Y). \tag{4.7.2}$$

Proof Let $a_i = \dim H_i(X)$, $b_i = \dim H_i(Y)$. The Künneth theorem implies that $\dim H_n(X \times Y) = \sum_{i+j=n} a_i b_j$ which proves (4.7.1). Equation (4.7.2) follows, since χ is the evaluation of $\mathfrak{P}_t$ at $t = -1$. Note that (4.7.2) also follows more elementarily from the cellular decomposition of $X \times Y$ (see Example 3.4.6). □

4.7.2 Slices

Let $y_0 \in Y$. The *slice inclusion* $s_X : X \to X \times Y$ at y_0 is the continuous map defined by $s_X(x) = (x, y_0)$. The slice inclusion $s_Y : Y \to X \times Y$ at $x_0 \in X$ is defined accordingly.

Using the bijection $Y \approx \mathcal{S}_0(Y)$, we see $y_0 \in Y$ as a 0-homology class $[y_0] \in H_0(Y)$. Hence, for $b \in H^0(Y)$ the number $\langle b, y_0 \rangle \in \mathbb{Z}_2$ is defined.

Lemma 4.7.2 *Let $s_X : X \to X \times Y$ be the slice inclusion at $y_0 \in Y$. Let $a \in H^m(X)$ and $b \in H^n(Y)$. Then,*

$$H^* s_X(a \times b) = \begin{cases} \langle b, y_0 \rangle\, a & \text{if } n = 0 \\ 0 & \text{otherwise.} \end{cases}$$

Proof One has $\pi_X \circ s_X = \mathrm{id}_X$, while $\pi_Y \circ s_X$ is the constant map c onto y_0. Thus, $H^* c\,(b) = 0$ if $n \neq 0$. When $n = 0$, $H^* c\,(b) = \langle b, y_0 \rangle \mathbf{1}$. Thus,

$$H^* s_X(a \times b) = H^* s_X(\pi_X^*(a) \smile \pi_Y^*(b)) = a \smile H^* c\,(b) = \langle b, y_0 \rangle\, a\,. \qquad \square$$

Here are two corollaries of Lemma 4.7.2 which enable us to detect cohomology classes via the slice homomorphisms.

Corollary 4.7.3 *Let X and Y be path-connected topological spaces such that $\tilde{H}^k(X) = 0$ for $k < n$. Then, the equation*

$$a = \mathbf{1} \times H^* s_Y(a) + H^* s_X(a) \times \mathbf{1}$$

is satisfied for all $a \in H^n(X \times Y)$.

Proof By the hypotheses and the Künneth theorem, the cross product provides an isomorphism

$$\times : H^0(X) \otimes H^n(Y) \oplus H^n(X) \otimes H^0(Y) \xrightarrow{\approx} H^n(X \times Y)\,.$$

and $H^0(X) \approx \mathbb{Z}_2 \approx H^0(Y)$. This implies that $a = \mathbf{1} \times v + u \times \mathbf{1}$ for some unique $u \in H^n(X)$ and $v \in H^n(Y)$. By Lemma 4.7.2, one has $H^* s_Y(a) = v$ and $H^* s_X(a) = u$, which proves the corollary. $\square$

The case $n = 1$ in Corollary 4.7.3 gives the following statement.

Corollary 4.7.4 *Let X and Y be path-connected spaces. Then, the equation*

$$a = \mathbf{1} \times H^* s_Y(a) + H^* s_X(a) \times \mathbf{1}$$

is satisfied for all $a \in H^1(X \times Y)$.

4.7.3 The Cohomology Ring of a Product of Spheres

We first note the associativity of the cross product.

Lemma 4.7.5 *Let X, Y and Z be three topological spaces. In $H^*(X \times Y \times Z)$, the cross product is associative: $(x \times y) \times z = x \times (y \times z)$ for all $x \in H^*(X)$, $y \in Y$ and $z \in Z$.*

Proof We have to consider the various projections $\pi_{12} : X \times Y \times Z \to X \times Y$, $\pi_{23} : X \times Y \times Z \to Y \times Z$, $\pi_1 : X \times Y \times Z \to X$, etc. Also, $\pi_1^{12} : X \times Y \to X$, etc. They satisfy $\pi_j^{ij} \circ \pi_{ij} = \pi_j$. Using the associativity and the functoriality of the cup product, we get

$$(x \times y) \times z = \pi_{12}^*(\pi_1^{12*}(x) \smile \pi_2^{12*}(y)) \smile \pi_3^*(z) = \pi_1^*(x) \smile \pi_2^*(y) \smile \pi_3^*(z)\,.$$

In the same way, $x \times (y \times z) = \pi_1^*(x) \smile \pi_2^*(y) \smile \pi_3^*(z)$. □

The cohomology of the sphere S^d being concentrated in dimension 0 and d, one has a **GrA**-isomorphism

$$\mathbb{Z}_2[x]/(x^2) \xrightarrow{\approx} H^*(S^d) \quad (x \text{ of degree } d)\,, \tag{4.7.3}$$

sending x to the generator $[S^d]^\sharp \in H^d(S^d)$. Here, $\mathbb{Z}_2[x]/(x^2)$ denotes the quotient of the polynomial ring $\mathbb{Z}_2[x]$, where x is a formal variable (here of degree d), by the ideal generated by x^2. The following proposition then follows directly from the Künneth theorem.

Proposition 4.7.6 *Let X be a topological space. The* **GrA**-*homomorphism*

$$H^*(X)[x]\big/(x^2) \to H^*(X \times S^d) \quad (x \text{ of degree } d)\,,$$

induced by $a \mapsto a \times \mathbf{1}$, for $a \in H^*(X)$, and $x \mapsto \mathbf{1} \times [S^d]^\sharp$, is a **GrA**-*isomorphism.*

Using Proposition 4.7.6 together with Lemma 4.7.5, we get the following proposition.

Proposition 4.7.7 *For $i = 1 \dots, m$, let x_i be a formal variable of degree d_i. Then, the* **GrA**-*homomorphism*

$$\mathbb{Z}_2[x_1, \dots, x_m]\big/(x_1^2, \dots, x_m^2) \to H^*(S^{d_1} \times \cdots \times S^{d_m})$$

induced by

$$x_i \mapsto \mathbf{1} \times \cdots \times \mathbf{1} \times [S^{d_i}]^\sharp \times \mathbf{1} \times \cdots \times \mathbf{1}$$

is a **GrA**-*isomorphism.*

4.7.4 Smash Products and Joins

Let (X, x) and (Y, y) be two pointed spaces. The base points provide an inclusion

$$X \vee Y \approx X \times \{y\} \cup \{x\} \times Y \hookrightarrow X \times Y\,.$$

The *smash product* $X \wedge Y$ of X and Y is the quotient space

$$X \wedge Y = X \times Y \big/ X \vee Y .$$

It is pointed by $x \wedge y$, the image of $X \vee Y$ in $X \wedge Y$.

Recall that (X, x) is *well pointed* if the pair $(X, \{x\})$ is well cofibrant. The following lemma is the first place where the full strength of this definition is used.

Lemma 4.7.8 *If (X, x) and (Y, y) are well pointed, so is $(X \wedge Y, x \wedge y)$.*

Proof By Lemma 3.1.40, the pair $(X \times Y, X \vee Y)$ is well cofibrant and, by Lemma 3.1.43, so is $(X \wedge Y, \{x \wedge y\})$. □

By Proposition 3.1.45 and (3.1.13), one has the isomorphisms

$$H^*(X \times Y, X \vee Y) \approx H^*(X \wedge Y, \{x \wedge y\}) \approx \tilde{H}^*(X \wedge Y) . \tag{4.7.4}$$

Proposition 4.7.9 *Let (X, x) and (Y, y) be well pointed spaces. Then, the homomorphisms induced by the inclusion $i : X \vee Y \to X \times Y$ and the projection $p : X \times Y \to X \wedge Y$ give rise to the short exact sequence*

$$0 \to \tilde{H}^*(X \wedge Y) \xrightarrow{\tilde{H}^* p} \tilde{H}^*(X \times Y) \xrightarrow{\tilde{H}^* i} \tilde{H}^*(X \vee Y) \to 0 .$$

Proof Using the isomorphism (4.7.4) and the exact sequence of Corollary 3.1.48, it is enough to prove that $\tilde{H}^* i$ is onto. Consider the commutative diagram

$$\begin{array}{ccccc}
\tilde{H}^*(X) & & \xrightarrow{\quad \mathrm{id} \quad} & & \tilde{H}^*(X) \\
& \searrow^{\tilde{H}^*\pi_X^*} & & \nearrow^{\tilde{H}^* j_X} & \\
& \tilde{H}^*(X \times Y) & \xrightarrow{\tilde{H}^* i} & \tilde{H}^*(X \vee Y) & \\
& \nearrow_{\tilde{H}^*\pi_Y^*} & & \searrow_{\tilde{H}^* j_Y^*} & \\
\tilde{H}^*(Y) & & \xrightarrow{\quad \mathrm{id} \quad} & & \tilde{H}^*(Y)
\end{array}$$

where π_X, π_Y are the projections and j_X, j_Y the inclusions. We note that $\pi_Y \circ j_X$ and $\pi_X \circ j_Y$ are constant maps. By Proposition 3.1.51, the homomorphism $\tilde{H}^*(X \vee Y) \xrightarrow{(\tilde{H}^* j_X, \tilde{H}^* j_Y^*)} \tilde{H}^*(X) \times \tilde{H}^*(Y)$ is an isomorphism. Hence $\tilde{H}^* i$ is onto. □

Remark 4.7.10 Using the relationship between the exact sequence of the pair $(X \times Y, X \vee Y)$ and that of Corollary 3.1.48, Proposition 4.7.9 implies that the homomorphism $H^* i : H^*(X \times Y) \to H^*(X \vee Y)$ is surjective, whence the short exact sequence

$$0 \to H^*(X \times Y, X \vee Y) \to H^*(X \times Y) \xrightarrow{H^* i} \tilde{H}^*(X \vee Y) \to 0 . \tag{4.7.5}$$

As $(X, \{x\})$ and $(Y, \{y\})$ are good pairs, the relative cross product

$$H^*(X, \{x\}) \otimes H^*(Y, \{y\}) \xrightarrow{\times} H^*(X \times Y, X \vee Y)$$

is defined by (4.6.6). Using the isomorphisms of (3.1.13), one constructs the commutative diagram

$$\begin{array}{ccc} H^*(X, \{x\}) \otimes H^*(Y, \{y\}) & \xrightarrow{\approx} & \tilde{H}^*(X) \otimes \tilde{H}^*(Y) \\ \downarrow \times & & \downarrow \tilde{\times} \\ H^*(X \times Y, X \vee Y) & \xrightarrow{\approx} & \tilde{H}^*(X \wedge Y) \end{array} \qquad (4.7.6)$$

which defines the *reduced cross product* $\tilde{\times}$. The relative Künneth theorem 4.6.10 gives the following *reduced Künneth theorem.*

Proposition 4.7.11 *Let (X, x) and (Y, y) be well pointed spaces, with Y of finite cohomology type. Then, the reduced cross product*

$$\tilde{\times} : \tilde{H}^*(X) \otimes \tilde{H}^*(Y) \xrightarrow{\approx} \tilde{H}^*(X \wedge Y)$$

is a multiplicative **GrV***-isomorphism.*

For a pointed space (Z, z), Diagram (3.1.14) provides an injective homomorphism $\tilde{H}^*(Z) \to H^*(Z)$. Using this together with Proposition 4.7.9 (or Remark 4.7.10), the Künneth theorem and its reduced form are summed up by the diagram

$$\begin{array}{ccc} \tilde{H}^*(X) \otimes \tilde{H}^*(Y) & \rightarrowtail & H^*(X) \otimes H^*(Y) \\ \approx \downarrow \tilde{\times} & & \approx \downarrow \times \\ \tilde{H}^*(X \wedge Y) & \rightarrowtail & H^*(X \times Y) \end{array} \quad . \qquad (4.7.7)$$

Example 4.7.12 Proposition 4.7.11 says that $\tilde{H}^k(S^p \wedge S^q) = 0$ for $k \neq p + q$ and $H^{p+q}(S^p \wedge S^q) = \mathbb{Z}_2$. Actually, $S^p \wedge S^q$ is homeomorphic to S^{p+q} by the following homeomorphisms. Let D^r be the compact unit disk of dimension r with boundary $\partial D^r = S^{r-1}$. Then $D^r/\partial D^r$ is homeomorphic to S^r and

$$\begin{aligned} S^p \wedge S^q &\xrightarrow{\approx} \left(D^p/\partial D^p \times D^q/\partial D^q\right) \Big/ [\partial D^p] \times D^q \cup D^p \times [\partial D^q] \\ &\xrightarrow{\approx} D^p \times D^q \Big/ \partial D^p \times D^q \cup D^p \times \partial D^q \\ &\xrightarrow{\approx} D^p \times D^q / \partial(D^p \times D^q) \approx S^{p+q} . \end{aligned}$$

Let (X, x) be a well pointed space. The smash product $X \wedge S^1$ is called the *reduced suspension* of X, which has the same homotopy type than the suspension ΣX. Indeed, let $\partial I = \{0, 1\}$. The map

$$F : \Sigma X = (X \times I)/(X \times \partial I) \to X \wedge S^1$$

given by $F(x, t) = [(x, e^{2i\pi t})]$ descends to a homeomorphism

$$\bar{F} : \bar{\Sigma} X = \Sigma X/(\{x\} \times I) \xrightarrow{\approx} X \wedge S^1 .$$

This homeomorphism preserves the base points, if we choose those to be $[x] \in \bar{\Sigma} X$ and $1 \in S^1$. The pair $(I, \partial I)$ is well cofibrant by Lemma 3.1.39. By Lemma 3.1.40, so is the product pair $(X \times I, X \times \partial I \cup \{x\} \times I)$. By Lemma 3.1.43, the pair $(\Sigma X, \{x\} \times I)$ is well cofibrant. As $\{x\} \times I$ is contractible, the projection $F : \Sigma X \twoheadrightarrow \bar{\Sigma} X \approx X \wedge S^1$ is a homotopy equivalence by Lemma 3.1.44.

Let b be the generator of $H^1(S^1) = \mathbb{Z}_2$. By Propositions 4.7.11 and 3.1.49 and the above, the three arrows in the following proposition are isomorphisms.

Lemma 4.7.13 *The diagram*

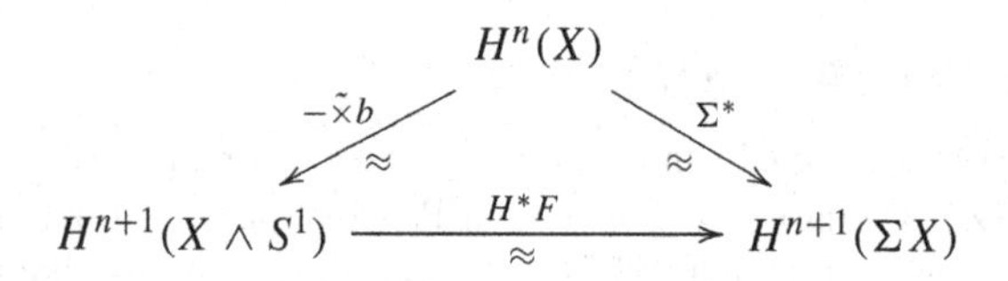

is commutative.

Proof As all these isomorphisms are functorial, it is enough to prove the lemma for $X = \mathcal{K}_n$. This is possible since the cellular decomposition of $\mathcal{K}_n$ given in Proposition 3.8.1 has 0-skeleton $\mathcal{K}_n^0 = \{x_0\}$, so $(\mathcal{K}_n, x_0)$ is a well pointed space by Proposition 3.4.1. In this particular case, the statement is obvious since the three groups are isomorphic to $\mathbb{Z}_2$. □

The smash product gives a geometric interpretation of the cup product. Let $a \in H^m(X)$ and $b \in H^m(X)$, given by maps $f_a : X \to \mathcal{K}_m$ and $f_m : X \to \mathcal{K}_n$ to Eilenberg-MacLane spaces. By Proposition 4.7.11, $H^{m+n}(\mathcal{K}_m \wedge \mathcal{K}_n) = \mathbb{Z}_2$, with generator corresponding to $g : \mathcal{K}_m \wedge \mathcal{K}_n \to \mathcal{K}_{m+n}$.

Proposition 4.7.14 *The composed map*

$$X \xrightarrow{(f_a, f_b)} \mathcal{K}_m \times \mathcal{K}_n \to \mathcal{K}_m \wedge \mathcal{K}_n \xrightarrow{g} \mathcal{K}_{m+n}$$

represents the class $a \smile b \in H^{m+n}(X)$.

Proof By Proposition 4.7.11, the generator of $H^{m+n}(\mathcal{K}_m \wedge \mathcal{K}_n) = \mathbb{Z}_2$ is the reduced cross product $\iota_m \tilde{\times} \iota_n$. By Diagram (4.7.7), it is send to $\iota_m \times \iota_n$ in $H^{m+n}(\mathcal{K}_m \times \mathcal{K}_n)$. Now, the composed map of Proposition 4.7.14 coincides with the composition

$$X \xrightarrow{\Delta} X \times X \xrightarrow{f_a \times f_b} \mathcal{K}_m \times \mathcal{K}_n \to \mathcal{K}_m \wedge \mathcal{K}_n \xrightarrow{g} \mathcal{K}_{m+n} .$$

Proposition 4.7.14 then follows from Remark 4.6.1. □

If we consider the composed map $f : X \xrightarrow{(f_a, f_b)} \mathcal{K}_m \times \mathcal{K}_n \to \mathcal{K}_m \wedge \mathcal{K}_n$, Proposition 4.7.14 gives the following corollary.

Corollary 4.7.15 *The diagram*

$$\begin{array}{ccc} H^m(\mathcal{K}_m) \otimes H^n(\mathcal{K}_n) & \xrightarrow[\approx]{\tilde{\times}} & H^{m+n}(\mathcal{K}_m \wedge \mathcal{K}_n) \\ \downarrow{\scriptstyle f_a^* \otimes f_b^*} & & \downarrow{\scriptstyle f^*} \\ H^m(X) \otimes H^n(X) & \xrightarrow{\smile} & H^{m+n}(X) \end{array}$$

is commutative.

Let X and Y be two topological spaces. Their *join* $X * Y$ is the quotient of $X \times Y \times I$ by the equivalence relation $(x, y, 0) \sim (x, y', 0)$ for $y, y' \in Y$ and $(x, y, 1) \sim (x', y, 1)$ for $x, x' \in X$. This topological join is related to the simplicial join in the following way: if K and L are locally finite simplicial complexes, then $|K * L|$ is homeomorphic to $|K| * |L|$ (see [155, Lemma 62.2]).

The two open subspaces $X \times Y \times [0, 1)$ and $X \times Y \times (0, 1]$ of $X \times Y \times I$ define open subspaces U_X and U_Y of $X * Y$. The space U_X retracts by deformation onto X and U_Y retracts by deformation onto Y. Moreover, $U_X \cap U_Y$ retracts by deformation onto $X \times Y \times \{\frac{1}{2}\}$. The following diagram is homotopy commutative,

$$\begin{array}{ccc} U_X \cap U_Y & \xrightarrow{\text{incl}} & U_X \\ \uparrow{\scriptstyle \simeq} & & \uparrow{\scriptstyle \simeq} \\ X \times Y & \xrightarrow{\pi_X} & X \end{array}$$

as well as the corresponding diagram for Y. Consider the homomorphism

$$H^k(X) \oplus H^k(Y) \xrightarrow{\pi_X^* + \pi_Y^*} H^k(X \times Y) .$$

If $k > 0$, then $(\pi_X^* + \pi_Y^*)(a, b) = a \times \mathbf{1} + \mathbf{1} \times b$ and, by the Künneth theorem, $\pi_X^* + \pi_Y^*$ is injective. As $X * Y$ is path-connected, the Mayer-Vietoris sequence for the data

$(X * Y, U_X, U_Y, U_X \cap U_Y)$ splits and gives, for all integers $k \geq 0$, the exact sequence

$$0 \to \tilde{H}^k(X) \oplus \tilde{H}^k(Y) \xrightarrow{\pi_X^* + \pi_Y^*} \tilde{H}^k(X \times Y) \to H^{k+1}(X * Y) \to 0. \qquad (4.7.8)$$

Example 4.7.16 The join $S^p * S^q$ is homeomorphic to S^{p+q+1}. Considering $S^{p+q+1} \subset \mathbb{R}^{p+1} \times \mathbb{R}^{q+1}$, a homeomorphism $S^{p+q+1} \xrightarrow{\approx} S^p * S^q$ is given by $(x, y) \mapsto [(x, y, |x|]$. The reader can check (4.7.8) on this example, including the case $p = q = 0$.

Observe that U_X and U_Y are contractible in $X * Y$. Hence, the Lusternik-Schnirelmann category of $X * Y$ is equal to 2. By Proposition 4.4.2, the cup product in $H^{>0}(X * Y)$ vanishes.

When the Künneth theorem is valid, one sees that the cohomology ring of $X * Y$ is isomorphic to that of $\Sigma(X \wedge Y)$. Actually, under some hypotheses, these two spaces have the same homotopy type (see [82, Ex. 24,p. 20]).

4.7.5 The Theorem of Leray-Hirsch

An important generalization of a product space is a locally trivial fiber bundle. A map $p : E \to B$ is a *locally trivial fiber bundle* with fiber F (in short: a *bundle*) if there exists an open covering $\mathcal{U}$ of B and, for each $U \in \mathcal{U}$, a homeomorphism $\psi_U : U \times F \xrightarrow{\approx} p^{-1}(U)$ such that $p \circ \psi(x, v) = x$ for all $(x, v) \in U \times F$. The space E is the *total space* and B is the *base space* of the bundle. If A is a subspace of B, we set $E_A = p^{-1}(A)$, getting a bundle $p : E_A \to A$. If $b \in B$, we set $E_b = E_{\{b\}}$ and denote by $i_b : E_b \hookrightarrow E$ the inclusion. A *fiber inclusion* is an embedding $i : F \to E$ which is a homeomorphism onto some fiber E_b. As elsewhere in the literature, we shall often speak about a *(locally trivial) bundle* $F \xrightarrow{i} E \xrightarrow{p} B$, meaning a locally trivial bundle $p : E \to B$ with fiber F together with a chosen fiber inclusion i.

If $p : E \to B$ is a bundle, then the homomorphism $p^* = H^*p : H^*(B) \to H^*(E)$ provides a structure of graded $H^*(B)$-module on $H^*(E)$.

A *cohomology extension of the fiber* is a **GrV**-morphism $\theta : H^*(F) \to H^*(E)$ such that, for each $b \in B$, the composite map

$$H^*(F) \xrightarrow{\theta} H^*(E) \xrightarrow{H^*i_b} H^*(E_b)$$

is a **GrV**-isomorphism. We do not require that θ is multiplicative. In the presentation of a bundle by a sequence $F \xrightarrow{i} E \xrightarrow{p} B$ with B path-connected, a cohomology extension θ of the fiber exists if and only if H^*i is surjective.

A cohomology extension θ of the fiber provides a morphism of graded $H^*(B)$-modules

$$H^*(B) \otimes H^*(F) \xrightarrow{\hat{\theta}} H^*(E)$$

given by $\hat{\theta}(a \otimes b) = p^*(a) \smile \theta(b)$.

Suppose that F is of finite cohomology type. As in Lemma 4.6.6, there is a unique **GrV**-homomorphism

$$\underline{\theta} : H_*(E) \to H_*(B) \otimes H_*(F)\,.$$

such that the formula

$$\langle \hat{\theta}(b \otimes u), \gamma \rangle = \langle b \otimes u, \underline{\theta}(\gamma) \rangle \tag{4.7.9}$$

holds true for all $b \in H^*(B)$, $u \in H^*(F)$ and $\gamma \in H_*(E)$. As in the proof of Lemma 4.6.6, we show that

$$\underline{\theta}(\gamma) = \sum_{m \in \mathcal{M}} H_* p\big(\theta(m^*) \frown \gamma\big) \otimes m$$

where $\mathcal{M}$ be a basis for $H_*(F)$ and $\mathcal{M}^* = \{m^* \in H^*(F) \mid m \in \mathcal{M}\}$ is the dual basis for the Kronecker pairing.

Theorem 4.7.17 (Leray-Hirsch) *Let $E \xrightarrow{p} B$ be a locally trivial fiber bundle with fiber F. Suppose that F is of finite cohomology type. Let $\theta : H^*(F) \to H^*(E)$ be a cohomology extension of the fiber. Then, $\hat{\theta}$ is an isomorphism of graded $H^*(B)$-modules and $\underline{\theta}$ is a* **GrV***-isomorphism.*

Proof By Kronecker duality, only the cohomology statement must be proven.

Let $A \subset B$ and let $h^*(A) = H^*(A) \otimes H^*(F)$. The composition

$$\theta_A : H^*(F) \xrightarrow{\theta} H^*(E) \to H^*(E_A)$$

is a cohomology extension of the fiber for the bundle $p : E_A \to A$, giving rise to $\hat{\theta}_A : h^*(A) \to H^*(E_A)$. We want to prove that $\hat{\theta}_B$ is an isomorphism. Considering the commutative diagram

$$\begin{array}{ccc}
h^*(B) & \xrightarrow{\hat{\theta}_B} & H^*(E) \\
\downarrow \approx & & \\
\big(\prod_{A\in\pi_0(B)} H^*(A)\big) \otimes H^*(F) & & \downarrow \approx \\
\Phi \downarrow \approx & & \\
\prod_{A\in\pi_0(B)} h^*(A) & \xrightarrow{\prod \hat{\theta}_A} & \prod_{A\in\pi_0(B)} H^*(E_A)
\end{array}$$

where Φ is the linear map of Lemma 4.6.8, which is an isomorphism since $H^k(F)$ is finite dimensional for all k, permits us to reduce to the case where the base is path-connected.

We now suppose that B is path-connected and that the bundle $E \to B$ is *trivial*, i.e. there exists a homeomorphism $\varphi : B \times F \to E$ such that $p \circ \varphi = \pi_B$, the projection to the factor B. Since F is of finite cohomology type, one may use $H^*\varphi$ together with the Künneth formula to identify $H^*(E)$ with $H^*(B) \otimes H^*(F)$.

Fix an integer n and consider the vector subspace P^n_k of $H^*(B) \otimes H^*(F)$ defined by

$$P^n_k = \{\bigoplus_{p,q} H^p(B) \otimes H^q(F) \mid p + q = n \text{ and } q \leq k\}.$$

These subspaces provide a filtration

$$0 = P^n_{-1} \subset P^n_0 \subset \cdots \subset P^n_n = H^n(E). \tag{4.7.10}$$

We denote by $\psi : H^*(F) \to H^*(F)$ the **GrV**-morphism defined by $\psi = H^*i \circ \theta$. The fiber inclusion i is a slice over one point. Let $u \in H^q(F)$. As B is path-connected, Lemma 4.7.2 implies that

$$\theta(u) = \mathbf{1} \otimes \psi(u) \mod P^q_{q-1}. \tag{4.7.11}$$

Hence, if $a \in H^p(B)$ with $p + q = n$, one has

$$\begin{aligned} \hat{\theta}(a \otimes u) &= (a \otimes \mathbf{1}) \smile \theta(u) \\ &= (a \otimes \mathbf{1}) \smile (\mathbf{1} \otimes \psi(u)) \quad \mod P^n_{n-1} \\ &= a \otimes \psi(u) \quad \mod P^n_{n-1} \end{aligned} \tag{4.7.12}$$

In particular, $\hat{\theta}$ preserves the filtration (4.7.10). It thus induces homomorphisms $\bar{\theta} : P^n_k/P^n_{k-1} \to P^n_k/P^n_{k-1}$. Moreover, one has a natural identification $P^n_k/P^n_{k-1} \approx H^{n-k}(B) \otimes H^k(F)$ under which $\bar{\theta}(a \otimes u) = a \otimes \psi(u)$. This enables us to prove by induction on k that $\hat{\theta} : P^n_k \to P^n_k$ is an isomorphism, using the five-lemma in the diagram

$$\begin{array}{ccccccccc} 0 & \longrightarrow & P^n_{k-1} & \longrightarrow & P^n_k & \longrightarrow & H^{n-k}(B) \otimes H^k(F) & \longrightarrow & 0 \\ & & \approx \downarrow \hat{\theta} & & \downarrow \hat{\theta} & & \approx \downarrow \mathrm{id} \otimes \psi & & \\ 0 & \longrightarrow & P^n_{k-1} & \longrightarrow & P^n_k & \longrightarrow & H^{n-k}(B) \otimes H^k(F) & \longrightarrow & 0 \end{array}$$

Indeed, the left vertical arrow is an isomorphism by induction hypothesis. Since $P^n_{-1} = 0$, the induction starts with $k = 0$, using (4.7.12). Therefore, the Leray-Hirsch theorem is true for a trivial bundle.

Let B_i, $i = 1, 2$, be two open sets of B with $B = B_1 \cup B_2$. Let $B_0 = B_1 \cap B_2$ and $E_i = p^{-1}(B_i)$. The Mayer-Vietoris cohomology sequence for (B, B_1, B_2, B_0)

may be tensored with $H^k(F)$ and remains exact, since we are dealing with $\mathbb{Z}_2$-vector spaces. The sum of these sequences provides the exact sequence of the top line of the commutative diagram

$$\begin{array}{ccccccc}
h^{k-1}(B_1) \oplus h^{k-1}(B_2) & \longrightarrow & h^{k-1}(B_0) & \longrightarrow & h^k(B) & \longrightarrow \\
\downarrow{\scriptstyle \hat{\theta}_1 \oplus \hat{\theta}_2} & & \downarrow{\scriptstyle \hat{\theta}_0} & & \downarrow{\scriptstyle \hat{\theta}} & \\
H^{k-1}(E_1) \oplus H^{k-1}(E_2) & \longrightarrow & H^{k-1}(E_0) & \longrightarrow & H^k(E) & \longrightarrow
\end{array}$$

$$\begin{array}{ccccc}
\longrightarrow & h^k(B_1) \oplus h^k(B_2) & \longrightarrow & h^k(B_0) \\
 & \downarrow{\scriptstyle \hat{\theta}_1 \oplus \hat{\theta}_2} & & \downarrow{\scriptstyle \hat{\theta}_0} \\
\longrightarrow & H^k(E_1) \oplus H^k(E_2) & \longrightarrow & h^k(E_0)
\end{array}$$

The bottom line is the Mayer-Vietoris sequence for the data (B, B_1, B_2, B_0). By the five-lemma, this shows that, if $\hat{\theta}_i$ are isomorphisms for $i = 0, 1, 2$, then $\hat{\theta}$ is an isomorphism.

What has been done so far implies that the $\hat{\theta}$ is an isomorphism for a bundle of *finite type*, i.e. admitting a finite covering $\mathcal{U}$ such that $E_U \xrightarrow{p} U$ is trivial for $U \in \mathcal{U}$. By Kronecker duality, $\underline{\theta}$ is an isomorphism in this case. As in Point (2) of the proof of the Künneth theorem (p. 161), $\underline{\theta}$ is the direct limit of $\underline{\theta}_A$ for $A \subset B$ such that $E_A \to A$ is of finite type. Therefore, $\underline{\theta}$ is an isomorphism and, by Kronecker duality, $\hat{\theta}$ is an isomorphism for any bundle. □

The Leray-Hirsch theorem also has the following version, in which the finite type hypothesis is on the base rather than on the fiber. The proof, involving the Serre spectral sequence, may be found in [141, Theorem 10].

Theorem 4.7.18 (Leray-Hirsch II) *Let $E \xrightarrow{p} B$ be a locally trivial fiber bundle with fiber F. Suppose that B is path-connected and of finite cohomology type. Let $\theta : H^*(F) \to H^*(E)$ be a cohomology extension of the fiber. Then, $\hat{\theta}$ is an isomorphism of graded $H^*(B)$-modules and $\underline{\theta}$ is a* **GrV***-isomorphism.*

Here are a few corollaries of the above Leray-Hirsch theorems.

Corollary 4.7.19 *Let $F \xrightarrow{i} E \xrightarrow{p} B$ be a locally trivial fiber bundle whose base B is path-connected and whose fiber F (or base B) is of finite cohomology type. Suppose that $H^*i : H^*(E) \to H^*(F)$ is surjective. Then*

(1) *$H^*p : H^*(B) \to H^*(E)$ is injective.*
(2) *$\ker H^*i$ is the ideal generated by the elements of positive degree in the image of H^*p.*

Proof Let $\theta : H^k(F) \to H^k(E)$ be a cohomology extension of the fiber such that $H^*i \circ \theta$ is the identity of $H(F)$. As $H^*p(b) = \hat{\theta}(b \otimes \mathbf{1})$, the homomorphism H^*p is injective.

To prove (2), we note that any element in $A \in H^*(E)$ may be written uniquely as a finite sum $A = \sum_{k\in\mathcal{K}} \hat{\theta}(b_k \otimes a_k)$. Let $\mathcal{K}_o = \{k \in \mathcal{K} \mid b_k = \mathbf{1}\}$. As $p \circ i$ is a constant map, one has $H^*(p\circ i)(\mathbf{1}) = \mathbf{1}$ and $H^*(p\circ i)(b_k) = 0$ if b_k has positive degree. Therefore,

$$\begin{aligned} H^*i(A) &= \textstyle\sum_{k\in\mathcal{K}} H^*i(H^*p(b_b k) \smile \theta(a_k)) \\ &= \textstyle\sum_{k\in\mathcal{K}} \big(H^*i \circ H^*p(b_k) \smile H^*i(\theta(a_k))\big) \\ &= \textstyle\sum_{k\in\mathcal{K}} \big(H^*(p\circ i)(b_k) \smile a_k\big) \\ &= \textstyle\sum_{k\in\mathcal{K}_0} a_k \,. \end{aligned}$$

Hence, $H^*i(A) = 0$ if and only if $\mathcal{K}_o = \emptyset$, which proves (2). □

As for Proposition 4.7.1, the Leray-Hirsch theorem implies the following result.

Corollary 4.7.20 *Let $F \xrightarrow{i} E \xrightarrow{p} B$ be a locally trivial fiber bundle whose whose base B is path-connected. Suppose that $H^*i : H^*(E) \to H^*(F)$ is surjective. Suppose that F and B are of finite cohomology type. Then, E is of finite cohomology type and the Poincaré series of F, E and B satisfy*

$$\mathfrak{P}_t(E) = \mathfrak{P}_t(F)\,\mathfrak{P}_t(B)\,. \tag{4.7.13}$$

Actually, Eq. (4.7.13) is equivalent to H^*i being surjective (see [15, Proposition 2.1]). Here is another kind of corollaries of the Leray-Hirsch theorem.

Corollary 4.7.21 *Let $p : E \to B$ be a locally trivial fiber bundle with fiber F. Suppose that $\tilde{H}^*(F) = 0$. Then, $H^*p : H^*(B) \to H^*(E)$ is a* **GrA**-*isomorphism.*

Remarks 4.7.22 (1) By Corollary 4.7.19, the existence of a cohomology extension of the fiber implies that $p^* : H^*(B) \to H^*(E)$ is injective. The converse is not true, even if the map p has a section (see e.g. [72]).
(2) In the Leray-Hirsch theorem the isomorphism $\hat{\theta}$ is not a morphism of algebras, unless θ is multiplicative. It is possible that there exists cohomology extensions of the fiber but that none of them is multiplicative (see Examples 4.7.45 or 7.1.16).
(3) The proof of the Leray-Hirsch theorem shows the following partial result. *Let $\theta : H^k(F) \to H^k(E)$ be a linear map defined for all $k \le n$. Suppose that, for each $b \in B$, the composition $H^*i_b \circ \theta : H^k(F) \to H^k(E_b)$ is an isomorphism for $k \le n$. Then, with the notation of the proof of Theorem* 4.7.17, *the linear map $\hat{\theta} : h^k(B) \to H^k(E)$ is an isomorphism for $k \le n$.* For instance, we get the following proposition.

Proposition 4.7.23 *Let $p : E \to B$ be a locally trivial fiber bundle with fiber F. Suppose that $\tilde{H}^k(F) = 0$ for all $k \le m$. Then, $H^*p : H^k(B) \to H^k(E)$ is an isomorphism for $k \le m$.*

The Leray-Hirsch theorem admits a version for bundle pairs. A *bundle pair* with fiber (F, F') is a topological pair (E, E') and a map $p : (E, E') \to (B, B)$ such that there exists an open covering $\mathcal{U}$ of B and, for each $U \in \mathcal{U}$, a homeomorphism $\psi_U : U \times (F, F') \xrightarrow{\approx} (p^{-1}(U), p^{-1}(U) \cap E')$ such that $p \circ \psi(x, v) = x$ for all $(x, v) \in U \times F$. In consequence, $p : E \to B$ is a bundle with fiber F and the restriction of p to E' is a bundle with fiber F'. A *cohomology extension of the fiber* is a **GrV**-morphism $\theta_{rel} : H^*(F, F') \to H^*(E, E')$ such that, for each $b \in B$, the composite

$$H^*(F, F') \xrightarrow{\theta_{rel}} H^*(E, E') \xrightarrow{H^* i_b} H^*(E_b, E'_b)$$

is a **GrV**-isomorphism. A cohomology extension θ of the fiber provides a morphism of graded $H^*(B)$-modules

$$H^*(B) \otimes H^*(F, F') \xrightarrow{\hat{\theta}_{rel}} H^*(E, E')$$

given by $\hat{\theta}_{rel}(a \otimes b) = p^*(a) \smile \theta_{rel}(b)$.

Suppose that F is of finite cohomology type. As in Lemma 4.6.6, there is a unique **GrV**-homomorphism

$$\underline{\theta}_{rel} : H_*(E, E') \to H_*(B) \otimes H_*(F, F') .$$

such that the formula

$$\langle \hat{\theta}_{rel}(b \otimes u), \gamma \rangle = \langle b \otimes u, \underline{\theta}_{rel}(\gamma) \rangle \tag{4.7.14}$$

holds true for all $b \in H^*(B)$, $u \in H^*(F, F')$ and $\gamma \in H_*(E, E')$. The formula

$$\underline{\theta}_{rel}(\gamma) = \sum_{m \in \mathcal{M}} H_* p\big(\theta_{rel}(m^*) \frown \gamma\big) \otimes m . \tag{4.7.15}$$

is satisfied for all $\gamma \in H_*(E, E')$, where $\mathcal{M}$ is a basis for $H_*(F, F')$ and $\mathcal{M}^* = \{m^* \in H^*(F, F') \mid m \in \mathcal{M}\}$ is the dual basis for the Kronecker pairing.

Theorem 4.7.24 (Leray-Hirsch relative) *Let $p : (E, E') \to (B, B)$ be a bundle pair with fiber (F, F'). Suppose that (F, F') is a well cofibrant pair and is of finite cohomology type. Let $\theta_{rel} : H^*(F, F') \to H^*(E, E')$ be a cohomology extension of the fiber. Suppose that (E, E') is a well cofibrant pair. Then, $\hat{\theta}_{rel}$ is an isomorphism of graded $H^*(B)$-modules and $\underline{\theta}_{rel}$ is a* **GrV***-isomorphism.*

The hypothesis that (E, E') is well cofibrant may be removed but this would necessitate some preliminary work. Besides, this hypothesis is easily fulfilled in our applications.

Proof By Kronecker duality, only the cohomology statement must be proven. We first reduce to the case where F' is a point. Let $\hat{E} = E/\!\sim$ where $\sim$ is the equivalence relation

$$x \sim y \iff p(x) = p(y) \text{ and } x, y \in E'$$

and let $\hat{E}' = E'/\!\sim$. Then the map p descends to a map $\hat{p} : (\hat{E}, \hat{E}') \to (B, B)$ which is a bundle pair with fiber $(F/F', y_0)$, where y_0 is the point given by F' in F/F'. In particular, $p : \hat{E}' \to B$ is a homeomorphism. Consider the commutative diagram:

$$\begin{array}{ccc} H^*(B) \otimes H^*(F/F', y^0) & \xrightarrow{\approx} & H^*(B) \otimes H^*(F, F') \\ \downarrow \hat{\theta}_{rel} & & \downarrow \hat{\theta}_{rel} \\ H^*(\hat{E}, \hat{E}') & \xrightarrow{\approx} & H^*(E, E') \end{array} \tag{4.7.16}$$

We shall show that the horizontal homomorphisms, induced by the quotient maps, are isomorphisms. Therefore, the right vertical arrow is bijective if and only if the left one is.

The top horizontal homomorphism of Diagram (4.7.16) is an isomorphism by Proposition 3.1.45 since (F, F') is a well cofibrant pair. To see that the bottom horizontal map is also bijective, consider the commutative diagram

$$\begin{array}{ccc} H^*(\hat{E}/\hat{E}', [\hat{E}']) & \xrightarrow{\approx} & H^*(E/E', [E']) \\ \downarrow \approx & & \downarrow \approx \\ H^*(\hat{E}, \hat{E}') & \longrightarrow & H^*(E, E') \end{array}$$

The top horizontal map is an isomorphism because the quotient spaces E/E' and $\hat{E}/\hat{E}'$ are equal. As (E, E') is well cofibrant, the right hand vertical map is bijective by Proposition 3.1.45. Also, Lemma 3.1.43 implies that $(\hat{E}, \hat{E}')$ is well cofibrant and thus, the left hand vertical map is an isomorphism by Proposition 3.1.45. Now, the diagram

$$\begin{array}{ccccc} H^*(F/F', [F']) & \overset{\theta_{rel}}{\dashrightarrow} & H^*(\hat{E}, \hat{E}') & \longrightarrow & H^*(\hat{E}_b, \hat{E}'_b) \\ \downarrow \approx & & \downarrow \approx & & \downarrow \approx \\ H^*(F, F') & \xrightarrow{\theta_{rel}} & H^*(E, E') & \longrightarrow & H^*(E_b, E'_b) \end{array}$$

(with the bottom composite $H^*(F, F') \to H^*(E_b, E'_b)$ marked $\approx$)

shows that the bundle pair $\hat{p}$ inherits a cohomology extension of the fiber θ_{rel}.

We are then reduced to the case $F' = \{y^0\}$ being a single point. Consider the commutative diagram:

$$\begin{array}{ccccc} H^*(B) \otimes H^*(F, y^0) & \rightarrowtail & H^*(B) \otimes H^*(F) & \twoheadrightarrow & H^*(B) \otimes H^*(y^0) \\ \downarrow \hat{\theta}_{rel} & & \approx \downarrow \hat{\theta} & & \approx \downarrow \hat{\theta} \\ H^*(E, E') & \rightarrowtail & H^*(E) & \twoheadrightarrow & H^*(E') \end{array}$$

The top line is the exact sequence of the pair (F, y^0) tensored by $H^*(B)$. It is exact since we are dealing with $\mathbb{Z}_2$-vector spaces and its splits since $\{y^0\}$ is a retract of F'. The bottom exact sequence of the pair (E, E') also splits since $p : E \to B \approx E'$ is a retraction of E onto E'. We shall check below the existence of a cohomology extension of the fiber $\theta : H^*(F) \to H^*(E)$, whence the middle vertical map $\hat{\theta}$. The two maps $\hat{\theta}$ are bijective by the absolute Leray-Hirsch theorem 4.7.17. By the five-lemma, θ_{rel} is then also an isomorphism.

The existence of a cohomology extension of the fiber $\theta : H^*(F) \to H^*(E)$ comes from $\theta_{rel} : H^*(F, y^0) \to H^*(E, E')$ when $* > 0$, since $H^k(F, y^0) \approx H^k(F)$ and $H^k(E_b, E'_b) \approx H^k(E_b)$ for $k > 0$. When $k = 0$, we consider the diagram

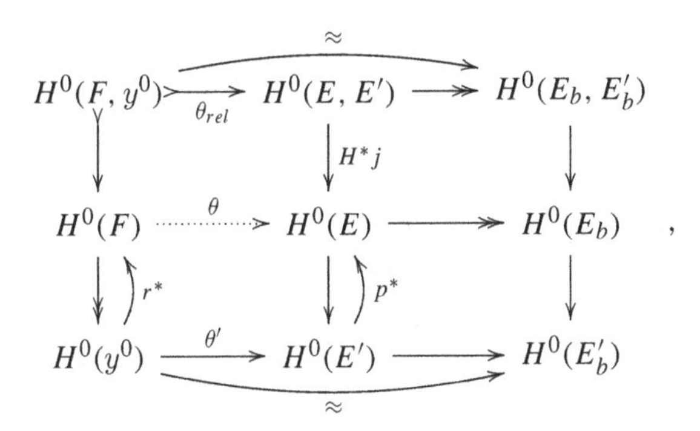

where $j : (E, \emptyset) \to (E, E')$ denotes the inclusion. The retraction $r : F \to \{y^0\}$ produces a section $r^* : H^0(y^0) \to H^0(F)$ of the homomorphism induced by the inclusion; this section provides an isomorphism $H^0(F) \approx H^0(F, y^0) \oplus H^0(y^0)$. As $p : E' \to B$ is a homeomorphism, one gets a section $p^* : H^0(E') \to H^0(E)$ of the homomorphism induced by the inclusion. The homomorphism $\theta : H^0(F) \approx H^0(F, y^0) \oplus H^0(y^0) \to H^0(E)$ given by $\theta(a, b) = H^* j \circ \theta_{rel}(a) + p^* \circ \theta'(b)$ completes the definition of the cohomology extension of the fiber θ in degree 0. □

Corollary 4.7.25 (Relative Künneth theorem II) *Let X be a topological space and (Y, C) be a well cofibrant pair which is of finite cohomology type. Then, the cross product*

$$\times : H^*(X) \otimes H^*(Y, C) \xrightarrow{\approx} H^*(X \times Y, X \times C)$$

is a **GrV**-*isomorphism.*

Proof We see the projection $\pi_1 : (X \times Y, X \times C) \to (X, X)$ as a trivial bundle pair with fiber (Y, C). Then $\theta = H^*\pi_2 : H^*(Y, C) \to H^*(X \times Y, X \times C)$ is a cohomology extension of the fiber. As (Y, C) is a well cofibrant pair, so is $(X \times Y, X \times C)$. The cohomological result then follows from the relative Leray-Hirsch theorem 4.7.24. □

4.7.6 The Thom Isomorphism

We start by some preliminary results.

Lemma 4.7.26 *Let $p : (\hat{E}, E) \to (B, B)$ be a bundle pair whose fiber $(\hat{F}, F)$ is a well cofibrant pair. Suppose that $H_k(\hat{F}, F) = 0$ for $k < r$ and that $H_r(\hat{F}, F) = \mathbb{Z}_2$. Then, $H_k(\hat{E}, E) = 0$ for $k < r$ and there is a unique isomorphism $\Phi_* : H_r(\hat{E}, E) \xrightarrow{\approx} H_0(B)$ such that, for each $b \in B$, the diagram*

$$\begin{array}{ccc} H_r(\hat{E}_b, E_b) & \rightarrowtail & H_r(\hat{E}, E) \\ \downarrow = & & \approx \downarrow \Phi_* \\ H_0(\{b\}) & \rightarrowtail & H_0(B) \end{array} \tag{4.7.17}$$

is commutative, where the horizontal homomorphisms are induced by the inclusions.

In Diagram (4.7.17), the left vertical isomorphism is abstract but well defined, since both $H_r(\hat{E}_b, E_b) \approx H_r(\hat{F}, F)$ and $H_0(\{b\})$ are equal to $\mathbb{Z}_2$.

Proof We first prove the uniqueness of Φ_*, if it exists. Indeed, for each path-connected component A of B, Diagram (4.7.17) implies that $\Phi_*(H^r(\hat{E}_A, E_A)) = H_0(A)$ and, as $H_0(A) = \mathbb{Z}_2$, the isomorphism Φ_* is unique.

If the bundle pair is trivial, the lemma follows from the relative Künneth theorem 4.7.25. Suppose that $B = B_1 \cup B_2$, where B_1 and B_2 are two open sets with $B_1 \cap B_2 = B_0$. Suppose that the conclusion of the lemma is satisfied for $(\hat{E}_i, E_i)$ for $i = 0, 1, 2$. Then, the Mayer-Vietoris sequence for the data $(\hat{E}_i, E_i)$ implies that $H_k(\hat{E}, E) = 0$ for $k < r$ and gives the diagram

$$\begin{array}{ccccccc} H_r(\hat{E}_0, E_0) & \longrightarrow & H_r(\hat{E}_1, E_1) \oplus H_r(\hat{E}_2, E_2) & \longrightarrow & H_r(\hat{E}, E) & \longrightarrow & 0 \\ \approx \downarrow \Phi_* & & \approx \downarrow \Phi_* & & \approx \vdots \bar{\Phi}_* & & \\ H_0(B_0) & \longrightarrow & H_0(B_1) \oplus H_0(B_2) & \longrightarrow & H_0(B) & \longrightarrow & 0 \end{array} .$$

Diagram (4.7.17) for each $b \in B_0$ implies that the left square is commutative. Therefore, the middle vertical isomorphism descends to a unique homomorphism

$\bar{\Phi}_* : H_r(\hat{E}, E) \to H_0(B)$ making the right square commutative, which an isomorphism by the five-lemma. It remains to prove the commutativity of Diagram (4.7.17) for $\bar{\Phi}_*$. Let $b \in B$. Without loss of generality, we may suppose that $b \in B_1$. Consider the diagram

$$\begin{array}{ccccc} H_r(\hat{E}_b, E_b) & \longrightarrow & H_r(\hat{E}_1, E_1) & \longrightarrow & H_r(\hat{E}, E) \\ \downarrow = & & \approx \downarrow \Phi_* & & \approx \downarrow \bar{\Phi}_* \\ H_0(\{b\}) & \longrightarrow & H_0(B_1) & \longrightarrow & H_0(B) \end{array} .$$

As both square commute, this gives the commutativity of Diagram (4.7.17) for $\bar{\Phi}_*$.

We have so far proven the lemma when the bundle pair $(\hat{E}, E) \to (B, B)$ is of finite type. Let $\mathcal{A}$ be the set of subspaces A of B such that the bundle pair $(\hat{E}_A, E_A) \to (A, A)$ is of finite type. Each compact of B is contained in some $A \in \mathcal{A}$ and each compact of E is contained in E_A for some $A \in \mathcal{A}$. By Proposition 3.1.29, this provides isomorphisms

$$\varinjlim_{A\in\mathcal{A}} H_r(\hat{E}_A, E_A) \approx H_r(\hat{E}, E) \text{ and } \varinjlim_{A\in\mathcal{A}} H_0(A) \approx H_0(B) . \tag{4.7.18}$$

Now, if $A, A' \in \mathcal{A}$ with $A \subset A'$, Diagram (4.7.17) for each $b \in A$ implies that the diagram

$$\begin{array}{ccc} H_r(\hat{E}_A, E_A) & \longrightarrow & H_r(\hat{E}_{A'}, E_{A'}) \\ \approx \downarrow \Phi_* & & \approx \downarrow \Phi_* \\ H_0(A) & \longrightarrow & H_0(A') \end{array} \tag{4.7.19}$$

is commutative. We therefore get isomorphisms

$$\varinjlim_{A\in\mathcal{A}} H_r(\hat{E}_A, E_A) \xrightarrow{\approx} \varinjlim_{A\in\mathcal{A}} H_0(A)$$

which, together with the isomorphisms of (4.7.18), produce the required isomorphism $\Phi_* : H_r(\hat{E}, E) \xrightarrow{\approx} H_0(B)$. □

By Kronecker duality, Lemma 4.7.26 gives the following lemma.

Lemma 4.7.27 *Let $p : (\hat{E}, E) \to (B, B)$ be a bundle pair whose fiber $(\hat{F}, F)$ is a well cofibrant pair. Suppose that $H^k(\hat{F}, F) = 0$ for $k < r$ and that $H^r(\hat{F}, F) = \mathbb{Z}_2$. Then, $H^k(\hat{E}, E) = 0$ for $k < r$ and there is a unique isomorphism $\Phi^* : H^0(B) \xrightarrow{\approx} H^r(\hat{E}, E)$ such that, for each $b \in B$, the diagram*

$$\begin{array}{ccc} H^0(B) & \longrightarrow\!\!\!\!\rightarrow & H^0(\{b\}) \\ \approx \Big\downarrow \Phi^* & & \Big\downarrow = \\ H^r(\hat{E}, E) & \longrightarrow\!\!\!\!\rightarrow & H^r(\hat{E}_b, E_b) \end{array} \tag{4.7.20}$$

is commutative, where the horizontal homomorphisms are induced by the inclusions.

Let $p : (\hat{E}, E) \to (B, B)$ a bundle pair satisfying the hypotheses of Lemma 4.7.27. The class $U = \Phi^*(\mathbf{1}) \in H^r(\hat{E}, E)$ is called the *Thom class* of the bundle pair p. If B is path-connected, the Thom class is just the non-zero element of $H^r(\hat{E}, E) = \mathbb{Z}_2$, whence the following characterization of the Thom class.

Lemma 4.7.28 *The Thom class of p is the unique class in $H^r(\hat{E}, E)$ which restricts to the generator of $H^r(\hat{E}_b, E_b)$ for all $b \in B$.*

Let Σ be a topological space having the same homology (mod 2) as the sphere S^{k-1}. For example, $\Sigma = S^{k-1}$ or a lens space with odd fundamental group. Let $p : E \to B$ be a bundle with fiber Σ. Let $\hat{E}$ be the mapping cylinder of p:

$$\hat{E} = (E \times I) \dot{\cup} B \Big/ \{(x, 1) \sim p(x)\}.$$

Let $C\Sigma$ be the cone over Σ. Then p extends to a bundle pair $p : (\hat{E}, E) \to (B, B)$ with fiber $(C\Sigma, \Sigma)$, called the *mapping cylinder bundle pair* of p. As $(C\Sigma, \Sigma)$ is a well cofibrant pair (by Lemma 3.1.39) and $H^k(C\Sigma, \Sigma) = 0$ for $k \neq r$ and $H^r(C\Sigma, \Sigma) = \mathbb{Z}_2$, the Thom class $U \in H^r(\hat{E}, E)$ is defined.

Theorem 4.7.29 (The Thom isomorphism theorem) *Let $p : E \to B$ be a bundle with fiber Σ, where Σ has the homology of the sphere S^{r-1}. Let $p : (\hat{E}, E) \to (B, B)$ be its mapping cylinder bundle pair. Let $U \in H^r(\hat{E}, E)$ be the Thom class. Then, the homomorphisms*

$$\Phi^* : H^k(B) \to H^{k+r}(\hat{E}, E) \text{ and } \Phi_* : H_k(\hat{E}, E) \to H_{k-r}(B)$$

given by

$$\Phi^*(a) = H^*p(a) \smile U \text{ and } \Phi_*(\gamma) = H_*p(U \frown \gamma)$$

are isomorphism for all $k \in \mathbb{Z}$.

Observe that Lemma 4.7.27 gives the result for $k \leq 0$.

Proof As $H^j(C\Sigma, \Sigma) = 0$ for $j \neq r$, the homomorphism $\theta_{rel} : H^*(C\Sigma, \Sigma) \to H^*(\hat{E}, E)$ sending the generator of $H^r(C\Sigma, \Sigma) = \mathbb{Z}_2$ onto the Thom class U is a cohomology extension of the fiber. Also, $(\hat{E}, E)$ and the fiber $(C\Sigma, \Sigma)$ are well cofibrant by Lemma 3.1.39. The relative Leray-Hirsch theorem 4.7.24 then provides a **GrV**-isomorphism $\hat{\theta}_{rel} : H^*(B) \otimes H^*(C\Sigma, \Sigma) \xrightarrow{\approx} H^*(\hat{E}, E)$. Let Φ^* be the composite isomorphism

$$\Phi^* : H^k(B) \approx H^k(B) \otimes H^r(C\Sigma, \Sigma) \xrightarrow{\hat{\theta}_{rel}} H^{k+r}(\hat{E}, E)$$

satisfy, by definition of θ_{rel}, the formula $\Phi^*(a) = H^*p(a) \smile U$. This proves the cohomology statement.

For the isomorphism Φ_*, let $0 \neq m \in H_r(C\Sigma, \Sigma)$. Then $\{m\}$ and $\{U\}$ are Kronecker dual bases for (co)homology of $(C\Sigma, \Sigma)$ in degree r. Theorem 4.7.24 and Formula (4.7.15) implies that the composite isomorphism

$$\Phi_* : H_{k+r}(\hat{E}, E) \xrightarrow{\underline{\theta}_{rel}} H_k(B) \otimes H_r(C\Sigma, \Sigma) \approx H_k(B)$$

satisfies $\Phi_*(\gamma) = H_*p(U \frown \gamma)$. □

Let $q : E \to B$ be a bundle with fiber F and let $f : A \to B$ be a continuous map. The *induced bundle* $f^*q : f^*E \to A$ is defined by

$$f^*E = \{(a, y) \in A \times E \mid f(a) = q(y)\}, \quad f^*q(a, y) = a ,$$

where f^*E is topologized as a subspace of $A \times E$. Then f^*q is a bundle over A with fiber F. The projection onto E gives a map $\tilde{f} : f^*E \to E$ and a commutative diagram

$$\begin{array}{ccc} f^*E & \xrightarrow{\tilde{f}} & E \\ {\scriptstyle f^*q}\downarrow & & \downarrow{\scriptstyle q} \\ A & \xrightarrow{f} & B \end{array} .$$

Let $p : E \to B$ be a bundle with fiber Σ, where Σ has the homology of the sphere S^{r-1} and let $p : (\hat{E}, E) \to (B, B)$ be its mapping cylinder bundle pair. Let $f : A \to B$ be a map. Then $(f^*\hat{E}, f^*E) \to (A; A)$ is the mapping cylinder bundle pair of the induced bundle f^*E. The following lemma states the *functoriality of the Thom class*.

Lemma 4.7.30 *If $U \in H^r(\hat{E}, E)$ is the Thom class of p, then $H^*\tilde{f}(U) \in H^r(f^*\hat{E}, f^*E)$ is the Thom class of f^*p.*

Proof For $a \in A$, let consider the commutative diagram

$$\begin{array}{ccc} H^r(\hat{E}, E) & \xrightarrow{H^*\tilde{f}} & H^r(f^*\hat{E}, f^*E) \\ \downarrow & & \downarrow \\ H^r(\hat{E}_{f(a)}, E_{f(a)}) & \xrightarrow[\approx]{H^*\tilde{f}} & H^r((f^*\hat{E})_a, (f^*E)_a) \end{array} .$$

Both cohomology groups downstairs are equal to $\mathbb{Z}_2$. The left vertical arrow sends the Thom class U to the non-zero element. Therefore, $H^*\tilde{f}(U)$ goes, by the right vertical arrow, to the non-zero element. As this is true for all $a \in A$, we deduce from Lemma 4.7.28 that $H^*\tilde{f}(U)$ is the Thom class of f^*p. □

Let $p : E \to B$ be a bundle with fiber Σ, where Σ has the homology of the sphere S^{r-1}. Let $p : (\hat{E}, E) \to (B, B)$ be its mapping cylinder bundle pair. As $C\Sigma$ is contractible, Proposition 4.7.23 implies that $H^*p : H^*(B) \to H^*(\hat{E})$ is a **GrA**-isomorphism. Therefore, there is a unique class $e \in H^r(B)$ such that

$$H^*j(U) = H^*p(e)\,, \tag{4.7.21}$$

where $U \in H^r(\hat{E}, E)$ is the Thom class and $j : (\hat{E}, \emptyset) \to (\hat{E}, E)$ is the pair inclusion. The class $e = e(p)$ is called the *Euler class* of the bundle p. If $\Phi^* : H^r(B) \to H^{2r}(\hat{E}, E)$ is the Thom isomorphism, one has the formula

$$\Phi^*(e) = U \smile U\,. \tag{4.7.22}$$

Indeed:

$$\Phi^*(e) = H^*p(e) \smile U = H^*j(U) \smile U = U \smile U\,,$$

the last equality coming from Lemma 4.1.7. The Euler class is functorial by the following lemma.

Lemma 4.7.31 *Let $p : E \to B$ be a bundle with fiber Σ, where Σ has the homology of the sphere S^{r-1}. Let $f : A \to B$ be a map. If $e \in H^r(B)$ is the Euler class of p, then $H^*f(e) \in H^r(A)$ is the Euler class of f^*p.*

Proof This follows from the definition of the Euler class, Lemma 4.7.30 and the commutativity of the diagram.

$$\begin{array}{ccccc}
H^r(\hat{E}, E) & \longrightarrow & H(\hat{E}) & \xleftarrow[\approx]{H^*p} & H^r(B) \\
\downarrow{\scriptstyle H^*\tilde{f}} & & \downarrow{\scriptstyle H^*\tilde{f}} & & \downarrow{\scriptstyle H^*\tilde{f}} \\
H^r(f^*\hat{E}, f^*E) & \longrightarrow & H^r(f^*\hat{E}) & \xleftarrow[\approx]{H^*f^*p} & H^r(A)
\end{array}\,.$$

The Euler class occurs in the *Gysin exact sequence*. □

Proposition 4.7.32 (Gysin exact sequence) *Let $p : E \to B$ be a bundle with fiber Σ, where Σ has the homology of the sphere S^{r-1}. Let $e \in H^r(B)$ be its Euler class. Then, there is a long exact sequence*

$$\cdots \to H^{k-1}(B) \xrightarrow{H^*p} H^{k-1}(E) \to H^{k-r}(B) \xrightarrow{-\smile e} H^k(B) \xrightarrow{H^*p} H^k(E) \to \cdots$$

which is functorial with respect to induced bundles.

Proof Let $p : (\hat{E}, E) \to (B, B)$ be the mapping cylinder pair of p. One uses the cohomology sequence of the pair $(\hat{E}, E)$ and the commutative diagram

$$\begin{array}{ccccccc} \cdots \longrightarrow & H^{k-1}(E) & \longrightarrow & H^k(\hat{E}, E) & \xrightarrow{H^*j} & H^k(\hat{E}) & \longrightarrow \cdots \\ & & & \Phi^* \uparrow \approx & & \approx \uparrow H^*p & \\ & & & H^{k-r}(B) & \xrightarrow{-\smile e} & H^k(B) & \end{array}$$

where $j : (\hat{E}, E) \to (\hat{E}, \emptyset)$ denotes the inclusion and Φ^* is the Thom isomorphism. The diagram is commutative since, for $a \in H^{k-r}(B)$,

$$\begin{aligned} H^*j \circ \Phi^*(a) &= H^*j\big(H^*p(a) \smile U\big) \\ &= H^*p(a) \smile H^*j(U) \\ &= H^*p(a) \smile H^*p(e) \\ &= H^*p(a \smile e) \,. \end{aligned}$$

(The second equality is the singular analogue of Lemma 4.1.8). The functoriality of the Gysin exact sequence comes from Lemma 4.7.30 and 4.7.31. □

Corollary 4.7.33 *Let $p : E \to B$ be a bundle with fiber Σ, where Σ has the homology of the sphere S^{r-1}. If p admits a continuous section, then the Euler class of p vanishes.*

Proof In the following segment of the Gysin sequence:

$$H^0(B) \xrightarrow{-\smile e} H^r(B) \xrightarrow{H^*p} H^r(E) \,,$$

the class $\mathbf{1} \in H^0(B)$ is sent to the Euler class e. If p admits a section, then H^*p is injective, which implies that $e = 0$. □

Remark 4.7.34 The vanishing of the Euler class of $p : E \to B$ does not imply that p admits a section. As an example, let $p : SO(3) \to S^2$ the map sending a matrix to its first column vector. Then p is an S^1-bundle, equivalent to the unit tangent bundle of S^2. The Gysin sequence gives the exact sequence

$$H^0(S^2) \xrightarrow{-\smile e} H^2(S^2) \xrightarrow{H^*p} H^2(SO(3)) \to 0$$

As $SO(3)$ is homeomorphic to $\mathbb{R}P^3$, $H^2(SO(3)) = \mathbb{Z}_2$ by Proposition 4.3.10 and all the cohomology groups in the above sequence are equal to $\mathbb{Z}_2$. Hence, $e = 0$. But it is classical that S^2 admits no nowhere zero vector field [82, Theorem 2.28].

Proposition 4.7.35 *Let $p : E \to B$ be a bundle with fiber Σ, where Σ has the homology of the sphere S^{r-1}. Let $e \in H^r(B)$ be its Euler class. Then, the following assertions are equivalent.*

(1) $e = 0$.
(2) *The restriction homomorphism* $H^*(E) \to H^*(\Sigma)$ *is surjective.*
(3) $H^{r-1}(E) \approx H^{r-1}(B) \oplus \mathbb{Z}_2$.

Proof Let $p : (\hat{E}, E) \to (B, B)$ be the mapping cylinder pair of p. Identifying Σ as the fiber over some point of B, we get a commutative diagram

$$\begin{array}{ccccccc} 0 \longrightarrow H^{r-1}(\hat{E}) & \longrightarrow & H^{r-1}(E) & \longrightarrow & H^r(\hat{E}, E) & \longrightarrow & H^r(\hat{E}) \\ \approx \uparrow H^*p & & \downarrow & & \downarrow \approx & & \\ H^{r-1}(B) & & H^{r-1}(\Sigma) & \xrightarrow{\approx} & H^r(C\Sigma, \Sigma) & & \end{array}$$

where the top line is the cohomology sequence of $(\hat{E}, E)$. But $H^r(\hat{E}, E) \approx \mathbb{Z}_2$ generated by the Thom class which, under the homomorphism $H^r(\hat{E}, E) \to H^r(\hat{E}) \approx H^r(B)$, goes to the Euler class. This proves the proposition. □

Let us consider the particular case of the Gysin sequence for an S^0-bundle. Such a bundle is simply a 2-fold covering $\xi = (p : \tilde{X} \to X)$. The Gysin sequence may thus be compared to the transfer exact sequence of Proposition 4.3.9.

Proposition 4.7.36 *Let* $\xi = (p : \tilde{X} \to X)$ *be a 2-fold covering (an* S^0*-bundle). Then, the Gysin and the transfer exact sequences of* ξ *coincide, i.e. the diagram*

$$\begin{array}{ccccccccc} \cdots \longrightarrow & H^k(X) & \xrightarrow{H^*p} & H^k(\tilde{X}) & \xrightarrow{tr^*} & H^k(X) & \xrightarrow{w(\xi)\smile -} & H^{k+1}(X) & \longrightarrow \cdots \\ & \updownarrow \text{id} & & \updownarrow \text{id} & & \updownarrow \text{id} & & \updownarrow \text{id} & \\ \cdots \longrightarrow & H^k(X) & \xrightarrow{H^*p} & H^k(\tilde{X}) & \longrightarrow & H^k(X) & \xrightarrow{e(p)\smile -} & H^{k+1}(X) & \longrightarrow \cdots \end{array}$$

is commutative. In particular, the Euler class $e(\xi) \in H^1(X)$ *and the characteristic class* $w(p) \in H^1(X)$ *are equal.*

Proof By Corollary 4.3.3, ξ is induced from $\xi_\infty = (p_\infty : S^\infty \to \mathbb{R}P^\infty)$ by a characteristic map $f : X \to \mathbb{R}P^\infty$. Both the Gysin and transfer exact sequences being functorial with respect to induced bundles, it suffices to prove the proposition for ξ_∞. This is trivial since the vector spaces occurring in the diagram are either equal to 0 or $\mathbb{Z}_2$. □

The Thom isomorphism is classically used for vector bundles. Recall that a (*real*) *vector bundle* ξ of *rank* r is a map $p : E \to B$ together with a $\mathbb{R}$-vector space structure on $E_b = p^{-1}(b)$ for each $b \in B$, satisfying the following local triviality condition: there is an open covering $\mathcal{U}$ of B and for each $U \in \mathcal{U}$, a homeomorphism $\psi_U : U \times \mathbb{R}^r \xrightarrow{\approx} p^{-1}(U)$ such that, for all $(b, v) \in U \times \mathbb{R}^r$, $p \circ \psi(b, v) = b$ and $\psi_U : \{b\} \times \mathbb{R}^r \to E_b$ is a $\mathbb{R}$-linear isomorphism. In consequence, p is a bundle with base $B = B(\xi)$, total space $E = E(\xi)$ and fiber $\mathbb{R}^r$. The map $\sigma_0 : B \to E$ sending $b \in B$ to the zero element of E_b is called the *zero section* of ξ (it satisfies $p \circ \sigma_0(b) = b$).

An *Euclidean vector bundle* is a vector bundle ξ together with a continuous map $v \mapsto |v| \in \mathbb{R}_{\geq 0}$ defined on $E(\xi)$ whose restriction to each fiber is quadratic and positive definite. Such a map is called an *Euclidean structure* (or *Riemannian metric*) on ξ. It is of course the same as defining a positive definite inner product on each fiber which varies continuously. Vector bundles with paracompact basis admit an Euclidean structure, [105, Chap. 3,Theorems 9.5 and 5.5]. If $\xi = (p : E \to B)$ is an Euclidean vector bundle, the restriction of p to

$$S(E) = \{v \in E \mid |v| = 1\} \text{ and } D(E) = \{v \in E \mid |v| \leq 1\}$$

gives the associated unit *sphere* and *disk bundles*. These bundles do not depend on the choice of the Euclidean structure on ξ. Indeed, using the map $(v, t) \mapsto tv$ from $S(E) \times I \to D(V)$ together with the zero section, the reader will easily construct a homeomorphism $(\widehat{S(E)}, S(E)) \xrightarrow{\approx} (D(E), S(E))$ over the identity of B, where $(\widehat{S(E)}, S(E)) \to (B, B)$ is the mapping cylinder bundle pair of $S(E) \to B$. Thus, the Thom class $U \in H^r(D(E), S(E))$ exists by Lemma 4.7.27 and, by Theorem 4.7.29, gives rise to the Thom isomorphisms $\Phi^* : H^k(B) \xrightarrow{\approx} H^{k+r}(D(E), S(E))$ and $\Phi_* : H_{k+r}(D(E), S(E)) \xrightarrow{\approx} H_k(B)$. Let $E_0 = E - \sigma_0(B)$ and $D(E)_0 = D(E) \cap E_0$. By excision and homotopy, one has

$$H^*(E, E_0) \xrightarrow{\approx} H^*(D(E), D(E)_0) \xrightarrow{\approx} H^*(D(E), S(E)) .$$

Hence the Thom class may be seen as an element $U(\xi) \in H^r(E, E_0)$ and one has the following theorem.

Theorem 4.7.37 (The Thom isomorphism theorem for vector bundles) *Let $\xi = (p : E \to B)$ be a vector bundle of rank r with B paracompact. Let $U(\xi) \in H^r(E, E_0)$ be the Thom class. Then, the homomorphisms*

$$\Phi^* : H^k(B) \to H^{k+r}(E, E_0) \text{ and } \Phi_* : H_k(E, E_0) \to H_{k-r}(B)$$

given by

$$\Phi^*(a) = H^* p(a) \smile U(\xi) \text{ and } \Phi_*(\gamma) = H_* p(U(\xi) \frown \gamma)$$

are isomorphism for all $k \in \mathbb{Z}$.

Let $\xi = (p : E \to B)$ be a vector bundle of rank r. The map $E \times I \to E$ given by $(v, t) \mapsto tv$ is a retraction by deformation of E onto the zero section of ξ. Hence, $H^* p : H^*(B) \to H^*(E)$ is a **GrA**-isomorphism. Therefore, there is a unique class $e(\xi) \in H^r(B)$ such that $H^* j(U(\xi)) = H^* p(e(\xi))$, where $j : (E, \emptyset) \to (E, E_0)$. The class $e(\xi)$ is called the *Euler class* of ξ (it coincides with the Euler $e(S(E))$ defined above). Lemma 4.7.31 and Corollary 4.7.33 imply the following two lemmas.

Lemma 4.7.38 *Let $\xi = (p : E \to B)$ be a vector bundle of rank r, with B paracompact. Let $f : A \to B$ be a continuous map. Then the equality $e(f^*\xi) = H^*f(e(\xi))$ holds in $H^r(A)$.*

Lemma 4.7.39 *Let $\xi = (p : E \to B)$ be a vector bundle of rank r, with B paracompact. If ξ admits a nowhere zero section, then $e(\xi) = 0$.*

Let $\xi_i = (p_i : E_i \to B_i)$ $(i = 1, 2)$ be two vector bundles of rank r_i, with B_i paracompact. The *product bundle* $\xi_1 \times \xi_2$ is the vector bundle of rank $r_1 + r_2$ given by $p_1 \times p_2 : E_1 \times E_2 \to B_1 \times B_2$. If $B_1 = B_2 = B$, the *Whitney sum* $\xi_1 \oplus \xi_2$ is the vector bundle of rank $r_1 + r_2$ over B given by $\xi_1 \oplus \xi_2 = \Delta^*(\xi_1 \times \xi_2)$ where $\Delta : B \to B \times B$ is the inclusion of the diagonal, $\Delta(x) = (x, x)$. The behavior of the Euler class under these constructions is as follows.

Proposition 4.7.40

(1) $e(\xi_1 \times \xi_2) = e(\xi_1) \times e(\xi_2)$.
(2) $e(\xi_1 \oplus \xi_2) = e(\xi_1) \smile e(\xi_2)$.

Proof Using Euclidean structures on ξ_i the Thom class $U(\xi_i)$ may be seen as an element of $H^{r_i}(D(E_i), S(E_i))$. Let $E = E_1 \times E_2$, $B = B_1 \times B_2$ and $r = r_1 + r_2$.

Let $j_i : (D(E_i), \emptyset) \to (D(E_i), S(E_i))$ and $j : (D(E), \emptyset) \to (D(E), S(E))$ denote the inclusions. There are homeomorphisms of pairs making the following diagram commutative

$$\begin{array}{ccc} (D(E), \emptyset) & \xrightarrow{\quad j \quad} & (D(E), S(E)) \\ \downarrow\approx & & \downarrow\approx \\ (D(E_1), \emptyset) \times (D(E_2), \emptyset) & \xrightarrow{j_1 \times j_2} & (D(E_1), S(E_1)) \times (D(E_2), S(E_2)) \end{array} .$$

In the same way, if $b = (b_1, b_2) \in B$, there is a homeomorphism of pairs

$$\big(D(E)_b, S(E)_b\big) \approx \big(D(E_1)_{b_1}, S(E_1)_{b_1}\big) \times \big(D(E_2)_{b_2}, S(E_2)_{b_2}\big) . \qquad (4.7.23)$$

By the relative Künneth theorem 4.6.10, the generator of $H^r(D(E)_b, S(E)_b) = \mathbb{Z}_2$ is the cross product of the generators of $H^{r_i}(D(E_i)_{b_i}, S(E_i)_{b_i})$. Using Lemma 4.7.28, we deduce that

$$U(\xi_1 \times \xi_2) = U(\xi_1) \times U(\xi_2) . \qquad (4.7.24)$$

Using Lemma 4.6.2, one has

$$\begin{aligned} H^*(p_1 \times p_2)(e(\xi)) &= H^*j(U(\xi)) \\ &= H^*j(U(\xi_1) \times U(\xi_2)) \\ &= H^*j_1(U(\xi_1)) \times H^*j_2(U(\xi_2)) \\ &= H^*p_1(e(\xi_1)) \times H^*p_2(e(\xi_2)) \\ &= H^*(p_1 \times p_2)(e(\xi_1) \times e(\xi_2)) . \end{aligned}$$

As $H^*(p_1 \times p_2)$ is an isomorphism, this proves (1). Point (2) is deduced from (1) using the definition of $\xi_1 \oplus \xi_2$ and Remark 4.6.1:

$$e(\xi_1 \oplus \xi_2) = H^*\Delta(e(\xi_1 \times \xi_2)) = H^*\Delta(e(\xi_1) \times e(\xi_2)) = e(\xi_1) \smile e(\xi_2). \qquad \square$$

The Thom class of a product bundle was computed in (4.7.24). For the Whitney sum, we use the projections $\pi_i : E(\xi_1 \oplus \xi_2) \to E(\xi_i)$.

Proposition 4.7.41 *Let ξ_1 and ξ_2 be two vector bundles over a paracompact basis. Let $U(\xi_i) \in H^{r_i}(D(E_i), S(E_i))$ be the Thom classes (for an Euclidean structure). Then*

$$U(\xi_1 \oplus \xi_2) = H^*\pi_1(U(\xi_1)) \smile H^*\pi_2(U(\xi_2)).$$

Proof Restricted to the fiber over $b \in B$, the right hand side of the formula gives the cross product of the generators of $H^{r_i}(D(E_i)_b, S(E_i)_b)$. The latter is the generator of $H^r(D(E)_b, S(E)_b)$. The proposition thus follows from Lemma 4.7.28. $\square$

4.7.7 Bundles Over Spheres

In this section, we study bundles $\xi = (p : E \to S^m)$ over the sphere S^m with fiber F. If $A \subset S^m$ we set $E_A = p^{-1}(A)$. Consider the cellular decomposition of S^m with one 0-cell b and one m-cell with characteristic map $\varphi : D^m \to S^m$ sending S^{m-1} onto b. We denote by $\phi : S^{m-1} \to \{b\}$ this constant map. We identify F with E_b, getting thus an inclusion $i : F \hookrightarrow E$. As D^m is contractible, any bundle over D^m is trivial [181, Corollary 11.6]. Therefore, there exists a trivialization $\varphi^*E \approx D^m \times F$ of the induced bundle $\varphi^*\xi$. The map (φ, ϕ) are covered by a bundle maps $\tilde{\varphi} : D^m \times F \to E$ and $\tilde{\phi} : S^{m-1} \times F \to F$. The latter satisfies, for each $x \in S^{m-1}$, that $\tilde{\phi} : \{x\} \times F \xrightarrow{\approx} F$ is a homeomorphism. Observe that

$$E = (D^m \times F) \cup_{\tilde{\varphi}} F. \tag{4.7.25}$$

Let $x_0 \in S^{m-1}$ be the base point corresponding to $1 \in S^0 \subset S^{m-1}$. By changing the trivialization of $\varphi^*\xi$ if necessary, we shall assume that $\tilde{\phi} : \{x_0\} \times F \to F$ is the projection onto F. The map $\tilde{\varphi}$ is called the *bundle characteristic map* and the map $\tilde{\phi}$ is called the *bundle gluing map* of the bundle ξ.

Lemma 4.7.42 *The bundle characteristic map $\tilde{\varphi} : D^m \times F \to E$ induces an isomorphism*

$$H^*\tilde{\varphi} : H^*(E, F) \xrightarrow{\approx} H^*(D^m \times F, S^{m-1} \times F).$$

Proof Consider the decomposition $D^m = B \cup C$, where B is the disk with center 0 and radius $1/2$ and C the closure of $D^m - B$; let $S = B \cap C$. As $\varphi(C)$ is a disk around

b, the bundle ξ is trivial above $\varphi(C)$: $E_{\varphi(C)} \approx \varphi(C) \times F$. As $\tilde{\varphi} : B \times F \to E_\varphi(B)$ is a homeomorphism, the lemma follows from the commutative diagram

$$\begin{array}{ccc} H^*(E, F) & \xrightarrow{H^*\tilde{\varphi}} & H^*(D^m \times F, S^{m-1} \times F) \\ \approx \uparrow \text{excision} & & \approx \uparrow \text{excision} \\ H^*(E, E_{\varphi(C)}) & & H^*(D^m \times F, C \times F) \\ \approx \downarrow \text{excision} & & \approx \downarrow \text{excision} \\ H^*(E_{\varphi(B)}, E_{\varphi(S)}) & \xrightarrow[\approx]{H^*\tilde{\varphi}} & H^*(B \times F, S \times F) \end{array}$$

□

Proposition 4.7.43 *Let $p : E \to S^m$ be a bundle with fiber F. There is a long exact sequence*

$$\cdots \to H^{k-1}(E) \xrightarrow{H^*i} H^{k-1}(F) \xrightarrow{\Theta} H^{k-m}(F) \xrightarrow{J} H^k(E) \xrightarrow{H^*i} H^k(F) \to \cdots .$$

The exact sequence of Proposition 4.7.43 is called the *Wang exact sequence.*

Proof We start with the exact sequence of the pair (E, F)

$$\cdots \to H^{k-1}(E) \xrightarrow{H^*i} H^{k-1}(F) \xrightarrow{\delta^*} H^k(E, F) \xrightarrow{H^*j} H^k(E) \to \cdots \quad (4.7.26)$$

where $j : (E, \emptyset) \to (E, F)$ denotes the pair inclusion. The following commutative diagram defines the homomorphism Θ and J.

$$\begin{array}{ccccc} H^{k-1}(F) & \xrightarrow{\delta^*} & H^k(E, F) & \xrightarrow{H^*j} & H^k(E) \\ & \searrow & \approx \downarrow H^*\tilde{\varphi} & \nearrow & \\ & \Theta & H^k(D^m \times F, S^{m-1} \times F) & J & \\ & & \approx \uparrow e\times - & & \\ & & H^{k-m}(F) & & \end{array} \quad (4.7.27)$$

Here, $H^*\tilde{\varphi}$ is an isomorphism by Lemma 4.7.42, $e \in H^m(D^m, S^{m-1}) = \mathbb{Z}_2$ is the generator and the map $e \times -$ is an isomorphism by the relative Künneth theorem 4.7.25. □

We now give some formulae satisfied by the homomorphism $\Theta : H^{k-1}(F) \to H^{k-m}(F)$. We start with the case $m = 1$ which deserves a special treatment. The bundle gluing map $\phi : S^0 \times F \to F$ satisfy $\phi(1, x) = x$ and $\phi(-1, x) = h(x)$ for some homeomorphism $h : F \to F$. The decomposition of (4.7.25) amounts to say that E is the *mapping torus* M_h of h:

$$E = M_h = \big([-1,1]\times F\big)/\{(1,x)\sim(-1,h(x))\}.$$

The bundle projection $p : M_h \to S^1$ is given by $p(t,x) = \exp(2i\pi t)$. The correspondence $x \to [(x,0)]$ gives an inclusion $j : F \hookrightarrow M_h$. Let $e \in H^1(S^1) = \mathbb{Z}_2$ be the generator. Proposition 4.7.43 may be rephrased and made more explicit in the following way.

Proposition 4.7.44 (Mapping torus exact sequence) *Let $h : F \to F$ be a homeomorphism. Then, there is a long exact sequence*

$$\cdots H^{k-1}(M_h) \xrightarrow{H^*i} H^{k-1}(F) \xrightarrow{\Theta} H^{k-1}(F) \xrightarrow{J} H^k(M_h) \xrightarrow{H^*i} H^k(F) \to \cdots,$$

with $\Theta = \mathrm{id} + H^*h$.

Proof We use the exact sequence (4.7.26) with $E = M_h$ and Diagram (4.7.27). It remains to identify Θ with $\mathrm{id} + H^*h$. Let $i_\pm : \{\pm 1\}\times F \to S^0 \times F$ denote the inclusions. Let $\alpha : H^{k-1}(F) \to H^{k-1}(S^0\times F)$ be the homomorphism such that $H^*i_+\circ\alpha(a) = a$ and $H^*i_-\circ\alpha(a) = H^*h(a)$. Consider the diagram.

$$\begin{array}{ccccc}
H^{k-1}(F) & \xrightarrow{\alpha} & H^{k-1}(S^0\times F) & \xrightarrow[\approx]{i^*} & H^{k-1}(\{1\}\times F)\oplus H^{k-1}(\{-1\}\times F) \\
\downarrow{\scriptstyle \delta^*} & & \downarrow{\scriptstyle \delta^*} & & \downarrow{\scriptstyle +} \\
H^k(M_h, F) & \xrightarrow[\approx]{H^*\tilde{\varphi}} & H^k(D^1\times F, S^0\times F) & \xleftarrow[\approx]{e\times -} & H^{k-1}(F)
\end{array}$$

where $i^* = (H^*i_+, H^*i_-)$. Let $\Psi_\pm : H^{k-1}(F) \to H^{k-1}(F)$ be the composed homomorphisms through the upper right or lower left corners. Then $\psi_+ = \mathrm{id} + H^*h$ and $\psi_- = \Theta$. The left square of the diagram being commutative by construction of M_h, it then suffices to prove the commutativity of the right square, that is $\delta^* = \psi$, where $\psi(a) = e\times\big(H^*i_+(a) + H^*i_-(a)\big)$. The homomorphisms $\delta*$ and ψ are both functorial. As, by Sect. 3.8, a class $a \in H^{k-1}(F)$ is represented by a map $F \to \mathcal{K}_{k-1}$, it suffices to prove that $\delta^* = \psi$ for $F = \mathcal{K}_{k-1}$. Observe that $\delta*$ and ψ are both surjective and have the same kernel, the image of $H^{k-1}(D^1\times F) \to H^{k-1}(S^0\times F)$. As $H^{k-1}(\mathcal{K}_{k-1}) = \mathbb{Z}_2$, this proves that $\delta^* = \psi$ when $F = \mathcal{K}_{k-1}$. □

Example 4.7.45 Let $h : S^1 \to S^1$ be the complex conjugation. Then, M_h is homeomorphic to the Klein bottle K and we get a bundle $S^1 \to K \xrightarrow{p} S^1$. The homomorphism Θ of Proposition 4.7.44 satisfies $\Theta = \mathrm{id} + H^*h = 0$. By the mapping torus exact sequence, we deduce that $H^*(K) \to H^*(S^1)$ is surjective (this can also be obtained using a triangulation like on p. 29 and computations like on p. 138). A cohomology extension of the fiber $\sigma : H^*(S^1) \to H^*(K)$ produces, by the Leray-Hirsch theorem 4.7.17, a **GrV**-isomorphism $\hat{\sigma} : H^*(S^1)\otimes H^*(S^1) \xrightarrow{\approx} H^*(K)$.

But $\hat{\sigma}$ is not a morphism of algebra. Indeed, the square map $x \mapsto x \bullet x$ vanishes in $H^*(S^1) \otimes H^*(S^1)$ while the cup-square map $x \mapsto x \smile x$ does not vanish in $H^*(K)$ (see, Proposition 4.2.3).

When $m > 1$, some information about the homomorphism $\Theta : H^{k-1}(F) \to H^{k-m}(F)$ may be obtained via the composition

$$H^{k-1}(F) \xrightarrow{\Theta} H^{k-m}(F) \xrightarrow{\dot{e}\times -} H^{k-1}(S^{m-1} \times F) \ ,$$

where $\dot{e} \in H^{m-1}(S^{m-1}) = \mathbb{Z}_2$ is the generator. The map $\dot{e} \times -$ is injective by the Künneth theorem.

Proposition 4.7.46 *Suppose that $m > 1$. Then*

$$\dot{e} \times \Theta(a) = H^*\tilde{\phi}(a) - H^* p_2(a) ,$$

where $\tilde{\phi}$, $p_2 : S^{m-1} \times F \to F$ are the bundle gluing map and the projection onto F.

Proof As F is a retract of $S^{m-1} \times F$, the cohomology sequence of the pair $(S^{m-1} \times F, F)$ splits into short exact sequences and, by the Künneth theorem and Lemma 4.7.2, there is a commutative diagram

$$\begin{array}{ccccccccc}
0 & \longrightarrow & H^{k-m}(F) & \longrightarrow & H^{k-m}(F) \oplus H^{k-1}(F) & \xrightarrow{+} & H^{k-1}(F) & \longrightarrow & 0 \\
 & & \downarrow \approx & \searrow^{\dot{e}\times -} & \downarrow \approx\ \alpha & & \downarrow \approx\ \mathrm{id} & & \\
0 & \longrightarrow & H^{k-1}(S^{m-1} \times F, F) & \longrightarrow & H^{k-1}(S^{m-1} \times F) & \xrightarrow{H^*i} & H^{k-1}(F) & \longrightarrow & 0
\end{array} \qquad (4.7.28)$$

where $i : F \to S^{m-1} \times F$ is the slice inclusion at the base point $x_0 \in S^{m-1}$ and $\alpha(a, b) = \dot{e} \times a + \mathbf{1} \times b$. Recall that we assume the restriction of $\tilde{\phi}$ to $\{x_0\} \times F$ to coincide with the projection p_2. Therefore, the composition

$$H^{k-1}(F) \xrightarrow{H^*\tilde{\phi} - H^* p_2} H^{k-1}(S^{m-1} \times F) \to H^{k-1}(F)$$

vanishes. Using Diagram (4.7.28), we get a factorization

$$\begin{array}{ccc}
H^{k-1}(F) & \xrightarrow{H^*\tilde{\phi} - H^* p_2} & H^{k-1}(S^{m-1} \times F) \\
 & \searrow^{\Theta'} \quad \nearrow_{\dot{e}\times -} & \\
 & H^{k-m}(F) &
\end{array}$$

which we introduce in the diagram

$$\begin{array}{ccc} H^{k-1}(F) & \xrightarrow{\delta^*} & H^k(E,F) \\ \big\downarrow {\scriptstyle H^*\tilde{\phi}+H^*p_2} & & \approx \big\downarrow {\scriptstyle H^*\tilde{\varphi}} \\ H^{k-1}(S^{m-1}\times F) & \xrightarrow{\delta^*} & H^k(D^m\times F, S^{m-1}\times F) \\ \big\uparrow {\scriptstyle \dot{e}\times -} & & \approx \big\uparrow {\scriptstyle e\times -} \\ H^{k-m}(F) & \xrightarrow{=} & H^{k-m}(F) \end{array} \quad (\Theta' : H^{k-1}(F) \to H^{k-m}(F)) \qquad (4.7.29)$$

We claim that the two square of Diagram (4.7.29) are commutative. By Diagram (4.7.27), this will imply that $\Theta' = \Theta$ and will prove the lemma.

As $\tilde{\phi}$ is the restriction of $\tilde{\varphi}$, the naturality of the connecting homomorphism δ^* implies that $\delta^* \circ H^*\tilde{\phi} = H^*\tilde{\varphi} \circ \delta^*$. Since p_2 extends to $D^n \times F$, the homomorphism $H^*p_2 : H^{k-1}(F) \to H^{k-1}(S^{m-1} \times F)$ factors through $H^{k-1}(D^m \times F)$ and thus $\delta^* \circ H^*p_2 = 0$. Hence, the top square is commutative. For the bottom one, let $a \in H^{k-1}(F)$. By Sect. 3.8, $a = H^*f(\iota)$ for some map f from F into the Eilenberg-MacLane space $\mathcal{K}_{k-1}$. The bottom square being functorial for the map f, it suffices to prove its commutativity for $F = \mathcal{K}_{k-1}$. As the source and range vector space are both then isomorphic to $\mathbb{Z}_2$, the commutativity holds trivially. □

As an exercise, the reader may adapt the proof of Proposition 4.7.46 to the case $m = 0$, thus getting an alternative proof of Proposition 4.7.44. The main point is to replace $\dot{e}$ (which has no meaning in $H^0(S^0)$ by the class of $\{-1\}$.

The family of homomorphisms $\Theta : H^{k-1}(F) \to H^{k-m}(F)$ forms an endomorphism of $H^*(F)$ of degree $m-1$ (it sends $H^q(F)$ to $H^{q-m+1}(F)$).

Proposition 4.7.47 *As an endomorphism of $H^*(F)$, Θ satisfies*

$$\Theta(a \smile b) = \Theta(a) \smile b + a \smile \Theta(b) .$$

Proof Proposition 4.7.46 may be rephrased as

$$H^*\tilde{\phi}(a) = H^*p_2(a) + \dot{e} \times \Theta(a) = \mathbf{1} \times a + \dot{e} \times \Theta(a) .$$

Therefore, if $a \in H^p(F)$ and $b \in H^q(F)$,

$$H^*\tilde{\phi}(a \smile b) = \mathbf{1} \times (a \smile b) + \dot{e} \times \Theta(a \smile b)$$

and, using Lemma 4.6.3,

$$\begin{aligned} H^*\tilde{\phi}(a) \smile H^*\tilde{\phi}(b) &= \big[\mathbf{1} \times a + \dot{e} \times \Theta(a)\big] \smile \big[\mathbf{1} \times b + \dot{e} \times \Theta(b)\big] \\ &= \mathbf{1} \times (a \smile b) + \dot{e} \times (\Theta(a) \smile b) + \dot{e} \times (a \smile \Theta(b)) \\ &= \mathbf{1} \times (a \smile b) + \dot{e} \times \big[\Theta(a) \smile b + a \smile \Theta(b)\big] . \end{aligned}$$

As $H^*\tilde{\phi}(a \smile b) = H^*\tilde{\phi}(a) \smile H^*\tilde{\phi}(b)$ and $e \times -$ is injective, this proves the proposition. □

Remark 4.7.48 The material of this section was inspired by [207, Sect. 1, Chap. VII]. As in this this reference, the following fact can also be proved:

(1) The Wang exact sequence holds for Serre fibrations. It also has a generalization to bundles over a suspension.
(2) A Wang exact sequence for homology exists.

Further properties of the Wang sequences are given in [207, Sect. 2, Chap. VII].

4.7.8 The Face Space of a Simplicial Complex

Let K be simplicial complex. Fix an integer $d > 0$. For each $v \in V(K)$, consider a copy S^d_v of the sphere S^d. It is pointed by $e_1 = (1, 0, \dots, 0) \in S^d_v$. For $\sigma \in \mathcal{S}(K)$, consider the space

$$\mathbf{F}_d(\sigma) = \{(z_v) \mid z_v = e_1 \text{if} v \notin \sigma\} \subset \prod_{v \in V(K)} S^d_v ,$$

which is homeomorphic to $\prod_{v\in\sigma} S^d_v$. The *face space* of K is the subset of $\prod_{v\in V(K)} S^d_v$ defined by

$$\mathbf{F}_d(K) = \bigcup_{\sigma\in\mathcal{S}(K)} \mathbf{F}_d(\sigma) \subset \prod_{v\in V(K)} S^d_v .$$

Remark 4.7.49 Let K be a flag simplicial complex (i.e. if K contains a graph L isomorphic to the 1-skeleton of an r-simplex, then L is contained in an r-simplex of K). Then the complex $\mathbf{F}_1(K)$ is the Salvetti complex of the right-angled Coxeter group determined by the 1-skeleton of K (see [30]).

The interest of the face space appears in the following proposition, based on a algebraic theorem of Gubeladze.

Proposition 4.7.50 *Let K and K' be two finite simplicial complexes. Let d be a positive integer. Then, K is isomorphic to K' if and only if $H^*(\mathbf{F}_d(K))$ and $H^*(\mathbf{F}_d(K'))$ are* **GrA**-*isomorphic.*

To explain the proof of Proposition 4.7.50, we compute the cohomology algebra of $\mathbf{F}_d(K)$ for a finite simplicial complex K. Let us number the vertices of K: $V(K) = \{1, \dots, m\}$. Consider the polynomial ring $\mathbb{Z}_2[x_1, \dots, x_m]$ with formal variables $x_1, \dots, x_m$ which are of degree d. If $J \subset \{1, \dots, m\}$, we denote by $x_J \in \mathbb{Z}_2[x_1, \dots, x_m]$ the monomial $\prod_{j\in J} x_j$. Let $\mathcal{I}(K)$ be the ideal of $\mathbb{Z}_2[x_1, \dots, x_m]$

generated by the squares x_i^2 of the variables and the monomials x_J for $J \notin \mathcal{S}(K)$ (non-face monomials). The quotient algebra

$$\Lambda_d(K) = \mathbb{Z}_2[x_1, \dots, x_m]/\mathcal{I}(K)$$

is called the *face exterior algebra* (because $u^2 = 0$ for all $u \in \Lambda_d(K)$; however, because the ground field is $\mathbb{Z}_2$, $\Lambda_d(K)$ is commutative).

Lemma 4.7.51 *The ring $H^*(\mathbf{F}_d(K))$ is isomorphic to $\Lambda_d(K)$.*

Proof (Compare [59, Proposition 4.3].) Let $\Delta_K = \mathcal{F}(V(K))$ be the full complex over the set $V(K) = \{1, \dots, m\}$. The simplicial inclusion $K \subset \Delta_K$ induces an inclusion

$$j : \mathbf{F}_d(K) \hookrightarrow \mathbf{F}_d(\Delta_K) = \prod_{v \in V(K)} S_v^d .$$

For $\sigma \subset \{1, \dots, m\}$, the fundamental class $[\mathbf{F}_d(\sigma)] \in H_{(\dim \sigma+1)d}(\mathbf{F}_d(\sigma))$ determines a class $[\sigma] \in H_{(\dim \sigma+1)d}(\mathbf{F}_d(\Delta_K))$ (by convention, $[\emptyset]$ is the generator of $H_0(\mathbf{F}_d(\Delta_K))$). If $\sigma \in \mathcal{S}(K)$, $[\sigma]$ is the image under H_*j of a class in $H_{(\dim \sigma+1)d}(\mathbf{F}_d(K))$, also called $[\sigma]$. Let

$$A = \{[\sigma] \in H_*(\mathbf{F}_d(K)) \mid \sigma \in \mathcal{S}(K) \cup \{\emptyset\}\} \subset H_*(\mathbf{F}_d(K))$$

and

$$B = \{[\sigma] \in H_*(\mathbf{F}_d(\Delta_K)) \mid \sigma \subset \{1, \dots, m\}\} \subset H_*(\mathbf{F}_d(\Delta_K)) .$$

By the Künneth theorem and Corollary 3.1.16, $H_*(\mathbf{F}_d(K))$ is generated by A and B is a basis of $H_*(\mathbf{F}_d(\Delta_K))$. It follows that A is a basis of $H_*(\mathbf{F}_d(K))$ and that H_*j is injective. By Kronecker duality, H^*j is surjective and the Kronecker-dual basis $B^\sharp$ of B is sent onto Kronecker-dual basis $A^\sharp$ of A by

$$H^*j([\sigma]^\sharp) = \begin{cases} [\sigma]^\sharp & \text{if } \sigma \in \mathcal{S}(K) \\ 0 & \text{otherwise.} \end{cases} \tag{4.7.30}$$

By the Künneth theorem again,

$$H^*(\mathbf{F}_d(\Delta_K)) \approx \mathbb{Z}_2[x_1, \dots, x_m]/(x_1^2, \dots, x_m^2) \tag{4.7.31}$$

and, if $\sigma \subset \{1, \dots, m\}$, then $[\sigma]^* = x_\sigma$. By (4.7.30), $\ker H^*j$ is the $\mathbb{Z}_2$-vector space in $H^*(\mathbf{F}_d(\Delta_K))$ with basis $\{x_\sigma \mid \sigma \notin \mathcal{S}(K)\}$. Using (4.7.31), we check that, under the epimorphism $\mathbb{Z}_2[x_1, \dots, x_m] \twoheadrightarrow H^*(\mathbf{F}_d(\Delta_K))$, $\ker H^*j$ is the image of $\mathcal{I}(K)$. □

The proof of Lemma 4.7.51 provides the following corollary.

Corollary 4.7.52 *The Poincaré polynomial of the algebra* $\Lambda_d(K)$ *is*

$$\mathfrak{P}_t(\Lambda_d(K)) = 1 + \sum_{\sigma \in \mathcal{S}(K)} t^{(\dim \sigma + 1)d} .$$

The proof of Proposition 4.7.50 follows from Lemma 4.7.51 and the following theorem of Gubeladze. For a proof, see [76, Theorem 3.1].

Theorem 4.7.53 (Gubeladze) *Let* K *and* K' *be two finite simplicial complexes. Suppose that* $\Lambda_d(K) = \mathbb{Z}_2[x_1, \dots, x_m]/\mathcal{I}(K)$ *and* $\Lambda_d(K') = \mathbb{Z}_2[y_1, \dots, y_{m'}]/\mathcal{I}(K')$ *are isomorphic as graded algebras. Then* $m = m'$ *and there is a bijection*

$$\phi : \{x_1, , \dots, x_m\} \xrightarrow{\approx} \{y_1, \dots, y_{m'}\}$$

such that $\phi(\mathcal{I}(K)) = \mathcal{I}(K')$.

4.7.9 Continuous Multiplications on $K(\mathbb{Z}_2, m)$

A continuous multiplication $\mu : X \times X \to X$ on a space X is *homotopy commutative* if the maps $(x, y) \mapsto \mu(x, y)$ and $(x, y) \mapsto \mu(y, x)$ are homotopic. A element $u \in X$ is a *homotopy unit* for μ if the maps $x \mapsto \mu(u, x)$ and $x \mapsto \mu(x, u)$ are homotopic to the identity of X. Note that, if $u_0 \in X$ is a homotopy unit for μ and if X is path-connected, then any $u \in X$ is also a homotopy unit.

Let $\mathcal{K} \approx K(\mathbb{Z}_2, m)$ be an Eilenberg-MacLane space in degree m, with its class $0 \neq \iota \in H^m(\mathcal{K})$. Recall from Sect. 3.8, the map $\phi : [X, \mathcal{K}] \to H^m(X)$ given by $\phi(f) = H^* f(\iota)$ is a bijection. In particular, if $\mathcal{K}$ and $\mathcal{K}'$ are two Eilenberg-MacLane spaces in degree m, there is a homotopy equivalence $g : \mathcal{K}' \to \mathcal{K}$ whose homotopy class is unique.

Proposition 4.7.54 *Let* $\mathcal{K}$ *be an Eilenberg-MacLane space in degree* m.

(1) *There exists a continuous multiplication on* $\mathcal{K}$ *admitting a homotopy unit and which is homotopy commutative.*
(2) *Any two continuous multiplications on* $\mathcal{K}$ *admitting a homotopy unit are homotopic.*
(3) *Let* $(\mathcal{K}, \mu)$ *and* $\mathcal{K}', \mu')$ *be two Eilenberg-MacLane spaces in degree* m *with continuous multiplications admitting homotopy units, Let* $g : \mathcal{K}'_m \to \mathcal{K}_m$ *a (unique up to homotopy) homotopy equivalence. Then, the diagram*

$$\begin{array}{ccc} \mathcal{K}' \times \mathcal{K}' & \xrightarrow{\mu'} & \mathcal{K}' \\ \downarrow{\scriptstyle g \times g} & & \downarrow{\scriptstyle g} \\ \mathcal{K} \times \mathcal{K} & \xrightarrow{\mu} & \mathcal{K} \end{array}$$

commutes up to homotopy.

Proof Consider the class

$$p = \iota \times \mathbf{1} + \mathbf{1} \times \iota \in H^m(\mathcal{K} \times \mathcal{K}) . \tag{4.7.32}$$

Since $[\mathcal{K} \times \mathcal{K}, \mathcal{K}]$ is in bijection with $H^m(\mathcal{K} \times \mathcal{K})$, one has $p = H^*\mu(\iota)$ for some continuous map $\mu : \mathcal{K} \times \mathcal{K} \to \mathcal{K}$, which we see as a continuous multiplication. The involution τ exchanging the coordinates on $\mathcal{K} \times \mathcal{K}$ satisfies $H^*\tau(p) = p$ and then $H^*(\mu \circ \tau) = H^*\mu$. Hence, $\mu \circ \tau$ is homotopic to μ, which says that μ is homotopy commutative.

Choose $u \in \mathcal{K}$ and let $i_1, i_2 : \mathcal{K} \to \mathcal{K} \times \mathcal{K}$ be the slice inclusions $i_1(x) = (x, u)$ and $i_2(x) = (u, x)$. By Lemma 4.7.2, $i_j^* \circ H^*\mu(\iota) = \iota$ for $j = 1, 2$. Hence, $\mu \circ i_j$ is homotopic to the identity, which proves that u is a homotopy unit. Point (1) is thus established.

For Point (2), let μ is continuous multiplication on $\mathcal{K}$ admitting a homotopy unit u. Let $i_1, i_2 : \mathcal{K} \to \mathcal{K} \times \mathcal{K}$ be the slice inclusions $i_1(x) = (x, u)$ and $i_2(x) = (u, x)$. As u is a homotopy unit, $h \circ i_j$ is homotopic to the identity for $j = 1, 2$, and thus $H^*i_j \circ H^*\mu(a) = a$ for all $a \in H^*(X)$. By Lemma 4.7.2, this implies that

$$H^*\mu(a) = a \times \mathbf{1} + \mathbf{1} \times a + \sum y \times y' , \tag{4.7.33}$$

where the degrees of y and y' are both positive. By the Künneth theorem, the cross product gives an isomorphism isomorphism $H^m(\mathcal{K}) \otimes H^0(\mathcal{K}) \oplus H^0(\mathcal{K}) \otimes H^m(\mathcal{K}) \approx H^m(\mathcal{K})$. Therefore, $H^*\mu(\iota) = p$, which says that the homotopy class of μ is well determined.

For Point (3), let $h : K \to K'$ be a homotopy inverse for g. Then, the formula $\mu''(x, y) = h \circ \mu(g(x), g(y))$ is a continuous multiplication of $\mathcal{K}'$ with a homotopy unit. By (2), μ'' is homotopic to μ', which proves (3). □

Examples 4.7.55 The following classical multiplications occur in Eilenberg-MacLane spaces $\mathcal{K}_m \approx K(\mathbb{Z}_2, m)$ (or more generally on $K(G, m)$ for an abelian group G).

- The loop space $\Omega\mathcal{K}_{m+1}$ is an Eilenberg-MacLane space in degree m [82, pp. 407 and ff.]. One can use the loop multiplication.
- Using semi-simplicial techniques, Milnor has shown that there exists an Eilenberg-MacLane space $\mathcal{K}_m$ which is an abelian topological group [146, Sect. 3].

The following property of the multiplication μ of Proposition 4.7.54 will be useful in Sect. 8.3.

Lemma 4.7.56 *Let $\mathcal{K}$ be an Eilenberg-MacLane space in degree m. Let $a \in H^k(\mathcal{K})$ for $m \leq k < 2m$. Then*

$$H^*\mu(a) = a \times \mathbf{1} + \mathbf{1} \times a . \tag{4.7.34}$$

Proof This comes from (4.7.33) since

$$H^k(\mathcal{K}) \otimes H^0(\mathcal{K}) \oplus H^0(\mathcal{K}) \otimes H^k(\mathcal{K}) \xrightarrow{\approx} H^k(\mathcal{K} \times \mathcal{K}) \tag{4.7.35}$$

for $m \leq k < 2m$ by the Künneth theorem. □

Remark 4.7.57 Together with the cup product, the map $H^*\mu$ makes $H^*(K(\mathbb{Z}_2, m))$ a *Hopf algebra* (see [82, Sect. 3.C]). In this setup, an element $a \in H^k(K(\mathbb{Z}_2, m))$ satisfying (4.7.34) is called *primitive*.

Let X be a CW-complex. The multiplication μ on $\mathcal{K} = K(\mathbb{Z}_2; m)$ induces a composition law

$$[X, \mathcal{K}] \times [X, \mathcal{K}] \xrightarrow{\star} [X, \mathcal{K}]$$

given by $f \star g\,(x) = \mu(f(x), g(x))$. It admits the following interpretation.

Proposition 4.7.58 *Let X be a CW-complex. Then, the bijection $\phi : H^m(X) \xrightarrow{\approx} [X, K]$ satisfies*

$$\phi(a) \star \phi(b) = \phi(a + b)\,.$$

for all $a, b, \in H^m(X)$.

Proof Let $f, g : X \to \mathcal{K}$ represent $\phi(a)$ and $\phi(b)$. Then $\phi(a) \star \phi(b)$ is represented by the composition

$$X \xrightarrow{(f,g)} \mathcal{K} \times \mathcal{K} \xrightarrow{\mu} \mathcal{K}\,.$$

The two projections $\pi_1, \pi_2 : \mathcal{K} \times \mathcal{K} \to \mathcal{K}$ satisfy $\pi_1 \circ (f, g) = f$ and $\pi_2 \circ (f, g) = g$. Using that $H^*\mu(\iota) = \iota \times \mathbf{1} + \mathbf{1} \times \iota$ (see the proof of Proposition 4.7.54), one has

$$\begin{aligned}
\phi(a) \star \phi(b) &= H^*(f, g) \circ H^*\mu(\iota) \\
&= H^*(f, g)(\iota \times \mathbf{1} + \mathbf{1} \times \iota) \\
&= H^*(f, g)(H^*\pi_1(\iota) + H^*\pi_2(\iota)) \\
&= H^* f(\iota) + H^* g(\iota) \\
&= \phi(a) + \phi(b)\,.
\end{aligned}$$

□

4.8 Exercises for Chapter 4

4.1. Write the proof of Lemma 4.1.8.

4.2. As $H^*(S^1 \vee S^1)$ has 4 elements, the bouquet of two circle has 4 inequivalent 2-fold coverings by the bijection (4.3.5). For each of them, describe the total space and the transfer exact sequence.

4.3. Same exercise as the previous one, replacing $S^1 \vee S^1$ by the Klein bottle. Compare with the discussion on p. 33.
4.4. Write the transfer exact sequence for a trivial 2-fold covering.
4.5. Let $p : \tilde{X} \to X$ be finite covering with an odd number of sheets. Prove that $H^* p$ is injective.
4.6. Let M and N be closed surfaces, with M orientable and N non-orientable. Prove that there is no continuous map $f : M \to N$ which is of degree one.
4.7. Show that there are no continuous map of degree one between the torus T and the Klein bottle K, in either direction. Same things for $S^1 \times S^2$ and $\mathbb{R}P^3$.
4.8. Let M be a closed topological manifold of dimension n. Let $h : D^n \to M$ be an embedding of the closed disk D^n into M. Form the manifold $\hat{M}$ as the quotient of $M - \operatorname{int} h(D^n)$ by the identification $h(x) \sim h(-x)$ for $x \in \operatorname{Bd} D^n$. Compute the ring $H^*(\hat{M})$. [Hint: express $\hat{M}$ as a connected sum.]
4.9. Show that the cohomology algebras of $(S^1 \times S^1) \sharp \mathbb{R}P^2$ and of $\mathbb{R}P^2 \sharp \mathbb{R}P^2 \sharp \mathbb{R}P^2$ are **GrA**-isomorphic. (It is classical that these two spaces are homeomorphic: see [136, Lemma 7.1]).
4.10. Using the triangulation of the Klein bottle given in Fig. 2.4, compute all the simplicial cap products. Check the formula $\langle a \smile b, \gamma \rangle = \langle a, b \frown \gamma \rangle$.
4.11. Show that the smash product and the join of two homology spheres is a homology sphere.
4.12. Compute the cohomology ring of (a) $X = \mathbb{R}P^\infty \times \cdots \times \mathbb{R}P^\infty$ (n times); (b) $Y = \mathbb{C}P^2 \wedge \mathbb{C}P^3$; (c) $Z = \mathbb{C}P^2 * \mathbb{C}P^3$.
4.13. Write the Mayer-Vietoris cohomology sequence for the decomposition

$$S^1 \times S^n = [(S^1 - \{1\})) \times S^n] \times [(S^1 - \{-1\})) \times S^n].$$

and describe its various homomorphisms. If $a \in H^1(S^1)$ and $b \in H^n(S^n)$ are the generators, describe how the elements $a \times 1$, $1 \times b$ and $a \times b$ behave with respect to the homomorphisms of the Mayer-Vietoris sequence.
4.14. Show that the product of two perfect CW-complexes is a perfect CW-complex.
4.15. What is the Lusternik-Schnirelmann category of $\mathbb{R}P^2 \times \mathbb{R}P^3$?
4.16. What is the Lusternik-Schnirelmann category of the n-dimensional torus $T^n = S^1 \times \cdots \times S^1$ (n times)?
4.17. Prove the relevant functoriality property for the homology cross product.
4.18. For m a positive integer, let $B(m)$ be a bouquet of m circles. Let $X = \prod_{i=1}^r B(a_i)$ and $Y = \prod_{j=1}^s B(b_j)$. Suppose that X and Y have the same Poincaré polynomial. Prove that $r = s$ and that $b_i = a_{\alpha(i)}$ for some permutation α.
4.19. *Cap product in the (co)homology of $X \times Y$.* Let X and Y be topological spaces, with Y being of finite cohomology type. Let $a \in H^*(X)$, $b \in H^*(Y)$, $\alpha \in H_*(X)$ and $\beta \in H_*(Y)$. Prove that the formula

$$\underline{\times}\big((a \times b) \frown \underline{\times}^{-1}(\alpha \otimes \beta)\big) = (a \frown \alpha) \otimes (b \frown \beta) \qquad (4.8.1)$$

holds in $H_*(X) \otimes H_*(Y)$, using the (co)homology cross products $\times$ and $\underline{\times}$ of Sect. 4.6.

4.20. *Slices in homology.* Let X and Y be topological spaces, with Y being of finite cohomology type. Let $y_0 \in Y$ and let $s_X : X \to X \times Y$ be the slice inclusion of X at y_0. Let $\alpha \in H_*(X)$. Prove that

$$\underline{\times}\big(H_* s_X(\alpha)\big) = \alpha \otimes y_0 \,,$$

where y_0 is seen as a 0-homology class of Y, using the bijection $Y \approx \mathcal{S}_0(Y)$.

4.21. Let $F \to E \xrightarrow{p} B$ be a locally trivial bundle containing a subbundle $F_0 \to E_0 \xrightarrow{p_0} B$. Prove that the cohomology sequence of (E, E_0) is a sequence of $H^*(B)$-modules.

4.22. Let K be a finite simplicial complex and let $\mathbf{F}_d(K)$ its face complex for an integer $d > 0$. What is the relationship between the Euler characteristic of $\mathbf{F}_d(K)$ and that of K?

4.23. Let $p : E \to S^n$ be a bundle with fiber Σ, where Σ has the homology of the sphere S^{n-1}. Let $e \in H^n(S^n)$ be its Euler class. Prove that

(a) $e \neq 0$ if and only if $H^*(E) \approx H^*(S^{2n-1})$.
(b) $e = 0$ if and only if there is a **GrV**-isomorphism $\phi : H^*(E) \approx H^*(S^n \times S^{n-1})$.
(c) if $n > 2$, prove that the **GrV**-isomorphism ϕ in (b) is a **GrA**-isomorphism.

4.24. Let $i : Q \to M$ be the inclusion of a smooth submanifold of codimension r in a smooth manifold M. Let ν be the normal bundle to Q. Suppose that $H^r(M) = 0$. Prove that the Euler class $e(\nu) \in H^r(Q)$ vanishes.

Chapter 5
Poincaré Duality

5.1 Algebraic Topology and Manifolds

Manifolds studied by algebraic topology tools occur in several categories: smooth, piecewise linear, topological, homology manifolds, etc. Below are a few words about this matter.

Henri Poincaré's paper *analysis situs* [161], published 1895, is considered as the historical start of algebraic topology (for the "prehistory" of the field, see [163]). The aim of Poincaré was to use tools of algebraic topology in order to distinguish *smooth manifolds up to diffeomorphism* (which he called "homeomorphism"). So, differential and algebraic topology were born together. The importance of studying smooth manifolds up to diffeomorphism was reaffirmed throughout the twentieth century by many great mathematicians (Thom, Smale, Novikov, Atiyah, etc.). It is based on the deep role played by global properties of smooth manifolds in analysis, differential geometry, dynamical systems and physics.

After the failure of defining homology using submanifolds (see [40, Sect. I.3]), Poincaré initiated a new approach [162], in which smooth manifolds are equipped with a triangulation. This permitted him to define what will later become simplicial homology. The existence and essential uniqueness of smooth triangulations were of course a problem, solved only in 1940 by Whitehead [208, Theorems 7 and 8]. Also, besides some developments in the twenties (Veblen, Morse), the real foundations of differential topology arose only after 1935 with the works of H. Whitney. As a result, homology was seen for 3 decades as combinatorial in nature and smooth manifolds were not considered as the right objects of study. In the prominent book written in 1934 by Seifert and Threlfall [174], smooth manifolds are not even mentioned, but replaced by a simplicial counterpart, i.e. *combinatorial* or *piecewise linear* (*PL*) *manifolds* (see definition in Sect. 5.2). Techniques analogous to those for smooth manifolds were later developed in the PL-framework (see [104]). Polyhedral homology manifolds were later introduced (see Sect. 5.2), whose importance may grow with the development of computational homology. For even more general objects, like ANR homology manifolds, see e.g. [206].

J.-C. Hausmann, *Mod Two Homology and Cohomology*, Universitext,
DOI 10.1007/978-3-319-09354-3_5

Topological manifolds have also long attracted the attention of topologists, mostly to know whether they carry smooth or piecewise linear structures (see, e.g. [8, p. 235], [132, p. 183]). Their status however remained mysterious until the 1960s. Kirby and Siebenmann produced examples in all dimension ≥5 of topological manifolds without PL-structures and developed many techniques to deal with these topological manifolds [116]. The field of topological versus smooth manifolds developed very much in dimension four, after 1980, with the work of M. Freedman and S. Donaldson.

Poincaré duality is one of the most remarkable properties of closed manifolds. In its strong form, it gives, for a compact n-manifold M, that $H^k(M)$ and $H_{n-k}(M)$ are isomorphic under the cap product with the fundamental class $[M]$. This result can be obtained in two contexts:

- by working with homology manifolds, using simplicial topology and dual cells. Taking its origin in the early work of Poincaré, this was achieved around 1930 in the work by Pontryagin et al. (see [40, Sect. II.4.C]). In the next sections, we follow this approach, akin to the presentation of [155, Chap. 8]. This proves Poincaré duality for triangulable topological manifold, whence for smooth manifolds. Observe that smooth manifolds techniques (Morse theory or handle presentations) give an isomorphism from $H^k(M)$ to $H_{n-k}(M)$ but not the identification of this isomorphism with a cap product (see e.g. [120, Sect. VII.6]).
- by working with topological manifolds, using Čech cohomology techniques (see, e.g. [179, Sect. 6.2] or [82, Sect. 3.3]). This is not done in this book.

5.2 Poincaré Duality in Polyhedral Homology Manifolds

A *polyhedral homology n-manifold* is a simplicial complex such that, for each $\sigma \in \mathcal{S}_k(M)$, the link $\mathrm{Lk}(\sigma)$ of σ in M is a simplicial complex of dimension $n-k-1$ which has the homology of the sphere S^{n-k-1}. (Recall that our homology is mod2 by default; thus, in a broader context, these objects may more accurately be called *polyhedral* $\mathbb{Z}_2$*-homology manifolds*).

Remark 5.2.1 (1) Let X be topological space satisfying the following local property: for any $x \in X$,

$$H^j(X, X - \{x\}) = \begin{cases} \mathbb{Z}_2 & j = n \\ 0 & j \neq n\,. \end{cases}$$

Such a space is called a *homology n-manifold*. For instance, an n-dimensional topological manifold is a homology n-manifold by (3.3.1). The following result is proven in e.g. [155, Theorem 63.2]: *if K is a simplicial complex such that* $|K|$ *is a homology n-manifold, then K is a polyhedral homology n-manifold.*

(2) Special kind of polyhedral homology n-manifolds are PL-manifolds. A simplicial complex M is a *PL-manifold*, or a *combinatorial manifold* if, for each

$\sigma \in \mathcal{S}_k(M)$, the link $\mathrm{Lk}(\sigma)$ of σ in M has a subdivision isomorphic to a subdivision of the boundary of the $(n-k)$-simplex. PL-manifolds were the combinatorial objects replacing smooth manifolds for algebraic topologists around 1930.

(3) A smooth manifold M admits a so-called $\mathcal{C}^1$*-triangulation*, making M a PL-manifold. Two $\mathcal{C}^1$-triangulations have isomorphic subdivisions. This was proven by Whitehead in [208, Theorems 7 and 8].

(4) By a result of Edwards (see [128]), any PL-manifold of dimension ≥ 5 admits non-PL triangulations (which are then polyhedral homology manifolds by (1) above). It is an open problem whether a closed topological manifold of dimension ≥ 5 admits a (possibly non-PL) triangulation. This is wrong in dimension 4 (see [168, Sect. 5]).

(5) There are polyhedral homological manifolds M such that $|M|$ is not a topological manifold. For instance, the suspension of a homology n-manifold N which has the mod 2 homology of S^n (a *homology sphere*) is an $(n+1)$-dimensional homology manifold. But there are many PL-homology sphere (even for integral homology) with non-trivial fundamental group [115]. More examples are given, for instance, by lens spaces with odd fundamental groups.

Here are two first consequences of the definition of a polyhedral homology n-manifold.

Lemma 5.2.2 *Let M be a polyhedral homology n-manifold. Then*

(1) *any simplex of M is contained in some n-simplex of M.*
(2) *any $(n-1)$-simplex of M is a face of exactly two n-simplexes of M.*

Proof If v is a vertex of M, then $\mathrm{Lk}(v)$ is $n-1$ dimensional, so M is n-dimensional. Let $\sigma \in \mathcal{S}_k(M)$. If $\mathrm{Lk}(\sigma) = \emptyset$, σ must be an n-simplex by the above. If $\mathrm{Lk}(\sigma)$ is not empty, it must contain a $(n-k-1)$-simplex τ. Then, σ is contained in the join $\sigma * \tau$ which is an n-simplex. This proves (1).

If $\sigma \in \mathcal{S}_{n-1}(M)$ then $\mathrm{Lk}(\sigma)$ is a 0-dimensional complex having the homology of S^0. Hence, $\mathrm{Lk}(\sigma)$ consists of two points, which proves (2). □

Let M be a finite polyhedral homology n-manifold. It follows from Point (2) of Lemma 5.2.2 that the n-chain $\mathcal{S}_n(M)$ is a cycle and represents a homology class $[M] \in H_n(M)$, called the *fundamental class of M*.

Theorem 5.2.3 (Poincaré Duality) *Let M be a finite polyhedral homology n-manifold. Then, for any integer k, the linear map*

$$-\frown[M] : H^k(M) \to H_{n-k}(M)$$

is an isomorphism.

The proof of this Poincaré duality theorem will start after Proposition 5.2.8. We first give some corollaries of Theorem 5.2.3. By Kronecker duality, we get

Corollary 5.2.4 (Poincaré Duality, weak form) *Let M be a finite polyhedral homology n-manifold. Then, for any integer k,*

$$\dim H_k(M) = \dim H_{n-k}(M)\,.$$

Thus, in the computation of the Euler characteristic of M, the Betti numbers essentially come in pairs, which gives the following corollary.

Corollary 5.2.5 *Let M be a finite polyhedral homology n-manifold. Then, the Euler characteristic $\chi(M)$ satisfies the following:*

(1) *if n is odd, then $\chi(M) = 0$.*
(2) *if $n = 2m$, then $\chi(M) \equiv \dim H_m(M) \pmod 2$.*

Expressed in terms of Poincaré polynomial, Corollary 5.2.4 has the following form.

Corollary 5.2.6 *Let M be a finite polyhedral homology n-manifold. Then,*

$$\mathfrak{P}_t(M) = t^n\,\mathfrak{P}_{1/t}(M)\,.$$

Another easy consequence of Poincaré duality is the following.

Corollary 5.2.7 *A finite polyhedral homology n-manifold which is connected is an n-dimensional pseudomanifold.*

Proof Let M be a finite polyhedral homology n-manifold. We may suppose that M is non-empty, otherwise there is nothing to prove. By Lemma 5.2.2, M satisfies Conditions (a) and (b) of the definition of an n-dimensional pseudomanifold. If M is connected, then $H^0(M) = \mathbb{Z}_2$. By Poincaré duality, this implies that $H_n(M) = \mathbb{Z}_2$. Using Proposition 2.4.5, we deduce that M is an n-dimensional pseudomanifold. □

By Corollary 5.2.7, a continuous map between connected finite polyhedral manifolds of the same dimension has a degree (see (2.5.4)).

Proposition 5.2.8 *Let $f : M' \to M$ be a continuous map of degree one between connected finite n-dimensional polyhedral manifolds. Then $H_*f : H_*(M') \to H_*(M)$ is surjective and $H^*f : H^*(M) \to H^*(M')$ is injective*

Proof By Kronecker duality, only the homology statement needs a proof. The hypotheses imply that $H^*f([M']) = [M]$. By Proposition 4.5.4, this implies that the diagram

$$\begin{array}{ccc} H^k(M') & \xleftarrow{H^*f} & H^k(M) \\ \downarrow{\scriptstyle -\frown[M']} & & \downarrow{\scriptstyle -\frown[M]} \\ H_{n-k}(M') & \xrightarrow{H_*f} & H_{n-k}(M) \end{array}$$

is commutative for all integer $k \geq 0$. This provides a section for H_*f. □

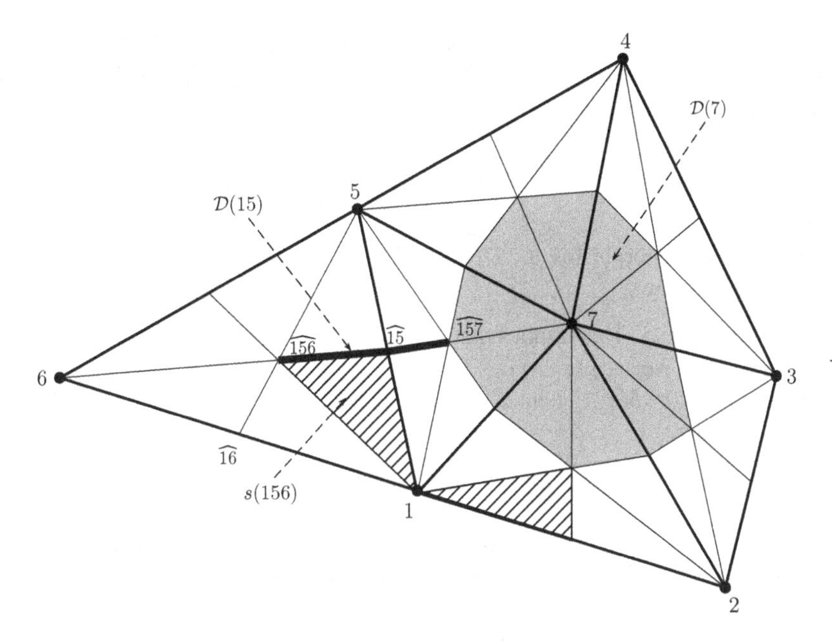

Fig. 5.1 *Dual cells and the map* $s : \mathcal{S}(M) \to \mathcal{S}(M')$ *of* (5.2.10). In the simplex notation, brackets and commas have been omitted: $156 = \{1, 5, 6\}$, etc.

The remainder of this section is devoted to the proof of Theorem 5.2.3. We shall introduce several simplicial or homology-cell complexes having all the homology of M. Let M' be the barycentric subdivision of M, with the notations introduced in p. 11. The simplicial complex M' is endowed with its natural simplicial order $\leq$ defined in (2.1.2). For $\sigma \in \mathcal{S}(M)$, define $\tilde{\mathcal{D}}(\sigma) \subset \mathcal{S}(M')$ by (Fig. 5.1)

$$\tilde{\mathcal{D}}(\sigma) = \{t \in \mathcal{S}(M') \mid \hat{\sigma} = \min t\}\,.$$

The simplicial subcomplex $\mathcal{D}(\sigma)$ of M' generated by $\tilde{\mathcal{D}}(\sigma)$ is called the *dual cell* of σ. Observe that $\dim \mathcal{D}(\sigma) = n - \dim(\sigma)$. The simplicial subcomplex $\dot{\mathcal{D}}(\sigma) = \mathrm{Lk}(\hat{\sigma}, \mathcal{D}(\sigma))$ is called the *boundary* of $\mathcal{D}(\sigma)$. Its dimension is one less than that of $\mathcal{D}(\sigma)$. We are interested in the topological spaces $\mathcal{E}(\sigma) = |\mathcal{D}(\sigma)|$ and $\dot{\mathcal{E}}(\sigma) = |\dot{\mathcal{D}}(\sigma)|$.

Lemma 5.2.9 *Let* $\sigma \in \mathcal{S}_k(M)$*, where* M *is a polyhedral homology* n*-manifold. Then,*

(a) *the space* $\mathcal{E}(\sigma)$ *is a homology* $(n-k)$*-cell with boundary* $\dot{\mathcal{E}}(\sigma)$*;*
(b) $\dot{\mathcal{D}}(\sigma)$ *is an* $(n-k-1)$*-dimensional pseudomanifold.*

Proof The space $\mathcal{E}(\sigma)$ is compact. Observe that $\mathcal{D}(\sigma)$ is the cone over $\dot{\mathcal{D}}(\sigma)$, with cone vertex $\hat{\sigma}$. Hence, $\mathcal{E}(\sigma)$ is the topological cone over $\dot{\mathcal{E}}(\sigma)$. Therefore, $(\mathcal{E}(\sigma), \dot{\mathcal{E}}(\sigma))$ is a good pair and $\tilde{H}_*(\mathcal{E}(\sigma)) = 0$. It then suffices to prove that

$H_*(\dot{\mathcal{E}}(\sigma)) \approx H_*\left(S^{n-k-1}\right)$. We shall see below that $\dot{\mathcal{D}}(\sigma)$ and $\mathrm{Lk}(\sigma, M)'$ are isomorphic simplicial complexes. As $|\mathrm{Lk}(\sigma, M)'| = |\mathrm{Lk}(\sigma, M)|$ and M is a polyhedral homology n-manifold, this implies that

$$H_*(\dot{\mathcal{E}}(\sigma)) \approx H_*\left(|\mathrm{Lk}(\sigma, M)'|\right) \approx H_*(\mathrm{Lk}(\sigma, M)) \approx H_*\left(S^{n-k-1}\right).$$

The simplicial isomorphisms $p : \dot{\mathcal{D}}(\sigma) \to \mathrm{Lk}(\sigma, M)'$ and $q : \mathrm{Lk}(\sigma, M)' \to \dot{\mathcal{D}}(\sigma)$ are defined as follows.

- Let $\hat{\tau} \in V\left(\dot{\mathcal{D}}(\sigma)\right)$. This implies that $\sigma \subset \tau$ and $\tau \in \mathcal{S}(M)$. Hence, $\kappa = \tau - \sigma \in \mathcal{S}(\mathrm{Lk}(\sigma, M))$. We set $p(\hat{\tau}) = \hat{\kappa}$.
- Let $\hat{\omega} \in V(\mathrm{Lk}(\sigma, M)')$. Then, $\omega \in \mathcal{S}(\mathrm{Lk}(\sigma, M))$, whence $\omega \cup \sigma \in \mathcal{S}(M)$. We set $q\left(\hat{\omega}\right) = \widehat{\omega \cup \sigma}$.

We check that p and q are simplicial maps and are inverse of each other. This proves Point (a).

To prove Point (b), let $\sigma \in \mathcal{S}_k(M)$. We leave as an exercise to the reader that a simplicial complex K is an m-dimensional pseudomanifold if and only if K' is so. Therefore, by Point (a) above and its proof, it is enough to prove that $L = \mathrm{Lk}(\sigma, M)$ is $(n-k-1)$-dimensional pseudomanifold.

Let $\tau \in \mathcal{S}(L)$. By Lemma 5.2.2, the simplex $\sigma * \tau$ is contained in some n-simplex of M, which is of the form $\sigma * \tau * \kappa$. Therefore, $\tau \subset \tau * \kappa \in \mathcal{S}_{n-k-1}(L)$. Now, if $\tau \in \mathcal{S}_{n-k-2}(L)$, then $\sigma * \tau$ is a common face of exactly two n-simplexes of M (by Lemma 5.2.2). Hence, τ is a common face of exactly two $(n-k-1)$-simplexes of L. We have proven that L satisfies Conditions (a) and (b) of the definition of an $(n-k-1)$-dimensional pseudomanifold. By Proposition 2.4.5, L is an $(n-k-1)$-dimensional pseudomanifold. □

Lemma 5.2.9 permits us to see $|M|$ as a homology-cell complex (see p. 108). The r-skeleton $|M|^r$ is defined by

$$|M|^r = \bigcup_{\substack{\sigma \in \mathcal{S}_s(M) \\ s \geq n-r}} \mathcal{E}(\sigma). \tag{5.2.1}$$

Indeed, the space $|M'|$ is the disjoint union of its geometric open simplexes

$$|M'| = \dot{\bigcup}_{t \in \mathcal{S}(M')} \left(|\bar{t}| - |\dot{t}|\right)$$

and each $|\bar{t}| - |\dot{t}|$ is contained in a single *open dual cell* $\mathcal{E}(\sigma) - \dot{\mathcal{E}}(\sigma)$, the one associated to σ for which $\hat{\sigma} = \min t$. This shows that $|M|^n = |M|$. If $\sigma \in \mathcal{S}_{n-r}(M)$, then $\dot{\mathcal{E}}(\sigma) = \mathcal{E}(\sigma) \cap |M|^{r-1}$; if $\sigma' \in \mathcal{S}_{n-r}(M)$ is distinct from σ, the open dual cells of σ and σ' are disjoint. This shows that $|M|^{r+1}$ is obtained by from $|M|^r$ by adjunction of the family of r-homology cells:

$$|M|^{r+1} = |M|^r \cup_\varphi \left(\dot{\bigcup}_{\sigma\in\mathcal{S}_{n-r}(M)} \mathcal{E}(\sigma)\right),$$

where φ is the attaching map

$$\varphi : \dot{\bigcup}_{\sigma\in\mathcal{S}_{n-r}(M)} \dot{\mathcal{E}}(\sigma) \twoheadrightarrow \bigcup_{\sigma\in\mathcal{S}_{n-r}(M)} \dot{\mathcal{E}}(\sigma) \subset |M|^r .$$

We denote by X the space $|M|$ endowed with this (regular) homology-cell structure. As noted in p. 108, the cellular homology $\dot{H}_*(X)$ (defined with the homology cells) is isomorphic to the singular homology $H_*(|M|)$ of $|M|$. If $\sigma \in \mathcal{S}_k(M)$, then $\dot{\mathcal{E}}(\sigma)$ is the union of those $\mathcal{E}(\tau)$ for which $\tau \in \mathcal{S}_{k+1}(M)$ has σ as a face. Using Formula (2.2.5), this amounts to

$$\dot{\mathcal{E}}(\sigma) = \bigcup_{\tau\in\delta(\sigma)} \mathcal{E}(\tau) . \tag{5.2.2}$$

On the other hand, since $\dot{\mathcal{D}}(\sigma)$ is a $(n-k-1)$-dimensional pseudomanifold by Lemma 5.2.9, Proposition 2.4.4 tells us that $H_{n-k-1}(\dot{\mathcal{D}}(\sigma)) = \mathbb{Z}_2$ is generated by $[\dot{\mathcal{D}}(\sigma)] = \mathcal{S}_{n-k-1}(\dot{\mathcal{D}}(\sigma))$ and the generator $H^{n-k-1}(\dot{\mathcal{D}}(\sigma)) = \mathbb{Z}_2$ is represented by any cochain formed by a single $(n-k-1)$-simplex. Hence, $H_{n-k}(\mathcal{D}(\sigma), \dot{\mathcal{D}}(\sigma)) = \mathbb{Z}_2$ is generated by $[\mathcal{D}(\sigma)] = \mathcal{S}_{n-k}(\mathcal{D}(\sigma))$ and the generator $H^{n-k}(\mathcal{D}(\sigma), \dot{\mathcal{D}}(\sigma)) = \mathbb{Z}_2$ is represented by any cochain formed by a single $(n-k)$-simplex of $\mathcal{D}(\sigma)$. The proof of Lemma 3.5.5 thus works and, using (5.2.2), $\dot{\partial} : \dot{C}_{n-k}(X) \to \dot{C}_{n-k-1}(X)$ satisfies

$$\dot{\partial}(\mathcal{D}(\sigma)) = \sum_{\tau\in\delta(\sigma)} [\mathcal{D}(\tau)] . \tag{5.2.3}$$

As M is a finite simplicial complex, $C^k(M)$ is isomorphic to the vector space generated by $\mathcal{S}_k(M)$. Therefore, the correspondence $\sigma \mapsto \mathcal{E}(\sigma)$ gives a linear map $\tilde{\Phi}_1 : C^k(M) \to \dot{C}_{n-k}(X)$ which, by (5.2.3), satisfies

$$\dot{\partial}\circ\tilde{\Phi}_1 = \Phi_1\circ\delta . \tag{5.2.4}$$

As $\tilde{\Phi}_1$ is bijective, the induced map

$$\Phi_1 : H^k(M) \xrightarrow{\approx} \dot{H}_{n-k}(X) \tag{5.2.5}$$

is an isomorphism. Observe that this proves the weak form of Poincaré duality of Corollary 5.2.5.

To prove Theorem 5.2.3, we now need to identify Φ_1 with a cap product. The correspondence $\mathcal{E}(\sigma) \mapsto [\mathcal{D}(\sigma)]$ provides a linear map $\tilde{\Phi}_2 : \dot{C}_{n-k}(X) \to C_{n-k}(M')$. By (5.2.2), $\tilde{\Phi}_2$ is a chain map, thus inducing a linear map $\Phi_2 : \dot{H}_{n-k}(X) \to H_{n-k}(M')$.

Lemma 5.2.10 $\Phi_2 : \dot{H}_{n-k}(X) \to H_{n-k}(M')$ *is an isomorphism.*

Proof The r-skeleton X^r of the homology-cell decomposition of X was given in (5.2.1). Note that $X^r = |K_r|$ where K_r is the subcomplex of M' given by

$$K_r = \bigcup_{\substack{\sigma \in \mathcal{S}_s(M) \\ s \geq n-r}} \mathcal{D}(\sigma) \,. \tag{5.2.6}$$

Thus, K_r is a simplicial complex of dimension r. We can use the simplicial pairs (K_r, K_{r-1}) to compute the simplicial homology of M'. Define $\ddot{C}_r(M') = H_r(K_r, K_{r-1})$ with the boundary $\ddot{\partial} : \ddot{C}_r(M') \to C_{r-1}(M')$ given by the composition

$$H_r(K_r, K_{r-1}) \to H_{r-1}(K_{r-1}) \to H_{r-1}(K_{r-1}, K_{r-2}) \,.$$

One has $\ddot{\partial} \circ \ddot{\partial} = 0$. Set $\ddot{H}_*(M') = \ker \ddot{\partial} / \mathrm{Image} \ddot{\partial}$. The correspondence $\mathcal{E}(\sigma) \mapsto [\mathcal{D}(\sigma)]$ gives an isomorphism $\Phi_2' : \dot{H}_r(X) \xrightarrow{\approx} \ddot{H}_r(M')$. Note that

$$\begin{aligned} \ddot{C}_r(M') &= H_r(K_r, K_{r-1}) \\ &= \ker \Big(\underbrace{C_r(K_r)/C_r(K_{r-1})}_{C_r(K_r)} \xrightarrow{\partial} C_{r-1}(K_r)/C_{r-1}(K_{r-1}) \Big) \,, \end{aligned}$$

whence

$$\ddot{C}_r(M') = \{\alpha \in C_r(K_r) \mid \partial\alpha \in C_{r-1}(K_{r-1})\} \subset C_r(K_r) \subset C_r(M') \,.$$

The inclusion $\tilde{\Phi}_2'' : \ddot{C}_*(M') \hookrightarrow C_*(M')$ is clearly a morphism of chain complexes. It induces a homomorphism $\Phi_2'' : \ddot{H}_*(M') \to H_*(M')$. As in the proof of Theorem 3.5.3, we have the commutative diagram

$$\begin{array}{ccccc}
H_{r+1}(K_{r+1}, K_r) & & \overset{0}{\dashrightarrow} & & H_r(K_{r+1}) \\
& \searrow^{\partial_{r+1}} & & \nearrow^{\mu} & \downarrow \approx \\
\downarrow \ddot{\partial}_{r+1} & & H_r(K_r) & \twoheadrightarrow & H_r(M') \\
& \swarrow^{j} & \vdots & & \\
H_r(K_r, K_{r-1}) & & \vdots\, 0 & & \\
& \searrow^{\partial_r} & \downarrow & & \\
\downarrow \ddot{\partial}_r & & H_{r-1}(K_{r-1}) & & \\
& \swarrow^{j} & & & \\
H_r(K_{r-1}, K_{r-2}) & & & &
\end{array} \tag{5.2.7}$$

which permits us to compute $\ddot{H}_*(M')$. As $\tilde{\Phi}_2''$ is just the inclusion, the diagram

$$\begin{array}{ccccc} H_r(K_r)/\mathrm{Im}\partial_{r+1} & \xrightarrow[\approx]{\mu} & \ker \partial_r/\mathrm{Im}\ddot{\partial}_{r+1} & \xrightarrow{=} & \ker \ddot{\partial}_r/\mathrm{Im}\ddot{\partial}_{r+1} \\ j \downarrow \approx & & & & \downarrow = \\ H_r(M') & & \xleftarrow{\quad \Phi_2'' \quad} & & \ddot{H}_r(M') \end{array}$$

is commutative, which proves that Φ_2'' is an isomorphism. Finally, the commutative diagram

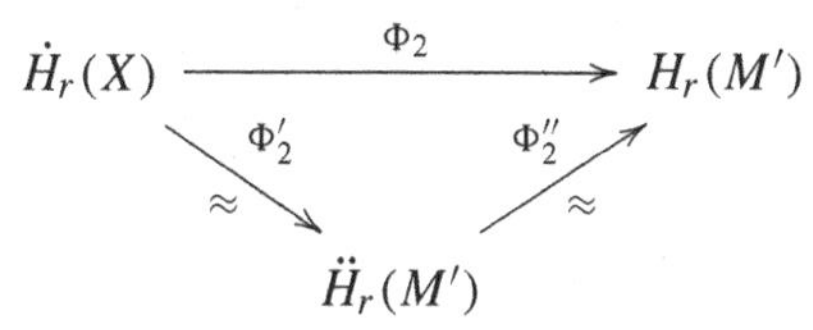

shows that Φ_2 is an isomorphism. □

We now need a good identification of the simplicial (co)homology of M with that of M'. Choose a simplicial order $\preceq$ on M. One has a simplicial map $g : M' \to M$ given, for $\sigma \in \mathcal{S}_m(M)$, by

$$g(\hat{\sigma}) = \max_{\preceq} \sigma \, . \tag{5.2.8}$$

In the other direction, one has a chain map $\mathrm{sd} : C_m(M) \to C_m(M')$ given, for $\sigma \in \mathcal{S}_m(M)$, by

$$\mathrm{sd}(\sigma) = \mathcal{S}_m(\sigma') \, . \tag{5.2.9}$$

(This chain map is in fact defined for any subdivision and is called the *subdivision operator*). Observe that, for any $\sigma \in \mathcal{S}_m(M)$, there exists a unique $\tau \in \mathcal{S}_m(M')$ such that $C_*g(\tau) = \sigma$. Indeed, if $\sigma = \{v_0, v_1, \dots, v_m\}$ with $v_0 \preceq v_1 \preceq \cdots \preceq v_m$, then $\tau = \{\hat{\sigma}_0, \hat{\sigma}_1, \dots, \hat{\sigma}_m\}$, where $\hat{\sigma}_i$ is the barycenter of $\{v_0, v_1, \dots, v_i\}$. The other m-simplexes of σ' are mapped to proper faces of σ. This defines a map

$$s : \mathcal{S}(M) \to \mathcal{S}(M') \tag{5.2.10}$$

by $s(\sigma) = \tau$. For $\sigma \in \mathcal{S}(M)$, one has

$$C_*g \circ C_*\mathrm{sd}(\sigma) = C_*g \circ s(\sigma) = \sigma \, , \tag{5.2.11}$$

which proves that $H_*g \circ H_*\mathrm{sd} = \mathrm{id}_{H_*(M)}$.

On the other hand, if $t = \{\hat{\sigma}_0, \hat{\sigma}_1, \dots, \hat{\sigma}_m\} \in \mathcal{S}_m(M')$, with $\sigma_0 \subset \cdots \subset \sigma_m$, then $\mathrm{sd} \circ g(t) \in C_*(\sigma_m')$. As, also $t \in C_*(\sigma_m')$, the correspondence $t \mapsto C_*(\sigma_m')$ is an acyclic carrier for both $\mathrm{sd} \circ g$ and $\mathrm{id}_{C_*(M')}$. By Proposition 2.9.1, this implies that

$H_* \mathrm{sd} \circ H_* g = \mathrm{id}_{H_*(M')}$. Therefore, g and sd induce isomorphisms in (co)homology which are inverse of each other. In particular, $H_* g$ et $H^* g$ do not depend on the order $\preceq$ since this is the case for sd.

It is straightforward that $\mathrm{sd}([M]) = [M']$. As $g : M' \to M$ is a simplicial map, Proposition 4.5.4 gives the formula

$$H_* g\big(H^* g(a) \frown [M']\big) = a \frown [M] ,$$

which is equivalent to the commutativity of the diagram.

$$\begin{array}{ccc} H^k(M) & \xrightarrow{\frown[M]} & H_{n-k}(M) \\ {\scriptstyle H^*g}\downarrow\approx & & \approx\uparrow{\scriptstyle H_* g} \\ H^k(M') & \xrightarrow{\frown[M']} & H_{n-k}(M') \end{array} \tag{5.2.12}$$

The identification of the isomorphism Φ_1 with the cap product with the fundamental class then follows from the following lemma.

Lemma 5.2.11 *The diagram*

$$\begin{array}{ccc} H^k(M) & \xrightarrow[\approx]{\Phi_1} & \dot{H}_{n-k}(X) \\ \approx\downarrow{\scriptstyle H^*g} & & \approx\downarrow{\scriptstyle \Phi_2} \\ H^k(M') & \xrightarrow{-\frown[M']} & H_{n-k}(M') \end{array}$$

is commutative.

Proof Let $\sigma \in \mathcal{S}_k(M)$. The properties of the map $s : \mathcal{S}(M) \to \mathcal{S}(M')$ defined in (5.2.10) imply that $C^* g(\sigma) = s(\sigma)$ and $\max s(\sigma) = \hat{\sigma}$ (for the natural simplicial order $\leq$ on M' defined in (2.1.2)). The isomorphism $\Psi^* = \Phi_2 \circ \Phi_1$ comes from the morphism of cochain-chains $\Psi : C^*(M) \to C_{n-*}(M')$ such that

$$\begin{aligned} \Psi(\sigma) = [D(\sigma)] &= \{t \in \mathcal{S}_{n-k}(M') \mid \min t = \hat{\sigma}\} \\ &= \{t \in \mathcal{S}_{n-k}(M') \mid s(\sigma) \cup t \in \mathcal{S}_n(M')\} . \end{aligned}$$

On the other hand, if $\tau = \{\hat{\sigma}_0, \dots \hat{\sigma}_n\} \in \mathcal{S}_n(M')$ with $\sigma_0 \subset \sigma_1 \subset \cdots \subset \sigma_n$, Formula (4.5.1) gives

$$C^* g(\sigma) \frown \tau = s(\sigma) \frown_{\leq} \tau = \langle s(\sigma), \{\hat{\sigma}_0, \dots, \hat{\sigma}_k\} \rangle \, \{\hat{\sigma}_k, \dots, \hat{\sigma}_n \, .$$

But

$$\langle s(\sigma), \{\hat{\sigma}_0, \ldots, \hat{\sigma}_k\}\rangle = \begin{cases} 1 & \text{if } s(\sigma) = \{\hat{\sigma}_0, \ldots, \hat{\sigma}_k\} \\ 0 & \text{otherwise} \end{cases}$$
$$= \begin{cases} 1 & \text{if } s(\sigma) \cup \{\hat{\sigma}_k, \ldots, \hat{\sigma}_n\} \in \mathcal{S}_n(M') \\ 0 & \text{otherwise.} \end{cases}$$

Therefore

$$C^* g(\sigma) \frown [M'] = s(\sigma) \frown [M'] = \Psi(\sigma) .$$ □

The proof of Poincaré Duality Theorem 5.2.3 is now complete.

5.3 Other Forms of Poincaré Duality

5.3.1 *Relative Manifolds*

A topological pair (X, Y) such that

$$H^j(X, X - \{x\}) = \begin{cases} \mathbb{Z}_2 & j = n \\ 0 & j \neq n . \end{cases}$$

for any $x \in X - Y$ is called a *relative homology n-manifold*. The condition is for instance fulfilled if $X - Y$ is n-dimensional topological manifold, by (3.3.1).

A simplicial pair (M, A) is a *relative polyhedral homology n-manifold* if, for each $\sigma \in \mathcal{S}_k(M) - \mathcal{S}_k(A)$, the link $\mathrm{Lk}(\sigma)$ of σ in M is a simplicial complex of dimension $n - k - 1$ which has the homology of the sphere S^{n-k-1}. For instance, $(M, \emptyset)$ is a relative polyhedral homology n-manifold if and only if M is a polyhedral homology n-manifold.

The following result is proven in e.g. [155, Theorem 63.2].

Proposition 5.3.1 *If (K, L) is a simplicial pair such that $(|K|, |L|)$ is a relative homology n-manifold, then K is a relative polyhedral homology n-manifold.*

A topological pair (X, Y) is *triangulable* if there exists a simplicial pair (K, L) and a homeomorphism of pair $h : (|K|, |L|) \to (X, Y)$. Such a homeomorphism h is called a *triangulation* of (X, Y).

Theorem 5.3.2 (Lefschetz duality) *Let (X, Y) be a compact relative homology n-manifold which is triangulable. Then, for any integer k, there is an isomorphism*

$$\Phi : H^k(X, Y) \approx H_{n-k}(X - Y) .$$

For a more general result, see [43, Proposition 7.2 in Chap. VII].

Proof Let (M, A) be a simplicial pair such that $(|M|, |A|)$ is homeomorphic to (X, Y). By Proposition 5.3.1, (M, A) is a relative polyhedral homology n-manifold. We shall construct an isomorphism

$$\Phi_0 : H^k(M, A) \xrightarrow{\approx} \dot{H}_{n-k}(|M| - |A|)\,, \tag{5.3.1}$$

where $H^k(M, A)$ is the simplicial cohomology. The proof is close to that of Theorem 5.2.3, so we just sketch the argument. For more details (see [155, Theorem 70.2]).

Let M^* be the subcomplex of the first barycentric subdivision of M consisting of all simplexes of M' that are disjoint from A. As in the proof of Theorem 5.2.3, consider the dual cell $\mathcal{D}(\sigma)$ for each $\sigma \in \mathcal{S}(M) - \mathcal{S}(A)$ and its geometric realization $\mathcal{E}(\sigma) = |\mathcal{D}(\sigma)|$. Lemma 5.2.9 holds for these dual cells and, as in the proof of Theorem 5.2.3, they provide a structure of a homology-cell complex on $|M^*|$. Call X^* the space $|M^*|$ endowed with this homology-cell decomposition. As M is a finite complex, then $C^k(M, A)$ is the vector space with basis $\mathcal{S}_k(M) - \mathcal{S}_k(A)$ (see p. 41). As in (5.2.5), the correspondence $\sigma \mapsto \mathcal{E}(\sigma)$ produces an isomorphism

$$\Phi_1 : H^k(M, A) \xrightarrow{\approx} \dot{H}_{n-k}(X^*)\,. \tag{5.3.2}$$

To get the isomorphism Φ_0 from Φ_1, we use that $|M^*| \approx X^*$ is a deformation retract of $|M| - |A|$ (see [155, Lemma 70.1]). □

Corollary 5.3.3 *Let (X, Y) be a connected compact relative homology n-manifold which is triangulable. If $Y \neq \emptyset$, then $H^n(X - Y) \approx H_n(X - Y) = 0$.*

Proof By Kronecker duality, it is enough to prove that $H_n(X - Y) = 0$. By Theorem 5.3.2, $H_n(X - Y) \approx H^0(X, Y)$ and, as X is path-connected, $H^0(X, Y) = 0$ if $Y \neq \emptyset$. (Corollary 5.3.3 may also be obtained using cohomology with compact supports: see [82, Theorem 3.35]). □

The following consequence of Corollary 5.3.3 is often referred to as the $\mathbb{Z}_2$-*orientability* of finite polyhedral homology n-manifolds (see e.g. [82, pp. 235–236]).

Corollary 5.3.4 *Let M be a finite polyhedral homology n-manifold and let $x \in M$. We denote by $j : (M, \emptyset) \to (M, M - \{x\})$ the pair inclusion. Then*

$$H_* j : H_m(M) \to H_m(M, M - \{x\})$$

sends $[M]$ onto the generator of $H_n(M, M - \{x\}) \approx \mathbb{Z}_2$. In particular, if M is connected, $H_* j$ is an isomorphism.

Proof The fundamental class of M being the sum of those of its connected components, it is enough to consider the case where M is connected. Corollary 5.3.4 then follows from the exact sequences

$$H_m(M - \{x\}) \to H_m(M) \xrightarrow{H_* j} H_m(M, M - \{x\}),$$

using that $H_m(M - \{x\}) = 0$ by Corollary 5.3.3. □

Let (X, Y) be a compact triangulable relative homology n-manifold. Choose a simplicial pair (M, A) such that $(|M|, |A|)$ is homeomorphic to (X, Y). Then, (M, A) is a finite relative polyhedral homology n-manifold by Proposition 5.3.1. Lemma 5.2.2 holds true for the simplexes of M which are not in A. As a consequence, the n-chain $\mathcal{S}_n(M) - \mathcal{S}_n(A)$ is a cycle relative to A and represent a homology class $[M] \in H_n(M, A)$ called the *fundamental class of* (M, A). Under the isomorphism between simplicial and singular homology of Theorem 3.6.3, the class $[M]$ corresponds to a singular class $[X] \in H_n(X, Y)$ called *fundamental class of* (X, Y). Let $i : X - Y \hookrightarrow X$ denote the inclusion. The isomorphism Φ of Theorem 5.3.2 is related to the cap product with $[X]$ in the following way.

Proposition 5.3.5 *Let (X, Y) be a compact relative homology n-manifold which is triangulable. Then the diagram*

$$\begin{array}{ccc} H^k(X, Y) & \xrightarrow[\approx]{\Phi} & H_{n-k}(X - Y) \\ & \searrow^{\frown [X]} & \downarrow H_* i \\ & & H_{n-k}(X) \end{array}$$

is commutative.

Proof As in the proof of Theorem 5.3.2, we choose a finite relative polyhedral homology n-manifold (M, A) such that such that $(|M|, |A|)$ is homeomorphic to (X, Y) and we use the same definitions and notations, such that $X^* \approx M^*$. The isomorphism $\Phi_2 : \dot{H}_{n-k}(X^*) \to H_{n-k}(M^*)$ may be established as in Lemma 5.2.10. The subdivision operator sd : $C_m(M) \to C_m(M')$ of (5.2.9) is defined, as well as the simplicial map $g : M' \to M$ of (5.2.8), choosing for the latter a simplicial order on M. They induced reciprocal isomorphisms on (co)homology. Ons has $\mathrm{sd}([M]) = [M']$, where $[M'] \in H_n(M', A')$ is the class of the relative cycle $\mathcal{S}_n(M') - \mathcal{S}_m(A')$. The commutative diagram (5.2.12) becomes

$$\begin{array}{ccc} H^k(M, A) & \xrightarrow{\frown [M]} & H_{n-k}(M) \\ H^* g \downarrow \approx & & \approx \uparrow H_* g \\ H^k(M', A') & \xrightarrow{\frown [M']} & H_{n-k}(M') \end{array} \tag{5.3.3}$$

If $i : M^* \to M'$ denotes the simplicial map given by the inclusion, the commutativity of the diagram

$$
\begin{array}{ccccc}
H^k(M,A) & \xrightarrow[\approx]{\Phi_1} & \dot{H}_{n-k}(X^*) & \xrightarrow[\approx]{\Phi_2} & H_{n-k}(M^*) \\
\approx\downarrow H^*g & & & & \approx\downarrow H^*i \\
H^k(M',A') & & \xrightarrow{-\frown[M']} & & H_{n-k}(M')
\end{array}
$$

is proven as in Lemma 5.2.11. Finally, as mentioned in the proof of Theorem 5.3.2, $|M^*| \approx X^*$ is a deformation retract of $|M| - |A|$, hence a commutative diagram involving simplicial and singular homology:

$$
\begin{array}{ccc}
H_*(M^*) & \xrightarrow{\approx} & H_*(|M|-|A|) \\
\downarrow H_*i & & \downarrow H_*j \\
H_*(M') & \xrightarrow{\approx} & H_*(|M|)
\end{array}
\quad . \tag{5.3.4}
$$

□

In the definition of a relative homology n-manifold (X, Y), it is not required that X itself is a homology n-manifold. If this is the case (and if X is compact and triangulable), the fundamental class $[X] \in H_n(X)$ is defined. To distinguish, call $[X]_{rel} \in H_n(X, Y)$ the class of Proposition 5.3.5. If (M, A) is a simplicial pair with $(|M|, |A|)$ homeomorphic to (X, Y), then $H_*j([M]) = [M]_{rel}$, where $j : (M, \emptyset) \to (M, A)$ (or $j : (X, \emptyset) \to (X, Y)$) denote the pair inclusion. Therefore $H^*j([X]) = [X]_{rel}$.

Proposition 5.3.6 *Let X be a compact homology n-manifold and let Y be a closed subset of X. Assume that the pair (X, Y) is triangulable. Then (X, Y) is a relative homology n-manifold and the diagram*

$$
\begin{array}{ccc}
H^k(X,Y) & \xrightarrow[\approx]{\Phi} & H_{n-k}(X-Y) \\
\downarrow H^*j & \searrow^{\frown[X]_{rel}} & \downarrow H_*i \\
H^k(X) & \xrightarrow{\frown[X]} & H_{n-k}(X)
\end{array}
$$

is commutative. Here, $j : (X, \emptyset) \to (X, Y)$ is the inclusion and ϕ is the Lefschetz duality isomorphism of Theorem 5.3.2.

Proof Only the commutativity of the diagram requires a proof. The commutativity of the upper triangle is established in Proposition 5.3.5. For the lower triangle, let $a \in H^k(X, Y)$ and $u \in H^{n-k}(X)$. One has

$$
\begin{aligned}
\langle u, a \frown [X]_{rel}\rangle &= \langle u \smile a, [X]_{rel}\rangle \\
&= \langle u \smile a, H_*j([X])\rangle \qquad \text{as } [X]_{rel} = H_*j([X]) \\
&= \langle H^*j(u \smile a), [X]\rangle \\
&= \langle u \smile H^*j(a), [X]\rangle \qquad \text{by Lemma 4.1.8} \\
&= \langle u, H^*j(a) \frown [X]\rangle ,
\end{aligned} \tag{5.3.5}
$$

which is, in formula, the commutativity of the lower triangle. □

5.3.2 Manifolds with Boundary

Let X be a compact topological n-manifold with boundary $Y = \mathrm{Bd}\, X$. Then (X, Y) is a compact relative homology n-manifold. As seen in the previous subsection, if the pair (X, Y) is triangulable, the fundamental class $[X] \in H_n(X, Y)$ is defined.

Theorem 5.3.7 *Let X be a compact topological n-manifold with boundary $Y = \mathrm{Bd}\, X$. Suppose that the pair (X, Y) is triangulable. Then, for any integer k, the linear maps*

$$-\frown[X] : H^k(X) \to H_{n-k}(X, Y)$$

and

$$-\frown[X] : H^k(X, Y) \to H_{n-k}(X)$$

given by the cap product with $[X] \in H_n(X, Y)$ are isomorphisms.

Theorem 5.3.7 is also true without the hypothesis of the triangulability of (X, Y), [82, Theorem 3.43].

Proof We first establish the isomorphism.

$$-\frown[X] : H^k(X, Y) \to H_{n-k}(X)\,.$$

As X is a topological manifold, its boundary admits a *collar neighbourhood*, i.e. there exists a embedding $h : Y \times [0, 1) \to X$, extending the identity on Y (see, e.g. [82, Proposition 3.42]). Then, $X - h(Y \times [0, 1/2])$ is a deformation retract of both X and $X - Y$. It follows that the inclusion $X - Y \hookrightarrow X$ is a homotopy equivalence. Hence, the result follows from Proposition 5.3.5.

The other isomorphism comes from the five lemma applied to the diagram

$$\begin{array}{ccccccccc}
\longrightarrow & H^k(X,Y) & \longrightarrow & H^k(X) & \longrightarrow & H^k(Y) & \xrightarrow{\delta^*} & H^{k+1}(X,Y) \\
 & \approx\downarrow{\frown[X]} & & \downarrow{\frown[X]} & & \approx\downarrow{\frown[Y]} & & \approx\downarrow{\frown[X]} \\
\longrightarrow & H_{n-k}(X) & \longrightarrow & H_{n-k}(X,Y) & \xrightarrow{\partial_*} & H_{n-k-1}(Y) & \longrightarrow & H_{n-k-1}(X)
\end{array}$$

The commutativity of the above diagram comes from Lemma 4.5.5, since

$$\partial_*([X]) = [Y]\,. \tag{5.3.6}$$

Indeed, if (M, N) be a finite a simplicial pair triangulating (X, Y), the fundamental class $[M]$ is represented by the chain $\mathcal{S}_n(M) \in C_*(M)$ and $\partial_*([M])$ is represented by $\partial(\mathcal{S}_n(M)) = \mathcal{S}_{n-1}(N)$. □

Corollary 5.3.8 *Let X be a compact triangulable topological n-manifold with boundary* $\mathrm{Bd}\, X = Y$. *Suppose that is $Y = Y_1 \cup Y_2$ the union of two compact $(n-1)$-manifolds with common boundary $Y_1 \cap Y_2 = \mathrm{Bd}\, Y_1 = \mathrm{Bd}\, Y_2$. Then, for any integer k, the linear map*

$$- \frown [X] : H^k(X, Y_1) \to H_{n-k}(X, Y_2)$$

given by the cap product with $[X] \in H_n(X, Y)$ is an isomorphism.

Again, Corollary 5.3.8 is true without the hypothesis of triangulability (see [82, Theorem 3.43]).

Proof Corollary 5.3.8 reduces to Theorem 5.3.7 by applying the five lemma to the diagram

$$\begin{array}{ccccccccc} \longrightarrow & H^k(X,Y) & \longrightarrow & H^k(X,Y_1) & \longrightarrow & H^k(Y,Y_1) & \longrightarrow & H^{k+1}(X,Y) & \longrightarrow \\ & \approx\downarrow \frown[X] & & \downarrow \frown[X] & & \approx\downarrow \mu & & \approx\downarrow \frown[X] & \\ \longrightarrow & H_{n-k}(X) & \longrightarrow & H_{n-k}(X,Y_2) & \longrightarrow & H_{n-k-1}(Y_2) & \longrightarrow & H_{n-k-1}(X) & \longrightarrow \end{array}$$

The top line is the cohomology sequence for the triple (X, Y, Y_1) and the bottom line is the homology sequence for the pair (X, Y_2). The isomorphism μ is the composition

$$\mu : H^k(Y, Y_1) \xrightarrow{\approx} H^k(Y_2, \mathrm{Bd}\, Y_2) \xrightarrow{\frown [Y_2]} H_{n-k-1}(Y_2)\,.$$

The commutativity of the above diagram is obtained as for those in the proofs in Sect. 5.3.1. □

Here are some applications of the Poincaré duality for compact manifolds with boundary.

Proposition 5.3.9 *Let X be a compact triangulable manifold of dimension $2n+1$, with boundary Y. Let $B = \mathrm{Image}\big(H^n(X) \to H^n(Y)\big)$. Then*

(1) *Let $u \in H^n(Y)$. Then*

$$u \in B \iff \langle u \smile B, [Y] \rangle = 0\,.$$

In particular, $\langle B \smile B, [Y] \rangle = 0$.

(2) $\dim H^n(Y) = 2 \dim B$.

For example, $\mathbb{R}P^{2n}$ is not the boundary of a compact manifold.

Proof We follow the idea of [133, Lemma 4.7 and Corollary 4.8]. Let $i : Y \to X$ denote the inclusion and let $a, b \in H^n(X)$. Then,

$$\langle H^*i(a) \smile H^*i(b), [Y] \rangle = \langle H^*i(a \smile b), [Y] \rangle = \langle a \smile b, H_*i([Y]) \rangle = 0\,,$$

since $H_*i([Y]) = 0$ by (5.3.6). This proves the implication $\Rightarrow$ of (1). Conversely, suppose that $\langle u \smile B, [Y]\rangle = 0$ for $u \in H^n(Y)$. Since $B = \ker(\delta : H^n(Y) \to H^{n+1}(X, Y)$, it suffices to prove that $\delta(u) = 0$. Let $v \in H^n(X)$. One has

$$\begin{aligned} 0 &= \langle u \smile H^*i(v), [Y]\rangle \\ &= \langle u \smile H^*i(v), \partial[X]\rangle \quad \text{by 5.3.6} \\ &= \langle \delta(u \smile H^*i(v)), [X]\rangle \\ &= \langle \delta(u) \smile v, [X]\rangle \qquad \text{by Lemma 4.1.9} \end{aligned}$$

This equality, holding for any $v \in H^n(X)$, implies, by Theorem 5.3.12, that $\delta(u) = 0$.

To prove (2), let us consider the linear map $\Phi : H^n(Y) \to H^n(Y)^\sharp$ given by $\Phi(a)(b) = \langle a \smile b, [Y]\rangle)$. Let Φ_B be the restriction of Φ to B. The map Φ is an isomorphism by Theorem 5.3.13. By (1), $\Phi_B(B) = A^\sharp$, where $A = H^n(Y)/B$. Thus, there is a quotient map $\hat{\Phi}$ fitting in the commutative diagram

$$\begin{array}{ccccccccc} 0 & \longrightarrow & B & \longrightarrow & H^n(Y) & \longrightarrow & A & \longrightarrow & 0 \\ & & \approx\downarrow\Phi_B & & \approx\downarrow\Phi & & \approx\downarrow\hat{\Phi} & & \\ 0 & \longrightarrow & A^\sharp & \longrightarrow & H^n(Y)^\sharp & \longrightarrow & B^\sharp & \longrightarrow & 0 \end{array} \tag{5.3.7}$$

(whose rows are exact) and $\hat{\Phi}$ is also an isomorphism. Therefore,

$$\dim H^n(Y) = \dim B + \dim A = 2\dim B\,.$$

(Remark: the proof does not use the map $\hat{\phi}$, only that Φ_B is an isomorphism; but Diagram (5.3.7) will be useful later). □

Proposition 5.3.9 and Corollary 5.2.4 have the following consequence on the Euler characteristic of bounding manifolds.

Corollary 5.3.10 *Let Y be a closed triangulable n-manifold. If Y is the boundary of a compact triangulable manifold, then $\chi(Y)$ is even.*

5.3.3 The Intersection Form

Let X be a compact topological n-manifold with boundary $Y = \mathrm{Bd}\, X$. We assume that the pair (X, Y) is triangulable. From Theorem 5.3.7, the cap product with $[X] \in H_n(X, Y)$ induces isomorphisms $H^q(X) \xrightarrow{\approx} H_{n-q}(X, Y)$ and $H^q(X, Y) \xrightarrow{\approx} H_{n-q}(X)$. We denote by PD the inverse of these isomorphisms. Thus, if $\alpha \in H_q(X)$ and $\beta \in H_q(X, Y)$, their *Poincaré dual* $\mathrm{PD}(\alpha) \in H^{n-q}(X, Y)$ and $\mathrm{PD}(\beta) \in H^{n-q}(X)$ are the classes determined by the equations

$$\mathrm{PD}(\alpha) \frown [X] = \alpha \;\text{ and }\; \mathrm{PD}(\beta) \frown [X] = \beta\,.$$

(The first equation uses the cap product of (4.5.16) and the second that of (4.5.14)). This permits us to define two *intersection forms* on the homology of X.

(1) If $\alpha \in H_q(X)$ and $\beta \in H_{n-q}(X)$, we set

$$\alpha \cdot_a \beta = \langle \mathrm{PD}(\alpha) \smile \mathrm{PD}(\beta), [X] \rangle .$$

This defines the *(absolute) intersection form* $H_*(X) \oplus H_{n-*}(X) \xrightarrow{\cdot_a} \mathbb{Z}_2$.

(2) Similarly, if $\alpha \in H_q(X)$ and $\beta \in H_{n-q}(X,Y)$, the same formula defines the *(relative) intersection form* $H_*(X) \oplus H_{n-*}(X,Y) \xrightarrow{\cdot_r} \mathbb{Z}_2$.

The name "intersection form" will be justified by Corollary 5.4.13.

Let $j : (X,\emptyset) \to (X,Y)$ denote the pair inclusion. For $\alpha \in H_q(X)$ and $\beta \in H_{n-q}(X)$, the absolute and relative intersection forms are related by the formula

$$\alpha \cdot_a \beta = H_* j(\alpha) \cdot_r \beta = H_* j(\beta) \cdot_r \alpha .$$

Indeed:

$$\begin{aligned} H_* j(\alpha) \cdot_r \beta &= \langle \mathrm{PD}(H_* j(\alpha)) \smile \mathrm{PD}(\beta), [X] \rangle \\ &= \langle (H^* j(\mathrm{PD}(\alpha)) \smile \mathrm{PD}(\beta), [X] \rangle && \text{by Lemma 4.5.5} \\ &= \langle \mathrm{PD}(\alpha) \smile \mathrm{PD}(\beta), [X] \rangle && \text{by Lemma 4.1.7} \\ &= \alpha \cdot_a \beta \end{aligned}$$

and the other equality is proven the same way.

The absolute and relative intersection forms coincide when Y is empty. Even when $Y \neq \emptyset$, we shall usually not distinguish between the two forms and just write $\alpha \cdot \beta$ when the context makes it clear. In both cases, since $\langle a \smile b, \gamma \rangle = \langle a, b \frown \gamma \rangle$ [see (4.5.2)], one has

$$\alpha \cdot \beta = \langle \mathrm{PD}(\alpha), \beta \rangle = \langle \mathrm{PD}(\beta), \alpha \rangle . \tag{5.3.8}$$

By Theorem 5.3.13, the relative intersection form is non-degenerate, i.e. induces an isomorphism $H_q(X) \xrightarrow{\approx} H_{n-q}(X,Y)^\sharp$ for all q. If $Y \neq \emptyset$, the absolute intersection form may be degenerate (example: $X = S^1 \times D^2$). In fact, if X is connected, it is always degenerate for $q = 0$, since $H_n(X) = 0$. However, one has the following proposition.

Proposition 5.3.11 *Suppose that X is connected and that Y is not empty. Then, the following conditions are equivalent.*

(a) *The absolute intersection form induces an isomorphism $H_q(X) \xrightarrow{\approx} H_{n-q}(X)^\sharp$ for $1 \le q \le n-1$.*
(b) *$Y = \mathrm{Bd}\, X$ is a $\mathbb{Z}_2$-homology sphere.*

Proof Let $j : (X,\emptyset) \to (X,Y)$ denote the pair inclusion. By (5.3.8), the composed homomorphism

$$H_q(X) \xrightarrow{H_q j} H_q(X, Y) \xrightarrow[\approx]{\text{PD}} H^{n-q}(X) \xrightarrow[\approx]{\mathbf{k}} H_{n-q}(X)^\sharp$$

is just the absolute intersection form of X. Thus, (a) is equivalent to $H_q j$ being an isomorphism for $1 \leq q \leq n-1$. By the exact homology sequence of (X, Y) this is equivalent to (b) if X is connected. □

5.3.4 Non Degeneracy of the Cup Product

Theorem 5.3.12 *Let M be a finite polyhedral homology n-manifold. Then, for any integer k, the bilinear map*

$$H^k(M) \times H^{n-k}(M) \xrightarrow{\smile} H^n(M) \xrightarrow{\langle -,[M]\rangle} \mathbb{Z}_2$$

induces an isomorphism $H^k(M) \xrightarrow{\approx} H^{n-k}(M)^\sharp$.

Proof By Corollary 5.2.4, it suffices to prove that the linear map $\Phi : H^k(M) \to H^{n-k}(M)^\sharp$ given by

$$a \overset{\Phi}{\mapsto} \Big(b \mapsto \langle a \smile b, [M]\rangle\Big)$$

is injective. Suppose that $a \in \ker \Phi$. Then

$$0 = \langle a \smile b, [M]\rangle = \langle b, a \frown [M]\rangle$$

for all $b \in H^{n-k}(M)$. By Point (a) of Lemma 2.2.3, we deduce that $a \frown [M] = 0$, which implies that $a = 0$ by Theorem 5.2.3. □

The same proof, using Corollary 5.3.8, gives the following result.

Theorem 5.3.13 *Let X be a compact triangulable topological n-manifold with boundary* $\mathrm{Bd}\, X = Y$. *Suppose that $Y = Y_1 \cup Y_2$ is the union of two compact $(n-1)$-manifolds Y_i with common boundary $Y_1 \cap Y_2 = \mathrm{Bd}\, Y_1 = \mathrm{Bd}\, Y_2$. Then, for any integer k, the bilinear map*

$$H^k(X, Y_1) \times H^{n-k}(X, Y_2) \xrightarrow{\smile} H^n(X, Y) \xrightarrow{\langle -,[X]\rangle} \mathbb{Z}_2$$

induces an isomorphism $H^k(X, Y_1) \xrightarrow{\approx} H^{n-k}(X, Y_2)^\sharp$.

5.3.5 Alexander Duality

The first version of Alexander Duality was proven in a paper [7] of Alexander II (1888–1971). This article pioneered several new methods and was very influential at the time (see [40, p. 56]). In his paper, Alexander used the mod 2 homology. Classical Alexander duality relates the cohomology of a closed subset A or S^n to the homology of $S^n - A$. We give below a version where S^n is replaced by a homology sphere (for instance a lens space with odd fundamental group).

Theorem 5.3.14 (Alexander Duality) *Let (X, A) be a compact triangulable pair with $\emptyset \neq A \neq X$. Suppose that X is a relative homology n-manifold and has its homology isomorphic to that of S^n. Then, for all integer k, there is an isomorphism*

$$\tilde{H}^k(A) \approx \tilde{H}_{n-k-1}(X - A)\,.$$

Particular case of Alexander duality were encountered in Proposition 3.3.6 and Corollary 3.3.7. For a version of Theorem 5.3.14 without the assumption of triangulability (see [82, Theorem 3.44]).

Proof The case $n = 0$ being trivial, we assume $n > 0$. The pair (X, A) satisfies the hypotheses of Lefschetz duality Theorem 5.3.2. This gives an isomorphism

$$\Phi : H^{k+1}(X, A) \approx H_{n-k-1}(X - A)\,.$$

Suppose that $k \neq n, n-1$. Since $H^*(X) \approx H^*(S^n)$, the connecting homomorphism $\delta^* : H^k(A) \to H^{k+1}(X, A)$ is an isomorphism and $H_{n-k-1}(X-A) \approx \tilde{H}_{n-k-1}(X-A)$. This proves the result in this case.

Let (M, L) be a simplicial pair such that $(|M|, |L|)$ is homeomorphic to (X, A). As M is a relative polyhedral homology n-manifold by Proposition 5.3.1. As L is a proper subcomplex of M, one has $H_n(L) = 0$, since $\mathcal{S}_n(M)$ is the only non-vanishing n-cycle of M. Hence, $\tilde{H}^n(A) \approx H^n(A) = 0$ by Kronecker duality. As, $\tilde{H}_{n-k-1}(X - A) = \tilde{H}_{-1}(X - A) = 0$, the theorem is true for $k = n$.

When $k = n - 1$, consider the diagram

$$\begin{array}{ccccccccc}
H^{n-1}(X) & \xrightarrow{H^*i} & H^{n-1}(A) & \longrightarrow & H^n(X, A) & \longrightarrow & H^n(X) & \longrightarrow & 0 \\
 & & \Big\downarrow{\scriptstyle \hat{\Phi}} & & \approx\Big\downarrow{\scriptstyle \Phi} & & \approx\Big\downarrow{\scriptstyle \frown[X]} & & \\
0 & \longrightarrow & \tilde{H}_0(X - A) & \longrightarrow & H_0(X - A) & \xrightarrow{H_*j} & H_0(X) & \longrightarrow & 0
\end{array}$$

where $i : A \to X$ and $j : (X - A) \to X$ denote the inclusions. The bottom line is the exact sequence of Lemma 3.1.10 and the commutativity of the right hand square is the contents of Proposition 5.3.6. Then the homomorphism $\hat{\phi} : H^{n-1}(A) \to \tilde{H}_0(X - A)$ exists, making the diagram commutative. If $n > 1$, $\tilde{H}^{n-1}(A) = H^{n-1}(A)$ and, as

$H^{n-1}(X) = 0$, the map $\hat{\phi}$ is an isomorphism by the five lemma. Finally, when $n = 1$, then coker $H^*i = \tilde{H}^0(A)$ by Lemma 3.1.10 and $\hat{\phi}$ induces an isomorphism from $\tilde{H}^{n-1}(A)$ to $\tilde{H}_0(X - A)$. □

5.4 Poincaré Duality and Submanifolds

In this section, we assume some familiarity of the reader with standard techniques of smooth manifolds, as exposed in e.g. [95].

5.4.1 The Poincaré Dual of a Submanifold

Let M be a smooth compact n-manifold and let $Q \subset M$ be a closed smooth submanifold of codimension r. Recall that smooth manifolds admit PL-triangulations [208], so the fundamental classes $[Q] \in H_{n-r}(Q)$ and $[M] \in H_n(M, \mathrm{Bd}\, M)$ do exist. We are interested in the Poincaré dual $\mathrm{PD}(H_*i([Q])) \in H^r(M, \mathrm{Bd}\, M)$ (see Sect. 5.3.3) of the class $H_*i([Q]) \in H_{n-r}(M)$, where $i : Q \to M$ denotes the inclusion. We write $\mathrm{PD}(Q)$ for $\mathrm{PD}(H_*i([Q]))$ and call it the *Poincaré dual* of Q. It is thus characterized by the equation

$$\mathrm{PD}(Q) \frown [M] = H_*i([Q]),$$

Two simple examples are given in Fig. 5.2.

Example 5.4.1 For a more elaborated example, let Q be a smooth closed connected manifold, seen as the diagonal submanifold of $M = Q \times Q$. Let $\mathcal{A} = \{a_1, a_2, \dots\} \subset H^*(Q)$ be an additive basis of $H^*(Q)$. By Theorem 5.3.12, there is a basis $\mathcal{B} = \{b_1, b_2, \dots\}$ of $H^*(Q)$ which is dual to $\mathcal{A}$ for the Poincaré duality, i.e.

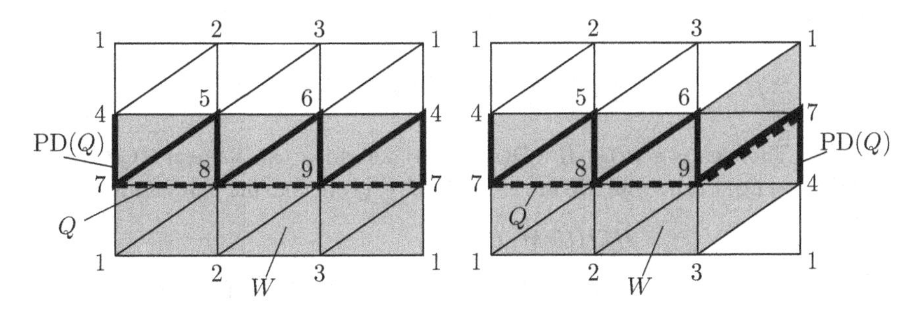

Fig. 5.2 The Poincaré dual $\mathrm{PD}(Q)$ *of a circle* Q in the torus (*left*) or the Klein bottle (*right*). This illustrates the localization principle of Remark 5.4.3: $\mathrm{PD}(Q)$ is supported in a tubular neighbourhood W of Q

$\langle a_i \smile b_j, [Q] \rangle = \delta_{ij}$. We claim that

$$\mathrm{PD}(Q) = \sum_i a_i \times b_i \,. \tag{5.4.1}$$

Indeed, by the Künneth theorem, $a_i \times b_j$ is a basis of $H^*(M)$, so there are unique coefficients $\gamma_{ij} \in \mathbb{Z}_2$ such that

$$\mathrm{PD}(Q) = \sum_{i,j} \gamma_{ij}\, a_i \times b_j \,.$$

Let $\Delta : Q \to M$ be the diagonal inclusion, $a_p \in \mathcal{A}$ and $b_q \in \mathcal{B}$. As $H^*\Delta(b_p \times a_q) = b_p \smile a_q$ (see Remark 4.6.1), one has

$$\langle b_p \times a_q, H_*\Delta([Q]) \rangle = \langle H^*\Delta(b_p \times a_q), [Q] \rangle = \langle b_p \smile a_q, [Q] \rangle = \delta_{pq} \,. \tag{5.4.2}$$

Without loss of generality, we may suppose that Q is connected. Let $[Q]^\sharp$ be the non zero element of $H^{\dim Q}(Q)$. One has $\langle [Q]^\sharp \times [Q]^\sharp, [M] \rangle = 1$ and

$$\begin{aligned} \langle b_p \times a_q, H_*\Delta([Q]) \rangle &= \langle (b_p \times a_q) \smile \mathrm{PD}(Q), [M] \rangle \\ &= \textstyle\sum_{i,j} \gamma_{ij} \, \langle (b_p \times a_q) \smile (a_i \times b_i), [M] \rangle \\ &= \textstyle\sum_{i,j} \gamma_{ij} \, \langle (b_p \smile a_i) \times (a_q \smile b_i), [M] \rangle \quad \text{by Remark 4.6.4} \\ &= \textstyle\sum_{i,j} \gamma_{ij} \, \langle \delta_{pi}[Q]^\sharp \times \delta_{qj}[Q]^\sharp, [M] \rangle \\ &= \gamma_{pq} \,. \end{aligned} \tag{5.4.3}$$

Thus, Equation (5.4.1) follows from (5.4.2) and (5.4.3).

The next two lemmas are recipes to compute $\mathrm{PD}(Q)$. Let us denote by $\nu = \nu(M, Q)$ the normal bundle of Q in M. A Riemannian metric provides a smooth bundle pair $(D(\nu), S(\nu))$ with fiber (D^r, S^{r-1}) and there is a diffeomorphism from $D(\nu)$ to a closed tubular neighbourhood W of Q in M. By excision,

$$H^*(M, M - Q) \xrightarrow{\approx} H^*(W, \mathrm{Bd}\, W) \approx H^*(D(\nu), S(\nu)) \,.$$

Hence, the Thom class $U(\nu) \in H^r(D(\nu), S(\nu))$ determines an element $U(M, Q) \in H^r(M, M - Q)$. Let $j : (M, \mathrm{Bd}\, M) \to (M, M - Q)$ denote the pair inclusion.

Lemma 5.4.2 $\mathrm{PD}(Q) = H^*j(U(M, Q))$.

Proof We first reduce to the case where Q is connected. Indeed, as Q is the finite union of components Q_i, with tubular neighbourhood W_i, then

$$H^r(M, M - Q) \xrightarrow{\approx} H^r(W, \mathrm{Bd}\, W) \approx \bigoplus H^r(W_i, \mathrm{Bd}\, W_i) \approx \bigoplus H^r(M, M - Q_i)$$

and $U(M,Q)=\sum U(M,Q_i)$. On the other hand, $\mathrm{PD}(Q)=\sum \mathrm{PD}(Q_i)$. Thus, we shall assume that Q is connected.

Let us consider the case $M = D(\nu)$ and Q is the image of the zero section. As $H_{n-r}(Q)=\mathbb{Z}_2=H_n(D(\nu),S(\nu))$, the Thom isomorphism of Theorem 4.7.29 says that $U(\nu)\frown[D(\nu)]=[Q]$. This proves the lemma for any tubular neighbourhood of Q, for instance W or a smaller tube W' contained in the interior of W.

Let us choose a triangulation of M for which W and W' are subcomplexes. The class $H^*j(U(M,Q))\in H^r(M)$ may then be represented by a simplicial cocycle $\tilde{q}\subset S_r(W')$. The n-simplexes of M involved in the computation of $H^*j(U(M,Q))\frown[M]$ are then all simplexes of W. Therefore

$$H^*j(U(M,Q))\frown[M]=H^*i\big(U(W,Q)\big)\frown[W]=H^*i([Q])\,.$$

.

□

Remark 5.4.3 We see in the proof of Lemma 5.4.2 that the Poincaré dual $\mathrm{PD}(Q)$ of a submanifold $Q\subset M$ is supported in an arbitrary small tubular neighbourhood of Q. This *localization principle* is illustrated in Fig. 5.2 for Q a circle in the torus or the Klein bottle. For the analogous localization principle in de Rham cohomology, see [19, Proposition 6.25].

Lemma 5.4.4 *The image of* $\mathrm{PD}(Q)$ *under the homomorphism* $H^r(M,\mathrm{Bd}\,M)\to H^r(M)\to H^r(Q)$ *is equal to the Euler class of the normal bundle* $\nu=\nu(M,Q)$.

Proof We use the notations of the proof of Lemma 5.4.2. Let $k:(M,\emptyset)\to(M,M-Q)$ denote the pair inclusion and let $\sigma_0:Q\to D(\nu)$ be the zero section. The various inclusions give rise to the commutative diagram

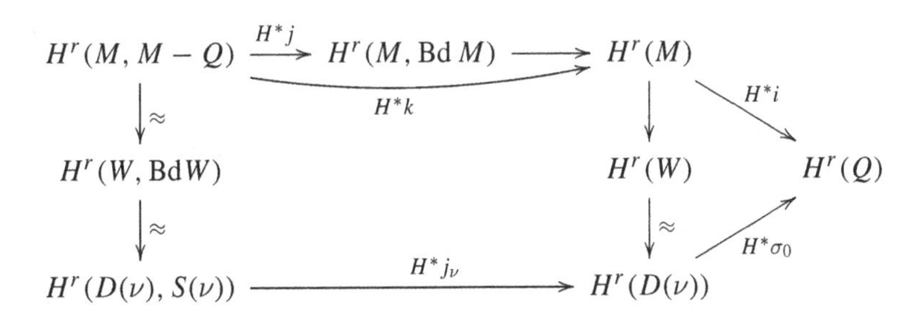

The Euler class $e(\nu)\in H^r(Q)$ is characterized by the equation $H^*j_\nu(U(\nu))=H^*p(e(\nu))$, where $p:D(\nu)\to Q$ is the bundle projection (see p. 184). Since $p\circ\sigma_0=\mathrm{id}_Q$, the previous diagram and Lemma 5.4.2 yield

$$\begin{aligned}H^*i(\mathrm{PD}(Q))&=H^*i\circ H^*k(U(M,Q))=H^*\sigma_0\circ H^*j_\nu(U(\nu))\\&=H^*\sigma_0\circ H^*p(e(\nu))=e(\nu)\,.\end{aligned}$$

□

We now compute $\ker\big(H^*i:H^*(M)\to H^*(Q)\big)$.

Proposition 5.4.5 *Let M be a smooth compact n-manifold and let $i : Q \hookrightarrow M$ be the inclusion of a closed smooth submanifold Q of codimension r. Then*

$$\ker H^*i \subset \mathrm{Ann}\,(\mathrm{PD}(Q)) = \{x \in H^*(M) \mid x \smile \mathrm{PD}(Q) = 0\}$$

(for a class b, the ideal Ann (b) *is called the annihilator of b). The above inclusion is an equality if and only if H^*i is surjective.*

Proof Let $a \in H^*(M)$ and let $q = \mathrm{PD}(Q)$. Then

$$\begin{aligned} H_*i\big(H^*i(a) \frown q\big) &= a \frown H_*i([Q]) && \text{by (4.5.5)} \\ &= a \frown (q \frown [M]) && \text{by definition of } q \\ &= (a \smile q) \frown [M] && \text{by Proposition 4.5.4} \end{aligned}$$

Therefore, if $H^*i(a) = 0$, then $a \in Ann\,(q)$ and the converse is true if and only if H_*i is injective. By Kronecker duality (see Corollary 2.3.11), the latter is equivalent to H^*i being surjective. □

Example 5.4.6 Let i be the standard inclusion of $Q = \mathbb{R}P^k$ in $M = \mathbb{R}P^{k+r}$. By Proposition 4.3.10, one has a commutative diagram

$$\begin{array}{ccc} \mathbb{Z}_2[a]/\left(a^{k+r+1}\right) & \twoheadrightarrow & \mathbb{Z}_2[a]/\left(a^{k+1}\right) \\ \downarrow \approx & & \downarrow \approx \\ H^*\left(\mathbb{R}P^{k+r}\right) & \xrightarrow{H^*i} & H^*\left(\mathbb{R}P^k\right) \end{array}$$

where a is of degree 1. Then, via the above vertical isomorphisms,

$$\ker H^*i = \left(a^{k+1}\right) = Ann\ \left(a^r\right) = Ann\ \left(\mathrm{PD}\left(\mathbb{R}P^k\right)\right).$$

The following proposition states the *functoriality of the Poincaré dual* for transversal maps.

Proposition 5.4.7 *Let $f : M \to N$ be a smooth map between smooth closed manifolds. Suppose that f is transversal to a closed submanifold Q of N. Then*

$$\mathrm{PD}\left(f^{-1}(Q)\right) = H^*f\,(\mathrm{PD}(Q))\,.$$

Proof Let $P = f^{-1}(Q)$. We consider the commutative diagram

$$\begin{array}{ccc} H^*(N, N-Q) & \xrightarrow{H^*f} & H^*(M, M-P) \\ \downarrow{\scriptstyle H^*i} & & \downarrow{\scriptstyle H^*j} \\ H^*(N) & \xrightarrow{H^*f} & H^*(M) \end{array}$$

where the vertical arrow are induced by the inclusions $i : (N, \emptyset) \to (N, N-Q)$ and $j : (M, \emptyset) \to (M, M-P)$. Then,

$$\begin{aligned} \mathrm{PD}(P) &= H^*j(U(M,P)) && \text{by Lemma 5.4.2} \\ &= H^*j \circ H^*f(U(N,Q)) && \text{by transversality and Lemma 4.7.30} \\ &= H^*f \circ H^*i(U(N,Q)) \\ &= H^*f(\mathrm{PD}(Q)) && \text{by Lemma 5.4.2.} \end{aligned}$$

□

5.4.2 The Gysin Homomorphism

Let (M, Q) be a pair of smooth compact manifolds, with Q closed. Let $i : Q \to M$ denote the inclusion. Set $q = \dim Q$ and $m = \dim M = q + r$. The *Gysin homomorphism* $\mathrm{Gys} : H^p(Q) \to H^{p+r}(M, \mathrm{Bd}\, M)$ is defined for all $p \in \mathbb{N}$ by the composed homomorphism

$$H^p(Q) \xrightarrow[\approx]{\frown[Q]} H_{q-p}(Q) \xrightarrow{H_*i} H_{q-p}(M) \xleftarrow[\approx]{\frown[M]} H^{p+r}(M, \mathrm{Bd}\, M).$$

The notation $i_!$ and the terminology *umkehr homomorphism* are also used in the literature.

For example, $\mathrm{Gys}(\mathbf{1}) = \mathrm{PD}(Q)$, the Poincaré dual of Q. More generally:

Lemma 5.4.8 *For $a \in H^p(M)$, one has* $\mathrm{Gys}\big(H^*i(a)\big) = a \smile \mathrm{PD}(Q)$.

Proof

$$\begin{aligned} (a \smile \mathrm{PD}(Q)) \frown [M] &= a \frown (\mathrm{PD}(Q)) \frown [M]) \\ &= a \frown H_*i([Q]) \\ &= H_*i\big(H^*i(a) \frown [Q]\big) && \text{by Proposition 4.5.4} \\ &= \mathrm{Gys}\big(H^*i(a)\big) \frown [M] \end{aligned}$$

As $- \frown [M]$ is an isomorphism, this proves the lemma. □

Example 5.4.9 Let M be the total space of an r-disk bundle $\pi : M \to Q$. We see Q as a submanifold of M via there 0-section $i : Q \to M$. Let $U \in H^r(M, \mathrm{Bd}\, M)$ be the Thom class. Since $U \frown [M] = H_*i([Q])$, we see that $U = \mathrm{PD}(Q)$. For $b \in H^p(Q)$, one has

$$\begin{aligned}
\mathrm{Gys}(b) \frown [M] &= \mathrm{Gys}\big(H^*i \circ H^*\pi(b)\big) \frown [M] && \text{since } \pi\circ i = \mathrm{id}_Q \\
&= \big(H^*\pi(b) \smile \mathrm{PD}(Q)\big) \frown [M] && \text{by Lemma 5.4.8} \\
&= \big(H^*\pi(b) \smile U\big) \frown [M] && \text{since } U = \mathrm{PD}(Q) \\
&= \mathrm{Thom}(b) \frown [M]\,.
\end{aligned}$$

Since $- \frown [M]$ is an isomorphism, we see in this example that *the Gysin homomorphism is identified with the Thom isomorphism.*

Proposition 5.4.10 *Let (M, Q) be a pair of smooth closed manifolds, with Q of codimension r. Let W be a closed tubular neighbourhood of Q in M. There is a commutative diagram*

$$\begin{array}{ccccccccc}
\to & H^{p-1}(M-Q) & \to & H^{p-r}(Q) & \xrightarrow{\mathrm{Gys}} & H^p(M) & \to & H^p(M-Q) & \to \\
& \downarrow & & \downarrow{=} & & \downarrow & & \downarrow & \\
\to & H^{p-1}(\mathrm{Bd}\,W) & \to & H^{p-r}(Q) & \to & H^p(Q) & \to & H^p(\mathrm{Bd}\,W) & \to
\end{array}$$

where the vertical arrows are induced by inclusions. The horizontal lines are exact sequences and the bottom one is the Gysin sequence of the sphere bundle $\mathrm{Bd}\,W \to Q$.

Proof We start with the commutative diagram

$$\begin{array}{ccccccccc}
\to & H^{p-1}(M-Q) & \to & H^p(M, M-Q) & \to & H^p(M) & \to & H^p(M-Q) & \to \\
& \downarrow & & \downarrow & & \downarrow & & \downarrow & \\
\to & H^{p-1}(\mathrm{Bd}\,W) & \to & H^p(W, \mathrm{Bd}\,W) & \to & H^p(W) & \to & H^p(\mathrm{Bd}\,W) & \to
\end{array}$$

using the cohomology sequences of the pairs $(M, M-Q)$ and $(W, \mathrm{Bd}\,W)$. To get the diagram of the proposition, we use the identification

$$H^{p-r}(Q) \underset{\approx}{\xrightarrow{\mathrm{Thom}}} H^p(W, \mathrm{Bd}\,W) \underset{\approx}{\longleftarrow} H^p(M, M-Q)$$

and $H^*(W) \approx H^*(Q)$. Thus the bottom line is the Gysin sequence of $\mathrm{Bd}\,W \to Q$. It remains to identify the homomorphism $H^{p-r}(Q) \to H^p(M)$ with the Gysin homomorphism. This amounts to the commutativity of the diagram.

$$\begin{array}{ccccccc}
H^{p-r}(Q) & \underset{\approx}{\xrightarrow{\mathrm{Thom}}} & H^p(W, \mathrm{Bd}\,W) & \underset{\approx}{\longleftarrow} & H^p(M, M - \mathrm{int}\,W) & \longrightarrow & H^p(M) \\
\approx\downarrow{-\frown[Q]} & & \approx\downarrow{-\frown[W]} & & & & \approx\downarrow{-\frown[M]} \\
H_{q-p+r}(Q) & \longrightarrow & H_{q-p+r}(W) & & \longrightarrow & & H_{q-p+r}(M)
\end{array}$$

The commutativity of the left square was observed in Example 5.4.9. That of the right square may be checked using simplicial (co)homology for a triangulation of M extending one of W. □

Proposition 5.4.11 *Let $f : M' \to M$ be a smooth map between closed manifolds. Let Q be a closed submanifold of codimension r in M. Suppose that f is transversal to Q. Then, for all $p \in \mathbb{N}$, the diagram*

$$\begin{array}{ccc} H^p(Q) & \xrightarrow{Gys} & H^{p+r}(M) \\ \downarrow{\scriptstyle H^*f} & & \downarrow{\scriptstyle H^*f} \\ H^p(f^{-1}(Q)) & \xrightarrow{Gys} & H^{p+r}(M') \end{array}$$

is commutative.

Proof Let $Q' = f^{-1}(Q)$. By transversality, $f : Q' \to Q$ is covered by a morphism of vector bundle $\tilde{f} : \nu(M', Q') \to \nu(M, Q)$. Put a Riemannian metric on $\nu(M, Q)$ and pull it back on $\nu(M', Q')$, so that $\tilde{f}$ is an isometry on each fiber. By standard technique of Riemannian geometry, one can find a tubular neighbourhood W of Q and a tubular neighbourhood W' of Q' and modify f by a homotopy relative to Q' so that $f(W') \subset W$, $f(\mathrm{Bd}W') \subset \mathrm{Bd}W$, $f(M' - \mathrm{int}\, W') \subset M - \mathrm{int}\, W$ and $f : W' \to W$ coincides with $\tilde{f}$ via the exponential maps of W and W'. We thus get a diagram.

$$\begin{array}{ccccccc} H^{p-r}(Q) & \xrightarrow[\approx]{\text{Thom}} & H^p(W, \mathrm{Bd}\, W) & \xleftarrow[\approx]{} & H^p(M, M - \mathrm{int}\, W) & \longrightarrow & H^p(M) \\ \downarrow{\scriptstyle H^*f} & & \downarrow{\scriptstyle H^*f} & & \downarrow{\scriptstyle H^*f} & & \downarrow{\scriptstyle H^*f} \\ H^{p-r}(Q') & \xrightarrow[\approx]{\text{Thom}} & H^p(W', \mathrm{Bd}\, W') & \xleftarrow[\approx]{} & H^p(M', M' - \mathrm{int}\, W') & \longrightarrow & H^p(M') \end{array} .$$

The left square is commutative by construction and the functoriality of the Thom isomorphism (coming from Lemma 4.7.30). The other squares are obviously commutative. But, as seen in the proof of Proposition 5.4.10, the compositions from the left end to the right end of the horizontal lines are the Gysin homomorphisms. □

5.4.3 Intersections of Submanifolds

Consider two closed submanifolds Q_i ($i = 1, 2$) of the compact smooth n-manifold M, Q_i being of codimension r_i. We suppose that Q_1 and Q_2 intersect transversally. Then, $Q = Q_1 \cap Q_2$ is a closed submanifold of codimension $r = r_1 + r_2$.

Proposition 5.4.12 *Under the above hypotheses*

$$\mathrm{PD}(Q) = \mathrm{PD}(Q_1) \smile \mathrm{PD}(Q_2) .$$

Proof As $(M - Q_1) \cup (M - Q_2) = M - Q$, the cup product provides a bilinear map

$$H^{r_1}(M, M - Q_1) \times H^{r_2}(M, M - Q_2) \xrightarrow{\smile} H^r(M, M - Q)\,.$$

In virtue of Lemma 5.4.2, it suffices to prove that

$$U(M, Q_1) \smile U(M, Q_2) = U(M, Q)\,. \tag{5.4.4}$$

We may suppose that $Q \neq \emptyset$ for, otherwise, the proposition is trivially true, since $q = 0$ and $r > n$.

If A is a submanifold of B, we denote by $\nu(B, A)$ the normal bundle of A in B. Choose an embedding $\mu : D(\nu(B, A)) \hookrightarrow B$ parameterizing a tubular neighbourhood $W(B, A)$. If $V \subset B$ the notation $W(A, B)_V$ means $\mu(D(\nu(B, A)_V))$. As Q_1 and Q_2 intersect transversally, one has

$$\nu(M, Q) = \nu(Q_1, Q)_{|Q} \oplus \nu(Q_2, Q)_{|Q}\,.$$

Let $b \in Q$. One may choose convenient tubular neighbourhood parameterizations so that $W(M, Q)_{\{b\}} \cap Q_j = W(Q_j, Q)_{\{b\}}$. Let $W_1 = W(Q_1, Q)_{\{b\}} \approx D^{r_2}$, $W_2 = W(Q_2, Q)_{\{b\}} \approx D^{r_1}$ and $W = W(M, Q)_{\{b\}} \approx W_1 \times W_2 \approx D^r$. Let $\pi_i : W \to W_i$ be the projection. By Lemma 5.4.2:

- the class $U(M, Q_1) \in H^{r_1}(M, M - Q_1)$ restricts to the non-zero element $a_1 \in H^{r_1}(W, W_1 \times \mathrm{Bd}\, W_2) = \mathbb{Z}_2$;
- the class $U(M, Q_2) \in H^{r_2}(M, M - Q_2)$ restricts to the non-zero element $a_2 \in H^{r_2}(W, \mathrm{Bd}\, W_1 \times W_2) = \mathbb{Z}_2$.

Hence, $U(M, Q_1) \smile U(M, Q_2)$ restricts to $a_1 \smile a_2 \in H^r(W, \mathrm{Bd}\, W) = \mathbb{Z}_2$. We have to prove that $a_1 \smile a_2 \neq 0$. Let $0 \neq \tilde{a}_i \in H^{r_i}(W_j, \mathrm{Bd}\, W_j)$ $(i \neq j)$. Then

$$a_1 \smile a_2 = H^*\pi_1(\tilde{a}_1) \smile H^*\pi_2(\tilde{a}_2) = \tilde{a}_1 \times \tilde{a}_2\,.$$

By the relative Künneth theorem 4.6.10, $\tilde{a}_1 \times \tilde{a}_2 \neq 0$ in $H^r(W, \mathrm{Bd}\, W)$. Hence, $U(M, Q_1) \smile U(M, Q_2)$ restricts to the non-zero element of $H^r(W(M, Q)_{\{b\}}, \mathrm{Bd}\, W(M, Q)_{\{b\}})$ for all $b \in Q$. By Lemma 5.4.2, this proves (5.4.4). □

An interesting case is when $\dim Q_1 + \dim Q_2 = \dim M$. If Q_1 and Q_2 intersect transversally, then $Q_1 \cap Q_2$ is a finite collection of points. Let $j_i : Q_i \to M$ denote the inclusion. The following result says that the parity of this number of points depends only on $[Q_i]_M = H_* j_i([Q_i])$ and justifies the terminology of *intersection form*.

Corollary 5.4.13 *Let Q_i $(i = 1, 2)$ be two closed submanifolds of the compact smooth n-manifold M, with* $\dim Q_1 + \dim Q_2 = n$. *Let* $q_i = \mathrm{PD}(Q_i)$. *Suppose that Q_1 and Q_2 intersect transversally. Then*

$$\sharp(Q_1 \cap Q_2) \equiv \langle q_1 \smile q_2, [M]\rangle \mod 2\,.$$

In other words,

$$\sharp\,(Q_1 \cap Q_2) \equiv H_* j_1([Q_1)] \cdot H_* j_2([Q_2)) \mod 2\,,$$

where "·" denotes the (absolute) intersection form (see Sect. 5.3.3).

Proof One has

$$\begin{aligned}\langle q_1 \smile q_2, [M]\rangle &= \langle \mathbf{1}, (q_1 \smile q_2) \frown [M]\rangle \\ &= \langle \mathbf{1}, [Q_1 \cap Q_2]\rangle \equiv \sharp\,(Q_1 \cap Q_2) \mod 2\,.\end{aligned}$$

□

Lemma 5.4.14 *Let $\xi = (p : E \to N)$ be a smooth vector bundle over a closed smooth manifold N. Let $\sigma, \sigma' : N \to E$ be two smooth sections of ξ which are transversal. Let Q be the submanifold of N defined by $Q = \sigma^{-1}\big(\sigma(N) \cap \sigma'(N)\big)$. Then, the Poincaré dual of Q in N is the Euler class $e(\xi)$ of ξ.*

Proof We us the following notation: if $\lambda : Y \to X$ is a continuous map and Y a closed manifold, we write $[Y]_X = H_*\lambda([Y]) \in H_{\dim Y}(X)$; the map λ is usually implicit, being an inclusion or an embedding obvious from the context.

Endow ξ with an Euclidean structure and consider the pair $(D, S) = (D(\xi), S(\xi))$ of the associated unit disk and sphere bundle. Using a homotopy in each fiber, we can assume that $\sigma(N)$ and $\sigma'(N)$ are contained in the interior of D. All the sections of a bundle are homotopic. By Lemma 5.4.2 and its proof, the Thom class U of ξ is the Poincaré dual in D of $[N]_D = H_*\sigma([N]) = H_*\sigma'([N])$. By Proposition 5.4.12, $U \smile U$ is the Poincaré dual in D of $[Q]_D$. Let $j : (D, \emptyset) \to (D, S)$ denote the pair inclusion. As $p \circ \sigma = \mathrm{id}_N$, one has

$$\begin{aligned}[Q]_N &= H_* p([Q]_D) \\ &= H_* p\big((U \smile U) \frown [D]\big) && \text{cap product of (4.5.9)} \\ &= H_* p\big(U \frown (U \frown [D])\big) && \text{by Formula (4.5.10)} \\ &= H_* p\big(U \frown [N]_D\big) \\ &= H_* p\big(H^* j(U) \frown [N]_D\big) && \text{by definition of the cap product (4.5.7)} \\ &= H_* p\big(H^* p(e(\xi)) \frown [N]_D\big) && \text{by definition of the Euler class} \\ &= e(\xi) \frown H_* p([N]_D) \\ &= e(\xi) \frown [N]\,.\end{aligned}$$

which proves the lemma. □

When, in Lemma 5.4.14, the rank of ξ is equal to the dimension of the manifold N, then $\sigma(N) \cap \sigma'(N)$ is a finite collection of point and one gets the following corollary.

Corollary 5.4.15 *Let $\xi = (p : E \to N)$ be a smooth vector bundle of rank n over a closed smooth n-manifold N. Let $\sigma, \sigma' : N \to E$ be two smooth sections of ξ which are transversal. Then*

$$\sharp\,\big(\sigma(N)\cap\sigma'(N)\big)\equiv\langle e(\xi),[N]\rangle \mod 2\,.$$

The following corollary is a justification for the name *Euler class.*

Corollary 5.4.16 *Let N be a smooth closed manifold. Then the following congruences* mod 2 *hold*:

$$\langle e(TN),[N]\rangle\equiv\chi(N)\equiv\dim H_*(N) \mod 2\,.$$

Proof As $\chi(N)=\sum_i(-1)^i\dim H_i(N)$, the second congruence is obvious. Let $\sigma_0: N\to D(TN)$ be the zero section and let $\sigma: N\to D(TN)$ be another smooth section (i.e. a vector field on N) which is transversal to σ_0. By Corollary 5.4.15, the number of zeros of σ is congruent mod 2 to $\langle e(TN),[N]\rangle$. It then suffices to find some vector field transversal to σ_0 for which we know that its number of zeros is congruent mod 2 to $\chi(N)$. Observe that, for a finite CW-complex X, the following congruence mod 2 holds

$$\chi(X)\equiv\dim H_*(X)\equiv\sharp\,\Lambda(X) \mod 2\,,$$

where $\Lambda(X)$ is set of cells of X. For the required vector field, one can take the gradient vector field $\sigma=\text{grad}\, f$ of a Morse function $f: N\to\mathbb{R}$. Then $\sigma_0^{-1}\big(\sigma(N)\cap\sigma_0(N)\big)=\text{Crit}\, f$, the set of critical points of f. The transversality of σ with σ_0 is equivalent to f being a Morse function (see [95, Chap. 6]). By Morse theory, N has then the homotopy type of a CW-complex X with $\sharp\,\Lambda(X)=\sharp\,\text{Crit}\, f$, [95, Chap. 6, Theorem 4.1]. One can also use the classical vector field associated to a C^1-triangulation of N, with one zero at the barycenter of each simplex (see, e.g. [180, pp. 611–612]). □

We give below a second proof of Corollary 5.4.16, using the following lemma.

Lemma 5.4.17 *Let Δ_N be the diagonal submanifold of $M=N\times N$, with normal bundle $\nu(M,\Delta_N)$. Then, there is a canonical isomorphism of vector bundles*

$$\nu(M,\Delta_N)\approx TN\,.$$

Proof Let $p_1,p_2: N\times N\to N$ be the projections onto the first and second factor. For $x\in N$, consider the commutative diagram in the category of real vector spaces

$$\begin{array}{ccccccccc}
0 & \longrightarrow & T_{(x,x)}(\Delta_N) & \longrightarrow & T_{(x,x)}(N\times N) & \longrightarrow & \nu_{(x,x)}(M,\Delta_N) & \longrightarrow & 0\\
 & & \downarrow & & \downarrow \phi & & \vdots\,\bar\phi & & \\
0 & \longrightarrow & \Delta(T_xN) & \longrightarrow & T_xN\times T_xN & \xrightarrow{\ -\ } & T_xN & \longrightarrow & 0
\end{array}$$

where the rows are exact and $\phi(v) = (Tp_1(v), Tp_2(v))$. The map ϕ is an isomorphism and sends $T_{(x,x)}(\Delta_N)$ onto $\Delta(T_x N)$. Hence, ϕ descends to the isomorphism $\bar{\phi} : \nu(M, \Delta_N) \xrightarrow{\approx} T_x N$ which, of course, depends continuously on x. □

Second proof of Corollary 5.4.16 We consider N as the diagonal submanifold of $M = N \times N$, with the diagonal inclusion $\Delta : N \to M$. The normal bundle $\nu(M, N)$ is isomorphic to the tangent bundle of M by Lemma 5.4.17.

By (5.4.1), the Poincaré dual of N is equal to $\sum_i a_i \times b_i$, where $\mathcal{A} = \{a_1, a_2, \dots\}$ and $\mathcal{B} = \{b_1, b_2, \dots\}$ are bases of $H^*(N)$ dual one to the other for the Poincaré duality.

$$\begin{aligned} \langle e(TM), [N]\rangle &= \langle e(\nu(M,N)), [N]\rangle \\ &= \langle H^*\Delta\left(\textstyle\sum_i a_i \times b_i\right), [N]\rangle && \text{by Lemma 5.4.4} \\ &= \langle \textstyle\sum_i a_i \smile b_i, [N]\rangle \equiv \dim H^*(N) \mod 2\,. \end{aligned}$$

Example 5.4.18 As $\chi(S^n) \equiv 0 \mod 2$, the Euler class of TS^n vanishes by Corollary 5.4.16. Let T^1S^n be the associated sphere bundle. By Proposition 4.7.35 and the Leray-Hirsch theorem, we get an isomorphism of $H^*(S^n)$-module

$$H^*\left(T^1S^n\right) \approx H^*\left(S^n\right) \otimes H^*\left(S^{n-1}\right).$$

If $n \geq 3$, Poincaré duality implies that this isomorphism is a ring-isomorphism. This is not true if $n = 2$ (see Remark 4.7.34).

Thus, for $n \geq 3$, T^1S^n has the same cohomology ring as $S^n \times S^{n-1}$. However, by [109, Theorem 1.12], these two spaces have the same homotopy type if and only if there exists a map $f : S^{2n+1} \to S^{n+1}$ with Hopf invariant one (see Sect. 6.3). By Theorem 8.6.7, such an f exists if and only if $n = 1, 3, 7$.

5.4.4 The Linking Number

Let Q and Q' be two disjoint closed submanifold of a closed manifold Σ (say, in the smooth category), with $q = \dim Q$, $q' = \dim Q'$ and $s = \dim \Sigma$. We assume that

(1) $q + q' = s - 1$.
(2) Σ is a $\mathbb{Z}_2$-homology sphere, i.e. $H_*(\Sigma) \approx H_*(S^s)$.
(3) $\langle \mathbf{1}_Q, [Q]\rangle = \langle \mathbf{1}_{Q'}, [Q']\rangle = 0$. This condition is always satisfied when q and q' are not zero. If, say, $q = 0$, it means that Q has an even number of points, so that $[Q] \in \tilde{H}_0(Q) = \ker\langle \mathbf{1}_Q, \,\rangle$.

Thanks to (2), Alexander duality (see Theorem 5.3.14) provides an isomorphism

$$A : \tilde{H}_q(Q) \xrightarrow{\approx} \tilde{H}^{s-q-1}(\Sigma - Q)\,.$$

Note that $s - q - 1 = q'$. By Condition (3), $[Q] \in \tilde{H}_q(Q)$ and $[Q'] \in \tilde{H}_{q'}(Q')$, so we can define the *linking number* (sometimes called the *linking coefficient*) $l(Q, Q')$ of Q and Q' in Σ by

$$l(Q, Q') = \langle A[Q], \tilde{H}_* i([Q']) \rangle \in \mathbb{Z}_2 , \tag{5.4.5}$$

where $i : Q' \to \Sigma$ denotes the inclusion. Although the asymmetry of the definition, the equality $l(Q, Q') = l(Q', Q)$ will be proven in Proposition 5.4.25.

The linking number $l(Q, Q')$ was introduced in 1911 by Lebesgue [131, pp. 173–175], with a definition in the spirit of Proposition 5.4.22. Lebesgue called Q and Q' "enlacées" if $l(Q, Q') = 1$. One year later, Brouwer [22, pp. 511–520] refined the idea when Σ, Q and Q' are oriented, defining an *integral linking number* whose reduction mod 2 is $l(Q, Q')$ (for the philosophy of Brouwer's definition, see Remark 2.5.10). More history and references about the linking numbers may be found in [40, pp. 176–179 and 185].

As Q is a submanifold of Σ, the isomorphism A may be described in the following way, which will be useful for computations. Let V be a closed tubular neighbourhood of Q in $\Sigma - Q'$ and let $X = \Sigma - \mathrm{int} V$. The cohomology sequence of (Σ, X)

$$H^{s-q-1}(\Sigma) \longrightarrow H^{s-q-1}(X) \xrightarrow{\delta^*} H^{s-q}(\Sigma, X) \longrightarrow H^{s-q}(\Sigma)$$

shows that the connecting homomorphism δ^* descends to an injection $\tilde{\delta}^*$ of

$$\mathrm{coker}\,\big(H^{s-q-1}(\Sigma) \to H^{s-q-1}(X)\big) \approx \tilde{H}^{s-q-1}(X)$$

into $H^{s-q}(\Sigma, X)$. Let $j : (V, \mathrm{Bd}\, V) \to (\Sigma, X)$ denote the pair inclusion. Identifying $H_*(V)$ with $H_*(Q)$, one has the diagram

$$\begin{array}{ccccc}
\tilde{H}^{s-q-1}(X) & & \xleftarrow[\approx]{A} & & \tilde{H}_q(Q) \\
\downarrow{\scriptstyle \tilde{\delta}^*} & & & & \downarrow \\
H^{s-q}(\Sigma, X) & \xrightarrow[\approx]{H^* j} & H^{s-q}(V, \mathrm{Bd} V) & \xrightarrow[\approx]{\frown [V]} & H_q(Q) \\
\downarrow & & & & \downarrow \\
H^{s-q}(\Sigma) & & \xrightarrow[\approx]{\frown [\Sigma]} & & H_q(\Sigma)
\end{array} \tag{5.4.6}$$

whose columns are exact (the right hand one by Lemma 3.1.10) and whose bottom part is commutative by Proposition 5.3.6. Then A is the unique isomorphism making the top rectangle commutative (compare the proof of Theorem 5.3.14).

Remark 5.4.19 Diagram (5.4.6) uses the singular (co)homology. But, via a triangulation of Σ, the various spaces may be the geometric realizations of simplicial complexes (by abuse of notations we use the same letters, i.e. $\Sigma = |\Sigma|$, etc.).

Then, Diagram (5.4.6) makes sense for simplicial (co)homology; the isomorphism $H^{s-q}(\Sigma, X) \approx H^{s-q}(V, \mathrm{Bd}V)$ is just the simplicial excision (see Exercise 2.17).

Remark 5.4.20 Suppose that Q is connected or consists of two points. Then $\tilde{H}_{q'}(\Sigma - Q) \approx \tilde{H}_q(Q) \approx \mathbb{Z}_2$. Therefore, $l(Q, Q') = 1$ if and only if $H_*i([Q']) \neq 0$ in $H_{q'}(\Sigma - Q)$. In this case, $l(Q, Q')$ determines $H_*i([Q'])$.

The following lemma shows that $l(Q, Q')$ is not always zero. We say that Q' is a *meridian sphere* for Q if Q' is the boundary of a $(s-q)$-disk Δ in Σ intersecting Q transversally in one point.

Lemma 5.4.21 *Let Q, Q' and Σ satisfying* (1)–(3) *above. Suppose that Q' is a meridian sphere for Q. Then $l(Q, Q') = 1$.*

Proof Let $\nu(Q, \Sigma)$ be the normal bundle of Q in Σ. A Riemannian metric provides a smooth bundle pair $(D(\nu), S(\nu))$ with fiber $\big(D^{m-q}, S^{m-q-1}\big)$ and a diffeomorphism $\phi : (D(\nu), S(\nu)) \xrightarrow{\approx} (V, \mathrm{Bd}V)$, where V is a tubular neighbourhood of Q in Σ. By choosing the Riemannian metric conveniently, we may assume that $\Delta \cap V$ is the image by ϕ of a fiber D^{m-q} of $D(\nu)$. One has $\tilde{H}_*i([Q']) = \tilde{H}_*j \circ \tilde{H}_*\phi([S^{m-q-1}])$, where $j : (V, \mathrm{Bd}V) \to (\Sigma, \Sigma - Q)$ denotes the pair inclusion, so $l(Q, Q') = l(Q, \phi(S^{m-q-1}))$ by (5.4.5).

In Diagram (5.4.6), one has $H^*\phi \circ H^*j \circ \tilde{\delta}^* \circ A([Q]) = U$, the Thom class of ν, as can be checked on each connected component of Q. Therefore,

$$\begin{aligned} l(Q, S) &= \langle A([Q]), \tilde{H}_*i([Q'])\rangle \\ &= \langle A([Q]), \tilde{H}_*j \circ \tilde{H}_*\phi([S^{m-q-1}])\rangle \\ &= \langle A([Q]), \partial_* \circ \tilde{H}_*j \circ \tilde{H}_*\phi([D^{m-q}])\rangle \\ &= \langle \tilde{H}^*\phi \circ \tilde{H}^*j \circ \tilde{\delta}^* \circ A([Q]), [D^{m-q}]\rangle \\ &= \langle U, [D^{m-q}]\rangle \end{aligned}$$

and $\langle U, [D^{m-q}]\rangle = 1$ by Lemma 4.7.28. □

The following proposition gives a common way to compute a linking number, related to the original definition of Lebesgue [131, pp. 173–175]. Let Q, Q' and Σ satisfying (1)–(3) above. Suppose that there exists a compact manifold W with $\mathrm{Bd}\, W = Q'$ so that the inclusion of Q' into Σ extends to a map $j : W \to \Sigma$ which is transverse to Q (j needs not to be an embedding). Then $j^{-1}(Q)$ is a finite number of points in W.

Proposition 5.4.22 $l(Q, Q') = \sharp\ \big(j^{-1}(Q)\big) \mod 2\,.$

Proof Let $k = \sharp\ \big(j^{-1}(Q)\big) \in \mathbb{N}$. Let W_0 be the manifold W minus an open tubular neighbourhood of $j^{-1}(Q)$. By (5.4.5), $l(Q, Q')$ depends only on the homology class $H_*i([Q']) \in H_{q'}(\Sigma - Q)$ which, thanks to the map j (see Exercise 5.6), is the same as that of k meridian spheres. The result then follows from Lemma 5.4.21. □

We now introduce some material for Lemma 5.4.23, which will enable us to compute linking numbers using convenient singular cochains. Let Q, Q' and Σ satisfying (1)–(3) above. Let W be a closed tubular neighbourhood of Q in $\Sigma - Q'$ and let V be a closed tubular neighbourhood of Q in int W. We also consider the symmetric data $Q' \subset V' \subset W' \subset \Sigma - Q$, assuming that $W \cap W' = \emptyset$. Let $\mathcal{B} = \{W, W', \Sigma - (V \cup V')\}$; note that the small simplex theorem 3.1.34 holds true for $\mathcal{B}$.

Let $c \in Z^{s-q}(W, W - \text{int}V)$ be a singular cocycle representing the Poincaré dual class of Q in $H^{s-q}(W, W - \text{int}V) \approx H^{s-q}(V, \text{Bd}V)$. We can see c as a cocycle of W and take its zero extension $\bar{c} \in C_{\mathcal{B}}^{s-q}(\Sigma)$, i.e.

$$\langle \bar{c}, \sigma \rangle = \begin{cases} \langle \bar{c}, \sigma \rangle & \text{if } \sigma \in \mathcal{S}_{s-q}(W) \\ 0 & \text{otherwise.} \end{cases}$$

Since c vanishes on $\mathcal{S}_{s-q}(W, W - \text{int}V)$, the cochain $\bar{c}$ is a $\mathcal{B}$-small cocycle, i.e. $\bar{c} \in Z_{\mathcal{B}}^{s-q}(\Sigma)$. We claim that we can choose $a \in C_{\mathcal{B}}^{s-q-1}(\Sigma)$ such that $\delta a = \bar{c}$. Indeed, by (2) and the small simplex theorem 3.1.34, $0 = H^{s-q}(\Sigma) \approx H_{\mathcal{B}}^{s-q}(\Sigma)$ when $q > 0$. When $q = 0$, $\bar{c}$ represents the Poincaré dual class of Q in $H_{\mathcal{B}}^{s}(\Sigma) \approx H^s(\Sigma)$. But, by (3), $[Q]$ represents 0 in $H_0(\Sigma)$, so $\bar{c}$ represents 0 in $H_{\mathcal{B}}^{s}(\Sigma)$. A cochain $a' \in C_{\mathcal{B}}^{s-q'-1}(\Sigma)$ with $\delta(a') = \bar{c}'$ may be also chosen, using the symmetric data $Q' \subset V' \subset W' \subset \Sigma - Q$.

Finally, let $\mu' \in Z_{q'}(\Sigma - \text{int } V)$ represent $H_*i([Q'])$ and let $\nu \in Z_s^{\mathcal{B}}(\Sigma)$ represent $[\Sigma]$ in $H_s^{\mathcal{B}}(\Sigma) \approx H_s(\Sigma)$.

Lemma 5.4.23 *The following equalities hold true.*

(a) $l(Q, Q') = \langle a, \mu' \rangle$.
(b) $l(Q, Q') = \langle a \smile \bar{c}', \nu \rangle$.

Proof We first establish some preliminary steps.
Step 1: If $a_1, a_2 \in C^{s-q-1}(\Sigma)$ *satisfy* $\delta(a_1) = \delta(a_2)$, *then* $\langle a_1, \mu' \rangle = \langle a_2, \mu' \rangle$. Indeed, one has then $\delta(a_1 + a_2) = 0$. If $s - q - 1 = q' > 0$, Condition (2) implies that there exists $b \in C^{s-q-2}(\Sigma)$ such that $\delta b = a_1 + a_2$. Hence,

$$\langle a_1, \mu' \rangle + \langle a_2, \mu' \rangle = \langle a_1 + a_2, \mu' \rangle = \langle \delta b, \mu' \rangle = \langle b, \partial \mu' \rangle = 0 .$$

If $q' = 0$, then $\delta(a_1 + a_2) = 0$ implies that $a_1 + a_2 = \mathbf{1}$ by Proposition 3.1.8, thus $\langle a_1 + a_2, \mu' \rangle = 0$ by Condition (3).
Step 2: Let $c_i \in Z^{s-q}(W, W - \text{int}V)$ $(i = 1, 2)$ *be singular cocycles both representing the Poincaré dual class of* Q *in* $H^{s-q}(W, W - \text{int}V)$. *Let* $a_i \in C_{\mathcal{B}}^{s-q-1}(\Sigma)$ *such that* $\delta a_i = \bar{c}_i$ *as above. Then* $\langle a_1, \mu' \rangle = \langle a_2, \mu' \rangle$. Indeed, there exists $b \in C^{s-q-1}(W, W - \text{int}V)$ such that $\delta(b) = c_1 + c_2$. Its zero extension $\bar{b} \in C_{\mathcal{B}}^{s-q-1}(\Sigma)$ then satisfies $\delta(\bar{b}) = \bar{c}_1 + \bar{c}_2$ and then $\delta(a_1 + \bar{b}) = \bar{c}_2$. Thus

$$\begin{aligned}\langle a_2, \mu' \rangle &= \langle a_1 + \bar{b}, \mu' \rangle \text{ by Step 1} \\ &= \langle a_1, \mu' \rangle \qquad \text{since } \langle \bar{b}, \mu' \rangle = 0 \,.\end{aligned}$$

We can now start the proof of Lemma 5.4.23. Given Steps 1 and 2, it is enough to prove (a) for a particular choice of c and a. We use Diagram (5.4.6) and see A as an isomorphism from $\tilde{H}_q(Q)$ onto $\tilde{H}^{s-q-1}(\Sigma - \operatorname{int} V)$. Let $a \in Z^{s-q-1}(\Sigma - \operatorname{int} V)$ represent $A([Q])$. Let $\bar{a} \in C_{\mathcal{B}}^{s-q-1}(\Sigma)$ be its zero extension and let $\bar{c} = \delta(\bar{a}) \in Z_{\mathcal{B}}^{s-q}(\Sigma)$. By Lemma 3.1.17, $\bar{c}$ represents $\tilde{\delta}^*(A([Q])$. Also, $\bar{c}$ is the zero extension of the cocycle $c \in Z^{s-q}(W, W - \operatorname{int} V)$ which, by definition of A and Diagram (5.4.6), represents the Poincaré dual class of Q in $H^{s-q}(W, W - \operatorname{int} V)$. Therefore, since a represents $A([Q])$ and $\mu' \in Z_{q'}(\Sigma - \operatorname{int} V)$ represents $H_*i([Q'])$, one has $l(Q, Q') = \langle a, \mu' \rangle = \langle \bar{a}, \mu' \rangle$.

To prove (b), consider the pair inclusions $j_1 : (\Sigma, \emptyset) \to (\Sigma, \Sigma - \operatorname{int} V')$ and $j_2 : (W', W' - \operatorname{int} V') \to (\Sigma, \Sigma - \operatorname{int} V')$. Since $\nu \in Z_s^{\mathcal{B}}(\Sigma)$, there exists a (unique) $\nu' \in Z_s(W', W' - \operatorname{int} V')$ such that $C_* j_2(\nu') = C_* j_1(\nu)$. As $H_* j_1$ and $H_* j_2$ are isomorphisms, ν' represents the generator of $H_s(W', W' - \operatorname{int} V') = \mathbb{Z}_2$. Therefore, $\bar{c}' \frown \nu = c' \frown \nu'$ represents $H_*i([Q'])$ and, by (a),

$$l(Q, Q') = \langle a, \bar{c}' \frown \nu \rangle = \langle a \smile \bar{c}', \nu \rangle \,.$$

□

Remark 5.4.24 The proof of Lemma 5.4.23 in the simplicial category (see Remark 5.4.19) is somewhat simpler. It uses only the tubular neighbourhoods V_i and not the W_i's, and, of course, does not require the use of small simplex techniques. Also, ν may be taken explicitly as $\mathcal{S}_s(\Sigma)$. Writing the details is left to the reader as an exercise.

Lemma 5.4.23 will be used for the Hopf invariant (see Sect. 6.3.3). For the moment, its main consequence is the following proposition.

Proposition 5.4.25 *Let Q, Q' and Σ satisfying* (1)–(3) *above. Then*

$$l(Q, Q') = l(Q', Q) \,.$$

Proof By Point (b) of Lemma 5.4.23, one has $l(Q, Q') = \langle a \smile \bar{c}', \nu \rangle$ and $l(Q', Q) = \langle a' \smile \bar{c}, \nu \rangle$. Then

$$\begin{aligned}l(Q, Q') + l(Q', Q) &= \langle a \smile \bar{c}' + a' \smile \bar{c}, \nu \rangle \\ &= \langle \delta(a \smile a'), \nu \rangle = \langle a \smile a', \partial(\nu) \rangle = 0 \,.\end{aligned}$$

□

5.5 Exercises for Chapter 5

5.1. Prove that the product of two homology manifolds is a homology manifold.

5.2. Let M be a compact manifold with boundary such that $\tilde{H}_*(M) = 0$. Show that the boundary of M is a homology sphere.

5.3. Check the Poincaré duality (Theorem 5.3.7) for the manifolds $S^1 \times I$ and the Möbius band.

5.4. Show that there is no continuous retraction of a non-empty compact manifold onto its boundary.

5.5. Let M be a closed triangulable manifold of dimension n. Prove that the homomorphism $H_k(M - \{pt\}) \to H_k(M)$ induced by the inclusion is an isomorphism for $k < n$.

5.6. Let M be a compact triangulable topological n-manifold with boundary $\mathrm{Bd}\, M = N$. Suppose that is $N = N_1 \dot{\cup} N_2$ the union of two closed $(n-1)$-manifolds. Let $f : M \to X$ be a continuous map. Show that $H_* f([N_1]) = H_* f([N_2])$ in $H_{n-1}(X)$. What happens if $N_2 = \emptyset$?

5.7. Let $f : M \to N$ be a map between closed n-dimensional manifolds of the same dimension. Show that the degree of f may be computed locally, using a topological regular value, like in Proposition 3.2.6.

5.8. Let Σ_m be the orientable surface of genus m and let $\bar{\Sigma}_n$ be the nonorientable surface of genus n. For which m and n does there exist a continuous map of degree one from Σ_m to $\bar{\Sigma}_n$ or from $\bar{\Sigma}_n$ to Σ_m?

5.9. Let M be a closed manifold of dimension m which is the products of two closed manifolds of positive dimensions. Does there exist a degree one map $f : M \to \mathbb{R}P^m$?

5.10. Let Q_1 and Q_2 be closed submanifolds of a closed manifold M (in the smooth category). Suppose that Q_1 and Q_2 intersect transversally in an odd number of points. Show that $[Q_1]$ and $[Q_2]$ represent non-zero classes in $H_*(M)$.

5.11. Let $i : Q \to M$ be the inclusion of a smooth closed submanifold of dimension q in a smooth closed manifold M. Suppose that $H_* i([Q]) = 0$ in $H_q(M)$. Prove that the Euler class of the normal bundle to Q vanishes.

5.12. For $A \subset \{0, 1, \dots, n\}$, let $P_A = \{[x_0 : \cdots : x_n] \in \mathbb{R}P^n \mid x_i = 0 \text{ when } i \notin A\}$. What is the diffeomorphism type of P_A? Show that, if $A \cup B = \{0, 1, \dots, n\}$, then P_A and P_B intersect transversally (what is the intersection?). How does Proposition 5.4.12 apply in this example?

5.13. *Poincaré dual classes in a product.* Let M_1 and M_2 be smooth compact manifolds. Let Q_i be a closed submanifold of M_i $(i = 1, 2)$. Then $Q_1 \times Q_2$ is a closed submanifold of $M_1 \times M_2$. Prove that

$$\mathrm{PD}(Q_1 \times Q_2) = \mathrm{PD}(Q_1) \times \mathrm{PD}(Q_2) \tag{5.5.1}$$

in $H^*(M_1 \times M_2)$.

5.14. *Poincaré dual classes in a product* II. Let M and M' be smooth closed manifolds and let $x \in M$ and $x' \in M'$. What are $\mathrm{PD}(\{x\} \times M')$ and $\mathrm{PD}(M \times \{x'\})$ in $H^*(M \times M')$? Check that $\mathrm{PD}(\{x\} \times M') \smile \mathrm{PD}(M \times \{x'\}) = \mathrm{PD}(\{(x, x')\})$.

5.15. Let Q and Q' be disjoint submanifolds of S^2, where Q consists of two circles and Q' of four points. Using Proposition 5.4.22, compute the linking numbers $l(Q, Q')$ and $l(Q', Q)$ for the various possibilities.

5.16. *Brouwer's definition of the linking number.* Let Q and Q' be two disjoint closed submanifolds of S^n, of dimension respectively q and q' satisfying $p + q = n - 1$. If Q (or Q') is of dimension 0, it should consist of an even number of points. See Q and Q' as submanifolds of $\mathbb{R}^n$ via a stereographic projection of $S^n - \{pt\}$ onto $\mathbb{R}^n$. Consider the *Gauss map* $\lambda : P \times Q \to S^{n-1}$ given by $\lambda(x, y) = \frac{x-y}{||x-y||}$. Show that the degree of λ is equal to the linking number $l(Q, Q')$ (see Sect. 5.4.4). [Hint: use Proposition 5.4.22.]

5.17. Write the proof of Lemma 5.4.23 in the simplicial category (see Remarks 5.4.19 and 5.4.24).

5.18. Let Σ be the unit sphere in $\mathbb{R}^{q+1} \times \mathbb{R}^{q'+1}$. Let $Q = S^q \times \{0\} \subset \Sigma$ and $Q' = \{0\} \times S^{q'} \subset \Sigma$. Compute the linking number $l(Q, Q')$ in Σ.

Chapter 6
Projective Spaces

Coming from algebraic geometry, projective spaces and their Hopf bundles play an important role in homotopy theory, as already seen in Sects. 3.8 and 4.3. The precise knowledge of their cohomology algebra has interesting applications, like the Borsuk-Ulam theorem, continuous multiplications in $\mathbb{R}^m$ and the Hopf invariant, which are presented in Sect. 6.2.

6.1 The Cohomology Ring of Projective Spaces—Hopf Bundles

The cohomology ring of $\mathbb{R}P^n$ for $n \leq \infty$ was established in Proposition 4.3.10, using the transfer (or Gysin) exact sequence for the double cover (S^0-bundle) $S^n \to \mathbb{R}P^n$. It gives a **GrA**-isomorphism $\mathbb{Z}_2[a]/(a^{n+1}) \to H^*(\mathbb{R}P^n)$. We give below a completely different proof of this fact, which is based on Poincaré duality (as $\mathbb{R}P^n$ is a smooth closed manifold, it can be triangulated as a polyhedral homology n-manifold: see p. 203). We shall also discuss the cases of complex and quaternionic projective spaces $\mathbb{C}P^n$ and $\mathbb{H}P^n$, and of the octonionic projective plane $\mathbb{O}P^2$.

Proposition 6.1.1 *The cohomology algebra of* $\mathbb{R}P^n$ *($n \leq \infty$) is given by*

$$H^*(\mathbb{R}P^n) \approx \mathbb{Z}_2[a]/(a^{n+1}), \quad H^*(\mathbb{R}P^\infty) \approx \mathbb{Z}_2[a]\,,$$

with $a \in H^1(\mathbb{R}P^n)$.

Proof We prove the first statement by induction on n. It is true for $n = 1$ since $\mathbb{R}P^1 = S^1$ (and also proven for $n = 2$ at p. 137). Suppose, by induction, that it is true in for $\mathbb{R}P^{n-1}$.

In Example 3.4.5 is given the standard CW-structure of $\mathbb{R}P^n$, with one k-cell for each $k = 0, 1, \ldots, n$. It follows that $H^k(\mathbb{R}P^n, \mathbb{R}P^{n-1}) = 0$ for $k \leq n-1$ and $H^n(\mathbb{R}P^n, \mathbb{R}P^{n-1}) = \mathbb{Z}_2$. By Poincaré duality, $H^n(\mathbb{R}P^n) = \mathbb{Z}_2$, so the exact sequence

J.-C. Hausmann, *Mod Two Homology and Cohomology*, Universitext,
DOI 10.1007/978-3-319-09354-3_6

for the pair $(\mathbb{R}P^n, \mathbb{R}P^{n-1})$ gives

$$0 \to H^{n-1}(\mathbb{R}P^n) \to \underbrace{H^{n-1}(\mathbb{R}P^{n-1})}_{\mathbb{Z}_2} \to \underbrace{H^n(\mathbb{R}P^n, \mathbb{R}P^{n-1})}_{\mathbb{Z}_2} \to \underbrace{H^n(\mathbb{R}P^n)}_{\mathbb{Z}_2} \to 0\,.$$

All that implies that the inclusion induces an isomorphism $H^k(\mathbb{R}P^n) \to H^k(\mathbb{R}P^{n-1})$ for $k \leq n-1$. By the induction hypothesis and functoriality of the cup product, $H^k(\mathbb{R}P^n) = \mathbb{Z}_2$ for $k \leq n-1$, generated by a^k.

By Poincaré duality, $H_n(\mathbb{R}P^n) = \mathbb{Z}_2$ with generator $[\mathbb{R}P^n]$ and, by Theorem 5.3.12, the bilinear map

$$H^1(\mathbb{R}P^n) \times H^{n-1}(\mathbb{R}P^n) \to H^n(\mathbb{R}P^n) \xrightarrow{\frown [\mathbb{R}P^n]} \mathbb{Z}_2$$

is not degenerated. Therefore, $a \smile a^{n-1} \neq 0$, which proves that $H^*(\mathbb{R}P^n) \approx \mathbb{Z}_2[a]/(a^{n+1})$.

Finally, by the standard CW-structure of $\mathbb{R}P^\infty$, one has, for all integer n, that $H^k(\mathbb{R}P^\infty, \mathbb{R}P^n) = 0$ for $k < n$. The first statement then implies the second. Note that we have also proven that the standard CW-structure of $\mathbb{R}P^n$ $(n \leq \infty)$ is perfect. □

Corollary 6.1.2 *The Poincaré series of* $\mathbb{R}P^n$ $(n \leq \infty)$ *are*

$$\mathfrak{P}_t(\mathbb{R}P^n) = 1 + t + \cdots + t^n = \frac{1 - t^{n+1}}{1-t} \quad \textit{and} \quad \mathfrak{P}_t(\mathbb{R}P^\infty) = \frac{1}{1-t}\,.$$

Remark 6.1.3 Proposition 6.1.1 and its proof show that the **GrA**-homomorphism $H^j(\mathbb{R}P^{n+k}) \to H^j(\mathbb{R}P^n)$ induced by the inclusion $\mathbb{R}P^n \hookrightarrow \mathbb{R}P^{n+k}$ is surjective $(k \leq \infty)$. In particular, it is an isomorphism for $j \leq n$.

Remark 6.1.4 The polynomial structure on $H^*(\mathbb{R}P^\infty)$ implies the following fact: if $f : \mathbb{R}P^\infty \to X$ is a continuous map with X a finite dimensional CW-complex, then $H^*f = 0$. Actually, *f is homotopic to a constant map*. This result is a weak version of the original *Sullivan conjecture* [188, p. 180], which lead to important researches in homotopy theory (see, e.g. [171]) and was finally proven, in a more general form, by Miller [143].

We now pass to the *complex projective space* $\mathbb{C}P^n$, the space of complex lines in $\mathbb{C}^{n+1}$. Such a line is represented by a non-zero vector $z = (z_0, \ldots, z_n) \in \mathbb{C}^{n+1} - \{0\}$, and two such vectors z and z' are in the same line if and only if $z' = \lambda z$ with $\lambda \in \mathbb{C}^* = \mathbb{C} - \{0\}$. If $|z| = |z'| = 1$, then $\lambda \in S^1$. Thus

$$\mathbb{C}P^n = (\mathbb{C}^{n+1} - \{0\})/\mathbb{C}^* = S^{2n+1}/S^1\,.$$

The image of $(z_0, \dots, z_n)$ in $\mathbb{C}P^n$ is denoted by $[z_0:z_1:\dots:z_n]$. As S^1 acts smoothly on S^{2n+1}, the quotient $\mathbb{C}P^n$ is a closed smooth manifold and the quotient map

$$p : S^{2n+1} \to \mathbb{C}P^n$$

is a principal S^1-bundle [82, Example 4.44], called the *Hopf bundle*. In this simple example, this can be proved directly. Consider the open set $V_k \subset \mathbb{C}^{n+1} - \{0\}$ given by $V_k = \{(z_0, \dots, z_n) \in \mathbb{C}^{n+1} \mid z_k \neq 0\}$. Its image in $\mathbb{C}P^n$ is an open set U_k, domain of the chart $\varphi_k : \mathbb{C}^n \xrightarrow{\approx} U_k$ given by

$$\varphi_k(z_0, \dots, z_{n-1}) = [z_0:z_1:\dots:z_{k-1}:1:z_k:\dots:z_{n-1}]. \tag{6.1.1}$$

On the other hand, a trivialization $\tilde{\varphi}_k : U_k \times S^1 \xrightarrow{\approx} p^{-1}(U_k)$ is given by

$$\tilde{\varphi}_k(\varphi_k(z_0, \dots, z_{n-1}), g) = \frac{1}{\sqrt{1 + \sum_{i=0}^{n-1} |z_i|^2}}(z_0, z_1, \dots, z_{k-1}, g, z_k, \dots, z_{n-1}). \tag{6.1.2}$$

It is also classical that $\mathbb{C}P^n$ is obtained from $\mathbb{C}P^{n-1}$ by attaching one cell of dimension $2n$,

$$\mathbb{C}P^n = \mathbb{C}P^{n-1} \cup_p D^{2n},$$

with the attaching map $p : S^{2n-1} \to \mathbb{C}P^{n-1}$ being the quotient map (see e.g. [82, Example 0.6] or [155, Theorem 40.2]). This gives a standard CW-structure on $\mathbb{C}P^n$ with one cell in each even dimension $\leq 2n$. For the direct limit $\mathbb{C}P^\infty$, we get CW-structure with one cell in each even dimension. For these CW-structure, the vector space of cellular chains vanish in odd degree, so the cellular boundary is identically zero. Therefore,

$$\mathfrak{P}_t(\mathbb{C}P^n) = 1 + t^2 + \cdots + t^{2n} = \frac{1 - t^{2(n+1)}}{1 - t^2} \quad \text{and} \quad \mathfrak{P}_t(\mathbb{C}P^\infty) = \frac{1}{1 - t^2}.$$

As $\mathbb{C}P^n$ is a smooth manifold, the same proof as for Proposition 6.1.1, using Poincaré duality, gives Proposition 6.1.5. One can also adapt the proof of Proposition 4.3.10, using the Gysin exact sequence of the Hopf bundle (see Exercise 6.2):

$$\cdots H^{k-1}(S^{2n+1}) \to H^{k-2}(\mathbb{C}P^n) \xrightarrow{-\smile e(\xi)} H^k(\mathbb{C}P^n) \xrightarrow{H^*p} H^{k-1}(S^{2n+1}) \to \cdots.$$

Proposition 6.1.5 *The cohomology algebra of $\mathbb{C}P^n$ ($n \leq \infty$) is given by*

$$H^*(\mathbb{C}P^n) \approx \mathbb{Z}_2[a]/(a^{n+1}), \quad H^*(\mathbb{C}P^\infty) \approx \mathbb{Z}_2[a],$$

with $a \in H^2(\mathbb{C}P^n)$. *The class* a *is the Euler class of the Hopf bundle* $S^{2n+1} \to \mathbb{C}P^n$.

If we replace the field of complex numbers by that of quaternions $\mathbb{H}$, we get *quaternionic projective space* $\mathbb{H}P^n$:

$$\mathbb{H}P^n = (\mathbb{H}^{n+1} - \{0\})/\mathbb{H}^* = S^{4n+3}/S^3$$

(it is usual to take the right $\mathbb{H}$-vector space structure on $\mathbb{H}^{n+1}$). The space $\mathbb{H}P^n$ is obtained from $\mathbb{H}P^{n-1}$ by attaching one cell of dimension $4n$, with the attaching map $p : S^{4n-1} \to \mathbb{H}P^{n-1}$ being the quotient map. The map p is an S^3-bundle called the *Hopf bundle*. This gives a standard CW-structure on $\mathbb{H}P^n$ with one cell in each dimension $4k \leq 4n$. For the direct limit $\mathbb{H}P^\infty$, we get CW-structure with one cell in each dimension $4k$ and

$$\mathfrak{P}_t(\mathbb{H}P^n) = 1 + t^4 + \cdots + t^{4n} = \frac{1 - t^{(4n+1)}}{1 - t^4} \quad \text{and} \quad \mathfrak{P}_t(\mathbb{H}P^\infty) = \frac{1}{1 - t^4}.$$

Proposition 6.1.6 is proven as Proposition 6.1.5, either using Poincaré duality ($\mathbb{H}P^n$ is a smooth $4n$-manifold), or the Gysin exact sequence of the Hopf bundle (see Exercise 6.2).

Proposition 6.1.6 *The cohomology algebra of* $\mathbb{H}P^n$ *(*$n \leq \infty$*) is given by*

$$H^*(\mathbb{H}P^n) \approx \mathbb{Z}_2[a]/(a^{n+1}), \quad H^*(\mathbb{H}P^\infty) \approx \mathbb{Z}_2[a],$$

with $a \in H^4(\mathbb{H}P^n)$. *The class* a *is the Euler class of the Hopf bundle* $S^{4n+3} \to \mathbb{H}P^n$.

Let $\mathbb{K} = \mathbb{R}$, $\mathbb{C}$ or $\mathbb{H}$ and let $d = d(\mathbb{K}) = \dim_{\mathbb{R}} \mathbb{K}$. The space $\mathbb{K}P^1$ has a CW-structure with one 0-cell and one d-cell and is then homeomorphic to S^d. The quotient maps $S^{2d-1} \twoheadrightarrow \mathbb{K}P^1$ thus give maps

$$h_{1,1} : S^1 \twoheadrightarrow S^1, \ h_{3,2} : S^3 \twoheadrightarrow S^2 \text{ and } h_{7,4} : S^7 \twoheadrightarrow S^4$$

called the *Hopf maps*. Note that $h_{1,1}$ is just a 2-covering. Using the homeomorphism $S^d \approx \hat{\mathbb{K}} = \mathbb{K} \cup \{\infty\}$ given by a stereographic projection, these Hopf maps admit the formula

$$h_{i,j}(v, w) = \begin{cases} vw^{-1} & \text{if } w \neq 0 \\ \infty & \text{otherwise.} \end{cases} \tag{6.1.3}$$

This formula also makes sense for $\mathbb{K} = \mathbb{O}$, the octonions, whose multiplication admits inverses for non zero elements. This gives one more Hopf map $h_{15,8}: S^{15} \to S^8$. One can prove that $h_{15,8}$ is an S^7-bundle (see [82, Example 4.47]), also called the *Hopf bundle*. Attaching a 16-cell to S^8 using $h_{15,8}$ produces the *octonionic projective plane* $\mathbb{O}P^2$ (because of non-associativity of the octonionic multiplication, there are no higher dimensional octonionic projective spaces).

Proposition 6.1.7 *The cohomology algebra of* $\mathbb{O}P^2$ *is given by*

$$H^*(\mathbb{O}P^2) \approx \mathbb{Z}_2[a]/(a^3)$$

with $a \in H^8(\mathbb{O}P^2) = \mathbb{Z}_2$. *In particular,* $\mathfrak{P}_t(\mathbb{O}P^2) = 1 + t^8 + t^{16}$.

Proof By its cellular decomposition, $H^k(\mathbb{O}P^2) = \mathbb{Z}_2$ for $k = 0, 8, 16$ and zero otherwise. Let $a \in H^8(\mathbb{O}P^2)$ and $b \in H^8(S^8)$ be the non-zero elements. The mapping cylinder $\hat{E}$ of $h_{15,8}$ is the disk bundle associated to the Hopf bundle and $\mathbb{O}P^2$ has the homotopy type of $\hat{E} \cup D^{16}$, with $\hat{E} \cap D^{16} = S^{15}$. The Thom class of the Hopf bundle $h_{15,8}$

$$U \in H^8(\hat{E}, S^{15}) \approx H^8(\mathbb{O}P^2, \operatorname{int} D^{16}) \approx H^8(\mathbb{O}P^2),$$

is not zero, so corresponds to $a \in H^8(\mathbb{O}P^2)$. The diagram

$$\begin{array}{ccccc} H^8(\hat{E}) & \xrightarrow[\approx]{-\smile U} & H^{16}(\hat{E}, S^{15}) & & \\ \uparrow \approx & & \uparrow \approx & & \\ H^8(\mathbb{O}P^2) & \longrightarrow & H^{16}(\mathbb{O}P^2, \operatorname{int} D^{16}) & \xrightarrow{\approx} & H^{16}(\mathbb{O}P^2) \end{array}$$

(with the bottom composite $H^8(\mathbb{O}P^2) \to H^{16}(\mathbb{O}P^2)$ labelled $-\smile a$)

is then commutative by the analogue in singular cohomology of Lemma 4.1.7. This proves that $- \smile a : H^8(\mathbb{O}P^2) \to H^{16}(\mathbb{O}P^2)$ is bijective. □

Remark 6.1.8 Proposition 6.1.7 may also be proved using Poincaré duality, since $\mathbb{O}P^2$ has the homotopy type of a closed smooth 16-manifold, in fact a homogeneous space of the exceptional Lie group F_4 (see [207, Theorem 7.21, p. 707]).

The computations of the cohomology algebra $H^*(\mathbb{K}P^2)$ have the following consequence.

Corollary 6.1.9 *The Hopf maps*

$$h_{1,1} : S^1 \to S^1\ ,\ h_{3,2} : S^3 \to S^2,\ h_{7,4} : S^7 \to S^4 \text{ and } h_{15,8} : S^{15} \to S^8$$

are not homotopic to constant maps.

Using the Steenrod squares, we shall prove in Chap. 8 that no suspension of these Hopf maps is homotopic to a constant map (see Proposition 8.6.1).

Proof One has $S^1 \cup_{h_{1,1}} D^2 \approx \mathbb{R}P^2$. If $h_{1,1}$ were null-homotopic, $\mathbb{R}P^2$ would have the homotopy type of $S^1 \vee S^2$ (see [82, Proposition 0.18]). But, in $H^*(S^1 \vee S^2)$, the cup-square map vanishes by (4.2.2), which is not the case in $H^*(\mathbb{R}P^2)$. The same proof works for the other Hopf maps. □

We now consider the vector bundle $\gamma_{\mathbb{K}}$ over $\mathbb{K}P^n$ associated to the Hopf bundle, where $\mathbb{K} = \mathbb{R}, \mathbb{C}, \mathbb{H}$ and $n \leq \infty$. For $n \in \mathbb{N}$, total space of $\gamma_{\mathbb{K}}$ is

$$E(\gamma_{\mathbb{K}}) = \{(a, v) \in \mathbb{K}P^n \times \mathbb{K}^{n+1} \mid v \in a\},$$

with bundle projection $(a, v) \mapsto a$. For instance, for $n = 1, E(\gamma_{\mathbb{R}})$ is the Möbius band. The correspondence $(a, v) \to V$ defines a map $E(\gamma_{\mathbb{K}}) \to \mathbb{K}^{n+1}$ whose restriction to each fiber is linear and injective. This endows $\gamma_{\mathbb{K}}$ with an Euclidean structure, whose unit sphere bundle is the Hopf bundle. The vector bundle $\gamma_{\mathbb{K}}$ is called the *Hopf vector bundle* or the *tautological bundle* over $\mathbb{K}P^n$. Its (real) rank is $d = d(\mathbb{K}) = \dim_{\mathbb{R}} \mathbb{K}$. By passing to the direct limit when $n \to \infty$, we get a tautological bundle $\gamma_{\mathbb{K}}$ over $\mathbb{K}P^\infty$. Propositions 6.1.5 and 6.1.6 (and 4.7.36 for $\mathbb{K} = \mathbb{R}$) gives the following result.

Proposition 6.1.10 *For $n \leq \infty$, the Euler class $e(\gamma_{\mathbb{K}})$ is the non-zero element $a_d \in H^d(\mathbb{K}P^n) = \mathbb{Z}_2$.*

The inclusions $\mathbb{R} \subset \mathbb{C} \subset \mathbb{H}$ induce inclusions

$$\mathbb{R}P^n \xrightarrow{j_1} \mathbb{C}P^n \xrightarrow{j_2} \mathbb{H}P^n_, \; n \leq \infty .$$

The above proposition permits us to determine the **GrA**-homomorphism induced in cohomology by these inclusions.

Proposition 6.1.11 $H^*j_d\,(a_{2d}) = a_d^2$.

Proof Observe that $\gamma_{\mathbb{C}}$ is a complex vector bundle, so the multiplication by $i \in \mathbb{C}$ is defined on each fiber. We notice that

$$i_1^*\gamma_{\mathbb{C}} = \gamma_{\mathbb{R}} \oplus i\,\gamma_{\mathbb{R}} \approx \gamma_{\mathbb{R}} \oplus \gamma_{\mathbb{R}} . \tag{6.1.4}$$

Then

$$\begin{aligned} H^*j_1\,(a_2) &= H^*j_1\,(e(\gamma_{\mathbb{C}})) && \text{by Proposition 6.1.11} \\ &= e(j_1^*\gamma_{\mathbb{C}}) && \text{by Lemma 4.7.31} \\ &= e(\gamma_{\mathbb{R}} \oplus \gamma_{\mathbb{R}}) && \text{by (6.1.4)} \\ &= e(\gamma_{\mathbb{R}}) \smile e(\gamma_{\mathbb{R}}) && \text{by Proposition 4.7.40} \\ &= a_1^2 && \text{by Proposition 6.1.11.} \end{aligned}$$

The proof that $H^*j_2\,(a_4) = a_2^2$ is the same, using the multiplication by $j \in \mathbb{H}$ on the fiber of $\gamma_{\mathbb{H}}$ which is a quaternionic vector bundle. □

The Hopf bundles are sphere bundles over S^p such that the total space is also a sphere. We shall see in Proposition 6.3.5 that $p = 1, 2, 4, 8$ are the only dimensions where such examples may occur.

6.2 Applications

6.2.1 The Borsuk-Ulam Theorem

A (continuous) map $f : \mathbb{R}^m \to \mathbb{R}^n$ or $f : S^m \to S^n$ such that $f(-x) = -f(x)$ is called an *odd map*.

Theorem 6.2.1 *Let $f : S^m \to S^n$ be an odd map. Then:*

(1) $n \geq m$.
(2) *if $m = n$, then* $\deg f = 1$.

Proof If f is odd, it descends to a map $\bar{f} : \mathbb{R}P^m \to \mathbb{R}P^n$ with a commutative diagram

$$\begin{array}{ccc} S^m & \xrightarrow{f} & S^n \\ \downarrow{\scriptstyle p_m} & & \downarrow{\scriptstyle p_n} \\ \mathbb{R}P^m & \xrightarrow{\bar{f}} & \mathbb{R}P^n \end{array} .$$

The two-fold covering p_n is induced from $p_\infty : S^\infty \to \mathbb{R}P^\infty$ by the inclusion $\mathbb{R}P^n \hookrightarrow \mathbb{R}P^\infty$. By Lemma 4.3.6 and Proposition 4.3.10, the characteristic classes $w(p_m) \in H^1(\mathbb{R}P^m)$ and $w(p_n) \in H^1(\mathbb{R}P^n)$ are the generators of these cohomology groups and $H^*\bar{f}(w(p_n)) = w(p_m)$. By Proposition 4.3.10 again, one has that $0 = H^*\bar{f}(w(p_n)^{n+1}) = w(p_m)^{n+1}$ which implies that $n \geq m$.

If $m = n$, observe that $H^*p_n : H^n(\mathbb{R}P^n) \to H^n(S^n)$ is the zero homomorphism since p_n is of local degree 2. The transfer exact sequence of (4.3.9), which is functorial, gives the commutative diagram

$$\begin{array}{ccc} H^n(S^n) & \underset{\approx}{\xrightarrow{\mathrm{tr}^*}} & H^n(\mathbb{R}P^n) \\ \downarrow{\scriptstyle H^*f} & & \approx\downarrow{\scriptstyle H^*\bar{f}} \\ H^n(S^n) & \underset{\approx}{\xrightarrow{\mathrm{tr}^*}} & H^n(\mathbb{R}P^n) \end{array} ,$$

proving that $\deg f = 1$. □

As a corollary, we get the theorem of Borsuk-Ulam.

Corollary 6.2.2 (Borsuk-Ulam theorem) *Let $g : S^n \to \mathbb{R}^n$ be a continuous map. Then, there exists $z \in S^n$ such that $g(z) = g(-z)$.*

Proof Otherwise, the map $f : S^n \to S^{n-1}$ defined by

$$f(z) = \frac{g(z) - g(-z)}{|g(z) - g(-z)|}$$

is continuous and odd, which contradicts Theorem 6.2.1. □

A famous consequence is the *ham sandwich theorem*. For an early history of this theorem, see [14].

Corollary 6.2.3 *Let $A_1, \dots, A_n$ be n bounded Lebesgue measurable subsets of $\mathbb{R}^n$. Then, there exists a hyperplane which bisects each A_i.*

Proof Identify $\mathbb{R}^n$ by an isometry with an affine n-subspace W of $\mathbb{R}^{n+1}$ not passing through the origin, and thus see $A_1, \dots, A_n \subset W$. For each unit vector $v \in \mathbb{R}^{n+1}$, consider the half-space $Q(v) = \{x \in \mathbb{R}^{n+1} \mid \langle v, x\rangle > 0\}$. Let $g_i : S^n \to \mathbb{R}$ defined by $g_i(v) = \text{measure}(A_i \cap Q(v))$. The maps g_i are the coordinates of a continuous map $g : S^n \to \mathbb{R}^n$. By Corollary 6.2.2, there is $z \in S^n$ such that $g(z) = g(-z)$, which means that $g_i(z) = \frac{1}{2}\,\text{measure}(A_i)$. Then, $P(z) \cap W$ is the desired bisecting hyperplane. □

6.2.2 Non-singular and Axial Maps

A continuous map $\mu : \mathbb{R}^m \times \mathbb{R}^m \to \mathbb{R}^k$ is called *non-singular* if

(1) $\mu(\alpha x, \beta y) = \alpha\beta\,\mu(x, y)$ for all $x, y \in \mathbb{R}^m$ and all $\alpha, \beta \in \mathbb{R}$, and
(2) $\mu(x, y) = 0$ implies that $x = 0$ or $y = 0$.

Non-singular maps generalize bilinear maps without zero divisors. They were introduced in [61], from where the results of this section are extracted. For references about the earlier literature, see also [61].

Non-singular maps are related to *axial* maps. A continuous map $g : \mathbb{R}P^m \times \mathbb{R}P^m \to \mathbb{R}P^\ell$, with $\ell \geq m$, is called *axial* if the restriction of g to each slice is not homotopic to a constant map. By Corollary 3.8.4, this is equivalent to ask that these restrictions $g_x^1 : \{x\} \times \mathbb{R}P^m \to \mathbb{R}P^\ell$ or $g_x^2 : \mathbb{R}P^m \times \{x\} \to \mathbb{R}P^\ell$ to be homotopic to the inclusion $\mathbb{R}P^m \hookrightarrow \mathbb{R}P^\ell$. Using Corollary 3.8.4, this is equivalent to ask that $H^* g_x^i(a_\ell) = a_m$, where a_j is the generator of $H^1(\mathbb{R}P^j)$. By Corollary 4.7.4, we deduce that a continuous map $g : \mathbb{R}P^m \times \mathbb{R}P^m \to \mathbb{R}P^\ell$ is axial if and only if

$$H^* g(a_\ell) = \mathbf{1} \times a_m + a_m \times \mathbf{1}\,. \tag{6.2.1}$$

The name of *axial map* appeared in [6] where references about the earlier literature on the subject may be found. It started with the work of Stiefel and Hopf [101].

Let $\mu : \mathbb{R}^m \times \mathbb{R}^m \to \mathbb{R}^k$ be a non-singular map. By Point (2) of the definition, we get a continuous map $\tilde{\mu} : S^{m-1} \times S^{m-1} \to S^{k-1}$ defined by

$$\tilde{\mu}(x, y) = \frac{\mu(x, y)}{|\mu(x, y)|}\,. \tag{6.2.2}$$

Point (1) above implies that $\tilde{\mu}$ descends to a map

$$\bar{\mu} : \mathbb{R}P^{m-1} \times \mathbb{R}P^{m-1} \to \mathbb{R}P^{k-1}\,. \tag{6.2.3}$$

For $x \in \mathbb{R}P^{m-1}$, the restriction of $\bar{\mu}_x$ to the slice $\{x\} \times \mathbb{R}P^{m-1}$ is covered by two-fold covering maps:

$$\begin{array}{ccccc}
\{x\} \times S^{m-1} & \longrightarrow & \mathbb{R}P^{m-1} \times S^{m-1} & \longrightarrow & S^{k-1} \\
\downarrow p_{m-1} & & \downarrow & & \downarrow p_{k-1} \\
\{x\} \times \mathbb{R}P^{m-1} & \longrightarrow & \mathbb{R}P^{m-1} \times \mathbb{R}P^{m-1} & \xrightarrow{\bar{\mu}} & \mathbb{R}P^{k-1}
\end{array}$$

(the bottom composite being $\bar{\mu}_x$).

By Lemma 4.3.6 and Proposition 4.3.10, the characteristic classes $w(p_{m-1}) \in H^1(\mathbb{R}P^{m-1})$ and $w(p_{k-1}) \in H^1(\mathbb{R}P^{k-1})$ are the generators of these cohomology groups and $H^*\bar{\mu}_x(w(p_{k-1})) = w(p_{m-1})$. Hence, $\bar{\mu}_x$ is not homotopic to a constant map. The same reasoning holds for the slices $\mathbb{R}P^{m-1} \times \{x\}$. Therefore, $\bar{\mu}$ is axial. Conversely, if $g : \mathbb{R}P^{m-1} \times \mathbb{R}P^{m-1} \to \mathbb{R}P^{k-1}$ is an axial map, it induces on universal covers a map $\tilde{g} : S^{m-1} \times S^{m-1} \to S^{k-1}$ satisfying $\tilde{g}(-x, y) = \tilde{g}(x, -y) = -\tilde{g}(x, y)$. The map $\mu : \mathbb{R}^m \times \mathbb{R}^m \to \mathbb{R}^k$ defined by

$$\mu(x, y) = |x| \cdot |y| \cdot \tilde{g}\left(\frac{x}{|x|}, \frac{y}{|y|}\right)$$

is a non-singular map. This proves the following lemma.

Lemma 6.2.4 *The correspondence $\mu \mapsto \bar{\mu}$ provides a bijection between non-singular maps $\mathbb{R}^m \times \mathbb{R}^m \to \mathbb{R}^k$ (up to multiplication by non-zero constants) and axial maps $\mathbb{R}P^{m-1} \times \mathbb{R}P^{m-1} \to \mathbb{R}P^{k-1}$.*

Let $\mu : \mathbb{R}^m \times \mathbb{R}^m \to \mathbb{R}^k$ be a non-singular map. The restriction of $\tilde{\mu}$ to each slice is odd. Hence, if a non-singular map $\mu : \mathbb{R}^m \times \mathbb{R}^m \to \mathbb{R}^k$ exists, it follows form Theorem 6.2.1 that $k \geq m$. When $m = k$, the following proposition is attributed to Stiefel. For other proofs (see [153, Theorem 4.7] or Remark 8.6.7).

Proposition 6.2.5 *Let $\mu : \mathbb{R}^m \times \mathbb{R}^m \to \mathbb{R}^m$ be a non-singular map. Then $m = 2^r$.*

In fact, by a famous result of J.F. Adams (see Remark 8.6.7), non-singular maps $\mathbb{R}^m \times \mathbb{R}^m \to \mathbb{R}^m$ exist only if $m = 1, 2, 4, 8$.

Proof We consider the associated axial map $\bar{\mu} : \mathbb{R}P^{m-1} \times \mathbb{R}P^{m-1} \to \mathbb{R}P^{m-1}$ and denote by a the generator of $H^{m-1}(\mathbb{R}P^{m-1})$. The Künneth theorem implies that the correspondence $x \mapsto \mathbf{1} \times a$ and $y \mapsto a \times \mathbf{1}$ provides a **GrA**-isomorphism

$$\mathbb{Z}_2[x, y]/(x^m, y^m) \xrightarrow{\approx} H^*(\mathbb{R}P^{m-1} \times \mathbb{R}P^{m-1}).$$

By (6.2.1), $H^*\bar{\mu}(a) = x + y$. Therefore, $(x + y)^m = 0$. As x^m and y^m also vanish, one has

$$(x + y)^m = \sum_{i=0}^{m} \binom{m}{i} x^i y^{m-i} = \sum_{i=1}^{m-1} \binom{m}{i} x^i y^{m-i}$$

This implies that $\binom{m}{i} \equiv 0 \bmod 2$ for all $i = 1, \ldots, m-1$ which, by Lemma 6.2.6, happens only if $m = 2^r$. □

For $n \in \mathbb{N}$, denote its dyadic expansion in the form $n = \sum_{j \in J(n)} 2^j$ where $J(n) \subset \mathbb{N}$.

Lemma 6.2.6 (Binomial coefficients mod 2) *Let $m, r \in \mathbb{N}$. Then*

$$\binom{m}{r} \equiv 1 \mod 2 \iff J(r) \subset J(m).$$

In other words, $\binom{m}{r} \equiv 1 \mod 2$ if and only if the dyadic expansion of r is a sub-sum of that of m.

Proof In $\mathbb{Z}_2[x]$, the equation $(1+x)^2 = 1 + x^2$ holds, whence $(1+x)^{2^j} = 1 + x^{2^j}$. Therefore

$$\sum_{r=0}^{m} \binom{m}{r} x^r = (1+x)^m = \prod_{j \in J(m)} (1+x)^{2^j} = \prod_{j \in J(m)} (1 + x^{2^j}).$$

The identification of the coefficient of x^r gives the lemma. □

The technique of the proof of Proposition 6.2.5 also gives a result of H. Hopf, [101, Satz I.e].

Proposition 6.2.7 *Let $\mu : \mathbb{R}^m \times \mathbb{R}^m \to \mathbb{R}^k$ be a non-singular map. If $m > 2^r$, then $k \geq 2^{r+1}$.*

Proof We already know that $k > m$. As in the proof of Proposition 6.2.5, we consider the associated axial map $\bar{\mu} : \mathbb{R}P^{m-1} \times \mathbb{R}P^{m-1} \to \mathbb{R}P^{k-1}$ and use the same notations. We get the equation $(x+y)^k = 0$ in $\mathbb{Z}_2[x, y]/(x^m, y^m)$, which, as $m > 2^r$, implies that $\binom{k}{i} = 0$ for all $1 \leq i \leq 2^r$. By Lemma 6.2.6, the dyadic expansion $k = \sum_j k_j 2^j$ must satisfy $k_j = 0$ for $j \leq r$, which is equivalent to $k \geq 2^{r+1}$. □

Remark 6.2.8 There always exists a non-singular map $\mu : \mathbb{R}^m \times \mathbb{R}^m \to \mathbb{R}^{2m-1}$ (see [61, Sect. 5]).

We finish this section by mentioning two results relating non-singular or axial maps to the immersion problem and the topological complexity of projective spaces. The following proposition was proven in [6].

Proposition 6.2.9 *There exists an axial map $g : \mathbb{R}P^n \times \mathbb{R}P^n \to \mathbb{R}P^k$ $(k > n)$ if and only if there exists an immersion of $\mathbb{R}P^n$ in $\mathbb{R}^k$.*

We shall not talk about the large literature and the many results on the problem immersing or embedding $\mathbb{R}P^m$ in $\mathbb{R}^q$ (see however Proposition 9.5.23). Tables and references are available in [36]. The existence of non-singular maps is also related to the topological complexity of the projective space. The following is proven in [61, Theorem 6.1].

Theorem 6.2.10 *The topological complexity* **TC** $(\mathbb{R}P^n)$ *is equal to the smallest integer k such that there is a non-singular map* $\mu : \mathbb{R}^{n+1} \times \mathbb{R}^{n+1} \to \mathbb{R}^k$.

Symmetric non-singular maps (i.e. $\mu(x, y) = \mu(y, x)$ are, in some range, related to embeddings of $\mathbb{R}P^n$ in Euclidean spaces or to the *symmetric topological complexity*. For results and references, see [68].

6.3 The Hopf Invariant

6.3.1 Definition

Let $f : S^{2m-1} \to S^m$ be a continuous map. The space $C_f = D^{2m} \cup_f S^m$ is a CW-complex with one cell in dimension 0, m and $2m$. Consider *the cup-square map* $\smile_m^2 : H^m(C_f) \to H^{2m}(C_f)$, given by $\smile_m^2 (x) = x \smile x$. The *Hopf invariant* Hopf $(f) \in \mathbb{Z}_2$ is defined by

$$\text{Hopf }(f) = \begin{cases} 1 & \text{if } \smile_m^2 \text{ is surjective for } C_f. \\ 0 & \text{otherwise.} \end{cases}$$

The space C_f depends only on the homotopy class of f (see, e.g. [82, Proposition 0.18]), then so does the Hopf invariant. A constant map has Hopf invariant 0. The computation of the cohomology ring of the various projective planes in Sect. 6.1 shows that the 2-fold cover $S^1 \to S^1$ as well as the other Hopf maps $h_{3,2} : S^3 \to S^2$, $h_{7,4} : S^7 \to S^4$ and $h_{15,8} : S^{15} \to S^8$ have Hopf invariant 1.

Our Hopf invariant is just the mod 2 reduction of the classical integral Hopf invariant defined in e.g. [82, Sect. 4.B]. The form of our definition is motivated by extending the statements to the case $m = 1$, usually not considered by authors.

Note that Hopf defined his invariant in 1931–35 [99, 100], before the invention of the cup product. He used linking numbers (see Sect. 6.3.4).

For $m = 1$, recall the bijection DEG: $[S^1, S^1] \xrightarrow{\approx} \mathbb{Z}$ given in (3.2.3).

Proposition 6.3.1 *Let* $f : S^1 \to S^1$. *Then*

$$\text{Hopf }(f) = \begin{cases} 0 & \text{if DEG (f)} \equiv 0 \mod 4 \\ 1 & \text{otherwise.} \end{cases}$$

Proof Let $C = C_f$. If DEG (f) is odd, then $\deg(f) = 1$ by Proposition 3.2.9. The computation of the cellular cohomology of C using Lemma 3.5.4 shows that $\tilde{H}^*(C) = 0$, so $\smile_1^2$ is surjective and Hopf $(f) = 1$. If DEG $(f) = 2k$, then, $H^1(C) \approx \mathbb{Z}_2 \approx H^2(C)$. Consider the 2-fold covering $p : \tilde{C} \to C$ whose characteristic class is the non zero element $a \in H^1(C)$. Its transfer exact sequence looks like

$$0 \to \underbrace{H^0(C)}_{\mathbb{Z}_2} \to \underbrace{H^1(C)}_{\mathbb{Z}_2} \xrightarrow{H^*p} H^1(\tilde{C}) \xrightarrow{tr^*} \underbrace{H^1(C)}_{\mathbb{Z}_2} \xrightarrow{-\smile a} \underbrace{H^2(C)}_{\mathbb{Z}_2}$$

By Van Kampen's theorem, $\pi_1(C)$ is cyclic of order $2k$. Thus, $\pi_1(\tilde{C})$ is cyclic of order k and $H^1(\tilde{C}) = \mathbb{Z}_2$ if k is even while it vanishes if k is odd. The proposition thus follows from the above transfer exact sequence. □

6.3.2 The Hopf Invariant and Continuous Multiplications

A classical construction associates a map $f_\kappa : S^{2m-1} \to S^m$ to a "continuous multiplication" $\kappa : S^{m-1} \times S^{m-1} \to S^{m-1}$. Let $D = D^m$ and $S = S^{m-1} = \partial D$. Divide S^m into the upper and lower hemisphere: $S^m = B_+ \cup B_- \subset \mathbb{R}^m \times \mathbb{R}$, with $B_+ \cap B_- = S^m \cap \mathbb{R}^m \times \{0\} = S$. Using the decomposition

$$\partial(D \times D) = \partial D \times D \cup D \times \partial D = S \times D \cup D \times S\,,$$

the map $f_\kappa : \partial(D \times D) \to S^m$ is defined, for $x, y \in S$ et $t \in [0, 1]$, by

$$f_\kappa(tx, y) = (t\,\kappa(x, y), \sqrt{1 - t^2}) \text{ and } f_\kappa(x, ty) = (t\,\kappa(x, y), -\sqrt{1 - t^2})\,.$$

For $u, v, \in S^{m-1}$, we consider the hypothesis $\mathcal{H}(u, v)$ on κ:

$$\mathcal{H}(u, v) : \kappa(u, -x) = -\kappa(u, x) \textit{ and } \kappa(-x, v) = -\kappa(x, v) \textit{ for all } x \in S^{m-1}.$$

For example, $\mathcal{H}(u, v)$ holds for all $u, v \in S^{m-1}$ if $\kappa = \tilde{\mu}$, the map associated using (6.2.2) to a non-singular map $\mu : \mathbb{R}^m \times \mathbb{R}^m \to \mathbb{R}^m$. Also, $\mathcal{H}(e, e)$ is satisfied if e is a neutral element for κ.

Example 6.3.2 Let $\kappa : S^0 \times S^0 \to S^0$ be the usual sign rule ($S^0 = S = \{\pm 1\}$). Then $D = [-1, 1]$ and the map $f_\kappa : \partial(D \times D) \to S^1$ is pictured in Fig. 6.1. One sees that f_κ has degree 2. By Proposition 6.3.1, this implies that Hopf $(f_\kappa) = 1$. Actually, the map f_κ is topologically conjugate to the projection $S^1 \twoheadrightarrow S^1/\{x \sim -x\}$, so C_{f_κ} is homeomorphic to $\mathbb{R}P^2$.

The same exercise may be done for the other possible multiplications on S^0. By changing the sign of κ if necessary, we may assume that $\kappa(1, 1) = 1$. There are then 8 cases (Table 6.1). One sees that Hopf $(f_\kappa) = 1$ if and only if $\mathcal{H}(u, v)$ is satisfied for some $u, v \in S^0$. This is partially generalized in the following result.

Proposition 6.3.3 *Let $\kappa : S^{m-1} \times S^{m-1} \to S^{m-1}$ be a continuous multiplication. Suppose that $\mathcal{H}(u, v)$ is satisfied for some $u, v \in S^{m-1}$. Then* Hopf $(f_\kappa) = 1$.

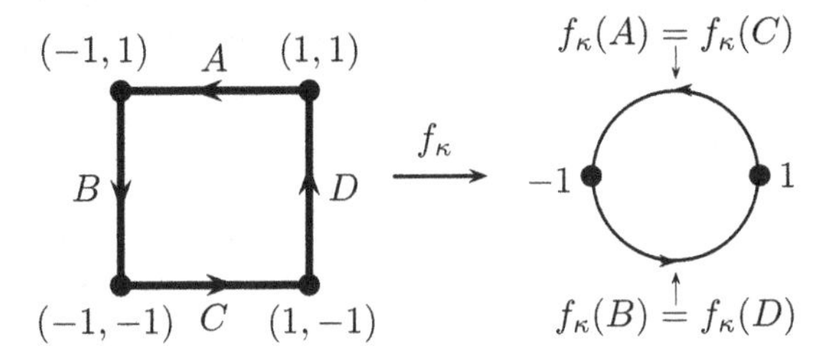

Fig. 6.1 The map f_κ for the usual sign rule

Table 6.1 The eight multiplications on S^0

	$\kappa(1,1)$	$\kappa(-1,1)$	$\kappa(-1,-1)$	$\kappa(1,-1)$	DEG (f_κ)	Hopf (f_κ)	Satisfies
1	1	1	1	1	0	0	
2	1	1	1	−1	1	1	$\mathcal{H}(1,-1)$
3	1	1	−1	1	−1	1	$\mathcal{H}(-1,-1)$
4	1	1	−1	−1	0	0	
5	1	−1	1	1	1	1	$\mathcal{H}(-1,1)$
6	1	−1	1	−1	2	1	$\mathcal{H}(u,v)\ \forall\, u,v$
7	1	−1	−1	1	0	0	
8	1	−1	−1	−1	1	1	$\mathcal{H}(1,1)$

Proof The case $m = 1$ was done in Example 6.3.2. We may thus assume that $m > 1$. The following proof is inspired by that of [81, Lemma 2.18]. Let $f = f_\kappa$. Consider the commutative diagram

$$\begin{array}{ccc}
H^m(C_f) \otimes H^m(C_f) & \xrightarrow{\quad\smile\quad} & H^{2m}(C_f) \\
\uparrow \approx & & \uparrow \approx \\
H^m(C_f, B_+) \otimes H^m(C_f, B_-) & \xrightarrow{\quad\smile\quad} & H^{2m}(C_f, S^m) \\
\downarrow \phi^*\otimes\phi^* & & \approx \downarrow \Phi^* \\
H^m(D\times D, S\times D) \otimes H^m(D\times D, D\times S) & \xrightarrow{\smile} & H^{2m}(D\times D, \partial(D\times D)) \\
\pi_1^*\otimes\pi_2^* \uparrow \approx & \nearrow{\scriptstyle \times}_{\approx} & \\
H^m(D,S) \otimes H^m(D,S) & &
\end{array}$$

where $\phi : D\times D \to C_f$ is the characteristic map for the $2m$-cell of C_f and $\phi^* = H^*\phi$. The cross-product map at the bottom of the diagram is an isomorphism by the relative Künneth theorem 4.6.10. Hence, Hopf $(f_\kappa) = 1$ if and only if the homomorphism $\phi^* \otimes \phi^*$ in the left column is an isomorphism. By symmetry, it is enough to prove that $\phi^* : H^m(C_f, B_+) \to H^m(D \times D, S \times D)$ is not zero. Consider the commutative diagram

$$\begin{array}{ccccc}
H^m(C_f,B_+) & \xrightarrow{\approx} & H^m(S^m,B_+) & \xrightarrow{\approx} & H^m(B_-,S)\\
\downarrow{\scriptstyle \phi^*} & & \downarrow{\scriptstyle f_\kappa^*} & & \downarrow{\scriptstyle f_\kappa^*}\\
H^m(D\times D, S\times D) & \xrightarrow{\approx} & H^m(\partial(D\times D),S\times D) & \xrightarrow{\approx} & H^m(D\times S,S\times S)\,.
\end{array}$$

The left horizontal maps are isomorphism since $m > 1$ and the right ones by excision. It then suffices to prove that $f_\kappa^* : H^m(B_-, S) \to H^m(D \times S, S \times S)$ is not zero.

As the restriction of f_κ to $S \times S$ is equal to κ, one has a commutative diagram

$$\begin{array}{ccccc}
 & & H^{m-1}(S) & \xrightarrow{\approx} & H^m(B_-,S)\\
 & & \downarrow{\scriptstyle \kappa^*} & & \downarrow{\scriptstyle f_\kappa^*}\\
H^{m-1}(D\times S) & \longrightarrow & H^{m-1}(S\times S) & \xrightarrow{\delta^*} & H^m(D\times S,S\times S)\\
{\scriptstyle\approx}\uparrow{\scriptstyle \pi_2^*} & \nearrow{\scriptstyle \pi_2^*} & & & \\
H^{m-1}(S) & & & &
\end{array}$$

where the second line is the cohomology sequence of the pair $(D \times S, S \times S)$. Let $u, v \in S^{m-1}$ such that $\mathcal{H}(u, v)$ is satisfied. Let $s^1, s^2 : S \to S \times S$ be the slice inclusions given by $s^1(x) = (x, v)$ and $s^2(x) = (u, x)$. The composition $\kappa \circ s^i$ is thus an odd map. Therefore, Theorem 6.2.1 implies that $H^*(\kappa \circ s^i)(a) = a$. By Corollary 4.7.3, one deduces that

$$H^*\kappa(a) = \mathbf{1} \times a + a \times \mathbf{1}\,. \tag{6.3.1}$$

On the other hand, $\ker \delta^* = \text{Image}\, \pi_2^* = \{0, \mathbf{1} \times a\}$. Therefore, f_κ^* does not vanish. □

In the proof of Proposition 6.3.3, the hypothesis on $\mathcal{H}(u, v)$ is only used to obtain Eq. (6.3.1). Therefore, one has the following proposition.

Proposition 6.3.4 *Let* $\kappa : S^{m-1} \times S^{m-1} \to S^{m-1}$ *be a continuous multiplication. Suppose that* $H^*\kappa(a) = \mathbf{1} \times a + a \times \mathbf{1}$ *for* $a \in H^{m-1}(S^{m-1})$. *Then* Hopf $(f_\kappa) = 1$.

6.3.3 Dimension Restrictions

We shall prove in Corollary 8.6.4 that, if there exists a map $f : S^{2m-1} \to S^m$ with Hopf invariant 1, then $m = 2^r$. Actually, $m = 1, 2, 4, 8$ by a famous theorem of Adams (see Theorem 8.6.6). This theorem implies the following result.

Proposition 6.3.5 *Let $S^q \xrightarrow{i} E \xrightarrow{\pi} S^p$ be a locally trivial bundle. Suppose that $H^*(E) \approx H^*(S^{p+q})$. Then $q = p - 1$ and $p = 1, 2, 4$ or 8.*

Proof If $p = 1$ and $q > 0$, then $H^*(E)$ is **GrV**-isomorphic to $H^*(S^1) \otimes H^*(S^q)$ by the argument of Example 4.7.45. Thus, we must have $q = 0$ and π is a non-trivial double cover of S^1.

Let us suppose that $p \geq 2$. If $H^*(E) \approx H^*(S^{p+q})$, then H^*i is not surjective; otherwise, $H^*(E)$ is **GrV**-isomorphic to $H^*(S^p) \otimes H^*(S^q)$ by the Leray-Hirsch theorem. The Wang exact sequences (see Proposition 4.7.43)

$$\cdots \to H^q(E) \xrightarrow{H^*i} H^q(S^q) \xrightarrow{\Theta} H^{q+1-p}(S^q) \to \cdots .$$

then implies that $q + 1 - p = 0$ (since $p > 1$). Therefore, $q = p - 1$.

The bundle gluing map $\tilde{\phi} : S^q \times S^q \to S^q$ (see p. 189) may thus be seen as a continuous multiplication, to which a map $f_{\tilde{\phi}} : S^{2p-1} \to S^p$ may be associated using (6.1.2). We shall prove that Hopf $(f_{\tilde{\phi}}) = 1$. By Theorem 8.6.6, this implies that $p = 1, 2, 4$ or 8.

Let $a \in H^q(S^q)$ be the generator. The restriction of $\tilde{\phi}$ to a slice $\{x\} \times S^q$ being a homeomorphism, one has, using Lemma 4.7.2, that

$$H^*\tilde{\phi}(a) = \mathbf{1} \times a + \lambda(\mathbf{1} \times a)$$

for some $\lambda \in \mathbb{Z}_2$. As $\Theta \neq 0$ and $p > 1$, one gets from Proposition 4.7.46 that

$$0 \neq \dot{e} \times \Theta(a) = H^*\tilde{\phi}(a) + \mathbf{1} \times a\,,$$

where $\dot{e} \in H^p(S^p)$ is the generator. Therefore, $H^*\tilde{\phi}(a) = a \times \mathbf{1} + \mathbf{1} \times a$. By Proposition 6.3.4, this implies that Hopf $(f_{\tilde{\phi}}) = 1$. □

Example 6.3.6 Consider the bundle $S^1 \to E \to S^2$ with gluing map $\tilde{\phi}(u, z) = u^k z$. The total space E, obtained by gluing two copies of $D^2 \times S^1$ using the map $\tilde{\phi}$, is then a lens space with fundamental group of order k. Thus, if k is odd, E satisfies the hypotheses of Proposition 6.3.5. Other famous examples are the bundles $S^3 \to E \to S^4$ which were used by J. Milnor to produce his exotic 7-spheres [145, Sect. 3]. Indeed, with a well chosen gluing map, the total space E is a smooth 7-manifold homeomorphic but not diffeomorphic to S^7.

6.3.4 Hopf Invariant and Linking Numbers

Let $f : S^{2m-1} \to S^m$ be a smooth map. Let $y, y' \in S^m$ be two distinct regular values of f. Then $Q = f^{-1}(\{y\})$ and $Q' = f^{-1}(\{y'\})$ are two disjoint closed submanifolds of S^{2m-1}, both of dimension $m - 1$. Therefore, their linking number $l(Q, Q') \in \mathbb{Z}_2$ (see Sect. 5.4.4) is defined, at least if $m > 1$ (see also Remark 6.3.8).

Proposition 6.3.7 *If $m > 1$, then* $\mathrm{Hopf}\,(f) = l(Q, Q')$.

Actually, $l(Q, Q')$ is the original definition by H. Hopf of his invariant [99, 100]. Proposition 6.3.7 goes back to the work of Steenrod [182], after which the definition of $\mathrm{Hopf}\,(f)$ with the cup product in C_f was gradually adopted.

Proof Let $\Sigma = S^{2m-1}$. We consider the mapping cylinder M_f of f

$$M_f = [(S^{2m-1} \times I) \dot{\cup} S^m]/\{(x, 0) \sim f(x) \mid x \in S^{2m-1}\}.$$

The correspondence $(x, t) \mapsto f(x)$ descends to a retraction by deformation ρ : $M_f \to S^m$. We identify Σ with the subspace $\Sigma \times \{1\}$ of M_f. The mapping cone $\hat{M}_f$ of f, defined as

$$\hat{M}_f = M_f \cup_\Sigma C\Sigma \approx M_f \cup_\Sigma D^{2m},$$

where $C\Sigma \approx D^{2m}$ is the cone over Σ, is homotopy equivalent to the CW-complex C_f.

We first introduce some material in order to compute $l(Q, Q')$ using Lemma 5.4.23. Let W_0 be a closed tubular neighbourhood of y in $S^m - \{y'\}$ and let V_0 be a closed tubular neighbourhood of y in $\mathrm{int}\, W_0$ (W_0 and V_0 are just m-disks). Let $y' \in V_0' \subset W_0'$ be a symmetric data for y' with $W_0 \cap W_0' = \emptyset$. Let $\mathcal{B}_0 = \{W_0, W_0', S^m - (V_0 \cup V_0')\}$. As y and y' are regular values of f, we may assume, provided W_0 and W_0' are small enough, that $W = f^{-1}(W_0)$ and $V = f^{-1}(V_0)$ are nested tubular neighbourhoods of Q and that $W' = f^{-1}(W_0')$ and $V' = f^{-1}(V_0')$ are nested tubular neighbourhoods of Q'. Let $\mathcal{B} = f^{-1}(\mathcal{B}_0) = \{W, W', \Sigma - (V \cup V')\}$.

Let us briefly repeat the preliminary constructions for Lemma 5.4.23 (see p. 234 for more details), in our context, with the dimensions $q = q' = m-1$ and $s = 2m-1$. Let $c_0 \in Z^m(W_0, W_0 - \mathrm{int}\, V_0)$ represent the Poincaré dual class of $\{y\}$ in $H^m(W_0, W_0 - \mathrm{int}\, V_0) \approx H^m(V_0, \mathrm{Bd}\, V_0)$ and let $c_0' \in Z^m(W_0', W_0' - \mathrm{int}\, V_0')$ represent the Poincaré dual class of $\{y'\}$ in $H^m(W_0', W_0' - \mathrm{int}\, V_0') \approx H^m(V_0', \mathrm{Bd}\, V_0')$. Let $\bar{c}_0, \bar{c}_0' \in Z^m_{\mathcal{B}_0}(S^m)$ be their zero extensions. Then $c = C^*f(c_0) \in Z^m(W, W - \mathrm{int}\, V)$ represent the Poincaré dual class of Q in $H^m(W, W - \mathrm{int}\, V) \approx H^m(V, \mathrm{Bd}\, V)$ and $c' = C^*f(c_0') \in Z^m(W', W' - \mathrm{int}\, V')$ represent the Poincaré dual class of Q' in $H^m(W', W' - \mathrm{int}\, V) \approx H^m(V', \mathrm{Bd}\, V')$. Also, $\bar{c} = C^*f(\bar{c}_0) \in Z^m_{\mathcal{B}}(\Sigma)$ and $\bar{c}' = C^*f(\bar{c}_0') \in Z^m_{\mathcal{B}}(\Sigma)$ are the zero extensions of c and c'. Choose $a \in C^{m-1}_{\mathcal{B}}(\Sigma)$ such that $\delta a = \bar{c}$. Let $\nu \in Z^{\mathcal{B}}_{2m-1}(\Sigma)$ represent $[\Sigma]_{\mathcal{B}}$ in $H^{\mathcal{B}}_{2m-1}(\Sigma) \approx H_{2m-1}(\Sigma)$. According to Lemma 5.4.23, one has $l(Q, Q') = \langle a \smile \bar{c}', \nu\rangle$. We note that $a \smile \bar{c}' \in Z^{2m-1}_{\mathcal{B}}(\Sigma)$. Indeed, for any $\sigma \in \mathcal{S}_{2m-1}(\Sigma)$, one has

$$\langle \delta(a \smile \bar{c}'), \sigma\rangle = \langle \bar{c} \smile \bar{c}', \sigma\rangle = 0 \tag{6.3.2}$$

since the support of $\bar{c}$ is in W and that of $\bar{c}'$ is in W'. Therefore, $\delta(a \smile \bar{c}') = 0$ and $a \smile \bar{c}'$ represents a cohomology classes $|a \smile \bar{c}'| \in H^{2m-1}_{\mathcal{B}}(\Sigma)$ (in this proof, we use the notation $|\;|$ for the cohomology class of a cocycle). The equality $l(Q, Q') = \langle a \smile \bar{c}', \nu\rangle$ is equivalent to

$$|a \smile \bar{c}'| = l(Q, Q')\,[\Sigma]^\sharp_{\mathcal{B}}, \tag{6.3.3}$$

an equality holding in $H^{2m-1}_{\mathcal{B}}(\Sigma)$, where $[\Sigma]^\sharp_{\mathcal{B}}$ is the generator of $H^{2m-1}_{\mathcal{B}}(\Sigma) \approx H^{2m-1}(\Sigma) = \mathbb{Z}_2$.

Let $\mathcal{B}_1 = \rho^{-1}(\mathcal{B})$. The inclusion $i : \Sigma \hookrightarrow M_f$ induces a morphism of cochain complexes $C^*i : C^*_{\mathcal{B}_1}(M_f) \to C^*_{\mathcal{B}}(\Sigma)$ whose kernel is denoted by $C^*_{\mathcal{B}_1}(M_f, \Sigma)$. Note that $C^*_{\mathcal{B}_1}(M_f, \emptyset) = C^*_{\mathcal{B}_1}(M_f)$ and so the inclusion $C^*_{\mathcal{B}_1}(M_f, \Sigma) \hookrightarrow C^*_{\mathcal{B}_1}(M_f)$ coincides with C^*j, the morphism induced by the pair inclusion $j : (M_f, \emptyset) \hookrightarrow (M_f, \Sigma)$ (see Remark 3.1.26). One has the commutative diagram

$$\begin{array}{ccccccccc} 0 & \longrightarrow & C^*_{\mathcal{B}_1}(M_f, \Sigma) & \xrightarrow{C^*j} & C^*_{\mathcal{B}_1}(M_f) & \xrightarrow{C^*i} & C^*_{\mathcal{B}}(\Sigma) & \longrightarrow & 0 \\ & & & & {\scriptstyle C^*\rho}\uparrow & \nearrow{\scriptstyle C^*f} & & & \\ & & & & C^*_{\mathcal{B}_0}(S^m) & & & & \end{array} \tag{6.3.4}$$

where the top row is an exact sequence of cochain complexes. This sequence gives rise to a connecting homomorphism δ^* sitting in the exact sequence

$$H^{2m-1}_{\mathcal{B}_1}(M_f) \xrightarrow{H^*i} H^{2m-1}_{\mathcal{B}}(\Sigma) \xrightarrow{\delta^*} H^{2m}_{\mathcal{B}_1}(M_f, \Sigma) \xrightarrow{H^*j} H^{2m}_{\mathcal{B}_1}(M_f)\,.$$

As $m > 1$, one has $H^{2m-1}_{\mathcal{B}_1}(M_f) \approx H^{2m-1}(M_f) = 0$ and $H^{2m}_{\mathcal{B}_1}(M_f) \approx H^{2m}(M_f) = 0$. Therefore $\delta^* : H^{2m-1}_{\mathcal{B}}(\Sigma) \to H^{2m}_{\mathcal{B}_1}(M_f, \Sigma)$ is an isomorphism. Let $b = \delta^*([\Sigma]^\sharp_{\mathcal{B}})$ be the generator of $H^{2m}_{\mathcal{B}_1}(M_f, \Sigma)$. By (6.3.3), the linking number $l(Q, Q')$ is then determined by the equation

$$\delta^*(|a \smile \bar{c}'|) = l(Q, Q')\, b\,. \tag{6.3.5}$$

To compute $\delta^*(|a \smile \bar{c}'|)$, write δ_M for the coboundary operator in $C^*_{\mathcal{B}_1}(M_f)$. Let $\bar{c}_1 = C^*\rho(\bar{c}_0)$ and $\bar{c}'_1 = C^*\rho(\bar{c}'_0)$, both in $Z^m_{\mathcal{B}_1}(M_f)$. Let $a_1 \in C^{m-1}(M_f)$ such that $C^*i(a_1) = a$. By the commutativity of diagram (6.3.4), one has $C^*i(a_1 \smile \bar{c}'_1) = a \smile \bar{c}'$. Then $\delta_M(a_1 \smile \bar{c}'_1)$ is a cocycle in $\ker C^*i$, so there is a unique $u \in Z^{2m}_{\mathcal{B}_1}(M_f, \Sigma)$ such that $C^*j(u) = \delta_M(a_1 \smile \bar{c}'_1)$. As in Lemma 2.7.1, one has

$$|u| = \delta^*(|a \smile \bar{c}'|)\,. \tag{6.3.6}$$

The cohomology class $|u|$ may be described in another way. As for (6.3.2), one has $\bar{c}_1 \smile \bar{c}'_1 = 0$ for support reasons. Therefore,

$$(\delta_M(a_1) + \bar{c}_1) \smile \bar{c}'_1 = \delta_M(a_1) \smile \bar{c}'_1 = \delta_M(a_1 \smile \bar{c}'_1)\,. \tag{6.3.7}$$

Now, $C^*i(\delta_M(a_1) + \bar{c}_1) = 0$, thus there is a unique $w \in Z^m_{\mathcal{B}_1}(M_f, \Sigma)$ with $C^*j(w) = \delta_M(a_1) + \bar{c}_1$. The first cup product of (6.3.7) may be understood as relative cochain cup product giving rise to a relative cohomology cup product

$$H^*_{\mathcal{B}_1}(M_f,\Sigma)\times H^*_{\mathcal{B}_1}(M_f)\xrightarrow{\smile} H^*_{\mathcal{B}_1}(M_f,\Sigma)$$

analogous to that of Lemma 4.1.14 (in the case $Y_2=\emptyset$). Equation (6.3.6) is equivalent to

$$|w|\smile|\bar{c}'_1|=\delta^*(|a\smile\bar{c}'|) \tag{6.3.8}$$

and, using (6.3.6), we get the equality

$$|w|\smile|\bar{c}'_1|=l(Q,Q')\,b \tag{6.3.9}$$

holding in $H^{2m}_{\mathcal{B}_1}(M_f,\Sigma)$.

Now, under the isomorphism $H^m(S^m)\xrightarrow{\approx} H^m_{\mathcal{B}_0}(S^m)$ (due to the small simplex theorem 3.1.34), the cohomology class $|\bar{c}_0|$ corresponds to the Poincaré dual PD($\{y\}$) $\in H^m(S^m)$, which is $[S^m]^\sharp$. Analogously, $|\bar{c}'_0|$ also corresponds to $[S^m]^\sharp$. Hence, under the isomorphism $H^m(M_f)\xrightarrow{\approx} H^m_{\mathcal{B}_1}(M_f)$, $|\bar{c}_1|$, $|\bar{c}'_1|$ and $H^*j(|w|)$ all correspond to $H^*\rho([S^m]^\sharp)$, that is the generator e of $H^m(M_f)\approx\mathbb{Z}_2$. The diagram

$$\begin{array}{ccc}
H^m_{\mathcal{B}_1}(M_f)\times H^m_{\mathcal{B}_1}(M_f,\Sigma) & \xrightarrow{\smile} & H^{2m}_{\mathcal{B}_1}(M_f,\Sigma)\\
\uparrow\approx & & \uparrow\approx\\
H^m(M_f)\times H^m(M_f,\Sigma) & \xrightarrow{\smile} & H^{2m}(M_f,\Sigma)\\
\uparrow\approx & & \uparrow\approx\\
H^m(\hat{M}_f)\times H^m(\hat{M}_f,C\Sigma) & \xrightarrow{\smile} & H^{2m}(\hat{M}_f,C\Sigma)\\
\downarrow\approx & & \downarrow\approx\\
H^m(\hat{M}_f)\times H^m(\hat{M}_f) & \xrightarrow{\smile} & H^{2m}(\hat{M}_f)
\end{array} \tag{6.3.10}$$

is commutative, where the vertical arrows are the obvious ones or induced by the inclusions (the commutativity of the bottom square is the content of the singular analogue of Lemma 4.1.8). Let $k:M_f\to\hat{M}_f$ denote the inclusion. Then $H^*_k(e)=\hat{e}$, the generator of $H^m(\hat{M}_f)\approx\mathbb{Z}_2$. Equation (6.3.9), obtained using the top line of (6.3.10), becomes, using the bottom line

$$\hat{e}\smile\hat{e}=l(Q,Q')\,\hat{b} \tag{6.3.11}$$

where $\hat{b}$ is the generator of $H^{2m}(\hat{M}_f) \approx \mathbb{Z}_2$. But, as $m > 1$, the equality $\hat{e} \smile \hat{e} = \text{Hopf}(f)\,\hat{b}$ holds true, by definition of the Hopf invariant. Therefore, $\text{Hopf}(f) = l(Q, Q')$.

Remark 6.3.8 For a map $f : S^1 \to S^1$ with even degree, the equality $\text{Hopf}(f) = l(Q, Q')$ holds true, using Proposition 6.3.1 (see Exercise 6.11). When $\deg f$ is odd, the linking number $l(Q, Q')$ is not defined. Indeed, both Q and Q' have an odd number of points and condition (3) of p. 231 is not satisfied.

6.4 Exercises for Chapter 6

6.1. What is the Lusternik-Schnirelmann category of $\mathbb{K}P^n$ for $\mathbb{K} = \mathbb{C}$ or $\mathbb{H}$.

6.2. Compute the cohomology ring of $\mathbb{C}P^n$ and $\mathbb{H}P^n$, using the Gysin exact sequence for the Hopf bundles. [Hint: like in Sect. 4.3.4.]

6.3. For $\mathbb{K} = \mathbb{C}$ or $\mathbb{H}$, prove that $H^*(\mathbb{K}P^\infty) \to H^*(\mathbb{R}P^\infty)$ is injective.

6.4. Let $f : X \to Y$ be a map. The *double mapping cylinder* CC_f of f is the union of two copies of the mapping cylinder C_f of f glued along X. Compute $H^*(CC_j)$ where j is the inclusion of $\mathbb{C}P^\infty \to \mathbb{H}P^\infty$.

6.5. Prove that $X = S^4 \times S^4$ and $Y = \mathbb{H}P^2 \sharp \mathbb{H}P^2$ have the same Poincaré polynomial. Are $H^*(X)$ and $H^*(Y)$ **GrA**-isomorphic?

6.6. For any positive integer n, construct a vector bundle ξ of rank n over a closed n-dimensional manifold such that $e(\xi) \neq 0$.

6.7. Let X be a CW-complex of dimension $n = 1, 2, 4, 8$ and let $a \in H^n(X)$. Prove that there exists a vector bundle ξ over X with $e(\xi) = a$.

6.8. Prove that there is no continuous injective map $f : \mathbb{R}^n \to \mathbb{R}^k$ if $n > k$. [Hint: use the Borsuk-Ulam theorem.]

6.9. Check the table of p. 251.

6.10. Show that the Hopf vector bundle over $\mathbb{K}P^1 \approx S^d$ ($d = \dim_{\mathbb{R}} \mathbb{K}$) cannot be the normal bundle of an embedding of S^d into a manifold M of dimension $2d$ with $H_d(M) = 0$.

6.11. Let $f : S^1 \to S^1$ be a smooth map with even degree. Show that the Hopf invariant of f is equal to the linking number of the inverse image of two regular values of f, as in Proposition 6.3.7. [Hint: use Proposition 6.3.1.]

6.12. Using the linking numbers and Proposition 6.3.7, show that the various Hopf maps have Hopf invariant 1. [Hint: use Formula (6.1.3) and Exercise 5.18.]

6.13. Let $g : S^{2m+1} \to S^m$ be a continuous map, as well as $f : S^m \to S^m$ and $h : S^{2m+1} \to S^{2m+1}$. Prove that $\text{Hopf}(f \circ g \circ h) = \deg(h) \deg(f)^2 \,\text{Hopf}(g)$. (Remark: of course, $\deg(f)^2 \equiv \deg(f) \mod 2$ but the formula is the one which is valid for the cohomology with any coefficients.)

6.14. Let $f : S^{2m-1} \to S^m$ be a smooth map and let $y \in S^m$ be a regular value for f. The closed $(m-1)$-manifold $Q = f^{-1}(\{y\})$ bounds a compact manifold W (see Exercise 9.14). As $S^{2m-1} - pt$ is contractible, the inclusion of Q into S^{2m-1} extends to a smooth map $j : W \to S^{2m-1}$ (see Exercise 3.2). We thus get a homomorphism $H_*(f \circ j) \ : H_m(W, Q) \to H_m(S^m, \{y\})$. As both these homology groups are isomorphic to $\mathbb{Z}_2$, this defines a degree for $f \circ j$, as in (2.5.4). Prove that Hopf $(f) = \deg(f \circ j)$.

Chapter 7
Equivariant Cohomology

In ordinary life, the symmetries of an object (like a ball or a cube) help us to apprehend it. The same should happen in topology when studying spaces with symmetries, i.e. endowed with actions of topological groups. Equivariant cohomology is one tool for such a purpose.

Our aim here is mostly to develop enough material needed in the forthcoming chapters. For instance, the definition and most properties of the Steenrod squares use equivariant cohomology for spaces with involution. This case is treated in detail in Sect. 7.1, at an elementary level and with *ad hoc* techniques. A second section deals with Γ-spaces for any topological group Γ (the proof of the Adem relations requires Γ-equivariant cohomology with Γ the symmetric group Sym_4). Equivariant cross products, treated in Sect. 7.4, will also be used. Only Sect. 7.3 is written uniquely for its own interest, devoted to some simple form of localization theorems and Smith theory. A final section presents the equivariant Morse-Bott theory, used in Sect. 9.5 to compute the cohomology of flag manifolds (see also Sect. 10.3.5). For further reading on equivariant cohomology, see e.g. [9, 37, 103].

7.1 Spaces with Involution

An *involution* on a topological space X is a continuous map $\tau : X \to X$ such that $\tau \circ \tau = \mathrm{id}$. The letter τ is usually used for all encountered involutions. We also use the symbol τ for the non-trivial element of the cyclic group $G = \{\mathrm{id}, \tau\}$ of order 2; an involution on X is thus equivalent to a continuous action of G on X and a space with involution is equivalent to a *G-space*, i.e. a space together with an action of G. We often pass from one language to the other. If X is a G-space, its *fixed point subspace* X^G is defined by

$$X^G = \{x \in X \mid \tau(x) = x\}.$$

As G has only two elements, the complement of X^G is the subspace where the action is free.

J.-C. Hausmann, *Mod Two Homology and Cohomology*, Universitext,
DOI 10.1007/978-3-319-09354-3_7

A continuous map $f : X \to Y$ between G-spaces is a *G-equivariant map*, or just a *G-map* if it commutes with the involutions: $f \circ \tau = \tau \circ f$. Two G-maps $f_0, f_1 : X \to Y$ are *G-homotopic* if there is a homotopy $F : X \times I \to Y$ connecting them which is a G-map. Here, the involution on $X \times I$ is $\tau(x, t) = (\tau(x), t)$. This permits us to define the notion of *G-homotopy equivalence* and of *G-homotopy type*. For instance, a G-space is *G-contractible* if it has the G-homotopy type of a point.

Let X be a G-space. A CW-structure on X with set of n-cells Λ_n is a *G-CW-structure* if the following condition is satisfied: *for each integer n, there is a G-action on Λ_n and a G-equivariant global characteristic map $\hat{\varphi}_n : \Lambda_n \times D^n \to X$, where the G-action on $\Lambda_n \times D^n$ is given by $\tau(\lambda, x) = (\tau(\lambda), x)$*. In particular, if $\lambda \in \Lambda_n$ satisfies $\tau(\lambda) = \lambda$, then τ restricted to $\{\lambda\} \times D^n$ is the identity. These cells are called the *isotropic cells*; they form a G-CW-structure for X^G. The other cells, the *free cells*, come in pairs $(\lambda, \tau(\lambda))$. A G-space endowed with a G-CW-structure is a *G-CW-complex*, or just a *G-complex*. Observe that, if X is a G-complex, then the quotient space X/G inherits a CW-structure (with set of n-cells equal to Λ_n/G) for which the quotient map is cellular. A smooth G-manifold admits a G-CW-structure, in fact a G-triangulation (see [106]).

Example 7.1.1 Let $X = S^n$ $(n \leq \infty)$ be the standard sphere endowed with the CW-structure where the m-skeleton is S^m and having two m-cells in each dimension $m \leq n$ (see Example 3.4.4). This is a G-CW-structure for the free involution given by the antipodal map $z \mapsto -z$. The quotient space X/G is $\mathbb{R}P^n$ with its standard CW-structure.

Let X be a space with an involution τ. The *Borel construction* X_G, also known as the *homotopy quotient*, is the quotient space

$$X_G = S^\infty \times_G X = (S^\infty \times X)/\sim \tag{7.1.1}$$

where $\sim$ is the equivalence relation $(z, \tau(x)) \sim (-z, x)$. If X and Y are G-spaces and if $f : Y \to X$ is a continuous G-equivariant map, the map $\mathrm{id} \times f : S^\infty \times Y \to S^\infty \times X$ descends to a map $f_G : Y_G \to X_G$. This makes the Borel construction a covariant functor from the category $\mathbf{Top}_G$ to $\mathbf{Top}$,$\mathbf{Top}_G$ where is the category of G-spaces and G-equivariant maps. Using the obvious homeomorphism between $(X \times I)_G$ and $X_G \times I$, a G-homotopy between two G-maps f^0 and $f^1 : X \to Y$ descends to a homotopy between f^0_G and f^1_G. Hence, X_G and Y_G have the same homotopy type if X and Y have the same G-homotopy type.

Let $\hat{p} : S^\infty \to \mathbb{R}P^\infty$ be the quotient map (this is a 2-fold covering projection). A map $p : X_G \to \mathbb{R}P^\infty$ is then given by $p([z, x]) = \hat{p}(z)$. Observe that p coincides with the map $f_G : X_G \to pt_G = \mathbb{R}P^\infty$ induced by the constant map $X \to pt$.

Example 7.1.2 Suppose that the involution τ is trivial, i.e. $\tau(x) = x$ for all $x \in X$. The projection $S^\infty \times X \to X$ then descends to $X_G \to X$. Together with the map p, this gives a homeomorphism $X_G \xrightarrow{\approx} \mathbb{R}P^\infty \times X$.

Lemma 7.1.3 (1) *The map $p : X_G \to \mathbb{R}P^\infty$ is a locally trivial fiber bundle with fiber homeomorphic to X.*

(2) *If $f : Y \to X$ is a G-equivariant map, then the diagram*

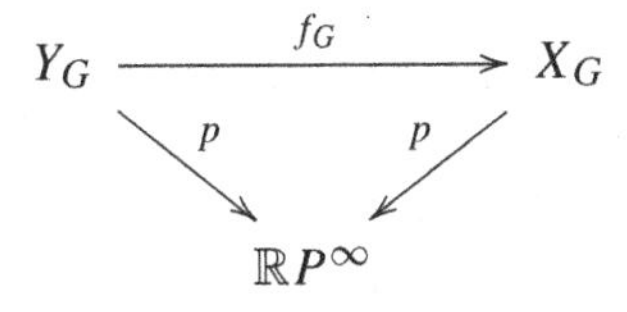

is commutative.

(3) *If τ has a fixed point, then p admits a section. More precisely, the choice of a point $v \in X^G$ provides a section $s_v : \mathbb{R}P^\infty \to X_G$ of p.*

(4) *The quotient map $S^\infty \times X \to X_G$ is a 2-fold covering admitting p as a characteristic map.*

Proof We use that $\hat{p} : S^\infty \to \mathbb{R}P^\infty$ is a principal G-bundle, i.e. a 2-fold covering. Denote by $z = (z_0, z_1, \dots)$ the elements of S^∞. The set $V_i = \{z \in S^\infty \mid z_i \neq 0\}$ is an open subspace of S^∞. As $\hat{p}$ is an open map, $U_i = \hat{p}(V_i)$ is an open set of $\mathbb{R}P^\infty$. A trivialization $\psi_i : V_i \to U_i \times \{\pm 1\}$ is given by $\psi_i(z) = (\hat{p}(z), z_i/|z_i|)$. Using the group isomorphism $\{\pm 1\} \xrightarrow{\approx} G$, this gives a trivialization $\psi_i : V_i \to U_i \times G$. Now, $\psi_i \times \mathrm{id} : V_i \times X \xrightarrow{\approx} U_i \times G \times X$ descends to a homeomorphism $p^{-1}(U_i) \xrightarrow{\approx} U_i \times (G \times_G X)$. Here, $G \times_G X$ denotes the quotient of $G \times X$ by the equivalence relation $(g, \tau(x)) \sim (g\tau, x)$. But the map $x \mapsto (\mathrm{id}, x)$ provides a homeomorphism from X onto $G \times_G X$. This proves Point (1). This also shows that, over $p^{-1}(U_i)$, the map $S^\infty \times X \to X_G$ looks like the projection $G \times p^{-1}(U_i) \to p^{-1}(U_i)$. Therefore, $S^\infty \times X \to X_G$ is a 2-fold covering, with the product involution $\tau_\times(z, x) = (-z, \tau(x))$ as deck transformation. The diagram

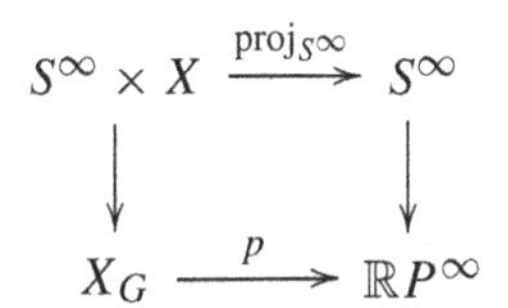

is commutative and $\mathrm{proj}_{S^\infty}(\tau_\times(y)) = -\mathrm{proj}_{S^\infty}(y)$. By Lemma 4.3.4, this implies that p is a characteristic map for the covering $S^\infty \times X \to X_G$. Point (4) is thus established.

Point (2) is obvious from the definitions. For Point (3), let $v \in X^G$. By Point (2), the inclusion $i : \{v\} \hookrightarrow X$ gives rise to a commutative diagram

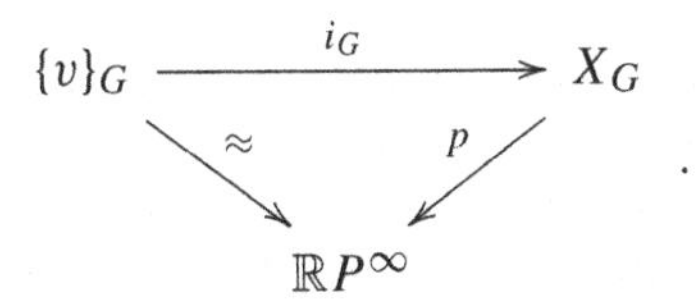

Hence, i_G provides a section $s_v : \mathbb{R}P^\infty \to X_G$ of p, depending on the choice of the fixed point v. □

The projection $\tilde{q} : S^\infty \times X \to X$ seen in Example 7.1.2 descends to $q : X_G \to X/G$.

Lemma 7.1.4 *Let X be a free G-space such that X is Hausdorff. Then, the* **GrA**-*morphism $H^*q : H^*(X/G) \to H^*(X_G)$ is an isomorphism. Moreover, if X is a free G-complex, then the map $q : X_G \to X/G$ is a homotopy equivalence and the map $p : X_G \to \mathbb{R}P^\infty$ is homotopic to the composition of q with a characteristic map for the covering $X \to X/G$.*

Proof If X is Hausdorff, such a projection $X \to X/G$ is a 2-fold covering and X/G is Hausdorff. Over a trivializing open set $U \subset X/G$, this covering is equivalent to $G \times U \to U$. Then $(G \times U)_G \approx S^\infty \times U$ since any class has a unique representative of the form $(z, \mathrm{id}, u) \in S^\infty \times G \times U$. Hence, $X_G \xrightarrow{q} X/G$ is a locally trivial bundle with fiber S^∞. As $\tilde{H}^*(S^\infty) = 0$, the map $q^* : H^*(X/G) \to H^*(X_G)$ is a **GrA**-isomorphism by Corollary 4.7.21. Actually as S^∞ is contractible [82, example 1.B.3 p. 88], the homotopy exact sequence of the bundle [82, Theorem 4.41 and Proposition 4.48] implies that q is a weak homotopy equivalence. If X is a G-complex, then X/G is a CW-complex. Also, $S^\infty \times X$ is a free G-complex and thus X_G is a CW-complex. Therefore, a weak homotopy equivalence is a homotopy equivalence by the Whitehead theorem [82, Theorem 4.5]. Also, again since S^∞ is contractible and X/G is a CW-complex, a direct proof that q is a homotopy equivalence is available using [42, Theorem 6.3].

Let $f : X/G \to \mathbb{R}P^\infty$ be a characteristic map for the covering $X \to X/G$. The diagram

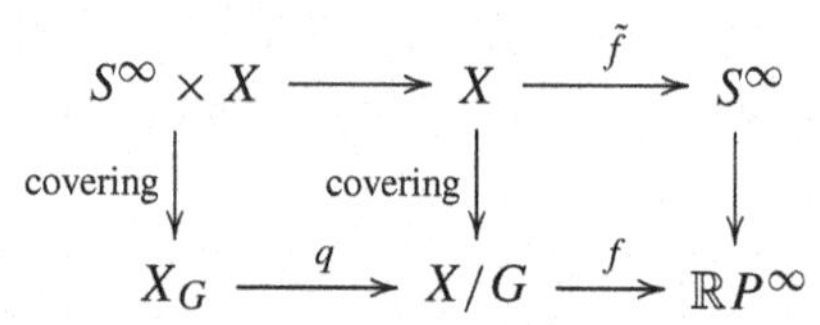

is commutative and the upper horizontal arrows commute with the deck involutions. By Lemma 4.3.4, this implies that $f \circ q$ is a characteristic map for the covering $S^\infty \times X \to X_G$. By Lemma 7.1.3, so is p. By Corollary 4.3.3, two characteristic maps of a covering are homotopic. Therefore, the maps p and $f \circ q$ are homotopic. □

Corollary 7.1.5 *Let X be a finite dimensional G-complex. Then, the following conditions are equivalent.*

(1) *X has a fixed point.*
(2) *The morphism $H^*p : H^*(\mathbb{R}P^\infty) \to H^*(X_G)$ is injective.*

Proof If X has a fixed point, then p admits a section by Lemma 7.1.3, so H^*p is injective. If X has no fixed point, then X is a free G-complex and, by Lemma 7.1.4,

$H^*(X_G) \approx H^*(X/G)$. Also, X/G is a finite dimensional CW-complex, so H^*p is not injective. □

Let X be a space with an involution τ and let $Y \subset X$ be an invariant subspace. Then $Y_G \subset X_G$. The (relative) *G-equivariant cohomology* $H^*_G(X, Y)$ is the cohomology algebra

$$H^*_G(X, Y) = H^*(X_G, Y_G).$$

We shall mostly concentrate on the absolute case $H^*_G(X) = H^*(X_G) = H^*_G(X, \emptyset)$. The map $p : X_G \to \mathbb{R}P^\infty$ induces a **GrA**-homomorphism $p^* : H^*(\mathbb{R}P^\infty) \to H_G(X)$. By Proposition 6.1.1, $H^*(\mathbb{R}P^\infty)$ is **GrA**-isomorphic to the polynomial ring $\mathbb{Z}_2[u]$, where u is a formal variable in degree 1. Hence, the **GrA**-homomorphism p^* gives on $H^*_G(X)$ a structure of $\mathbb{Z}_2[u]$-algebra. In particular, $H^*_G(pt) = \mathbb{Z}_2[u]$.

As an important example, let us consider the case of a G-space Y with $Y = Y^G$, i.e. the involution τ is trivial. As seen in Example 7.1.2, we get an identification $Y_G = \mathbb{R}P^\infty \times Y$. By the Künneth theorem,

$$H^*_G(Y) \approx \mathbb{Z}_2[u] \otimes H^*(Y) \approx H^*(Y)[u]. \tag{7.1.2}$$

The **GrA**-homomorphism $H^*(Y) \to H^*_G(Y)$ induced by the projection $\mathbb{R}P^\infty \times Y \to Y$ corresponds to the inclusion of the "ring of constants" $H^*(Y)$ into $H^*(Y)[u]$.

The functoriality of the Borel construction and of the cohomology algebra, together with Point (2) of Lemma 7.1.3, says that, if $f : Y \to X$ is a G-equivariant map between G-spaces, then $H^*f_G : H^*_G(X) \to H^*_G(Y)$ is a **GrA**-homomorphism commuting with the multiplication by u. We are then driven to consider the category **GrA**$[u]$ whose objects are graded $\mathbb{Z}_2[u]$-algebras and whose morphisms are **GrA**-homomorphism commuting with the multiplication by u. Hence, the correspondence $X \mapsto H^*_G(X)$ is a contravariant functor from $\mathbf{Top}_G$ to **GrA**$[u]$. If $f : Y \to \bar{Y}$ is a G-equivariant map between *trivial* G-spaces (i.e., any continuous map), the following diagram is commutative:

$$\begin{array}{ccc} H^*(\bar{Y})[u] & \xrightarrow{\approx} & H^*_G(\bar{Y}) \\ \Big\downarrow{\scriptstyle H^*f[u]} & & \Big\downarrow{\scriptstyle H^*_G f} \\ H^*(Y)[u] & \xrightarrow{\approx} & H^*_G(Y) \end{array} \quad . \tag{7.1.3}$$

Choosing a point $z \in S^\infty$ provides, for each G-space X, a map $i_z : X \to X_G$ defined by $i_z(x) = [z, x]$. As S^∞ is path-connected, the homotopy class of i_z does not depend on z. Therefore, we get a well defined **GrA**-homomorphism

$$\rho : H^*_G(X) \to H^*(X) \tag{7.1.4}$$

given by $\rho = H^*i_z$ for some $z \in S^\infty$. We can call ρ the *forgetful homomorphism* (it forgets the G-action). Observe that ρ is functorial. Indeed, if $f : X \to \bar{X}$ is a G-equivariant map, the diagram

$$\begin{array}{ccc} X & \xrightarrow{i_z} & X_G \\ \downarrow f & & \downarrow f_G \\ \bar{X} & \xrightarrow{i_{f(z)}} & \bar{X}_G \end{array}$$

is commutative, and so is the diagram

$$\begin{array}{ccc} H^*(\bar{X}) & \xleftarrow{\rho_{\bar{X}}} & H^*_G(\bar{X}) \\ \downarrow H^*f & & \downarrow H^*_G f \\ H^*(X) & \xleftarrow{\rho_X} & H^*_G(X) \end{array} \quad . \tag{7.1.5}$$

If Y is a G-space with $Y = Y^G$, we get, using the $\mathbf{GrA}[u]$-isomorphism of (7.1.2), the commutative diagram

$$\begin{array}{ccc} H^*(Y)[u] & \xrightarrow{\approx} & H^*_G(Y) \\ & {\scriptstyle \mathrm{ev}_0} \searrow \quad \swarrow {\scriptstyle \rho} & \\ & H^*(Y) & \end{array} \tag{7.1.6}$$

where ev_0 is the *evaluation* of a polynomial at $u = 0$, i.e. the unique algebra homomorphism extending the identity on $H^*(Y)$ and sending u to 0.

We now explain how some information on $H^*_G(X)$ may be obtained form transfer exact sequences. Observe that $\rho = H^*\bar{\rho}$ where $\bar{\rho} : X \to X_G$ is the composition

$$\bar{\rho} : X \xrightarrow{\text{slice}} X \times S^\infty \to X_G .$$

As S^∞ is contractible [82, example 1.B.3 p. 88], the slice inclusion is a homotopy equivalence. By Lemma 7.1.3, the map $X \times S^\infty \to X_G$ is then a 2-fold covering. Its characteristic map is $p : X_G \to \mathbb{R}P^\infty$ and, by Lemma 4.3.6, its characteristic class coincides with $u \in H^1_G(X)$. Therefore, the transfer exact sequence of the covering (see Proposition 4.3.9) gives the exact sequence

$$\cdots \to H^{m-1}_G(X) \xrightarrow{-\cdot u} H^m_G(X) \xrightarrow{\rho} H^m(X) \xrightarrow{\mathrm{tr}^*} H^m_G(X) \xrightarrow{-\cdot u} H^{m+1}_G(X) \to \cdots . \tag{7.1.7}$$

Denote by (u) the ideal of $H_G^*(X)$ generated by u and by

$$\mathrm{Ann}\,(u) = \{x \in H_G^*(X) \mid ux = 0\}$$

the annihilator of u. The information carried by Sequence (7.1.7) may be concentrated in the following short exact sequence of graded $\mathbb{Z}_2[u]$-modules:

$$0 \to H_G^*(X)/(u) \xrightarrow{\rho} H^*(X) \xrightarrow{\mathrm{tr}^*} \mathrm{Ann}\,(u) \to 0. \tag{7.1.8}$$

A G-space X is called *equivariantly formal* if $\rho : H_G^*(X) \to H^*(X)$ is surjective. For instance, X is equivariantly formal if the G-action is trivial. See 7.2.9 for a discussion of this definition in a more general setting.

Proposition 7.1.6 *For a G-space X, the following conditions are equivalent.*

(1) *X is equivariantly formal.*
(2) *$H_G^*(X)$ is a free $\mathbb{Z}_2[u]$-module.*
(3) $\mathrm{Ann}\,(u) = 0$.

Proof That (2) $\Rightarrow$ (3) is obvious and (3) $\Leftrightarrow$ (1) follows from (7.1.8). For (1) $\Rightarrow$ (2), choose a **GrV**-section $\theta : H^*(X) \to H_G^*(X)$ of ρ (as ρ is surjective). Then θ is a cohomology extension of the fiber for the fiber bundle $X \to X_G \xrightarrow{p} \mathbb{R}P^\infty$. As $\mathbb{R}P^\infty$ is path-connected and of finite cohomology type, the Leray-Hirsch theorem II (Theorem 4.7.18) implies that $H_G^*(X)$ is a free $\mathbb{Z}_2[u]$-module generated by $\theta(\mathcal{B})$, where $\mathcal{B}$ is a $\mathbb{Z}_2$-basis of $H^*(X)$. □

Remark 7.1.7 As noted before, $H^*(X) \otimes \mathbb{Z}_2[u]$ is isomorphic, as a $\mathbb{Z}_2[u]$-algebra, to $H^*(X)[u]$. If X is equivariantly formal, the Leray-Hirsch theorem II (Theorem 4.7.18) thus provides an isomorphism of $\mathbb{Z}_2[u]$-modules between $H_G^*(X)$ and $H^*(X)[u]$. This isomorphism depends on the choice of a **GrV**-section $\theta : H^*(X) \to H_G^*(X)$ of $\rho : H_G^*(X) \to H^*(X)$ and is not, in general, an isomorphism of algebras. However, as in the case of a trivial G-action, Diagram (7.1.6) is commutative.

Corollary 7.1.8 *Let X be a G-space. Suppose that $r : H_G^*(X) \to H_G^*(X^G)$ is injective. Then X is equivariantly formal.*

Proof Since $H_G^*(X^G)$ is a free $\mathbb{Z}_2[u]$-module, the hypothesis implies that Ann $(u) = 0$. □

The converse of Corollary 7.1.8 is true in some cases (see Proposition 7.3.9), but not in general. For example, S^∞ is equivariantly formal since $H^*(S^\infty) \approx H^*(pt)$. But $H_G^*(S^\infty) = H^*(\mathbb{R}P^\infty)$ by Lemma 7.1.4 and $H_G^*((S^\infty)^G) = H_G^*(\emptyset) = 0$.

For X a G-space, let $r : H_G^*(X) \to H_G^*(X^G)$ and $\bar{r} : H^*(X) \to H^*(X^G)$ be the **GrA**$[u]$-homomorphism induced by the inclusion $X^G \hookrightarrow X$. One can compose $\mathrm{tr}^* : H^*(X) \to H_G^*(X)$ with r.

Proposition 7.1.9 $r \circ \mathrm{tr}^* = 0$.

Proof As G acts trivially on X^G, one has $H^*_G(X^G) \approx H^*(\mathbb{R}P^\infty) \otimes H^*(X^G)$. The commutative diagram

$$\begin{array}{ccccc} H^*(S^\infty \times X) & \xrightarrow{\bar{r}} & H^*(S^\infty \times X^G) & \xleftarrow{\approx} & H^*(S^\infty) \otimes H^*(X^G) \\ \downarrow{\scriptstyle \mathrm{tr}^*} & & \downarrow{\scriptstyle \mathrm{tr}^*} & & \downarrow{\scriptstyle \mathrm{tr}^* \otimes \mathrm{id}} \\ H^*(S^\infty \times_G X) & \xrightarrow{r} & H^*(\mathbb{R}P^\infty \times X^G) & \xleftarrow{\approx} & H^*(\mathbb{R}P^\infty) \otimes H^*(X^G) \end{array}$$

proves the proposition since $\mathrm{tr}^* : H^*(S^\infty) \to H^*(\mathbb{R}P^\infty)$ is the zero map (even in degree 0, by the transfer exact sequence). □

To say more about the image of $\rho : H^*_G(X) \to H^*(X)$, define

$$H^*(X)^G = \{a \in H^*(X) \mid H^*\tau(a) = a\} \subset H^*(X).$$

As $H^*\tau$ is a **GrA**-morphism, $H^*(X)^G$ is a $\mathbb{Z}_2$-graded subalgebra of $H^*(X)$.

Lemma 7.1.10 $\rho(H^*_G(X)) \subset H^*(X)^G$.

Proof Let $z \in S^\infty$ and $b \in H^*_G(X)$. Then,

$$H^*\tau \circ \rho(b) = H^*\tau \circ H^* i_z(b) = H^*(i_z \circ \tau)(b) = H^* i_{-z}(b) = \rho(b).$$ □

The reverse inclusion in Lemma 7.1.10 may be wrong, as shown by the following example.

Example 7.1.11 Let τ be the antipodal map on the sphere $X = S^n$, making X a free G-complex, as seen in Example 7.1.1. The equality $H^n(X) = H^n(X)^G$ holds true since $H^n(X) = \mathbb{Z}_2$. For any $z \in S^\infty$, the composition $X \xrightarrow{i_z} X_G \xrightarrow{q} X/G$ coincides with the quotient map $X \twoheadrightarrow X/G = \mathbb{R}P^n$. By Lemma 7.1.4, the map q is a homotopy equivalence. But, since $S^n \twoheadrightarrow \mathbb{R}P^n$ is of local degree 2, the homomorphism $H^n(\mathbb{R}P^n) \to H^n(S^n)$ vanishes. Thus $\rho = 0$. In this example, the non-existence of fixed points is important (see Proposition 7.1.12).

Let X be a G-space. The *reduced equivariant cohomology* $\tilde{H}^*_G(X)$ is the **GrA**$[u]$-algebra defined by

$$\tilde{H}^*_G(X) = \mathrm{coker}\left(H^* p_G : H^*_G(pt) \to H^*_G(X)\right) \tag{7.1.9}$$

where $p_G : X \to pt$ denotes the constant map to a point (which is G-equivariant). Warning: $\tilde{H}^*_G(X) \neq \tilde{H}^*(X_G)$. Here are some examples.

(1) $\tilde{H}^*_G(pt) = 0$.
(2) Let $X = S^n$ with the antipodal involution. By Lemma 7.1.4, X_G has the homotopy type of $\mathbb{R}P^n$ and $H^*p : H^*(\mathbb{R}P^\infty) \to H^*_G(X)$ is surjective. Therefore, $\tilde{H}^*_G(X) = 0$.

(3) If Y is a space with trivial G-action, one has a natural $\mathbf{GrA}[u]$-isomorphism

$$\tilde{H}^*_G(Y) = H^*(Y)[u]/\mathbb{Z}_2[u] \approx \tilde{H}^*(Y)[u] \tag{7.1.10}$$

(4) If X is equivariantly formal, we get, as in Remark 7.1.7, an isomorphism of $\mathbb{Z}_2[u]$-modules between $\tilde{H}^*_G(X)$ and $\tilde{H}^*(X)[u]$. This isomorphism depends on the choice of a section of $\rho : H^*_G(X) \to H^*(X)$ and is not, in general, an isomorphism of algebra.

Any G-equivariant map $f : Y \to X$ satisfies $p \circ f = p$, so $\tilde{H}^*_G$ is a contravariant functor from $\mathbf{Top}_G$ to $\mathbf{GrA}[u]$. One checks that the homomorphisms $\rho : H^*_G(X) \to H^*(X)^G$ and $\mathrm{tr}^* : H^*(X) \to H^*_G(X)$ descend to $\tilde{\rho} : \tilde{H}^*_G(X) \to \tilde{H}^*(X)^G$ and to $\tilde{\mathrm{tr}}^* : \tilde{H}^*(X) \to \tilde{H}^*_G(X)$.

The equivariant reduced cohomology will be further developed in a more general setting (see 7.2.10 in the next section). Here, we shall only prove the following proposition, which plays an important role in the construction of the Steenrod squares in Chap. 8. Note that $\tilde{H}^*(X)^G$ contains the classes $a + \tau^*(a)$ for all $a \in H^*(X)$.

Proposition 7.1.12 *Let X be a G-space with $X^G \neq \emptyset$. Suppose that $\tilde{H}^j(X) = 0$ for $0 \leq j < r$. Then, $\tilde{\rho} : \tilde{H}^r_G(X) \to \tilde{H}^r(X)^G$ is an isomorphism. Moreover, $\tilde{\rho}^{-1}(a + \tau^*(a)) = \tilde{\mathrm{tr}}^*(a)$ for all $a \in H^r(X)$.*

Proof The last assertion follows from the main one since $\tilde{\rho} \circ \tilde{\mathrm{tr}}^*(a) = a + \tau^*(a)$ by definition of the transfer. We shall prove that the sequence

$$0 \to H^r_G(pt) \xrightarrow{H^*p} H^r_G(X) \xrightarrow{\rho} H^r(X)^G \to 0 \tag{7.1.11}$$

is exact. This will prove the main assertion.

For $0 \leq k \leq \infty$, let

$$Z_k = S^k \times_G X.$$

Thus, $X_G = Z_\infty$ and, as in Lemma 7.1.3, there is a natural locally trivial bundle $p : Z_k \to \mathbb{R}P^k$ with fiber X.

Choosing a point $z \in S^1$ provides a map $i_z : X \to Z_1 \subset Z_k$, defined by $i_z(x) = [z, x]$, which induces a $\mathbf{GrA}$-homomorphism $\rho : H^*(Z_k) \to H^*(X)$ (independent of z) given by $\rho = H^*i_z$. As in Lemma 7.1.10, one proves that $\rho(H^*(Z_k)) \subset H^*(X)^G$. We shall prove, by induction on k, that the sequence

$$0 \to H^r(\mathbb{R}P^k) \xrightarrow{H^*p} H^r(Z_k) \xrightarrow{\rho} H^r(X)^G \to 0 \tag{7.1.12}$$

is exact for each $k \geq 1$. Since any compact subset of $X_G = Y_\infty$ is contained in Z_k for some k, the exactness of (7.1.11) will follow, using Corollary 3.1.16.

Observe that, in Sequence (7.1.12), the homomorphism H^*p is injective since the choice of a G-fixed point in X provides a section of p. It is also clear that $\rho \circ H^*p = 0$.

We start with $k = 1$. The space Z_1 is the mapping torus of τ. The mapping torus exact sequence of Proposition 4.7.44 is of the form

$$\cdots \to H^{r-1}(X) \xrightarrow{\Theta} H^{r-1}(X) \xrightarrow{J} H^r(Z_1) \xrightarrow{\rho} H^r(X) \xrightarrow{\Theta} H^r(X) \to \cdots ,$$

where $\Theta = \mathrm{id} + H^*\tau$. Hence, $\ker(H^r(X) \xrightarrow{\Theta} H^r(X)) = H^r(X)^G$. If $r \geq 2$, then $H^1(X) = 0$, which proves that $\rho : H^r(Z_1) \to H^r(X)^G$ is an isomorphism and thus Sequence (7.1.12) is exact (for $k = 1$). If $r = 1$, then $\Theta : H^0(X) \to H^0(X)$ is the null-homomorphism, since X is path-connected. Then, $\ker(H^1(Z_1) \xrightarrow{\rho} H^1(X)^G) \approx \mathbb{Z}_2$ which implies that Sequence (7.1.12) is also exact when $k = r = 1$.

Take, as induction hypothesis, that Sequence (7.1.12) is exact for $k = \ell - 1 \geq 1$. We have to prove that it is exact for $k = \ell$. The space Z_ℓ is obtained from $Z_{\ell-1}$ by gluing $D^\ell \times X$ using the projection $S^{\ell-1} \times X \twoheadrightarrow Z_{\ell-1}$. Let e be the generator of $H^\ell(D^\ell, S^{\ell-1}) = \mathbb{Z}_2$. Using excision and the relative Künneth theorem, we get the commutative diagram

$$\begin{array}{ccccc} H^*(\mathbb{R}P^\ell, \mathbb{R}P^{\ell-1}) & \xrightarrow[\approx]{} & H^*(D^\ell, S^{\ell-1}) & \xleftarrow[\approx]{e\times -} & H^{*-\ell}(pt) \\ \downarrow & & \downarrow & & \downarrow \\ H^*(Z_\ell, Z_{\ell-1}) & \xrightarrow[\approx]{} & H^*(D^\ell \times X, S^{\ell-1} \times X) & \xleftarrow[\approx]{e\times -} & H^{*-\ell}(X) \end{array}$$

This diagram, together with the cohomology sequences for the pairs $(\mathbb{R}P^\ell, \mathbb{R}P^{\ell-1})$ and $(Z_\ell, Z_{\ell-1})$ gives the commutative diagram:

$$\begin{array}{ccccccc} H^{r-\ell}(pt) & \longrightarrow & H^r(\mathbb{R}P^\ell) & \longrightarrow & H^r(\mathbb{R}P^{\ell-1}) & \longrightarrow & H^{r+1-\ell}(pt) \\ \downarrow & & \downarrow{\scriptstyle H^*p} & & \downarrow{\scriptstyle H^*p} & & \downarrow \\ H^{r-\ell}(X) & \longrightarrow & H^r(Z_\ell) & \longrightarrow & H^r(Z_{\ell-1}) & \longrightarrow & H^{r+1-\ell}(X) \\ & & \downarrow{\scriptstyle \rho} & & \downarrow{\scriptstyle \rho} & & \\ & & H^r(X)^G & \xrightarrow{=} & H^r(X)^G & & \end{array} \quad . \quad (7.1.13)$$

where the two long lines are exact. The induction step follows by comparing the two middle columns. The argument divides into four cases.

Case 1 $\ell < r$. As $\ell \geq 2$, one has $0 < r - \ell \geq r - 2$. By hypothesis, $H^j(X) = 0$ for $1 \leq j < r$. Therefore the left and right columns vanish and the two middle columns are isomorphic.

Case 2 $\ell = r$. Since $\ell \geq 2$, one has $H^1(X) = 0$ and the right column vanishes. Also, $H^r(\mathbb{R}P^{\ell-1}) = 0$ and the left vertical arrow is an isomorphism (since $\tilde{H}^0(X) = 0$). The induction step follows. Observe that $H^0(X) \to H^r(Z_\ell)$ is injective.

Case 3 $\ell = r+1$. The left column vanishes. By step 2, Diagram (7.1.13) continues on the right by injections

$$\begin{array}{ccc} H^0(pt) & \rightarrowtail & H^{r+1}(\mathbb{R}P^\ell) \\ \downarrow{\approx} & & \downarrow{H^*p} \\ H^0(X) & \rightarrowtail & H^{r+1}(Z_\ell) \end{array} .$$

Hence, the two middle columns are isomorphic.

Case 4 $\ell > r+1$. The left and right columns vanish for dimensional reasons, so the two middle columns are isomorphic.

Remark 7.1.13 The Serre spectral sequence for the bundle $X \to X_G \to \mathbb{R}P^\infty$ provides a shorter proof of the exactness of sequences (7.1.11) and (7.1.12). This will be used to prove the more general Proposition 7.2.17 in the next section.

Example 7.1.14 Linear involution on spheres. Let S^n be the standard sphere equipped with an involution $\tau \in O(n+1)$. In $\mathbb{R}^{n+1}$, the equality

$$x = \frac{x+\tau(x)}{2} + \frac{x-\tau(x)}{2}$$

gives the decomposition $\mathbb{R}^{n+1} = V_+ \oplus V_-$ with $V_\pm$ being the eigenspace for the eigenvalue ± 1. As τ is an isometry, the vector spaces V_+ and V_- are orthogonal. Therefore, two elements $\tau, \tau' \in O(n+1)$ of order 2 are conjugate in $O(n+1)$ if and only if $\dim(S^n)^\tau = \dim(S^n)^{\tau'}$. We write S^n_p, $(-1 \le p \le n)$ for the sphere S^n equipped with an involution $\tau \in O(n+1)$ such that $\dim(S^n)^\tau = p$. Hence, $(S^n_p)^\tau \approx S^p$. The equivariant CW-structure on S^n (see Example 3.4.5) provides a G-CW-structure on S^n_p for all $p \le n$.

The involution on S^n_{-1} is just the antipodal map and, by Lemma 7.1.4, $(S^n_{-1})_G \approx \mathbb{R}P^n$. For $p \ge 0$, the inclusion $S^p = (S^n_p)^G \hookrightarrow S^n_p$ gives rise to **GrA**$[u]$-morphisms

$$r : H^*_G(S^n_p) \to H^*_G((S^n_p)^G) = H^*_G(S^p) \approx H^*(S^p)[u]$$

and

$$\tilde r : \tilde H^*_G(S^n_p) \to \tilde H^*_G((S^n_p)^G) = \tilde H^*_G(S^p) \approx \tilde H^*(S^p)[u].$$

If $n \ge 1$, then $\tilde H^j(S^n_p) = 0$ for $0 \le j < n$ and Proposition 7.1.12 asserts that $\tilde\rho : \tilde H^n_G(S^n_p) \to \tilde H^n(S^n_p)^G = H^n(S^n_p)$ is an isomorphism (this is also true if $n = p = 0$). Let $a \in \tilde H^n(S^n)$ and $b \in \tilde H^p(S^p)$ be the generators.

Proposition 7.1.15 *When* $p \ge 0$ *the* **GrA**$[u]$*-morphisms* r *and* $\tilde r$ *are injective. Moreover,* $\tilde r \circ \tilde\rho^{-1}(a) = b\, u^{n-p}$.

Proof The proposition is trivial if $n = p$, so we can suppose that $n > p \geq 0$. Using the commutative diagram

$$\begin{array}{ccccccccc} 0 & \longrightarrow & H_G^*(pt) & \longrightarrow & H_G^*(S_p^n) & \longrightarrow & \tilde{H}_G^*(S_p^n) & \longrightarrow & 0 \\ & & \downarrow = & & \downarrow r & & \downarrow \tilde{r} & & \\ 0 & \longrightarrow & H_G^*(pt) & \longrightarrow & H_G^*(S^p) & \longrightarrow & \tilde{H}_G^*(S^p) & \longrightarrow & 0 \end{array},$$

the five-lemma technique show that r is injective if and only if $\tilde{r}$ is injective. Thus we shall prove that r is injective.

We first prove that r is injective when $p = 0$. One can see S_0^n as the suspension of S_{-1}^n, with $(S_0^n)^G = \{\omega_+, \omega_-\} \approx S^0$. Then, $X = S_0^n$ is the union of the G-equivariantly contractible open sets $X^+ = S_0^n - \{\omega_-\}$ and $X^- = S_0^n - \{\omega_+\}$, with intersection X^0 having the G-equivariant homotopy type of S_{-1}^n. Hence, $X_G^{\pm}$ has the homotopy type of $\{\omega_\pm\}_G$ and

$$H_G^*(S^0) \approx H_G^k(\{\omega_-\}) \oplus H_G^k(\{\omega_+\}) \approx H_G^k(X^-) \oplus H_G^k(X^+).$$

By Lemma 7.1.4, X_G^0 has the homotopy type of $\mathbb{R}P^{n-1}$. By Proposition 3.1.53, the Mayer-Vietoris data $(X_G, X_G^+, X_G^-, X_G^0)$ gives rise to the long exact sequence

$$H^{k-1}(\mathbb{R}P^n) \xrightarrow{\delta_{MV}} H_G^k(S_0^n) \xrightarrow{r} H_G^*(S^0) \xrightarrow{J} H^k(\mathbb{R}P^n) \xrightarrow{\delta_{MV}} \cdots \tag{7.1.14}$$

with $J = H_G^* j^+ + H_G^* j^-$, where $j^\pm : X^0 \hookrightarrow X^\pm$ denotes the inclusion. The map $j^\pm$ is G-homotopy equivalent to the constant map $S_{-1}^n \mapsto \omega_\pm$. As noted before Example 7.1.2, the induced map $H_G^*(\omega_\pm) \to H_G^*(X^0)$ is the **GrA**-morphism $H^*p : H^*(\mathbb{R}P^\infty) \to H^*(\mathbb{R}P^{n-1})$. By Lemma 7.1.4, p is the characteristic map for the covering $S_{-1}^{n-1} \to (S_{-1}^{n-1})_G = \mathbb{R}P^{n-1}$. As noticed in Example 4.3.5, this map is just the inclusion $\mathbb{R}P^{n-1} \hookrightarrow \mathbb{R}P^\infty$ and, then, $H_G^* j^\pm$ is surjective by Proposition 4.3.10. Hence, J is surjective and the exact sequence (7.1.14) splits. This proves that $r : H_G^*(S_0^n) \to H_G^*(S^0)$ is injective. For a more precise analysis of $H_G^*(S_0^n)$ see Examples 7.1.16 or 7.6.9.

Suppose, by induction on $p \geq 1$, that $r : H_G^i(S_{p-1}^m) \to H_G^i(S^{p-1})$ is injective for all $i \in \mathbb{N}$ and all $m \geq p-1$. We have to prove that $r : H_G^k(S_p^n) \to H_G^k(S^p)$ is injective for all $k \in \mathbb{N}$ and all $n \geq p$. As $(S_p^n)_G$ and $(S^p)_G$ are path-connected, the required assertion is true for $k = 0$ by Lemma 3.1.9. Thus, we may suppose that $k \geq 1$. As $n \geq p \geq 1$, the G-sphere S_{p-1}^{n-1} exists and S_p^n is the suspension of S_{p-1}^{n-1}, with $(S_p^n)^G$ being the suspension of $(S_{p-1}^{n-1})^G$. Let $Y = Y^G = \{\omega_-, \omega_+\}$ be the suspension points. As above, we decompose the $X = S_p^n = X^- \cup X^+$ with $X^- \cap X^+ = X - Y$. The maps r sit in a commutative diagram

$$\begin{array}{ccccccc}
H_G^{k-1}(Y) & \longrightarrow & H_G^{k-1}(S^{n-1}_{p-1}) & \xrightarrow{\delta_{MV}} & H_G^k(S^n_p) & \longrightarrow & H_G^k(Y) \\
\approx\downarrow r & & \downarrow r & & \downarrow r & & \approx\downarrow r \\
H_G^{k-1}(Y^G) & \longrightarrow & H_G^{k-1}(S^{p-1}) & \xrightarrow{\delta_{MV}} & H_G^k(S^p) & \longrightarrow & H_G^k(Y^G)
\end{array} \tag{7.1.15}$$

where the horizontal line are the Mayer-Vietoris sequences for the data $(X, X^+, X^-, X - Y)$ and $(X^G, (X^+)^G, (X^-)^G, (X - Y)^G)$. Hence, $r : H_G^k(S^n_p) \to H_G^k(S^p)$ is injective by the proof of the five lemma (see [82, p. 129]).

The last assertion is now obvious, since $\tilde{r} \circ \tilde{\rho}^{-1}$ is injective and, as $\tilde{H}_G^*(S^p) \approx \tilde{H}^*(S^p)[u]$, one has $\tilde{H}_G^n(S^p) = \mathbb{Z}_2 u^{n-p}$. □

Example 7.1.16 We use the notations of the proof of Proposition 7.1.15 in the case $p = 0$, with $(S^n_0)^G = S^0 = \{\omega_\pm\}$. The isomorphism $\sigma_- : H^n(S^n_0) \to H_G^n(S^n_0)$ defined by the commutative diagram

$$\begin{array}{ccccccc}
\tilde{H}^n(S^n_0) & \xrightarrow[\approx]{\tilde{\rho}^{-1}} & \tilde{H}_G^n(S^n_0) & \xrightarrow{\tilde{r}} & \tilde{H}_G^n(S^0) & & \\
\uparrow & & \approx\uparrow & & \approx\uparrow & & \\
\approx & & H_G^n(S^n_0, \omega_-) & \xrightarrow{r} & H_G^n(S^0, \omega_-) & & \\
| & & \downarrow & & \downarrow & & \\
H^n(S^n_0) & \xrightarrow{\sigma_-} & H_G^n(S^n_0) & \xrightarrow{r} & H_G^n(S^0) & \xleftrightarrow{\approx} & H_G^n(\omega_-) \oplus H_G^n(\omega_+)
\end{array}$$

is an extension of the fiber for the bundle $S^n_0 \to (S^n_0)_G \to \mathbb{R}P^\infty$. Another one, σ_+, is obtained using ω_+ (there are two of them by the exact sequence (7.1.11) and $\sigma_+(a) = \sigma_-(a) + u^n$). Then, $r \circ \sigma_-(a) = (u^n, 0)$ and $r \circ \sigma_+(a) = (0, u^n)$. Hence, neither σ_- nor σ_+ is multiplicative. We see the relation $r \circ \sigma_\pm(a)^2 = u^n\, r \circ \sigma_\pm(a)$. Hence, as r is a monomorphism of $\mathbb{Z}_2[u]$-module, the relation $\sigma_\pm(a)^2 = u^n\, \sigma_\pm(a)$ holds in $H_G^*(S^n_0)$. By the Leray-Hirsch Theorem 4.7.17, $H_G^*(S^n_0)$ is a free $\mathbb{Z}_2[u]$-module generated by $A = \sigma_+(a)$ (or, by $B = \sigma_-(a)$). By dimension counting, we check that $H_G^*(S^n_0)$ admits, as a $\mathbb{Z}_2[u]$-algebra, the presentation

$$H_G^*(S^n_0) \approx \mathbb{Z}_2[u][A] \,/\, (A^2 + u^n A)\,. \tag{7.1.16}$$

As $\sigma_-(a)\sigma_+(a) = 0$, a more symmetric presentation is obtained using the two generators A and B:

$$H_G^*(S^n_0) \approx \mathbb{Z}_2[u][A, B]/\mathcal{I}$$

where $\mathcal{I}$ is the ideal generated by

$$A + B + u^n \ , \ \text{and } A^2 + u^n A.$$

Note that AB and $B^2 + u^n B$ are in $\mathcal{I}$. Indeed, mod $\mathcal{I}$, one has

$$AB = A(A + u^n) = A^2 + u^n A = 0$$

and

$$B^2 = (A + u^n)^2 = A^2 + u^{2n} = u^n A + u^{2n} = u^n(A + u^n) = u^n B.$$

Corollary 7.1.17 *If $p \geq 0$, then S^n_p is equivariantly formal. There is a section $\sigma : H^n(S^n_p) \to H^n_G(S^n_p)$ of ρ such that $r \circ \sigma : H^n(S^n_p) \to H^n_G((S^n_p)^G) \to H^*(S^p)[u]$ satisfies*

$$r \circ \sigma(a) = b\, u^{n-p}. \tag{7.1.17}$$

Proof By Proposition 7.1.15 $\tilde{\rho} : \tilde{H}^n_G(S^n_p) \to \tilde{H}^n(S^n_p)$ is an isomorphism, so the commutative diagram

$$\begin{array}{ccc} H^n_G(S^n_p) & \xrightarrow{\rho} & H^n(S^n_p) \\ \downarrow & & \approx\downarrow \\ \tilde{H}^n_G(S^n_p) & \xrightarrow[\approx]{\tilde{\rho}} & \tilde{H}^n(S^n_p) \end{array}$$

shows that ρ is surjective and thus S^n_p is equivariantly formal. Choose a section σ of ρ. By Proposition 7.1.15, Eq. (7.1.17) holds true modulo $\ker\big(H^n_G(S^n_p) \to \tilde{H}^n_G(S^n_p)\big) = \mathbb{Z}_2 u^n$. By changing $\sigma(a)$ by $\sigma(a)+u^n$ if necessary, (7.1.17) will hold true strictly. □

As another example, we consider $\mathbb{C}P^n$ as a G-space with the involution τ being the complex conjugation. Thus, $(\mathbb{C}P^n)^G = \mathbb{R}P^n$. Let $0 \neq a \in H^2(\mathbb{C}P^n)$ and $0 \neq b \in H^1(\mathbb{R}P^n)$.

Proposition 7.1.18 *For $n \leq \infty$, $\mathbb{C}P^n$ is equivariantly formal. Moreover, there is a section $\sigma : H^*(\mathbb{C}P^n) \to H^*_G(\mathbb{C}P^n)$ which is multiplicative and satisfies $r \circ \sigma(a) = bu + b^2$.*

Proof As $\tilde{H}^i(\mathbb{C}P^n) = 0$ for $i \leq 1$, Proposition 7.1.12 implies that $\tilde{\rho} : \tilde{H}^2_G(\mathbb{C}P^n) \to \tilde{H}^2(\mathbb{C}P^n)$ is an isomorphism. As in the proof of Corollary 7.1.17, this implies that $\rho : H^2_G(\mathbb{C}P^n) \to H^2(\mathbb{C}P^n)$ is surjective. As $H^*(\mathbb{C}P^n)$ is generated by a as an algebra and ρ is a **GrA**-morphism, we deduce that $\rho : H^*_G(\mathbb{C}P^n) \to H^*(\mathbb{C}P^n)$ is surjective. Thus $\mathbb{C}P^n$ is equivariantly formal.

Choose a section $\sigma_2 : H^2(\mathbb{C}P^n) \to H^2_G(\mathbb{C}P^n)$ of ρ. As $H^*_G(\mathbb{R}P^n) \approx H^*(\mathbb{R}P^n)[u]$, there exists λ, μ and ν in $\mathbb{Z}_2$ such that

$$r \circ \sigma_2 = \lambda u^2 + \mu b u + \nu b^2.$$

By changing $\sigma_2(a)$ by $\sigma_2'(a) = \sigma(a) + \lambda u^2$, we may assume that $\lambda = 0$. We must prove that $\mu = \nu = 1$. The inclusions i
$\mathbb{R}P^n \to \mathbb{C}P^n$ and $j : \mathbb{C}P^1 \to \mathbb{C}P^n$ provide commutative diagrams

$$\begin{array}{ccc} H^2_G(\mathbb{C}P^n) & \xrightarrow{r} & H^2_G(\mathbb{R}P^n) \\ \downarrow \rho & & \downarrow \rho^G \\ H^2(\mathbb{C}P^n) & \xrightarrow{H^*i} & H^2(\mathbb{R}P^n) \end{array} \qquad \begin{array}{ccc} H^2_G(\mathbb{C}P^n) & \underset{\sigma}{\overset{\rho}{\rightleftarrows}} & H^2(\mathbb{C}P^n) \\ \downarrow H^*_G j & & \approx \downarrow H^* j \\ H^2_G(\mathbb{C}P^1) & \underset{\sigma_1}{\overset{\rho_1}{\rightleftarrows}} & H^2(\mathbb{C}P^1) \end{array}$$

with $r = H^*_G i$. By Proposition 6.1.11, $H^* i(a) = b^2$, so $\nu = 1$. Note that $\sigma_1 = H^*_G j \circ \sigma \circ (H^* j)^{-1}$ is a section of ρ_1. As $\mathbb{C}P^1$ with the complex conjugation is G-diffeomorphic to S^2_1 (via the stereographic projection of S^2 onto $\mathbb{C} \cup \{\infty\} \approx \mathbb{C}P^1$), Corollary 7.1.17 implies that $r_1 \circ \sigma_1(a) = au$, which proves that $\mu = 1$.

For $n \leq \infty$, we now define $\sigma[n] : H^*(\mathbb{C}P^n) \to H^*_G(\mathbb{C}P^n)$ by $\sigma[n](a^k) = \sigma_2(a)^k$. This is a section of ρ and $\sigma[\infty]$ is clearly multiplicative. As $\sigma[n]$ is the composition of $\sigma[\infty]$ with the morphism $H^*_G(\mathbb{R}P^\infty) \to H^*_G(\mathbb{R}P^n)$ induced by the inclusion, the section $\sigma = \sigma_n$ of ρ is multiplicative and satisfies the requirements of Proposition 7.1.18. □

Corollary 7.1.19 *For $n \leq \infty$, let $\sigma : H^*(\mathbb{C}P^n) \to H^*_G(\mathbb{C}P^n)$ be the section of Proposition 7.1.18. Then, the correspondence $a \mapsto \sigma(a)$ provides a* **GrA**$[u]$-*isomorphism*

$$\mathbb{Z}_2[u,a]/(a^{n+1}) \xrightarrow{\approx} H^*_G(\mathbb{C}P^n).$$

Corollary 7.1.20 *For $n \leq \infty$, the restriction morphism $r : H_G(\mathbb{C}P^n) \to H_G(\mathbb{R}P^n)$ is injective.*

Proof Let $x \in H^m(\mathbb{C}P^n)$ with $x \in \ker r$. Write x under the form $x = \sigma(a^k)u^r + lt_r$ $(k + r = m)$, where σ is given by Corollary 7.1.19 and lt_r denotes some polynomial in the variable u of degree less than r. Then, the equation

$$0 = r(x) = b^k u^{r+k} + lt_{r+k} \tag{7.1.18}$$

holds in $H^*_G(\mathbb{R}P^n) = H^*(\mathbb{R}P^n)[u]$. This first proves that $k > 0$. Choose x so that k is minimal. Then, (7.1.18) again implies that $b^k = 0$. Hence, $n < \infty$ and $k > n$. As σ is multiplicative, one has $\sigma(a^k) = 0$ and $x = lt_r$, contradicting the minimality of k. □

The proof of Corollary 7.1.20 generalizes for conjugation spaces (see Lemma 10.2.8). For $n < \infty$, Corollary 7.1.20 is a consequence of the equivariant formality of $\mathbb{C}P^n$ (see Proposition 7.3.9).

Remark 7.1.21 As an exercise, the reader may develop the analogous of Proposition 7.1.18 and Corollaries 7.1.19 and 7.1.20 for the G-space $X = \mathbb{H}P^n$, where G acts via the involutions on $\mathbb{H}$ defined by $\tau(x + iy + jz + kt) = x + iy - jz - kt$ (thus $X^G \approx \mathbb{C}P^n$), or $\tau(x + iy + jz + kt) = x - iy - jz - kt$ (thus $X^G \approx \mathbb{R}P^n$). The same work may be done with $X = \mathbb{O}P^2$ with various $\mathbb{R}$-linear involutions on $\mathbb{O}$ so that $X^G \approx \mathbb{H}P^2$, $\mathbb{C}P^2$ or $\mathbb{R}P^2$.

t

7.2 The General Case

Let Γ be a topological group. Let $\hat{p} : E\Gamma \to B\Gamma$ be the universal principal Γ-bundle constructed by Milnor [144]. The space $E\Gamma$ is contractible, being the join of infinitely many copies of Γ with a convenient topology. An element of $E\Gamma$ is represented by a sequence $(t_i\gamma_i)$ $(i \in \mathbb{N})$ with $(t_i) \in \Delta^\infty$ and $\gamma_i \in \Gamma$; two such sequences $(t_i\gamma_i)$ and $(t_i'\gamma_i')$ represent the same class in $E\Gamma$ if $t_i = t_i'$ and $\gamma_i = \gamma_i'$ whenever $t_i \neq 0$. There is a free right action of Γ on $E\Gamma$ given by $(t_i\gamma_i)\,g = (t_i\gamma_i\,g)$. One defines $B\Gamma = E\Gamma/\Gamma$. The quotient map $\hat{p} : E\Gamma \to B\Gamma$ enjoys local triviality, in other words is a principal Γ-bundle. These constructions are functorial: a continuous homomorphism $\alpha : \Gamma' \to \Gamma$ induces a continuous map $E\alpha : E\Gamma' \to E\Gamma$, defined by $E\alpha(t_i\gamma_i) = (t_i\alpha(\gamma_i))$, which descends to a continuous map $B\alpha : B\Gamma' \to B\Gamma$.

Example 7.2.1 Consider the case $\Gamma = G = \{I, \tau\}$. Then, $EG \to BG$ is homotopy equivalent to $S^\infty \to \mathbb{R}P^\infty$. This is because the join of a space Y with $G \approx S^0$ is homeomorphic to the suspension of Y. In the same way, $ES^1 \to BS^1$ is homotopy equivalent to $S^\infty \to \mathbb{C}P^\infty$.

Lemma 7.2.2 *Let Γ be a finite group of odd order. Then $H^*(B\Gamma) \approx H^*(pt)$.*

Proof By Kronecker duality, it is equivalent to prove that $H_*(B\Gamma) \approx H_*(pt)$. When Γ is a discrete group, the principal Γ-bundle $\hat{p} : E\Gamma \to B\Gamma$ is the universal covering of $B\Gamma$. One has the transfer chain map $\mathrm{tr}_* : C_m(B\Gamma) \to C_m(E\Gamma)$ as in Sect. 4.3.3, sending a singular simplex $\sigma : \Delta^m \to B\Gamma$ to the set of its liftings in $E\Gamma$. If the number of sheets is odd, the composition

$$H_*(B\Gamma) \xrightarrow{\mathrm{tr}} H_*(E\Gamma) \xrightarrow{H_*\hat{p}} H_*(B\Gamma)$$

is the identity. Since $E\Gamma$ is contractible, this proves the lemma. □

When Γ is a discrete group, the cohomology of $B\Gamma$ is isomorphic to the *cohomology $H^*(\Gamma; \mathbb{Z}_2)$ of the group* Γ in the sense of [3, 26]. The isomorphism

$H^*(B\Gamma) \approx H^*(\Gamma;\mathbb{Z}_2)$ is proven in e.g. [3, § II.2]. The following proposition, proven in [26, Proposition III.8.3] will be useful.

Proposition 7.2.3 *Let Γ be a discrete group. Let α be an inner automorphism of Γ, i.e. $\alpha(g) = g_0 g g_0^{-1}$ for some $g_0 \in \Gamma$. Then $H^* B\alpha(a) = a$ for all $a \in H^*(B\Gamma)$.*

Let X be a left Γ-space. The *Borel construction* X_Γ or *homotopy quotient*, is the quotient space

$$X_\Gamma = E\Gamma \times_\Gamma X = (E\Gamma \times X)/\sim$$

where $\sim$ is the equivalence relation $(z, \gamma x) \sim (z\gamma, x)$ for all $x \in X$, $z \in E\Gamma$ and $\gamma \in \Gamma$. A map $p : X_\Gamma \to B\Gamma$ is then given by $p(z,x) = \hat{p}(z)$. If $\Gamma = G = \{I, \tau\}$, this Borel construction coincides with that defined in (7.1.1).

As in Sect. 7.1, one proves the following statements.

(1) The Borel construction is a covariant functor from the category $\mathbf{Top}_\Gamma$ to **Top**, where $\mathbf{Top}_\Gamma$ is the category of Γ-spaces and Γ-equivariant maps. The map $p : X_\Gamma \to B\Gamma$ coincides with the map $X_\Gamma \to pt_\Gamma = B\Gamma$ induced by the constant map $X \to pt$.
(2) X_Γ and Y_Γ have the same homotopy type if X and Y have the same Γ-homotopy type.
(3) The map $p : X_\Gamma \to B\Gamma$ is a locally trivial fiber bundle with fiber homeomorphic to X.
(4) If $f : Y \to X$ is a G-equivariant map, then the diagram

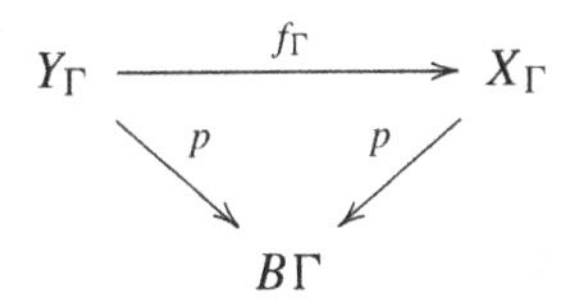

is commutative.
(5) If the Γ action on X has a fixed point, then p admits a section. More precisely, the choice of a point $v \in X^\Gamma$ provides a section $s_v : B\Gamma \to X_\Gamma$ of p.
(6) If Γ acts trivially on X, then X_Γ has the homotopy type of $B\Gamma \times X$ (see the proof of Point (3) of Lemma 7.1.3).
(7) The projection $E\Gamma \times X \to X_\Gamma$ is a Γ-principal bundle induced from the universal bundle by p.

Example 7.2.4 Let $i : \Gamma_0 \to \Gamma$ denote the inclusion of a closed subgroup Γ_0 of Γ. We consider the homogeneous Γ-space $X = \Gamma/\Gamma_0$. Then, the map $h : E\Gamma \times X \to E\Gamma/\Gamma_0$ given by $h(z, [\gamma]) = z\gamma$ descends to a homeomorphism $\bar{h} : E\Gamma \times_\Gamma X \xrightarrow{\approx} E\Gamma/\Gamma_0$. Consider the commutative diagram

$$\begin{array}{ccccccc}
\Gamma_0 & \longrightarrow & E\Gamma_0 & \longrightarrow & E\Gamma_0/\Gamma_0 & \stackrel{\equiv}{\longleftrightarrow} & B\Gamma_0 \\
\downarrow{=} & & \downarrow{Ei} & & \downarrow{\bar{E}i} \;\swarrow{g} & & \\
\Gamma_0 & \longrightarrow & E\Gamma & \longrightarrow & E\Gamma/\Gamma_0 & & \downarrow{Bi} \\
 & & \updownarrow{=} & & \downarrow & & \\
 & & E\Gamma & \longrightarrow & E\Gamma/\Gamma & \stackrel{\equiv}{\longleftrightarrow} & B\Gamma
\end{array}$$

The two upper lines are Γ_0-principal bundles. As both $E\Gamma_0$ and $E\Gamma$ have vanishing homotopy groups, the map $\bar{E}i$ is a weak homotopy equivalence, and so is g. Hence X_Γ has the weak homotopy type of $B\Gamma_0$. In addition, the map $Bi : B\Gamma_0 \to B\Gamma$ is weakly homotopy equivalent to the locally trivial bundle $X_\Gamma \to B\Gamma$ with fiber X. More generally, let Y is a Γ-space and consider $\Gamma/\Gamma_0 \times Y$ endowed with the diagonal Γ-action; then $H^*_\Gamma(\Gamma/\Gamma_0 \times Y) \approx H^*_{\Gamma_0}(Y)$ (see the proof of Theorem 7.4.3).

For a general Γ-space Y, the quotient map $q : Y \to \Gamma\backslash Y$ descends to a surjective map $\bar{q} : Y_\Gamma \to \Gamma\backslash Y$ such that $q^{-1}([y])$ has the weak homotopy type of $B\Gamma_y$ for all $y \in Y$, where Γ_y is the stabilizer of y.

Let (X, Y) be a Γ-pair, i.e. a Γ-space X with an Γ-invariant subspace Y. The *Γ-equivariant cohomology* $H^*_\Gamma(X, Y)$ is the cohomology algebra

$$H^*_\Gamma(X, Y) = H^*(X_\Gamma, Y_\Gamma) \text{ and } H^*_\Gamma(X) = H^*(X_\Gamma) = H_\Gamma(X, \emptyset).$$

In particular, $H^*_\Gamma(pt) = H^*(B\Gamma)$. The map $p : X_\Gamma \to B\Gamma$ induces an **GrA**-homomorphism $H^*p : H^*(B\Gamma) \to H^*_\Gamma(X)$, endowing the latter with a structure of $H^*_\Gamma(pt)$-algebra.

7.2.5 *Changing spaces and groups.* Let $\alpha : \Gamma' \to \Gamma$ be a continuous homomorphism. Let X be a Γ-space and X' be a Γ'-space. A continuous map $f : X' \to X$ satisfying

$$f(\gamma x) = \alpha(\gamma) f(x)$$

is called *equivariant with respect to* α. The continuous map $E\Gamma' \times X' \xrightarrow{E\alpha \times f} E\Gamma \times X$ then descends to a continuous map $f_{\Gamma',\Gamma} : X_{\Gamma'} \to X_\Gamma$ (depending on α). There is a commutative diagram

$$\begin{array}{ccc}
X_{\Gamma'} & \xrightarrow{f_{\Gamma',\Gamma}} & X_\Gamma \\
\downarrow & & \downarrow \\
B\Gamma' & \xrightarrow{B\alpha} & B\Gamma
\end{array} \quad . \tag{7.2.1}$$

The map $f_{\Gamma',\Gamma}$ induces a **GrA**-homomorphism

$$f^*_{\Gamma',\Gamma} : H^*_\Gamma(X) \to H^*_{\Gamma'}(X'). \tag{7.2.2}$$

By commutativity of the Diagram (7.2.1), one has

$$f^*_{\Gamma',\Gamma}(av) = f^*_{\Gamma',\Gamma}(a)(B\alpha)^*(v) \quad \forall\, a \in H^*_\Gamma(X) \text{ and } v \in H^*_\Gamma(pt)$$

for all $a \in H^*_\Gamma(X)$ and $v \in H^*_\Gamma(pt)$. We say that $f^*_{\Gamma',\Gamma}$ *preserves the module structures via* α. More simply, the Γ-space X becomes a Γ'-space via α, thus $H^*_\Gamma(X)$ becomes a $H^*(B\Gamma')$-algebra and $f^*_{\Gamma',\Gamma}$ is a morphism of $H^*(B\Gamma')$-algebras. An important case is given by $f = \mathrm{id} : X \to X$, Setting $\mathrm{id}^*_{\Gamma,\Gamma'} = \alpha^*$, we get a map $\alpha^* : H^*_\Gamma(X) \to H^*_{\Gamma'}(X)$ which is a morphism $H^*(B\Gamma')$-algebras.

7.2.6 *Free actions.* Let Γ_0 be a closed normal subgroup of Γ and let X be a Γ-space. For $x \in X$, $\gamma \in \Gamma$ and $\gamma_0, \gamma_0' \in \Gamma_0$, the equation

$$(\gamma\gamma_0)\,(\gamma_0' x) = (\gamma\gamma_0\gamma_0'\gamma^{-1})\,\gamma x$$

shows that the Γ-action on X descends to a (Γ/Γ_0)-action on $\Gamma_0\backslash X$. By the functoriality of equivariant cohomology (see 7.2.5), we get a map

$$H^*_{\Gamma/\Gamma_0}(\Gamma_0\backslash X) \to H^*_\Gamma(X) \tag{7.2.3}$$

which is a homomorphism of $H^*(B(\Gamma/\Gamma_0))$-algebras. The following lemma generalizes Lemma 7.1.4. To avoid point-set topology complications, we restrict ourselves to the smooth action of a Lie group.

Lemma 7.2.7 *Let Γ_0 be a compact normal subgroup of a Lie group Γ. Let X be a smooth Γ-manifold on which Γ_0 acts freely. Then, the map* (7.2.3) *is an isomorphism of $H^*(B(\Gamma/\Gamma_0))$-algebras.*

Proof Let $Y = E(\Gamma/\Gamma_0) \times_{\Gamma/\Gamma_0} (\Gamma_0\backslash X)$. Consider the commutative diagram

$$\begin{array}{ccc}
E\Gamma \times X & \xrightarrow{\ q\ } & E\Gamma \times_\Gamma X \\
\downarrow p & & \downarrow \bar{p} \\
E(\Gamma/\Gamma_0) \times (\Gamma_0\backslash X) & \xrightarrow{\ \bar{q}\ } & Y
\end{array}\ .$$

Let $a \in Y$ represented by $((t_i\gamma_i), x)$ in $E\Gamma \times X$. Then,

$$(\bar{q}\circ p)^{-1}(a) = \{\big((t_i\gamma_i\tilde{\delta}_i), \delta x\big) \mid \tilde{\delta}_i, \delta \in \Gamma_0\}.$$

Therefore,

$$\bar{p}^{-1}(a) = q((\bar{q}\circ p)^{-1}(a)) \approx \{\big((t_i\gamma_i\delta_i), x\big) \mid \delta_i \in \Gamma_0\} \xrightarrow{\approx} E\Gamma_0, \tag{7.2.4}$$

the last homeomorphism being given by $\big((t_i\gamma_i\delta_i), x\big) \mapsto (t_i\delta_i)$.

As Γ_0 acts smoothly and freely on X, the quotient map $X \to \Gamma_0\backslash X$ is a locally trivial bundle (this follows from the slice theorem: see [12, Theorem 2.2.1]). Hence, p is homotopy equivalent to a locally trivial bundle, which is numerable ([179, p. 94]) since $\Gamma_0\backslash X$ is paracompact. The map $\bar{q}$ is also a numerable locally trivial bundle (see (7) on p. 275). Therefore, $\bar{q} \circ p$ is a fibration (i.e. satisfies the homotopy covering property for any space: see [179, Theorem 12, p. 95]), and so does $\bar{p}$. As $\Gamma_0\backslash X$ is a manifold and $E(\Gamma/\Gamma_0)$ is contractible, the space Y admits a numerable covering $\{V_\lambda\}_{\lambda\in\Lambda}$ such that each inclusion $V_\lambda \to Y$ is null-homotopic. As each fiber of $\bar{p}$ is contractible by (7.2.4), [42, Theorem 6.3] implies that $\bar{p}$ is a homotopy equivalence, which proves the lemma. □

7.2.8 *The forgetful homomorphism.* Choosing a point $\zeta \in E\Gamma$ provides a map $i_\zeta : X \to X_\Gamma$ defined by $i_\zeta(x) = [\zeta, x]$. As $E\Gamma$ is path-connected, the homotopy class of i_ζ does not depend on ζ. For instance, we can take $\zeta = \zeta^0 = (1e, 0, 0, \dots)$ where $e \in \Gamma$ is the unit element. Therefore, we get a well defined **GrA**-homomorphism

$$\rho : H^*_\Gamma(X) \to H^*(X) \tag{7.2.5}$$

given by $\rho = H^* i_\zeta$ for some $\zeta \in E\Gamma$. As in (7.1.5), one proves that ρ is functorial. In fact, using 7.2.5, the homomorphism ρ coincides with the homomorphism $\mathrm{id}_{\{e\},\Gamma}$ induced by inclusion of the trivial group $\{e\}$ into Γ:

$$\rho = \mathrm{id}_{\{e\},\Gamma} : H^*_\Gamma(X) \to H^*_{\{e\}}(X) = H^*(X). \tag{7.2.6}$$

Indeed, i_{ζ^0} factors through $X \to X_{\{e\}} \approx X$. Hence, ρ may be seen as a *forgetful homomorphism* (one forgets the Γ-action).

A consequence of (7.2.5) is that ρ is functorial for the changing of groups: if $\alpha : \Gamma' \to \Gamma$ is a continuous homomorphism and X a Γ-space, the diagram

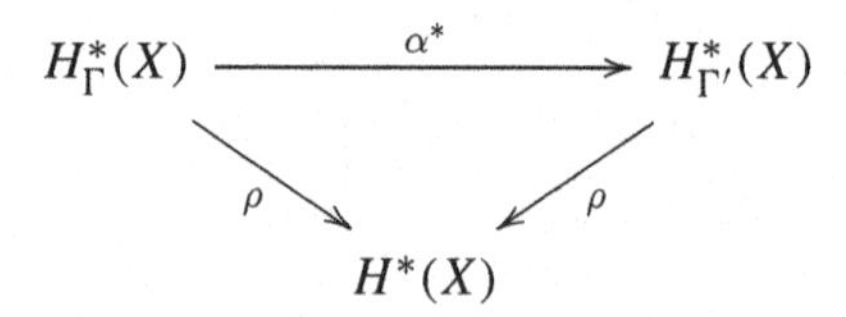

is commutative.

7.2.9 *Equivariant formality.* A Γ-space X is called *equivariantly formal* if $\rho : H_\Gamma(X) \to H^*(X)$ is surjective. For instance, X is equivariantly formal if the Γ-action is trivial. For relationships with other kind of "formal" spaces, see [173]. If X is equivariantly formal, one can choose, as in the proof of Proposition 7.1.6, a **GrV**-section $\theta : H^*(X) \to H^*_\Gamma(X)$ of ρ. Then θ is a cohomology extension of the fiber for the fiber bundle $X_\Gamma \xrightarrow{p} B\Gamma$. If X is of finite cohomology type, the Leray-Hirsch

Theorem 4.7.17 then gives an map (depending on θ)

$$H_\Gamma^*(pt) \otimes H^*(X) \xrightarrow{\approx} H_\Gamma^*(X) \tag{7.2.7}$$

which is an isomorphism of $H_\Gamma^*(pt)$-modules.

7.2.10 *Reduced cohomology.* Let X be a Γ-space. The *reduced equivariant cohomology* $\tilde{H}_\Gamma^*(X)$ is the $H_\Gamma^*(pt)$-algebra defined by

$$\tilde{H}_\Gamma^*(X) = \operatorname{coker}\big(H^* p_\Gamma : H_\Gamma^*(pt) \to H_\Gamma^*(X)\big) \tag{7.2.8}$$

where $p : X \to pt$ denotes the constant map to a point (which is Γ-equivariant). Warning: $H_\Gamma^*(X) \neq \tilde{H}^*(X_\Gamma)$. Examples:

(1) $\tilde{H}_\Gamma^*(pt) = 0$.
(2) If Y is a space with trivial Γ-action, there is an isomorphism of $H_\Gamma^*(pt)$-algebras

$$\tilde{H}_\Gamma^*(Y) = \big(H^*(Y) \otimes H_\Gamma^*(pt)\big)/(\mathbf{1} \otimes H_\Gamma^*(pt)) \approx \tilde{H}^*(Y) \otimes H_\Gamma^*(pt). \tag{7.2.9}$$

(3) If X is equivariantly formal and is of finite cohomology type, one uses (7.2.7) to provides an isomorphism of $H_\Gamma^*(pt)$-modules between $\tilde{H}_\Gamma^*(X)$ and $\tilde{H}^*(X) \otimes H_\Gamma^*(pt)$. This isomorphism depends on the choice of a section of $\rho : H^*(X) \to H_\Gamma^*(X)$ and is not, in-general, an isomorphism of algebras.

Any Γ-equivariant map $f : Y \to X$ satisfies $p \circ f = p$, so $\tilde{H}_\Gamma^*$ is a contravariant functor from $\mathbf{Top}_\Gamma$ to the category of $H_\Gamma^*(pt)$-algebra.

Let $v \in X^\Gamma$. As for 3.1.14, one has the following diagram.

$$\begin{array}{ccccc}
 & & H_\Gamma^*(pt) & & \\
 & & \big\downarrow{\scriptstyle p_\Gamma^*} & \searrow{\scriptstyle \approx} & \\
H^*(X_\Gamma, \{v\}_\Gamma) & \xrightarrow{j^*} & H_\Gamma^*(X) & \xrightarrow{i_\Gamma^*} & H_\Gamma^*(v) \\
 & \searrow{\scriptstyle \approx} & \big\downarrow & & \\
 & & \tilde{H}_\Gamma^*(X) & &
\end{array} \tag{7.2.10}$$

where the line and the column are exact. This proves that

$$H^*(X_\Gamma, \{v\}_\Gamma) \xrightarrow{\approx} \tilde{H}_\Gamma^*(X). \tag{7.2.11}$$

Observe that, in (7.2.10), i_Γ^* coincides with the section s_v of p_Γ^*. We see that the choice of $v \in X^\Gamma$ produces a supplementary vector subspace to $p_\Gamma^*(H_\Gamma^*(pt))$ in $H_\Gamma^*(X)$.

A pair (X, A) of Γ-spaces is called *equivariantly well cofibrant* if it admits a presentation (u, h) as a well cofibrant pair which is Γ-equivariant, i.e. $u(\gamma x) = u(x)$ and $h(\gamma x, t) = \gamma h(x, t)$ for all $\gamma \in \Gamma$, $x \in X$ and $t \in I$.

Lemma 7.2.11 *Let (X, A) be a pair of Γ-spaces which is equivariantly well cofibrant. Then (X_Γ, A_Γ) is well cofibrant.*

Proof Let (u, h) be a presentation of (X, A) as an equivariantly well cofibrant pair. Define $\tilde{u} : E\Gamma \times X \to I$ and $\tilde{h} : E\Gamma \times X \times I \to E\Gamma \times X$ by $\tilde{u}(z, x) = u(x)$ and $\tilde{h}(z, x, t) = (z, h(x, t))$. We check that these maps descend to $u_\Gamma : X_\Gamma \to I$ and $h_\Gamma : X_\Gamma \times I \to X_\Gamma$ and that (u_Γ, h_Γ) is a presentation of (X_Γ, A_Γ) as a well cofibrant pair. □

Lemma 7.2.12 *Let (X, A) be a Γ_1-equivariantly well cofibrant pair of Γ_1-spaces. Let (Y, B) be a Γ_2-equivariantly well cofibrant pair of Γ_2-spaces. Then, $(X \times Y, A \times Y \cup X \times B)$ is a Γ_{12}-equivariantly well cofibrant pair of Γ_{12}-spaces, where $\Gamma_{12} = \Gamma_1 \times \Gamma_2$.*

Proof One checks that the proof of Lemma 3.1.40 works equivariantly. □

If (X, A) is a pair of Γ-spaces, the quotient space X/A inherits a Γ-action, with $[A] \in (X/A)^\Gamma$, where $[A]$ denotes the set A as a class in X/A. The proof of the following lemma is the same as that of Lemma 3.1.43.

Lemma 7.2.13 *If (X, A) is an equivariantly well cofibrant pair of Γ-spaces, so is the pair $(X/A, A/A)$.*

Example 7.2.14 Let (X, x) be a pointed space. The group $G = \{I, \tau\}$ acts on $X \times X$ by exchanging the coordinates and this action descends to $X \wedge X$. If (X, x) is well pointed, the proof of Lemma 3.1.40 shows that the pair $(X \times X, X \vee X)$ is G-equivariantly well cofibrant. By Lemma 7.2.13, $(X \wedge X, x \wedge x)$ is G-equivariantly well pointed.

The quotient map $\pi : (X, A) \to (X/A, A/A)$ is a Γ-equivariant map of pairs which induces $\pi_\Gamma : (X_\Gamma, A_\Gamma) \to ((X/A)_\Gamma, (A/A)_\Gamma)$.

Proposition 7.2.15 *Let (X, A) be a pair of Γ-spaces which is equivariantly well cofibrant. Then,*

$$\pi_\Gamma^* : H^*((X/A)_\Gamma, (A/A)_\Gamma) \xrightarrow{\approx} H^*(X_\Gamma, A_\Gamma)$$

is an isomorphism.

Proof Let $(K, L) = (X/A, A/A)$. Let (u, h) be a presentation of (X, A) as an equivariant well cofibrant pair and let $(\bar{u}, \bar{h})$ be the induced presentation of (K, L). Let $V = u^{-1}([0, 1/2])$ and $W = \bar{u}^{-1}([0, 1/2]) = \pi(V)$. As noticed in the proof of Lemma 3.1.41, the condition $u(h(x)) \le u(x)$ implies that h and $\bar{h}$ restrict to Γ-equivariant deformation retractions from V to A and from W to L. The tautological homeomorphism from $E\Gamma \times (V \times I)$ onto $(E\Gamma \times V) \times I$ descends to a homeomorphism $(V \times I)_\Gamma \xrightarrow{\approx} V_\Gamma \times I$. Using this, h and $\bar{h}$ descend to a deformation retractions $h_\Gamma : V_\Gamma \times I \to V_\Gamma$, and $\bar{h}_\Gamma : W_\Gamma \times I \to W_\Gamma$ onto A and L, making (X_Γ, A_Γ) and (K_Γ, L_Γ) good pairs.

The inclusion $(K, L) \to (K, W)$ gives rise to a morphism of exact sequences

$$\begin{array}{ccccccccc}
H^{k-1}(K_\Gamma) & \longrightarrow & H^{k-1}(L_\Gamma) & \longrightarrow & H^k(K_\Gamma, L_\Gamma) & \longrightarrow & H^k(K_\Gamma) & \longrightarrow & H^k(L_\Gamma) \\
\downarrow{\scriptstyle =} & & \downarrow{\scriptstyle \approx} & & \downarrow & & \downarrow{\scriptstyle =} & & \downarrow{\scriptstyle \approx} \\
H^{k-1}(K_\Gamma) & \longrightarrow & H^{k-1}(W_\Gamma) & \longrightarrow & H^k(K_\Gamma, W_\Gamma) & \longrightarrow & H^k(K_\Gamma) & \longrightarrow & H^k(W_\Gamma)
\end{array}$$

which, by the five lemma, implies that $H^k(K_\Gamma, L_\Gamma) \xrightarrow{\approx} H^k(K_\Gamma, W_\Gamma)$ is an isomorphism. The same proof gives the isomorphism $H^k(X_\Gamma, A_\Gamma) \xrightarrow{\approx} H^k(X_\Gamma, V_\Gamma)$. Proposition 7.2.15 then comes from the commutativity of the following diagram (where the vertical arrows are induced by inclusions)

$$\begin{array}{ccc}
H_*(K_\Gamma, L_\Gamma) & \xrightarrow{\pi_\Gamma^*} & H_*(X_\Gamma, A_\Gamma) \\
{\scriptstyle \approx}\uparrow & & {\scriptstyle \approx}\uparrow \\
H_*(K_\Gamma, W_\Gamma) & \xrightarrow{\pi_\Gamma^*} & H_*(X_\Gamma, V_\Gamma) \\
{\scriptstyle \approx}\downarrow{\scriptstyle \text{excision}} & & {\scriptstyle \approx}\downarrow{\scriptstyle \text{excision}} \\
H_*(K_\Gamma - L_\Gamma, W_\Gamma - L_\Gamma) & \underset{\approx}{\xrightarrow{\pi_\Gamma^*}} & H_*(X_\Gamma - A_\Gamma, V_\Gamma - A_\Gamma).
\end{array}$$

The bottom horizontal arrow is indeed an isomorphism since $\pi : (X - A, V - A) \to (K - L, W - L)$ is a Γ-equivariant homeomorphism. □

Corollary 7.2.16 *Let (X, A) be a pair of Γ-spaces which is equivariantly well cofibrant. If A is non-empty, there is a functorial isomorphism of $H_\Gamma^*(pt)$-algebras*

$$\tilde{H}^*(X_\Gamma / A_\Gamma) \xrightarrow{\approx} \tilde{H}_\Gamma^*(X/A).$$

The hypothesis $A \neq \emptyset$ is necessary since $\tilde{H}^*(X_\Gamma)$ is not isomorphic to $\tilde{H}_\Gamma^*(X)$.

Proof This follows from the following diagram

$$\begin{array}{ccc}
H^*((X/A)_\Gamma, (A/A)_\Gamma) & \xrightarrow{\approx} & \tilde{H}_\Gamma^*(X/A) \\
{\scriptstyle \approx}\downarrow{\scriptstyle \pi_\Gamma^*} & & \\
H^*(X_\Gamma, A_\Gamma) & \xrightarrow{\approx} & \tilde{H}^*(X_\Gamma / A_\Gamma)
\end{array} .$$

The bijectivities come

- from (7.2.11) for the top horizontal arrow, since, as $A \neq \emptyset$, A/A is a point.
- from Lemma 7.2.11 and Proposition 3.1.45 for the bottom horizontal arrow.
- from Proposition 7.2.15 for the vertical arrow. □

As in 7.2.5, the reduced equivariant cohomology is functorial for changing groups. In particular, as in 7.2.8, the inclusion of the trivial group $\{e\}$ into Γ provides the *forgetful homomorphism*

$$\tilde{\rho} : \tilde{H}^*_\Gamma(X) \to \tilde{H}^*_{\{e\}}(X) \approx \tilde{H}^*(X)$$

which is functorial.

As in Lemma 7.1.10, one proves that $\rho(H^*_\Gamma(X)) \subset H^*(X)^\Gamma$. The following statement generalizes Proposition 7.1.12.

Proposition 7.2.17 *Let X be a Γ-space with $X^\Gamma \neq \emptyset$. Suppose that $\tilde{H}^j(X) = 0$ for $0 \leq j < r$. Then, $\tilde{\rho} : \tilde{H}^r_\Gamma(X) \to \tilde{H}^r(X)^\Gamma$ is an isomorphism.*

Proof (Using a spectral sequence). In the E_2-term of the Serre spectral sequence of the bundle $X \to X_\Gamma \to B\Gamma$, the lines from 1 to $n-1$ vanish:

$$\begin{array}{cccc}
H^r(X)^\Gamma & \bullet & \bullet \\
0 & & \\
\vdots & 0 & \\
0 & & \\
H^0(B\Gamma) & H^1(B\Gamma) & H^2(B\Gamma)
\end{array}$$

Therefore, it gives rise to the *edge exact sequence*:

$$0 \to H^r(B\Gamma) \xrightarrow{H^*p} H^r_\Gamma(X) \xrightarrow{\rho} H^r(X)^\Gamma \to H^{r+1}(B\Gamma) \xrightarrow{H^*p} H^{r+1}_\Gamma(X). \tag{7.2.12}$$

The choice of a fixed point $v \in X^\Gamma$ provides a section $s_v : B\Gamma \to X_\Gamma$, as seen in Lemma 7.1.3. Therefore, H^*p is injective and $H^r_\Gamma(X) \xrightarrow{\rho} H^r(X)^\Gamma$ is surjective. Proposition 7.2.17 follows form this, as in the proof of Proposition 7.1.12. □

Remark 7.2.18 When Γ is discrete, a proof of Proposition 7.2.17 without spectral sequence is possible, following the pattern of that of Proposition 7.1.12. The role of $\mathbb{R}P^k$ is played by $B_k\Gamma$, the quotient by Γ of the join $E_k\Gamma = \Gamma * \cdots * \Gamma$ ($k+1$ times) (see [144, Sect. 3]). The space Y_k is defined to be $E_k\Gamma \times_\Gamma X$. We leave this proof as an exercise to the reader.

7.3 Localization Theorems and Smith Theory

As in Sect. 7.1, we consider in this section the group $G = \{I, \tau\}$ of order 2, so $BG \approx \mathbb{R}P^{\infty}$ and $H_G^*(pt) = \mathbb{Z}_2[u]$ with u of degree one. Strong results come out if we invert u, namely if we tensor the $\mathbb{Z}_2[u]$-modules with the ring of Laurent polynomials $\mathbb{Z}_2[u, u^{-1}]$. For a pair (X, Y) of G-spaces, we thus define

$$h_G^*(X, Y) = \mathbb{Z}_2[u, u^{-1}] \otimes_{\mathbb{Z}_2[u]} H_G^*(X, Y),$$

with the notation $h_G^*(X) = h_G^*(X, \emptyset)$. Note that $\mathbb{Z}_2[u, u^{-1}]$ is $\mathbb{Z}$-graded, with $\mathbb{Z}_2[u, u^{-1}]^k = \mathbb{Z}_2 u^k$ and we use the graded tensor product. Hence, $h_G^*(X, Y)$ is a $\mathbb{Z}$-graded $\mathbb{Z}_2[u, u^{-1}]$-algebra, with

$$h_G^k(X, Y) = \bigoplus_{i+j=k} \mathbb{Z}_2 u^i \otimes H_G^j(X, Y) \approx \bigoplus_{\ell \in \mathbb{Z}} H_G^{k-\ell}(X, Y). \qquad (7.3.1)$$

The theorem below is an example of the so called *localization theorems*. For more general statements, see e.g. [37, Chap. 3] or [9, Chap. 3].

Theorem 7.3.1 *Let X be a finite dimensional G-complex. Then, the inclusion $X^G \subset X$ induces an isomorphism*

$$h_G^*(X) \xrightarrow{\approx} h_G^*(X^G)$$

of $\mathbb{Z}$-graded $\mathbb{Z}_2[u, u^{-1}]$-algebras.

Before proving Theorem 7.3.1, we discuss a few examples.

Example 7.3.2 Suppose, in Theorem 7.3.1, that X is a free G-complex. By Lemma 7.1.4, $H_G^*(X) \approx H^*(X/G)$. As X/G is a finite dimensional CW-complex, there exists an integer m such that $u^m \cdot H_G^*(X) = 0$. As u is invertible in $\mathbb{Z}_2[u, u^{-1}]$, this proves that $h_G^*(X) = 0$, as predicted by Theorem 7.3.1, since $X^G = \emptyset$. We see here that the finite dimensional hypothesis is necessary in Theorem 7.3.1. Indeed, the free G-complex $S^{\infty} = EG$ satisfies $H_G^*(EG) = H^*(BG) = \mathbb{Z}_2[u]$, so $h_G^*(EG) = \mathbb{Z}_2[u, u^{-1}]$.

Example 7.3.3 Consider the G-space S_p^n of Example 7.1.14, i.e. the sphere S^n endowed with a linear G-action with $(S_p^n)^G \approx S^p$. We assume that $0 \le p \le n$. Using Corollary 7.1.17, there are elements $\bar{a} \in H_G^n(S_p^n)$ and $\bar{b} \in H_G^p((S_p^n)^G)$ generating respectively $H_G^*(S_p^n)$ and $H_G^*((S_p^n)^G)$ as free $\mathbb{Z}_2[u]$-modules and $r(a) = bu^{n-p}$. Then, as predicted by Theorem 7.3.1, $r : h_G^*(S_p^n) \to h_G^*((S_p^n)^G)$ admits an inverse, sending b to au^{p-n}.

Example 7.3.4 Consider the G-space $\mathbb{C}P^n$ ($n \ge 1$), where G acts via the complex conjugation, with $(\mathbb{C}P^n)^G = \mathbb{R}P^n$. By Proposition 7.1.18 and Corollary 7.1.19, there is a commutative diagram

$$\begin{array}{ccc} \mathbb{Z}_2[u,u^{-1},a]/(a^{n+1}) & \xrightarrow{\hat{r}} & \mathbb{Z}_2[u,u^{-1},b]/(b^{n+1}) \\ \downarrow \approx & & \downarrow \approx \\ h_G^*(\mathbb{C}P^n) & \xrightarrow{r} & h_G^*(\mathbb{R}P^n) \end{array}$$

with a of degree 2, b of degree 1 and $\hat{r}(a) = bu + b^2$. If $n < \infty$, the correspondence

$$b \mapsto au^{-1} + a^2u^{-3} + a^4u^{-7} + \cdots = \sum_{i\geq 0} a^{2^i} u^{2^{i+1}-1} \tag{7.3.2}$$

extends to a $\mathbf{GrA}[u]$-isomorphism $\hat{r}^{-1} : \mathbb{Z}_2[u,u^{-1},b]/(b^{n+1}) \to \mathbb{Z}_2[u,u^{-1},a]/(a^{n+1})$ which is the inverse of $\hat{r}$.

Example 7.3.5 If $n = \infty$, the right hand member of (7.3.2) is not a polynomial and no inverse of $\hat{r}$ may be defined this way. In fact, $\hat{r}$ (and then r) is not an isomorphism. Indeed, the composition of $\hat{r}$ with the epimorphism $\mathbb{Z}_2[u,u^{-1},b] \to \mathbb{Z}_2$ sending both b and u to 1 is the zero map. Of course, $\mathbb{C}P^\infty$ violates the finite dimensional hypothesis in Theorem 7.3.1.

Proof of Thorem 7.3.1 The proof is by induction on the dimension of X, which starts trivially with $X = \emptyset$ (dimension -1). The induction step reduces to proving that, if the theorem is true for X, it is then true for $Z = X \cup C$ where C is a finite family of G-cells. We consider the commutative diagram.

$$\begin{array}{ccccccccc} h_G^{k-1}(X) & \longrightarrow & h_G^k(Z,X) & \longrightarrow & h_G^k(Z) & \longrightarrow & h_G^k(X) & \longrightarrow & h_G^{k+1}(Z,X) \\ \approx\downarrow r_X & & \downarrow r_{Z,X} & & \downarrow r_Z & & \approx\downarrow r_X & & \downarrow r_{Z,X} \\ h_G^{k-1}(X^G) & \longrightarrow & h_G^k(Z^G,X^G) & \longrightarrow & h_G^k(Z^G) & \longrightarrow & h_G^k(X^G) & \longrightarrow & h_G^{k+1}(Z^G,X^G) \end{array}$$

The two lines are exact sequences, obtained by tensoring with $\mathbb{Z}_2[u,u^{-1}]$ the exact sequence of (Z,X) for H_G^* (as in the proof of Lemma 4.6.9, we use that a direct sum of exact sequences is exact and that, over a field, tensoring with a vector space preserves exactness). If r_X is an isomorphism by induction hypothesis, it is enough, using the five lemma, to prove that $r_{Z,X}$ is an isomorphism. Note that C is a disjoint union of free G-cells C_f and of isotropic G-cells G_i. By excision, one has the commutative diagram

$$\begin{array}{ccccc} h_G^*(Z,X) & \xrightarrow{\approx} & h_G^*(C,\mathrm{BdC}) & \xrightarrow{\approx} & h_G^*(C_f,\mathrm{BdC_f}) \times h_G^*(C_i,\mathrm{BdC_i}) \\ \downarrow r_{Z,X} & & & & \downarrow r \\ h_G^*(Z^G,X^G) & & \xrightarrow{\approx} & & h_G^*(C_i,\mathrm{BdC_i}) \end{array}$$

where $r(a, b) = b$. It is then enough to prove that $h^*_G(C_f, \mathrm{BdC_f}) = 0$. But this follows from the exact sequence of $(C_f, \mathrm{BdC_f})$ for h^*_G and, since C_f and $\mathrm{BdC_f}$ are free G-space, from Example 7.3.2.

We are now leading toward the Smith inequalities. Let us extend our ground ring $\mathbb{Z}_2[u, u^{-1}]$ to the fraction field $\mathbb{Z}_2(u)$ of $\mathbb{Z}_2[u]$ (this is just a field of characteristic 2, the grading is lost). For a space X, the *total Betti number* $b(X)$ of X is defined by

$$b(X) = \sum_{i=0}^{\infty} \dim H^*(X) \in \mathbb{N} \cup \{\infty\}.$$

Lemma 7.3.6 *Let X be a finite dimensional G-complex with $b(X) < \infty$. Then, as a vector space over $\mathbb{Z}_2(u)$,*

$$\dim \mathbb{Z}_2(u) \otimes_{\mathbb{Z}_2[u]} H^*_G(X) \leq b(X)$$

with equality if and only if X is equivariantly formal.

Proof From the transfer exact sequence (7.1.7), we extract the exact sequence

$$H^{k-1}_G(X) \xrightarrow{\smile u} H^k_G(X) \xrightarrow{\rho} H^k(X) \to H^k_G(X) \xrightarrow{\smile u} H^{k+1}_G(X).$$

We deduce that $H^k_G(X)$ is generated by $u \cdot H^{k-1}_G(X)$ and a number of elements $\leq \dim H^k(X)$, which proves the first assertion. Moreover, $\dim \mathbb{Z}_2(u) \otimes_{\mathbb{Z}_2[u]} H^*_G(X) = b(X)$ if and only if $\rho : H^*_G(X) \to H^*(X)$ is surjective, that is X is equivariantly formal. □

Proposition 7.3.7 *Let X be a finite dimensional G-complex with $b(X) < \infty$. Then*

$$b(X^G) \leq b(X) \tag{7.3.3}$$

with equality if and only if X is equivariantly formal.

Proof

$$\begin{aligned} b(X) &\geq \dim \mathbb{Z}_2(u) \otimes_{\mathbb{Z}_2[u]} H^*_G(X) && \text{by Lemma 7.3.6} \\ &= \dim \mathbb{Z}_2(u) \otimes_{\mathbb{Z}_2[u]} H^*_G(X^G) && \text{by Theorem 7.3.1} \\ &= b(X^G), \end{aligned}$$

the last equality coming from Lemma 7.3.6, since X^G is equivariantly formal. From Lemma 7.3.6 again, the above inequality is an equality if and only if X is equivariantly formal. □

Formula (7.3.3) is an example of *Smith inequalities*, a development of the work of P. Smith started in 1938 [178]. The following corollary is a classical result in the theory.

Corollary 7.3.8 *Let X be a finite dimensional G-complex. Then,*

(1) *If $H^*(X) \approx H^*(pt)$, then $H^*(X^G) \approx H^*(pt)$.*
(2) *If X has the cohomology of a sphere, so does X^G.*

Proof If X has the cohomology of a point, it is equivariantly formal and, by Proposition 7.3.7, $b(X^G) = 1$ which proves (1). For Point (2), Proposition 7.3.7 implies that $b(X^G) \leq 2$. Statement (2) is true if $b(X^G) = 2$ or if $b(X^G) = 0$ (since $\emptyset = S^{-1}$). It remains to prove that $b(X^G) = 1$ is impossible if $H^*(X) \approx H^*(S^n)$.

If $b(X^G) = 1$, then X is not equivariantly formal. Using Exact sequence (7.1.7), this implies that $H_G^*(X) \approx H^*(\mathbb{R}P^n)$. As in Example 7.3.2, we deduce that $h_G^*(X) = 0$, contradicting Theorem 7.3.1 ($h_G^*(X^G) = \mathbb{Z}_2[u, u^{-1}]$ if $b(X^G) = 1$). □

Here is another consequence of Theorem 7.3.1.

Proposition 7.3.9 *Let X be a finite dimensional G-complex. Then, the following statements are equivalent.*

(1) *X is equivariantly formal.*
(2) *$r : H_G^*(X) \to H_G^*(X^G)$ is injective.*

Proof From Corollary 7.1.8, we already know that (2) implies (1). For the converse, suppose that X is equivariantly formal. Then $H_G^*(X)$ is a free $\mathbb{Z}_2[u]$-module by Proposition 7.1.6 and thus $j : H_G^*(X) \to \mathbb{Z}_2[u, u^{-1}] \otimes_{\mathbb{Z}_2[u]} H_G^*(X) = h_G^*(X)$ is injective. Therefore, in the commutative diagram

$$\begin{array}{ccc} H_G^*(X) & \xrightarrow{r} & H_G^*(X^G) \\ \downarrow j & & \downarrow j^G \\ h_G^*(X) & \xrightarrow{\tilde{r}} & h_G^*(X^G) \end{array}$$

the left vertical arrow is injective. When X is finite dimensional G-complex, the bottom arrow is an isomorphism by Theorem 7.3.1. Hence j is injective. □

We shall now prove a localization theorem analogous to Theorem 7.3.1 for S^1-spaces. Since we are working with $\mathbb{Z}_2$-cohomology, an important role is played by the subgroup $\{\pm 1\} = S^0$ of S^1. We also need the notion of a Γ-CW-complex for Γ a topological group. If Γ_0 is a closed subgroup of Γ, the Γ-space $\Gamma/\Gamma_0 \times D^n$ is called a *Γ-cell of dimension n* (of *type* Γ_0), with *boundary* $\Gamma/\Gamma_0 \times S^{n-1}$ (the group Γ acts on the left on Γ/Γ_0 and trivially on D^n). One can attach a Γ-cell to a Γ-space Y via a G-equivariant map $\varphi : \Gamma/\Gamma_0 \times S^{n-1} \to Y$. A Γ-CW-structure on a Γ-space X is a filtration

$$\emptyset = X^{-1} \subset X^0 \subset X^1 \subset \cdots \subset X = \bigcup_{n \in \mathbb{N}} X^n \tag{7.3.4}$$

by Γ-subspaces, such that, for each n, the space X^n (the *n-skeleton*) is Γ-homeomorphic to a Γ-space obtained from X^{n-1} by attachment of a family Γ-cells of dimension n (of various type). A Γ-space endowed with a Γ-CW-structure is a *Γ-CW-complex* (or just a *Γ-complex*). The topology of X is the weak topology with respect to the filtration (7.3.4).

If X is a Γ-complex, then X/Γ admits a CW-structure so that the projection $X \to X/\Gamma$ is cellular. For $\Gamma = G$ of order 2, the above definition is easily made equivalent to that of p. 260 (compare also [37, pp. 101–102]). If Γ is a compact Lie group acting smoothly on a smooth manifold X, then X admits a Γ-CW-structure (see [107]).

The Milnor classifying space BS^1 for principal S^1-bundles is homotopy equivalent to $\mathbb{C}P^\infty$. Then, by Proposition 6.1.5, $H^*_{S^1}(pt) = \mathbb{Z}_2[v]$ with v of degree 2. For a pair (X, Y) of S^1-spaces, we thus define

$$h^*_{S^1}(X, Y) = \mathbb{Z}_2[v, v^{-1}] \otimes_{\mathbb{Z}_2[v]} H^*_{S^1}(X, Y),$$

with the notation $h^*_{S^1}(X) = h^*_{S^1}(X, \emptyset)$. As in (7.3.1), $h^*_{S^1}(X, Y)$ is a $\mathbb{Z}$-graded $\mathbb{Z}_2[v, v^{-1}]$-algebra.

Theorem 7.3.10 *Let X be a finite dimensional S^1-complex such that $X^{S^1} = X^{S^0}$. Then, the inclusion $X^{S^1} \subset X$ induces an isomorphism*

$$h^*_{S^1}(X) \xrightarrow{\approx} h^*_{S^1}(X^{S^1})$$

of $\mathbb{Z}$-graded $\mathbb{Z}_2[v, v^{-1}]$-algebras.

The hypothesis $X^{S^1} = X^{S^0}$ is necessary in the above localization theorem Theorem 7.3.10. For example, let $X = S^1$ with S^1-action $g \cdot z = g^2 z$. Then $X_{S^1} \approx BS^0 \approx \mathbb{R}P^\infty$ by Example 7.2.4, so $h^*_{S^1}(X) \approx \mathbb{Z}_2[v, v^{-1}, u]$ while $X^{S^1} = \emptyset$.

Proof The proof follows the plan of that of Theorem 7.3.1, by induction on the skeleton of X, starting trivially with the (-1)-skeleton which is the empty set. The induction step reduces to proving that, if the theorem is true for X, it is then true for $Z = X \cup C$ where C is a family of S^1-cells. As for Theorem 7.3.1, this eventually reduces to proving that $h^*_G(C, \mathrm{BdC}) = 0$ when C is not an isotropy cell. As $X^{S^1} = X^{S^0}$, the isotropy group Γ of C is then a finite group of odd order. The pair $(C, \mathrm{Bd}C)$ is of the form $(S^1/\Gamma \times D^n, S^1/\Gamma \times S^{n-1})$ and, as seen in Example 7.2.4, $C_{S^1} \approx B\Gamma$ and $(\mathrm{Bd}C)_{S^1} \approx B\Gamma \times S^{n-1}$. By Lemma 7.2.2, $H^*_{S^1}(C) = H^*(pt)$ and $H^*_{S^1}(\mathrm{BdC}) = H^*(S^{n-1})$. In particular, the multiplication by u is the zero map and thus $h^*_{S^1}(C) = h^*_{S^1}(\mathrm{Bd}C) = 0$. From the exact sequence of (C, BdC) for h^*_G, it follows that $h^*_G(C, \mathrm{BdC}) = 0$. □

The Smith theory for S^1-complexes with $X^{S^1} = X^{S^0}$ is very similar to that of S^0-spaces. Let $\mathbb{Z}_2(v)$ be the fraction field of $\mathbb{Z}_2[v]$.

Lemma 7.3.11 *Let X be a finite dimensional S^1-complex with $b(X) < \infty$ and $X^{S^1} = X^{S^0}$. Then, as a vector space over $\mathbb{Z}_2(v)$,*

$$\dim \mathbb{Z}_2(v) \otimes_{\mathbb{Z}_2[v]} H^*_{S^1}(X) \leq b(X)$$

with equality if and only if X is equivariantly formal.

Proof The proof is the same as that of Lemma 7.3.6. The transfer exact sequence is replaced by the Gysin exact sequence of the S^1-bundle $X \times S^\infty \to X_{S^1}$ which, as indicated in (7) p. 275, is induced from the universal bundle by $p : X_{S^1} \to BS^1 \approx \mathbb{C}P^\infty$. Therefore, this Gysin sequence looks like

$$H^{k-1}_{S^1}(X) \xrightarrow{\smile v} H^{k+1}_{S^1}(X) \xrightarrow{\rho} H^{k+1}(X) \to H^k_{S^1}(X) \xrightarrow{\smile v} H^{k+2}_{S^1}(X)$$

and permits us the same arguments as for Lemma 7.3.6. □

The proofs of 7.3.12–7.3.14 below are then the same as those of 7.3.7–7.3.9, replacing Theorem 7.3.1 by Theorem 7.3.10.

Proposition 7.3.12 *Let X be a finite dimensional S^1-complex with $b(X) < \infty$ and $X^{S^1} = X^{S^0}$. Then*

$$b(X^{S^1}) \leq b(X) \tag{7.3.5}$$

with equality if and only if X is equivariantly formal.

Corollary 7.3.13 *Let X be a finite dimensional S^1-complex with $X^{S^1} = X^{S^0}$. Then,*

(1) *If $H^*(X) \approx H^*(pt)$, then $H^*(X^{S^1}) \approx H^*(pt)$.*
(2) *If X has the cohomology of a sphere, so does X^{S^1}.*

Proposition 7.3.14 *Let X be finite dimensional S^1-complex with $X^{S^1} = X^{S^0}$. Then, the following statements are equivalent.*

(1) *X is equivariantly formal.*
(2) *$r : H^*_{S^1}(X) \to H^*_{S^1}(X^{S^1})$ is injective.*

7.4 Equivariant Cross Products and Künneth Theorems

Let Γ_1 and Γ_2 be two topological groups; we set $\Gamma_{12} = \Gamma_1 \times \Gamma_2$. Let X be a Γ_1-space and Y be a Γ_2-space. Then $X \times Y$ is a Γ_{12}-space by the product action $(\gamma_1, \gamma_2)\cdot(x, y) = (\gamma_1 x, \gamma_2 y)$. The projections $P_1 : X \times Y \to X$ and $P_2 : X \times Y \to Y$ are equivariant with respect to the projection homomorphisms $\Gamma_{12} \to \Gamma_i$. Passing to the Borel construction gives a map

$$(X \times Y)_{\Gamma_{12}} \xrightarrow{P} X_{\Gamma_1} \times Y_{\Gamma_2} \tag{7.4.1}$$

The map P is a homotopy equivalence, being induced by the homotopy equivalence

$$\tilde{P} : E\Gamma_{12} \times (X \times Y) \to (E\Gamma_1 \times X) \times (E\Gamma_2 \times Y) \tag{7.4.2}$$

given by

$$\tilde{P}\big((t_i(a_i, b_i), (x, y)\big) = \big((t_i a_i, x), (t_i b_i, y)\big),$$

where $(t_i) \in \Delta^\infty$, $(a_i, b_i) \in \Gamma_{12}$ and $(x, y) \in X \times Y$. The case $X = Y = pt$ provides a homotopy equivalence $P_0 : B(\Gamma_{12}) \to B\Gamma_1 \times B\Gamma_2$ and a commutative diagram

$$\begin{array}{ccc} (X \times Y)_{\Gamma_{12}} & \xrightarrow{P} & X_{\Gamma_1} \times Y_{\Gamma_2} \\ \downarrow & & \downarrow \\ B(\Gamma_{12}) & \xrightarrow{P_0} & B\Gamma_1 \times B\Gamma_2 \end{array} . \tag{7.4.3}$$

The cross product $H^*(B\Gamma_1) \otimes H^*(B\Gamma_2) \xrightarrow{\times} H^*(B\Gamma_1 \times B\Gamma_2)$ post-composed with $H^* P_0$ gives a ring homomorphism

$$h : H^*_{\Gamma_1}(pt) \otimes H^*_{\Gamma_2}(pt) \to H^*_{\Gamma_{12}}(pt).$$

Note that, if $B\Gamma_1$ or $B\Gamma_2$ is of finite cohomology type, the Künneth theorem implies that h is an isomorphism. The homotopy equivalence (7.4.1) together with (7.4.3) and the Künneth theorem gives the following lemma.

Lemma 7.4.1 *The composed map*

$$\times_{\Gamma_{12}} : H^*_{\Gamma_1}(X) \otimes H^*_{\Gamma_2}(Y) \xrightarrow{\times} H^*(X_{\Gamma_1} \times Y_{\Gamma_2}) \xrightarrow[\approx]{H^*P} H^*_{\Gamma_{12}}(X \times Y).$$

is an homomorphism of algebras. The $(H^*_{\Gamma_1}(pt) \otimes H^*_{\Gamma_2}(pt))$-module structure on $H^*_{\Gamma_1}(X) \otimes H^*_{\Gamma_2}(Y)$ and the $H^*_{\Gamma_{12}}(pt)$-module structure on $H^*_{\Gamma_{12}}(X \times Y)$ are preserved via h. If Y_{Γ_2} is of finite cohomology type, then $\times_{\Gamma_{12}}$ is an isomorphism.

Example 7.4.2 Let $\Gamma_1 = \Gamma_2 = G = \{\pm 1\}$. We let Γ_1 act on the linear sphere $X = S^m_0$ with $X^{\Gamma_1} = \{\omega^1_\pm\}$, and let Γ_2 act on $Y = S^n_0$ with $Y^{\Gamma_2} = \{\omega^2_\pm\}$ (see Example 7.1.16). Set $H^*_{\Gamma_1}(pt) = \mathbb{Z}_2[u_1]$ and $H^*_{\Gamma_2}(pt) = \mathbb{Z}_2[u_2]$ (u_i of degree 1). As seen in Example 7.1.16, $H^*_{\Gamma_1}(X)$ and $H^*_{\Gamma_2}(Y)$ admit the following presentations

$$H^*_{\Gamma_1}(X) \approx \mathbb{Z}_2[u_1, A_1, B_1] \big/ (A_1 + B_1 + u_1^m, A_1^2 + u_1^m A_1)$$

and

$$H^*_{\Gamma_2}(Y) \approx \mathbb{Z}_2[u_2, A_2, B_2] \,/\, (A_2 + B_2 + u_2^n, A_2^2 + u_2^n A_2),$$

where A_1, B_1 are of degree m and A_2, B_2 are of degree n. To shorten the formulae, we also denote by A_1 the element $A_1 \times_{\Gamma_{12}} \mathbf{1} \in H^*_{\Gamma_{12}}(X \times Y)$, by A_2 the element $\mathbf{1} \times_{\Gamma_{12}} A_2 \in H^*_{\Gamma_{12}}(X \times Y)$, etc. By Lemma 7.4.1. we thus get the presentation

$$H^*_{\Gamma_{12}}(X \times Y) \approx \mathbb{Z}_2[u_1, u_2, A_1, B_1, A_2, B_2] \,/\, \mathcal{I}, \tag{7.4.4}$$

where $\mathcal{I}$ is the ideal generated by

$$A_1 + B_1 + u_1^m \ , \ A_1^2 + u_1^m A_1 \ , \ A_2 + B_2 + u_2^n \text{ and } A_2^2 + u_2^n A_2.$$

One can of course eliminate the B_i's and get the shorter presentation

$$H^*_{\Gamma_{12}}(X \times Y) \approx \mathbb{Z}_2[u_1, u_2, A_1, A_2] \,/\, (A_1^2 + u_1^m A_1, A_2^2 + u_2^n A_2).$$

The commutative diagram

$$\begin{array}{ccc} H^*_{\Gamma_1}(X) \otimes H^*_{\Gamma_2}(Y) & \xrightarrow[\approx]{\times_{\Gamma_{12}}} & H^*_{\Gamma_{12}}(X \times Y) \\ \Big\downarrow{\scriptstyle r_X \otimes r_Y} & & \Big\downarrow{\scriptstyle r} \\ H^*_{\Gamma_1}(X^{\Gamma_1}) \otimes H^*_{\Gamma_2}(Y^{\Gamma_2}) & \xrightarrow[\approx]{\times_{\Gamma_{12}}} & H^*_{\Gamma_{12}}((X \times Y)^{\Gamma_{12}}) \end{array} \tag{7.4.5}$$

permits us to compute the image under r of the various classes of $H^*_{\Gamma_{12}}(X \times Y)$. Set

$$H^*_{\Gamma_1}(X^{\Gamma_1}) = \mathbb{Z}_2[u_1]\omega_-^1 \oplus \mathbb{Z}_2[u_1]\omega_+^1 \text{ and } H^*_{\Gamma_2}(Y^{\Gamma_2}) = \mathbb{Z}_2[u_2]\omega_-^2 \oplus \mathbb{Z}_2[u_2]\omega_+^2.$$

Denote the four points of $(X \times Y)^{\Gamma_{12}} = X^{\Gamma_1} \times Y^{\Gamma_2}$ by $\omega_{--} = (\omega_-^1, \omega_-^2)$, $\omega_{-+} = (\omega_-^1, \omega_+^2)$, etc. With the notation $R = \mathbb{Z}_2[u_1, u_2]$, one has

$$H^*_{\Gamma_{12}}((X \times Y)^{\Gamma_{12}}) \approx R\,\omega_{--} \oplus R\,\omega_{+-} \oplus R\,\omega_{-+} \oplus R\,\omega_{++} \tag{7.4.6}$$

One has

$$r(A_1) = r_X(A_1) \times_{\Gamma_{12}} r_Y(\mathbf{1}) = u_1^m \omega_+^1 \times_{\Gamma_{12}} (\mathbf{1}\omega_-^2 + \mathbf{1}\,\omega_+^2) = u_1^m\, \omega_{+-} + u_1^m\, \omega_{++}.$$

Hence, the coordinates of $r(A_1)$ using (7.4.6) are $(0, u_1^m, 0, u_1^m)$. Similar computations provide the following table.

We now concentrate our interest to the case where $\Gamma_1 = \Gamma_2 = \Gamma$ and see $X \times Y$ as a Γ-space using the diagonal homomorphism $\Delta : \Gamma \to \Gamma \times \Gamma$. We get a

x	Coord. of $r(x)$ in (7.2.6)			
$\mathbf{1}$	$\mathbf{1}$	$\mathbf{1}$	$\mathbf{1}$	$\mathbf{1}$
u_i	u_i	u_i	u_i	u_i
A_1	0	u_1^m	0	u_1^m
A_2	0	0	u_2^n	u_2^n
$A_1 A_2$	0	0	0	$u_1^m u_2^n$
B_1	u_1^m	0	u_1^m	0
B_2	u_2^n	u_2^n	0	0
$B_1 B_2$	$u_1^m u_2^n$	0	0	0

homomorphism $\Delta^* : H^*_{\Gamma\times\Gamma}(X \times Y) \to H^*_\Gamma(X \times Y)$. The composed map

$$H^*_\Gamma(X) \otimes H^*_\Gamma(Y) \xrightarrow{\times_{\Gamma\times\Gamma}} H^*(X_\Gamma \times Y_\Gamma) \underset{\approx}{\xrightarrow{H^*P}} H^*_{\Gamma\times\Gamma}(X \times Y) \xrightarrow{\Delta^*} H^*_\Gamma(X \times Y) \qquad (7.4.7)$$

(the dotted composite arrow being labelled $\times_\Gamma$)

is called the *equivariant cross product*. For $X = Y = pt$, one has the commutative diagram

$$\begin{array}{ccc} H^*_\Gamma(pt) \otimes H^*_\Gamma(pt) & \xrightarrow{\times_\Gamma} & H^*_\Gamma(pt) \\ \downarrow \approx & & \downarrow \approx \\ H^*(B\Gamma) \otimes H^*(B\Gamma) & \xrightarrow{\smile} & H^*(B\Gamma) \end{array} . \qquad (7.4.8)$$

Indeed, one has

$$B\Gamma \xrightarrow{B\Delta} B(\Gamma \times \Gamma) \underset{\approx}{\xrightarrow{P}} B\Gamma \times B\Gamma \qquad (7.4.9)$$

(the composite arrow $B\Gamma \to B\Gamma \times B\Gamma$ being labelled $\Delta_{B\Gamma}$)

and $H^*\Delta_{B\Gamma}(a \times b) = a \smile b$ by (4.6.5).

The equivariant cross product $\times_\Gamma$ will be useful in Sect. 8.3 but one may wish to get some Künneth theorem. As this is not even the case for $X = Y = pt$, some adaptation is needed. Lemma 7.4.1 together with diagram (7.4.8) implies that

$$(w \cdot a) \times_\Gamma b = a \times_\Gamma (w \cdot b) = w \cdot (a \times_\Gamma b). \qquad (7.4.10)$$

for all $a \in H^*_\Gamma(X)$, $b \in H^*_\Gamma(Y)$ and $w \in H^*\Gamma(pt)$. Therefore, $\times_\Gamma$ descend to the *strong equivariant cross product*

$$\bar{\times}_\Gamma : H^*_\Gamma(X) \otimes_{H^*_\Gamma(pt)} H^*_\Gamma(Y) \to H^*_\Gamma(X \times Y).$$

The tensor product $H^*_\Gamma(X) \otimes_{H^*_\Gamma(pt)} H^*_\Gamma(Y)$ still carries an $H^*_\Gamma(pt)$-action, defined by $w \cdot (a \otimes b) = (w \cdot a) \otimes b = a \otimes (w \cdot b)$. Lemma 7.4.1 together with (7.4.10) implies that $\bar{\times}_\Gamma$ is a morphism of $H^*_\Gamma(pt)$-algebras.

Theorem 7.4.3 (Equivariant Künneth theorem) *Let Γ be a topological group such that $B\Gamma_0$ is of finite cohomology type for any closed subgroup Γ_0 of Γ. Let X and Y be Γ-spaces, where X is a finite dimensional Γ-CW-complex. Suppose that Y is of finite cohomology type and is equivariantly formal. Then, the strong equivariant $\bar{\times}_\Gamma$ cross product is an isomorphism of $H^*_\Gamma(pt)$-algebras.*

Proof As $\bar{\times}_\Gamma$ is a morphism of $H^*_\Gamma(pt)$-algebra, it suffices to prove that it is a **GrV**-isomorphism. We follow the idea of the proof of the ordinary Künneth Theorem 4.6.7, fixing the Γ-space Y and comparing the "equivariant cohomology theories"

$$h^*(X, A) = H^*_\Gamma(X, A) \otimes_{H^*_\Gamma(pt)} H^*_\Gamma(Y) \text{ and } k^*(X, A) = H^*_\Gamma(X \times Y, A \times Y)$$

defined for a Γ-pair (X, A). The definition of the strong equivariant cross product extends to pairs and we get a morphism of $H^*_\Gamma(pt)$-algebras

$$\bar{\times}_\Gamma : h^*(X, A) \to k^*(X, A)$$

One gets a commutative diagram

$$\begin{array}{ccccccccc}
h^*(X) & \longrightarrow & h^*(A) & \xrightarrow{\delta_*} & h^{*+1}(X, A) & \longrightarrow & h^{*+1}(X) & \longrightarrow & h^{*+1}(A) \\
\downarrow \bar{\times}_\Gamma & & \downarrow \bar{\times}_\Gamma & & \downarrow \bar{\times}_\Gamma & & \downarrow \bar{\times}_\Gamma & & \downarrow \bar{\times}_\Gamma \\
h^*(X) & \longrightarrow & k^*(A) & \xrightarrow{\delta_*} & k^{*+1}(X, A) & \longrightarrow & k^{*+1}(X) & \longrightarrow & k^{*+1}(A)
\end{array} \tag{7.4.11}$$

where the lines are exact. That the square diagram with the δ^*'s commutes comes from the definition of $\times_\Gamma$, using the commutativity of Diagram (4.6.13).

The theorem is proven by induction on the dimension of X. If X is 0-dimensional, it is a disjoint union of homogeneous Γ-spaces. As the disjoint union axiom holds for our theories, the induction starts by proving the theorem for $X = \Gamma/\Gamma_0$, where Γ_0 is a closed subgroup of Γ. As Y is Γ-equivariantly formal, it is also Γ_0-equivariantly formal and one has

$$\begin{aligned}
h^*(X) &= H^*_\Gamma(\Gamma/\Gamma_0) \otimes_{H^*_\Gamma(pt)} H^*_\Gamma(Y) \\
&\approx H^*(B\Gamma_0) \otimes_{H^*(B\Gamma)} \big(H^*(B\Gamma) \otimes H^*(Y)\big) \\
&\approx H^*(B\Gamma_0) \otimes H^*(Y) \\
&\approx H^*_{\Gamma_0}(Y).
\end{aligned}$$

On the other hand, consider the map $\tilde{\alpha} : E_\Gamma \times (\Gamma \times Y) \to E_\Gamma \times Y$ given by

$$\tilde{\alpha}(z, (\gamma, y)) = (z\gamma, \gamma^{-1}y).$$

It satisfies $\tilde{\alpha}(z, (\delta\gamma, \delta y)) = \tilde{\alpha}(z\delta, (\gamma, y))$ and $\tilde{\alpha}(z, (\gamma\gamma_0, y)) = (z\gamma\gamma_0, \gamma_0^{-1}\gamma^{-1}y)$; it thus descends to a map

$$\alpha : E_\Gamma \times_\Gamma (\Gamma/\Gamma_0 \times Y) \xrightarrow{\approx} E\Gamma \times_{\Gamma_0} Y$$

which is a homeomorphism: its inverse is induced by the map $\beta(z, y) = (z, ([e], y))$, where $e \in \Gamma$ is the unit element. Hence, $k^*(X)$ is also isomorphic, as an $H^*_\Gamma(pt)$-algebra, to $H^*_{\Gamma_0}(Y)$. It remains to show that $\bar{\times}_\Gamma$ is a **GrV**-isomorphism. As Y and $B\Gamma_0$ are both of finite cohomology type, the graded vector space $H_{\Gamma_0}(Y) \approx H^*(B\Gamma_0) \otimes H^*(Y)$ is finite dimensional in each degree. Therefore, it suffices to prove that $\bar{\times}_\Gamma$ is surjective.

If Z is a Γ-space, we denote by $i : Z \to Z_\Gamma$ the inclusion $i(z) = [(1e, 0, ...), z]$ (it induces the forgetful homomorphism $H^*i = \rho : H^*_\Gamma(Z) \to H^*(Z)$). One has a commutative diagram

$$\begin{array}{ccc}
\Gamma/\Gamma_0 \times Y & \xleftarrow{\;\;s\;\;} & Y \\
\downarrow{\scriptstyle i} & & \downarrow{\scriptstyle i} \\
E_\Gamma \times_\Gamma (\Gamma/\Gamma_0 \times Y) & \xleftarrow{\;\;\beta\;\;} & E\Gamma \times_{\Gamma_0} Y
\end{array}$$

where s is the slice inclusion $s(y) = ([e], y])$. We thus get a commutative diagram

$$\begin{array}{ccccc}
H^*(\Gamma/\Gamma_0) \otimes H^*(Y) & \xrightarrow{\;\times\;} & H^*(X \times Y) & \xrightarrow{H^*s} & H^*(Y) \\
\uparrow{\scriptstyle \rho\otimes\rho} & & \uparrow{\scriptstyle \rho} & & \uparrow{\scriptstyle \rho} \\
H^*_\Gamma(\Gamma/\Gamma_0) \otimes_\Gamma H^*(Y)_\Gamma & \xrightarrow{\;\bar{\times}_\Gamma\;} & H^*_\Gamma(X \times Y) & \xrightarrow[\approx]{H^*\beta} & H^*_{\Gamma_0}(Y)
\end{array}$$

Let $\mathcal{B}$ be a **GrV**-basis of $H^*(Y)$. Let $\sigma : H^k(Y) \to H^k_\Gamma(Y)$ be a section of ρ. For $b \in \mathcal{B}$, one has

$$H^*s(\rho(\mathbf{1}) \times \rho(\sigma(b)) = H^*s(\mathbf{1}\,\bar{\times}_\Gamma\, b) = b,$$

the last equality coming from Lemma 4.7.2. Therefore, $H^*s \circ \times \circ (\rho\otimes\rho)$ is surjective and the formula

$$\hat{\sigma}(a) = H^*\beta(\mathbf{1}\,\bar{\times}_\Gamma\, \sigma(a))$$

defines a section $\hat{\sigma} : H^*(Y) \to H^*_{\Gamma_0}(Y)$ of ρ. The Leray-Hirsch theorem then implies that $H^*_{\Gamma_0}(Y)$ is generated, as a $H^*_\Gamma(pt)$-module, by $\hat{\sigma}(\mathcal{B})$. Hence, $\bar{\times}_\Gamma$ is surjective.

The induction step reduces to proving that, if the theorem is true for A, it is then true for $X = A \cup C$ where C is a family of Γ-cells. By the five lemma in Diagram (7.4.11), it suffices to prove that $\bar{\times}_\Gamma : h^*(X, A) \to k^*(X, A)$ is an isomorphism. By excision and the disjoint union axiom one can restrict ourselves to the case of a pair $(X_0, A_0) = \Gamma/\Gamma_0 \times (D^n, S^{n-1})$ (a Γ-cell). By the five lemma in Diagram (7.4.11) for the pair (X_0, A_0), it suffices to prove the theorem for X_0 and for A_0. The former is covered by the 0-dimensional case (sinceX_0 is Γ-homotopy equivalent to Γ/Γ_0) and the latter is $(n-1)$-dimensional, thus covered by the induction hypothesis. □

Remark 7.4.4 If $\Gamma = \{e\}$, Theorem 7.4.3 reduces to the ordinary Künneth Theorem 4.6.7. Therefore, the hypotheses that Y is of finite cohomology type is essential. Theorem 7.4.3 is also wrong if Y is not equivariantly formal. For example, set $\Gamma = \{\pm 1\}$, $X = S^1$ and $Y = S^2$, with the antipodal involution. These are free Γ-spaces and, by Lemma 7.1.4,

$$H^*_\Gamma(X) \approx H^*(S^1/\pm 1) \approx \mathbb{Z}_2[u]/(u^2) \text{ and } H^*_\Gamma(Y) \approx H^*(S^2/\pm 1) \approx \mathbb{Z}_2[u]/(u^3).$$

Moreover, $H^*_\Gamma(pt) = H^*(\mathbb{R}P^\infty) \approx \mathbb{Z}_2[u]$ and, using Lemma 7.1.4 again together with Proposition 4.3.10, the $\mathbb{Z}[u]$-morphisms $H^*_\Gamma(pt) \to H^*_\Gamma(X)$ and $H^*_\Gamma(pt) \to H^*_\Gamma(Y)$ are surjective. Therefore,

$$H^*_\Gamma(X) \otimes_{H^*_\Gamma(pt)} H^*_\Gamma(Y) \approx \mathbb{Z}_2[u]/(u^2) \times_{\mathbb{Z}_2[u]} \mathbb{Z}_2[u]/(u^3) \approx \mathbb{Z}_2[u]/(u^2).$$

In particular, $H^*_\Gamma(X) \otimes_{H^*_\Gamma(pt)} H^*_\Gamma(Y)$ vanishes in degree 3, while $H^3_\Gamma(X \times Y) = \mathbb{Z}_2$. Indeed $H^*_\Gamma(X \times Y) \approx H^*((X \times Y)/\pm 1)$ and $(X \times Y)/\pm 1$ is a closed manifold of dimension 3.

The hypothesis that $B\Gamma_0$ is of finite cohomology type is fulfilled if Γ_0 is a compact Lie group. Note that it is only used in the proof for the stabilizers of points of X. For other kind of equivariant Künneth theorems (see [176]).

Example 7.4.5 Consider the diagonal action of the group $G = \{\pm 1\}$ on the product of linear spheres $S^m_0 \times S^n_0$. Set $H^*_G(pt) = \mathbb{Z}_2[u]$, with u of degree 1. By Example 7.4.2 and Theorem 7.4.3, one has

$$H^*_G(S^m_0 \times S^n_0) \approx \mathbb{Z}_2[u, A_1, B_1, A_2, B_2]/\mathcal{I},$$

where $\mathcal{I}$ is the ideal generated by

$$A_1 + B_1 + u^m \ , \ A_1^2 + u^m A_1 \ , \ A_2 + B_2 + u^n \text{ and } A_2^2 + u^n A_2.$$

Using the notations of Example 7.4.2 for the fixed points, one has

$$H^*_G((S^m_0 \times S^n_0)^G) \approx \mathbb{Z}_2[u]\,\omega_{--} \oplus \mathbb{Z}_2[u]\,\omega_{+-} \oplus \mathbb{Z}_2[u]\,\omega_{-+} \oplus \mathbb{Z}_2[u]\,\omega_{++} \quad (7.4.12)$$

and one has the following table for $r : H^*_G(S^m_0 \times S^n_0) \to H^*_G((S^m_0 \times S^n_0)^G)$

For a generalization of this example, see Proposition 10.3.5.

We now define the equivariant *reduced* cross product, related to the equivariant cohomology of a smash product. Let X be a Γ_1-space and Y be a Γ_2-space, pointed

x	coord. of $r(x)$ in (7.4.12)			
$\mathbf{1}$	$\mathbf{1}$	$\mathbf{1}$	$\mathbf{1}$	$\mathbf{1}$
u	u	u	u	u
A_1	0	u^m	0	u^m
A_2	0	0	u^n	u^n
$A_1 A_2$	0	0	0	u^{m+n}
B_1	u^m	0	u^m	0
B_2	u^n	u^n	0	0
$B_1 B_2$	u^{m+n}	0	0	0

by $x \in X^{\Gamma_1}$ and $y \in Y^{\Gamma_2}$. Then, $X \vee Y$ is a Γ_{12}-invariant subspace of $X \times Y$. Consider the space

$$X_{\Gamma_1} \bar{\vee} Y_{\Gamma_2} = (X_{\Gamma_1} \times \{y\}_{\Gamma_2}) \cup (\{x\}_{\Gamma_1} \times Y_{\Gamma_2}) \subset X_{\Gamma_1} \times Y_{\Gamma_2}.$$

If the pairs $(X, \{x\})$ and $(Y, \{y\})$ are equivariant well cofibrant pairs, we say that (X, x) and (Y, y) are *equivariantly well pointed.*

Lemma 7.4.6 *Let* (X, x) *be an equivariantly well pointed* Γ_1*-space and* (Y, y) *be an equivariantly well pointed* Γ_2*-space. Then, the map* $P : (X \times Y)_{\Gamma_{12}} \to X_{\Gamma_1} \times Y_{\Gamma_2}$ *of* (7.4.1) *sends* $(X \vee Y)_{\Gamma_{12}}$ *onto* $X_{\Gamma_1} \bar{\vee} Y_{\Gamma_2}$ *and induces an isomorphism*

$$H^*P : H^*(X_{\Gamma_1} \bar{\vee} Y_{\Gamma_2}) \xrightarrow{\approx} H^*_{\Gamma_{12}}(X \vee Y)$$

Proof That $P((X \vee Y)_{\Gamma_{12}}) = X_{\Gamma_1} \bar{\vee} Y_{\Gamma_2}$ follows directly from the definition of P, using (7.4.2). This gives a commutative diagram

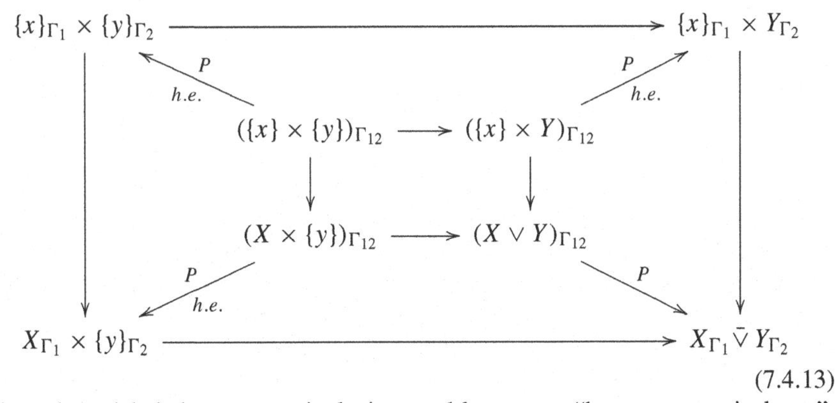

(7.4.13)

where the unlabeled arrows are inclusions and *h.e.* means "homotopy equivalence". Our hypotheses and Lemma 7.2.11 imply that pairs like $((X \times \{y\})_{\Gamma_{12}}, (\{x\} \times \{y\})_{\Gamma_{12}})$, etc., are good. Hence, the hypotheses of Proposition 3.1.54 to get Mayer-Vietoris sequences are fulfilled. We thus get a morphism from the Mayer-Vietoris sequence for the outer square of (7.4.13) to that of the inner square, and the proposition follows from the five-lemma. □

Remark 7.4.7 The map $P : (X \vee Y)_{\Gamma_{12}} \to X_{\Gamma_1} \bar{\vee} Y_{\Gamma_2}$ of Lemma 7.4.6 is actually a weak homotopy equivalence, since the squares in (7.4.13) are homotopy co-Cartesian diagrams (see [38, Prop. 5.3.3]).

As $X \vee Y$ is a Γ_{12}-invariant subspace of $X \times Y$, the wedge product $X \wedge Y$ inherits a Γ_{12}-action.

Lemma 7.4.8 *Let (X, x) be an equivariantly well pointed Γ_1-space and (Y, y) be an equivariantly well pointed Γ_2-space. Then, there is a natural isomorphism*

$$H^*(X_{\Gamma_1} \times Y_{\Gamma_2}, X_{\Gamma_1} \bar{\vee} Y_{\Gamma_2}) \xrightarrow{\approx} \tilde{H}^*_{\Gamma_{12}}(X \wedge Y)).$$

Proof By Lemma 7.4.6, the map P produces a morphism from the cohomology sequence of the pair $(X_{\Gamma_1} \times Y_{\Gamma_2}, X_{\Gamma_1} \bar{\vee} Y_{\Gamma_2})$ to that of the pair $\big((X \times Y)_{\Gamma_{12}}, (X \vee Y)_{\Gamma_{12}}\big)$. By Lemma 7.4.6 again and the fact that the map P of (7.4.1) is a homotopy equivalence, the five lemma implies that

$$H^*P : H^*(X_{\Gamma_1} \times Y_{\Gamma_2}, X_{\Gamma_1} \bar{\vee} Y_{\Gamma_2}) \to H^*((X \times Y)_{\Gamma_{12}}, (X \vee Y)_{\Gamma_{12}}\big)$$

is an isomorphism. By Lemma 7.2.12, the pair $(X \times Y, X \vee Y)$ is Γ_{12}-equivariantly well cofibrant. As $X \wedge Y)$ is not empty, Corollary 7.2.16 provides a natural isomorphism between $H^*((X \times Y)_{\Gamma_{12}}, (X \vee Y)_{\Gamma_{12}}\big)$ and $\tilde{H}^*_{\Gamma_{12}}(X \wedge Y)$.

Using the isomorphism of Lemma 7.4.8 as well as those of (7.2.11). one constructs the commutative diagram

$$\begin{array}{ccc} H^*(X_{\Gamma_1}, \{x\}_{\Gamma_1}) \otimes H^*(Y_{\Gamma_2}, \{y\}_{\Gamma_2}) & \xrightarrow{\approx} & \tilde{H}^*_{\Gamma_1}(X) \otimes \tilde{H}^*_{\Gamma_2}(Y) \\ \downarrow \times & & \vdots\, \tilde{\times}_{\Gamma_{12}} \\ H^*(X_{\Gamma_1} \times Y_{\Gamma_2}, X_{\Gamma_1} \bar{\vee} Y_{\Gamma_2}) & \xrightarrow{\approx} & \tilde{H}^*_{\Gamma_{12}}(X \wedge Y) \end{array} \tag{7.4.14}$$

which defines the *equivariant reduced cross product* $\tilde{\times}_{\Gamma_{12}}$. The relative cross product (left vertical arrow) is indeed defined as in (4.6.6), since, as (Y, y) is equivariantly well pointed, the couple $(Y_{\Gamma_2}, \{y\}_{\Gamma_2})$ is a good pair by Lemma 7.2.11.

In the case where $\Gamma_1 = \Gamma_2 = \Gamma$, one can see $X \wedge Y$ as a Γ-space via the diagonal homomorphism $\Delta : \Gamma \to \Gamma \times \Gamma$. Composing $\tilde{\times}_{\Gamma_{12}}$ with $\tilde{\Delta}^* : \tilde{H}^*_{\Gamma \times \Gamma}(X \wedge Y) \to \tilde{H}^*_{\Gamma}(X \wedge Y)$, we get the equivariant reduced cross product

$$\tilde{H}^*_{\Gamma}(X) \otimes \tilde{H}^*_{\Gamma}(Y) \xrightarrow{\tilde{\times}_{\Gamma}} \tilde{H}^*_{\Gamma}(X \wedge Y). \tag{7.4.15}$$

Lemma 7.4.9 *Let (X, x) and (Y, y) be equivariantly well pointed Γ-spaces. Then,*

(1) *there is an equivariant reduced cross product*

$$\tilde{H}^*_\Gamma(X) \otimes \tilde{H}^*_\Gamma(Y) \xrightarrow{\tilde{\times}_\Gamma} \tilde{H}^*_\Gamma(X \wedge Y)$$

which is a bilinear map.

(2) *the diagram*

$$\begin{array}{ccc} \tilde{H}^*_\Gamma(X) \otimes \tilde{H}^*_\Gamma(Y) & \xrightarrow{\tilde{\times}_\Gamma} & \tilde{H}^*_\Gamma(X \wedge Y) \\ \downarrow{\scriptstyle \rho\otimes\rho} & & \downarrow{\scriptstyle \rho} \\ \tilde{H}^*(X) \otimes \tilde{H}^*(Y) & \xrightarrow{\tilde{\times}} & \tilde{H}^*(X \wedge Y) \end{array} \qquad (7.4.16)$$

is commutative, where ρ is the forgetful homomorphism.

(3) *the hypotheses on X, Y are inherited by X^Γ, Y^Γ and there is a commutative diagram*

$$\begin{array}{ccc} \tilde{H}^*_\Gamma(X) \otimes \tilde{H}^*_\Gamma(Y) & \xrightarrow{\tilde{\times}_\Gamma} & \tilde{H}^*_\Gamma(X \wedge Y) \\ \downarrow{\scriptstyle r\otimes r} & & \downarrow{\scriptstyle r} \\ \tilde{H}^*_\Gamma(X^\Gamma) \otimes \tilde{H}^*_\Gamma(Y^\Gamma) & \xrightarrow{\tilde{\times}_\Gamma} & \tilde{H}^*_\Gamma(X^\Gamma \wedge Y^\Gamma) \\ \updownarrow{\scriptstyle \approx} & & \updownarrow{\scriptstyle \approx} \\ [\tilde{H}^*(X^\Gamma) \otimes H^*(B\Gamma)] \otimes [\tilde{H}^*(Y^\Gamma) \otimes H^*(B\Gamma)] & \xrightarrow{\tilde{\times}\otimes\smile} & \tilde{H}^*(X^\Gamma \wedge Y^\Gamma) \otimes H^*(B\Gamma) \end{array}$$
(7.4.17)

Proof The equivariant reduced cross product of (1) is obtained by post-composing $\tilde{\times}_{\Gamma_{12}}$ of (7.4.14) (with $\Gamma_1 = \Gamma_2 = \Gamma$) with $\tilde{\Delta}^* : \tilde{H}^*_{\Gamma\times\Gamma}(X \wedge Y) \to \tilde{H}^*_\Gamma(X \wedge Y)$.

Let $\alpha : \Gamma' \to \Gamma$ is a continuous homomorphism. Then (X, x) and (Y, y) are Γ'-equivariantly well cofibrant. Our constructions are natural enough so that there is a commutative diagram

$$\begin{array}{ccc} \tilde{H}^*_\Gamma(X) \otimes \tilde{H}^*_\Gamma(Y) & \xrightarrow{\tilde{\times}_\Gamma} & \tilde{H}^*_\Gamma(X \wedge Y) \\ \downarrow{\scriptstyle \alpha^*\otimes\alpha^*} & & \downarrow{\scriptstyle \alpha^*} \\ \tilde{H}^*_{\Gamma'}(X) \otimes \tilde{H}^*_{\Gamma'}(Y) & \xrightarrow{\tilde{\times}_{\Gamma'}} & \tilde{H}^*_{\Gamma'}(X \wedge Y) \end{array} . \qquad (7.4.18)$$

For $\Gamma' = \{I\}$, the homomorphism α^* coincides with the forgetful homomorphism ρ (see 7.2.8), which proves (2).

To prove (3), we note that the upper square of (7.4.16) commutes by obvious naturality of the equivariant reduced cross product with respect to equivariant maps. The commutativity of the lower square is obtained using the considerations of (7.4.8) and (7.4.9). □

Example 7.4.10 Let (Z, z) be a well pointed space, considered with the trivial action of $G = \{I, \tau\}$. Then, $\tilde{H}^*_G(Z) \approx \tilde{H}^*(Z)[u]$ and the bottom square in (7.4.17) becomes

$$\begin{array}{ccc} \tilde{H}^*_G(Z) \otimes \tilde{H}^*_G(Z) & \xrightarrow{\tilde{\times}_G} & \tilde{H}^*_G(Z \wedge Z) \\ \updownarrow \approx & & \updownarrow \approx \\ \tilde{H}^*(Z)[u] \otimes \tilde{H}^*(Z)[u] & \xrightarrow{\tilde{\times}_{[u]}} & \tilde{H}^*(Z \wedge Z)[u] \end{array},$$

where, for $a, b \in \tilde{H}^*(Z)$, $\tilde{\times}_{[u]}$ is defined by

$$au^m \tilde{\times}_{[u]} bu^n = (a \tilde{\times} b)\, u^{m+n}.$$

7.5 Equivariant Bundles and Euler Classes

Although we are mostly interested in equivariant vector bundles, passing through equivariant principal bundles is easier and more powerful. Let A be a topological group. A *principal A-bundle* ζ over a space Y consists of a continuous map $p : P \to Y$, a continuous right action of A on P such that $p(u\alpha) = p(u)$ for all $u \in P$ and all $\alpha \in A$; in addition, the following local triviality should hold: for each $x \in X$ there is a neighbourhood U of x and a homeomorphism $\psi : U \times A \to p^{-1}(U)$ such that $p \circ \psi(x, \alpha) = x$ and $\psi(x, \alpha\beta) = \psi(x, \alpha)\beta$. In consequence, A acts freely on P and transitively on each fiber. Also, p is a surjective open map, descending to a homeomorphism $P/A \xrightarrow{\approx} X$ (use [44, Sect. I Chap. VI]). Two principal A-bundles $\zeta = (P \xrightarrow{p} X)$ and $\bar{\zeta} = (\bar{P} \xrightarrow{\bar{p}} X)$ are *isomorphic* if there exists an A-equivariant homeomorphism $h : P \to \bar{P}$ such that $\bar{p} \circ h = p$.

Let Γ be a topological group and let X be a (left) Γ-space. An A-principal bundle $\zeta : P \xrightarrow{p} X$ is called a *Γ-equivariant principal A-bundle* if it is given a left action $\Gamma \times P \to P$ commuting with the free right action of A and such that the projection p is Γ-equivariant (a more general setting is considered in e.g. [37, 126, 139]). Two Γ-equivariant principal A-bundles $\zeta = (P \xrightarrow{p} X)$ and $\bar{\zeta}' = (\bar{P} \xrightarrow{\bar{p}} X)$ are *isomorphic* if there exists an (Γ, A)-equivariant homeomorphism $h : P \to \bar{P}$ such that $\bar{p} \circ h = p$.

Example 7.5.1 Let $p : P \to pt$ be a Γ-equivariant principal A-bundle over a point. The A-action on P is free and transitive. Hence, choosing a point $s \in P$ provides a continuous map $\mu : \Gamma \to A$ by the equation $\gamma s = s\mu(\gamma)$. For $\gamma, \gamma' \in \Gamma$, one has

$$s\mu(\gamma\gamma') = (\gamma\gamma')s = \gamma(\gamma' s) = (\gamma s)\mu(\gamma') = (s\mu(\gamma))\mu(\gamma') = s(\mu(\gamma)\mu(\gamma')),$$

which proves that μ is a homomorphism. Another point $\tilde{s} \in P$ is of the form $\tilde{s} = s\alpha$ for some $\alpha \in A$. The map $\tilde{\mu}$ obtained from $\tilde{s}$ is related to μ by

$$s\alpha\tilde{\mu}(\gamma) = \tilde{s}\tilde{\mu}(\gamma) = \gamma\tilde{s} = \gamma s\alpha = s\mu(\gamma)\alpha$$

and hence $\tilde{\mu}(\gamma) = \alpha^{-1}\mu(\gamma)\alpha$. If $\bar{p} : \bar{P} \to pt$ is another Γ-equivariant principal A-bundle and if $h : P \to \bar{P}$ is a (Γ, A)-equivariant homeomorphism, then $\gamma h(s) = h(s)\mu(\gamma)$. This provides a map from the isomorphism classes of Γ-equivariant principal A-bundles over a point and the set $\hom(\Gamma, A)/A$ of the conjugation classes of continuous homomorphisms from Γ to A. This map is a bijection. A homomorphism $\mu : \Gamma \to A$ is realized by the bundle $A \to pt$ with the Γ-action $\gamma \cdot \alpha = \mu(\gamma)\alpha$ (hence, if e is the unit element of A, one has indeed $\gamma \cdot e = e\mu(\gamma)$). This proves the surjectivity. The proof of the injectivity is left to the reader.

Let $\zeta : P \xrightarrow{p} X$ be a Γ-equivariant principal A-bundle. Being Γ-equivariant, the map p induces a map $p_\Gamma : P_\Gamma \to X_\Gamma$. Let $i : X \to X_\Gamma$ be an inclusion as in (7.2.8).

Lemma 7.5.2 *The map $p_\Gamma : P_\Gamma \to X_\Gamma$ is a principal A-bundle, denoted by ζ_Γ. Moreover, ζ is isomorphic to the induced principal A-bundle $i^*\zeta_\Gamma$.*

Example 7.5.3 Let ξ be a Γ-equivariant principal A-bundle over a point, corresponding to $[\mu] \in \hom(\Gamma, A)/A$ (see Example 7.5.1). It is then isomorphic to $A \to pt$ with the Γ-action $\gamma \cdot \alpha = \mu(\gamma)\alpha$. Then ξ_Γ is the principal A-bundle over $B\Gamma$ induced by the map $B\mu : B\Gamma \to BA$. Indeed, the map $f : E\Gamma \times_\Gamma A \to EA$ given by $f([(t_i\gamma_i), \alpha]) = (t_i\mu(\gamma_i)\alpha)$ is A-equivariant and covers the map B_μ.

Before proving Lemma 7.5.2, let us recall the standard local cross-sections for the Milnor construction of the universal Γ-bundle $\hat{p} : E\Gamma \to B\Gamma$. For $i \in \mathbb{N}$, let $(E\Gamma)_i = \{(t_j\gamma_j) \in E\Gamma \mid t_i \neq 0\}$ and let $(B\Gamma)_i = \hat{p}((E\Gamma)_i)$. There is a cross-section s_i of $\hat{p}$ over $(B\Gamma)_i$ sending $b \in (B\Gamma)_i$ to the unique element in $(t_j\Gamma_j) \in \hat{p}^{-1}(b)$ with $\gamma_i = 1$. If Z is a Γ-space, the map $\psi_i : (B\Gamma)_i \times Z \xrightarrow{\approx} (E\Gamma)_i \times_\Gamma Z$ given by $\psi_i(b, u) = [s_i(b), u]$ is a homeomorphism: its inverse is induced by the correspondence $[(t_j\gamma_j), u] \mapsto (\hat{p}(t_j\gamma_j), \gamma_i z)$.

Proof of Lemma 7.5.2 The right A-action on $P_\Gamma = E\Gamma \times_\Gamma P$ is defined by $[z, u]\alpha = [z, u\alpha]$. For $i \in \mathbb{N}$, one has the commutative diagram

$$\begin{array}{ccc}
(B\Gamma)_i \times P & \xrightarrow[\approx]{\psi_i} & (E\Gamma)_i \times_\Gamma P \\
\Big\downarrow {\scriptstyle \mathrm{id}\times p} & & \Big\downarrow {\scriptstyle p_\Gamma} \\
(B\Gamma)_i \times X & \xrightarrow[\approx]{\psi_i} & (E\Gamma)_i \times_\Gamma X
\end{array}.$$

The upper homeomorphism is A-equivariant and $\mathrm{id} \times p$ is a principal A-bundle. As $\{(E\Gamma)_i \times_\Gamma X\}_{i\in\mathbb{N}}$ is an open covering of X_Γ, the map p_Γ admits local trivializations of a principal A-bundle.

We have proven that p_Γ is a principal A-bundle. If $z \in E\Gamma$, let $i_z : X \to X_\Gamma$ be the inclusion defined in (7.2.8). Then, $\tilde{i}_z : P \to P_\Gamma$ is an A-equivariant map covering i_z, inducing an isomorphism of principal A-bundles $\zeta \approx i_z^*\zeta_\Gamma$.

Remark 7.5.4 When A is abelian, the last assertion of Lemma 7.5.2 may be strengthened: a principal A-bundle over a Γ-space X admits a structure of a Γ-equivariant bundle if and only if it is induced from a principal A-bundle over X_Γ (see [127]).

The construction $\zeta \mapsto \zeta_\Gamma$ enjoys some functorial properties. Let $\mu : \Gamma' \to \Gamma$ be a continuous homomorphism between topological groups. Let X' be a Γ'-space, X a Γ-space and let $f : X' \to X$ be a continuous map which is Γ'-equivariant with respect to μ. Recall from (7.2.1) that f induces a map $f_{\Gamma',\Gamma} : X_{\Gamma'} \to X_\Gamma$. If ζ is a Γ-equivariant principal A-bundle over X, then $f^*\zeta$ is a Γ'-equivariant principal A-bundle over X'.

Lemma 7.5.5 $(f^*\zeta)_{\Gamma'} \approx f^*_{\Gamma',\Gamma}\zeta_\Gamma$.

Proof The map f is covered by a Γ'-equivariant map of principal A-bundle $\tilde{f} : P(f^*\zeta) \to P$, where $P(f^*\zeta)$ denotes the total space of $f^*\zeta$. By functoriality of the Borel construction (see 7.2.5), there is a commutative diagram

$$\begin{array}{ccccc} P(f^*\zeta)_{\Gamma'} & \xrightarrow{\tilde{f}_{\Gamma',\Gamma}} & P_\Gamma & \xleftrightarrow{=} & P(\zeta_\Gamma) \\ \downarrow & & \downarrow & & \downarrow \\ X'_{\Gamma'} & \xrightarrow{f_{\Gamma',\Gamma}} & X_\Gamma & \xleftrightarrow{=} & X_\Gamma \end{array}$$

Thanks to the description of the A-actions (see the proof of Lemma 7.5.2), the map $\tilde{f}_{\Gamma',\Gamma}$ A-equivariant. Hence, $\tilde{f}_{\Gamma',\Gamma}$ factor through an isomorphism $(f^*\zeta)_{\Gamma'} \approx f^*_{\Gamma',\Gamma}\zeta_\Gamma$. □

For another functoriality of ζ_Γ let $\varphi : A \to A'$ be a continuous homomorphism between topological groups. It makes A' a left A-space ($\alpha \cdot \alpha' = \varphi(\alpha)\alpha'$). If $\zeta : P \to X$ is a Γ-equivariant A-principal bundle, $P \times_A A'$ is, in an obvious way, the total space of a Γ-equivariant A'-principal bundle $\varphi_*\zeta$. The tautological homeomorphism $E\Gamma \times_\Gamma (P \times_A A') \approx (E\Gamma \times_\Gamma P) \times_A A'$ gives an isomorphism of A'-principal bundles

$$(\varphi_*\zeta)_\Gamma \approx \varphi_*\zeta_\Gamma. \tag{7.5.1}$$

By a *Γ-equivariant $\mathbb{K}$-vector bundle* ξ over X ($\mathbb{K} = \mathbb{R}$ or $\mathbb{C}$), we mean a Γ-equivariant map $p : E = E(\xi) \to X$ which is a $\mathbb{K}$-vector bundle, such that, for each $\gamma \in \Gamma$ and each $x \in X$, the map $y \mapsto \gamma y$ is a $\mathbb{K}$-linear map from $p^{-1}(x)$ to $p^{-1}(\gamma x)$. The tangent bundle to a smooth Γ-manifold is an example for $\mathbb{K} = \mathbb{R}$.

It is convenient here to see a $\mathbb{K}$-vector bundle ξ of rank r as associated to a principal $GL(\mathbb{K}^r)$-bundle, the bundle $\mathrm{Fra}(\xi)$ of *frames* of ξ. Its total space is

$$\mathrm{Fra}(\xi) = \{\nu : \mathbb{K}^r \to E(\xi) \mid \nu \text{ is a } \mathbb{K} - \text{linear isomorphism onto some fiber of } \xi\}.$$

with the map $p_{\mathrm{Fra}} : \mathrm{Fra}(\xi) \to X$ given by $p_{\mathrm{Fra}}(\nu) = p \circ \nu(0)$. The right $GL(\mathbb{K}^r)$-action on $\mathrm{Fra}(\xi)$ is by precomposition (we use the same notation for the bundle $\mathrm{Fra}(\xi)$ and for its total space). The evaluation map sending $[\nu, t] \in \mathrm{Fra}(\xi) \times_{GL(\mathbb{K}^r)} \mathbb{K}^r$ to $\nu(t) \in E(\xi)$ defines an isomorphism of $\mathbb{K}$-vector bundles

$$\mathrm{Fra}(\xi) \times_{GL(\mathbb{K}^r)} \mathbb{K}^r \xrightarrow{\approx} E(\xi). \tag{7.5.2}$$

For more details and developments, see 9.1.9 If ξ is a Γ-equivariant $\mathbb{K}$-vector bundle, then Γ acts on $\mathrm{Fra}(\xi)$ by $(\gamma\nu)(t) = \gamma \cdot \nu(t)$. Hence, $\mathrm{Fra}(\xi)$ is a Γ-equivariant principal $GL(\mathbb{K}^r)$-bundle and (7.5.2) is an isomorphism of Γ-equivariant $\mathbb{K}$-vector bundles. The tautological homeomorphism

$$E\Gamma \times_\Gamma (\mathrm{Fra}(\xi) \times_{GL(\mathbb{K}^r)} \mathbb{K}^r) \approx (E\Gamma \times_\Gamma \mathrm{Fra}(\xi)) \times_{GL(\mathbb{K}^r)} \mathbb{K}^r$$

implies that

$$E(\xi)_\Gamma \approx \mathrm{Fra}(\xi)_\Gamma \times_{GL(\mathbb{K}^r)} \mathbb{K}^r. \tag{7.5.3}$$

Using Lemma 7.5.2, this proves the following lemma.

Lemma 7.5.6 *Let $\xi = (p : E(\xi) \to X)$ be a Γ-equivariant $\mathbb{K}$-vector bundle of rank r. Then, the map $p_\Gamma : E(\xi)_\Gamma \to X_\Gamma$ is a $\mathbb{K}$-vector bundle of rank r, denoted by ξ_Γ. Moreover, ξ is isomorphic to the induced vector bundle $i^*\xi_\Gamma$.*

Let $\mu : \Gamma' \to \Gamma$ and $f : X' \to X$ as for Lemma 7.5.5. If ξ is a Γ-equivariant $\mathbb{K}$-vector bundle of rank r over X, then $f^*\xi$ is a Γ'-equivariant $\mathbb{K}$-vector bundle over X' of the same rank. One has an isomorphism of Γ'-equivariant $GL(\mathbb{K}^r)$-principal bundles $f^*_{\Gamma',\Gamma}\mathrm{Fra}(\xi) \approx \mathrm{Fra}(f^*\xi)$. Therefore, Lemma 7.5.5 gives an isomorphism of $\mathbb{K}$-vector bundles

$$(f^*\xi)_{\Gamma'} \approx f^*_{\Gamma',\Gamma}\xi_\Gamma. \tag{7.5.4}$$

The correspondence $\xi \mapsto \xi_\Gamma$ commutes with some operations on vector bundles, like the Whitney sum or the tensor product. We first define these operations in the category of Γ-equivariant $\mathbb{K}$-vector bundles. Let ξ (respectively ξ') be two such bundles over a Γ-space X, of ranks r (respectively r'). Set $F = \mathrm{Fra}(\xi)$, $F' = \mathrm{Fra}(\xi')$, $G = GL(\mathbb{K}^r)$ and $G' = GL(\mathbb{K}^{r'})$. The diagonal inclusion $\Delta_X : X \to X \times X$ is Γ-equivariant with respect to the diagonal homomorphism $\Delta_\Gamma : \Gamma \to \Gamma \times \Gamma$. Hence, $\Delta_X^*(F \times F')$ is a Γ-equivariant principal $(G \times G')$-bundle. The linear $(G \times G')$-action on $\mathbb{K}^r \oplus \mathbb{K}^{r'}$ given by $(R, R') \cdot (v, v') = (Rv, R'v')$ defines a continuous homomorphism $\varphi^\oplus : G \times G' \to GL(\mathbb{K}^r \oplus \mathbb{K}^{r'})$. This permits us to define the Whitney sum

$$\xi \oplus \xi' = \varphi^\oplus_* \Delta_X^*(F \times F') \times_{GL(\mathbb{K}^r \oplus \mathbb{K}^{r'})} (\mathbb{K}^r \oplus \mathbb{K}^{r'})$$

as a Γ-equivariant $\mathbb{K}$-vector space. The Γ-equivariant tensor product is defined accordingly

$$\xi \otimes \xi' = \varphi_*^{\otimes} \Delta_X^*(F \times F') \times_{GL(\mathbb{K}^r \otimes \mathbb{K}^{r'})} (\mathbb{K}^r \otimes \mathbb{K}^{r'}), \tag{7.5.5}$$

using the homomorphism $\varphi^{\otimes} : G \times G' \to GL(\mathbb{K}^r \otimes \mathbb{K}^{r'})$ induced by the unique linear action of $G \times G$ on $\mathbb{K}^r \otimes \mathbb{K}^{r'}$ satisfying $(R, R') \cdot (v \otimes v') = (Rv \otimes R'v')$.

Lemma 7.5.7 $(\xi \oplus \xi')_\Gamma \approx \xi_\Gamma \oplus \xi'_\Gamma$ *and* $(\xi \otimes \xi')_\Gamma \approx \xi_\Gamma \otimes \xi'_\Gamma$.

Proof One has

$$\begin{aligned}(\xi \oplus \xi')_\Gamma &= \big[\varphi_*^{\oplus} \Delta_X^*(F \times F') \times_{GL(\mathbb{K}^r \oplus \mathbb{K}^{r'})} (\mathbb{K}^r \oplus \mathbb{K}^{r'})\big]_\Gamma \\ &\approx \big[\varphi_*^{\oplus} \Delta_X^*(F \times F')\big]_\Gamma \times_{GL(\mathbb{K}^r \oplus \mathbb{K}^{r'})} (\mathbb{K}^r \oplus \mathbb{K}^{r'}) \quad \text{by (7.5.3)} \\ &\approx \varphi_*^{\oplus} \big[\Delta_X^*(F \times F')\big]_\Gamma \times_{GL(\mathbb{K}^r \oplus \mathbb{K}^{r'})} (\mathbb{K}^r \oplus \mathbb{K}^{r'}) \quad \text{by (7.5.1)}\end{aligned}$$

while

$$\xi_\Gamma \oplus \xi'_\Gamma = \varphi_*^{\oplus} \Delta_{X_\Gamma}^*(F_\Gamma \times F'_\Gamma) \times_{GL(\mathbb{K}^r \oplus \mathbb{K}^{r'})} (\mathbb{K}^r \oplus \mathbb{K}^{r'}).$$

Therefore, it is enough to construct an isomorphism of principal $(G \times G')$-bundles

$$\Delta_{X_\Gamma}^*(F_\Gamma \times F'_\Gamma) \approx \big[\Delta_X^*(F \times F')\big]_\Gamma. \tag{7.5.6}$$

This will prove the lemma for Whitney sum, and also for the tensor product, using $\varphi^{\otimes}$ instead of $\varphi^{\oplus}$.

As $\Delta_X : X \to X \times X$ is Γ-equivariant with respect to the diagonal homomorphism $\Delta_\Gamma : \Gamma \to \Gamma \times \Gamma$, it induces a map $\Phi = (\Delta_X)_{\Gamma,\Gamma\times\Gamma} : X_\Gamma \to (X \times X)_{\Gamma\times\Gamma}$. By Lemma 7.5.5, one has

$$\big[\Delta_X^*(F \times F')\big]_\Gamma \approx \Phi^*(F \times F')_\Gamma. \tag{7.5.7}$$

For Γ-space Z and Z', a natural homotopy equivalence $P : (Z \times Z')_\Gamma \xrightarrow{\simeq} Z_\Gamma \times Z'_\Gamma$ was constructed in (7.4.1). For $Z = F$ and $Z' = F'$, we thus get a homotopy equivalence $\tilde{P} : (F \times F')_\Gamma \xrightarrow{\simeq} F_\Gamma \times F'_\Gamma$ which is $(G \times G')$-equivariant. The diagram

$$\begin{array}{ccccc} \Phi^*(F \times F')_\Gamma & \xrightarrow{\tilde{\Phi}} & (F \times F')_\Gamma & \xrightarrow[\simeq]{\tilde{P}} & F_\Gamma \times F'_\Gamma \\ \downarrow & & \downarrow & & \downarrow \\ X_\Gamma & \xrightarrow{\Phi} & (X \times X)_{\Gamma\times\Gamma} & \xrightarrow[\simeq]{P} & X_\Gamma \times X_\Gamma \end{array}$$

is commutative, thus each square is a morphism of principal $(G \times G')$-bundles. By the definition of P in Sect. 7.4, one has $P \circ \Phi = \Delta_{X_\Gamma}$. Therefore, $\Phi^*(F \times$

$F')_\Gamma \approx \Delta^*_{X_\Gamma}(F_\Gamma \times F'_\Gamma)$. This together with (7.5.7) gives the required isomorphism of (7.5.6).

Let ξ be a Γ-vector bundle of rank r over X. The *equivariant Euler class* $e_\Gamma(\xi)$ is the Euler class of ξ_Γ:

$$e_\Gamma(\xi) = e(\xi_\Gamma) \in H^r_\Gamma(X).$$

Example 7.5.8 Let $\chi : \Gamma \to GL(V)$ be a representation of Γ on a vector space V of dimension r. This makes V a Γ-space. One can also see V as a vector bundle with basis a point. This gives a Γ-vector bundle $\check{\chi}$ of rank r over a point and then a vector bundle $\check{\chi}_\Gamma = (V_\Gamma \to B\Gamma)$ of rank r over $B\Gamma$. Its equivariant Euler class $e_\Gamma(\check{\chi})$ is an element of $H^r_\Gamma(pt) = H^r(B\Gamma)$. Its vanishing is related to the existence of a non-zero fixed vector in V, as seen in the following lemma.

Lemma 7.5.9 *If* $V^\Gamma \neq \{0\}$, *then* $e_\Gamma(\check{\chi}) = 0$.

Proof A non-zero fixed vector $0 \neq v \in V^\Gamma$ determines a nowhere-zero section of $p : V_\Gamma \to B\Gamma$ (see (5) p. 275). This implies that $e_\Gamma(\check{\chi}) = 0$ by Lemma 4.7.39. □

We now give a few recipes to compute an equivariant Euler class. A Γ-equivariant vector bundle $p : E \to X$ is called *rigid* if the Γ-action on X is trivial.

Lemma 7.5.10 *Let* $\xi = (p : E \to X)$ *be a rigid* Γ*-vector bundle of rank* r. *Let* $\chi : \Gamma \to GL(E_x)$ *be the representation of* Γ *on the fiber* E_x *over* $x \in X$. *Suppose that* $\tilde{H}^k(X) = 0$ *for* $k < r$. *Then, the equation*

$$e_\Gamma(\xi) = \mathbf{1} \times e(\xi) + e_\Gamma(\check{\chi}) \times \mathbf{1}$$

holds in $H^r_\Gamma(X) = H^r(B\Gamma \times X)$.

Proof The inclusion $j : \{x\} \to X$ satisfies $j^*\xi = \check{\chi}$. It induces $j_\Gamma : B\Gamma = \{x\}_\Gamma \to X_\Gamma$ satisfying $j^*_\Gamma \xi_\Gamma = \check{\chi}_\Gamma$. Hence, $H^* j_\Gamma(e_\Gamma(\xi)) = e_\Gamma(\check{\chi})$. By construction of ξ_Γ, one has $i^*\xi_\Gamma = \xi$, where $i : X \to X_\Gamma$ denotes the inclusion. Hence, $H^*i(e_\Gamma(\xi)) = e(\xi)$. Using the homeomorphism $X_\Gamma \approx B\Gamma \times X$, the maps j_Γ and i are slice inclusions. The lemma then follows from Corollary 4.7.3. □

Let $\chi : \Gamma \to O(1)$ be a continuous homomorphism, permitting Γ to act on $\mathbb{R}$. This gives a Γ-line bundle $\check{\chi}$ over a point (see Example 7.5.8). Its equivariant Euler class lives in $H^1_\Gamma(pt) = H^1(B\Gamma)$. As χ is continuous, it factors through the homomorphism $\pi_0\chi : \pi_0(\Gamma) \to O(1) \approx \mathbb{Z}_2$. As $E\Gamma$ is contractible, the homotopy exact sequence of $\Gamma \to E\Gamma \to B\Gamma$ identifies $\pi_0(\Gamma)$ with $\pi_1(B\Gamma)$. One thus gets (using Lemma 4.3.1) the isomorphism

$$\hom_{cont}(\Gamma, O(1)) \approx \hom(\pi_0(\Gamma), \mathbb{Z}_2) \approx \hom(\pi_1(B\Gamma), \mathbb{Z}_2) \approx H^1(B\Gamma) = H^1_\Gamma(pt). \tag{7.5.8}$$

Lemma 7.5.11 *Under the isomorphism of* (7.5.8), *one has* $e_\Gamma(\check{\chi}) = \chi$.

Proof Note that $E\Gamma \times_\Gamma O(1) \to B\Gamma$ is a 2-fold covering and that

$$E(\check{\chi}) = E\Gamma \times_\Gamma \mathbb{R} = (E\Gamma \times_\Gamma O(1)) \times_{O(1)} \mathbb{R}.$$

Then, $E\Gamma \times_\Gamma O(1) \to B\Gamma$ is the sphere bundle $S(\check{\chi}_\Gamma)$ for the Euclidean structure on $\check{\chi}_\Gamma$ given by the standard Euclidean structure on $\mathbb{R}$. By Proposition 4.7.36, the Euler class $e_\Gamma(\check{\chi})$ coincides with the characteristic class $w(S(\check{\chi}_\Gamma))$ of the two-covering $S(\check{\chi}_\Gamma) \to B\Gamma$. But $E\Gamma \times_\Gamma O(1) \approx B\ker\chi$ and thus

$$\pi_1(S(\check{\chi}_\Gamma)) = \pi_1(B\ker\chi) = \pi_0(\ker\chi) = \ker \pi_0\chi = \ker \pi_1 B\chi.$$

Therefore, $S(\check{\chi}_\Gamma) \to B\Gamma$ is the 2-covering with fundamental group to $\ker \pi_1 B\chi \subset \pi_1(B\Gamma)$. Diagram (4.3.4) then implies that $w(S(\check{\chi}_\Gamma)) = \chi$. □

A discrete group Γ is a 2-*torus* if it is finitely generated and if every element has order 2. It follows that Γ is isomorphic to $\{\pm 1\}^m$, the integer m being called the *rank* of Γ. As seen in Sect. 7.4,

$$B\Gamma \simeq B(\{\pm 1\}^m) \simeq (B\{\pm 1\})^m \simeq (\mathbb{R}P^\infty)^m. \tag{7.5.9}$$

Hence, $H^*_\Gamma(pt)$ is isomorphic to a polynomial algebra

$$H^*_\Gamma(pt) \approx H^*((\mathbb{R}P^\infty)^m) \approx H^*(\mathbb{R}P^\infty) \otimes \cdots \otimes H^*(\mathbb{R}P^\infty) \approx \mathbb{Z}_2[u_1, \dots, u_m], \tag{7.5.10}$$

where degree$(u_i) = 1$. Under the identifications of (7.5.8), $u_i \in H^1\Gamma(pt)$ corresponds to the homomorphism $\{\pm 1\}^n \to \{\pm 1\}$ which is the projection onto the ith factor.

Example 7.5.12 Let Γ be the 2-torus formed by the diagonal matrices of $O(n)$. Then $H^*(B\Gamma) \approx \mathbb{Z}_2[u_1, \dots, u_n]$, where $u_i \in H^1(B\Gamma)$ corresponds to the homomorphism $\pi_i : \Gamma \to \{\pm 1\}$ given by the i-th diagonal entry. The inclusion $\chi : \Gamma \to O(n)$ provides a Γ-equivariant vector bundle $\check{\chi}$ of rank n over a point. Note that χ is a direct sum of 1-dimensional representations π_i. Using Lemmas 7.5.7 and 7.5.11, we get that

$$w(\check{\chi}) = \prod_{i=1}^{n} (1 + u_i).$$

Lemma 7.5.13 *Let* $\chi : \Gamma \to GL(V)$ *be a representation of a* 2-*torus* Γ *on a finite dimensional vector space* V. *Then the following two conditions are equivalent.*

(1) $V^\Gamma = \{0\}$.
(2) $e_\Gamma(\check{\chi}) \neq 0$.

Proof That (2) implies (1) is given by Lemma 7.5.9. To prove the converse, we use the fact that χ is diagonalizable, with eigenvalues ± 1: indeed, this is true for a linear involution (see Example 7.1.14) and, if $a, b \in GL(V)$ commute, then b preserves the eigenspaces of a. Thus, $V = V_1 \oplus \cdots \oplus V_r$ and Γ acts on V_j through a homomorphism $\chi_j : \Gamma \to \{\pm 1\} = O(1)$. Hence, $\check{\chi}$ is the Whitney sum $\check{\chi}_1 \oplus \cdots \oplus \check{\chi}_r$. Therefore,

$$\begin{aligned} e_\Gamma(\check{\chi}) &= e(\check{\chi}_\Gamma) \\ &= e((\check{\chi}_1)_\Gamma \oplus \cdots \oplus (\check{\chi}_r)_\Gamma) && \text{by Lemma 7.5.7} \\ &= e((\check{\chi}_1)_\Gamma) \smile \cdots \smile e((\check{\chi}_r)_\Gamma) && \text{by Proposition 4.7.40} \\ &= \chi_1 \smile \cdots \smile \chi_r && \text{by Lemma 7.5.11.} \end{aligned}$$

The condition $V^\Gamma = \{0\}$ implies that none of the χ_j vanishes. As $H^*(B\Gamma)$ is a polynomial algebra, this implies that $e_\Gamma(\check{\chi}) \neq 0$. □

Proposition 7.5.14 *Let $\xi = (E \xrightarrow{p} X)$ be a rigid Γ-vector bundle of rank r, where Γ is a 2-torus. Suppose that E^Γ consists only of the image of the zero section. Then, the cup-product with the equivariant Euler class*

$$H^*_\Gamma(X) \xrightarrow{-\smile e_\Gamma(\xi)} H^{*+r}_\Gamma(X)$$

is injective.

Proof Without loss of generality, we may suppose that X is path-connected and non-empty. Let $x \in X$. Consider the slice inclusion $s : B\Gamma \to B\Gamma \times X$ with image $B\Gamma \times \{x\}$. Then, $H^*_\Gamma s(e_\Gamma(\xi)) = e_\Gamma(\xi_x)$, where $\xi_x = (E_x \to \{x\})$ is the restriction of ξ over the point x. As X is path-connected, Lemma 4.7.2 implies that the component of $e_\Gamma(\xi) \in H^r(B\Gamma \otimes X)$ in $H^r(B\Gamma) \times H^0(X)$ is equal to $e_\Gamma(\xi_x) \times \mathbf{1}$ (as $B\Gamma$ is of finite cohomology type, we identify $H^r(B\Gamma \times X)$ with $H^*(B\Gamma) \otimes H^*(X)$ by the Künneth theorem).

Now let $0 \neq a \in H^k(B\Gamma \times X)$. We isolate its *minimal component* $a_{\min}$ by the formula

$$a = a_{\min} + A$$

with

$$0 \neq a_{\min} \in H^{k-p}(B\Gamma) \otimes H^p(X) \quad \text{and} \quad A \in \bigoplus_{q>p} H^{k-q}(B\Gamma) \otimes H^q(X).$$

Then

$$a \smile e_\Gamma(\xi) = a_{\min} \smile (e_\Gamma(\xi_x) \otimes \mathbf{1}) + A'$$

with

$$a_{\min} \smile (e_\Gamma(\xi_x) \otimes \mathbf{1}) \in H^{k-p+r}(B\Gamma) \otimes H^p(X) \text{ and } A' \in \bigoplus_{q>p} H^{k-q+r}(B\Gamma) \otimes H^q(X).$$

Therefore it suffices to prove that $a_{\min} \smile (e_\Gamma(\xi_x) \otimes \mathbf{1}) \neq 0$. The condition on E^Γ implies that $E_x^\Gamma = \{0\}$ and thus, by Lemma 7.5.13, $e_\Gamma(\xi_x) \neq 0$. As $H^*(B\Gamma)$ is a polynomial algebra, this implies that $a \smile e_\Gamma(\xi) \neq 0$. Let $\mathcal{B}$ be a basis of $H^{k-p}(B\Gamma)$ and $\mathcal{C}$ be a basis of $H^p(X)$. Then, $\{b \otimes c \mid (b,c) \in \mathcal{B} \times \mathcal{C}\}$ is a basis of $H^{k-p}(B\Gamma) \otimes H^p(X)$. As $H^*(B\Gamma)$ is a polynomial algebra, the family $\{b \smile e_\Gamma(\xi_x) \mid b \in \mathcal{B}\}$ is free in $H^{k-p+r}(B\Gamma)$. Hence, if

$$0 \neq a_{\min} = \sum_{(b,c)\in\mathcal{B}\times\mathcal{C}} \lambda_{bc}\,(b \otimes c) \qquad (\lambda_{bc} \in \mathbb{Z}_2),$$

then

$$a_{\min} \smile (e_\Gamma(\xi_x) \otimes \mathbf{1}) = \sum_{(b,c)\in\mathcal{B}\times\mathcal{C}} \lambda_{bc}\,\big((b \smile e_\Gamma(\xi_x)) \otimes c\big) \neq 0\,. \qquad \square$$

Statements 7.5.11, 7.5.13 and 7.5.14 have analogues, replacing $O(1)$ by $SO(2)$ and 2-tori by tori. Let Γ be a topological group and let $\chi : \Gamma \to SO(2)$ be a continuous homomorphism, making Γ to act on $\mathbb{R}^2$. This gives a Γ-vector bundle $\check{\chi}$ of rank 2 over a point (see Example 7.5.8). Its equivariant Euler class lives in $H^2_\Gamma(pt) = H^2(B\Gamma)$. As $SO(2) \approx S^1$, one has $BSO(2) \approx \mathbb{C}P^\infty$ (see Example 7.2.1). Define

$$\kappa(\chi) = H^* B\chi(\iota) \in H^2_\Gamma(pt) = H^2(B\Gamma), \tag{7.5.11}$$

where ι is the non-zero element of $H^2(BSO(2)) = \mathbb{Z}_2$.

Lemma 7.5.15 $e_\Gamma(\check{\chi}) = \kappa(\chi)$.

Proof The homomorphism $\chi : \Gamma \to SO(2)$ makes $SO(2)$ a Γ-space. The map $E\chi : E\Gamma \to ESO(2)$ descends to continuous maps $E\Gamma \times_\Gamma SO(2) \to ESO(2)$ and there is a commutative diagram

$$\begin{array}{ccccc} E\Gamma \times_\Gamma SO(2) & \longrightarrow & ESO(2) & \overset{\approx}{\longleftrightarrow} & S^\infty \\ \downarrow & & \downarrow & & \downarrow \\ B\Gamma & \xrightarrow{B\chi} & BSO(2) & \overset{\approx}{\longleftrightarrow} & \mathbb{C}P^\infty \end{array} \tag{7.5.12}$$

Note that $E\Gamma \times_\Gamma SO(2) \to B\Gamma$ is an $SO(2)$-principal bundle and that

$$E(\check{\chi}) = E\Gamma \times_\Gamma \mathbb{R}^2 = (E\Gamma \times_\Gamma SO(2)) \times_{SO(2)} \mathbb{R}^2.$$

As $SO(2) \approx S^1$, $E\Gamma \times_\Gamma SO(2)$ is the sphere bundle $S(\check{\chi}_\Gamma)$ for the Euclidean structure on $\check{\chi}_\Gamma$ given by the standard Euclidean structure on $\mathbb{R}^2$. Diagram (7.5.12) implies that $S(\check{\chi}_\Gamma)$ is induced by $B\chi$ from the Hopf bundle $S^\infty \to \mathbb{C}P^\infty$, whose Euler class in $\iota \in H^2(\mathbb{C}P^\infty)$ (see Proposition 6.1.6). Hence,

$$e_\Gamma(\check{\chi}) = e(S(\check{\chi}_\Gamma)) = H^* B\chi(\iota) = \kappa(\chi). \qquad \square$$

A *torus* Γ is a Lie group isomorphic to $(S^1)^m$, the integer m being called the *rank of* Γ. For instance, $SO(2)$ is a torus of rank 1. As seen in Sect. 7.4,

$$B\Gamma \simeq B((S^1)^m) \simeq (BS^1)^m \simeq (\mathbb{C}P^\infty)^m. \qquad (7.5.13)$$

Hence, $H^*_\Gamma(pt)$ is isomorphic to a polynomial algebra

$$H^*_\Gamma(pt) \approx H^*((\mathbb{C}P^\infty)^m) \approx H^*(\mathbb{C}P^\infty) \otimes \cdots \otimes H^*(\mathbb{C}P^\infty) \approx \mathbb{Z}_2[v_1, \ldots, v_m], \qquad (7.5.14)$$

where degree$(v_i) = 2$. One has $v_i = \kappa(\chi_i)$ where $\chi_i : (S^1)^n \to S^1 \approx SO(2)$ is the projection onto the ith factor. If Γ is a torus, its *associated* 2*-torus* Γ_2 is the subgroup of elements of order 2 in Γ.

Lemma 7.5.16 *Let Γ be a torus and Γ_2 be its associated 2-torus. Let $\chi : \Gamma \to GL(V)$ be a representation of Γ on a finite dimensional vector space V. Then the following two conditions are equivalent.*

(1) $V^{\Gamma_2} = \{0\}$.
(2) $e_\Gamma(\check{\chi}) \neq 0$.

Moreover, if (1) or (2) holds true, then dim V *is even.*

Proof As Γ is a torus, $\chi(\Gamma)$ is contained in a maximal torus $\mathcal{T}$ of $GL(V)$. Those are all conjugate (see [21, Sect. IV.1]). If dim $V = 2s + 1$, there is an isomorphism $V \approx \mathbb{R}^2 \oplus \cdots \oplus \mathbb{R}^2 \oplus \mathbb{R}$ intertwining $\mathcal{T}$ with $SO(2) \times \cdots SO(2) \times 1$ (see [21, Chap. IV, Theorem 3.4]). This contradicts the condition $V^\Gamma \neq \{0\}$. We can then suppose that dim $V = 2s$, in which case there is an isomorphism $V \approx \mathbb{R}^2 \oplus \cdots \oplus \mathbb{R}^2$ conjugating $\mathcal{T}$ with $SO(2) \times \cdots SO(2)$. The homomorphism χ takes the form $\chi = (\chi_1, \ldots, \chi_s)$ where $\chi_j : \Gamma \to SO(2)$. Hence $\check{\chi} = \check{\chi}_1 \oplus \cdots \oplus \check{\chi}_s$ and, using Lemma 7.5.7 and Proposition 4.7.40,

$$e_\Gamma(\check{\chi}) = e_\Gamma(\check{\chi}_1) \smile \cdots \smile e_\Gamma(\check{\chi}_s) = \kappa(\chi_1) \smile \cdots \smile \kappa(\chi_s),$$

the last equality coming from Lemma 7.5.15. Since $H^*(B\Gamma)$ is a polynomial algebra, the condition $e_\Gamma(\check{\chi}) \neq 0$ is equivalent to $\kappa(\chi_j) \neq 0$ for all j. The condition $V^\Gamma = \{0\} = V^{\Gamma_2}$ is equivalent to $V_j^\Gamma = \{0\} = V_j^{\Gamma_2}$ for all j, where V_j is the 2-dimensional vector space corresponding to the jth factor $\mathbb{R}^2$ in the decomposition of V.

We are thus reduced to the case dim $V = 2$ and $\chi : \Gamma \to SO(2)$. We start with preliminaries. Choose isomorphisms $\Gamma \approx (S^1)^m$, $SO(2) \approx S^1$ and $S^1 \approx \mathbb{R}/\mathbb{Z}$. We get a commutative diagram

$$\begin{array}{ccccc} \mathbb{Z}^m & \rightarrowtail & \mathbb{R}^m & \twoheadrightarrow & (S^1)^m \\ \downarrow{\scriptstyle \pi_1\chi} & & \downarrow{\scriptstyle \tilde{\chi}} & & \downarrow{\scriptstyle \chi} \\ \mathbb{Z} & \rightarrowtail & \mathbb{R} & \twoheadrightarrow & S^1 \end{array} \qquad (7.5.15)$$

where the vertical arrows are homomorphisms. Therefore, $\tilde{\chi}(x_1, \dots, x_m) = \sum_i b_i x_i$ with $b_i \in \mathbb{Z}$ and

$$\chi(\gamma_1, \dots, \gamma_m) = \gamma_1^{b_1} \cdots \gamma_m^{b_m}. \tag{7.5.16}$$

We deduce that

$$V^\Gamma = \{0\} \Longleftrightarrow \pi_1\chi \text{ non-trivial } \Longleftrightarrow \chi \text{ surjective} \Longleftrightarrow b_j \neq 0 \,\forall\, j. \tag{7.5.17}$$

If χ is surjective, one gets a fiber bundle $\ker\chi \to \Gamma \to S^1$ and, using its homotopy exact sequence and (7.5.16), we get

$$V^\Gamma = \{0\} \Longrightarrow \text{coker}\,\pi_1\chi \approx \pi_0(\ker\chi) \approx \mathbb{Z}/\gcd(b_1, \dots, b_m)\mathbb{Z}. \tag{7.5.18}$$

Since $0 \in V^\Gamma \subset V^{\Gamma_2}$, the condition $V^{\Gamma_2} = \{0\}$ implies that $V^\Gamma = \{0\}$. Hence, using (7.5.15)–(7.5.18),

$$\begin{aligned} V^{\Gamma_2} = \{0\} &\Longleftrightarrow \Gamma_2 \subset \ker\chi \\ &\Longleftrightarrow \tfrac{1}{2}\mathbb{Z} \subset \ker\pi_1\chi \\ &\Longleftrightarrow 2 \mid \gcd(b_1, \dots, b_m) \\ &\Longleftrightarrow \hom(\pi_0(\ker\chi); \mathbb{Z}_2) = 0 \end{aligned}. \tag{7.5.19}$$

As in the proof of Lemma 7.5.15, we consider $S = E\Gamma \times_\Gamma S^1$, which is the total space for the sphere bundle $S(\check{\chi})$. One has a commutative diagram

$$\begin{array}{ccccc} \Gamma & \xrightarrow{\chi} & E\Gamma & \longrightarrow & B\Gamma \\ \downarrow & & \downarrow & & \downarrow{=} \\ S^1 & \longrightarrow & S & \longrightarrow & B\Gamma \end{array}$$

whose rows are fiber bundles. Passing to the homotopy exact sequences, we get a commutative diagram

$$\begin{array}{ccccccc} \pi_2(B\Gamma) & \xrightarrow{\approx} & \pi_1(\Gamma) & \longrightarrow & 0 & & \\ \downarrow{=} & & \downarrow{\pi_1\chi} & & \downarrow & & \\ \pi_2(B\Gamma) & \xrightarrow{\approx} & \pi_1(S^1) & \longrightarrow & \pi_1(S) & \longrightarrow & 0 \end{array}$$

whose rows are exact sequences. Hence,

$$\pi_1(S) \approx \text{coker}\,\pi_1\chi \tag{7.5.20}$$

Now, the Gysin sequence for $S \to B\Gamma$ gives

$$\underbrace{H^1(B\Gamma)}_{0} \to H^1(S) \to \underbrace{H^0(B\Gamma)}_{\mathbb{Z}_2} \xrightarrow{\smile e_\Gamma(\check{\chi})} H^2(B\Gamma). \tag{7.5.21}$$

By Lemma 7.5.9, one knows that $e_\Gamma(\check{\chi}) \neq 0$ implies $V^\Gamma = \{0\}$. Therefore, using (7.5.18)–(7.5.21),

$$\begin{aligned} e_\Gamma(\check{\chi}) \neq 0 &\Longleftrightarrow H^1(S) = 0 \\ &\Longleftrightarrow \hom(\pi_1(S); \mathbb{Z}_2) = 0 \\ &\Longleftrightarrow \hom(\text{coker}\, \pi_1\chi; \mathbb{Z}_2) = 0\,. \\ &\Longleftrightarrow \hom(\pi_0(\ker \chi); \mathbb{Z}_2) = 0 \\ &\Longleftrightarrow V^{\Gamma_2} = \{0\} \end{aligned} \tag{7.5.22}$$

□

Proposition 7.5.17 *Let Γ be a torus and Γ_2 be its associated 2-torus. Let $\xi = (p : E \to X)$ be a rigid Γ-vector bundle of rank r. Suppose that E^{Γ_2} consists only of the image of the zero section. Then r is even and the cup-product with the equivariant Euler class*

$$H^*_\Gamma(X) \xrightarrow{-\smile e_\Gamma(\xi)} H^{*+r}_\Gamma(X)$$

is injective.

Proof The proof is the same as that of Proposition 7.5.14, using Lemma 7.5.16 instead of Lemma 7.5.13. □

7.6 Equivariant Morse-Bott Theory

Let $f : M \to \mathbb{R}$ be a smooth function defined on a smooth manifold M. A point $x \in M$ is *critical* for f if $df(x) = 0$. Let Crit $f \subset M$ be the subspace of critical points for f. Then, $f(\text{Crit}\, f) \subset \mathbb{R}$ is the set of *critical values of f*. We say that f is *Morse-Bott* if the following two conditions hold

- Crit f is a disjoint union of submanifolds. A connected component of Crit f is called a *critical manifold of f*.
- the kernel of the Hessian $\mathcal{H}_x$ at a critical point x equals the tangent space to the critical manifold N containing x.

This definition coincides with that of a *Morse function* when Crit f is a discrete set. See e.g. [13, 18, 95, 149, 151] for presentations of Morse and Morse-Bott theory. The *index* of $x \in$ Crit f is the number of negative eigenvalues of $\mathcal{H}_x$. This number is constant over a critical manifold and thus defines a function ind : $\pi_0(\text{Crit}\, f) \to \mathbb{N}$. Also, the normal bundle νN to a critical manifold N decomposes into a Whitney sum $\nu N \approx \nu^- N \oplus \nu^+ N$ of the *negative* and *positive normal bundles*, i.e. the bundles spanned at each $x \in N$ respectively by the negative and positive eigenspaces of $\mathcal{H}_x$. Note that rank $\nu^-(N) = \text{ind}\, N$.

A (continuous) map $g : X \to Y$ is called *proper* if the pre-image of any compact set is compact. For instance, if X is compact, then any map g is proper. Let $f : M \to \mathbb{R}$ be a proper Morse-Bott function and let $a < b$ be two regular values. Define $M_{a,b} = f^{-1}([a, b])$, a compact manifold whose boundary is the union of $M_a = f^{-1}(a)$ and $M_b = f^{-1}(b)$. Denote by $f_{a,b}$ the restriction of f to $M_{a,b}$. The *Morse-Bott polynomial* $\mathfrak{M}_t(M_{a,b})$ is defined by

$$\mathfrak{M}_t(f_{a,b}) = \sum_{N \in \pi_0(\text{Crit } f_{a,b})} t^{\,\text{ind} N} P_t(N)$$

(the sum is finite since, as f is proper, Crit $f_{a,b}$ is compact).

Proposition 7.6.1 (Morse-Bott inequalities) *There is a polynomial R_t, with positive coefficients, such that*

$$\mathfrak{M}_t(f_{a,b}) = \mathfrak{P}_t(M_{a,b}, M_a) + (1 + t)R_t. \tag{7.6.1}$$

Equation (7.6.1) implies that the coefficients of $\mathfrak{M}_t(f_{a,b})$ are greater or equal to those of $\mathfrak{P}_t(M_{a,b}, M_a)$ (whence the name of *Morse-Bott inequalities*). For the equivalence of (7.6.1) with other classical and more subtle forms of the Morse-Bott inequalities (see [13, Sect. 3.4]).

Proof The map $f_{a,b}$ has a finite number of critical values, all in the interior of $[a, b]$. Let $a = a_0 < a_1 < \cdots < a_r = b$ be regular values such that $[a_i, a_i + 1]$ contains a single critical value. We shall prove by induction on i that (7.6.1) holds true for f_{a,a_i}. The induction starts trivially for $i = 0$, with the three terms of (7.6.1) being zero.

As $[a_i, a_{i+1}]$ contains a single critical level, there is a homotopy equivalence

$$M_{a_i,a_{i+1}} \simeq M_{a_i} \cup_{S_i} D_i \tag{7.6.2}$$

where (D_i, S_i) is the disjoint union over $N \in \pi_0(\text{Crit } f_{a_i,a_i+1})$ of the pairs formed by the disk and sphere bundles of $\nu^-(N)$ (see [18, pp. 339–344]). By excision and the Thom isomorphism,

$$H^*(M_{a_i,a_{i+1}}, M_a) \approx H^*(D_i, S_i) \approx \prod_{N \in \pi_0(\text{Crit } f_{a_i,a_i+1})} H^{*-\text{ind } N}(N). \tag{7.6.3}$$

Therefore,

$$\mathfrak{M}_t(f_{a_i,a_i+1}) = \mathfrak{P}_t(M_{a_i,a_i+1}, M_{a_i}). \tag{7.6.4}$$

On the other hand, Corollary 3.1.27 applies to the triple $(M_{a,a_{i+1}}, M_{a,a_i}, M_a)$ gives the equality

$$\mathfrak{P}_t(M_{a,a_{i+1}}, M_a) + (1 + t)\, Q_t = \mathfrak{P}_t(M_{a,a_{i+1}}, M_{a,a_i}) + \mathfrak{P}_t(M_{a,a_i}, M_a), \tag{7.6.5}$$

for some $Q_t \in \mathbb{N}[t]$. By excision and (7.6.4), one gets

$$\mathfrak{P}_t(M_{a,a_{i+1}}, M_{a,a_i}) = \mathfrak{P}_t(M_{a_i,a_i+1}, M_{a_i}) = \mathfrak{M}_t(f_{a_i,a_i+1}). \tag{7.6.6}$$

Thus, (7.6.5) and (7.6.6) provide the induction step. □

A proper Morse-Bott function $f : M \to \mathbb{R}$ is called *perfect* if for any two regular values $a < b$, Eq. (7.6.1) reduces to

$$\mathfrak{M}_t(f_{a,b}) = \mathfrak{P}_t(M_{a,b}, M_a). \tag{7.6.7}$$

The easiest occurrence of perfectness is the following *lacunary principle*.

Lemma 7.6.2 *Suppose that no consecutive powers of f occur in $\mathfrak{M}_t(f)$. Then, f is perfect.*

Proof Suppose that $R_t \neq 0$ in (7.6.1). Then, two successive powers of t occur in $(1+t)R_r$. The same happens then in $\mathfrak{M}_t(f_{a,b})$, and then in $\mathfrak{M}_t(f)$. □

Other simple criteria for perfectness are given by the following three results. For a regular value x of $f : M \to \mathbb{R}$, set $W_x = f^{-1}(-\infty, x]$.

Lemma 7.6.3 *Let $f : M \to \mathbb{R}$ be a proper Morse-Bott function. Then, the following two conditions are equivalent.*

(1) *f is perfect.*
(2) *For any regular values $a < b < c$ of f, the cohomology sequence of the triple (W_c, W_b, W_a) cuts into a global short exact sequence*

$$0 \to H^*(W_c, W_b) \to H^*(W_c, W_a) \to H^*(W_b, W_a) \to 0.$$

Proof Suppose that f is perfect. Then, by excision,

$$\mathfrak{P}_t(W_c, W_a) = \mathfrak{P}_t(M_{a,c}, M_a) = \mathfrak{M}_t(f_{a,c})$$

and analogously for $\mathfrak{P}_t(W_b, W_a)$ and $\mathfrak{P}_t(W_c, W_b)$. As

$$\mathfrak{M}_t(f_{c,a}) = \mathfrak{M}_t(f_{c,b}) + \mathfrak{M}_t(f_{b,a}),$$

one has

$$\mathfrak{P}_t(W_c, W_a) = \mathfrak{P}_t(W_c, W_b) + \mathfrak{P}_t(W_b, W_a).$$

By Corollary 3.1.27 and its proof, this implies that $H^*(W_b, W_a) \to H^*(W_c, W_b)$ is surjective, whence (2).

Conversely, suppose that (2) holds true. For two regular values $a < c$, we prove that $\mathfrak{P}_t(M_{a,c}, M_a) = \mathfrak{P}_t(W_c, W_a) = \mathfrak{M}_t(f_{a,c})$, by induction on the number $n_{a,c}$ of critical values in the segment $[a, c]$. This is trivial for $n_{a,c} = 0$, since then $M_{a,c}$ is then diffeomorphic to $M_a \times [a, c]$ (see [95, Chap. 6, Theorem 2.2]). When $n_{a,c} = 1$, one uses (7.6.4). For the induction step, when $n_{a,c} \geq 2$, choose a regular value $b \in (a, c)$ such that $n_{a,b} = n_{a,c} - 1$. Then,

$$\begin{aligned}\mathfrak{P}_t(W_c, W_a) &= \mathfrak{P}_t(W_c, W_b) + \mathfrak{P}_t(W_b, W_a) && \text{by (2)}\\ &= \mathfrak{M}_t(f_{c,b}) + \mathfrak{M}_t(f_{b,a}) && \text{by induction hypothesis}\\ &= \mathfrak{M}_t(f_{a,c}),\end{aligned}$$

which proves the induction step. □

Lemma 7.6.4 *Let $f : M \to \mathbb{R}$ be a proper Morse-Bott function. Then if for any two regular values $a < b$, one has*

$$\dim H^*(M_{a,b}, M_a) \leq \dim H^*(\mathrm{Crit}\, f_{a,b}) \tag{7.6.8}$$

and f is perfect if and only if (7.6.8) is an equality.

Proof The evaluation of (7.6.1) at $t = 1$ implies (7.6.8) and the equality is equivalent to $R_t = 0$. □

In the case where M is a closed manifold, one has the following result.

Proposition 7.6.5 *Let $f : M \to \mathbb{R}$ be a Morse-Bott function, where M is a closed manifold. Then f is perfect if and only if*

$$\dim H^*(M) = \dim H^*(\mathrm{Crit}\, f). \tag{7.6.9}$$

Proof Equation (7.6.8) implies (7.6.9) when $f(M) \subset (a, b)$. Conversely, let $a < b$ be two regular values of f. Let $a' < a$ and $b' > b$ such that $f(M) \subset (a', b')$. Using Corollary 3.1.28 and excision, we get

$$\begin{aligned}\dim H^*(M) &= \dim H^*(M_{a',b'}, M_{a'})\\ &\leq \dim H^*(M_{a',b}, M'_a) + \dim H^*(M_{a',b'}, M_{a',b})\\ &= \dim H^*(M_{a',b}, M'_a) + \dim H^*(M_{b,b'}, M_b).\end{aligned} \tag{7.6.10}$$

Doing the same for $\dim H^*(M_{a',b}, M'_a)$ and using (7.6.1) gives

$$\begin{aligned}\dim H^*(M) &\leq \dim H^*(M_{a',a}, M'_a) + \dim H^*(M_{a,b}, M_a) + \dim H^*(M_{b,b'}, M_b)\\ &\leq \dim H^*(\mathrm{Crit}\, f_{a',a}) + \dim H^*(\mathrm{Crit}\, f_{a,b}) + \dim H^*(\mathrm{Crit}\, f_{b,b'})\\ &= \dim H^*(\mathrm{Crit}\, f).\end{aligned} \tag{7.6.11}$$

Now, if (7.6.9) holds true, then all the inequalities occurring in (7.6.10) are equalities, including $\dim H^*(M_{a,b}, M_a) = \dim H^*(\mathrm{Crit}\, f_{a,b})$. □

Theorem 7.6.6 *Let M be a smooth Γ-manifold, where Γ is a 2-torus. Let $f : M \to \mathbb{R}$ be a proper Γ-invariant Morse-Bott function which is bounded below. Suppose that* $\mathrm{Crit}\, f = M^{\Gamma}$. *Then*

(1) *f is perfect.*
(2) *M is Γ-equivariantly formal.*
(3) *the restriction morphism $H^*_{\Gamma}(M) \to H^*_{\Gamma}(M^{\Gamma})$ is injective.*

Remark 7.6.7 When $\Gamma = \{\pm 1\}$, Theorem 7.6.6 follows from Smith theory. Indeed, for any regular values $a < b$ of f,

$$\begin{aligned} \dim \mathrm{Crit}\, f_{a,b} &= \dim H^*(M^{\Gamma}_{a,b}) && \text{by hypothesis} \\ &\leq \dim H^*(M_{a,b}) && \text{by Proposition 7.3.7} \\ &\leq \dim \mathrm{Crit}\, f_{a,b} && \text{by Lemma 7.6.4.} \end{aligned}$$

Therefore, the above inequalities are equalities and f is perfect by Lemma 7.6.4 and equivariantly formal by Proposition 7.3.7. Point (3) then follows from Proposition 7.3.9.

Remark 7.6.8 Under the hypotheses of Theorem 7.6.6, when $\Gamma = \{\pm 1\}$ and f is a Morse function, M. Farber and D. Schütz have proven that each integral homology group $H_i(M; \mathbb{Z})$ is free abelian with rank equal to the number of critical points of index i [60, Theorem 4]. By the universal coefficients theorem [82, Theorem 3.2], such a function is perfect.

Proof of Theorem 7.6.6 For a regular value x of f, set $W_x = f^{-1}(-\infty, x]$. We first prove that

$$H^*_{\Gamma}(W_x) \to H^*_{\Gamma}(W^{\Gamma}_x) \text{ is injective for all regular value } x. \tag{7.6.12}$$

This is proven by induction on the number n_x of critical values in the interval $(-\infty, x]$, following the argument of [198, proof of Proposition 2.1]. If $n_x = 0$, then $W_x = \emptyset$ and $H^*_{\Gamma}(W_x) = 0$, which starts the induction as f is bounded below.

Suppose that $n \geq 1$ and that (7.6.12) holds true when $n_x < n$. Let y be a regular value of f with $n_y = n$. Choose $z < y$ such that $n_z = n - 1$ (this is possible since the set of critical values is discrete Morse-Bott function is discrete). As in (7.6.2), one has a homotopy equivalence

$$M_{z,y} \simeq M_y \cup_S D,$$

where (D, S) is the disjoint union over $N \in \pi_0(\mathrm{Crit}\, f_{z,x})$ of the pairs formed by the disk and sphere bundles of $\nu^-(N)$. Using (7.6.3) and the proof of Proposition 4.7.32, we get the commutative diagram

$$\begin{array}{ccccccc}
H^*_\Gamma(W_y, W_z) & \xleftarrow[\text{excision}]{\approx} & H^*_\Gamma(M_{z,y}, M_z) & \xleftarrow[\text{excision}]{\approx} & H^*_\Gamma(D, S) & \xleftarrow[\text{Thom}]{\approx} & \prod_N H_\Gamma^{*-\mathrm{ind}\, N}(N) \\
\downarrow & & \downarrow & & & & \downarrow {\scriptstyle -\smile(e_\Gamma(\nu^-(N)))} \\
H^*_\Gamma(W_y) & \longrightarrow & H^*_\Gamma(M_{z,y}) & \longrightarrow & H^*_\Gamma(M^\Gamma_{z,y}) & \xrightarrow{\approx} & \prod_N H^*_\Gamma(N)
\end{array} \tag{7.6.13}$$

where N runs over $\pi_0(\mathrm{Crit}\, f_{z,y}) = \pi_0(M^\Gamma_{z,y})$. As $M^\Gamma = \mathrm{Crit}\, f$, the linear Γ-action of $\nu^-(N)$ has fixed point set consisting only of the image of the zero section. By Lemma 7.5.13, the right vertical arrow of (7.6.13) is injective. Thus, we deduce from (7.6.13) that $H^*_\Gamma(W_y, W_z) \to H^*_\Gamma(W_y)$ is injective. This cuts the Γ-equivariant cohomology sequence of (W_y, W_z) into short exact sequences. The same cutting occurs for the pair (W^Γ_y, W^Γ_z) using Proposition 3.1.21, and one has a commutative diagram

$$\begin{array}{ccccccccc}
0 & \longrightarrow & H^*_\Gamma(W_y, W_z) & \longrightarrow & H^*_\Gamma(W_y) & \longrightarrow & H^*_\Gamma(W_z) & \longrightarrow & 0 \\
 & & \downarrow {\scriptstyle r_{z,y}} & & \downarrow {\scriptstyle r_y} & & \downarrow {\scriptstyle r_z} & & \\
0 & \longrightarrow & H^*_\Gamma(M^\Gamma_{z,y}) & \longrightarrow & H^*_\Gamma(W^\Gamma_y) & \longrightarrow & H^*_\Gamma(W^\Gamma_z) & \longrightarrow & 0
\end{array} \tag{7.6.14}$$

where the vertical arrows are induced by the inclusions. The left vertical arrow is injective by (7.6.13). Since $n_z = n - 1$, the right one is injective by induction hypothesis. By diagram-chasing, we deduce that the middle vertical arrow is injective, which proves (7.6.12).

Warning: As W^Γ_y *is the disjoint union of* $M^\Gamma_{z,y}$ *and* W^Γ_z*, one has* $H^*_\Gamma(W^\Gamma_y) \approx H^*_\Gamma(M^\Gamma_{z,y}) \oplus H^*_\Gamma(W^\Gamma_z)$. *Consider the image* $\mathrm{Im}\, r_y$ *of* r_y *under this decomposition. The above arguments imply that* $\mathrm{Im}\, r_{z,y} \times 0 \subset \mathrm{Im}\, r_y$. *But, in general* $0 \times \mathrm{Im}\, r_z \not\subset \mathrm{Im}\, r_y$ (*see* Example 7.6.9).

We deduce Point (3) from (7.6.12). Indeed as $M = \bigcup_x W_x$ Corollary 3.1.16 provides a commutative diagram

$$\begin{array}{ccc}
H^*_\Gamma(M) & \xrightarrow{\approx} & \varprojlim_x H^*_\Gamma(W_x) \\
\downarrow & & \downarrow \\
H^*_\Gamma(M^\Gamma) & \xrightarrow{\approx} & \varprojlim_x H^*_\Gamma(W^\Gamma_x)
\end{array} . \tag{7.6.15}$$

As the right vertical arrow is injective by (7.6.12), so is the left one.

For Point (2), we first prove that $\rho_x : H^*_\Gamma(W_x) \to H^*(W_x)$ is surjective for all regular value x. This is also done by induction on n_x, starting trivially when $n_x = 0$. For the induction step, consider as above two regular values $z < y$ such that $n_y = n_z + 1$. The cohomology sequences of the pair (W_y, W_z) give the commutative diagram

$$\begin{array}{ccccccccc} 0 & \longrightarrow & H^k_\Gamma(W_y, W_z) & \longrightarrow & H^k_\Gamma(W_y) & \longrightarrow & H^k_\Gamma(W_z) & \longrightarrow & 0 \\ & & \downarrow \rho_{y,z} & & \downarrow \rho_y & & \downarrow \rho_z & & \\ H^{k-1}(W_z) & \longrightarrow & H^k(W_y, W_z) & \longrightarrow & H^k(W_y) & \longrightarrow & H^k(W_z) & \longrightarrow & H^{k+1}(W_y, W_z) \end{array}, \tag{7.6.16}$$

the top sequence being cut as seen above. Similarly to (7.6.13), we get a commutative diagram

$$\begin{array}{ccccccc} H^*_\Gamma(W_y, W_z) & \xleftarrow[\text{excision}]{\approx} & H^*_\Gamma(M_{z,y}, M_z) & \xleftarrow[\text{excision}]{\approx} & H^*_\Gamma(D, S) & \xleftarrow[\text{Thom}]{\approx} & \prod_N H_\Gamma^{*-\operatorname{ind} N}(N) \\ \downarrow \rho_{y,z} & & \downarrow & & \downarrow & & \downarrow \rho_{\mathrm{Crit}} \\ H^*(W_y, W_z) & \xleftarrow[\text{excision}]{\approx} & H^*(M_{z,y}, M_z) & \xleftarrow[\text{excision}]{\approx} & H^*(D, S) & \xleftarrow[\text{Thom}]{\approx} & \prod_N H_\Gamma^{*-\operatorname{ind} N}(N) \end{array}. \tag{7.6.17}$$

Since $\mathrm{Crit}\, f \subset M^\Gamma$, the map ρ_{Crit} is surjective and so is $\rho_{y,z}$. If ρ_z is surjective by induction hypothesis, a diagram-chase proves that ρ_y is surjective. Now, recall that, for a Γ-space X, $\rho: H^*_\Gamma(X) \to H^*(X)$ is equal to $H^* i$ for some fiber inclusion $i: X \to X_\Gamma$. One can thus consider the Kronecker dual $\rho_*: H_*(X) \to H_*(X_\Gamma)$. One has a commutative diagram

$$\begin{array}{ccc} \varinjlim_x H_*(W_x) & \xrightarrow{\approx} & H_*(M) \\ \varinjlim \rho_{x,*} \downarrow & & \downarrow \rho_* \\ \varinjlim_x H_*((W_x)_\Gamma) & \xrightarrow{\approx} & H_*(M_\Gamma) \end{array}. \tag{7.6.18}$$

As ρ_x is surjective, $\rho_{x,*}$ is injective and thus $\varinjlim \rho_{x,*}$ is injective. By Diagram (7.6.18) and Kronecker duality, $\rho: H^*_\Gamma(M) \to H^*(M)$ is surjective and M is equivariantly formal.

Let us finally prove Point (1). Consider two regular values $z < y$ such that $n_y = n_z + 1$. As in (7.6.16) the vertical maps are surjective, the cohomology sequence of (W_y, W_z) cuts into a global short exact sequence. By the proof that (2) implies (1) in Lemma 7.6.3, this implies that f is perfect.

Example 7.6.9 Consider the action of $\Gamma = \{\pm 1\}$ on $M = S^n \subset \mathbb{R} \times \mathbb{R}^n$ given by $\gamma \cdot (t, x) = (t, \gamma x)$, with fixed points $p_\pm = (\pm 1, 0)$. Note that M is a sphere with linear involution S^n_0 in the sense of Example 7.1.14. The formula $f(t, x) = t$ defines a Morse function M satisfying the hypotheses of Theorem 7.6.6. Taking $z = 0$ and $y = 2$ as regular values of f, one has $W_y = M$ and Diagram (7.6.14) becomes

$$\begin{array}{ccccccccc} 0 & \longrightarrow & H^*_\Gamma(M, W_0) & \xrightarrow{j} & H^*_\Gamma(M) & \longrightarrow & H^*_\Gamma(W_0) & \longrightarrow & 0 \\ & & \downarrow r_+ & & \downarrow r & & \approx \downarrow r_- & & \\ 0 & \longrightarrow & H^*_\Gamma(p_+) & \longrightarrow & H^*_\Gamma(p_+) \oplus H^*_\Gamma(p_-) & \longrightarrow & H^*_\Gamma(p_-) & \longrightarrow & 0 \end{array} .$$

Set $H^*(B\Gamma) = \mathbb{Z}_2[u]$. By Lemma 7.5.13, $e_\Gamma(\nu^-(p_+)) = u^n$. Together with Diagram (7.6.13), this shows that $r \circ j(U) = (u^n, 0)$, where

$$U \in H^n_\Gamma(M, W_0) \approx H^n_\Gamma(M_{0,2}, M_0) \approx H^0_\Gamma(p_+)$$

is the Thom class of $\nu^-(p_+)$. Let $B = j(U)$. Using the diagram

$$\begin{array}{ccccc} H^0_\Gamma(p_+) & \xrightarrow[\approx]{\text{Thom}} & H^n_\Gamma(M, W_0) & \xrightarrow{j} & H^n_\Gamma(M) \\ \downarrow \approx & & \downarrow & & \downarrow \rho \\ H^0(p_+) & \xrightarrow[\approx]{\text{Thom}} & H^n(M, W_0) & \xrightarrow{\approx} & H^n(M) \end{array}$$

one sees that $\rho(B)$ is the generator of $H^n(S^n) = \mathbb{Z}_2$. Hence, ρ is surjective, as expected by Theorem 7.6.6. By the Leray-Hirsch theorem, $H^*_\Gamma(M)$ is the free $\mathbb{Z}[u]$-module generated by B. Now, $r(B) = (u^n, 0)$ and $r(u) = (u, u)$. As is r injective by Theorem 7.6.6, the relation $B^2 = u^n B$ holds true in $H^*_\Gamma(M)$. Given the dimension of $H^k_\Gamma(M)$, this establishes the **GrA**$[u]$-isomorphism

$$\mathbb{Z}_2[B, u]/(B^2 + u^n B) \approx H^*_\Gamma(M)$$

Taking the image by ρ adds the relation $u = 0$ and we recover that $H^*(S^n) \approx \mathbb{Z}_2[B]/(B^2)$. Note that $(0, u)$ is not in the image of r, confirming the warning in the proof of Theorem 7.6.6. But $(0, u^n) = r(B + u^n)$ is in the image of r, corresponding to the generator $A = B + u^n$ of $H^*_\Gamma(M)$ (see Example 7.1.16). Had we considered $-f$ instead of f, the above discussion would have selected the generator A first. The relation $A^2 = u^n A$ also holds true and recall from Example 7.1.16 that $H^*_\Gamma(M)$ admits the presentation

$$\mathbb{Z}_2[u, A, B]/(A^2 + u^n A, A + B + u^n) \approx H^*_\Gamma(M).$$

Example 7.6.10 Let Γ be a 2-torus and let $\chi : \Gamma \to \{\pm 1\}$ be a homomorphism, identified with $\chi \in H^1(B\Gamma)$ under the bijection (7.5.8). Consider the Γ-action on $M = S^1 \subset \mathbb{R}^2$ given by $\gamma \cdot (t, x) = (t, \chi(\gamma)\, x)$, with fixed points $p_\pm = (\pm 1, 0)$. We call M a *χ-circle*. As in Example 7.6.9, one sees that the image of

$$r : H^*_\Gamma(M) \to H^*_\Gamma(p_\pm) \approx H^*_\Gamma(p_-) \oplus H^*_\Gamma(p_+) \approx H^*(B\Gamma) \oplus H^*(B\Gamma)$$

is the $H^*(B\Gamma)$-module generated by $B = (\chi, 0)$ and that $H^*_\Gamma(M)$ admits the presentation

$$H^*_\Gamma(M) \approx H^*(B\Gamma)[B]/(B^2 + \chi\, B).$$

Moreover, the image of r is the set of classes (a, b) such that $b - a$ is a multiple of χ.

Theorem 7.6.6 admits the following analogue for torus actions.

Theorem 7.6.11 *Let Γ be a torus and Γ_2 be its associated 2-torus. Let M be a smooth Γ-manifold. Let $f : M \to \mathbb{R}$ be a proper Γ-invariant Morse-Bott function which is bounded below. Suppose that* Crit $f = M^\Gamma = M^{\Gamma_2}$. *Then*

(1) *f has only critical manifolds of even index. In particular, if f is a Morse function, then M is of even dimension.*
(2) *f is perfect.*
(3) *M is Γ-equivariantly formal.*
(4) *the restriction morphism $H^*_\Gamma(M) \to H^*_\Gamma(M^\Gamma)$ is injective.*

Proof The proof is the same as that of Theorem 7.6.6. The hypothesis Crit $f = M^\Gamma$ implies that the negative normal bundles are Γ-vector bundles and the hypothesis Crit $f = M^{\Gamma_2}$ permits us to use Proposition 7.5.17 instead of Proposition 7.5.14. □

In Theorem 7.6.11, note that the perfectness of f is implied by (1). When $\Gamma = S^1$, Points (3) and (4) follows from Smith theory, in the same way as in Remark 7.6.7.

Example 7.6.12 Let Γ be a torus with associated 2-torus Γ_2. Let $\chi : \Gamma \to S^1$ be a continuous homomorphism. Consider the Γ-action on $M = S^2 \subset \mathbb{R} \times \mathbb{C}$ given by $\gamma \cdot (t, x) = (t, \chi(\gamma)\, x)$, with fixed points $p_\pm = (\pm 1, 0)$. We call M a *χ-sphere*. Let us assume that the restriction of χ to the associated 2-torus Γ_2 of Γ is not trivial. This implies that $M^\Gamma = M^{\Gamma_2}$, so we can apply Theorem 7.6.11 and, as in Example 7.6.9, one sees that the image of

$$r : H^*_\Gamma(M) \to H^*_\Gamma(p_\pm) \approx H^*_\Gamma(p_-) \oplus H^*_\Gamma(p_+) \approx H^*(B\Gamma) \oplus H^*(B\Gamma)$$

is the $H^*(B\Gamma)$-module generated by $B = (\kappa(\chi), 0)$, where $\kappa(\chi) \in H^2(BT)$ is defined in (7.5.11). Also, $H^*_\Gamma(M)$ admits the presentation

$$H^*_\Gamma(M) \approx H^*(B\Gamma)[B]/(B^2 + \kappa(\chi)\, B).$$

Moreover, the image of r is the set of classes (a, b) such that $b - a$ is a multiple of $\kappa(\chi)$.

A consequences of Theorem 7.6.6 and 7.6.11 are the *surjectivity theorems à la Kirwan* (see Remark 7.6.16 below). For $f : M \to \mathbb{R}$ is a continuous map, we set

$M_- = f^{-1}((-\infty, 0])$, $M_+ = f^{-1}([0 - \infty))$ and $M_0 = M_- \cap M_+ = f^{-1}(0)$. The inclusions form a commutative diagram.

$$\begin{array}{ccc} M_0 & \xrightarrow{i_+} & M_+ \\ \downarrow{\scriptstyle i_-} & \searrow^{i} & \downarrow{\scriptstyle j_+} \\ M_- & \xrightarrow{j_-} & M \end{array} . \qquad (7.6.19)$$

Proposition 7.6.13 *Let M be a closed smooth Γ-manifold, where Γ is a 2-torus. Let $f : M \to \mathbb{R}$ be a Γ-invariant Morse-Bott function satisfying* Crit $f = M^\Gamma$. *Suppose that 0 is a regular value of f. Then*

$$H_\Gamma^* i : H_\Gamma^*(M) \to H_\Gamma^*(M_0)$$

is surjective and its kernel is the ideal $\ker H_\Gamma^* j_- + \ker H_\Gamma^* j_+$, *generated by* $\ker H_\Gamma^* j_-$ *and* $\ker H_\Gamma^* j_+$.

For applications of this proposition, see Sect. 10.3.2.

Proof We use the abbreviations $i_\pm^* = H_\Gamma^* i_\pm$, $j_\pm^* = H_\Gamma^* j_\pm$, etc. As M is compact, f is proper and bounded. By Theorem 7.6.6, the restriction homomorphism $H_\Gamma^*(M) \to H_\Gamma^*(M^\Gamma) = H_\Gamma^*(\text{Crit } f)$ is injective. The commutative diagram

$$\begin{array}{ccc} H_\Gamma^*(M) & \xrightarrow{(j_-^*, j_+^*)} & H_\Gamma^*(M_-) \oplus H_\Gamma^*(M_+) \\ \downarrow & & \downarrow \\ H_\Gamma^*(M^\Gamma) & \xrightarrow{\approx} & H_\Gamma^*(M_-^\Gamma) \oplus H_\Gamma^*(M_+^\Gamma) \end{array}$$

shows that the Mayer-Vietoris sequence in equivariant cohomology for Diagram (7.6.19) splits into a global short exact sequence

$$0 \to H_\Gamma^*(M) \xrightarrow{(j_-^*, j_+^*)} H_\Gamma^*(M_-) \oplus H_\Gamma^*(M_+) \xrightarrow{i_-^* + i_+^*} H_\Gamma^*(M_0) \to 0. \qquad (7.6.20)$$

Suppose that $x_1 < x_2 < x_3 < \dots$ are regular values of f such that $f_{x_i, x_{i+1}}$ has only one critical level (we use the notations of Theorem 7.6.6 and its proof). Then, by (7.6.14), $H_\Gamma^*(W_{i+1}) \to H_\Gamma^*(W_i)$ is surjective. As $W_0 = M_-$ and M is compact, this argument shows that j_-^* is surjective. By symmetry, replacing f by $-f$ (using again that M is compact), one also has that j_+^* is surjective.

Let $a \in H_\Gamma^*(M_0)$. Using (7.6.20), choose $a_\pm \in H^*\Gamma(M_\pm)$ such that $a = i_-^*(a_-) + i_-^*(a_+)$. As $j_\pm^*$ is surjective, there exist $b_\pm \in H_\Gamma^*(M)$ with $i_\pm^*(b_\pm) = a_\pm$. Then

$$i^*(b_- + b_+) = i^*(b_-) + i^*(b_+) = i_-^* \circ j_-^*(b_-) + i_+^* \circ j_+^*(b_+) = i_-^*(a_-) + i_+^*(a_+) = a,$$

which proves that $i^* = H^*_\Gamma i$ is surjective. As $i^* = i^*_\pm \circ j_\pm$, we have also proven that $i^*_\pm$ is surjective. Therefore, one has a commutative diagram

$$\begin{array}{ccccc} H^*_\Gamma(M, M_+) & \rightarrowtail & H^*_\Gamma(M) & \xrightarrow{j^*_+} & H^*_\Gamma(M_+) \\ \downarrow \approx & & \downarrow j^*_- & \searrow^{i^*} & \downarrow i^*_+ \\ H^*_\Gamma(M_-, M_0) & \rightarrowtail & H^*_\Gamma(M_-) & \xrightarrow{i^*_-} & H^*_\Gamma(M_0) \end{array} \tag{7.6.21}$$

where the horizontal and vertical sequences are exact and the left hand vertical arrow is an isomorphism by excision. Hence, $\ker i^* = (j^*_-)^{-1}(\ker i^*_-)$, which yields to an exact sequence

$$0 \to \ker j^*_- \to \ker i^* \to \ker i^*_- \to 0.$$

But Diagram (7.6.21) provides a section of $\ker i^* \to \ker i^*_-$, whose image is equal to $\ker j^*_+$. This proves the assertion on $\ker i^*$ (which is actually **GrV**-isomorphic to $\ker j^*_- \oplus \ker j^*_+$). □

Example 7.6.14 Consider the action of $\Gamma = \{\pm 1\}$ on $M = \mathbb{R}P^n$ given by

$$\gamma \cdot [x_0, \ldots, x_n] = [x_0, \ldots, x_{n-1}, \gamma x_n].$$

The Morse-Bott function defined on M by $f([x_0, \ldots, x_n]) = 1 - 2x_n^2$ satisfies the hypotheses of Theorem 7.6.6 and Proposition 7.6.13. Set $M_c = f^{-1}(c)$. The critical submanifolds are $M_{\pm 1}$ and one has $M_- \simeq M_{-1} = pt$ and $M_+ \simeq M_1 = \mathbb{R}P^{n-1}$. Let $u \in H^1(B\Gamma) = \mathbb{Z}_2$ be the generator.

The bundle projection $p : M_\Gamma \to B\Gamma$ and its restriction p_c to $(M_c)_\Gamma$ give elements $H^* p_c(u) = u_c \in H^*_\Gamma(M_c)$ and $H^* p(u) = v \in H^*_\Gamma(M)$. One has

$$H^*_\Gamma(M_-) \approx \mathbb{Z}_2[u_{-1}] \text{ and } H^*_\Gamma(M_+) \approx H^*(B\Gamma \times M_1) \approx \mathbb{Z}_2[b, u_1]/(b^n),$$

with the degree of b equal to 1. Consider the commutative diagram

$$\begin{array}{ccccccc} H^0_\Gamma(M_1) & \xrightarrow[\approx]{\text{Thom}} & H^1_\Gamma(M, M_-) & \longrightarrow & H^1_\Gamma(M) & \xrightarrow{\rho} & H^1(M) \\ & & \alpha \text{ (composite)} & & \downarrow r_1 & & \downarrow \approx \\ & & & & H^1_\Gamma(M_1) & \xrightarrow{\rho_1} & H^1(M_1) \end{array} .$$

By Diagram (7.6.13), $r_1 \circ \alpha(\mathbf{1}) = e_\Gamma(\nu^-)$, the equivariant Euler class of the (negative) normal bundle $\nu^-(M_1)$. By Lemma 7.5.10,

$$e_\Gamma(\nu^-) = \mathbf{1} \times e(\nu^-(M_1)) + e(\hat{\chi}) \times \mathbf{1} \in H^1(B\Gamma \times M_1),$$

where $\hat{\chi}$ is the representation of Γ on a fiber of $\nu^-(M_1)$. Note that $\nu^-(M_1) = \nu(M_1)$ is the canonical line bundle over $\mathbb{R}P^{n-1}$ (since $M = \mathbb{R}P^n$ is obtained by attaching an n-cell to $M_1 = \mathbb{R}P^{n-1}$ by the Hopf map (two fold covering) $S^{n-1} \to \mathbb{R}P^{n-1}$). Then, by Proposition 4.7.36 and its proof, $e(\nu^-(M_1)) = b$. As χ has no non-zero fixed vector, $e_\Gamma(\hat{\chi}) = u$ by Lemma 7.5.13. Therefore,

$$e_\Gamma(\nu^-) = \mathbf{1} \times b + u \times \mathbf{1} = b + u_1,$$

the last formula making sense in the presentation $H^*_\Gamma(M_1) \approx \mathbb{Z}_2[b, u_1]/(b^n)$. As $\rho_1(b + u_1) = b$, this proves that ρ is surjective, as already known by Theorem 7.6.6. Let $a = \alpha(\mathbf{1}) + v$. One also has $\rho_1 \circ r_1(a) = b$ so, by the Leray-Hirsch theorem, $H^*_\Gamma(M)$ is the free $\mathbb{Z}_2[v]$-module generated by $a, a^2, \ldots, a^{n-1}$ and the Poincaré series of M_Γ is

$$\mathfrak{P}_t(M_\Gamma) = \mathfrak{P}_t(M) \cdot \mathfrak{P}_t(B\Gamma) = \frac{1 - t^{n+1}}{(1-t)^2}. \tag{7.6.22}$$

By Diagram (7.6.14), $r_{-1} \circ \alpha(\mathbf{1}) = 0$ in $H^1_\Gamma(M_-) \approx H^1_\Gamma(M_{-1})$. Therefore, the homomorphism $r : H^*_\Gamma(M) \to H^*_\Gamma(M^\Gamma) \approx H^*_\Gamma(M_{-1}) \oplus H^*_\Gamma(M_1)$ satisfies $r(a) = (u_{-1}, b)$ and $r(v) = (u_{-1}, u_1)$. Finally, we claim that there is a **GrA**-isomorphism

$$\mathbb{Z}_2[v, a]/(a^{n+1} + va) \xrightarrow{\approx} H^*_\Gamma(M). \tag{7.6.23}$$

Indeed, one already knows that $H^*_\Gamma(M)$ is **GrA**-generated by v and a, and, using the injective homomorphism r, one checks that the relation $a^{n+1} = va$ holds true. This gives the **GrA**-morphism of (7.6.23) which is surjective, and hence bijective since both sides of (7.6.23) have the same Poincaré series, computed in (7.6.22).

Replacing Theorem 7.6.6 by Theorem 7.6.11 in the proof of Proposition 7.6.13 gives the following result.

Proposition 7.6.15 *Let M be a closed smooth Γ-manifold, where Γ is a torus with associated 2-torus Γ_2. Let $f : M \to \mathbb{R}$ be a Γ-invariant Morse-Bott function satisfying* Crit $f = M^\Gamma = M^{\Gamma_2}$. *Suppose that* 0 *is a regular value of f. Then*

$$H^*_\Gamma i : H^*_\Gamma(M) \to H^*_\Gamma(M_0)$$

is surjective and its kernel is the ideal $\ker H^*_\Gamma j_- + \ker H^*_\Gamma j_+$*, generated by* $\ker H^*_\Gamma j_-$ *and* $\ker H^*_\Gamma j_+$.

As an example, one can take the complex analogue of Example 7.6.14, i.e. $\Gamma = S^1$ acting on $M = \mathbb{C}P^n$ given by $\gamma \cdot [x_0, \ldots, x_n] = [x_0, \ldots, x_{n-1}, \gamma x_n]$ and the Morse-Bott function $f([x_0, \ldots, x_n]) = 1 - 2x_n^2$. All the formulae of Example 7.6.14 hold true, with the degrees of all the classes multiplied by 2.

Remark 7.6.16 For $\Gamma = S^1$, the hypotheses of Proposition 7.6.15 are realized when f is the moment map of a Hamiltonian circle action (see [12]). In this case, it follows from F. Kirwan's thesis [117, Sect. 5] that $H^*_\Gamma(M;\mathbb{Q}) \to H^*_\Gamma(M_0;\mathbb{Q})$ is surjective (see e.g. [198, Theorem 2]). This justifies the terminology of *surjectivity theorem à la Kirwan* used above to introduce Propositions 7.6.13 and 7.6.15. For the assertion on $\ker H_\Gamma i$ in these propositions, compare [199, Theorem 2]; our proofs followed the hint of [199, Remark 3.5].

7.7 Exercises for Chapter 7

Notations. As in Sect. 7.1, G denotes the group with 2 elements $G = \{\mathrm{id}, \tau\}$. A G-space is thus a space endowed with an involution τ. The notation S^n_p stands for the G-linear sphere S^n with $(S^n_p)^G = S^p$, as in Example 7.1.14.

7.1. Let G acting on $X = S^1$ with $\tau(z) = \bar{z}$. Prove that X_G is homeomorphic to the double mapping cylinder CC_q, where $q : S^\infty \to \mathbb{R}P^\infty$ is the covering projection. Prove that X_G has the homotopy type of $\mathbb{R}P^\infty \vee \mathbb{R}P^\infty$ (use [82, Proposition 0.17] and that S^∞ is contractible).

7.2. If $p \geq 1$, prove that $H^*_G(S^n_p)$ admits, as a $\mathbb{Z}_2[u]$-algebra, the presentation $H^*_G(S^n_p) \approx \mathbb{Z}_2[u][A] \,/\, (A^2)$, where A is of degree n.

7.3. Let X be an equivariantly formal G-space which is of finite cohomology type. Find a formula giving $\mathfrak{P}_t(X_G)$ and $\mathfrak{P}_t(u \cdot H^*(X_G))$ in terms of $\mathfrak{P}_t(X)$.

7.4. Write the details for Remark 7.1.21.

7.5. What is $\tilde{H}^*_G(S^n_p)$?

7.6. Let $X = S^d \vee S^d$ with $d \geq 1$, endowed with the G action intertwining the two spheres. Prove that $H^*_G(X)$ is, as a **GrA**$[u]$-algebra, isomorphic to $\mathbb{Z}_2[u,a]/(ua)$, with a of degree d. Prove that $a^2 = 0$.

7.7. Let Δ be the subgroup of $SU(2)$ formed by the diagonal matrices. Prove that the map $B\Delta \to BSU(2)$ induced on the Milnor classifying spaces by the inclusion is, up to homotopy type, equivalent to the inclusion $\mathbb{C}P^\infty \to \mathbb{H}P^\infty$.

7.8. Let Γ be a topological group acting on a space X. Let Y be a space of finite cohomology type, considered as a Γ-space with trivial Γ-action ($Y = Y^\Gamma$). Prove that $H^*_\Gamma(X \times Y) \approx H^*_\Gamma(X) \otimes H^*(Y)$ (tensor product over $\mathbb{Z}_2$).

7.9. Let Γ be a compact Lie group. Let X and Y be two Γ-space which are equivariantly formal. We suppose that X is a finite dimensional Γ-complex and that Y is of finite cohomology type. Prove that $X \times Y$ (with the diagonal Γ-action) is equivariantly formal.

7.10. Let Y be a G-space with $Y^G = Y$. We suppose that Y is of finite cohomology type. Let $X = S^1_0 \times Y$, with the diagonal G-action. Give a presentation of $H^*_G(X)$ as a $\mathbb{Z}_2[u]$-algebra. Describe, for H^*_G, the Mayer-Vietoris sequence analogous to that of Exercise 4.13. Describe the injective restriction homomorphism $r : H^*_G(X) \to H^*_G(X^G)$.

7.11. Let $X = S^1_0 \times S^1_0$, with the diagonal G-action. Give a presentation of $H^*_G(X)$ as a $\mathbb{Z}_2[u]$-algebra. Prove that the map $f : X \to \mathbb{R}$ given by $f(e^{i\alpha}, e^{i\beta}) = \cos\alpha + 2\cos\beta$ is an equivariant Morse function satisfying the hypotheses of Theorem 7.6.6 and, with the help of this theorem, describe the injective restriction homomorphism $H^*_G(X) \to H^*_G(X^G)$.

7.12. Find a connected equivariantly formal G-space X such that X^G is the disjoint union of a point and of a sphere.

7.13. For $0 \le p \le n$, let P^n_p denote the projective space $\mathbb{R}P^n$ endowed with the involution

$$\tau(x_0 : x_1 : \cdots : x_n) = (-x_0 : \cdots : -x_p : x_{p+1} : \cdots : x_n).$$

(a) Prove that P^n_p is G-equivariantly formal.

(b) Describe the restriction homomorphism $r : H_G(P^n_p) \to H_G((P^n_p)^G)$ (it is injective by (a) and Proposition 7.3.9).

(c) Prove that $H^*_G(P^n_0)$ admits, as $\mathbb{Z}_2[u]$-algebra, the presentations

$$H^*_G(P^n_0) \approx \mathbb{Z}_2[u][A]/(A^{n+1}+uA^n) \text{ or } \mathbb{Z}_2[u][B]/((B+u)^{n+1}+u(B+u)^n)$$

(d) Prove the **GrA**$[u]$-isomorphisms

(d.1) $\mathbb{Z}_2[u][A]/(A^{n+1} + u^2A^{n-1}) \xrightarrow{\approx} H^*_G(P^n_1)$ and

(d.2) $\mathbb{Z}_2[u][A]/(A^{12} + uA^{11} + u^4A^8 + u^5A^7) \xrightarrow{\approx} H^*_G(P^{11}_4)$.

7.14. Prove that the algebras

$$R = \mathbb{Z}_2[u, A]/(A^3 + uA^2) \text{ and } S = \mathbb{Z}_2[u, B]/(B^3 + u^2B)$$

are **GrA**$[u]$-isomorphic (A and B of the same degree). Find a G-space X such that $H^*_G(X)$ is **GrA**$[u]$-isomorphic to R, with A of degree 4.

7.15. Let P^n_p be the G-space of Exercise 7.13. One checks that the function $f : P^n_p \to \mathbb{R}$ given by

$$f(x_0 : x_1 : \cdots : x_n) = x_0^2 + \cdots x_p^2$$

is a Morse-Bott function. Prove that it satisfies the hypotheses of Theorem 7.6.6. With the help of the proof of this theorem (as in Example 7.6.9), describe the restriction homomorphism $r : H_G(P^n_p) \to H_G((P^n_p)^G)$. Compare Exercise 7.13 (b).

7.16. What would be the analogue of Exercises 7.13 and 7.15 for S^1-actions?

7.17. Let Γ be a topological group. Prove that any functor J from **Top** to **Top** extends to a functor J_Γ from **Top**$_\Gamma$ to **Top**$_\Gamma$. What is here the meaning of "extends"?

7.18. We apply Exercise 7.17 to the functor suspension $X \mapsto \Sigma X$. Let X be a Γ-space.

(a) Prove that there exists a *suspension homomorphism*

$$\Sigma_\Gamma^* : \tilde{H}_\Gamma^*(X) \to \tilde{H}_\Gamma^{*+1}(\Sigma_\Gamma X)$$

which is a morphism of $H_\Gamma^*(pt)$-module and which is injective. Discuss its functoriality.

(b) Find an example where Σ_Γ^* is not surjective.

(c) Suppose that X is Γ-equivariantly formal. Prove that $\Sigma_\Gamma X$ is Γ-equivariantly formal and that Σ_Γ^* is an isomorphism.

7.19. Let X be a G-space. The G-action on X may be extended to a G-action on ΣX, permuting the suspension points, giving rise to a G-space $\check{\Sigma} X$ (note that $\check{\Sigma} X \approx X * G$). Suppose that X is a connected finite dimensional G-complex satisfying $b(X) = b(X^G) < \infty$. Let $i : X \to \Sigma X$ denote the inclusion. Prove that the sequence

$$0 \longrightarrow \tilde{H}_G^*(\check{\Sigma} X) \xrightarrow{\tilde{H}_G^* i} \tilde{H}_G^*(X) \xrightarrow{\rho} \tilde{H}^*(X) \longrightarrow 0$$

is exact.

7.20. Let $\Gamma = SU(2)$ acting on $X = SU(2)$ by conjugation.

(a) Show that X_Γ has the homotopy type of the double mapping cylinder CC_j where j is the inclusion of $\mathbb{C}P^\infty \to \mathbb{H}P^\infty$ (see Exercise 6.4). [Hint: use that X/Γ is homeomorphic to a segment]

(b) Deduce from Exercise 6.4 that X is Γ-equivariantly formal.

(c) Prove that there is a (unique) isomorphism of $H_\Gamma^*(pt)$-algebras

$$H_\Gamma^*(X) \approx H_\Gamma^*(pt)[b]/(b^2) \approx \mathbb{Z}_2[a, b]/(b^2),$$

where a is of degree 4 and b of degree 3.

7.21. Let (X, X_1, X_2, X_0) be a Mayer-Vietoris data. Suppose that X is a Γ-space (Γ a topological group) and that X_i are closed Γ-invariant subspaces of X. Suppose that $X = X_1 \cup X_2$ and that (X_i, X_0) is a Γ-equivariantly well cofibrant pair for $i = 1, 2$. Prove that there is a Mayer-Vietoris sequence for the Γ-equivariant cohomology.

7.22. Let X and Y be Γ-spaces, equivariantly well pointed by $x \in X^\Gamma$ and $y \in Y^\Gamma$. Thus, $X \wedge Y$ (using these base points) is a Γ-space. Prove that there is a H_Γ^*-algebra isomorphism $\tilde{H}_\Gamma^*(X \vee Y) \approx \tilde{H}_\Gamma^*(X) \oplus \tilde{H}_\Gamma^*(Y)$.

Chapter 8
Steenrod Squares

In Chap. 4, the power of cohomology was much increased by the introduction of the cup product, making $H^*(X)$ a graded algebra. Another rich structure on $H^*(X)$ comes from *cohomology operations*, i.e. the natural self-transformations of the mod 2 cohomology functor (see Sect. 8.1). The basic examples of such operations, the *Steenrod squares* $\mathrm{Sq}^i\colon H^*(X) \to H^{*+i}(X)$, were discovered by Norman Steenrod and Henri Cartan in the late 1940s (see, e.g. [40, pp. 510–523] for historical details). The **GrA**-morphism induced by any continuous map must then commute with all the Steenrod squares, which imposes strong restrictions. For instance, the spaces $Y = S^2 \vee S^3$ and $Y' = \Sigma\mathbb{R}P^2$ do not have the same homotopy type, although their cohomology are **GrA**-isomorphic. Indeed, Sq^1 vanishes on $H^*(Y)$ but not on $H^*(Y')$. In the same way, we show that all suspensions of Hopf maps are essential, i.e. not homotopic to a constant map (see Sect. 8.6).

After an introductory section on cohomology operations, we state in Sect. 8.2 the basic properties and make some computations of Steenrod squares. Their constructions and the proof of Adem relations are given in Sects. 8.3 and 8.4. Based on equivariant cohomology, these two technical sections may be skipped on first reading, since the applications of Steenrod squares are consequences of the properties presented in Sect. 8.2.

The last two sections of this chapter treat applications of Steenrod squares. Prominent among them are Adams theorem on "the Hopf invariant one problem" and Serre's computation of the cohomology algebra of Eilenberg-MacLane spaces $K(\mathbb{Z}_2, n)$. The latter implies that mod 2 cohomology operations are, in some sense, generated by sums, cup products and iterations of Steenrod squares (see Remark 8.5.7). More applications will appear in Chap. 9, for instance Thom's definition of Stiefel-Whitney classes and Wu's formula.

J.-C. Hausmann, *Mod Two Homology and Cohomology*, Universitext,
DOI 10.1007/978-3-319-09354-3_8

8.1 Cohomology Operations

This section contains some generalities on mod 2 cohomology operations, in order to present Steenrod squares. We take a global approach which may shed a new light with respect to existing texts on the subject.

A *cohomology operation* is a map

$$Q = Q_{(X,Y)} \colon H^*(X, Y) \to H^*(X, Y)$$

defined for any topological pair (X, Y), satisfying the following two conditions:

(1) Q is functorial, i.e. if $g\colon (X', Y') \to (X, Y)$ is a continuous map of pairs, then

$$H^*g \circ Q = Q \circ H^*g\,. \tag{8.1.1}$$

(2) $Q_{(X,Y)} = \sum Q_{[i](X,Y)}$ where $Q_{[i](X,Y)}$ is the restriction of $Q_{(X,Y)}$ to $H^i(X, Y)$.

We may restrict the definition to some classes of pairs, like CW-pairs, etc. For instance, restricting to pairs $(X, \emptyset)$ gives operations on absolute cohomology, since $H^*(X, \emptyset) \xrightarrow{\approx} H^*(X)$. Point (2) is a partial linearity (Q is not supposed to be linear) and permits us to define Q via its restrictions $Q_{[i]}$.

Examples of cohomology operations are given by $Q = 0$ or $Q = \mathrm{id}$. A less trivial example is the cohomology operation Q such that by $Q_{[n]}(a) = a^n$ for all $n \in \mathbb{N}$, where $a^n = a \smile \cdots \smile a$ (n times). Cohomology operations may be added, multiplied by cup products and composed, giving rise to more examples.

Here are a few remarks about cohomology operations. They are used throughout this section, without always an explicit mention.

8.1.1 By Theorem 3.7.1, a topological pair has, in a functorial way, the same cohomology as a CW-pair. Hence, when studying cohomology operations, we do not lose generality by restricting to CW-pairs. For instance, a cohomology operation defined for CW-pairs extends in a unique way to a cohomology operation defined for all topological pairs.

8.1.2 Let (X, Y) be a CW-pair with Y non-empty. The quotient map $(X, Y) \to (X/Y, [Y])$ induces an isomorphism $H^*(X/Y, [Y]) \xrightarrow{\approx} H^*(X, Y)$ (Proposition 3.1.45). Most questions on cohomology operations may thus be settled by considering the CW-pairs of type $(X, \emptyset)$ and (X, pt). In particular, a cohomology operation defined for these pairs extends to a unique cohomology operation. Moreover, a cohomology operation Q defined on absolute cohomology for CW-complexes extends to a unique cohomology operation for CW-pairs, using the commutative diagram

$$\begin{array}{ccccccccc} 0 & \longrightarrow & H^*(X,pt) & \longrightarrow & H^*(X) & \longrightarrow & H^*(pt) & \longrightarrow & 0 \\ & & \Big\downarrow Q & & \Big\downarrow Q & & \Big\downarrow Q & & \\ 0 & \longrightarrow & H^*(X,pt) & \longrightarrow & H^*(X) & \longrightarrow & H^*(pt) & \longrightarrow & 0 \end{array}, \tag{8.1.2}$$

where (X, pt) is a CW-pair.

8.1.3 Let (X, Y) be a CW-pair. By Corollary 3.1.12, the family of inclusions $i_A : A \to X$ for $A \in \pi_0(X)$ gives rise to the commutative diagram

$$\begin{array}{ccc} H^*(X,\emptyset) & \xrightarrow[\approx]{(H^*i_A)} & \prod_{A\in\pi_0(X)} H^*(A,\emptyset) \\ \Big\downarrow Q & & \Big\downarrow \prod Q \\ H^*(X,\emptyset) & \xrightarrow[\approx]{(H^*i_A)} & \prod_{A\in\pi_0(X)} H^*(A, emptyset) \end{array}$$

or, if $pt \in A_0 \in \pi_0(X)$,

$$\begin{array}{ccc} H^*(X,pt) & \xrightarrow[\approx]{(H^*i_A)} & H^*(A_0,pt) \times \prod_{A\in\pi_0(X)-A_0} H^*(A,\emptyset) \\ \Big\downarrow Q & & \Big\downarrow \prod Q \\ H^*(X,pt) & \xrightarrow[\approx]{(H^*i_A)} & H^*(A_0,pt) \times \prod_{A\in\pi_0(X)-A_0} H^*(A,\emptyset) \end{array}$$

In other words, a cohomology operation preserves the connected components. Together with 8.1.2, this permits us often to restrict, without loss of generality, a cohomology operation to pairs (X, Y) where X is path-connected.

Lemma 8.1.4 *If Q be a cohomology operation, then $Q(0) = 0$.*

Proof The class $0 \in H^*(X, Y)$ is in the image of $H^*(X, X) \to H^*(X, Y)$. As, $H^*(X, X) = 0$, the lemma follows from functoriality. □

An important property of cohomology operations is that it does not decrease dimensions.

Lemma 8.1.5 *Let Q be a cohomology operation. Then, there is a function $N : \mathbb{N} \to \mathbb{N}$, satisfying $N(0) = 0$ and $N(m) \geq m$, such that*

$$Q(H^m(X, Y)) \subset \bigoplus_{k=m}^{N(m)} H^k(X, Y) \tag{8.1.3}$$

for all topological pairs (X, Y).

Proof By 8.1.1–8.1.3 above, it is enough to prove the lemma for CW-pairs (X, Y) with X connected and $Y = pt$ or $\emptyset$. As $H^0(X, pt) = 0$, $Q(H^0(X, pt)) = 0$ by Lemma 8.1.4. The constant map $X \to pt$ induces an isomorphism $H^0(pt, \emptyset) \to H^0(X, \emptyset)$. As $H^{>0}(pt, \emptyset) = 0$, the functoriality implies (8.1.3) for $m = 0$, with $N(0) = 0$.

If $m > 0$, then $H^m(X, pt) \xrightarrow{\approx} H^m(X, \emptyset) \xrightarrow{\approx} H^m(X)$, so it is enough to prove (8.1.3) in the absolute case. By functoriality, the following diagram is commutative.

$$\begin{array}{ccc} H^m(X, X^{m-1}) & \twoheadrightarrow & H^m(X) \\ \downarrow Q & & \downarrow Q \\ H^*(X, X^{m-1}) & \longrightarrow & H^*(X) \end{array}$$

As $H^k(X, X^{m-1}) = 0$ for $k < m$, this proves that the direct sum in (8.1.3) starts at $k = m$. Also, any class $a \in H^m(X)$ is of the form $a = H^*f(\iota)$ for some map $f\colon X \to \mathcal{K}_m$. Thus, $N(m)$ is the maximal degree of $Q(\iota) \in H^*(\mathcal{K}_m)$. □

A cohomology operation Q restricts to a cohomology operation Q' on absolute cohomology by $Q'_X = Q_{(X,\emptyset)}$. Not every absolute cohomology operation Q' is such restriction because it may not satisfy $Q'(0) = 0$, contradicting Lemma 8.1.4. As an example, we can, for X path-connected and non-empty, define Q' by $Q'(a) = 0$ for $a \in H^m(X)$ $(m \neq 1)$ and $Q'(H^1(X)) = \mathbf{1}$ (the functoriality of Q' coming from Lemma 2.5.4). In fact, we have the following lemma.

Lemma 8.1.6 *Let Q' be a cohomology operation defined on absolute cohomology for connected CW-complexes. Then, there exists a cohomology operation Q such that $Q'_X = Q_{(X,\emptyset)}$ if and only if $Q'(0) = 0$. The cohomology operation Q is unique.*

Proof The condition is necessary by Lemma 8.1.4. For the converse, using 8.1.1–8.1.3 above, it suffices to define $Q_{(X,pt)}$ for an path-connected non-empty CW-complex X. On $H^0(X, pt) = 0$, Q is defined by $Q(0) = 0$ (this is compulsory by Lemma 8.1.4). The functoriality for the inclusion $(X, \emptyset) \hookrightarrow (X, pt)$ on H^0 is guaranteed by the condition $Q'(0) = 0$. Let $P\colon H^*(X) \to H^0(X)$ be the projection onto the component of degree 0. Let $j\colon pt \to X$ be an inclusion of a point in X. As $Q'(0) = 0$, the commutative diagram

$$\begin{array}{ccccc} H^{>0}(X) & \xrightarrow{Q'} & H^*(X) & \xrightarrow{P} & H^0(X) \\ \downarrow H^*j & & \downarrow H^*j & & \approx\downarrow H^*j \\ H^{>0}(pt) & \xrightarrow{Q'=0} & H^*(pt) & \xrightarrow{P} & H^0(pt) \end{array}$$

shows that $Q'(H^{>0}(X)) \subset H^{>0}(X)$. Hence, the commutative diagram

$$\begin{array}{ccc} H^{>0}(X,pt) & \xrightarrow{\approx} & H^{>0}(X) \\ \Big\downarrow{\scriptstyle Q} & & \Big\downarrow{\scriptstyle Q'} \\ H^{>0}(X,pt) & \xrightarrow{\approx} & H^{>0}(X) \end{array}$$

defines Q and shows its uniqueness. □

The notion of cohomology operation makes sense for the reduced cohomology, with the same definition.

Lemma 8.1.7 *A cohomology operation Q descends to a unique cohomology operation on reduced cohomology, also called Q.*

Proof Let $p: X \to pt$ be the unique map from X to a point. Consider the diagram

$$\begin{array}{ccccc} H^*(pt) & \xrightarrow{H^*p} & H^*(X) & \twoheadrightarrow & \tilde{H}^*(X) \\ \Big\downarrow{\scriptstyle Q} & & \Big\downarrow{\scriptstyle Q} & & \Big\downarrow{\scriptstyle Q} \\ H^*(pt) & \xrightarrow{H^*p} & H^*(X) & \twoheadrightarrow & \tilde{H}^*(X) \end{array}$$

where the line are exact. As Q is a cohomology operation, the left square is commutative, so there is a unique $Q: \tilde{H}^*(X) \to \tilde{H}^*(X)$ so that the right square commutes and this construction is functorial. (Recall that, if X is path-connected and Y is non-empty, then $\tilde{H}^*(X, Y) = H^*(X, Y)$). □

We now study the multiplicativity of a cohomology operation Q. Note that, by Lemma 4.1.14, the relative cup product $H^*(X, Y) \otimes H^*(X, Y) \to H^*(X, Y)$ is defined for all topological pairs (X, Y). Consider the following four statements.

(a) $Q(a \smile b) = Q(a) \smile Q(b)$ for all $a, b \in H^*(X)$ and all spaces X.
(b) $Q(a \smile b) = Q(a) \smile Q(b)$ for all $a, b \in H^*(X, Y)$ and all topological pairs (X, Y).
(c) $Q(a \times b) = Q(a) \times Q(b)$ for all $a \in H^*(X_1)$, $b \in H^*(X_2)$ and all spaces X_1 and X_2.
(d) $Q(a \tilde{\times} b) = Q(a) \tilde{\times} Q(b)$ for all $a \in \tilde{H}^*(X_1)$, $b \in \tilde{H}^*(X_2)$ and all pointed spaces X_1 and X_2.

Proposition 8.1.8 *For a cohomology operation Q, Conditions* (a), (b) *and* (c) *are equivalent and* (a) *implies* (d). *If $Q(\mathbf{1}) = \mathbf{1}$, then* (d) *implies* (a).

Proof Without loss of generality, we may suppose that the spaces X and X_i are connected CW-complexes. Statement (b) is stronger than (a) since $H^*(X) = H^*(X, \emptyset)$. To prove that (a) implies (b), it suffices to consider the case $Y = pt$, which is obvious.

Using the functoriality of Q, (a)$\Rightarrow$(c) follows from the definition of the cross product and (c)$\Rightarrow$ (a) from the formula $a \smile b = \Delta^*(a \times b)$ (see Remark 4.6.1).

That (c)⇒(d) is obvious, so (a)⇒(d). Now, (d) implies (c) for classes of positive degree. As (c)⇒(a), property (d) implies that $Q(a \smile b) = Q(a) \smile Q(b)$ except possibly for a or b equal to $\mathbf{1}$. If, say $a = \mathbf{1}$, then

$$Q(\mathbf{1} \smile b) = Q(b) = \mathbf{1} \smile Q(b) = Q(\mathbf{1}) \smile Q(b)\,,$$

since $Q(\mathbf{1}) = \mathbf{1}$. Thus, *(d)*⇒*(a)* if $Q(\mathbf{1})) = \mathbf{1}$. □

Corollary 8.1.9 *Let Q be a cohomology operation with $Q(\mathbf{1}) = \mathbf{1}$. Then, (a) is equivalent to*

(d') the diagram

$$\begin{array}{ccc}
\tilde{H}^m(\mathcal{K}_m) \otimes \tilde{H}^n(\mathcal{K}_n) & \xrightarrow{\tilde{\times}} & \tilde{H}^{m+n}(\mathcal{K}_m \wedge \mathcal{K}_n) \\
\big\downarrow {\scriptstyle Q\otimes Q} & & \big\downarrow {\scriptstyle Q} \\
\tilde{H}^*(\mathcal{K}_m) \otimes \tilde{H}^*(\mathcal{K}_n) & \xrightarrow{\tilde{\times}} & \tilde{H}^*(\mathcal{K}_m \wedge \mathcal{K}_n)
\end{array}$$

is commutative for all positive integers m and n.

Proof It is clear that *(d)*⇒*(d')*. The corollary will then follow from Proposition 8.1.8 if we prove that *(d')*⇒*(d)*.

It suffices to prove *(d)* for $a \in \tilde{H}^m(X_1)$ and $b \in \tilde{H}^n(X_2)$ where X_1 and X_2 are connected CW-complexes, so $m, n > 0$. Then $a = f_a^*(\iota_m)$ and $a = f_b * (\iota_n)$ for maps $f_a\colon X_1 \to \mathcal{K}_m$ and $f_b\colon X_2 \to \mathcal{K}_n$. Condition *(d')* says that $Q(\iota_m \tilde{\times} \iota_n) = Q(\iota_m) \tilde{\times} Q(\iota_n)$ and $Q(a \tilde{\times} b) = Q(a) \tilde{\times} Q(b)$ from this special case, using the functoriality of $\tilde{\times}$ and of Q. □

8.2 Properties of Steenrod Squares

One of the most remarkable feature of mod 2 cohomology is the existence of cohomology operations, introduced by N. Steenrod and H. Cartan in the late 1940s (see, e.g. [40, pp. 510–523]), called the *Steenrod squares* $\mathrm{Sq}^i\colon H^*(X, Y) \to H^*(X, Y)$ ($i \in \mathbb{N}$). For $a \in H^m(X, Y)$, one has $\mathrm{Sq}^i(a) = 0$ for $i > m$ (see 2a in Theorem 8.2.1). Hence, the sum $\sum_{i\in\mathbb{N}} \mathrm{Sq}^i(a)$ has only a finite number of non-zero terms and thus defines the *total Steenrod square*

$$\mathrm{Sq}\colon H^*(X, Y) \to H^*(X, Y)\,, \quad \mathrm{Sq}(a) = \sum_{i\in\mathbb{N}} \mathrm{Sq}^i(a)\,. \tag{8.2.1}$$

Here is the main theorem of this section.

Theorem 8.2.1 *There exists a cohomology operation* Sq *and* Sq^i *as in* (8.2.1), *which enjoys the following properties:*

(1) Sq *is* $\mathbb{Z}_2$*-linear.*
(2) *if* $a \in H^n(X,Y)$ *then* $\mathrm{Sq}^i(a) \in H^{n+i}(X,Y)$ *and*
   (a) $\mathrm{Sq}^i(a) = 0$ *for* $i < 0$ *and* $i > n$.
   (b) $\mathrm{Sq}^0(a) = a$.
   (c) $\mathrm{Sq}^n(a) = a \smile a$.
(3) $\mathrm{Sq}(a \smile b) = \mathrm{Sq}(a) \smile \mathrm{Sq}(b)$. *This is equivalent to the formula*

$$\mathrm{Sq}^k(a \smile b) = \sum_{i+j=k} \mathrm{Sq}^i(a) \smile \mathrm{Sq}^j(b) \text{ (Cartan's formula)}.$$

(4) $\Sigma^* \circ \mathrm{Sq} = \mathrm{Sq} \circ \Sigma^*$, *where* $\Sigma^* : \tilde{H}^*(X) \xrightarrow{\approx} \tilde{H}^{*+1}(\Sigma X)$ *is the suspension isomorphism of Proposition* 3.1.49.
(5) *The Adem relations:*

$$\mathrm{Sq}^i\mathrm{Sq}^j = \sum_{k=0}^{[i/2]} \binom{j-k-1}{i-2k} \mathrm{Sq}^{i+j-k}\mathrm{Sq}^k \quad (0 < i < 2j).$$

The Steenrod squares are characterized amongst cohomology operations by some of these properties (see Proposition 8.5.12). Also, the Adem relations generate all the polynomial relations amongst the compositions of Sq^i's which hold true for any space (see Corollary 8.5.11).

Example 8.2.2 Theorem 8.2.1 permits us to compute Sq^i easily for the projective spaces $\mathbb{R}P^n$, $\mathbb{C}P^n$ and $\mathbb{H}P^n$. Indeed, one has the following results.

(a) Let $a \in H^1(X)$. Then (2) implies that $\mathrm{Sq}(a) = a + a^2$ (we write a^n for the cup product of n copies of a). Then, (3) implies that

$$\mathrm{Sq}(a^n) = (a + a^2)^n = a^n \smile (\mathbf{1} + a)^n = a^n \smile \sum_{i=1}^{n} \binom{n}{i} a^i .$$

Therefore

$$\mathrm{Sq}^i(a^n) = \binom{n}{i} a^{n+i} . \tag{8.2.2}$$

(b) If $a \in H^2(X)$ satisfies $\mathrm{Sq}^1(a) = 0$, then $\mathrm{Sq}(a) = a + a^2$ and, as in (a), one has

$$\mathrm{Sq}^{2i}(a^n) = \binom{n}{i} a^{n+i} \text{ and } \mathrm{Sq}^{2i+1}(a^n) = 0 . \tag{8.2.3}$$

(c) If $a \in H^4(X)$ satisfies $\mathrm{Sq}(a) = a + a^2$, then

$$\mathrm{Sq}^{4i+k}(a^n) = \begin{cases} \binom{n}{i} a^{n+i} & \text{if } k \equiv 0 \mod 4. \\ 0 & \text{otherwise.} \end{cases} \tag{8.2.4}$$

Besides the trivial case of $\mathbb{O}P^2$, there are no more such examples. Indeed, by Corollary 8.6.3 and Theorem 8.6.6, if $a \in H^m(X)$ satisfies $\mathrm{Sq}(a) = a + a^2$ with $a^2 \neq 0$, then $m = 1, 2, 4$ or 8.

We finish by two more properties of Steenrod squares. As $\mathrm{Sq}(a \smile b) = \mathrm{Sq}(a) \smile \mathrm{Sq}(b)$, Proposition 8.1.8 implies the following result.

Proposition 8.2.3 *Let X_1 and X_2 be topological spaces [pointed for (b)]. Then,*

(1) $\mathrm{Sq}(a \times b) = \mathrm{Sq}(a) \times \mathrm{Sq}(b)$ *for all* $a \in H^*(X_1)$, $b \in H^*(X_2)$.
(2) $\mathrm{Sq}(a \mathbin{\tilde{\times}} b) = \mathrm{Sq}(a) \mathbin{\tilde{\times}} \mathrm{Sq}(b)$ *for all* $a \in \tilde{H}^*(X_1)$, $b \in \tilde{H}^*(X_2)$.

Proposition 8.2.4 *Let (X, Y) be a topological pair. Then*

$$\mathrm{Sq} \circ \delta^* = \delta^* \circ \mathrm{Sq}\,,$$

where δ^* is the connecting homomorphism $\delta^* \colon H^*(Y) \to H^{*+1}(X, Y)$.

Proof By 8.1.1 , we may suppose that (X, Y) is a CW-pairs. Let $Z = X \cup [-2, 1] \times Y$ and $Z' = X \cup [-2, -1] \times Y \cup \{1\} \times Y$. The projection $p \colon [-2, 1] \times Y \to \{-2\} \times Y$ extends to a homotopy equivalence of pairs $p \colon (Z, \{1\} \times Y) \to (X, Y)$. The commutative diagram

$$\begin{array}{ccccc}
H^*(\{1\} \times Y) & & \xleftarrow{\quad p^* \quad} & & H^*(S^0 \times Y) \\
\downarrow{\scriptstyle \delta^*} & & & & \downarrow{\scriptstyle \delta^*} \\
H^{*+1}(Z, \{1\} \times Y) & \longleftarrow & H^{*+1}(Z, Z') & \underset{\text{excision}}{\xleftrightarrow{\approx}} & H^{*+1}(D^1, S^0 \times Y)
\end{array}$$

shows that it is enough to prove that $\mathrm{Sq} \circ \delta^* = \delta^* \circ \mathrm{Sq}$ for a CW-pair of the type $(D^1 \times Y, S^0 \times Y)$. In the proof of Proposition 4.7.44, we showed that

$$\delta^*(a) = e \times \big(H^* i_+(a) + H^* i_-(a)\big)\,,$$

where $i_\pm \colon \{\pm 1\} \times F \to S^0 \times F$ denote the inclusions and $0 \neq e \in H^1(D^1, S^0) = \mathbb{Z}_2$. One has $\mathrm{Sq}(e) = \mathrm{Sq}^0(e) = e$. Using the linearity of Sq and Proposition 8.2.3, we get

$$\begin{aligned}
\mathrm{Sq} \circ \delta^*(a) &= \mathrm{Sq}\big(e \times \big(H^* i_+(a) + H^* i_-(a)\big)\big) \\
&= \mathrm{Sq}(e) \times \mathrm{Sq}(H^* i_+(a) + H^* i_-(a)) \\
&= e \times \big(\mathrm{Sq} \circ H^* i_+(a) + \mathrm{Sq} \circ H^* i_-(a)\big) \\
&= e \times \big(H^* i_+ \circ \mathrm{Sq}(a) + H^* i_- \circ \mathrm{Sq}(a)\big) \\
&= \delta^* \circ \mathrm{Sq}(a)\,.
\end{aligned}$$

□

Sections 8.3 and 8.4 contain the proof of Theorem 8.2.1. They are based on the ideas of Steenrod (see [184, VII.1 and VIII.1]) using the equivariant cohomology. Other treatments of similar ideas are developed in [3, VI.7] and [82, Sect. 4.L].

8.3 Construction of Steenrod Squares

The involution τ on $X \times X$ given by $\tau(x, y) = (y, x)$ makes $X \times X$ a G-space for $G = \{\mathrm{id}, \tau\}$. We consider the *cross-square map* $\beta\colon H^n(X) \to H^{2n}(X \times X)$ defined by $\beta(a) = a \times a$. Its image is obviously contained in $H^{2n}(X \times X)^G$. By Lemma 7.1.10, the image of $\rho\colon H^*_G(X \times X) \to H^*(X \times X)$ is also contained in $H^*(X \times X)^G$. The same considerations are valid for the reduced cross-square map $\tilde\beta\colon \tilde H^n(X) \to \tilde H^{2n}(X \wedge X)^G$, defined for a space X which is well pointed by $x \in X$. The maps β and $\tilde\beta$ are not linear but they are functorial: if $f\colon X' \to X$ is a continuous maps, then $H^*(f \times f) \circ \beta = \beta \circ H^* f$ and $\tilde H^*(f \wedge f) \circ \tilde\beta = \tilde\beta \circ \tilde H^* f$. Using Diagram (4.7.7), the use of base points $x \in X$ and $(x, x) \in X \wedge X$ provide a commutative diagram

$$
\begin{array}{ccc}
\tilde H^*(X \wedge X) & \rightarrowtail & H^*(X \times X) \\
\uparrow{\scriptstyle\tilde\beta} & & \uparrow{\scriptstyle\beta} \\
\tilde H^*(X) & \rightarrowtail & H^*(X)
\end{array}
\qquad (8.3.1)
$$

Lemma 8.3.1 *Let X be a connected CW-complex, pointed by $x \in X^0$. Then, the cross-square maps $\tilde\beta$ and β admit liftings $\tilde\beta_G$ and β_G so that the diagram*

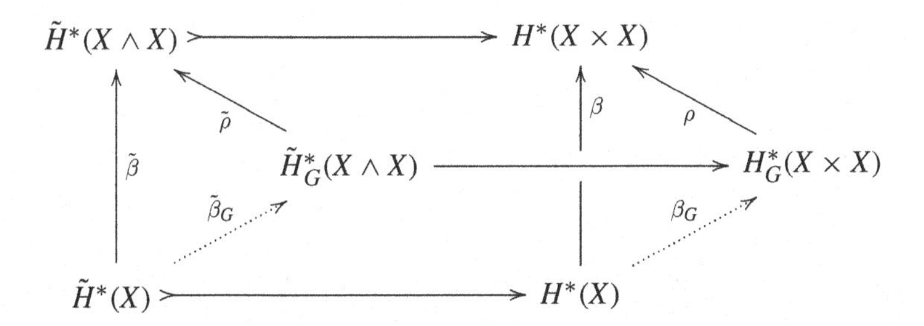

is commutative. These liftings are functorial and satisfy $\tilde\beta_G(0) = 0$ and $\beta_G(0) = 0$. Such liftings are unique.

Proof The maps β, ρ, $\tilde\beta$ and $\tilde\rho$ preserving the connected components, one may suppose that X is connected. Lemma 8.3.1 is obvious when $n = 0$, giving $\beta_G(0) = 0$ and $\beta_G(\mathbf{1}) = \mathbf{1}$. We can then assume $n > 0$. In this case, $\tilde H^n(X) \to H^n(X)$ is an isomorphism, so it is enough to define $\tilde\beta_G$.

We first define $\tilde{\beta}_G$ when $X = \mathcal{K}_n = K(\mathbb{Z}_2, n)$ with its CW-structure given in the proof of Proposition 3.8.1, whose 0-skeleton consists in a single point x. By Proposition 4.7.11, the G-space $\mathcal{K}_n \wedge \mathcal{K}_n$ satisfies $\tilde{H}^k(\mathcal{K}_n \wedge \mathcal{K}_n) = 0$ for $k < 2n$. Hence, Proposition 7.1.12 implies that $\tilde{\rho}\colon \tilde{H}^{2n}_G(\mathcal{K}_n \times \mathcal{K}_n) \to \tilde{H}^{2n}(\mathcal{K}_n \times \mathcal{K}_n)^G$ is an isomorphism. We define $\tilde{\beta}_G = \tilde{\rho}^{-1} \circ \tilde{\beta}$.

Now, a cohomology class $a \in \tilde{H}^n(X)$ is of the form $a = \tilde{H}^* f_a(\iota)$ for a map $f_a\colon X \to \mathcal{K}_n$, well defined up to homotopy. We define

$$\tilde{\beta}_G(a) = \tilde{H}^*_G(f_a \times f_a) \circ \tilde{\beta}_G(\iota)\,. \tag{8.3.2}$$

This definition makes $\tilde{\beta}_G$ functorial. Indeed, let $g\colon Y \to X$ be a continuous map and $a = \tilde{H}^* f_a(\iota) \in H^n(X)$. If $\tilde{b} = H^* g(a)$, then $b = \tilde{H}^* f_b(\iota)$ with $f_b = f_a \circ g$. By definition of $\tilde{\beta}_G$, one has

$$\tilde{H}^*_G(g \times g) \circ \tilde{\beta}_G(a) = \tilde{H}^*_G(g \times g) \circ \tilde{H}^*_G(f_a \times f_a) \circ \beta_G(\iota) = \tilde{H}^*_G(f_b \times f_b) \beta_G(\iota) = \tilde{\beta}_G(b)\,.$$

The functoriality of β_G follows from that of $\tilde{\beta}_G$. The uniqueness of $\tilde{\beta}_G$ (then, that of β_G) is obvious, since there was no choice for $X = \mathcal{K}_n$ and Definition (8.3.2) is compulsory by the required functoriality. □

Remark 8.3.2 When $\tilde{H}^i(X) = 0$ for $i < n$, then $H^j(X \wedge X) = 0$ for $j < 2n$ by Proposition 4.7.11. The homomorphism $\tilde{\rho}\colon H^{2n}(X \wedge X) \to H^{2n}(X \wedge X)^G$ is an isomorphism which is natural. Then, by functoriality, the formula $\tilde{\beta}_G = \tilde{\rho}^{-1} \circ \tilde{\beta}$ holds true.

The map $\beta_G\colon H^n(X) \to H^{2n}_G(X \times X)$ extends to $\beta_G\colon H^*(X) \to H^*_G(X \times X)$ by

$$\beta_G\Big(\sum_{j\in\mathbb{N}} a_j\Big) = \sum_{j\in\mathbb{N}} \beta_G(a_j)\,, \text{ where } a_j \in H^j(X)\,. \tag{8.3.3}$$

The inclusion of $(X \times X)^G$ into $X \times X$ induces a $\mathbf{GrA}[u]$-morphism $r\colon H^*_G(X \times X) \to H^*_G((X \times X)^G)$. Observe that $(X \times X)^G$ is the diagonal subspace $\{(x, x)\}$ of $X \times X$, hence homeomorphic to X. We thus write $r\colon H^*_G(X \times X) \to H^*_G(X)$, considering X as a G-space with trivial G-action. Using (7.1.2), we thus get a $\mathbf{GrA}[u]$-isomorphism $H^*_G(X) \approx H^*(X)[u]$. We also consider the (non-graded) ring homomorphism $\mathrm{ev}_{\mathbf{1}}\colon H^*(X)[u] \to H^*(X)$ which extends the identity on $H^*(X)$ by sending u to $\mathbf{1}$ (*evaluation of a polynomial at* $\mathbf{1}$).

Let X be a CW-complex. By definition, the *Steenrod square* $\mathrm{Sq}\colon H^*(X) \to H^*(X)$ is the composition

$$\mathrm{Sq} = \mathrm{ev}_{\mathbf{1}} \circ r \circ \beta_G \tag{8.3.4}$$

making the following diagram commutative.

$$\begin{array}{ccccc} H_G^*(X\times X) & \xrightarrow{r} & H_G^*(X) & \overset{\approx}{\longleftrightarrow} & H^*(X)[u] \\ \beta_G\uparrow & & & & \downarrow \mathrm{ev}_1 \\ H^*(X) & & \xrightarrow{\mathrm{Sq}} & & H^*(X) \end{array} \quad . \tag{8.3.5}$$

By (8.3.3) and Lemma 8.3.1, the map Sq is a cohomology operation, defined so far on absolute cohomology for connected CW-complexes. As $\beta_G(0)=0$, Lemma 8.1.6 implies that this partial definition of Sq extends to a unique cohomology operation $\mathrm{Sq}\colon H^*(X,Y)\to H^*(X,Y)$ defined for all topological pairs (X,Y).

For $a\in H^n(X,Y)$ let $\mathrm{Sq}^i(a)$ be the component of $\mathrm{Sq}(a)$ in $H^{n+i}(X,Y)$. Again, defining $\mathrm{Sq}^i\colon H^*(X,Y)\to H^*(X,Y)$ by

$$\mathrm{Sq}^i(\sum_{j\in\mathbb{N}} a_j)=\sum_{j\in\mathbb{N}}\mathrm{Sq}^i(a_j)\ ,\ \text{ where } a_j\in H^j(X) \tag{8.3.6}$$

provides a family of cohomology operation Sq^i. *A priori*, $i\in\mathbb{Z}$ but, by Lemma 8.1.5, $\mathrm{Sq}^i=0$ if $i<0$. It follows from these definitions that $\mathrm{Sq}=\sum_{i\in\mathbb{N}}\mathrm{Sq}^i$.

By Lemma 8.1.7, the Steenrod squares Sq and Sq^i are also cohomology operations on reduced cohomology. In this case, the following diagram is commutative.

$$\begin{array}{ccccc} \tilde H_G^*(X\wedge X) & \xrightarrow{r} & \tilde H_G^*(X) & \overset{\approx}{\longleftrightarrow} & \tilde H^*(X)[u] \\ \tilde\beta_G\uparrow & & & & \downarrow \mathrm{ev}_1 \\ \tilde H^*(X) & & \xrightarrow{\mathrm{Sq}} & & \tilde H^*(X) \end{array} \tag{8.3.7}$$

In the important case where $\tilde H^j(X)=0$ for $j<n$ (e.g. $X=\mathcal{K}_n$), one ca use Remark 8.3.2 to get the following commutative diagram

$$\begin{array}{ccccccc} \tilde H^{2n}(X\wedge X)^G & \underset{\approx}{\xrightarrow{\tilde\rho^{-1}}} & \tilde H_G^{2n}(X\wedge X) & \xrightarrow{r} & \tilde H_G^*(X) & \overset{\approx}{\longleftrightarrow} & \tilde H^*(X)[u] \\ \tilde\beta\uparrow & \nearrow \tilde\beta_G & & & & & \downarrow \mathrm{ev}_1 \\ \tilde H^n(X) & & & \xrightarrow{\mathrm{Sq}} & & & \tilde H^*(X) \end{array} \tag{8.3.8}$$

We now prove Properties (1)–(4) of Theorem 8.2.1, for the absolute cohomology $H^*(X)$ of a connected CW-complex X.

Proof of (2). That Sq^i sends $H^n(X)$ to $H^{n+i}(X)$ is by definition and we already noticed that $\mathrm{Sq}^i=0$ for $i<0$. If $a\in H^n(X)$, then $r\circ\beta_G(a)\in H_G^{2n}(X)$, which implies that $\mathrm{Sq}^i=0$ for $i>n$. Note that, from (8.3.4) and the definition of the Sq^i, one has, for $a\in H^n(X)$:

$$r\circ\beta_G(a)=\sum_{i=0}^{n}\mathrm{Sq}^i(a)\,u^{n-i}\,. \tag{8.3.9}$$

Let us prove that $\mathrm{Sq}^n(a) = a \smile a$. From the above equation, we deduce that $\mathrm{Sq}^n(a) = \mathrm{ev}_0\circ r\circ\beta_G(a)$. Using Diagrams (7.1.5) and (7.1.6), we get the diagram

$$\begin{array}{ccccccc}
H^*(X) & \xrightarrow{\beta_G} & H^*_G(X\times X) & \xrightarrow{r} & H^*_G(X) & \xleftrightarrow{\approx} & H^*(X)[u] \\
\updownarrow{=} & & \downarrow{\rho} & & & & \downarrow{\mathrm{ev}_0} \\
H^*(X) & \xrightarrow{\beta} & H^*(X\times X) & & \xrightarrow{\Delta^*} & & H^*(X)
\end{array}$$

where Δ^* is induced by the diagonal map $\Delta\colon X\to X\times X$. By (4.6.5) p. 157, one has

$$\mathrm{Sq}^n(a)=\mathrm{ev}_0\circ r\circ\beta_G(a)=\Delta^*\circ\beta(a)=\Delta^*(a\times a)=a\smile a\,.$$

It remains to prove that $\mathrm{Sq}^0(a) = a$. By naturality and since $H^n(\mathcal{K}_n) = \mathbb{Z}_2$, it suffices to find, for each integer n, some space X with a class $a\in H^n(X)$ such that $\mathrm{Sq}^0(a)\neq 0$. The space X will be the sphere S^n. Indeed, there is a homeomorphism $h\colon S^n\wedge S^n \xrightarrow{\approx} S^{2n}$ (see Example 4.7.12). We leave as an exercise to the reader to construct such a homeomorphism h which conjugates the G-action on $S^n\wedge S^n$ to a linear involution on S^{2n}. Therefore, $S^n\wedge S^n$ is G-homeomorphic to the sphere S^{2n}_n of Example 7.1.14. We now use Diagram (8.3.8). By the reduced Künneth theorem, $\tilde\beta(a)=\tilde a$, the generator of $H^{2n}(S^n\wedge S^n)=\mathbb{Z}_2$. By Proposition 7.1.15, $r\circ\tilde\rho^{-1}(\tilde a)= a\,u^n$, whence $\mathrm{Sq}^0(a)=a$.

Linearity of Sq. We have to prove that for each $n\in\mathbb{N}$, the map $\mathrm{Sq}\colon H^n(X)\to H^*(X)$ is linear. By Point (2) already proven, the restriction of Sq to H^0 is $\mathrm{Sq}^0=\mathrm{id}$, so we may assume that $n\geq 1$. By functoriality of Sq, the diagram

$$\begin{array}{ccc}
H^n(X/X^{n-1}) & \twoheadrightarrow & H^n(X) \\
\downarrow{\mathrm{Sq}} & & \downarrow{\mathrm{Sq}} \\
H^*(X/X^{n-1}) & \longrightarrow & H^*(X)
\end{array}$$

is commutative, where the horizontal maps are induced by the projection $X\to X/X^{n-1}$. Therefore, it is enough to prove the linearity of $\mathrm{Sq}\colon H^n(X/X^{n-1})\to H^*(X/X^{n-1})$. We may thus assume that $\tilde H^k(X) = 0$ for $k < n$ and use Diagram (8.3.8) to define Sq. It is then enough to show that $r\circ\tilde\rho^{-1}\circ\tilde\beta\colon H^n(X)\to H^{2n}_G(X)$ is linear. One has

$$\begin{aligned}
\tilde\beta(a+b) &= \tilde\beta(a)+\tilde\beta(b)+a\mathbin{\tilde\times} b+b\mathbin{\tilde\times} a\\
&= \tilde\beta(a)+\tilde\beta(b)+a\mathbin{\tilde\times} b+\tau^*(a\mathbin{\tilde\times} b)\,.
\end{aligned}$$

Using Proposition (7.1.12), we get

$$\tilde{\rho}^{-1} \circ \tilde{\beta}(a+b) = \tilde{\beta}(a) + \tilde{\beta}(b) + \tilde{\text{tr}}^*(a \tilde{\times} b)\,.$$

As $r \circ \text{tr}^* = 0$ by Proposition 7.1.9, this prove that $r \circ \tilde{\rho}^{-1} \circ \tilde{\beta}$ is linear.

From the already proven properties of Sq, we now deduce a structure result about $H^*_G(X \times X)$ (Proposition 8.3.3) which will be used to prove the multiplicativity of Sq. Let $N\colon H^*(X \times X) \to H^*(X \times X)$ be the **GrV**-morphism defined by $N(x \times y) = x \times y + y \times x$ and let $\mathcal{N}$ be the image of N. Note that

$$\ker N = H^*(X \times X)^G = \mathcal{D} \oplus \mathcal{N}$$

where $\mathcal{D}$ is the subgroup generated by $\{x \times x \mid x \in H^*(X)\}$. By definition of the transfer map $\text{tr}^*\colon H^*(X \times X) \to H^*_G(X \times X)$ [see (7.1.7)], one has

$$\rho \circ \text{tr}(x \times y) = x \times y + \tau^*(x \times y) = N(x \times y)\,.$$

The correspondence $(x \times x, N(y \times z)) \mapsto \beta_G(x) + \text{tr}^*(y \times z)$ produces a section $\sigma\colon \mathcal{D} \oplus \mathcal{N} \to H^*_G(X \times X)$ of $\rho\colon H^*_G(X \times X) \to H^*(X \times X)^G$. We identify $\mathcal{D} \oplus \mathcal{N}$ with $\sigma(\mathcal{D} \oplus \mathcal{N})$ and thus see $\mathcal{D} \oplus \mathcal{N}$ as a subgroup of $H^*_G(X \times X)$. Let $\hat{\mathcal{D}}$ be the $\mathbb{Z}[u]$-module generated in $H^*_G(X \times X)$ by $\mathcal{D}$. The following result is due to Steenrod (unpublished, but compare [78, Sect. 2]).

Proposition 8.3.3 *With the above identifications, the following properties hold true.*

(a) *The restriction of ρ to $\mathcal{D} \oplus \mathcal{N}$ coincides with the identity $\mathcal{D} \oplus \mathcal{N} \xrightarrow{\text{id}} H^*(X \times X)^G$.*
(b) $\mathcal{N} = \text{Ann}\,(u)$.
(c) *$\hat{\mathcal{D}}$ is a free $\mathbb{Z}_2[u]$-module with basis $\mathcal{D} - \{0\}$. In particular, $\hat{\mathcal{D}}$ is isomorphic to $\mathbb{Z}_2[u] \otimes \mathcal{D}$.*
(d) *As a $\mathbb{Z}_2[u]$-module, $H^*_G(X \times X) = \hat{\mathcal{D}} \oplus \mathcal{N}$.*

Proof Point (a) is obvious from the identification via σ. Hence, $H^*(X \times X)^G$ is the image of ρ and one has the commutative diagram

$$\begin{array}{ccccccccc}
0 & \longrightarrow & H^*_G(X \times X)/(u) & \longrightarrow & H^*(X \times X) & \xrightarrow{\text{tr}^*} & \text{Ann}\,(u) & \longrightarrow & 0 \\
 & & \downarrow{\scriptstyle \rho} & & \updownarrow{\scriptstyle =} & & \downarrow{\scriptstyle \rho} & & \\
0 & \longrightarrow & H^*(X \times X)^G & \longrightarrow & H^*(X \times X) & \longrightarrow & \mathcal{N} & \longrightarrow & 0
\end{array}$$

where the lines are exact (the upper line is the transfer exact sequence (7.1.8)). By the techniques of the five lemma, we deduce that $\rho\colon \text{Ann}\,(u) \to \mathcal{N}$ is an isomorphism. As $\sigma(N)$ is contained in the image Ann (u) of the transfer map, this proves (b). Also, we get the isomorphism

$$\rho \colon H^*_G(X \times X)/(u) \xrightarrow{\approx} H^*(X \times X)^G \,. \tag{8.3.10}$$

Let $\mathcal{B}$ be $\mathbb{Z}_2$-basis $H^*(X)$ formed by homogeneous classes. To prove (c), one has to show that

$$\sum_{a \in \mathcal{B}^0} \beta_G(a) u^{k(a)} \neq 0 \,. \tag{8.3.11}$$

for any non-empty finite subset $\mathcal{B}^0$ of $\mathcal{B}$ and any function $k \colon \mathcal{B}^0 \to \mathbb{N}$. Let $\mathcal{B}^0_{min}$ be the subset formed by the elements in $\mathcal{B}^0$ which are of minimal degree. Then

$$\mathrm{ev}_{\mathbf{1}} \circ r\Big(\sum_{a \in \mathcal{B}^0} \beta_G(a) u^{k(a)}\Big) = \sum_{a \in \mathcal{B}^0} \mathrm{Sq}(a) = \sum_{a \in \mathcal{B}^0_{min}} a + \textit{terms of higher degrees} \neq 0 \,.$$

To prove (d), let $A = H^*_G(X \times X)$ and $B = \hat{\mathcal{D}} \oplus \mathcal{N}$. If $B \neq A$, let $a \in A - B$ be of minimal degree. By the above and Sequence (7.1.8), one has $B/uB = A/uA \approx H^*(X \times X)^G$. Hence, there exists $b \in B$ and $c \in A$ such that $a = b + uc$. By the minimality hypotheses on a, one has $c \in B$ and thus $a \in B$ (contradiction). □

Proof that $\mathrm{Sq}(a \smile b) = \mathrm{Sq}(a) \smile \mathrm{Sq}(b)$ (*multiplicativity*). One has

$$\beta(a \smile b) = (a \smile b) \times (a \smile b) = (a \smile a) \times (b \smile b) = \beta(a) \smile \beta(b) \,.$$

Hence,

$$\beta_G(a \smile b) = \beta_G(a) \smile \beta_G(b) + x$$

with $x \in \ker \rho = (u)$ [the last equality was established in (8.3.10)]. We may suppose that $a \in H^m(X)$ and $b \in H^n(X)$. Let $\mathcal{V}$ be $\mathbb{Z}_2$-basis $H^{<m+n}(X)$ formed by homogeneous classes. By Proposition 8.3.3, $x = \sum_{v \in \mathcal{V}^0} \beta_G(a) u^{k(v)}$ for some finite subset $\mathcal{V}^0$ of $\mathcal{V}$, with $k(v) > 0$. Let $\mathcal{V}^0_{min}$ be the subset formed by the elements in $\mathcal{V}^0$ which are of minimal degree. Then,

$$\mathrm{Sq}(a \smile b) = \mathrm{ev}_{\mathbf{1}} \circ r \circ \beta_G(a \smile b) = \sum_{v \in \mathcal{V}^0_{min}} a + \textit{terms of higher degrees} \,.$$

Since $\mathrm{Sq}^i(a \smile b) = 0$ for $i < 0$, one has $\mathcal{V}^0_{min} = \emptyset$ and therefore $\mathcal{V}^0 = \emptyset$. This implies that $\beta_G(a \smile b) = \beta_G(a) \smile \beta_G(b)$ and thus $\mathrm{Sq}(a \smile b) = \mathrm{Sq}(a) \smile \mathrm{Sq}(b)$.

Proof that $\Sigma^* \circ \mathrm{Sq} = \mathrm{Sq} \circ \Sigma^*$. By Lemma 4.7.13, using the reduced suspension $S^1 \wedge X$, the relation $\Sigma^* \circ \mathrm{Sq} = \mathrm{Sq} \circ \Sigma^*$ is equivalent to the following equation in $H^*(S^1 \wedge X)$:

$$b \mathbin{\tilde{\times}} \mathrm{Sq}(c) = \mathrm{Sq}(b \mathbin{\tilde{\times}} c) \quad \forall c \in H^n(X) \,, \tag{8.3.12}$$

where b is the generator of $H^1(S^1) = \mathbb{Z}_2$. As noticed in Proposition 8.2.3, the formula $\mathrm{Sq}(b \smile c) = \mathrm{Sq}(b) \smile \mathrm{Sq}(c)$ already proven implies that $\mathrm{Sq}(b \,\tilde{\times}\, c) = \mathrm{Sq}(b) \,\tilde{\times}\, \mathrm{Sq}(c)$. Also, by (2) already established, $\mathrm{Sq}(b) = b$. Hence, Eq. (8.3.12) holds true.

The proof of Points (1)–(4) in Theorem 8.2.1 is now complete for pairs $(X, \emptyset)$ with X a connected CW-complex. We check easily that the extension of Sq and Sq^i to topological pairs given by Lemma 8.1.6 satisfies the same properties.

8.4 Adem Relations

The Adem relations are relations amongst the compositions $\mathrm{Sq}^i\mathrm{Sq}^j$. They were conjectured by Wu wen-Tsün around 1950 and first proved in 1952 by Adem in his thesis at Princeton University (summary in [4] and full proofs in [5]). We present below a proof based on the idea of Steenrod [184, Chap. VIII], using the equivariant cohomology for the symmetric group Sym_4. The proof in [82, Sect. 4.L] is another adaptation of the same idea. For different proofs, see [33] and Remark 8.5.10.

Let X be a topological space. Consider the map

$$\widetilde{\mathrm{Sq}} : H^*(X) \xrightarrow{\beta_G} H^*_G(X \times X) \xrightarrow{r} H^*_G((X \times X)^G) \approx H^*(BG \times X)$$

(recall that $BG \approx \mathbb{R}P^\infty$: see Example 7.2.1). The map $\widetilde{\mathrm{Sq}}$ would sit diagonally in (8.3.5), the diagram used to define Steenrod squares. By Sect. 8.3, the map $\widetilde{\mathrm{Sq}}$ is functorial, $\mathbb{Z}_2$-linear and multiplicative.

Consider now the iterated map

$$\widetilde{\mathrm{Sq}} \circ \widetilde{\mathrm{Sq}} : H^*(X) \to H^*(BG \times BG \times X) \approx H^*(X)[u, v] ,$$

using the Künneth theorem and that $H^*(BG \times BG) \approx \mathbb{Z}_2[u, v]$ with u and v in degree 1.

Proposition 8.4.1 *For any $a \in H^*(X)$, the polynomial $\widetilde{\mathrm{Sq}} \circ \widetilde{\mathrm{Sq}}(a)$ is symmetric in the variables u and v.*

Before proving Proposition 8.4.1, we do some preliminaries. Let Y be a K-space for a topological group K. The equivariant cross product $\times_K$ of Sect. 7.4 gives rise to an *equivariant cross square map* $\beta^K : H^n_K(Y) \to H^{2n}_K(Y \times Y)^G$, defined by $\beta^K(a) = a \times_K a$, where $G = \{I, \tau\}$ acting on $Y \times Y$ by exchanging the factors. A map $\tilde{\beta}^K : \tilde{H}^n_K(Y) \to \tilde{H}^{2n}_K(Y \wedge Y)^G$ is similarly defined, using the reduced equivariant cross product $\tilde{\times}_K$. The following is a generalization of Lemma 8.3.1.

Lemma 8.4.2 *Let Y be a K-space, equivariantly well pointed by $y \in Y^K$. Then,*

(1) *The cross-square maps $\tilde{\beta}^K$ and β^K admit liftings $\tilde{\beta}^K_G$ and β^K_G so that the diagram*

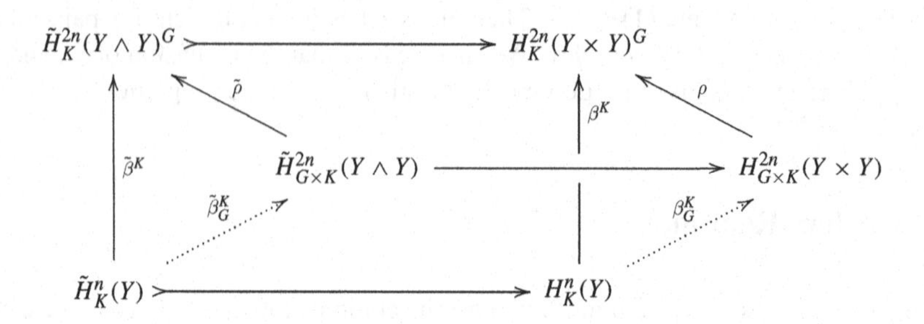

is commutative, where $\tilde{\rho}$ and ρ are induced by the homomorphism $K \to G \times K$. When K is the trivial group, these lifting coincide with those of Lemma 8.3.1.

(2) *The lifting β^K_G is functorial in Y and G, i.e. if Y' is a K'-space, $f: Y \to Y'$ is equivariant with respect to a continuous homomorphism $\varphi: K \to K'$, then the diagram*

$$\begin{array}{ccc} H^{2n}_{G\times K'}(Y' \times Y') & \xrightarrow{f^* \times f^*} & H^{2n}_{G\times K}(Y \times Y) \\ {\scriptstyle \beta^{K'}_G}\uparrow & & {\scriptstyle \beta^K_G}\uparrow \\ H^n_{K'}(Y') & \xrightarrow{f^*} & H^n_K(Y) \end{array}$$

is commutative. The analogous property holds for $\tilde{\beta}_G$.

(3) *Suppose that the K-action on Y is trivial. Then, the following diagram is commutative*

$$\begin{array}{ccccc} H^{2n}_{G\times K}(Y \times Y) & \xrightarrow{r} & H^{2n}_{G\times K}(Y) & \xrightarrow{\approx} & H^{2n}(B(G \times K) \times Y) \\ {\scriptstyle \beta^G_K}\uparrow & & & & \uparrow{\scriptstyle P}\, \approx \\ H^n_K(Y) & \xleftrightarrow{\approx} & H^n(BK \times Y) & \xrightarrow{\widetilde{\mathrm{Sq}}} & H^{2n}(BG \times BK \times Y) \end{array}$$

Proof The lifting β^K_G is defined using the lifting β_G of Lemma 8.3.1 and the commutative diagram

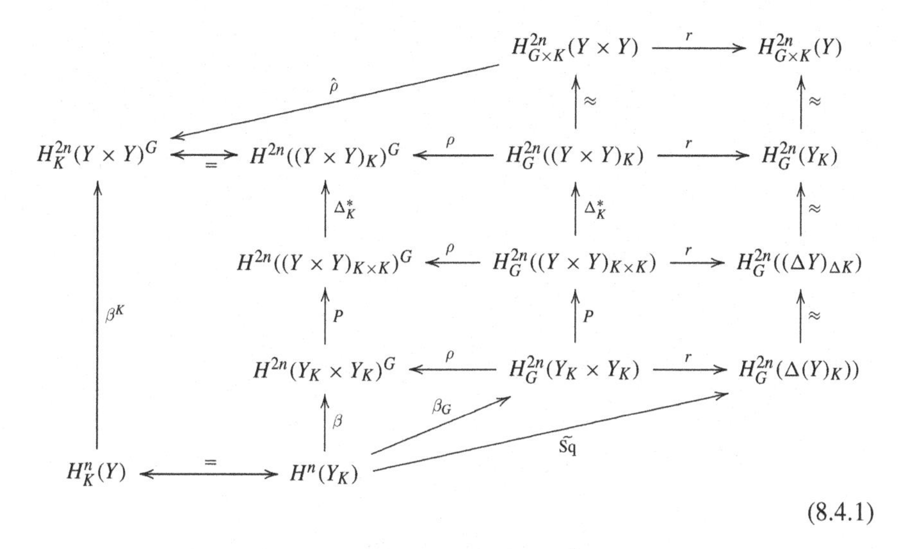

(8.4.1)

(we do not need the last column for the definition of β^K_G but it will be used later). The commutativity of the right rectangle is the definition of β^K, using the formula (7.4.7) for the equivariant cross product. The top vertical isomorphisms come from the following fact: if Z is a $(G\times K)$-space, the homotopy equivalence $E(G\times K)\times Z\to EG\times EK\times Z$ given by $\big((t_i(g_i,k_i)),z\big)\mapsto\big((t_ig_i,z),(t_ik_i,z)\big)$ descends to a homotopy equivalence $E(G\times K)_{G\times K}\times Z\to EG\times_G(EK\times_K Z)$. The same homotopy equivalence is used to get the map

$$Z_K=EK\times_K Z\to EG\times_G(EK\times_K Z)\approx E(G\times K)\times_{G\times K}Z=Z_{G\times K} \quad (8.4.2)$$

giving rise (for $Z=Y\times Y$) to the homomorphism $\hat{\rho}$. For $\tilde{\beta}^K_G$, we use the commutative diagram

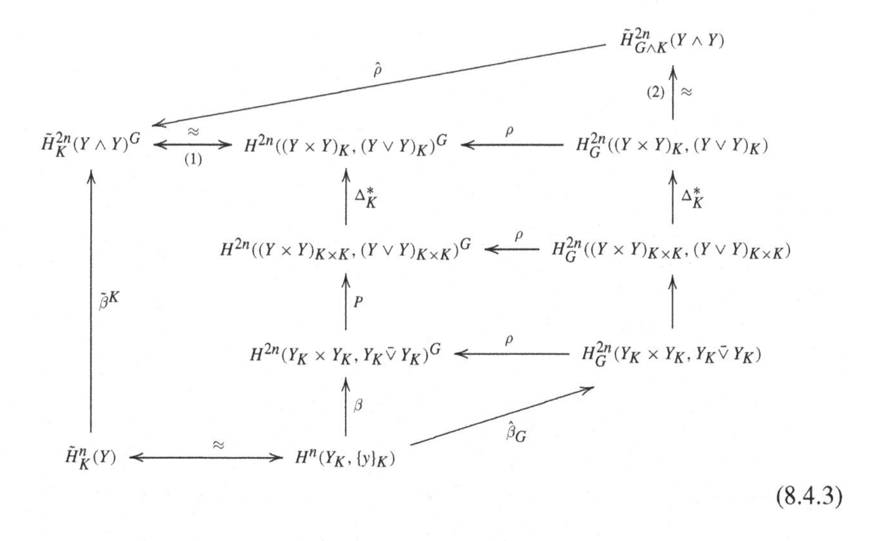

(8.4.3)

As Y is equivariantly well pointed by y, the pair $(Y_K, \{y\}_K)$ is well cofibrant, so the cross-square map β is defined. Also, the pair $(Y \times Y, Y \vee Y)$ is K-equivariantly well cofibrant (the proof is the same as for Lemma 7.2.12). Hence, Identification (1) then comes from Corollary 7.2.16. The same argument gives Identification (2), using (8.4.2) for $Z = Y \times Y$ and $Z = Y \vee Y$. The lifting $\hat{\beta}_G$ is defined using the following diagram.

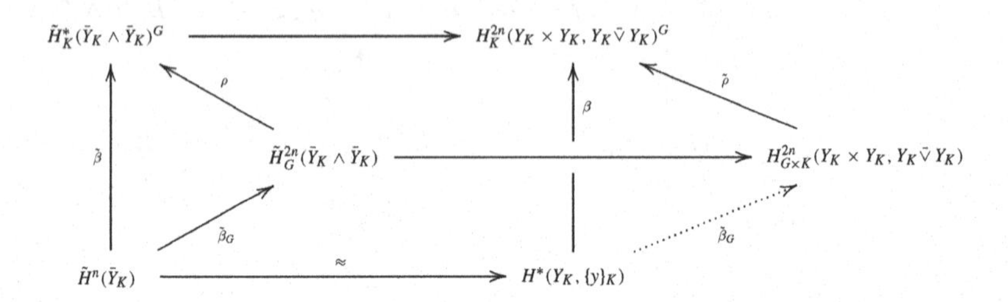

With these definitions, Diagram (8.4.3) is mapped into Diagram (8.4.1), giving rise to the diagram of Point (1) of our lemma. That $\tilde{\beta}_G^K$ and β_G^K coincide with $\tilde{\beta}_G$ and β_G when K is the trivial group follows from the definitions, as well as Point (2). Point (3) comes from (8.4.1), where the occurrence of $\widetilde{\mathrm{Sq}}$ is noticed. □

Proof of Proposition 8.4.1 By naturality, it is enough to prove the proposition for $X = \mathcal{K}_n$ and a the generator of $H^n(\mathcal{K}_n) = \mathbb{Z}_2$ $(n \geq 1)$. We consider X as pointed by $x \in X^0$, using the CW-structure given in the proof of Proposition 3.8.1, whose 0-skeleton consists in the single point x.

Let the symmetric group $\Sigma = \mathrm{Sym}_4$ act on $X^{\wedge 4} = X \wedge X \wedge X \wedge X$ and $X^4 = X \times X \times X \times X$ by permutation of the factors. This action may be restricted to the subgroup of Γ of Σ generated by $s = (1,2)(3,4)$ and $t = (1,3)(2,4)$. As for (8.3.1), the use the base points $x \in X$ and $(x,x,x,x) \in X \wedge X$ provide a commutative diagram

$$\begin{array}{ccc} \tilde{H}^*(X^{\wedge 4}) & \rightarrowtail & H^*(X^4) \\ \uparrow \tilde{\beta}\circ\tilde{\beta} & & \uparrow \beta\circ\beta \\ \tilde{H}^*(X) & \rightarrowtail & H^*(X) \end{array} \quad . \tag{8.4.4}$$

By Proposition 4.7.11, $H^k(X^{\wedge 4}) = 0$ for $k < 4n$ and $\tilde{H}^{4n}(X^{\wedge 4})^{\Sigma} = \tilde{H}^{4n}(X^{\wedge 4}) = \mathbb{Z}_2$. As in Lemma 8.3.1, using Proposition 7.2.17, we get liftings

$$(\tilde{\beta}\circ\tilde{\beta})_\Sigma : \tilde{H}^n(X) \to \tilde{H}^{4n}_\Sigma(X^{\wedge 4}) \text{ and } (\beta\circ\beta)_\Sigma : H^n(X) \to H^{4n}_\Sigma(X^4)$$

of $\tilde{\beta}\circ\tilde{\beta}$ and $\beta\circ\beta$. Consider the composite map.

$$\Phi : H^n(X) \xrightarrow{(\beta\circ\beta)_\Sigma} H^{4n}_\Sigma(X^4) \xrightarrow{r} H^*_\Sigma((X^4)^\Sigma) \approx H^*(X) \otimes H^*(B\Sigma)\,.$$

Let Γ be the subgroup of Γ generated by $s = (1,2)(3,4)$ and $t = (1,3)(2,4)$. As s and t commute, Γ is isomorphic to $G \times G$. Let $i \colon \Gamma \to \Sigma$ denote the inclusion. We shall prove that the diagram

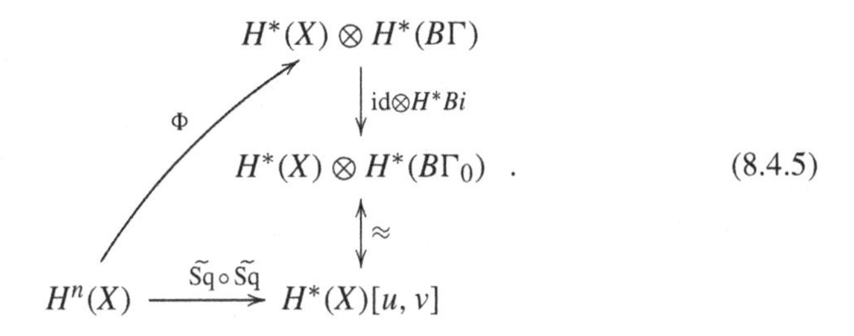

(8.4.5)

is commutative. For the moment, let us show that this property implies Proposition 8.4.1. It is enough to prove that the image of $H^*(B\Sigma) \to \mathbb{Z}_2[u,v]$ consists of symmetric polynomials. Under the isomorphism $\mathbb{Z}_2[u,v] \approx H^*(B\Gamma)$ the automorphism exchanging u and v corresponds to that induced by exchanging s and t. But in Σ, exchanging s and t is achieved by the conjugation by the transposition $(2,4)$. Such an inner automorphism of Σ induces the identity on $H^*(B\Sigma)$ by Proposition 7.2.3.

It remains to prove that Diagram (8.4.5) is commutative. Let G_1 and G_2 be the subgroups of Γ generated by s and t respectively, so $\Gamma \approx G_1 \times G_2$. The commutativity of Diagram (8.4.5) comes from that of the diagram

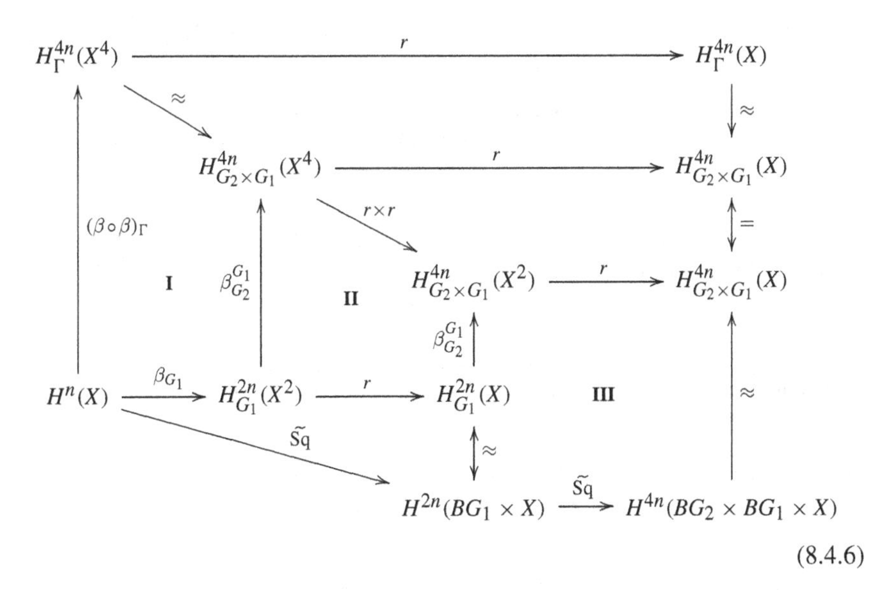

(8.4.6)

together with the obvious commutativity of the diagram

$$
\begin{array}{ccccccc}
H^{4n}_{\Sigma}(X^4) & & \xrightarrow{\quad r \quad} & & H^{4n}_{\Sigma}(X) & \xrightarrow{\approx} & H^{4n}(B\Sigma \times X) \\
\uparrow{\scriptstyle \beta^2_{\Sigma}} & \searrow{\scriptstyle j^*} & & & \downarrow{\scriptstyle j^*} & & \downarrow{\scriptstyle (Bj\times \mathrm{id})^*} \\
H^n(X) & \xrightarrow{\beta^2_{\Gamma}} & H^{4n}_{\Gamma}(X^4) & \xrightarrow{r} & H^{4n}_{\Gamma}(X) & \xrightarrow{\approx} & H^{4n}(B\Gamma \times X)
\end{array}\; .
$$

The commutativity of Diagram (8.4.6) comes from that of its subdiagram I, II and III (the commutativity of the other diagrams is obvious). Diagrams II and III commute because of Points (2) and (3) of Lemma 8.4.2. To verify the commutativity of Diagram (I), we put it as the inner square of the following diagram.

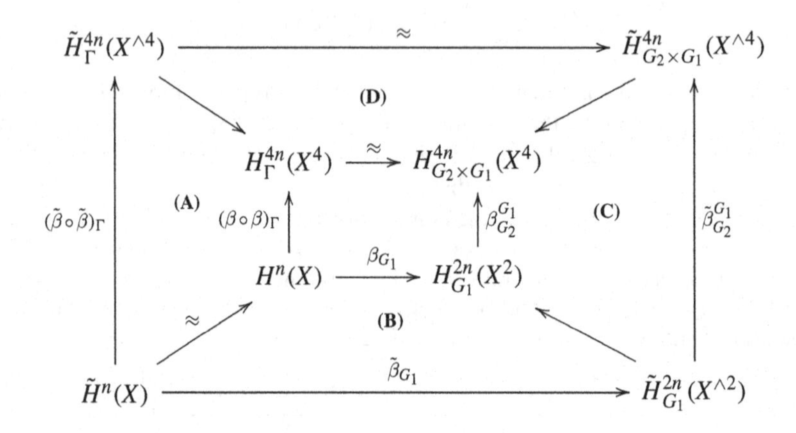

Diagrams (A), (B) and (C) commute because of Lemmas 8.3.1 and 8.4.2, and Diagram (A) is obviously commutative. As $\tilde H^n(X) \to H^n(X)$ is an isomorphism, the inner inner square will commute if the outer does. But the outer square commutes by default, since, using Proposition 7.2.17, all the groups are equal to $\mathbb{Z}_2$ and the maps are non trivial (for $\tilde\beta_{G_1}$ and $\tilde\beta^{G_1}_{G_2}$, this is checked using the restrictions to the trivial group and Point(2) of Lemma 8.4.2).

From Proposition 8.4.1, we deduce relations between $\mathrm{Sq}^i\mathrm{Sq}^j$ called the *Adem relations*.

Theorem 8.4.3 (Adem relations) *Let X be a topological space. For each $i, j \in \mathbb{N}$ with $i < 2j$, the relation*

$$
\mathrm{Sq}^i\mathrm{Sq}^j = \sum_{k=0}^{[i/2]} \binom{j-k-1}{i-2k} \mathrm{Sq}^{i+j-k}\mathrm{Sq}^k \quad (0 < i < 2j)
$$

holds amongst the non-graded endomorphisms of $H^*(X)$.

Example 8.4.4 (1) When $i = 1$, the right hand member in the Adem relation reduces to $\binom{j-1}{1}\mathrm{Sq}^{j+1}$. Hence

$$\mathrm{Sq}^1\mathrm{Sq}^j = \begin{cases} \mathrm{Sq}^{j+1} & \text{if } j \text{ is even} \\ 0 & \text{if } j \text{ is odd} \end{cases}.$$

For instance, $\mathrm{Sq}^1\mathrm{Sq}^1 = 0$ and $\mathrm{Sq}^1\mathrm{Sq}^2 = \mathrm{Sq}^3$.
(2) In the limit case $i = 2j-1$ with $j > 0$, the binomial coefficient $\binom{j-k-1}{2j-1-2k}$ vanish if $k \leq j-1$, since $2j-1-2k > j-k-1$. Then

$$\mathrm{Sq}^{2j-1}\mathrm{Sq}^j = 0 \text{ if } j > 0\,.$$

Proof of the Adem Relations Let $a \in H^n(X)$. The homomorphism $\widetilde{\mathrm{Sq}}\colon H^n(X) \to H^*(\mathbb{R}P^\infty \times X) \xrightarrow{\approx} H^*(X)[u]$ satisfies

$$\widetilde{\mathrm{Sq}}(a) = \sum_{\mu=0}^{n} \mathrm{Sq}^\mu(a)u^{n-\mu} = \sum_{\mu\in\mathbb{Z}} \mathrm{Sq}^\mu(a)u^{n-\mu}\,.$$

The range extension $\mu \in \mathbb{Z}$ is possible since the other summands vanish. We shall do that repeatedly in the computations below and, without other specification, the summations will be over the integers (with only a finitely many non-zero terms). This permits us to exchange the summation symbols.

Observe that $\widetilde{\mathrm{Sq}}(u) = u^2 + uv = u(u+v)$ thus, by multiplicativity of $\widetilde{\mathrm{Sq}}$, one has $\widetilde{\mathrm{Sq}}(u^k) = \widetilde{\mathrm{Sq}}(u)^k = u^k(u+v)^k$. Therefore

$$\begin{aligned}
\widetilde{\mathrm{Sq}}\circ\widetilde{\mathrm{Sq}}(a) &= \textstyle\sum_\mu \widetilde{\mathrm{Sq}}(u^{n-\mu})\widetilde{\mathrm{Sq}}(\mathrm{Sq}^\mu(a)) \\
&= \textstyle\sum_\mu u^{n-\mu}(u+v)^{n-\mu}\sum_\lambda \mathrm{Sq}^\lambda\mathrm{Sq}^\mu(a)v^{n+\mu-\lambda} \\
&= \textstyle\sum_{\mu,\lambda} u^{n-\mu}v^{n+\mu-\lambda}(u+v)^{n-\mu}\mathrm{Sq}^\lambda\mathrm{Sq}^\mu(a) \\
&= \textstyle\sum_{\mu,\lambda,\nu} \binom{n-\mu}{\nu} u^{n-\mu+\nu}v^{2n-\lambda-\nu}\mathrm{Sq}^\lambda\mathrm{Sq}^\mu(a)
\end{aligned}$$

Setting $\lambda + \nu = i$ yields

$$\widetilde{\mathrm{Sq}}\circ\widetilde{\mathrm{Sq}}(a) = \textstyle\sum_{\nu,\mu,i} \binom{n-\mu}{\nu} u^{n-\mu+\nu}v^{2n-i}\mathrm{Sq}^{i-\nu}\mathrm{Sq}^\mu(a)\,.$$

Setting $n - \mu + \nu = 2n - q$ yields

$$\widetilde{\mathrm{Sq}}\circ\widetilde{\mathrm{Sq}}(a) = \textstyle\sum_{\nu,q,i} \binom{2n-q-\nu}{\nu} u^{2n-q}v^{2n-i}\mathrm{Sq}^{i-\nu}\mathrm{Sq}^{q+\nu-n}(a)\,. \tag{8.4.7}$$

By Proposition 8.4.1, for each i and q, the coefficient of $u^{2n-q}v^{2n-i}$ in (8.4.7) must be equal to that of $u^{2n-i}v^{2n-q}$. This leads to the equation

$$\textstyle\sum_\nu \binom{2n-q-\nu}{\nu}\mathrm{Sq}^{i-\nu}\mathrm{Sq}^{q+\nu-n}(a) = \sum_\nu \binom{2n-i-\nu}{\nu}\mathrm{Sq}^{q-\nu}\mathrm{Sq}^{i+\nu-n}(a)\,.$$

In the left hand member, set $j = q - n$. In the right hand member, set $k = i + \nu - n$ and use the relation $\binom{x}{y} = \binom{x}{x-y}$. In both side, restrict the range of summation so that the summands are not zero for obvious reasons. This gives

$$\sum_{\nu=0}^{i} \binom{n-j-\nu}{\nu} \mathrm{Sq}^{i-\nu} \mathrm{Sq}^{j+\nu}(a) = \sum_{k=0}^{[i/2]} \binom{n-k}{i-2k} \mathrm{Sq}^{j+i-k} \mathrm{Sq}^{k}(a) \,. \tag{8.4.8}$$

Now, i and j being fixed, suppose that $n = 2^r - 1 + j$ for r large. Then, by Lemma 6.2.6,

$$\binom{n-j-\nu}{\nu} = \binom{2^r-1-\nu}{\nu} = 0 \text{ if } \nu \neq 0$$

for, the dyadic expansion of $2^r - 1 - \nu$ has a zero where that of ν has a one. Hence, the left hand member of (8.4.8) reduces to the single term $\mathrm{Sq}^i\mathrm{Sq}^j(a)$. Also,

$$\binom{n-k}{i-2k} = \binom{2^r+j-k-1}{i-2k} = \binom{j-k-1}{i-2k} \text{ if} i < 2j$$

since the length of the dyadic expansions of $i - 2k$ is not more than that of $j - 1$ and adding 2^r to $j - k - 1$ only puts a single 1 far to the left.

Thus, (8.4.8) proves the Adem relations classes of degree $2^r - 1 + j$ with r large ($2^r > \max\{i, j\}$). As $\Sigma^* \circ \mathrm{Sq} = \mathrm{Sq} \circ \Sigma^*$, Eq. (8.4.8) holds for a if and only if it holds for $\Sigma^*(a)$. But the suspension isomorphism may be iterated on a class $a \in H^n(X)$ till its degree becomes of the form $2^r - 1 + j$ with r large. This proves the Adem relations.

8.5 The Steenrod Algebra

The *Steenrod algebra* $\mathcal{A}$ is the graded $\mathbb{Z}_2$-algebra generated by indeterminates Sq^i (in degree i) and subject to the Adem relations and to $\mathrm{Sq}^0 = 1$. The properties of Steenrod squares imply that the cohomology $H^*(X)$ of a space X is an $\mathcal{A}$-module. The algebraic study of $\mathcal{A}$-modules is a rich subject (see, e.g. [212] for a survey).

Lemma 8.5.1 *As an algebra, $\mathcal{A}$ is generated by* $\{\mathrm{Sq}^n \mid n = 2^r\}$.

Proof Let $m = 2^r + s$ with $s < 2^r$ ($r \geq 1$). As $\binom{2^r-1}{s} \equiv 1 \mod 2$, the Adem relation

$$\mathrm{Sq}^s Sq^{2^r} = \mathrm{Sq}^m + \sum_{k=1}^{[s/2]} \binom{2^r-k-1}{s-2k} \mathrm{Sq}^{m-k}\mathrm{Sq}^k \quad (0 < i < 2j)$$

expresses Sq^m as a sum of $\mathrm{Sq}^i\mathrm{Sq}^j$. If $s > 0$ then $i, j < m$, which permits us to prove the lemma by induction on m. □

Here are a few examples of decompositions of Sq^i according to Lemma 8.5.1 and its proof:

$$\begin{aligned} \mathrm{Sq}^3 &= \mathrm{Sq}^1\mathrm{Sq}^2 \\ \mathrm{Sq}^5 &= \mathrm{Sq}^1\mathrm{Sq}^4 \\ \mathrm{Sq}^6 &= \mathrm{Sq}^2\mathrm{Sq}^4 + \mathrm{Sq}^5\mathrm{Sq}^1 = \mathrm{Sq}^2\mathrm{Sq}^4 + \mathrm{Sq}^1\mathrm{Sq}^4\mathrm{Sq}^1\,. \\ \mathrm{Sq}^7 &= \mathrm{Sq}^1\mathrm{Sq}^6 = \mathrm{Sq}^1\mathrm{Sq}^2\mathrm{Sq}^4 \end{aligned} \tag{8.5.1}$$

For $0 \neq a \in H^1(\mathbb{R}P^\infty)$, the formula $\mathrm{Sq}(a^{2^n}) = a^{2^n} + a^{2^{n+1}}$ shows that Sq^{2^n} is not a sum of $\mathrm{Sq}^i\mathrm{Sq}^j$ with $i, j < 2^n$. On the other hand, the Adem relations imply that

$$\mathrm{Sq}^2\mathrm{Sq}^2 = \mathrm{Sq}^3\mathrm{Sq}^1 = \mathrm{Sq}^1\mathrm{Sq}^2\mathrm{Sq}^1\,.$$

Therefore, the Sq^{2^r} do not generate $\mathcal{A}$ freely. In order to achieve that, we shall take another system of generators.

For a sequence $I = (i_1, \dots, i_k)$ of positive integers, we set $\mathrm{Sq}^I = \mathrm{Sq}^{i_1} \cdots \mathrm{Sq}^{i_k} \in \mathcal{A}$. The *degree* of I is $i_1 + \cdots + i_k$. The sequence I is called *admissible* if $i_j \geq 2i_{j+1}$. Let $\mathrm{Adm_n}$ be the (finite) set of admissible sequences of degree n. A monomial Sq^I is called *admissible* if I is admissible.

Proposition 8.5.2 *$\mathcal{A}$ is the polynomial algebra over the admissible monomials.*

The family of admissible monomials is sometimes called the *Cartan-Serre basis* of $\mathcal{A}$.

Before proving Proposition 8.5.2, we develop some preliminaries. Fix an integer n and consider

$$w_n = x_1 \cdots x_n \in H^*((\mathbb{R}P^\infty)^n) \approx \mathbb{Z}_2[x_1, \dots, x_n]\,. \tag{8.5.2}$$

As $\mathrm{Sq}(x_i) = x_i + x_i^2 = x_i(1 + x_i)$, one has

$$\mathrm{Sq}(w_n) = \prod_{i=1}^{n} \mathrm{Sq}(x_i) = w_n \prod_{i=1}^{n}(1 + x_i)\,.$$

Hence,

$$\mathrm{Sq}^k(w_n) = w_n\sigma_k \tag{8.5.3}$$

where σ_k is the k-th elementary symmetric polynomial:

$$\begin{aligned} \sigma_1 &= x_1 + \cdots + x_n \\ \sigma_2 &= x_1x_2 + \cdots x_{n-1}x_n \\ \sigma_k &= \textstyle\sum_{i_1 < \cdots < i_k} x_{i_1} \cdots x_{i_k}\,. \end{aligned} \tag{8.5.4}$$

For an integer p, we use the joker notation L_p for any polynomial in $\sigma_1, \dots, \sigma_p$. For instance, the equations $L_p = L_{p+q}$ and $\mathrm{Sq}^i(L_p) = L_{p+i}$ hold true but, as $\mathrm{Sq}^p(\sigma_p) = \sigma_p^2$ and $\mathrm{Sq}^i(\sigma_p) = 0$ for $i > p$, one has $\mathrm{Sq}^i(L_p) = L_{2p-1}$ for all i.

Lemma 8.5.3 *Let $I = (i_1, \dots, i_r)$ be an admissible sequence. Then, for $w = w_n$ with $n \geq i_1$, one has*

$$\mathrm{Sq}^I(w) = w(\sigma_{i_1} L_{i_1-1} + L_{i_1-1}) = wL_{i_1} \,.$$

Proof Only the first equation has to be proven. We proceed by induction on r. For r=1, the lemma follows from (8.5.3). Suppose that $k \geq 2$.

Let $\bar{I} = (i_2, \dots, i_r)$. By induction hypothesis, Eq. (8.5.3) and the Cartan formula, one has

$$\begin{aligned} \mathrm{Sq}^I(w) &= \mathrm{Sq}^{i_1}\mathrm{Sq}^{\bar{I}}(w) \\ &= \mathrm{Sq}^{i_1}(wL_{i_2}) \\ &= \mathrm{Sq}^{i_1}(w)L_{i_2} + \textstyle\sum_{j\geq 1} \mathrm{Sq}^{i_1-j}(w)\mathrm{Sq}^j(L_{i_2}) \\ &= w\sigma_{i_1} L_{i_2} + w \textstyle\sum_{j\geq 1} \sigma_{i_1-j} L_{2i_2-1} \\ &= w(\sigma_{i_1} L_{i_1-1} + L_{i_1-1}) \end{aligned}$$

since $i_1 \geq 2i_2$. □

Proof of Proposition 8.5.2 The Adem relations imply that any monomial Sq^I is a sum of admissible ones. It remains to see that the admissible monomials are linearly independent.

Suppose that $\sum_{I\in\mathcal{I}} \lambda_I \mathrm{Sq}^I = 0$ for $\mathcal{I} \subset \mathrm{Adm_n}$ and $\lambda_I \in \mathbb{Z}_2$. We must prove that $\lambda_I = 0$ for all I. We proceed by induction on the cardinality of $\mathcal{I}$. Equation (8.5.3) for $n \geq i_1$ proves the assertion for $\mathcal{I} = \{(i_1)\}$.

Let $\max_j \mathcal{I} = \max\{0, i_j \mid (i_1, \dots, i_k) \in \mathcal{I}\}$ and let $w = w_n$ as in (8.5.2) with $n \geq \max \mathcal{I}$. Set $\mathcal{I} = \mathcal{I}_0 \dot{\cup} \mathcal{I}_1$ where $\mathcal{I}_0 = \{(i_1, \dots, i_k) \in \mathcal{I} \mid i_1 = \max_1 \mathcal{I}\}$. By Lemma 8.5.3, one has

$$0 = \sum_{I\in\mathcal{I}} \lambda_I \mathrm{Sq}^I(w) = \sum_{I\in\mathcal{I}_0} \lambda_I \mathrm{Sq}^I(w) + \sum_{I\in\mathcal{I}_1} \lambda_I \mathrm{Sq}^I(w) = w(\sigma^{\max \mathcal{I}} L_{\max \mathcal{I}-1} + L_{\max \mathcal{I}-1}) \,.$$

Therefore,

$$\sum_{I\in\mathcal{I}_0} \lambda_I \mathrm{Sq}^I(w) = 0 \text{ and } \sum_{I\in\mathcal{I}_1} \lambda_I \mathrm{Sq}^I(w) = 0 \,. \tag{8.5.5}$$

(This uses a very easy case of the uniqueness of the expression of a symmetric polynomial as a polynomial in the σ_i's).

If the decomposition $\mathcal{I} = \mathcal{I}_0 \cup \mathcal{I}_1$ is non-trivial (i.e. $\mathcal{I} \neq \mathcal{I}_0$), (8.5.5) permits us to apply the induction hypothesis. Otherwise, we decompose $\mathcal{I}_0$ with respect to

$\max_2(\mathcal{I}_0)$: $\mathcal{I}_0 = \mathcal{I}_{00}\dot{\cup}\mathcal{I}_{01}$, where $\mathcal{I}_0 = \{(i_1, \dots, i_k) \in \mathcal{I} \mid i_2 = \max_2 \mathcal{I}_0\}$ and, as in (8.5.5), obtain that

$$\sum_{I\in\mathcal{I}_{00}} \lambda_I \mathrm{Sq}^I(w) = 0 \text{ and } \sum_{I\in\mathcal{I}_{01}} \lambda_I \mathrm{Sq}^I(w) = 0 . \tag{8.5.6}$$

If $\sharp\mathcal{I} > 1$, iterating this process will once produce a non-trivial decomposition, enabling us to use the induction hypothesis.

The proof of Proposition 8.5.2 shows that the map $A \mapsto A(w)$ sends $\{\mathrm{Sq}^I \mid I \in \mathrm{Adm_n}\}$ into a free family of $H^*((\mathbb{R}P^\infty)^n)$. This proves the following result.

Proposition 8.5.4 *Let $0 \neq a \in H^1(\mathbb{R}P^\infty)$. The evaluation map $\mathcal{A} \to H^*((\mathbb{R}P^\infty)^n)$ given by $A \mapsto A(a \times \cdots \times a)$ (n times) is injective in degree $\leq n$.*

We no turn our interest to the cohomology ring $H^*(\mathcal{K}_m)$ of the Eilenberg-MacLane complex $\mathcal{K}_m$. It contains the classes $\mathrm{Sq}^I(\iota)$ $(0 \neq \iota \in H^m(\mathcal{K}_m) = \mathbb{Z}_2)$ and the admissible monomials play an important role. Define the *excess* $e(I)$ of an admissible sequence $I = \{i_1, \dots, i_k\}$ by

$$e(I) = (i_1 - 2i_2) + (i_2 - 2i_3) + \cdots + (i_{k-1} - 2i_k) + i_k = i_1 - i_2 - \cdots - i_k .$$

The excess of an admissible monomial Sq^I is the excess of I. Here is a famous theorem of Serre [175, Sect. 2].

Theorem 8.5.5 *$H^*(\mathcal{K}_m)$ is the polynomial algebra generated by* $\mathrm{Sq}^I(\iota)$ *for I admissible of excess $< m$.*

The proof of this theorem uses spectral sequences and will not be given here. The condition $e(I) < m$ is natural: $\mathrm{Sq}^I(\iota) = 0$ is $e(I) > n$ since $i_1 = e(I)+i_2+\cdots+i_k > n + i_2 + \cdots + i_k$. If $e(I) = n$, then

$$\mathrm{Sq}^I(\iota) = (\mathrm{Sq}^{i_2} \cdots \mathrm{Sq}^{i_k})^2 = \cdots = (\mathrm{Sq}^{i_r} \cdots \mathrm{Sq}^{i_k})^{2^{r-1}}$$

where $e(i_r, \dots, i_k) < n$.

Example 8.5.6 (1) Only the empty sequence has excess 0. Then $H^*(\mathcal{K}_1)$ is the polynomial algebra generated by $\iota \in H^1(\mathcal{K}_1)$. This is not a surprise since $\mathcal{K}_1 \approx \mathbb{R}P^\infty$ by Proposition 3.8.3.
In order to formulate the other examples, observe that if $I = (i_1, \dots, i_k)$ is admissible, so is $I^+ = (2i_1, i_1, \dots, i_k)$ and $e(I^+) = e(I)$. We denote by $\mathcal{F}(I)$ the family of admissible sequences obtained from I by iterating this construction.
(2) The family of admissible monomials with excess 1 is $\mathcal{F}(1)$. Thus $H^*(\mathcal{K}_2)$ is a polynomial algebra with one generator degree $2^i + 1$, $i \in \mathbb{N}$. Its Poincaré series is

$$\mathfrak{P}_t(\mathcal{K}_2) = \prod_{i\in\mathbb{N}} \frac{1}{1 - t^{2^i+1}} .$$

For the Poincaré series of $\mathcal{K}_m$, see Lemma 8.5.13.
(3) The set of admissible monomials with excess 2 is the union of the families $\mathcal{F}(2^r+1, 2^r, \dots, 2, 1)$ for $r \geq 0$.
(4) The Poincaré series of $\mathcal{K}_m$ is computed in [175, Sect. 17].

Remark 8.5.7 The coefficient exact sequence $0 \to \mathbb{Z}_2 \to \mathbb{Z}_4 \to \mathbb{Z}_2 \to 0$ gives rise to a *Bockstein homomorphism* $\beta \colon H^*(X) \to H^{*+1}(X)$ (see [82, Sect. 3.E]). As β is functorial and not trivial, one has $\beta(\iota) \neq 0$ in $H^{n+1}(\mathcal{K}_n)$. But, by Theorem 8.5.5, the only non-trivial element in $H^{n+1}(\mathcal{K}_n)$ is $\mathrm{Sq}^1(\iota)$. By naturality of β and Sq^1, this proves that $\beta = \mathrm{Sq}^1$. This argument illustrates the following corollary of Theorem 8.5.5, saying that the actions of the Steenrod algebra on $H^n(-)$ for all $n \in \mathbb{N}$ generate all the mod 2 cohomology operations.

Corollary 8.5.8 *Let Q be a cohomology operation, and let $Q_{[n]}$ its restriction to $H^n(-)$. Then, there exists $A_n \in \mathcal{A}$ such that $Q_{[n]}(x) = A_n x$ for all $x \in H^n(X)$ and all spaces X.*

Proof By functoriality, it suffices to prove the statement for $X = \mathcal{K} = K(\mathbb{Z}_2, n)$ and $x = \iota$, the generator of $H^n(\mathcal{K})$. But $Q_{[n]}(\iota) \in H^{\geq n}(\mathcal{K})$ by Lemma 8.1.5 and $H^{\geq i}(\mathcal{K}) = \mathcal{A} \cdot \iota$ by Theorem 8.5.5. □

We now list other corollaries of Theorem 8.5.5. The following one comes from Proposition 8.5.4.

Corollary 8.5.9 *Let $0 \neq a \in H^1(\mathbb{R}P^\infty)$ and let $y = a \times \cdots \times a \in H^n((\mathbb{R}P^\infty)^n)$. Let $f_y \colon (\mathbb{R}P^\infty)^n \to \mathcal{K}_n$ such that $Hf_w(\iota) = y$. Then, $H^*f \colon H^i(\mathcal{K}_n) \to H^i((\mathbb{R}P^\infty)^n)$ is injective for $i \leq 2n$.*

Remark 8.5.10 The proofs of both Theorem 8.5.5 and Proposition 8.5.4, and then that of Corollary 8.5.9, do not use the Adem relations. Thus, one can use Corollary 8.5.9 to give an alternative proof of the Adem relations, as in [175, Sect. 33], [154, pp. 29–31] or [27].

Corollary 8.5.11 *The Adem relations are the only relations amongst the* Sq^I*'s which hold true for all spaces.*

Proof A relation amongst the Sq^I's would be of the form $P(\mathrm{Sq}^{I_1}, \dots, \mathrm{Sq}^{I_r}) = 0$, where P is a $\mathbb{Z}_2$-polynomial in r variables. The Adem relations imply that any monomial Sq^I is a sum of admissible ones. Therefore, there is a relation of the form $\bar{P}(\mathrm{Sq}^{J_1}, \dots, \mathrm{Sq}^{J_s}) = 0$, where P is a $\mathbb{Z}_2$-polynomial in s variables and $J_1, \dots, J_s$ are admissible sequences. Let m be the maximal excess of $J_1, \dots, J_s$. For ι the generator of $H^m(K(\mathbb{Z}_2, m))$, the equation $\bar{P}(\mathrm{Sq}^{J_1}(\iota), \dots, \mathrm{Sq}^{J_s}(\iota)) = 0$ implies, by Theorem 8.5.5, that $\bar{P} = 0$. Hence, the original relation $P(\mathrm{Sq}^{I_1}, \dots, \mathrm{Sq}^{I_r}) = 0$ was a consequence of the Adem relations. □

Another consequence of Theorem 8.5.5 is that Steenrod squares are characterized by some of their properties listed in Theorem 8.2.1.

Proposition 8.5.12 *Suppose that for each CW-complex X, there exists a map $P: H^*(X) \to H^*(X)$ satisfying the following properties.*

(a) *If $g: Y \to X$ is a continuous map, then $H^*g \circ P = P \circ H^*g$.*
(b) $P(H^n(X)) \subset H^{\leq 2n}(X)$.
(c) *If $a \in H^1(\mathbb{R}P^\infty)$ then $P(a) = a + a \smile a$.*
(d) *$P(x \smile y) = P(x) \smile P(y)$ for all $x, y \in H(X)$.*

Then $P = \mathrm{Sq}$.

Proof Using (a) and (d) together with the definition of the cross product, we get

(d') $P(x \times y) = P(x) \times P(y)$ for all $x \in H(X)$ and $y \in H(Y)$.

Let $w = x_1 \dots x_n \in H^*((\mathbb{R}P^\infty)^n) \approx \mathbb{Z}_2[x_1, \dots, x_n]$. Using (c) and (d') we prove, as for (8.5.3), that $P(w) = w\,\sigma_k$. But, by Formula (8.5.3) again, this shows that $P(w) = \mathrm{Sq}(w)$. Using (a), (b) and Corollary 8.5.9, we deduce that $P = \mathrm{Sq}$ on $H^n(\mathcal{K}_n)$. By (a), this proves that $P = \mathrm{Sq}$ in general. □

As a last application of Theorem 8.5.5, we compute the Poincaré series of $\mathcal{K}_m$, following [175, Sect. 17]. By Theorem 8.5.5, one has

$$\mathfrak{P}_t(\mathcal{K}_m) = \prod_{r=0}^{\infty} \frac{1}{1 - t^{m+a(r)}},$$

where

$$a(r) = \sharp\,\{I \mid I \text{admissible}, e(I) < m \quad\text{and}\quad \deg(I) = r\}.$$

To compute $a(r)$, we note that an admissible sequence $I = (i_1, \dots, i_k)$ is determined by its *excess components* $\alpha_1 = i_1 - 2i_2, \dots, \alpha_{k-1} = i_{k-1} - 2i_k$, $\alpha_k = i_k$. Therefore,

$$a(r) = \sharp\,\{(\alpha_1, \dots, \alpha_k) \mid \sum_{i=1}^{k} \alpha_i < m \quad\text{and}\quad \sum_{i=1}^{k} \alpha_i(2^i - 1) = r\}. \tag{8.5.7}$$

Set $\alpha_0 = m - 1 - \sum_{i=1}^k \alpha_i$. Then

$$m + r = 1 + \sum_{i=0}^{k} \alpha_i 2^i = 1 + \underbrace{2^0 + \dots + 2^0}_{\alpha_0} + \dots + \underbrace{2^k + \dots + 2^k}_{\alpha_k}. \tag{8.5.8}$$

Using that $\sum_{i=0}^k \alpha_i = m - 1$ and writing the power of 2 in (8.5.8) in decreasing order $h_1 \geq \dots \geq h_{m-1}$, we get

$$m + a(r) = \sharp\,\{(h_1, \dots, h_{m-1}) \in \mathbb{N}^{r-1}) \mid h_1 \geq \dots \geq h_{m-1} \text{and} 2^{h_1} + \dots + 2^{h_r} + 1 = m + r\}.$$

This proves the following result of [175, Sect. 17].

Lemma 8.5.13 *The Poincaré series of $\mathcal{K}_m$ is*

$$\mathfrak{P}_t(\mathcal{K}_m) = \prod_{h_1 \geq \cdots \geq h_{m-1} \geq 0} \frac{1}{1 - t^{2^{h_1} + \cdots + 2^{h_{m-1}} + 1}} .$$

Let $r < m$. If I is an admissible sequence with $\deg(I) = r$, the condition $e(I) < m$ is automatic since $e(I) \leq \deg(I)$. Using (8.5.7), we see that $a(r)$ is equal to the number of partitions of r into integers of the form $2^i - 1$. Also, $H^{m+r}(\mathcal{K}_m)$ only contains classes of the form $\mathrm{Sq}^I(\iota)$ (products like $\mathrm{Sq}^I(\iota)\mathrm{Sq}^J(\iota)$ have higher degree). This proves the following result of [191, p. 37].

Lemma 8.5.14 *If $r < m$, then* $\dim H^{m+r}(\mathcal{K}_m)$ *is equal to the number of partitions of r into integers of the form $2^i - 1$.*

8.6 Applications

Suspension of the Hopf maps. Recall that the non triviality of the cup-square map $\alpha(a) = a \smile a$ is $H^*(\mathbb{K}P^2)$ for $\mathbb{K} = \mathbb{R}, \mathbb{C}, \mathbb{H}$ or $\mathbb{O}$ implies that the Hopf maps

$$h_{1,1} \colon S^1 \to S^1 \ , \ h_{3,2} \colon S^3 \to S^2 \ , \ h_{7,4} \colon S^7 \to S^4 \text{ and } h_{15,8} \colon S^{15} \to S^8$$

are not homotopic to a constant maps (see Corollary 6.1.9). This argument cannot be applied to the suspensions of the Hopf maps $\Sigma^k h_{p,q} \colon S^{p+k} \to S^{q+k}$ since the cup product in $H^{>0}(\Sigma^k \mathbb{K}P^2)$ vanish by dimensional reasons (also by Corollary 4.4.4). But, for instance in $\mathbb{R}P^2$, $\alpha(a) = \mathrm{Sq}^1(a)$. As $\Sigma^* \circ \mathrm{Sq} = \mathrm{Sq} \circ \Sigma^*$, one deduces that Sq^1 is not trivial on $\Sigma^k \mathbb{R}P^2$ and therefore $\Sigma^k h_{1,1}$ is not homotopic to a constant map for all $k \in \mathbb{N}$ [though $H^{k+1}\Sigma^k h_{1,1}$ vanishes on $H^{k+1}(\Sigma^k \mathbb{R}P^2)$]. The same argument applies for the other Hopf maps, so we get the following proposition.

Proposition 8.6.1 *For all $k \geq 0$, the k-th suspension of the Hopf maps*

$$\Sigma^k h_{1,1} \colon S^{k+1} \to S^{k+1} \ , \ \Sigma^k h_{3,2} \colon S^{k+3} \to S^{k+2} \ ,$$

$$\Sigma^k h_{7,4} \colon S^{k+7} \to S^{k+4} \text{ and } \Sigma^k h_{15,8} \colon S^{k+15} \to S^{k+8}$$

are not homotopic to a constant maps.

Actually, for $k \geq 1$, $\Sigma^k h_{3,2}$ represents the generator of $\pi_{k+3}(S^{k+2}) \approx \mathbb{Z}_2$ (see [197, Proposition 5.1]).

Restrictions on cup-squares. The action of the Steenrod algebra on the cohomology imposes strong restrictions for the existence of classes with non-vanishing cup-square. Let $\mathcal{A}_{<n}$ be the subalgebra of $\mathcal{A}$ generated by $\{\mathrm{Sq}^i \mid i < n\}$.

Proposition 8.6.2 *Let X be a topological space and let $a \in H^m(X)$. If m is not a power of 2, then $a \smile a \in \mathcal{A}_{<m}(a)$.*

Proof By Lemma 8.5.1, $\mathcal{A}$ is generated by $\{\mathrm{Sq}^k \mid k = 2^i\}$, so $\mathrm{Sq}^m \in \mathcal{A}_{<m}$ if m is not a power of 2. As $a \smile a = \mathrm{Sq}^m(a)$, this proves the proposition. □

Corollary 8.6.3 *Let X be a topological space. Let $a \in H^m(X)$ such that $a \smile a \neq 0$. Then, there exists $k \leq m$ with $k = 2^i$ such that $\mathrm{Sq}^k(a) \neq 0$.*

As a consequence of Proposition 8.6.2, if $a \in H^m(X)$ satisfies $a \smile a \neq 0$ with m not a power of 2, there must be a non-zero group $H^{m+k}(X)$ with $0 < k < m$. This is not the case for the space $X = C_f = D^{2m} \cup_f S^m$ used in Sect. 6.3 to define the Hopf invariant of $f: S^{2m-1} \to S^m$. Therefore, Proposition 8.6.2 has the following corollary, a result due to Adem [4].

Corollary 8.6.4 *Suppose that $f: S^{2m-1} \to S^m$ satisfies* Hopf $(f) = 1$. *Then $m = 2^r$.*

When m is a power of 2, Proposition 8.6.2 is wrong, as seen with the projective spaces $\mathbb{K}P^2$. Using secondary cohomology operations, Adams proved deeper results [2, Theorem 4.6.1 and Sect. 1.2] implying the following theorem.

Theorem 8.6.5 *Let X be a space such that $H^{m+k}(X) = 0$ for $1 \leq k \leq 2^r$. If $r \geq 3$, then $\mathrm{Sq}^{2^{r+1}}(H^m(X)) = 0$.*

Combining this theorem with Corollary 8.6.4, Adams got his famous result [2, Theorem 1.1.1]:

Theorem 8.6.6 *Continuous maps $f: S^{2m-1} \to S^m$ with Hopf invariant 1 exist only for $m = 1, 2, 4, 8$.*

Remark 8.6.7 Recall that a non-singular map $\mu: \mathbb{R}^m \times \mathbb{R}^m \to \mathbb{R}^m$ (see Sect. 6.2.2) determines, using (6.2.2), a continuous multiplication $\tilde{\mu}: S^{m-1} \times S^{m-1} \to S^{m-1}$, giving rise to a map $f_{\tilde{\mu}}: S^{2m-1} \to S^m$ with Hopf $(f_{\tilde{\mu}}) = 1$ (Proposition 6.3.3). Thus, by Theorem 8.6.6, non-singular maps $\mathbb{R}^m \times \mathbb{R}^m \to \mathbb{R}^m$ exit only for $m = 1, 2, 4, 8$. Note that Corollary 8.6.4 gives another proof of Proposition 6.2.5.

Somehow related to Theorem 8.6.6 are the results of E. Thomas on the cohomology of H-spaces (see e.g. [194]). They also heavily use Steenrod square techniques.

A variant of the Sullivan conjecture. If $f: R \to S$ is a continuous map, then H^*f is a morphism of algebras over the Steenrod algebra $\mathcal{A}$. The following result, conjectured by H. Miller, was proven by Lannes [124, Theorem 0.4].

Theorem 8.6.8 *Let Y be a simply connected space of finite cohomology type. Let $B = (\mathbb{R}P^\infty)^n$. Then the map*

$$[B, Y] \to hom_{\mathcal{A}}(H^*(Y), H^*(B))$$

is a bijection.

In particular, if Y is a finite dimensional CW-complex, then $[B, Y]$ is a singleton. This is a weak version of the Sullivan conjecture (see Remark 6.1.4).

8.7 Exercises for Chapter 8

8.1. Let $T = S^1 \times S^1$ and let K be the Klein bottle. Is ΣT homotopy equivalent to ΣK?

8.2. In $H^*(\mathbb{R}P^\infty \times \mathbb{R}P^\infty) \approx \mathbb{Z}_2[a, b]$, compute $\mathrm{Sq}^6(a^5 b^7)$.

8.3. Let $a \in H^m(X)$ such that $\mathrm{Sq}(a) \neq a$. Prove that the smallest integer i such that $\mathrm{Sq}^i(a) \neq 0$ is a power of 2.

8.4. Verify the relations in the Steenrod algebra given in (8.5.1).

8.5. Check that $\mathrm{Sq}^2\mathrm{Sq}^2\mathrm{Sq}^2\mathrm{Sq}^2 = 0$ and that $\mathrm{Sq}^3\mathrm{Sq}^3\mathrm{Sq}^3 = 0$

8.6. Find a minimal set of generators of $H^*(\mathbb{R}P^\infty)$ as a module over the Steenrod algebra.

8.7. Let X be a space such that there is an **GrA**-epimorphism $\mathbb{Z}_2[a] \twoheadrightarrow H^*(X)$, with a of degree r. Prove that $r = 2^k$ with $k \leq 3$.

8.8. Let $S^{r-1} \to S^{n-1} \xrightarrow{p} M$ be a locally trivial fiber bundle, where M is a closed manifold of positive dimension. Prove that $n = kr$ and $r = 1, 2, 4, 8$.

8.9. Let M be a closed connected manifold with total Betti number 3. What can be said about its dimension?

8.10. Let M be a smooth closed G-manifold (G of order 2). Let $f : M \to \mathbb{R}$ be a G-invariant Morse-Bott function satisfying the hypotheses of Theorem 7.6.6. Suppose that f has only two critical values ± 1, with $M_{-1} = pt$ and $\dim M_1 > 0$. Suppose that $n = \dim M$ is odd. Prove that the pair (M, M_1) is homeomorphic to $(\mathbb{R}P^n, \mathbb{R}P^{n-1})$. [Hint: this uses Exercise 8.8.]

8.11. Let Sq_* be the cohomology operation defined as follows: for $a \in H^n(X)$, we set $\mathrm{Sq}_i(a) = \mathrm{Sq}^{n-i}(a)$ (these Sq_i's were the first operations defined by Steenrod: see [40, Chap. 6]).

(a) Prove that $\mathrm{Image}(\mathrm{Sq}_0) \subset \ker \mathrm{Sq}_1$.
(b) Prove that $\mathrm{Image}(\mathrm{Sq}_0) = \ker \mathrm{Sq}_1$ for $X = \mathbb{R}P^\infty$.
(c) Is (b) true for $X = \mathbb{C}P^\infty$?

Chapter 9
Stiefel-Whitney Classes

In Sect. 4.7.6, we encountered the Euler class $e(\xi) \in H^n(B)$ of a vector bundle ξ of rank n over B, which sometimes distinguishes between non-isomorphic vector bundles. This is an example of a *characteristic class*, meaning that it is natural with respect to induced bundles: $e(f^*\xi) = H^*f(e(\xi))$ for any continuous map $f : B' \to B$. Characteristic classes play a major role in the study of vector bundles.

In Sect. 9.4, we shall see that the Euler class belongs to a family $w_i(\xi) \in H^i(B)$ of characteristic classes called the *Stiefel-Whitney classes*. Historically, these were the first characteristic classes discovered, in the simultaneous work of Eduard Stiefel and Hassler Whitney starting around 1935 (see [40, pp. 421–426]). Using Morse theory, we shall see in Sect. 9.5 that the cohomology of the Grassmannians is **GrA**-generated by the Stiefel-Whitney classes of their tautological bundles. This implies that any characteristic class in mod 2 cohomology is a polynomial in the Stiefel-Whitney classes (see Corollary 9.5.10). The Chern classes of a complex vector bundle, when reduced mod 2, are also Stiefel-Whitney classes (see Remark 9.7.6).

Geometric interpretations of w_1 and w_2 are given in Sects. 9.2 and 9.3, related to orientations and spin structures. Stiefel-Whitney classes have important applications in the topology of manifolds, for instance via the Wu formula (see Sect. 9.8) or in the work by R. Thom on cobordism, surveyed in Sect. 9.9.2.

9.1 Trivializations and Structures on Vector Bundles

We recall below some classical facts about vector bundles (defined on p. 186). Two vector bundles $\xi = (p : E \to X)$ and $\xi' = (p' : E' \to X)$ over the same space X are *isomorphic* if there exists a homeomorphism $h : E \to E'$ such that $p' \circ h = p$ and such that the restriction of h to each fiber is linear. Two such isomorphisms $h_0, h_1 : E \to E'$ are *isotopic* if there exists a family of isomorphisms $h_t : E \to E'$ depending continuously on $t \in [0, 1]$ and joining h_0 to h_1. Unless otherwise mentioned, the total space of a vector bundle ξ is denoted by $E(\xi)$ and the bundle projection by p.

J.-C. Hausmann, *Mod Two Homology and Cohomology*, Universitext,
DOI 10.1007/978-3-319-09354-3_9

Let ξ be a vector bundle of rank r over Y and let $f : X \to Y$ be a continuous map. The *induced vector bundle* $f^*\xi$ is the vector bundle of rank r over X defined by

$$E(f^*\xi) = \{(x, z) \in X \times E(\xi) \mid f(x) = p(z)\},$$

with the projection onto X as the bundle projection. The projection to $E(\xi)$ gives a map $\tilde{f} : E(f^*\xi) \to E(\xi)$ which is a linear isomorphism on each fiber and such that the diagram

$$\begin{array}{ccc} E(f^*\xi) & \xrightarrow{\tilde{f}} & E(\xi) \\ \Big\downarrow{\scriptstyle \mathrm{pr}_X} & & \Big\downarrow{\scriptstyle p} \\ X & \xrightarrow{f} & Y \end{array} \tag{9.1.1}$$

is commutative. Note that if f and g coincide over $A \subset X$, the induced bundles $f^*\xi$ and $g^*\xi$ restrict to the same bundle over A. Two such maps are said *homotopic relative to* A if there is a homotopy $F : X \times I \to X$ between f and g such that $F(a, t) = f(a) = g(a)$ for all $a \in A$. The next proposition is proven in [181, Theorem 11.3].

Proposition 9.1.1 *Let ξ be a vector bundle over a space B. Let $f, g : X \to B$ be two maps, where X is a paracompact space. Suppose that f and g coincide over $A \subset X$. If f and g are homotopic relative to A, then there exists an isomorphism between $f^*\xi$ and $g^*\xi$ which is the identity over A.*

A *trivialization* of a vector bundle ξ over X is an isomorphism of ξ with the *product bundle* $\eta_r = (pr_X : X \times \mathbb{R}^r \to X)$. A vector bundle admitting a trivialization is called *trivial*. Denote by $\tilde{\mathcal{T}}(\xi)$ the space of trivializations of ξ (endowed with the compact-open topology). Then, $\mathcal{T}(\xi) = \pi_0(\tilde{\mathcal{T}}(\xi))$ is the set of isotopy classes of trivializations of ξ is $\pi_0(\tilde{\mathcal{T}}(\xi))$.

By post-composition, the topological group Aut (η_r) of the automorphisms of the product bundle η_r acts continuously on the left on $\tilde{\mathcal{T}}(\xi)$. As usual we denote by $GL(r, \mathbb{R})$ the *general linear group*, i.e. the topological group of automorphisms of $\mathbb{R}^r$. Alternatively, $GL(r, \mathbb{R})$ is the group of invertible $(r \times r)$-matrices with real coefficients, topologized as an open set of $\mathbb{R}^{r^2}$. Note that Aut $(\eta_r) \approx \mathrm{Map}(X, GL(r, \mathbb{R}))$, so the group $\pi_0(\mathrm{Map}(X, GL(r, \mathbb{R}))) = [X, GL(r, \mathbb{R})]$ acts on $\mathcal{T}(\xi)$.

Lemma 9.1.2 *Let ξ be a trivial vector bundle over a topological space X. Then the action*

$$[X, GL(r, \mathbb{R})] \times \mathcal{T}(\xi) \to \mathcal{T}(\xi) \tag{9.1.2}$$

is simply transitive. In particular, given a trivialization T_0 of ξ, the correspondence $\lambda \mapsto \lambda \cdot T_0$ induces a bijection $[X, GL(r, \mathbb{R})] \xrightarrow{\approx} \mathcal{T}(\xi)$.

Proof Let $h, h' \in \tilde{\mathcal{T}}(\xi)$. Then $h' = (h' \circ h^{-1}) \circ h$ and $h' \circ h^{-1} \in \text{Aut}\,(\eta_r)$. This shows that $\text{Aut}\,(\eta_r)$ acts transitively on $\mathcal{T}(\xi)$, whence the action (9.1.2) is transitive.

On the other hand, if $h_1 = \lambda \cdot h_0 \in \mathcal{T}(\xi)$ is isotopic to h_0 via a homotopy $h_t \in \mathcal{T}(\xi)$, the formula $\lambda_t = h_t \circ h_0^{-1}$ defines a continuous path $\lambda_t \in \text{Aut}\,(\eta_r)$ joining $\lambda_0 = \text{id}$ to $\lambda_1 = \lambda$. This shows that $[X, GL(r, \mathbb{R})]$ acts simply on $\mathcal{T}(\xi)$. □

Remark 9.1.3 The bijection of Lemma 9.1.2 depends on the choice of $[h_0] \in \mathcal{T}(\xi)$. In general, there is no natural choice, except for the product bundle (like in the next example).

Example 9.1.4 We see the projection $S^1 \times \mathbb{C} \to S^1$ as the product vector bundle ξ of rank 2 over the circle, identifying $\mathbb{C}$ with $\mathbb{R}^2$. By Lemma 9.1.2

$$\mathcal{T}(\xi) \approx [S^1, GL(2, \mathbb{R})] \approx [S^1, O(2)] \approx \{\pm 1\} \times \mathbb{Z}$$

(by Gram-Schmidt orthonormalization, $GL(2, \mathbb{R})$ has the homotopy type of $O(2)$ which is homeomorphic to $\{\pm 1\} \times S^1$). The fiber linear map corresponding to $(1, k)$ is given by $(x, z) \mapsto x^k z$ and that corresponding to $(-1, k)$ is given by $(x, z) \mapsto x^k \bar{z}$.

Example 9.1.5 Let ξ be a vector bundle over Y and let $f : X \to Y$ be a continuous map. If ξ is trivial, so is $f^*\xi$. More precisely, let $h : E(\xi) \to Y \times \mathbb{R}^r$ be a trivialization of ξ. Write h under the form $h(z) = (p(z), \hat{h}(z))$ where $\hat{h} : E(\xi) \to \mathbb{R}^d$ is a map whose restriction to each fiber is a linear isomorphism. Then the map $f^*h : E(f^*\xi) \to X \times \mathbb{R}^r$ given by $f^*h(x, z) = (x, \hat{h}(z))$ is a trivialization of $f^*\xi$. This process descends to a map

$$f^* : \mathcal{T}(\xi) \to \mathcal{T}(f^*\xi)$$

which is equivariant for the actions defined in (9.1.2), via the homomorphism $f^* : [Y, GL(r, \mathbb{R})] \to [X, GL(r, \mathbb{R})]$ induced by $\lambda \mapsto \lambda \circ f$. Indeed, for $\lambda : Y \to GL(r, \mathbb{R})$ and $T \in \tilde{\mathcal{T}}(\xi)$, the following formula holds true in $\tilde{\mathcal{T}}(f^*\xi)$.

$$f^*(\lambda\, T) = (f \circ \lambda)\, f^*T. \tag{9.1.3}$$

Proposition 9.1.6 *Let ξ be a vector bundle over a space X. If X is paracompact and contractible, then ξ is trivial. Moreover, any trivialization of the restriction of ξ over a point $x_0 \in X$ extends to a trivialization of ξ which is unique up to isotopy.*

Proof Let $\xi = (p : E \to X)$ and let $\xi_0 = (p : E_0 \to \{x_0\})$ be the restriction of ξ over the point x_0. The vector bundle ξ is induced by the identity of X: $\xi = \text{id}_X^*\xi$. As X is contractible, id_X is homotopic to a constant map c onto x_0 and, by Proposition 9.1.1, $\xi \approx c^*\xi_0$. The contractability of X also implies that $c^* : [\{x_0\}, GL(r, \mathbb{R})] \to [X, GL(r, \mathbb{R})]$ is an isomorphism. Proposition 9.1.6 then follows from Lemma 9.1.2 and the considerations of Example 9.1.5. □

We now consider a vector bundle ξ over a space $X = Z \cup_\varphi Y$ obtained by attaching a pair (Z, A) to a space Y using an attaching map $\varphi : A \to Y$ (see p. 97). Let $j_Y : Y \to X$ and $j_Z : Z \to X$ be the natural maps. Let $\xi_Y = j_Y^* \xi$, $\xi_Z = j_Z^* \xi$ and denote by ξ_A the restriction of ξ_Z over A.

Lemma 9.1.7 *Let ξ be a vector bundle over $X = Z \cup_\varphi Y$ as above. Suppose that there exists trivializations $h_Y \in \tilde{\mathcal{T}}(\xi_Y)$ and $h_Z \in \tilde{\mathcal{T}}(\xi_Z)$ such that the restriction of h_Z to ξ_A is isòtopic to $\varphi^* h_Y$. Suppose that Z is paracompact and that the pair (Z, A) is cofibrant. Then, the bundle ξ is trivial.*

Proof Let $h_A \in \mathcal{T}(\xi_A)$ be the restriction of h_Z to ξ_A. An isotopy from h_A to $\varphi^* h_Y$ produces a trivialization $\bar{h} \in \mathcal{T}(\xi_A \times I)$ (where $\xi_A \times I$ is the vector bundle $p \times \mathrm{id} : E(\xi_A) \times I \to A \times I$) restricting to h_A on $\xi_A \times \{0\}$ and to $\varphi^* h_Y$ on $\xi_A \times \{1\}$. The pair (Z, A) being cofibrant, there exists a retraction $r : Z \times I \to Z \times \{0\} \cup A \times I$ of $Z \times I$ onto $Z \times \{0\} \cup A \times I$ (see Lemma 3.1.37). Let ξ_1 be the restriction of $r^* \xi_Z$ above $Z \times \{1\} \approx Z$. As Z is paracompact and r is actually a strong deformation retraction (see Remark 3.1.42), ξ_1 is isomorphic to ξ_Z relative to A by Proposition 9.1.1. Then, $r^* \bar{h}$ produces at level 1 a trivialization $\bar{h}_Z$ of ξ_Z which is *equal* to $\varphi^* h_Y$ on $E(\xi_A)$. This condition implies that the formula

$$H(u) = \begin{cases} \bar{h}_Z(z, u) & \text{if } p(u) = i_Z(z) \\ h_Y(y, u) & \text{if } p(u) = i_Y(y) \end{cases}$$

defines a trivialization $H : E(\xi) \to X \times \mathbb{R}^r$ of ξ. □

The above lemmas have analogues for principal bundles. Let A be a topological group. Recall that a A-*principal bundle* over X consists of a continuous map $p : P \to X$, a continuous right action of A on P such that $p(z\alpha) = p(z)$ for all $z \in P$ and all $\alpha \in A$. In addition, the following local triviality should hold: for each $x \in X$ there is a neighbourhood U of x and a homeomorphism $h : U \times A \to p^{-1}(U)$ such that $p \circ h(x, a) = x$ and $h(x, a\alpha) = h(x, a)\alpha$. In consequence, A acts simply transitively on each fiber and p is a surjective open map, thus descending to a homeomorphism $P/A \xrightarrow{\approx} X$ (use [44, Sect. I of Chap. VI]). An isomorphism of A-principal bundles from $p : P \to X$ to $p' : P' \to X$ is a A-equivariant homeomorphism $h : P \to P'$ such that $p' \circ h = p$. Two such isomorphisms $h_0, h_2 : P \to P'$ are *isotopic* if there exists a family of isomorphisms $h_t : P \to P'$ depending continuously on $t \in [0, 1]$ and joining h_0 to h_1. The notion of induced A-principal bundle works as for vector bundles and Proposition 9.1.1 also follows from [181, Theorem 11.3].

A *trivialization* of an A-principal bundle is an isomorphism with the product bundle $X \times A \to X$. An A-principal bundle is trivial if and only if it admits a section (see, e.g. [181, I.8]). This gives a bijection between sections and trivializations and between homotopy classes of sections and isotopy classes of trivialization. If $\sigma_1, \sigma_2 : X \to P$ are two sections of p, then $\sigma_2(x) = \sigma_1(x)\alpha(x)$ for a unique $\alpha : \mathrm{Map}(X, A)$ whence the analogue of Lemma 9.1.2: *the action of* $\mathrm{Map}(X, A)$ *on the trivializations of P is simply transitive*. Also, Proposition 9.1.6 holds true for principal bundles.

Let $p : P \to X$ be an A-principal bundle, $p' : P' \to X$ be an A'-principal bundle and let $\varphi : A \to A'$ be a continuous homomorphism. A continuous map $g : P \to P'$ is called a *φ-map* if $p' \circ g = p$ and $g(z\alpha) = g(z)\,\varphi(\alpha)$ for all $z \in P$ and $\alpha \in A$.

Let $\rho : A \to GL(r, \mathbb{R})$ be a representation (i.e. a continuous homomorphism) of the topological group A. Let $P \to X$ be an A-principal bundle. Then the $\mathbb{R}^r$-associated bundle with total space

$$P \times_{(A,\rho)} \mathbb{R}^r = P \times \mathbb{R}^r / \{(z\alpha, u) \sim (z, \rho(\alpha)u)\}$$

is a vector bundle of rank r over X. Most of the time, the representation ρ is implicit and we just write $P \times_A \mathbb{R}^r$ for $P \times_{(A,\rho)} \mathbb{R}^r$. A (A, ρ)-*structure* (or just A-*structure*) for a vector bundle ξ of rank r over X is an A-principal bundle $P \to X$ together with a vector bundle isomorphism $f : P \times_A \mathbb{R}^r \to E(\xi)$. Two A-structures (P, f) and (P', f') are

- *strongly equivalent* if there exists an isomorphism of A-principal bundles $h : P \to P'$ such that $f' \circ (h \times \mathrm{id}_{\mathbb{R}^r}) = f$.
- *weakly equivalent* if there exists an isomorphism of A-principal bundles $h : P \to P'$ such that $f' \circ (h \times \mathrm{id}_{\mathbb{R}^r})$ is isotopic to f.

Here are a few examples.

9.1.8 $A = \{1\}$, the trivial group. An $\{1\}$-structure on a vector bundle ξ is just a trivialization of ξ. Strong equivalence coincides here with equality. Weak equivalence classes of $\{1\}$-structures correspond to isotopy classes of trivializations.

9.1.9 Each vector bundle ξ of rank r admits a $GL(r, \mathbb{R})$-structure which is unique up to strong equivalence. Indeed, consider the *space of frames* of ξ:

$$\mathrm{Fra}(\xi) = \{\nu : \mathbb{R}^r \to E(\xi) \mid \nu \text{ is a linear isomorphism onto some fiber of } \xi\}.$$

with the map $p_{\mathrm{Fra}} : \mathrm{Fra}(\xi) \to X$ given by $p_{\mathrm{Fra}}(\nu) = p \circ \nu(0)$ (The image by $\nu \in \mathrm{Fra}(\xi)$ of the standard basis of $\mathbb{R}^r$ is a frame (basis) of $\nu(\mathbb{R}^r)$, whence the name *space of frames*). By precomposition, the topological group $GL(r, \mathbb{R})$ acts continuously and freely on the right upon $\mathrm{Fra}(\xi)$. The map p_{Fra} descends to a continuous bijection $\mathrm{Fra}(\xi)/GL(r, \mathbb{R}) \to X$. Using local trivializations, one checks that this map is a homeomorphism. Also, a local trivialization of ξ gives rise to a local section of p_{Fra}. Hence, p_{Fra} is a $GL(r, \mathbb{R})$-principal bundle, called the *framed bundle* of ξ.

Consider the evaluation map $f_{\mathrm{Fra}} : \mathrm{Fra}(\xi) \times_{GL(r,\mathbb{R})} \mathbb{R}^r \to E(\xi)$ sending $[\nu, t]$ to $\nu(t)$. This map is a continuous bijection which is linear on each fiber and, using local trivializations of ξ again, one checks that it is a homeomorphism. Hence, f_{Fra} is a $GL(r, \mathbb{R})$-structure on ξ. We claim that f_{Fra} is a *universal structure* in the following sense. Each A-structure (for a representation $\rho : A \to GL(r, \mathbb{R})$)

$f : P \times \mathbb{R}^r \to E(\xi)$ determines a ρ-map $\hat{f} : P \to \mathrm{Fra}(\xi)$. The map $\hat{f}$ sends $z \in P$ to the map $f(z, -) \in \mathrm{Fra}(\xi)$. We check that two A-structure (P, f) and (P', f') are

(a) *strongly equivalent* if there exists an isomorphism of A-principal bundles $h : P \to P'$ such that $\hat{f}' \circ h = \hat{f}$.
(b) *weakly equivalent* if there exists an isomorphism of A-principal bundles $h : P \to P'$ such that $\hat{f}' \circ h$ is homotopic to $\hat{f}$.

The case $A = GL(r, \mathbb{R})$ give the uniqueness claimed above: any $GL(r, \mathbb{R})$-structure on ξ (for $\rho = \mathrm{id}$) is strongly equivalent to f_{Fra}.

9.1.10 $GL^+(r, \mathbb{R})$-structures and orientations (*see also* Sect. 9.2). Recall that an *orientation* of a finite dimensional real vector space is an equivalence class of (ordered) basis, where two basis are equivalent if their change-of-basis matrix is in $GL^+(r, \mathbb{R})$, i.e. has positive determinant. An *orientation* of a vector bundle ξ is an orientation of each fiber which varies continuously, i.e. there are local trivializations $p^{-1}(U) \xrightarrow{\approx} U \times \mathbb{R}^r$ whose restriction to each fiber is orientation-preserving (for the standard orientation of $\mathbb{R}^r$). A vector bundle admitting an orientation is called *orientable* and the choice of an orientation makes it *oriented.*

If P is a $GL^+(r, \mathbb{R})$-principal bundle, then $P \times_{GL^+(r,\mathbb{R})} \mathbb{R}^r$ is oriented, using the standard orientation of $\mathbb{R}^r$. Hence, a $GL^+(r, \mathbb{R})$-structure (P, f) on a vector bundle ξ makes it oriented. Conversely, an oriented vector bundle ξ admits a $GL^+(r, \mathbb{R})$-structure: one just restricts the canonical $(\mathrm{Fra}(\xi), f_{\mathrm{Fra}})$ of 9.1.9 to $\mathrm{Fra}^+(\xi)$, where

$$\mathrm{Fra}^+(\xi) = \{\nu \in \mathrm{Fra}(\xi) \mid \nu \text{ preserves the orientation}\}. \tag{9.1.4}$$

(for the standard orientation of $\mathbb{R}^r$). As in 9.1.9, we check that $(\mathrm{Fra}^+(\xi), f_{\mathrm{Fra}})$ is universal for the A-structures on ξ with a representation $\rho : A \to GL^+(r, \mathbb{R})$.

9.1.11 $O(r)$-structures. Consider the orthogonal group $O(r)$ with its standard representation $O(r) \hookrightarrow GL(r, \mathbb{R})$. Let $f : P \times_{O(r)} \mathbb{R}^r \xrightarrow{\approx} E(\xi)$ be a $O(r)$-structure on ξ. Then, the standard inner product on $\mathbb{R}^r$ gives, via f, an Euclidean structure on ξ (see p. 187) for which f is an isometry on each fiber. On the other hand, an Euclidean bundle ξ of rank r admits an $O(r)$-structure: one restricts f_{Fra} to the subbundle of $\mathrm{Fra}(\xi)$ formed by orthonormal frames:

$$\mathrm{Fra}_\perp(\xi) = \{\nu : \mathbb{R}^r \to E(\xi) \mid \nu \text{ is a linear isometry onto some fiber of } \xi\}.$$

As in (9.1.9), one shows that such an $O(r)$-structure compatible with a given Euclidean structure on ξ is unique up to strong equivalence. This process provides a bijection between Euclidean structures on ξ and strong equivalences of $O(r)$-structures.

On the other hand, a vector bundle over a paracompact space admits Euclidean structures which form a convex space. Let (ξ, e_t) be the vector bundle ξ endowed with an Euclidean structure e_t depending continuously on $t \in I$. Then

$$\mathrm{Fra}_e(\xi) = \{\nu_t : \mathbb{R}^r \to E(\xi) \mid t \in I,\ \nu_t \text{ is a linear isometry onto some fiber of } (\xi, e_t)\}$$

is the total space of an $O(r)$-principal bundle over $X \times I$. Note that $\mathrm{Fra}_e(\xi)$ is the union indexed by I of $\mathrm{Fra}_\perp(\xi, e_t)$, the bundle of orthonormal frames for the Euclidean structure e_t. If X is paracompact, $\mathrm{Fra}_e(\xi) \to X \times I$ is isomorphic to the principal bundle $\mathrm{Fra}_\perp(\xi, e_0) \times I \to X \times I$ [105, Chap. 4, Theorem 9.8]. Hence, there is an isomorphism $h : \mathrm{Fra}_\perp(\xi, e_0) \to \mathrm{Fra}_\perp(\xi, e_1)$ such that $i_1 \circ h$ is homotopic to i_0 (where $i_t : \mathrm{Fra}_\perp(\xi, e_t) \to \mathrm{Fra}(\xi)$ denotes the inclusion). This proves the following statement: *a vector bundle over a paracompact space admits an $O(r)$-structure which is unique up to weak equivalence.*

9.1.12 The case of orthogonal representations. Let $\rho : A \to O(r)$ be an orthogonal representation of a topological group A. Let $f : P \times_A \mathbb{R}^r \to E(\xi)$ be an A-structure on the vector bundle ξ (for the representation ρ). As ρ is orthogonal, $P \times_A \mathbb{R}^r$ inherits a natural Euclidean structure which is transported on ξ via f (see 9.1.11). As in 9.1.9, there is an $O(r)$-structure (P_O, f_O) on ξ such that $f = f_O \circ \hat{f}$ for a ρ-map $\hat{f} : P \to P_O$. If $h : P \to P'$ induces a strong equivalence between the A-structures (P, f) and (P', f'), then $P_O = P'_O$ and $\hat{f}' \circ h = \hat{f}$. If h only induces a weak equivalence, it descends to an isomorphism $h_O : P_O \to P'_O$ making the following diagram commutative

$$\begin{array}{ccc} P & \xrightarrow{h} & P' \\ \downarrow{\scriptstyle \hat{f}} & & \downarrow{\scriptstyle \hat{f}'} \\ P_O & \xrightarrow{h_O} & P'_O \end{array} .$$

We can pull-back $\hat{f}'$ over P_O via h_O, getting another representative of the weak equivalence class of (P', f'). This permits us to assume that $P_O = P'_O$ and $h_O = \mathrm{id}$; this also means that the Euclidean structure induced by f and f' coincide. Now, using the Gram-Schmidt orthonormalization process in each fiber of ξ, the isotopy between $f' \circ (h \times \mathrm{id}_{\mathbb{R}^r})$ and f may be deformed into an isotopy of isometries. This implies that $\hat{f}$ and $\hat{f}' \circ h$ are homotopic.

These considerations drive us to the following point of view for an A-structure on ξ, in the case of an orthogonal representation. One first fix an Euclidean structure on ξ and consider only A-structures (P, f) on ξ for which f is an isometry. In other words, an A-structure may be seen as a ρ-map $\hat{f} : P \to \mathrm{Fra}_\perp(\xi)$ and strong or weak equivalences are described as in (a) and (b) of 9.1.9.

Note that, if ξ is an *oriented* Euclidean bundle, one can consider the oriented orthonormal frames

$$\mathrm{Fra}_{\perp}^{+}(\xi) = \mathrm{Fra}_{\perp}(\xi) \cap \mathrm{Fra}^{+}(\xi) \tag{9.1.5}$$

which is the total space of an $SO(r)$-principal bundle over X. Then $(\mathrm{Fra}_{\perp}^{+}(\xi), f_{\mathrm{Fra}})$ is an $SO(r)$-structure which is universal for the A-structures associated to a representation $A \to SO(r)$. The special case of the representation $Spin(r) \to SO(r)$ (spin structures) is treated in Sect. 9.3 (compare [130, Sect. II.1]).

The best known example of $rep_{\perp}^{+}(\xi)$ is for $\xi = TS^n$, the tangent bundle to the standard unit sphere S^n. One sees TS^n as the space of couples $(v, w) \in \mathbb{R}^{n+1} \times \mathbb{R}^{n+1}$ such that $|v| = 1$ and $\langle v, w\rangle = 0$ (up to translation, w is tangent to $v \in S^n$). Then the map $q : SO(n+1) \to S^n$ defined by $q(A) = Ae_1$ (first column vector) is the oriented frame bundle for TS^n. The bundle isomorphism $SO(n+1) \xrightarrow{\approx} \mathrm{Fra}_{\perp}^{+}(TS^n)$ sends A to the map $\nu_A : \mathbb{R}^n \to TS^n$ defined by

$$\nu_A(t_1, \dots, t_n) = \Big(Ae_1, \sum_{i=1}^{n} t_i\, Ae_{i+1}\Big).$$

9.1.13 Complex vector bundles. By replacing $\mathbb{R}$ by $\mathbb{C}$ in the definition of a vector bundle, we get the notion of a *complex vector bundle*. The notion of a *A-structure* is defined accordingly, using a complex representation $\rho : A \to GL(r, \mathbb{C})$. As in 9.1.9, a complex vector bundle ξ of rank r admits a $GL(r, \mathbb{C})$-structure which is unique up to strong equivalence. Also, as in 9.1.11, using a Hermitian structure on ξ (those form a contractible space), ξ admits an $U(r)$ structure which is unique up to weak equivalence.

9.1.14 Classifying spaces and structures. Let $EGL(r, \mathbb{R}) \to BGL(r, \mathbb{R})$ be the universal bundle for the principal $GL(r, \mathbb{R})$-bundles and let ζ its associated vector bundle. Recall that, for X paracompact, the correspondence $[c] \to c^*\zeta$ provides a bijection from $[X, BGL(r, \mathbb{R})]$ and the set of isomorphism classes of vector bundles over X of rank r (using (9.1.9)). We shall use the following consequence of this result.

Lemma 9.1.15 *Let (X, A) be a cofibrant pair with X paracompact. Let ξ be a vector bundle over X whose restriction over A is trivial. Then, there exists a vector bundle $\bar{\xi}$ over X/A such that $\xi \approx \pi^*\bar{\xi}$, where $\pi : X \to X/A$ is the quotient map.*

Proof Let $c : X \to BGL(r, \mathbb{R})$ be a classifying map for ξ (r = rank of ξ), i.e. $\xi \approx c^*(\zeta)$. In the definition of a cofibrant pair, A is supposed to be closed, so A is paracompact. The restriction of c to A is then null-homotopic. As (X, A) is cofibrant, there is a homotopy from c to $\tilde{c}$ such that $\tilde{c}_{|A}$ is a constant map. Therefore, $\tilde{c}$ descends to $\bar{c} : X/A \to BGL(r, \mathbb{R})$. Hence, $\xi \approx \tilde{c}^*\zeta \approx \pi^*(\bar{c}^*\zeta)$. □

Structures on vector bundles and the classifying spaces are related as follows. Let ξ be a vector bundle of rank r over a paracompact space X. Fix a characteristic map $c : X \to BGL(r, \mathbb{R})$ for $\mathrm{Fra}(\xi)$. A representation $\rho : A \to GL(r, \mathbb{R})$ of A induces a continuous map $B\rho : BA \to BGL(r, \mathbb{R})$ which may be made a Serre fibration.

A *lifting of* c is a continuous map $\tilde{c} : X \to BA$ such that $B\rho \circ \tilde{c} = c$ and two such liftings $\tilde{c}_0$ and $\tilde{c}_1$ are homotopic if there exists a homotopy $h : X \times I \to BA$ between $\tilde{c}_0$ and $\tilde{c}_1$ such that $B\rho \circ h(x, t) = c(x)$ for all $(x, t) \in X \times I$. Then, the set of weak equivalence of (A, ρ)-structures on ξ is in bijection with the homotopy classes of liftings of c. For details (For details, see [23, Sect. 4]).

9.2 The Class w_1—Orientability

Let $\mathcal{O}(V)$ be the set of the two orientations of a finite dimensional vector space V. Let ξ be a vector bundle of rank r over a topological space X. Using the canonical $GL(r, \mathbb{R})$-structure $(\mathrm{Fra}(\xi), f_{\mathrm{Fra}})$ of 9.1.9, we define

$$\mathcal{O}(\xi) = \mathrm{Fra}(\xi) \times_{GL(r,\mathbb{R})} \mathcal{O}(\mathbb{R}^r).$$

The projection $\mathcal{O}(\xi) \to X$ is a locally trivial bundle whose fiber over $x \in X$ is, via f_{Fra}, in bijection with $\mathcal{O}(p^{-1}(x))$. In consequence, $\mathcal{O}(\xi) \to X$ is a two fold covering. An orientation of ξ (see 9.1.10) is clearly a continuous section of $\mathcal{O}(\xi) \to X$.

The characteristic class $w(\mathcal{O}(\xi) \to X) \in H^1(X)$ (see Sect. 4.3.2), is called the *first Stiefel-Whitney class* of ξ and is denoted by $w_1(\xi)$.

Proposition 9.2.1 *A vector bundle ξ over a CW-complex X is orientable if and only if $w_1(\xi) = 0$ in $H^1(X)$. The set of orientations of an orientable bundle is in bijection with $H^0(X)$.*

Proof The 2-covering $\mathcal{O}(\xi)$ is trivial if and only if it admits a continuous section, that is to say if and only if ξ is orientable. On the other hand, if X is a CW-complex, the 2-covering $\mathcal{O}(\xi) \to X$ is trivial if and only if its characteristic class vanish (see Lemma 4.3.6). There are two orientation for the restriction of ξ over each connected component, whence the last assertion. □

Remark 9.2.2 If η is a trivial vector bundle, then $w_1(\xi \oplus \eta) = w_1(\xi)$. Indeed, there exists a natural map $\mathcal{O}(\xi) \to \mathcal{O}(\xi \oplus \eta)$ which is an isomorphism of 2-fold coverings.

Proposition 9.2.3 *Vector bundles over an 1-dimensional CW-complex are classified by their rank and their first Stiefel-Whitney class.*

Proof Let ξ and ξ' be two vector bundles over a CW-complex X. If ξ and ξ' are isomorphic, they have the same rank and the 2-coverings $\mathcal{O}(\xi)$ and $\mathcal{O}(\xi')$ are isomorphic, which implies that $w_1(\xi) = w_1(\xi')$.

To prove the converse, note that, when X is 1-dimensional, any vector bundle ζ over X is the Whitney sum of a line bundle λ with the trivial vector bundle [105, Chap. 8, Theorem 1.2]. By Remark 9.2.2, $w_1(\zeta) = w_1(\lambda)$. We are thus reduced to ξ and ξ' being both of rank 1, in which case we use Proposition 9.2.4. □

Let $\mathcal{L}(X)$ be the set of isomorphism classes of real lines bundles over a space X. The tensor product (see (7.5.5)) of two line bundles is again a line bundle. This provides an operation $\otimes$ on $\mathcal{L}(X)$.

Proposition 9.2.4 *Let X be a CW-complex. Then the first Stiefel-Whitney class provides an isomorphism*

$$w_1 : (\mathcal{L}(X), \otimes) \xrightarrow{\approx} (H^1(X), +).$$

In particular, $(\mathcal{L}(X), \otimes)$ is an elementary abelian 2-group.

Proof Let $\xi = (p : E \to B)$ be a line bundle over X. Endow ξ with a Euclidean structure. The unit sphere bundle $S(E) \to X$ is then a 2-fold covering which is clearly isomorphic to $\mathcal{O}(\xi)$. By (4.3.5), $w(\mathcal{O}(\xi))$ determines $\mathcal{O}(\xi)$ and then $S(E)$. But $S(E)$ determines ξ by the isomorphism $S(E) \times_{O(1)} \mathbb{R} \xrightarrow{\approx} E$ (an $O(1)$-structure on ξ). Hence, w_1 is injective. For the surjectivity, let $a \in H^1(X)$. By (4.3.5), there is a 2-covering $\tilde{X} \to X$ with characteristic class a. We see it as an $O(1)$-principal bundle. Then, $\tilde{X} \times_{O(1)} \mathbb{R} \to X$ is a Euclidean line bundle ξ whose sphere bundle is $\tilde{X}$. By the above, $w_1(\xi) = a$.

It remains to show that $w_1(\xi \otimes \xi') = w_1(\xi) + w_1(\xi')$ for two line bundles ξ and ξ' over X. We start with some preliminaries. The linear group $GL(\mathbb{R})$ is isomorphic the multiplicative group $\mathbb{R}^\times = \mathbb{R} - \{0\}$. Set $K = \mathbb{R}^\times \times \mathbb{R}^\times$. Using the $\mathbb{R}$-vector space isomorphism $\mathbb{R} \otimes \mathbb{R} \xrightarrow{\approx} \mathbb{R}$ such that $x \otimes y \mapsto xy$, the diagram

$$\begin{array}{ccc} GL(\mathbb{R}) \times GL(\mathbb{R}) & \xrightarrow{\varphi^{\otimes}} & GL(\mathbb{R} \otimes \mathbb{R}) \\ \downarrow \approx & & \downarrow \approx \\ K & \xrightarrow{\varphi} & \mathbb{R}^\times \end{array}$$

is commutative, where $\varphi^{\otimes}$ is the natural homomorphism (see p. 302) and φ is just the standard product, which is a continuous homomorphism.

Let $F = \mathrm{Fra}(\xi)$, $F' = \mathrm{Fra}(\xi')$ and $F^{\otimes} = \mathrm{Fra}(\xi \otimes \xi')$ seen, using the above, as principal $\mathbb{R}^\times$-bundles. Let $\Delta : X \to X \times X$ be the diagonal inclusion. The definition of $F^{\otimes}$ in (7.5.5) becomes

$$F^{\otimes} = \varphi_*^{\otimes} \Delta^*(F \times F') \approx \Delta^*(\varphi_*^{\otimes}(F \times F')).$$

Let $\beta, \beta' : X \to B\mathbb{R}^\times$ be characteristic maps for F and F'. One has morphisms of $\mathbb{R}^\times$-principal bundles, i.e. commutative diagrams

$$\begin{array}{ccccc} E(F^{\otimes}) & \longrightarrow & E(F\times F')\times_K \mathbb{R}^\times & \longrightarrow & (E\mathbb{R}^\times \times E\mathbb{R}^\times)\times_K \mathbb{R}^\times \\ \downarrow & & \downarrow & & \downarrow \\ X & \xrightarrow{\Delta} & X\times X & \xrightarrow{\beta\times\beta'} & B\mathbb{R}^\times \times B\mathbb{R}^\times \end{array} \qquad X \xrightarrow{(\beta,\beta')} B\mathbb{R}^\times \times B\mathbb{R}^\times \tag{9.2.1}$$

where the maps on the top line are $\mathbb{R}^\times$-equivariant.

Recall from (7.4.3) that the two projections of K onto $\mathbb{R}^\times$ induce a homotopy equivalence $P : BK \xrightarrow{\simeq} B\mathbb{R}^\times \times B\mathbb{R}^\times$. Let P' be a homotopy inverse of P. One has more morphisms of $\mathbb{R}^\times$-principal bundles:

$$\begin{array}{ccccc} (E\mathbb{R}^\times \times E\mathbb{R}^\times)\times_K \mathbb{R}^\times & \longrightarrow & EK\times_K B\mathbb{R}^\times & \xrightarrow{\kappa} & E\mathbb{R}^\times \\ \downarrow & & \downarrow & & \downarrow \\ B\mathbb{R}^\times \times B\mathbb{R}^\times & \underset{\simeq}{\xrightarrow{P'}} & BK & \xrightarrow{B\varphi} & B\mathbb{R}^\times \end{array} \tag{9.2.2}$$

where κ is defined by $\kappa([(t_i(g_i, g_i'), \lambda]) = (t_i, g_i g_i'\lambda)$. Diagrams (9.2.1) and (9.2.2) imply that $B\varphi \circ P' \circ (\beta, \beta')$ is a characteristic map for the $\mathbb{R}^\times$-principal bundle $F^{\otimes}$.

As the inclusion $\{\pm 1\} \hookrightarrow \mathbb{R}^\times$ is a homomorphism and a homotopy equivalence, one has $B\mathbb{R}^\times \simeq B\{\pm 1\} \simeq \mathbb{R}P^\infty \simeq K(\mathbb{Z}_2, 1)$. The map $B\varphi \circ P'$ thus defines continuous multiplication on $K(\mathbb{Z}_2, 1)$. One checks that it is homotopy commutative and admits a homotopy unit. Let u be the generator of $H^1(B\mathbb{R}^\times) = \mathbb{Z}_2$. By Proposition 4.7.54 and Lemma 4.7.56, one has

$$H^*(B\varphi \circ P')(u) = u \times \mathbf{1} + \mathbf{1} \times u. \tag{9.2.3}$$

Hence,

$$\begin{aligned} w_1(\xi \otimes \xi') = H^*(\beta, \beta')(u \times \mathbf{1} + \mathbf{1} \times u) &= H^*\beta(u) + H^*\beta'(u) \\ &= w_1(\xi) + w_1(\xi'). \end{aligned}$$

□

The above shows that the correspondence $[\tilde{X} \to X] \mapsto [\tilde{X} \times_{O(1)} \mathbb{R} \to X]$ induces a bijection between the set $\mathrm{Cov}_2(X)$ of equivalence classes of 2-fold coverings of X and $\mathcal{L}(X)$, with a commutative diagram

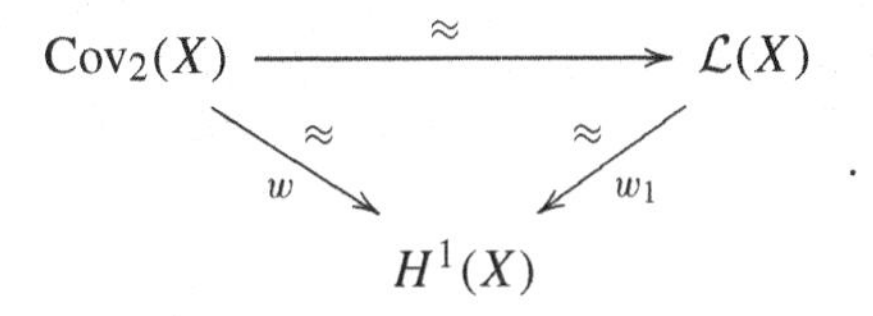

By Proposition 4.7.36, we get the following corollary.

Corollary 9.2.5 *Let λ be a line bundle over a CW-complex X. Then $w_1(\lambda) \in H^1(X)$ coincides with the Euler class $e(\lambda)$ of λ.*

Example 9.2.6 The tautological line bundle over $\mathbb{R}P^n$. The 2-covering $\zeta = (S^n \to \mathbb{R}P^n)$ $(1 \leq n \leq \infty)$ is an $O(1)$-principal bundle. Its associated line bundle $\lambda = (S^n \times_{O(1)} \mathbb{R} \to \mathbb{R}P^n)$ satisfies $w_1(\lambda) = w(\zeta) \neq 0$ in $H^1(\mathbb{R}P^n) = \mathbb{Z}_2$ (see the proof of Proposition 4.3.10). Seeing $\mathbb{R}P^n$ as the space of vector lines in $\mathbb{R}^{n+1}$, λ identifies itself with the *tautological line bundle over $\mathbb{R}P^n$* , i.e.

$$E(\lambda) = \{(a, v) \in \mathbb{R}P^n \times \mathbb{R}^{n+1} \mid v \in a\}\,,\ p(a, v) = a.$$

The identification $S^n \times_{O(1)} \mathbb{R} \xrightarrow{\approx} E(\lambda)$ is given by $[z, t] \mapsto (\mathbb{R}z, tz)$.

Proposition 9.2.7 *Let ξ be a vector bundle over a CW-complex X. Then, the following conditions are equivalent.*

(1) *The restriction of ξ over the 1-skeleton of X is trivial.*
(2) $w_1(\xi) = 0$.

Proof Let $i : X^1 \to X$ denote the inclusion and let $\xi_1 \approx i^*\xi$ be the restriction of ξ over X_1. By Lemma 4.3.6, $w_1(\xi_1) = i^*(w_1(\xi))$ and $w_1(\xi_1) = 0$ if ξ_1 is trivial. As $i^* : H^1(X) \to H^1(X^1)$ is injective, this proves the implication (1) $\Rightarrow$ (2). Conversely, if $w_1(\xi) = 0$, then $w_1(\xi_1) = 0$ and ξ_1 is trivial by Proposition 9.2.3. □

We finish this section by describing a singular cocycle representing $w_1(\xi)$ in terms of transporting orientations. Let $c : I \to X$ be a path and let $\alpha \in \mathcal{O}(E_{c(0)})$ (where E_x is the fiber of ξ over $x \in X$). We see α as an element of the fiber of $\text{Fra}(\xi)$ over $c(0)$. The unique lifting $\tilde{c} : I \to \text{Fra}(\xi)$ of c with $\tilde{c}(0) = \alpha$ provides an orientation $c_*\alpha = \tilde{c}(1) \in \mathcal{O}(E_{c(1)})$. We say that $c_*\alpha$ is obtained from α by *transport along c*.

Choose a set-theoretic section $\alpha : X \to \text{Fra}(\xi)$, i.e. an assignation of an orientation of E_x for each $x \in X$ (which has not to vary continuously). We see a singular simplex $\sigma \in \mathcal{S}_1(X)$ as a path $\sigma : I \to X$ via the identification $I \approx \Delta^1$ sending t to $(t, 1-t)$. The cochain $\tilde{w}_1(\xi, \alpha)$ defined by

$$\langle \tilde{w}_1(\xi, \alpha), \sigma \rangle = \begin{cases} 1 & \text{if } \sigma_*\alpha(\sigma(0)) \neq \alpha(\sigma(1)) \\ 0 & \text{otherwise.} \end{cases}$$

The cochains $\tilde{w}_1(\xi, \alpha)$ and $\tilde{w}(\text{Fra}(\xi), \alpha)$ of Sect. 4.3.2 clearly coincide. Hence, by Proposition 4.3.7, $\tilde{w}_1(\xi, \alpha)$ is a cocycle representing $w_1(\xi)$.

Let $x_0 \in X$ and $\alpha_0 \in \mathcal{O}(E_{x_0})$. We say that a loop $c : I \to X$ at x_0 *preserves the orientation* if $c_*\alpha_0 = \alpha_0$ (this condition does not depend on the choice of α_0). Note that, if d is another loop at x_0, then $(dc)_*\alpha_0 = d_*(c_*\alpha_0)$. Also, $c_*\alpha_0$ depends only on the homotopy class of the loop c, by the homotopy lifting property of the covering $\text{Fra}(\xi) \to X$. Hence the correspondence

$$c \mapsto \begin{cases} 0 & \text{if } c \text{ preserves the orientation} \\ 1 & \text{otherwise.} \end{cases}$$

provides a homomorphism $w : \pi_1(X, x_0) \to \mathbb{Z}_2$ which corresponds to $w_1(\xi) \in H^1(X)$ via the isomorphism $H^1(X) \xrightarrow{\approx} \hom(\pi_1(X, x_0), \mathbb{Z}_2)$ of Lemma 4.3.1.

9.3 The Class $\dot{w}_2$—Spin Structures

In this section, we define a *cellular* second Stiefel-Whitney class $\dot{w}_2(\xi) \in \dot{H}^2(X)$, when ξ is an orientable vector bundle over a *regular CW-complex* X (recall that $\dot{H}^*$ denotes the cellular cohomology introduced in Sect. 3.5). A more general $w_2(\xi) \in H^2(X)$ (X any space) is considered in Sect. 9.4, but the results below are used to establish the relationship between $w_2(\xi)$ and the existence of spin structures on ξ.

Let ξ be an orientable vector bundle of rank $r \geq 2$ over a CW-complex X. Denote by ξ_i the restriction of ξ over the i-skeleton X^i of X. By Propositions 9.2.1 and 9.2.7, ξ_1 is trivial. Fix an orientation of ξ. Choose a trivialization T_1 of ξ_1 which is compatible with this orientation. The restriction T_0 of T_1 over X^0 is thus uniquely determined up to isotopy.

Let e be a 2-cell of X with characteristic map $\varphi : (D^2, S^1) \to (X, X^1)$. As D^2 is contractible, there is a unique (up to isotopy) trivialization T_e of $\varphi^*\xi$ over D^2 which is compatible with the orientation (see Proposition 9.1.6). Thus, φ^*T_1 and T_e provides two trivialization of $\varphi^*\xi$ restricted to the boundary S^1 of D^2. By Lemma 9.1.2, these two trivializations differ by the action of a map from S^1 to $GL(r, \mathbb{R})$, whose range is $GL^+(r, \mathbb{R})$ since φ^*T_1 and T_e are both compatible with the chosen orientation. Let

$$\tilde{w}_2(e) \in [S^1, GL^+(r, \mathbb{R})] \approx \pi_1(GL^+(r, \mathbb{R})) \approx \pi_1(SO(r)) \tag{9.3.1}$$

be the homotopy class of this map. Here, the isomorphism $[S^1, GL^+(r, \mathbb{R})] \approx \pi_1(GL^+(r, \mathbb{R}))$ holds true since $GL^+(r, \mathbb{R})$ is a topological group and $GL^+(r, \mathbb{R})$ has the homotopy type of $SO(r)$ by the Gram-Schmidt orthonormalization process. If $r \geq 3$, then $\pi_1(SO(r)) = \mathbb{Z}_2$. If $r = 2$ then $\pi_1(SO(r)) \approx \mathbb{Z}$ and, by convention, we take $\tilde{w}_2(e) \bmod 2$. The correspondence $e \mapsto \tilde{w}_2(e)$ thus defines a cellular 2-cochain $\tilde{w}_2 = \tilde{w}_2(\xi, T_1) \in \dot{C}^2(X)$.

Lemma 9.3.1 *Suppose that X has no 3-cells or that X^3 is regular complex. Then, the cochain $\tilde{w}_2(\xi, T_1)$ is a cellular cocycle. Its cohomology class $\dot{w}_2(\xi) \in \dot{H}^2(X)$ depends only of the isomorphism class of ξ.*

The class $\dot{w}_2(\xi) \in \dot{H}^2(X)$ is called the *second Stiefel-Whitney class* of the vector bundle ξ. If the rank of ξ is ≤ 1, we set by convention that $\dot{w}_2(\xi) = 0$.

Proof We may assume that X is connected. Observe first that the cochain $\tilde{w}_2(\xi, T_1)$ does not depend on the orientation of ξ, since the other choice would just change all the orientations under consideration. Let us see how $\tilde{w}_2(\xi, T_1)$ depends on the isotopy class of the trivialization $[T_1] \in \mathcal{T}(\xi_1)$. Let $T_1' \in \tilde{\mathcal{T}}(\xi_1)$ be another trivialization compatible with the orientation. As seen above, this compatibility implies that the restriction T_0' of T_1' to ξ_0 coincides with T_0 up to isotopy. Such an isotopy may be realized in a neighbourhood of X_0 in X_1, so one may assume that $T_0 = T_0'$. By Lemma 9.1.2, there is a unique map $\tilde{a} : X^1 \to GL^+(r, \mathbb{R})$ such that $T_1' = \tilde{a} \cdot T_1$. As $T_0 = T_0'$, the restriction of $\tilde{a}$ to X^0 is constant to the identity of $\mathbb{R}^r$. Hence, each 1-cell ε with characteristic map $\varphi_\varepsilon : D^1 \to X^1$ gives rise to a homotopy class

$$a(\varepsilon) = [\tilde{a} \circ \varphi_\varepsilon] \in [(D^1, S^0), (GL^+(r, \mathbb{R}), \mathrm{id})] \approx \pi_1(GL^+(r, \mathbb{R}), \mathrm{id}) \approx \pi_1(SO(r), \mathrm{id}) = \mathbb{Z}_2$$

which does not depend on the choice of φ_ε (since $\pi_1(SO(r), \mathrm{id}) = \mathbb{Z}_2$; again, if $r = 2$, one takes by convention the value mod 2 of $a \in \pi_1(SO(2)) \approx \mathbb{Z}$). The correspondence $\varepsilon \mapsto a(\varepsilon)$ determines a cellular 1-cochain $a \in \dot{C}^1(X)$. Let e be a 2-cell of X with characteristic map $\varphi : (D^2, S^1) \to (X, X^1)$. Using the cellular boundary formula (3.5.4) and writing the abelian group $\pi_1(GL^+(r, \mathbb{R})) = \mathbb{Z}_2$ additively, we get that $[\tilde{a} \circ \varphi] = \delta(a)$. In the same way, using (9.1.3), one gets the following formula.

$$\tilde{w}_2(\xi, T_1') = \tilde{w}_2(\xi, \tilde{a}\, T_1) = \tilde{w}_2(\xi, T_1) + \delta(a). \tag{9.3.2}$$

We now prove that $\tilde{w}_2$ is a cellular cocycle. If X has no 3-cells, there is nothing to prove. Let ϵ be a 3-cell of X which, as X^3 is now supposed to be regular, can be identified with a subcomplex (also called ϵ) of X. As ϵ is contractible, there is a unique (up to isotopy) trivialization T^ϵ of $\xi_{|\epsilon}$, compatible with the orientation. As above, one may assume that T^ϵ coincides with T_0 over ϵ^0. A trivialization $T_1^\epsilon \in \tilde{\mathcal{T}}(\xi_1)$ may be thus defined by

$$T_1^\epsilon(z) = \begin{cases} T^\epsilon(z) & \text{if } \ p(z) \in \epsilon \\ T_1(z) & \text{otherwise.} \end{cases}$$

By (9.3.2) and the construction of T_1^ε, one has

$$\delta(\tilde{w}_2(\xi, T_1))(\epsilon) = \delta(\tilde{w}_2(\xi, T_1^\epsilon))(\epsilon) = 0. \tag{9.3.3}$$

Equation (9.3.3) shows that $\delta(\tilde{w}_2(\xi, T_1))(\epsilon) = 0$ for all 3-cell ϵ, proving that $\delta(\tilde{w}_2(\xi, T_1)) = 0$. Also, formula (9.3.2) show that $[\tilde{w}_2(\xi, T_1)] \in \dot{H}^2(X)$ does not depend on the choice of T_1 and thus depends only on ξ. Finally, if ξ' isomorphic to ξ via a homeomorphism $h : E(\xi') \xrightarrow{\approx} E(\xi)$ over the identity of X, the trivialization $T_1' = T_1 \circ h$ satisfies $\tilde{w}_2(\xi', T_1') = \tilde{w}_2(\xi, T_1)$. Therefore, $\dot{w}_2(\xi') = \dot{w}_2(\xi)$. □

Lemma 9.3.2 *Let ξ be an orientable vector bundle of rank $r \geq 2$ over a CW-complex X satisfying the hypotheses of* Lemma 9.3.1. *If η is a trivial vector bundle over X, then $\dot{w}_2(\xi \oplus \eta) = \dot{w}_2(\xi)$.*

Proof Note that $\xi \oplus \eta$ is orientable since ξ is so, thus $\dot{w}_2(\xi \oplus \eta)$ is defined. Let us represent $\dot{w}_2(\xi)$ by a cocycle $\tilde{w}_2(\xi, T_1)$ as above. If $T_\eta : X \times \mathbb{R}^s \xrightarrow{\approx} E(\eta)$ is a fixed trivialization of η, one checks that $\tilde{w}(\xi, T_1 \oplus T_\eta) = \pi_1 j \circ \tilde{w}(\xi, T_1)$, where $j : SO(r) \to SO(r+s)$ is the inclusion. But $\pi_1 j$ is an isomorphism if $r \geq 3$ or the reduction mod 2 if $r = 2$, which proves the lemma in these cases.

If $r = 1$, then $\dot{w}_2(\xi) = 0$ by convention. Also, as ξ is orientable, it is trivial by Proposition 9.2.4. Thus, $\xi \oplus \eta$ is trivial and $\dot{w}_2(\xi \oplus \eta) = 0$ by the proof that (2) implies (3) in the next proposition (this part of the proof is valid for any r). □

Example 9.3.3 Let $X = S^2$, with its cellular decomposition with one 0-cell and one 2-cell: $X = D^2 \cup_{S^1} pt$. Let $\alpha : S^1 \to GL^+(r, \mathbb{R})$, representing $[\alpha] \in \pi_1(GL^+(r, \mathbb{R}))$. Then

$$E(\xi_{[\alpha]}) = D^2 \times \mathbb{R}^r \cup_\alpha pt \times \mathbb{R}^r$$

is the total space of a vector bundle $\xi_{[\alpha]}$ of rank r over X. This process gives a bijection between $\pi_1(GL^+(r, \mathbb{R}))$ and the isomorphism classes of vector bundles of rank r over S^2 (compare [181, Sect. 18]). The two cochain associating to the 2-cell of X the element $[\alpha] \in \pi_1(GL^+(r, \mathbb{R})) = \mathbb{Z}_2$ if $r \geq 3$ (or its reduction mod 2 if $r = 2$) represents $\dot{w}_2(\xi_{[\alpha]}) \in \dot{H}^2(X) = \mathbb{Z}_2$. Summing up, there are two vector bundles (up to isomorphism) of rank ≥ 3 over S^2: the trivial bundle η, satisfying $\dot{w}_2(\eta) = 0$, and the non-trivial bundle ξ, characterized by the property that $w_2(\xi) \neq 0$. This is an example of Proposition 9.3.4.

Proposition 9.3.4 *Let ξ be a vector bundle of rank $r \geq 3$ over a CW-complex X. Suppose that X has no 3-cell or that X^3 is a regular complex. Let ξ_i be the restriction of ξ over the X^i. Then, the following conditions are equivalent.*

(1) *ξ_3 is trivial.*
(2) *ξ_2 is trivial.*
(3) *$w_1(\xi) = 0$ and $\dot{w}_2(\xi) = 0$.*

In the next section, Proposition 9.3.4 will be generalized, using the singular second Stiefel-Whitney class (see Proposition 9.4.6).

Proof By Proposition 9.2.7, $w_1(\xi) = 0$ if ξ_2 is trivial. Also, ξ is orientable by Proposition 9.2.1, so $\dot{w}_2(\xi)$ is defined. We can represent $\dot{w}_2(\xi)$ by a cocycle $\tilde{w}_2(\xi, T_1)$ where T_1 is the restriction of $\Gamma \in \tilde{\mathcal{T}}(\xi_2)$. For each 2-cell e, we thus have $T^e = \Gamma_{|e}$, thus $\tilde{w}_2(\xi, T_1) = 0$, proving that $\dot{w}_2(\xi) = 0$. Thus, (2) implies (3).

Conversely, suppose that $w_1(\xi) = 0$ and $\dot{w}_2(\xi) = 0$. Choose a trivialization T_1 of ξ_1 and let $\varphi : \Lambda_2 \times (D^2, S^1) \to (X, X^1)$ be a global characteristic map for the 2-cells of X. Then, $\tilde{w}_2(\xi, T_1) = \delta(a)$ for $a \in \dot{C}^1(X)$. As in the proof of Lemma 9.3.1, the cochain a may be used to modify (relative to ξ_0) T_1 into a trivialization T_1' such that $\tilde{w}_2(\xi, T_1') = 0$ (this uses that $r \geq 3$). This means that, over $\Lambda_2 \times S^1$, the trivialization $\varphi^* T_1'$ coincides up to isotopy with the unique (up to isotopy) trivialization over $\Lambda_2 \times D^2$ compatible with the orientation. By Lemma 9.1.7, this implies that ξ_2 is trivial.

We have so far proven that (2) $\Leftrightarrow$ (3). We now prove the equivalence (1) $\Leftrightarrow$ (2), which is true for any CW-complex. The implication (1) $\Rightarrow$ (2) is trivial. Conversely, let $\Gamma \in \mathcal{T}(\xi_2)$. Let $\varphi : \Lambda_3 \times (D^3, S^2) \to (X, X^2)$ be a global characteristic map for the 3-cells of X. As above, one compares the trivialization $\varphi^*\Gamma$ over $\Lambda_3 \times S^2$ with the unique (upto isotopy) trivialization over $\Lambda_3 \times D^3$ which is compatible with the orientation. Their isotopy classes differ by the action of an element of $[\Lambda_3 \times S^2, GL^+(r, \mathbb{R})]$. But $\pi_2(SO(r)) = 0$ (see, e.g. [181, 22.10]), thus

$$[S^2, GL^+(r, \mathbb{R})] \approx [S^2, SO(r)] \approx \pi_2(SO(r)) = 0.$$

As above, using Lemma 9.1.7, this implies that ξ_3 is trivial. □

We now see the second Stiefel-Whitney class $\dot{w}_2(\xi)$ as the obstruction to the existence of the existence of a *spin structure* on ξ. This refers to the standard orthogonal representation $\rho_0 : Spin(r) \to SO(r)$ which is a 2-covering. As in 9.1.12, it is natural to fix an Euclidean structure and an orientation on ξ, in which case, a spin structure is seen a ρ_0-map $\hat{f} : P \to \mathrm{Fra}_\perp^+(\xi)$, where P is a $Spin(r)$-principal bundle. Note that $\hat{f}$ is a 2-covering whose restriction to each fiber is modeled by ρ_0, and any such covering would be a spin structure. Alternative equivalent definitions of spin structures are to be found in [150] or [130, Sect. 2.1].

Strong and weak equivalences for spin structures may be expressed as in (a) and (b) in 9.1.9 but then, by the homotopy lifting property for coverings, these two notions of equivalence coincide. Since $\pi_1(SO(r))$ is cyclic, ρ_0 is, upto equivalence, the only 2-covering of $SO(r)$ which is non-trivial (see Sect. 4.3.1). These considerations prove the following lemma.

Lemma 9.3.5 *Let ξ be an oriented Euclidean bundle ξ of rank r over a CW-complex X. Then, there is a bijection between*

- *the strong (or weak) equivalence classes of spin structures on ξ.*
- *the isomorphism classes of 2-coverings of $\mathrm{Fra}_\perp^+(\xi)$ whose restriction to the fibers of p_{Fra} is non-trivial.*

In general, the above sets are empty. For the existence of a spin structure, one has the following proposition (see Proposition 9.4.6 for a more general framework).

Proposition 9.3.6 *Let ξ be a vector bundle of rank $r \geq 2$ over a regular CW-complex X. Then, the following conditions are equivalent.*

(a) *ξ admits a spin structure.*
(b) *$w_1(\xi) = 0$ and $\dot{w}_2(\xi) = 0$.*

Moreover, if (a) *or* (b) *holds true, then the set of strong (or weak) equivalence classes of spin structures on ξ is in bijection with $H^1(X)$.*

Proof Let (P, f) be a spin structure on ξ and let $P_2 \to X^2$ be the restriction of P onto the 2-skeleton of X. Suppose first that $r \geq 3$. As $\pi_1(Spin(r)) = 0$, the principal bundle $P_2 \to X^2$ admits a section by [181, Corollary 34.4] and is therefore trivial.

This implies that ξ_2 is trivial and (b) holds by Proposition 9.3.4. When $r = 2$, we apply the above argument to $\xi \oplus \eta$ where η is a trivial bundle and use Remark 9.2.2 and Lemma 9.3.2.

Conversely, suppose that $w_1(\xi)$ and $\dot{w}_2(\xi)$ vanish. Fix an orientation and an Euclidean structure on ξ (thus inducing an orientation and an Euclidean structure on ξ_k). Suppose first that $r \geq 3$. By Proposition 9.3.4, ξ_2 is trivial and thus admits a spin structure $\hat{f}_2 : P_2 \to \mathrm{Fra}_\perp^+(\xi_2)$. Consider the commutative diagram

$$\begin{array}{ccccccccc} \pi_2(X^2) & \longrightarrow & \pi_1(SO(r)) & \xrightarrow{i_*} & \pi_1(\mathrm{Fra}_\perp^+(\xi_2)) & \longrightarrow & \pi_1(X^2) & \longrightarrow & \pi_0(SO(r)) \\ \downarrow & & \downarrow\approx & & \downarrow & & \downarrow\approx & & \downarrow\approx \\ \pi_2(X) & \longrightarrow & \pi_1(SO(r)) & \xrightarrow{i_*} & \pi_1(\mathrm{Fra}_\perp^+(\xi)) & \longrightarrow & \pi_1(X) & \longrightarrow & \pi_0(SO(r)) \end{array} \tag{9.3.4}$$

where the rows are the homotopy exact sequences of the bundles $\mathrm{Fra}_\perp^+(\zeta)$ for $\zeta = \xi_2$ or ξ. By Lemma 9.3.5, the set of strong equivalence classes of spin structures on ζ is in bijection with the set

$$\mathcal{E}(\zeta) = \{\kappa \in \hom(\pi_1(\mathrm{Fra}_\perp^+(\zeta)), \mathbb{Z}_2) \mid \kappa \circ i_* \neq 0\}.$$

Diagram (9.3.4) implies that $\pi_1(\mathrm{Fra}_\perp^+(\xi_2)) \to \pi_1(\mathrm{Fra}_\perp^+(\xi))$ is an isomorphism (five-lemma) and that $\mathcal{E}(\xi) = \mathcal{E}(\xi_2)$. Hence, the spin structure on ξ_2 extends to ξ.

In the case $r = 2$, we apply the above argument to $\xi \oplus \eta$ where η is a trivial bundle of rank 1. Note that $w_1(\xi \oplus \eta)$ and $\dot{w}_2(\xi \oplus \eta)$ vanish by Remark 9.2.2 and Lemma 9.3.2. Hence, $\xi \oplus \eta$ admits a spin structure. Taking the sum with a fixed frame of η gives a $(SO(2) \to SO(3))$-map $\mathrm{Fra}_\perp^+(\xi) \to \mathrm{Fra}_\perp^+(\xi \oplus \eta)$ and thus a commutative diagram

$$\begin{array}{ccccccccc} \pi_2(X) & \longrightarrow & \pi_1(SO(2)) & \xrightarrow{i_*} & \pi_1(\mathrm{Fra}_\perp^+(\xi)) & \longrightarrow & \pi_1(X) & \longrightarrow & \pi_0(SO(2)) \\ \downarrow\approx & & \downarrow & & \downarrow & & \downarrow\approx & & \downarrow\approx \\ \pi_2(X) & \longrightarrow & \pi_1(SO(3)) & \xrightarrow{i_*} & \pi_1(\mathrm{Fra}_\perp^+(\xi \oplus \eta)) & \longrightarrow & \pi_1(X) & \longrightarrow & \pi_0(SO(3)) \end{array} \tag{9.3.5}$$

As $\mathcal{E}(\xi \oplus \eta) \neq \emptyset$, Diagram (9.3.5) shows that $\mathcal{E}(\xi) \neq \emptyset$ and hence ξ admits a $Spin(2)$-structure.

Finally, note that, in (9.3.4), the last horizontal map vanish ($\pi_0(SO(r)) \to \pi_0(\mathrm{Fra}_\perp^+(\xi))$ is injective). Also, $\hom(\pi_1(\mathrm{Fra}_\perp^+(\zeta)), \mathbb{Z}_2)$ is an Abelian group. Hence, if $\mathcal{E}(\xi) \neq \emptyset$, it is in bijection with

$$\{\kappa \in \hom(\pi_1(\mathrm{Fra}_\perp^+(\zeta)), \mathbb{Z}_2) \mid \kappa \circ i_* = 0\} \approx \hom(\pi_1(X), \mathbb{Z}_2) \approx H^1(X),$$

the last isomorphism being established in Lemma 4.3.1. □

9.4 Definition and Properties of Stiefel-Whitney Classes

Let $\xi = (p : E \to X)$ be a vector bundle of rank r over a paracompact space X. In Theorem 4.7.37 was established the Thom isomorphism

$$\Phi^* : H^k(X) \xrightarrow{\approx} H^{k+r}(E, E_0),$$

where $E_0 \subset E$ is the complement of the zero section. Using the Steenrod squaring $\mathrm{Sq} : H^*(E, E_0) \to H^*(E, E_0)$, the *(total) Stiefel-Whitney class* $w(\xi) \in H^*(X)$ of ξ is defined by

$$w(\xi) = \Phi^{-1} \circ \mathrm{Sq} \circ \Phi(\mathbf{1}), \tag{9.4.1}$$

where $\mathbf{1} \in H^0(X)$ is the unit class. The component of $w(\xi)$ in $H^i(X)$ is denoted by $w_i(\xi)$ and is called the i-th Stiefel-Whitney class of ξ.

Equation (9.4.1) is one of the multiple definitions of the Stiefel-Whitney class and it is due to Thom [190, Sects. II and III]. For a history of Stiefel-Whitney classes, see [153, p. 38] and [40, Chap. IV, Sect. 1]. The main properties of Stiefel-Whitney class are given in the following proposition.

Theorem 9.4.1 *Let $\xi = (p : E \to X)$ be a vector bundle of rank r over a paracompact space X.*

(1) $w(\xi) = \mathbf{1} + w_1(\xi) \cdots + w_r(\xi)$. *In particular, $w_i(\xi) = 0$ if $i > r$.*
(2) *If $f : Y \to X$ be a continuous map, then $w(f^*\xi) = H^* f(w(\xi))$. In particular, if ξ is isomorphic to ξ', then $w(\xi) = w(\xi')$.*
(3) *If ξ is trivial, then $w(\xi) = \mathbf{1}$.*
(4) *Let ξ' be a vector bundle over a paracompact space X'. Then,*

$$w(\xi \times \xi') = w(\xi) \times w(\xi') \in H^*(X \times X'). \tag{9.4.2}$$

If $X' = X$, then

$$w(\xi \oplus \xi') = w(\xi) \smile w(\xi') \in H^*(X). \tag{9.4.3}$$

(5) *If η is a trivial vector bundle over X, then $w(\xi) = w(\xi \oplus \eta)$.*
(6) *$w_r(\xi)$ is the Euler class of ξ.*

Property (2) says that the total Stiefel-Whitney and its homogeneous components are *characteristic classes*. Historically, they were the first characteristic classes defined, in the simultaneous work of Eduard Stiefel and Hassler Whitney starting around 1935 (see [40, pp. 421–426]).

Proof Recall that $\Phi(\mathbf{1})$ is the Thom class $U(\xi) \in H^r(E, E_0)$; thus, (9.4.1) is equivalent to

$$w(\xi) = \Phi^{-1} \circ \mathrm{Sq}(U(\xi)). \tag{9.4.4}$$

Now, Theorem 9.4.1 comes from the properties of Sq established in Theorem 8.2.1. Since $\mathrm{Sq}^0 = \mathrm{id}$, one has $w_0(\xi) = \phi^{-1} \circ \mathrm{Sq}^0(U(\xi)) = \mathbf{1}$. As $\mathrm{Sq}^i(U(\xi)) = 0$ for $i > r$, this proves (1). The naturality (2) comes from the naturality of all the ingredients of (9.4.4): the Thom class is natural (Lemma 4.7.30), and so is Φ, and Sq is also natural, being a cohomology operation. Now, (3) is a consequence of (2) since a trivial bundle is induced by a map to a point.

To prove (4), one has

$$\begin{aligned} w(\xi \times \xi') &= \phi^{-1} \circ \mathrm{Sq}(U(\xi \times \xi')) \\ &= \phi^{-1} \circ \mathrm{Sq}\big(U(\xi) \times U(\xi')\big) && \text{using (4.7.24)} \\ &= \phi^{-1}\big(\mathrm{Sq}(U(\xi)) \times \mathrm{Sq}(U(\xi'))\big) && \text{by (3) of Theorem 8.2.1} \end{aligned} \tag{9.4.5}$$

On the other hand, if $a \in H^*(X)$ and $a' \in H^*(X')$, one has

$$\begin{aligned} \phi(a \times a') &= H^*(p \times p')(a \times a') \smile U(\xi \times \xi') \\ &= [H^*p(a) \times H^*p'(a')] \smile [U(\xi) \times U(\xi')] \\ &= [H^*p(a) \smile U(\xi)] \times [H^*p(a') \smile U(\xi')] \\ &= \Phi(a) \times \Phi(a'). \end{aligned} \tag{9.4.6}$$

Thus, (9.4.5) together with (9.4.6) proves (9.4.2). If $X' = X$, then $\xi \oplus \xi' = \Delta^*(\xi \times \xi')$ where $\Delta : X \to X \times X$ is the diagonal inclusion. Therefore, (9.4.3) comes from (2) already proven, (9.4.2) and Remark 4.6.1:

$$w(\xi \oplus \xi') = H^*\Delta(w(\xi \times \xi')) = H^*\Delta(w(\xi) \times (\xi')) = w(\xi) \smile w(\xi').$$

Property (5) is a consequence of (3) and (4). Finally, (6) follows from

$$w_r(\xi) = \Phi^{-1} \circ \mathrm{Sq}^r(U(\xi)) = \Phi^{-1}(U(\xi) \smile U(\xi)) = e(\xi),$$

the last equality coming from (4.7.22). □

Remark 9.4.2 Versions of Properties (1), (2), (9.4.3) *and* (6) *uniquely characterize the total Stiefel-Whitney class.* See Proposition 9.6.4, [153, Theorem 7.3] or [105, Chap. 16, Sect. 5]. This is the philosophy of the axiomatic presentation of Stiefel-Whitney class (see [153]), inspired by that of the Chern classes introduced by Hirzebruch [96, p. 58].

Remark 9.4.3 As the Steenrod squares are used for Definition (9.4.1), the Adem relations provide constraints amongst Stiefel-Whitney classes. For instance, the relation $\mathrm{Sq}^{2i+1} = \mathrm{Sq}^1\mathrm{Sq}^{2i}$ (see Example 8.4.4) implies that $w_{2i+1}(\xi) = 0$ if $w_{2i}(\xi) = 0$. Also, if $w_{2^k}(\xi) = 0$ for $k = 1 \dots, r$, then, by Lemma 8.5.1, $w_j(\xi) = 0$ for $0 < j < 2^{r+1}$.

We now discuss the relationship with the classes w_1 and w_2 defined in Sects. 9.2 and 9.3.

Proposition 9.4.4 *Let ξ be a vector bundle over a CW-complex X. Then, the first Stiefel-Whitney class $w_1(\xi) \in H^1(X)$ defined above coincides with that defined in Sect. 9.2. In particular, $w_1(\xi) = 0$ if and only if ξ is orientable.*

Proof Both definitions enjoy naturality for induced bundles. We can then restrict ourselves to X being 1-dimensional, since $H^1(X) \to H^1(X^1)$ is injective. In this case, $\xi \approx \lambda \oplus \eta$ where λ is a line bundle and η a trivial vector bundle (see e.g. [105, Chap. 8, Theorem 1.2]). By Remark 9.2.2 and (5) of Theorem 9.4.1, we are reduced to the case of a line bundle. Then, both definitions coincide with the Euler class by Corollary 9.2.5 and Point (6) of Theorem 9.4.1. □

A similar result holds for the second Stiefel-Whitney class.

Proposition 9.4.5 *Let ξ be an orientable vector bundle over a CW-complex X. Suppose that X has no 3-cells or that X^3 is a regular complex. Then, the second Stiefel-Whitney class $w_2(\xi) \in H^2(X)$ defined above coincides with the cellular one $\dot{w}_2(\xi) \in \dot{H}^2(X)$ defined in* Sect. 9.3.

Proof Recall that the condition on X (and the orientability of ξ) was necessary for us to define $\dot{w}_2(\xi)$. The coincidence between $w_2(\xi) \in H^2(X)$ and $\dot{w}_2(\xi) \in \dot{H}^2(X)$ holds under the identification of $H^2(X)$ and $\dot{H}^2(X)$ as the same subgroup of $H^2(X^2)$ (see (3.5.5)). The class w_2 is natural by Point (2) of Theorem 9.4.1 and, by construction, $\dot{w}_2$ is natural for the restriction to a subcomplex. We can thus suppose that $X = X^2$ and that X is connected.

As ξ is orientable, its restriction over X^1 is trivial. By Lemma 9.1.15, $\xi \approx p^*\bar{\xi}$, where $p : X \to \bar{X} = X/X_1$. Again, $w_2(\xi) = H^*p(w_2(\bar{\xi}))$ and, by construction of $\dot{w}_2$, $\dot{w}_2(\xi) = \dot{H}^*p(\dot{w}_2(\bar{\xi}))$. We can thus suppose that X is a bouquet of 2-sphere, or even that $X = S^2$ with its minimal cell decomposition.

If η is a trivial bundle, both equations $w_2(\xi \oplus \eta) = w_2(\xi)$ and $\dot{w}_2(\xi \oplus \eta) = \dot{w}_2(\xi)$ hold true, by Point (5) of Theorem 9.4.1 and Lemma 9.3.2. We can thus suppose that ξ has rank ≥ 3. As seen in Example 9.3.3, there is only one non-trivial such bundle over S^2, characterized by $\dot{w}_2(\xi) \neq 0$. Let $\gamma_{\mathbb{C}}$ be the tautological bundle over $\mathbb{C}P^1 \approx S^2$. By Proposition 6.1.10, one has

$$0 \neq e(\gamma_{\mathbb{C}}) = w_2(\gamma_{\mathbb{C}}) = w_2(\gamma_{\mathbb{C}} \oplus \eta).$$

which finishes the proof of our proposition. Incidentally, we have proven that $\gamma_{\mathbb{C}}$ is stably non-trivial. □

Proposition 9.4.5 permits us to generalize the framework of Propositions 9.3.4 and 9.3.6.

Proposition 9.4.6 *Let ξ be a vector bundle of rank $r \geq 3$ over a CW-complex X. Then, the following conditions are equivalent.*

(1) $w_1(\xi) = 0$ *and* $w_2(\xi) = 0$.
(2) *the restriction* ξ_3 *of* ξ *over* X^3 *is trivial.*

Proof By Theorem 9.4.1, (1) implies $w_1(\xi_2) = 0$ and $w_2(\xi_2) = 0$. As ξ_2 has no 3-cells, $\dot{w}_2(\xi_2)$ is defined and, by Proposition 9.4.5, $\dot{w}_2(\xi_2) = 0$. By Proposition 9.3.4, ξ_2 is trivial which, as seen in the proof of Proposition 9.3.4, implies that ξ_3 is trivial. Thus, (1) implies (2). To prove that (2) implies (1), let $j : X^3 \to X$ denote the inclusion. Then $j^*(w_i(\xi)) = w_i(\xi_3)) = 0$ for and $j^* : H^k(X) \to H^k(X^3)$ is injective for $k \leq 3$ (We have also proven that (1) implies $w_3(\xi) = 0$, but this is already known by Remark 9.4.3). □

Proposition 9.4.7 *Let* ξ *be a vector bundle of rank* $r \geq 2$ *over a CW-complex* X. *Then, the following conditions are equivalent.*

(1) $w_1(\xi) = 0$ *and* $w_2(\xi) = 0$.
(2) ξ *admits a spin structure*

Moreover, if (2) *holds true, then the set of strong (or weak) equivalence classes of spin structures on* ξ *is in bijection with* $H^1(X)$.

Proof Suppose first that $r \geq 3$. If ξ admits a spin structure, then ξ_2 is trivial (see the proof of Proposition 9.3.6), which implies (1) by Proposition 9.4.6. Conversely, if (1) holds true, then ξ_2 is trivial by Proposition 9.4.6 and thus ξ_2 admits a spin-structure. That this structure extends to ξ is established as in the proof of Proposition 9.3.6. For the case $r = 2$ as well as for the last assertion of the proposition, the proofs are the same as those for Proposition 9.3.6. □

9.5 Real Flag Manifolds

Most of the results of this section come from [15], but we do not use spectral sequences. The Leray-Hirsch Theorem 4.7.17 for locally trivial bundles, together with some perfect Morse theory, is sufficient for our needs. We shall deal with *homogeneous spaces* of the form Γ/Γ_0, where Γ is a Lie group and Γ_0 a compact subgroup (therefore, a Lie subgroup). Then Γ/Γ_0 inherits a smooth manifold structure [37, Chap. 1, Proposition 5.3]. More generally, [20, Chap. II, Theorem 5.8] implies the following lemma.

Lemma 9.5.1 *Let* Γ *be a Lie group and* $H \subset G$ *be compact subgroups of* Γ. *Then, the quotient map* $\Gamma/H \to \Gamma/G$ *is a smooth locally trivial fiber bundle with fiber* G/H. *If* $H = \{1\}$, *then the quotient map* $\Gamma \to \Gamma/G$ *is a smooth* G*-principal bundle.*

9.5.1 Definitions and Morse Theory

Let $n_1, \dots n_r$ be positive integers and let $n = n_1 + n_2 + \cdots n_r$. By the *flag manifold* $\mathrm{Fl}(n_1, \dots, n_r)$, we mean any smooth manifold diffeomorphic to the homogeneous space

$$\mathrm{Fl}(n_1, \dots, n_r) \approx O(n)\big/O(n_1) \times O(n_2) \times \cdots \times O(n_r). \tag{9.5.1}$$

Here are some examples.

(1) *Nested subspaces.* $\mathrm{Fl}(n_1, \dots, n_r)$ is the set of nested vector subspaces $V_1 \subset \cdots \subset V_r \subset \mathbb{R}^n$ with $\dim V_i = \sum_{j=1}^{i} n_j$.
(2) *Mutually orthogonal subspaces.* $\mathrm{Fl}(n_1, \dots, n_r)$ is the set of r-tuples $(W_1, \dots, W_r)$ of vector subspaces $\mathbb{R}^n$ which are mutually orthogonal and satisfy $\dim W_i = n_i$. The correspondence from this definition to Definition (1) associates to $(W_1, \dots, W_r)$ the nested family $\{V_i\}$ where V_i is the vector space generated by $W_1 \cup \cdots \cup W_i$.
(3) *Isospectral symmetric matrices.* Let $\lambda_1 > \cdots > \lambda_r$ be real numbers. Consider the manifold $SM(n)$ of all symmetric real $(n \times n)$-matrices, on which $O(n)$ acts by conjugation. Then $\mathrm{Fl}(n_1, \dots, n_r)$ occurs as the orbit of the diagonal matrix having entries λ_i with multiplicity n_i.

$$\mathrm{Fl}(n_1, \dots, n_r) = \big\{R\,\mathrm{dia}\big(\underbrace{\lambda_1, \dots, \lambda_1}_{n_1}, \cdots, \underbrace{\lambda_r, \dots, \lambda_r}_{n_r}\big)\,R^{-1} \mid R \in O(n)\big\}. \tag{9.5.2}$$

In other words, $\mathrm{Fl}(n_1, \dots, n_r)$ is here the space of symmetric real $(n \times n)$-matrices with characteristic polynomial equal to $\sum_{i=1}^{r}(x - \lambda_i)^{n_i}$. Indeed, elementary linear algebra teaches us that two matrices in $SM(n)$ are in the same $O(n)$-orbit if and only if they have the same characteristic polynomial. The correspondence from this definition to Definition (2) associates, to a matrix M, its eigenspaces for the various eigenvalues.

Concrete definition (3) will be our working definition for $\mathrm{Fl}(n_1, \dots, n_r)$ throughout this section. Special classes of flag manifolds are given by the Grassmannians

$$\mathrm{Gr}(k; \mathbb{R}^n) = \mathrm{Fl}(k, n-k) \approx O(n)\big/O(k) \times O(n-k)$$

of k-planes in $\mathbb{R}^n$. This is a closed manifold of dimension

$$\dim \mathrm{Gr}(k; \mathbb{R}^n) = \dim O(n) - \dim O(k) - \dim O(n-k) = k(n-k).$$

For example, $\mathrm{Gr}(1; \mathbb{R}^n) \approx \mathbb{R}P^{n-1}$, of dimension $n - 1$. Using Definition (3) above, our "concrete Grassmannian" will be

$$\mathrm{Gr}(k;\mathbb{R}^n) = \big\{R\,\mathrm{dia}\big(\underbrace{1,\ldots,1}_{k},\underbrace{0,\ldots,0}_{n-k}\big)\,R^{-1} \mid R \in O(n)\big\}. \tag{9.5.3}$$

In other words, $\mathrm{Gr}(k;\mathbb{R}^n)$ is the space of orthogonal projectors on $\mathbb{R}^n$ of rank k. Another interesting flag manifold is the *complete flag manifold*

$$\mathrm{Fl}(1,\ldots,1) \approx O(n)\big/O(1)\times\cdots\times O(1)$$

with $\dim \mathrm{Fl}(1,\ldots,1) = \dim O(n) = \frac{n(n-1)}{2}$.

We now define real functions on the flag manifolds by restriction of the *weighted trace* on $f : SM(n) \to \mathbb{R}$ defined by

$$f(M) = \sum_{j=1}^{n} j\,M_{jj}$$

where M_{ij} denotes the (i,j)-entry of M.

Proposition 9.5.2 *Let* $\mathrm{Fl}(n_1,\ldots,n_r) \subset SM(n)$ *be the flag manifold as presented in* (9.5.2)*. Then, the restriction* $f : \mathrm{Fl}(n_1,\ldots,n_r) \to \mathbb{R}$ *of the weighted trace is a perfect Morse function whose critical points are the diagonal matrices in* $\mathrm{Fl}(n_1,\ldots,n_r)$*. The index of the critical point* $\mathrm{dia}(x_1,\ldots,x_n)$ *is the number of pairs* (i,j) *with* $i<j$ *and* $x_i < x_j$.

For a general discussion about such Morse functions on flag manifolds, see [13, Chap. 8].

Example 9.5.3 For $\mathrm{Gr}(2;\mathbb{R}^5) = \mathrm{Fl}(2,3)$, we get the following $\binom{5}{2} = 10$ critical points, with their index and value by f.

Critical point	Index	Value
dia(1, 1, 0, 0, 0)	0	3
dia(1, 0, 1, 0, 0)	1	4
dia(1, 0, 0, 1, 0), dia(0, 1, 1, 0, 0)	2	5
dia(1, 0, 0, 0, 1), dia(0, 1, 0, 1, 0)	3	6
dia(0, 0, 1, 1, 0), dia(0, 1, 0, 0, 1)	4	7
dia(0, 0, 1, 0, 1)	5	8
dia(0, 0, 0, 1, 1)	6	9

Remark 9.5.4 The function $\bar{f} : \mathrm{Gr}(k;\mathbb{R}^n) \to \mathbb{R}$ given by

$$\bar{f}(M) = -\frac{k(k+1)}{2} + f(M)$$

is a Morse function which is *self-indexed*, i.e. $\bar{f}(M) = j$ if M is a critical point of index j.

Proof of Proposition 9.5.2 We introduce precise notations which will be used later. For $1 \leq i < j \leq n$, let $r^{ij} : \mathcal{M}_2(\mathbb{C}) \to \mathcal{M}_n(\mathbb{C})$ defined by requiring that the entries of $r^{ij}(N)$ are those of the identity matrix I_n, except for

$$r^{ij}(N)_{ii} = N_{11} \,,\; r^{ij}(N)_{ij} = N_{12} \,,\; r^{ij}(N)_{ji} = N_{21} \,,\; r^{ij}(N)_{jj} = N_{22}.$$

The restriction of r^{ij} to $SO(2)$ gives an injective homomorphism $r^{ij} : SO(2) \to SO(n)$ whose image is formed by the matrices

$$R_t^{ij} = r^{ij}\begin{pmatrix} \cos t & -\sin t \\ \sin t & \cos t \end{pmatrix} \quad (t \in \mathbb{R}).$$

The action of R_t^{ij} on $\mathrm{Fl}(n_1, \dots, n_r) \subset SM(n)$ by conjugation produces a flow and thus a vector field V^{ij} on $\mathrm{Fl}(n_1, \dots, n_r)$, whose value V_M^{ij} at $M \in \mathrm{Fl}(n_1, \dots, n_r)$ is $V_M^{ij} = \frac{d}{dt}(R_t^{ij} M R_{-t}^{ij})_{|t=0}$ (we identify $T_M\mathrm{Fl}(n_1, \dots, n_r)$ as a subspace of $SM(n)$). A direct computation gives that

$$\begin{aligned}
(R_t^{ij} M R_{-t}^{ij})_{ii} &= M_{ii} \cos^2 t - M_{ij} \sin 2t + M_{jj} \sin^2 t \\
(R_t^{ij} M R_{-t}^{ij})_{jj} &= M_{ii} \sin^2 t + M_{ij} \sin 2t + M_{jj} \cos^2 t \\
(R_t^{ij} M R_{-t}^{ij})_{ij} &= M_{ij} \cos 2t + (M_{ii} - M_{jj}) \sin t \cos t.
\end{aligned} \tag{9.5.4}$$

Moreover,

$$\begin{aligned}
(R_t^{ij} M R_{-t}^{ij})_{ik} &= M_{ik} \cos t - M_{jk} \sin t && \text{if } i \neq k \neq j \\
(R_t^{ij} M R_{-t}^{ij})_{kj} &= M_{ki} \sin t + M_{kj} \cos t && \text{if } i \neq k \neq j \\
(R_t^{ij} M R_{-t}^{ij})_{kl} &= M_{kl} && \text{if } i \neq k \text{ and } j \neq l.
\end{aligned} \tag{9.5.5}$$

Let $g^{ij}(t) = f(R_t^{ij} M R_{-t}^{ij})$. The first derivative $\dot{g}^{ij}(t)$ satisfies

$$\dot{g}^{ij}(t) = (j-i)(M_{ii} - M_{jj}) \sin 2t + 2(j-i) M_{ij} \cos 2t.$$

Hence,

$$V_M^{ij} f = \dot{g}^{ij}(0) = 2(j-i) M_{ij}, \tag{9.5.6}$$

which proves that only the diagonal matrices in $\mathrm{Fl}(n_1, \dots, n_r)$ may be critical points of the weighted trace.

Suppose that $\Delta \in \mathrm{Fl}(n_1, \dots, n_r)$ is a diagonal matrix. Let

$$\mathcal{J}_\Delta = \{(i, j) \mid 1 \leq i < j \leq n \text{ and } \Delta_{ii} \neq \Delta_{jj}\}.$$

and let $\mathcal{V}_\Delta = \{V_\Delta^{ij} \mid (i,j) \in \mathcal{J}_\Delta\} \subset T_\Delta \mathrm{Fl}(n_1, \dots, n_r)$. By (9.5.4) and (9.5.5), $\frac{d}{dt}(R_t^{ij} \Delta R_{-t}^{ij})(0)$ has only non-zero term away from the diagonal, namely $\frac{d}{dt}$ $(R_t^{ij} \Delta R_{-t}^{ij})_{ij}(0) = \Delta_{ii} - \Delta_{jj}$. Hence, vectors of $\mathcal{V}_\Delta$ are linearly independent. But

$$\sharp \mathcal{J}_\Delta = \frac{n(n-1)}{2} - \sum_{k=1}^{r} \frac{n_k(n_k-1)}{2} = \dim O(n) - \sum_{k=1}^{r} \dim O(n_k) = \dim \mathrm{Fl}(n_1, \dots, n_r).$$

Therefore, $\mathcal{V}_\Delta$ is a basis of $T_\Delta \mathrm{Fl}(n_1, \dots, n_r)$. Using (9.5.6), this proves that the diagonal matrices in $\mathrm{Fl}(n_1, \dots, n_r)$ are exactly the critical points of the weighted trace. The matrix of the Hessian form $\mathcal{H}f$ on $T_\Delta \mathrm{Fl}(n_1, \dots, n_r)$ is

$$\begin{aligned} \mathcal{H}f(V_\Delta^{kl}, V_\Delta^{ij}) &= V_\Delta^{kl}(V^{ij} f) \\ &= V_\Delta^{kl}\big(M \mapsto 2(j-i)M_{ij}\big) \qquad \text{by (9.5.6)} \\ &= 2(j-i)\big[\tfrac{d}{dt}(R_t^{kl} \Delta R_{-t}^{kl})_{|t=0}\big]_{ij}\,. \end{aligned}$$

Using (9.5.4) and (9.5.5), we see that the matrix of $\mathcal{H}f$ in the basis $\mathcal{V}_\Delta$ is diagonal, with diagonal term

$$\mathcal{H}f(V_\Delta^{ij}, V_\Delta^{ij}) = 2(j-i)(\Delta_{ii} - \Delta_{jj}).$$

As $(i,j) \in \mathcal{J}_\Delta$, this proves that f is a Morse function as well as the assertion on the Morse index of Δ.

It remains to prove that f is perfect. Let Γ be the subgroup of $O(n)$ formed by the diagonal matrices (with coefficients ± 1). The $O(n)$-action on $SM(n)$ by conjugation may be restricted to Γ and f is Γ-invariant. Moreover, the diagonal matrices in $\mathrm{Fl}(n_1, \dots, n_r)$ are exactly the fixed points of the Γ-action. The perfectness of f then follows from Theorem 7.6.6.

Here is a first consequence of Proposition 9.5.2.

Corollary 9.5.5

$$\dim H^*(\mathrm{Fl}(n_1, \dots, n_r)) = \frac{n!}{n_1! \cdots n_r!}.$$

In particular,

$$\dim H^*(\mathrm{Fl}(k, n-k)) = \mathrm{Gr}(k; \mathbb{R}^n) = \tbinom{n}{k} \ \textit{and} \ \dim H^*(\mathrm{Fl}(1, \cdots, 1)) = n!.$$

Proof By Proposition 9.5.2, the weighted trace $f : \mathrm{Fl}(n_1, \dots, n_r) \to \mathbb{R}$ is a perfect Morse function. Hence, by Proposition 7.6.4, $\dim H^*(\mathrm{Fl}(n_1, \dots, n_r)) = \sharp \mathrm{Crit}\, f$. But $\mathrm{Crit}\, f$ consists of the diagonal matrices in $\mathrm{Fl}(n_1, \dots, n_r)$, which are all conjugate to

$$\mathrm{dia}\big(\underbrace{\lambda_1, \dots, \lambda_1}_{n_1}, \cdots, \underbrace{\lambda_r, \dots, \lambda_r}_{n_r}\big)$$

by a permutation matrix. Hence, Crit f is an orbit of the symmetric group Sym_n, with isotropy group $\mathrm{Sym}_{n_1} \times \cdots \times \mathrm{Sym}_{n_r}$, whence the formulae. □

Remark 9.5.6 The critical points of f in Proposition 9.5.2 are related to the Schubert cells (see Sect. 9.5.3).

Consider the inclusion $SM(n) \subset SM(n+1)$ with image the matrices with vanishing last row and column. Seeing $\mathrm{Gr}(k;\mathbb{R}^n) \subset SM(n)$ as in (9.5.3), this gives an inclusion $\mathrm{Gr}(k;\mathbb{R}^n) \subset \mathrm{Gr}(k;\mathbb{R}^{n+1})$.

Lemma 9.5.7 *The homomorphism $H^j(\mathrm{Gr}(k;\mathbb{R}^{n+1})) \to H^j(\mathrm{Gr}(k;\mathbb{R}^n))$ induced by the inclusion is surjective for all j and is an isomorphism for $j \le n-k$.*

Proof Let us use the Morse function $\bar f : \mathrm{Gr}(k;\mathbb{R}^{n+1}) \to \mathbb{R}$ of Remark 9.5.4 and let $\bar f'$ be its restriction to $\mathrm{Gr}(k;\mathbb{R}^n)$. Then, $\bar f'$ and $\bar f$ are self-indexed and $\mathrm{Crit}\,\bar f' \subset \mathrm{Crit}\,\bar f \subset \mathbb{N}$. For $m \in \mathbb{N}$, let $W_m = \bar f^{-1}((\infty, m+1/2])$ and $W'_m = (\bar f')^{-1}((\infty, m+1/2])$. For the first assertion, we prove, by induction on m that $H^*(W_m) \to H^*(W'_m)$ is surjective for all $m \in \mathbb{N}$. The induction starts with $m=0$, since $W_0 \simeq W'_0 \simeq pt$. The induction step involves the cohomology sequences

$$\begin{array}{ccccccccc} 0 & \longrightarrow & H^*(W_m, W_{m-1}) & \longrightarrow & H^*(W_m) & \longrightarrow & H^*(W_{m-1}) & \longrightarrow & 0 \\ & & \big\downarrow {\scriptstyle i^*_{m,m-1}} & & \big\downarrow {\scriptstyle i^*_m} & & \big\downarrow {\scriptstyle i^*_{m-1}} & & \\ 0 & \longrightarrow & H^*(W'_m, W'_{m-1}) & \longrightarrow & H^*(W'_m) & \longrightarrow & H^*(W'_{m-1}) & \longrightarrow & 0 \end{array} \tag{9.5.7}$$

obtained by Lemma 7.6.3, since $\bar f$ and $\bar f'$ are perfect by Proposition 9.5.2. From Proposition 9.5.2 again and its proof, the critical points of $\bar f'$ have the negative normal directions in W_{m-1} or in W'_{m-1}. Hence, using excision, the Morse lemma and Thom isomorphisms, we get the commutative diagram

$$\begin{array}{ccc} H^*(W_m, W_{m-1}) & \xrightarrow{\approx} & \prod_{C\in \mathrm{Crit}\,\bar f \cap \bar f^{-1}(m)} H^{*-m}(C) \\ \big\downarrow {\scriptstyle i^*_{m,m-1}} & & \big\downarrow {\scriptstyle proj} \\ H^*(W'_m, W'_{m-1}) & \xrightarrow{\approx} & \prod_{C\in \mathrm{Crit}\,\bar f' \cap \bar f^{-1}(m)} H^{*-m}(C) \end{array}$$

which proves that $i^*_{m,m-1}$ is onto. If i^*_{m-1} is surjective by induction hypothesis, we get that i^*_m is surjective by diagram-chasing.

Note that the point $D \in \mathrm{Crit}\,\bar f - \mathrm{Crit}\,\bar f'$ of lowest index is

$$D = \mathrm{dia}(1,\ldots,1,0,\ldots,0,1) \in SM(n+1)$$

satisfies $\bar f(D) = \mathrm{index}(D) = n-k+1$ (the number of zeros in D). Hence, $\mathrm{Crit}\,\bar f \cap W_{n-k} = \mathrm{Crit}\,\bar f' \cap W'_{n-k}$. The same induction argument as above shows that $H^j(\mathrm{Gr}(k;\mathbb{R}^{n+1})) \to H^j(\mathrm{Gr}(k;\mathbb{R}^n))$ is an isomorphism for $j \le n-k$. □

9.5.2 Cohomology Rings

The cohomology ring of a flag manifold V will be generated by Stiefel-Whitney classes of some tautological bundles over V. Consider a flag manifold $\mathrm{Fl}(n_1,\dots,n_r)$, with $n = n_1 + \dots + n_r$. Consider the following closed subgroups of $O(n)$.

$$B_i = O(n_1) \times \dots \times \{1\} \times \dots \times O(n_r) \subset O(n_1) \times \dots \times O(n_r) \subset O(n),$$

where $\{1\}$ sits at the i-th place. Then

$$P_i = O(n)/B_i \twoheadrightarrow O(n)\big/O(n_1) \times \dots \times O(n_r) = \mathrm{Fl}(n_1,\dots,n_r)$$

is an $O(n_i)$-principal bundle over $\mathrm{Fl}(n_1,\dots,n_r)$. Indeed, if K is a compact subgroup of a Lie group G, then $G \to G/K$ is a principal K-bundle (see, e.g. [12, Theorem 2.1.1, Chap. I]). Let ξ_i be the vector bundle of rank n_i associated to P_i, i.e. $E(\xi_i) = P_i \times_{O(n_i)} \mathbb{R}^{n_i}$. The vector bundle ξ_i is called the *i-th-tautological vector bundle* over $\mathrm{Fl}(n_1,\dots,n_r)$. Being associated to an $O(n_i)$-principal bundle, ξ_i is endowed with an Euclidean structure and its space of orthogonal frames $\mathrm{Fra}_\perp(\xi_i)$ is equal to P_i.

In the mutually orthogonal subspaces description (presentation (2), p. 376) of $\mathrm{Fl}(n_1,\dots,n_r)$, we see that

$$E(\xi_i) = \{(W_1 \dots, W_r, v) \in \mathrm{Fl}(n_1,\dots,n_r) \times \mathbb{R}^n \mid v \in W_i\}.$$

Note that $\xi_1 \oplus \dots \oplus \xi_r$ is trivial. Indeed,

$$E(\xi_1 \oplus \dots \oplus \xi_r) = \{((W_1,\dots,W_r),(v_1,\dots,v_r)) \in \mathrm{Fl}(n_1,\dots,n_r) \times (\mathbb{R}^n)^r \mid v_i \in W_i\}$$

and the correspondence

$$((W_1,\dots,W_r),(v_1,\dots,v_r)) \mapsto v_1 + \dots + v_r \tag{9.5.8}$$

restricts to a linear isomorphism on each fiber. Such a map thus provides a trivialization of $\xi_1 \oplus \dots \oplus \xi_r$.

If one sees $\mathrm{Fl}(n_1,\dots,n_r)$ as the space of matrices $M \in SM(n)$ with characteristic polynomial equal to $\sum_{i=1}^r (x-\lambda_i)^{n_i}$ (presentation (3), p. 376), then

$$E(\xi_i) = \{(M, v) \in \mathrm{Fl}(n_1,\dots,n_r) \times \mathbb{R}^n \mid Mv = \lambda_i v\}. \tag{9.5.9}$$

The vector bundle ξ_1 over $\mathrm{Fl}(k, n-k) = \mathrm{Gr}(k;\mathbb{R}^n)$ is called the *tautological vector bundle* over the Grassmannian $\mathrm{Gr}(k;\mathbb{R}^n)$; it is of rank k and is denoted by ζ, ζ_k or $\zeta_{k,n}$. The space of $\mathrm{Fra}_\perp(\zeta_k)$ is the *Stiefel manifold* $\mathrm{Stief}(k,\mathbb{R}^n)$ of orthonormal k-frames in $\mathbb{R}^n$.

The inclusion $\mathbb{R}^n \approx \mathbb{R}^n \times \{0\} \hookrightarrow \mathbb{R}^{n+1}$ induces an inclusion $\mathrm{Gr}(k;\mathbb{R}^n) \hookrightarrow \mathrm{Gr}(k;\mathbb{R}^{n+1})$ and we may consider the inductive limit

$$\mathrm{Gr}(k;\mathbb{R}^\infty) = \lim_{n} \mathrm{Gr}(k;\mathbb{R}^n)$$

which is a CW-space. The tautological vector bundle ζ_k is also defined over $\mathrm{Gr}(k;\mathbb{R}^\infty)$ and induces that over $\mathrm{Gr}(k;\mathbb{R}^n)$ by the inclusion $\mathrm{Gr}(k;\mathbb{R}^n) \hookrightarrow \mathrm{Gr}(k;\mathbb{R}^\infty)$. It is classical that $\pi_i(\mathrm{Stief}(k,\mathbb{R}^n)) = 0$ for $i < n-k$ (see [181, Theorem 25.6]), thus $\mathrm{Stief}(k,\mathbb{R}^\infty) = \mathrm{Fra}(\zeta_k)$ is contractible. Hence, the $O(k)$-principal bundle $\mathrm{Stief}(k,\mathbb{R}^\infty) \to \mathrm{Gr}(k;\mathbb{R}^\infty)$ is a universal $O(k)$-principal bundle (see [181, Sect. 19.4]) and thus homotopy equivalent to the Milnor universal bundle $EO(k) \to BO(k)$. In particular, $\mathrm{Gr}(k;\mathbb{R}^\infty)$ has the homotopy type of $BO(k)$. As a consequence, any vector bundle of rank k over a paracompact space X is induced from ζ_k by a map $X \to \mathrm{Gr}(k;\mathbb{R}^\infty)$ (for a direct proof of that (see [153, Theorem 5.6]).

Theorem 9.5.8 *The cohomology ring of $BO(k)$ is* **GrA***-isomorphic to the polynomial ring*

$$H^*(BO(k)) \approx H^*(\mathrm{Gr}(k;\mathbb{R}^\infty)) \approx \mathbb{Z}_2[w_1,\dots,w_k]$$

generated by the Stiefel-Whitney classes $w_i = w_i(\zeta_k)$ of the tautological bundle ζ_k.

Proof Slightly more formally, we consider the polynomial ring $\mathbb{Z}_2[w_1,\dots,w_k]$ with formal variables w_i of degree i. The correspondence $w_i \mapsto w_i(\zeta_k)$ provides a **GrA**-morphism $\psi : \mathbb{Z}_2[w_1,\dots,w_k] \to H^*(BO(k))$ which we shall show that it is bijective.

For the injectivity, we consider the tautological line bundle γ over $\mathbb{R}P^\infty$ and its n-times product γ^n over $(\mathbb{R}P^\infty)^n$. As seen above, ζ_k is universal so γ^n is induced by a map $f : (\mathbb{R}P^\infty)^n \to BO(n)$. Recall from Proposition 4.3.10 that $H^*(\mathbb{R}P^\infty) = \mathbb{Z}_2[a]$ with a of degree 1 and, by Theorem 9.4.1, $w(\gamma) = 1 + a$. By the Künneth theorem, there is a **GrA**-isomorphism $\mathbb{Z}_2[a_1,\dots,a_n] \xrightarrow{\approx} H^*((\mathbb{R}P^\infty)^n)$ and, by Theorem 9.4.1, $w(\gamma^n) = (1+a_1) \smile \cdots \smile (1+a_n)$. As $H^*f(w(\zeta_k)) = w(\gamma^n)$, there is a commutative diagram

$$\begin{array}{ccc}
\mathbb{Z}_2[w_1,\dots,w_k] & \xrightarrow{\psi} & H^*(BO(k)) \\
\big\downarrow \phi & & \big\downarrow H^*f \\
\mathbb{Z}_2[a_1,\dots,a_n] & \xrightarrow{\approx} & H^*((\mathbb{R}P^\infty)^n)
\end{array}$$

with

$$\phi(w(\zeta_k)) = (1+a_1)\cdots(1+a_n) = 1 + \sigma_1 + \cdots \sigma_n,$$

where $\sigma_i = \sigma_i(a_1, \ldots, a_n)$ is the i-th elementary symmetric polynomial in the variables a_j (see (8.5.4)). Thus, $\phi(w_i) = \sigma_i$. Now, if $0 \neq A \in \mathbb{Z}_2[w_1, \ldots, w_k]$ satisfies $\psi(A) = 0$, then $\phi(A) = 0$ would be a non-trivial polynomial relation between the σ_i's. But the elementary symmetric polynomials are algebraically independent (see e.g. [122]). Thus, ψ is injective.

For $d \in \mathbb{N}$, let

$$\mathcal{B}_d = \{(d_1, \ldots, d_k) \in \mathbb{N}^k \mid \sum_{j=1}^{k} j\, d_j = d\}.$$

The correspondence $(d_1, \ldots, d_k) \mapsto w_1^{d_1} \cdots w_k^{d_k}$ is a bijection from $\mathcal{B}_d$ onto a basis of the vector subspace $\mathbb{Z}_2[w_1, \ldots, w_k]^{[d]}$ formed by the elements in $\mathbb{Z}_2[w_1, \ldots, w_k]$ which are of degree d. On the other hand, consider $\mathrm{Gr}(k; \mathbb{R}^n) \subset SM(n)$ as in (9.5.3), with n large. Let $\mathrm{Crit}_d f \subset \mathrm{Gr}(k; \mathbb{R}^n)$ be the set of critical points of index d for the weighted trace. Then the correspondence

$$(d_1, \ldots, d_k) \mapsto \mathrm{dia}(\underbrace{0, \ldots, 0}_{d_k}, 1, \underbrace{0, \ldots, 0}_{d_{k-1}}, 1, \ldots, \underbrace{0, \ldots, 0}_{d_1}, 1, 0, \ldots, 0)$$

provides a bijection $\mathcal{B}_d \xrightarrow{\approx} \mathrm{Crit}_d f$. As f is a perfect Morse function by Proposition 9.5.2, one has

$$\sharp\, \mathcal{B}_d = \sharp\, \mathrm{Crit}_d f = \dim H^d(\mathrm{Gr}(k; \mathbb{R}^n)) = \dim H^d(BO(k)),$$

the last equality coming from Lemma 9.5.7 when n is large enough. Therefore,

$$\dim \mathbb{Z}_2[w_1, \ldots, w_k]^{[d]} = \dim H^d(BO(k)).$$

As ψ is injective, it is then bijective. □

Define

$$Q_r(t) = \frac{1}{1 - t^r} = 1 + t^r + t^{2r} + \cdots \in \mathbb{Z}[[t]], \tag{9.5.10}$$

which is the Poincaré series of $\mathbb{Z}_2[x]$ if x is of degree r. Here is a direct consequence of Theorem 9.5.8.

Corollary 9.5.9 *The Poincaré series of $BO(k)$ is*

$$\mathfrak{P}_t(BO(k)) = Q_1(t) \cdots Q_k(t).$$

As any vector bundle of rank k over a paracompact space is induced from the universal bundle ζ_k [153, Theorem 5.6], Theorem 9.5.8 has the following corollary.

Corollary 9.5.10 *Any characteristic class in* mod 2 *cohomology for vector bundles of finite rank over paracompact spaces is a polynomial in the Stiefel-Whitney classes* w_i.

Also, Theorem 9.5.8 together with Lemma 9.5.7 gives the following corollary.

Corollary 9.5.11 *The cohomology ring* $H^*(\mathrm{Gr}(k;\mathbb{R}^n))$ *is generated, as a ring, by the Stiefel-Whitney classes* $w_1(\zeta_k), \dots, w_k(\zeta_k)$ *of the tautological bundle* ζ_k.

Theorem 9.5.8 permits us to compute the cohomology of $BSO(k)$. The latter also has a tautological bundle $\tilde{\zeta}_k = (ESO(k) \times_{SO(k)} \mathbb{R}^k \to BSO(k))$ which is orientable.

Corollary 9.5.12 *The cohomology ring of* $BSO(k)$ *is* **GrA**-*isomorphic to the polynomial ring*

$$H^*(BSO(k)) \approx \mathbb{Z}_2[w_2, \dots, w_k]$$

generated by the Stiefel-Whitney classes $w_i = w_i(\tilde{\zeta}_k)$ *of the tautological bundle* $\tilde{\zeta}_k$.

Proof Let $i : SO(k) \to O(k)$ denote the inclusion. By Example 7.2.4, the map $Bi : BSO(k) \to BO(k)$ is homotopy equivalent to a two fold covering, which is non-trivial since $BSO(k)$ is connected. By Lemma 4.3.6, its characteristic class $w(Bi) \in H^1(BO(k))$ is not trivial. By Theorem 9.5.8, the only non-zero element in $H^1(BO(k))$ is $w_1(\xi_k)$, so $w(Bi) = w_1(\xi_k)$.

By Theorem 9.5.8 and the transfer exact sequence (Proposition 4.3.9), the ring homomorphism $H^*Bi : H^*(BO(k)) \to H^*(BSO(k))$ is surjective with kernel the ideal generated by $w_1(\zeta_k)$. As Bi is covered by a bundle map from $\tilde{\zeta}_k$ to ζ_k, one has $H^*Bi(w_i(\zeta_k) = w_i(\tilde{\zeta}_k)$. The corollary follows. □

Remark 9.5.13 In contrast with the simplicity of $H^*(BSO(k))$, the cohomology ring $H^*(BSpin(k))$ is complicated and its computation requires spectral sequences (see [167]). The stable case $BSpin = \lim_k BSpin(k)$ is slightly simpler (see [193]).

We are now in position to give a **GrA**-presentation of $H^*(\mathrm{Fl}(n_1, \dots, n_r))$. Let

$$w(\xi_j) = \mathbf{1} + w_1(\xi_j) + \cdots + w_{n_j}(\xi_j) \in H^*(\mathrm{Fl}(n_1, \dots, n_r)) \tag{9.5.11}$$

be the Stiefel-Whitney class of the tautological vector bundle ξ_j. As seen in (9.5.8), $\xi_1 \oplus \cdots \oplus \xi_r$ is trivial. By Theorem 9.4.1, the equation

$$w(\xi_1) \smile \cdots \smile w(\xi_r) = \mathbf{1} \tag{9.5.12}$$

holds true. Hence, the homogeneous components of $w(\xi_1) \smile \cdots \smile w(\xi_r)$ in positive degrees vanish, giving rise to n equations.

Theorem 9.5.14 *The cohomology algebra* $H^*(\mathrm{Fl}(n_1, \dots, n_r))$ *is* **GrA**-*isomorphic to the quotient of the polynomial ring*

$$\mathbb{Z}_2[w_i(\xi_j)]\,,\ 1 \le i \le r_j\,,\ j = 1, \ldots, r$$

by the ideal generated by the homogeneous components of $w(\xi_1)\cdots w(\xi_r)$ *in positive degrees.*

Proof We first prove that $H^*(\mathrm{Fl}(n_1, \ldots, n_r))$ is, as a ring, generated by the Stiefel-Whitney classes $w_i(\xi_j)$ ($1 \le i \le r_j$, $j = 1, \ldots, r$). This is done by induction on r (note that $r \ge 2$ in order for the definition of $\mathrm{Fl}(n_1, \ldots, n_r)$ to make sense). For $r = 2$, as $\mathrm{Fl}(n_1, n_2) = \mathrm{Gr}(n_1; \mathbb{R}^{n_1+n_2})$, the result comes from Corollary 9.5.11. For the induction step, let us define a map $\pi : \mathrm{Fl}(n_1, \ldots, n_r) \to \mathrm{Fl}(n - n_r, n_r)$ by $\pi(W_1 \ldots, W_r) = (W_1 \oplus \cdots \oplus W_{r-1}, W_r)$ (using the mutually orthogonal definition (2) of the flag manifolds). By Lemma 9.5.1, this gives a locally trivial bundle

$$\mathrm{Fl}(n_1, \ldots, n_{r-1}) \xrightarrow{\iota} \mathrm{Fl}(n_1, \ldots, n_r) \xrightarrow{\pi} \mathrm{Fl}(n - n_r, n_r).$$

By induction hypothesis, $H^*(\mathrm{Fl}(n_1, \ldots, n_{r-1}))$ is generated, as a ring, by the Stiefel-Whitney classes of its tautological bundles, say $w_i(\bar{\xi}_j)$. Note that these bundles are induced by the tautological bundles (called ξ_j) over $\mathrm{Fl}(n_1, \ldots, n_r)$: $\bar{\xi}_j = \iota^*\xi_j$. Then, $H^*\iota$ is surjective and $w_i(\bar{\xi}_j) \mapsto w_i(\xi_j)$ is a cohomology extension of the fiber (see p. 172). On the other hand,

$$\mathrm{Fl}(n - n_r, n_r) \approx \mathrm{Gr}(n - n_r; \mathbb{R}^n) \approx \mathrm{Gr}(n_r; \mathbb{R}^n),$$

the last isomorphism sending an $(n - n_r)$-dimensional subspace of $\mathbb{R}^n$ to its orthogonal complement. By Corollary 9.5.11, $H^*(\mathrm{Gr}(n_r; \mathbb{R}^n))$ is **GrA**-generated by $w_1(\zeta_{n_r}), \ldots, w_{n_r}(\zeta_{n_r})$ and $H^*\pi(w_i(\zeta_{n_r})) = w_i(\xi_r)$. By the Leray-Hirsch Theorem 4.7.17, $H^*(\mathrm{Fl}(n_1, \ldots, n_r))$ is then **GrA**-generated by $w_i(\xi_j)$ ($1 \le i \le n_j$ and $j = 1 \ldots r$).

Let $\Gamma = O(n_1) \times \cdots \times O(n_r) \subset O(n)$ and consider the commutative diagram

$$\begin{array}{ccccc} O(n) & \longrightarrow & EO(n) & \longrightarrow & BO(n) \\ \downarrow & & \downarrow & & \downarrow = \\ O(n)/\Gamma & \longrightarrow & EO(n)/\Gamma & \longrightarrow & BO(n) \end{array} \tag{9.5.13}$$

where the top line is the $O(n)$-universal bundle. Hence, the bottom line is a locally trivial bundle with fiber equal to $O(n)/\Gamma = \mathrm{Fl}(n_1, \ldots, n_r)$; as $EO(n)$ is contractible, there are homotopy equivalences

$$EO(n)/\Gamma \simeq B\Gamma \simeq BO(n_1) \times \cdots \times BO(n_r),$$

the last homotopy equivalence coming from (7.4.3). Hence, Diagram (9.5.13) may be rewritten in the following way.

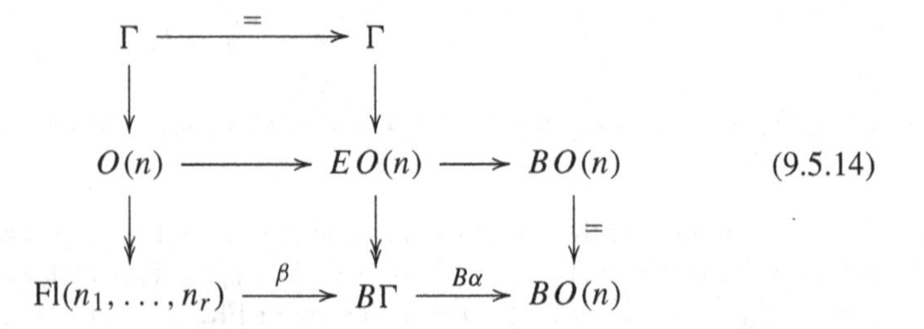

(9.5.14)

where α denotes the inclusion of Γ in $O(n)$. The left column is a Γ-principal bundle which is a Γ-structure on $\xi = \xi_1 \oplus \cdots \oplus \xi_r$. The central column is the Γ-principal bundle associated to the vector bundle $\zeta = \zeta_{n_1} \times \cdots \times \zeta_{n_r}$ over $B\Gamma \simeq BO(n_1) \times \cdots \times BO(n_r)$. Thus, the map β is a classifying map for the Γ-structure on ξ: it lifts the map $B\alpha \circ \beta$, which is classifying for ξ as a vector bundle (that $B\alpha \circ \beta$ is null-homotopic is coherent with the triviality of ξ, seen in (9.5.8)). Hence

$$\beta^* \zeta_{n_j} \approx \xi_j \ , \ \beta^* \zeta \approx \xi$$

and thus

$$H^*\beta(w_i(\zeta_{n_j})) = w_i(\xi_j) \ , \ H^*\beta(w_i(\zeta)) = w_i(\xi).$$

As $H^*(\mathrm{Fl}(n_1, \dots, n_r))$ is **GrA**-generated by the classes $w_i(\xi_j)$, $H^*\beta$ is surjective and one may apply the Leray-Hirsch Theorem 4.7.17 and its corollaries. By Theorem 9.5.8 and the Künneth theorem, there is a **GrA**-isomorphism

$$\mathbb{Z}_2[w_i(\zeta_{n_j})] \xrightarrow{\approx} H^*(B\Gamma) \ , \ (1 \le i \le r_j \ , \ j = 1, \dots, r).$$

On the other hand, $H^*(BO(n)) \approx \mathbb{Z}_2[w_i(\zeta_n)]$ and $H^*B\alpha(w_i(\zeta_n)) = w_i(\zeta)$. By Corollary 4.7.19, $H^*(\mathrm{Fl}(n_1, \dots, n_r))$ is then **GrA**-isomorphic to the quotient of $H^*(B\Gamma)$ by the ideal generated by $w_i(\zeta)$ $(i > 0)$. Hence, one has the commutative diagram

$$\begin{array}{ccc} \mathbb{Z}_2[w_i(\zeta_j)]/\big(w_i(\zeta), i > 0\big) & \xrightarrow{\approx} & \mathbb{Z}_2[w_i(\xi_j)]/\big(w_i(\xi), i > 0\big) \\ & \searrow^{\approx} \quad \swarrow & \\ & H^*(\mathrm{Fl}(n_1, \dots, n_r)) & \end{array}$$

which proves Theorem 9.5.14. □

In the following corollary, we use the notations Q_i of (9.5.10).

Corollary 9.5.15 *The Poincaré polynomial of* $\mathrm{Fl}(n_1, \dots, n_r)$ *is given by the formula*

$$\mathfrak{P}_t(\mathrm{Fl}(n_1, \dots, n_r)) = \frac{\prod_{j=1}^r [Q_1(t) \cdots Q_{n_j}(t)]}{Q_1(t) \cdots Q_n(t)}.$$

In particular,

$$\mathfrak{P}_t(\mathrm{Gr}(k;\mathbb{R}^n)) = \mathfrak{P}_t(\mathrm{Fl}(k,n-k)) = \frac{Q_1(t)\cdots Q_k(t)}{Q_{n-k+1}(t)\cdots Q_n(t)}$$

and

$$\mathfrak{P}_t(\mathrm{Fl}(1,\dots,1)) = \frac{Q_1(t)^n}{Q_1(t)\cdots Q_n(t)} = \frac{(1-t)(1-t^2)\cdots(1-t^n)}{(1-t)^n}.$$

Remark 9.5.16 The above formulae, evaluated at $t=1$ using L'Hospital's rule, give dim $H^*(\mathrm{Fl}(n_1,\dots,n_r))$, etc., giving again the formulae of Corollary 9.5.5.

Proof of Corollary 9.5.15 We have seen in the proof of Theorem 9.5.14 that

$$\mathrm{Fl}(n_1,\dots,n_r) \xrightarrow{\beta} B\Gamma \xrightarrow{B\alpha} BO(n)$$

is a locally trivial bundle satisfying the hypotheses of the Leray-Hirsch theorem. We know by Corollary 9.5.9 that $\mathfrak{P}_t(BO(k)) = Q_1(t)\cdots Q_k(t)$. By the Künneth formula, we get that

$$\mathfrak{P}_t(B\Gamma) = \mathfrak{P}_t(BO(n_1)\times\cdots\times BO(n_r)) = \prod_{j=1}^{r}[Q_1(t)\cdots Q_{n_j}(t)].$$

The first formula then comes from Corollary 4.7.20. The other formulae are consequences of the first one.

We now give some illustrations of Theorem 9.5.14.

Example 9.5.17 Consider the case of the complete flag manifold $\mathrm{Fl}(1,\dots,1)$. Theorem 9.5.14 says that $H^*(\mathrm{Fl}(1,\dots,1))$ is generated by $x_i = w_1(\xi_i)$ for $i = 1\dots,n$. In this generating system, $w_i(\xi_1\oplus\cdots\oplus\xi_n) = \sigma_i$, the ith elementary symmetric polynomial in the variables x_i. Hence, by Theorem 9.5.14,

$$H^*(\mathrm{Fl}(1,\dots,1)) \approx \mathbb{Z}_2[x_1,\dots,x_n]\big/\big(\sigma_1,\dots,\sigma_n\big).$$

Example 9.5.18 Consider the case of the Grassmannian

$$\mathrm{Gr}(k;\mathbb{R}^n) = \mathrm{Fl}(k,n-k).$$

Set $\zeta = \xi_1$, with Stiefel-Whitney class $w(\zeta) = w = \mathbf{1} + w_1 + w_2 + \cdots$, and $\zeta^\perp = \xi_2$, with $w(\zeta^\perp) = \bar{w} = \mathbf{1} + \bar{w}_1 + \bar{w}_2 + \cdots$. Note that the fiber of the vector bundle $\zeta^\perp$ over $P \in \mathrm{Gr}(k;\mathbb{R}^n)$ is the set of vectors in $\mathbb{R}^n$ which are orthogonal to P. Equation (9.5.12) becomes

$$w \smile \bar{w} = \mathbf{1}, \tag{9.5.15}$$

which is equivalent to the following system of equations:

$$\bar{w}_i = \sum_{r=1}^{k} w_r \bar{w}_{i-r} \ (i = 1, \dots, n-k) \ \text{and} \ \bar{w}_i = 0 \text{ if } i > n-k. \tag{9.5.16}$$

This system has the following unique solution.

Lemma 9.5.19 *With the convention $\bar{w}_i = w_i = 0$ for $i < 0$, the equation*

$$\bar{w}_r = \begin{vmatrix} w_1 & 1 & & & 0 \\ w_2 & \ddots & \ddots & & \\ \vdots & & \ddots & \ddots & \\ w_{r-1} & & & \ddots & 1 \\ w_r & w_{r-1} & \cdots & w_2 & w_1 \end{vmatrix} = \det\left(w_{i+1-j}\right)_{1 \le i,j \le r}.$$

holds true in $H^r(\mathrm{Gr}(k;\mathbb{R}^n))$. The symmetric formula $w_r = \det\left(\bar{w}_{i+1-j}\right)_{1\le i,j\le r}$ holds true as well. These equalities are both equivalent to Eq. (9.5.15).

Proof The first equation is proved by induction on r, starting, for $r = 1$, with $\bar{w}_1 = w_1$ (this also gives the uniqueness of the solution). The induction step is achieved by expanding the determinant with respect to the first column: the $(s, 1)$-th minor is equal to $\bar{w}_{r-s}$ by induction hypothesis and the result follows from (9.5.16). The symmetric equation follows from the symmetry in w_i and $\bar{w}_i$ of (9.5.16) (coming from the symmetry of (9.5.15)). □

Below are two special case of Example 9.5.18.

Example 9.5.20 Consider the case of $\mathrm{Gr}(1;\mathbb{R}^n) = \mathrm{Fl}(1, n-1) \approx \mathbb{R}P^{n-1}$. The relation $w \smile \bar{w} = \mathbf{1}$ gives rise to the system of equations

$$\begin{aligned} w_1 + \bar{w}_1 &= 0 \\ \bar{w}_i + \bar{w}_{i-1} w_1 &= 0 \quad (i = 2, \dots, n-1) \\ \bar{w}_{n-1} w_1 &= 0 \end{aligned}$$

from which we deduce the usual presentation $H^*(\mathbb{R}P^{n-1}) \approx \mathbb{Z}_2[w_1]/(w_1^n)$.

Example 9.5.21 In the case $\mathrm{Gr}(2;\mathbb{R}^4) = \mathrm{Fl}(2,2)$, the relation $w \smile \bar{w} = \mathbf{1}$ gives rise to four equations

$$\begin{aligned} \bar{w}_1 &= w_1 \\ \bar{w}_2 &= w_1^2 + w_2 \\ w_1^3 &= 0 \\ w_2 w_1^2 + w_2^2 &= 0. \end{aligned} \tag{9.5.17}$$

and Theorem 9.5.14 says that $H^*(\mathrm{Gr}(2;\mathbb{R}^4))$ is generated by $w_1, w_2, \bar{w}_1$ and $\bar{w}_2$, subject to Relations (9.5.17). The first two equations imply that $H^*(\mathrm{Gr}(2;\mathbb{R}^4))$ is generated by w_1 and w_2, as known since Corollary 9.5.11. We check that an additive basis of $H^*(\mathrm{Gr}(2;\mathbb{R}^4))$ is given by $\mathbf{1}$, w_1, w_2, w_1^2, w_2w_1 and $w_2w_1^2 = w_2^2$. The Poincaré polynomial of $\mathrm{Gr}(2;\mathbb{R}^4)$ is given by Corollary 9.5.15:

$$\mathfrak{P}_t(\mathrm{Gr}(2;\mathbb{R}^4)) = \frac{Q_1(t)Q_2(t)}{Q_3(t)Q_4(t)} = \frac{(1-t^3)(1-t^4)}{(1-t)(1-t^2)} = 1+t+2t^2+t^3+t^4.$$

For any bundle of ξ rank k over a space X the *dual (or normal) Stiefel-Whitney class* $\bar{w}_r(\xi)$ are defined by the equation of Lemma 9.5.19. Set $\bar{w}(\xi) = \mathbf{1}+\bar{w}_1(\xi)+\cdots$ for the total dual Stiefel-Whitney class. Equations (9.5.15) and (9.5.16) are satisfied. If there exists a vector bundle η over X such that $\xi \oplus \eta$ is trivial, then $w(\eta) = \bar{w}(\xi)$. Thus, if η is of rank r, then $\bar{w}_i(\xi) = 0$ for $i > r$. The same condition is necessary for ξ being induced from the tautological bundle ζ by a map $f : X \to \mathrm{Gr}(k;\mathbb{R}^{k+r})$.

For example, let M be a smooth manifold of dimension k which admits an immersion $\beta : M \to \mathbb{R}^{k+r}$. Let $x \in M$. By identifying $T_{\beta(x)}\mathbb{R}^{k+r}$ with $\mathbb{R}^{k+r}$, the k-vector space $T_x\beta(TM)$ becomes an element of $\mathrm{Gr}(k;\mathbb{R}^{k+r})$. This produces a map $\tilde{\beta} : M \to \mathrm{Gr}(k;\mathbb{R}^{k+r})$ and $TM = \tilde{\beta}^*\zeta$. We thus get the following result.

Proposition 9.5.22 *If a smooth manifold M of dimension k admits an immersion into $\mathbb{R}^{k+r}$, then $\bar{w}_i(TM) = 0$ for $i > r$.*

For improvements of Proposition 9.5.22 concerning also smooth embeddings, see Proposition 9.8.23 and Corollary 9.8.24. Usually, Proposition 9.5.22 does not give the smallest integer r for which M immerses into $\mathbb{R}^{k+r}$. This is however the case in the following example, taken from [153, Theorem 4.8].

Proposition 9.5.23 *For $k = 2^j$ ($j \geq 1$), the projective space $\mathbb{R}P^k$ immerses into $\mathbb{R}^N$ if and only if $N \geq 2k-1$.*

Proof That a manifold of dimension $k \geq 2$ immerses into $\mathbb{R}^{2k-1}$ is a classical theorem of Whitney [210]. Conversely, we shall see in Proposition 9.8.10 that

$$w(T\mathbb{R}P^{2^j}) = (1+a)^{2^j+1} = 1+a+a^{2^j},$$

where $0 \neq a \in H^1(\mathbb{R}P^{2^j}) = \mathbb{Z}_2$. Hence,

$$\bar{w}(T\mathbb{R}P^{2^j}) = 1+a+a^2+\cdots+a^{2^j-1},$$

which, using Proposition 9.5.22 implies that $\mathbb{R}P^{2^j}$ does not immerses into $\mathbb{R}^{2^{j+1}-2}$. □

9.5.3 Schubert Cells and Stiefel-Whitney Classes

Let $f : M \to \mathbb{R}$ be a Morse function on a manifold M. It is classical that M has the homotopy type of a CW-complex whose r-cells are in bijection with the critical points of index r of f (see, e.g. [13, Theorem 3.28]). For the weighted trace f (or $\bar{f}$) defined on $M = \mathrm{Gr}(k; \mathbb{R}^n)$ in Proposition 9.5.2 (or Remark 9.5.4), a very explicit such CW-structure is given, using the *Schubert cells* (there are generalizations for flag manifolds). Inspired by works of H. Schubert on enumerative geometry in the XIXth century (see e.g. [169]), Schubert cells were introduced in 1934 (for complex Grassmannians) by Ch. Ehresmann [47] ([48] for the real Grassmannians). See [40, 224–25] for a history. We restrict ourselves here to a very elementary point of view, Schubert calculus being a huge subject in algebraic geometry.

Recall that Crit $\bar{f}$ are diagonal matrices in $SM(n)$. We write $\mathrm{dia}(\lambda_1, \dots, \lambda_n) = \mathrm{dia}(\lambda)$, where $\lambda = \lambda_1 \cdots \lambda_n$ is a binary word of length n. Let $[{n \atop k}]$ be the set or such words with $\sum \lambda_i = k$ (they are $\binom{n}{k}$ in number). The correspondence $\lambda \mapsto \lambda 0$ identifies $[{n \atop k}]$ with a subset of $[{n+1 \atop k}]$, permitting us to define $[{\infty \atop k}]$ as the direct limit of $[{n \atop k}]$.

Let $F = (F_1 \subset \cdots \subset F_n)$ be a complete flag in $\mathbb{R}^n$ (adding the convention that $F_0 = \{0\}$). For $\lambda \in [{n \atop k}]$, the *Schubert cell* C_λ^F with respect to F is defined by

$$C_\lambda^F = \{P \in \mathrm{Gr}(k; \mathbb{R}^n) \mid \dim(P \cap F_i) = \sum_{j=1}^{i} \lambda_j\} \subset \mathrm{Gr}(k; \mathbb{R}^n).$$

(This convention is close to that of [119], except for the binary words being written in the reverse order, so it works for $n = \infty$, in the spirit of [153, Sect. 6]). The following facts may be proven.

(1) The Schubert cells $\{C_\lambda^F \mid \lambda \in [{n \atop k}]\}$ are the open cells of a CW-structure $\mathfrak{X}_F$ on $\mathrm{Gr}(k; \mathbb{R}^n)$ (see [153, Sect. 6]). The dimension of C_λ^F is

$$d(\lambda) = \mathrm{index}\,(\mathrm{dia}(\lambda)) = \bar{f}(\mathrm{dia}(\lambda)) = -\frac{k(k+1)}{2} + \sum_{i \geq 1} \lambda_i.$$

By Proposition 9.5.2, the cellular chains have then the same Poincaré polynomial as the homology. Therefore, $\mathfrak{X}_F$ is a perfect CW-structure.

(2) The closure $\bar{C}_\lambda^F$, called the *Schubert variety*, satisfies

$$\bar{C}_\lambda^F = \{P \in \mathrm{Gr}(k; \mathbb{R}^n) \mid \dim(P \cap F_i) \geq \sum_{j=1}^{i} \lambda_j\} \subset \mathrm{Gr}(k; \mathbb{R}^n)$$

and is a subcomplex of $\mathfrak{X}_F$ (see e.g. [47, Sect. 10]). As $\mathfrak{X}_F$ is perfect, so is $\bar{C}_\lambda^F$ and thus $\bar{C}_\lambda^F$ defines a homology class

$$[\lambda] = [\bar{C}_\lambda^F] \in H_{d(\lambda)}(\mathrm{Gr}(k; \mathbb{R}^n)) \quad (n \leq \infty)$$

which does not depend on F since, by Proposition 9.5.2, the complete flag manifold is path-connected. It corresponds, under the isomorphism (3.5.6) between cellular and singular homology, to the cellular homology class for $\mathfrak{X}_F$ indexed by λ. It follows that the $\mathcal{S} = \{[\lambda] \in H_*(\mathrm{Gr}(k;\mathbb{R}^n)) \mid \lambda \in [{n \atop k}]\}$ is a basis of $H_*(\mathrm{Gr}(k;\mathbb{R}^n))$ $(n \leq \infty)$.

(3) Let $P \in \mathrm{Gr}(k;\mathbb{R}^n)$. Using a basis of $\mathbb{R}^n$ compatible with the flag F, let M_P be the matrix of a linear epimorphism $\mathbb{R}^n \to \mathbb{R}^{n-k}$ with kernel P. The condition $P \in \bar{C}_\lambda^F$ is equivalent to the vanishing of various minors of M_P. Therefore, $P \in \bar{C}_\lambda^F$ is a compact real algebraic variety. This is another proof of the existence of the class $[\lambda]$, since such a variety carries a fundamental class (see [192, p. 67] or [16, Theorem 3.7 and Sect. 3.8]).

(4) Suppose that F is the standard flag ($F_i = \mathbb{R}^i \times 0$). Then $\bar{f}(\bar{C}_\lambda^F) = [0, d(\lambda)]$ and $\bar{C}_\lambda^F \cap \bar{f}^{-1}(d(\lambda)) = \mathrm{dia}(\lambda)$. Recall from Proposition 9.5.2 that $d(\lambda)$ is equal to the number of pairs (i, j) with $1 \leq i < j \leq n$ such that $\lambda_i < \lambda_j$. For such a pair (i, j), let R_t^{ij} be the one-parameter subgroup of $SO(n)$ considered in the proof of Proposition 9.5.2. Then, the curve $R_t^{ij}\,\mathrm{dia}(\lambda) R_{-t}^{ij}$ is contained in $\bar{C}_\lambda^F$ for $t \in \mathbb{R}$ and stays in C_λ^F when $|t| < \pi/2$. By the proof of Proposition 9.5.2, these curves generate the negative normal bundle for $\bar{f}$ at $\mathrm{dia}(\lambda)$.

Example 9.5.24 Consider the case of $\mathrm{Gr}(1;\mathbb{R}^{n+1}) \approx \mathbb{R}P^n$. For F the standard flag in $\mathbb{R}^{n+1}$, the Schubert cells give the standard CW-structure on $\mathbb{R}P^n$, the cell C_λ^F for $\lambda = 0^r 1 0^{n-r}$ being of dimension r. The Schubert variety $\bar{C}_\lambda^F$ is equal to $\mathbb{R}P^r$ (a rare case where it is a smooth manifold).

Note that $H_*\alpha([\lambda]) = [\lambda 0]$ where $\alpha : \mathrm{Gr}(k;\mathbb{R}^n) \hookrightarrow \mathrm{Gr}(k;\mathbb{R}^{n+1})$ is induced by the inclusion $\mathbb{R}^n \approx \mathbb{R}^n \oplus 0 \hookrightarrow \mathbb{R}^n \oplus \mathbb{R}$. We often identify $[\lambda]$ with $[\lambda 0]$. In this way, for instance, $[100101]$ determines a class in $H_5(\mathrm{Gr}(k;\mathbb{R}^n))$ for $n \geq 6$.

Let $\mathcal{S}^\sharp = \{[\lambda]^\sharp \mid \lambda \in [{n \atop k}]\}$ $(n \leq \infty)$ be the additive basis of $H^*(\mathrm{Gr}(k;\mathbb{R}^n))$ which is dual for the Kronecker pairing to the basis $\mathcal{S}$ (see (2) above): the class $[\lambda]^\sharp$ is defined by

$$\langle [\lambda]^\sharp, [\mu] \rangle = \delta_{\lambda\mu},$$

where $\delta_{\lambda\mu}$ is the Kronecker symbol. The basis $\mathcal{S}^\sharp$ was studied in [31, 32]. Because of intersection theory, a more widely used additive basis for $H^*(\mathrm{Gr}(k;\mathbb{R}^n))$ (defined only or $n < \infty$) is $\mathcal{S}^{PD}$, is formed by the Poincaré duals $[\lambda]^{PD}$ for all $\lambda \in [{n \atop k}]$. Though some intersection theory in used in the proof of Proposition 9.5.29, we shall not use $\mathcal{S}^{PD}$. We just note the following result.

Lemma 9.5.25 *For any $k \leq n < \infty$, the two sets $\mathcal{S}^\sharp$ and $\mathcal{S}^{PD}$ in $H^*(\mathrm{Gr}(k;\mathbb{R}^n))$ are equal.*

Proof Let F be the standard flag ($F_i = \mathbb{R}^i \times 0$) and let F^- be the anti-standard one ($F_i^- = 0 \times \mathbb{R}^i$). For $\lambda \in [{n \atop k}]$, define $\lambda^- \in [{n \atop k}]$ by $\lambda_i^- = \lambda_{n+1-i}$. The cycles $\bar{C}_\lambda^F$ and $\bar{C}_{\lambda^-}^{F^-}$ are of complementary dimensions and, by (4) above, they intersect transversally in a single point (the k-plane generated by $\lambda_i e_i$ for $i = 1, 2, \ldots, n$). In the same way, if $\mu \in [{n \atop k}]$ satisfies $d(\lambda) = d(\mu)$ but $\mu \neq \lambda$, then $\bar{C}_\lambda^F \cap \bar{C}_{\mu^-}^{F^-} = \emptyset$. Analogously to Proposition 5.4.12, one has

$$[\mu^-]^{PD} \smile [\lambda]^{PD} = [\bar{C}_\lambda^F \cap \bar{C}_{\lambda^-}^{F^-}] \tag{9.5.18}$$

(see Remark 9.5.26). This implies that

$$\langle [\mu^-]^{PD} \smile [\lambda]^{PD}, [\mathrm{Gr}(k; \mathbb{R}^n)] \rangle = \delta_{\mu\lambda}$$

But, using (4.5.13),

$$\begin{aligned}\langle [\mu^-]^{PD} \smile [\lambda]^{PD}, [\mathrm{Gr}(k; \mathbb{R}^n)] \rangle &= \langle [\mu^-]^{PD}, [\lambda]^{PD} \frown [\mathrm{Gr}(k; \mathbb{R}^n)] \rangle \\ &= \langle [\mu^-]^{PD}, [\lambda] \rangle.\end{aligned}$$

This proves that $[\lambda]^\sharp = [\lambda^-]^{PD}$ (or $[\lambda]^{PD} = [\lambda^-]^\sharp$). □

Remark 9.5.26 In the above proof, (9.5.18) is not a consequence of Proposition 5.4.12, which would require that $\bar{C}_\lambda^F$ and $\bar{C}_{\lambda^-}^{F^-}$ are submanifolds of $\mathrm{Gr}(k; \mathbb{R}^n)$. In this simple situation, one could use the Morse function $\bar{f}$ to isolate the intersection point around the critical level $d(\lambda)$ and deal with an intersection of submanifolds (with boundaries). For more general situation (see the proof of Proposition 9.5.29), one must rely on the intersection theory for real algebraic varieties (see, e.g. [16, (1.12) and Sect. 5]).

In addition to the above *ambient inclusion* $\alpha : \mathrm{Gr}(k; \mathbb{R}^n) \hookrightarrow \mathrm{Gr}(k; \mathbb{R}^{n+1})$ we also consider the *fattening inclusion* $\beta : \mathrm{Gr}(k; \mathbb{R}^n) \to \mathrm{Gr}(k+1; \mathbb{R}^{n+1})$ sending P to $\mathbb{R} \oplus P \subset \mathbb{R} \oplus \mathbb{R}^n$. Then $H_*j([\lambda]) = [1\lambda]$ for all $\lambda \in [{}^k_n]$. This drives us to decompose a word $\lambda \in [{}^n_k]$ into its *prefix*, *stem* and *suffix*, delimited by the first 0 and the last 1 of λ:

$$\lambda = \underbrace{111111111}_{\text{prefix}}\underbrace{00101101}_{\text{stem}}\underbrace{0000}_{\text{suffix}}.$$

Given n and k, a word $\lambda \in [{}^n_k]$ (and then a class $[\lambda] \in H_*(\mathrm{Gr}(k; \mathbb{R}^n))$ or $[\lambda]^\sharp \in H^*(\mathrm{Gr}(k; \mathbb{R}^n))$) is determined by its stem. For example, 0101 is, for $k = 4$ and $n = 8$, the stem of the unique class $[11010100]^\sharp \in H^3(\mathrm{Gr}(4; \mathbb{R}^8))$. The stem of $\mathbf{1} \in H^0(\mathrm{Gr}(k; \mathbb{R}^n))$ is just 0. Here is a first use of the prefix-stem-suffix decomposition.

Proposition 9.5.27 *Let n, k and i be integers with $0 \le i \le k$. Then, for $k+1 \le n \le \infty$, the Stiefel-Whitney class $w_i = w_i(\zeta_k)$ is the class in $H^i(\mathrm{Gr}(k; \mathbb{R}^n))$ with stem 01^i. For example, $w_3(\zeta_4) = [1011100]^\sharp \in H^3(\mathrm{Gr}(4; \mathbb{R}^7))$.*

Proof The proposition is true if $i = 0$, since $w_0(\zeta_k) = \mathbf{1}$. Let us assume that $i \ge 1$. We first prove that $w_i(\zeta_k) = [01^i]^\sharp$ in $H^i(\mathrm{Gr}(i; \mathbb{R}^\infty))$. Recall $H^*(\mathrm{Gr}(i; \mathbb{R}^\infty)) \approx \mathbb{Z}_2[w_1, \ldots, w_i]$ (where $w_j = w_j(\zeta_i)$) so

$$K^* = \big(\ker H^*\beta : H^*(\mathrm{Gr}(i; \mathbb{R}^\infty)) \to H^*(\mathrm{Gr}(i-1; \mathbb{R}^\infty))\big)$$

is the ideal generated by w_i. Hence, K^i is one-dimensional generated by w_i. As $H_*\beta([\lambda]) = [1\lambda]$, one has

$$H^*\beta([\mu]^\sharp) = \begin{cases} [\lambda]^\sharp & \text{if } \mu = 1\lambda \\ 0 & \text{otherwise.} \end{cases} \tag{9.5.19}$$

Hence $0 \neq [01^i]^\sharp \in K^i$. Therefore, $w_i(\zeta_k) = [01^i]^\sharp$. Proposition 9.5.27 follows from the above particular case since $H^*\alpha(w_i(\zeta_k)) = w_i(\zeta_k)$ and $H^*\beta(w_i(\zeta_k)) = w_i(\zeta_{k-1})$. □

Let $\lambda, \mu \in [{n \atop k}]$. As $\mathcal{S}$ is a basis for $H_*(\mathrm{Gr}(k; \mathbb{R}^n))$ and $\mathcal{S}^\sharp$ is the Kronecker dual basis for $H^*(\mathrm{Gr}(k; \mathbb{R}^n))$, we can write

$$[\lambda]^\sharp \smile [\mu]^\sharp = \sum_{\nu \in [{n \atop k}]} \Gamma^\nu_{\lambda\mu} \, [\nu]^\sharp$$

where

$$\Gamma^\nu_{\lambda\mu} = \langle [\lambda]^\sharp \smile [\mu]^\sharp, [\nu] \rangle \in \mathbb{Z}_2.$$

Computing the "structure constants" $\Gamma^\nu_{\lambda\mu}$ is a version of the *Schubert calculus* (mod 2). The usual Schubert calculus deals with the structure constants $C^\nu_{\lambda\mu}$ for the basis $\mathcal{S}^{PD}$, defined by

$$[\lambda]^{PD} \smile [\mu]^{PD} = \sum_{\nu \in [{n \atop k}]} C^\nu_{\lambda\mu} \, [\nu]^{PD}.$$

By Lemma 9.5.25 and its proof, one has $\Gamma^\nu_{\lambda\mu} = C^{\nu^-}_{\lambda^-\mu^-}$. Again, Schubert calculus was initiated by Ehresmann in [47, 48] and further developed in e.g. [31, 32, 66, 75]. For a more recent as well as an equivariant version, see [119]. Note that $\Gamma^\nu_{\lambda\mu} = 0$ unless $d(\lambda) + d(\mu) = d(\nu)$.

A binary word $\lambda \in [{n \atop k}]$ is determined by its *Schubert symbol*, i.e. the k-tuple of integers indicating the positions of the 1's in λ. For instance, the Schubert symbol of 0100101 is $(2, 5, 7)$. We use the Schubert symbol of λ for all the cohomology classes $[\lambda 0^j]^\sharp$ (λ and $\lambda 0$ having the same symbol). For the reverse correspondence, we decorate the Schubert symbol by a flat sign $\flat$. Example:

$$\begin{aligned} [01001010^r]^\sharp &= (2,5,7) && \text{in } H^7(\mathrm{Gr}(3; \mathbb{R}^{7+r})) \\ (2,5,7)^\flat &= [01001010^r] && \text{in } H_7(\mathrm{Gr}(3; \mathbb{R}^{7+r})) \end{aligned}$$

Our notation for Schubert symbols are that of [153], close to the original one of [47]. Other conventions are used in e.g. [32, 75].

Remark 9.5.28 Fix the integers $k \leq n$ and let $a = (a_1, \ldots, a_k)$ be a k-tuple of integers. In order for a to be a Schubert symbol determining a class in $H^*(\mathrm{Gr}(k; \mathbb{R}^n))$, it should satisfy

$$1 \leq a_1 < a_2 < \cdots < a_n \leq n. \tag{9.5.20}$$

When this is not the case, we decide by convention, that a represents the class 0.

A Schubert cell C_λ^F will be also labeled by the Schubert symbol of λ: $C_\lambda^F = C_a^F$ if $a = [\lambda]^\sharp$. For the Poincaré duality (see Lemma 9.5.25), we set

$$a^- = [\lambda_-]^\sharp = [\lambda]^{PD}.$$

If $a = (a_1, \ldots, a_k)$ then $a^- = (n+1-a_k, \ldots, n+1-a_1)$. The definition of $\Gamma_{\lambda\mu}^\nu$ is also transposed for Schubert symbols:

$$a \smile b = \sum_c \Gamma_{ab}^c\, c$$

where the sum runs over all Schubert symbols c and

$$\Gamma_{ab}^c = \langle a \smile b, c^\flat \rangle \in \mathbb{Z}_2.$$

The following proposition and its proof is a variant, in our language, of Reduction Formula I of [75, p. 202].

Proposition 9.5.29 (Reduction formula) *Let $k \geq 2$ be an integer. Let r, s and t be positive integers $\leq k$ satisfying $t = r + s - 1$. Let a, b and c be Schubert k-symbols. Then*

$$\Gamma_{ab}^c = \begin{cases} 0 & \text{if } c_t < a_r + b_s - 1 \\ \Gamma_{\bar a \bar b}^{\bar c} & \text{if } c_t = a_r + b_s - 1, \end{cases}$$

where $\bar a$, $\bar b$ and $\bar c$ are the Schubert $(k-1)$-symbols

$$\begin{aligned} \bar a &= (a_1, \ldots, a_{r-1}, a_{r+1} - 1, \ldots, a_k - 1) \\ \bar b &= (b_1, \ldots, b_{s-1}, b_{s+1} - 1, \ldots, b_k - 1) \\ \bar c &= (c_1, \ldots, c_{t-1}, c_{t+1} - 1, \ldots, c_k - 1) \end{aligned}$$

Example 9.5.30 Let us use the formula for $s = 1$ and suppose that $b_s = 1$. Thus $b^\flat = [\mu] = [1\bar\mu]$ with $\bar\mu \in [{n-1 \atop k-1}]$. The condition $t = r + s - 1$ reduces to $t = r$ and $c_t = a_r + b_s - 1$ becomes $c_r = a_r$. Writing it in terms of $a^\flat = [\lambda]$ and $c^\flat = [\nu]$ this means that if $\lambda_r = \nu_r = 1$ for some r, one can remove λ_r from λ and ν_r from ν and replace μ by $\bar\mu$. For instance,

$$\begin{aligned}
\Gamma^{01101}_{10101,11010} &= \Gamma^{0110}_{1010,1010} && s = 1 \text{ and } r = t = 5 \\
&= \Gamma^{010}_{100,010} && s = 1 \text{ and } r = t = 3 \\
&= 1 && \text{since } [100]^\sharp = \mathbf{1} \text{ in } H^0(\mathrm{Gr}(1; \mathbb{R}^3)).
\end{aligned}$$

Proof of Proposition 9.5.29 Let F, F' and F'' be three complete flags in $\mathbb{R}^n$. If chosen generically, then $\bar{C}^F_x$, $\bar{C}^{F'}_y$ and $\bar{C}^{F''}_z$ are pairwise transverse for any Schubert symbols x, y and z. Therefore, if $d(a) + d(b) = d(c)$, $\bar{C}_{a^-} \cap \bar{C}_{b^-} \cap \bar{C}_c$ is 0-dimensional and

$$\begin{aligned}
\Gamma^c_{ab} &= \langle a \smile b, c^\flat \rangle \\
&= \langle a \smile b, (c^\flat)^{PD} \frown [\mathrm{Gr}(k; \mathbb{R}^n)] \rangle \\
&= \langle a \smile b \smile (c^\flat)^{PD}, [\mathrm{Gr}(k; \mathbb{R}^n)] \rangle \\
&= \sharp(\bar{C}_{a^-} \cap \bar{C}_{b^-} \cap \bar{C}_c) \mod 2,
\end{aligned} \tag{9.5.21}$$

the last equality coming from the intersection theory analogous to Proposition 5.4.12 but for algebraic cycles (see Remark 9.5.26).

Let $P \in \mathrm{Gr}(k; \mathbb{R}^n)$. If $P \in \bar{C}_{a^-} \cap \bar{C}_{b^-} \cap \bar{C}_c$ then

$$\begin{aligned}
\dim(P \cap F_{n+1-a_r}) &\geq k + 1 - r \\
\dim(P \cap F'_{n+1-b_s}) &\geq k + 1 - s \\
\dim(P \cap F''_{c_t}) &\geq t.
\end{aligned}$$

Therefore, the condition $t = r + s - 1$ implies that

$$\dim(P \cap F_{n+1-a_r} \cap F'_{n+1-b_s} \cap F''_{c_t}) \geq 1.$$

On the other hand, as F, F' and F'' are transverse flags,

$$\dim(F_{n+1-a_r} \cap F'_{n+1-b_s} \cap F''_{c_t}) = t - r - s + 2.$$

Thus, $\Gamma^c_{ab} = 0$ if $c_t < a_r + b_s - 1$. If $c_t = a_r + b_s - 1$, then $F_{n+1-a_r} \cap F'_{n+1-b_s} \cap F''_{c_t}$ is a line L, which must be contained in any $P \in \bar{C}_{a^-} \cap \bar{C}_{b^-} \cap \bar{C}_c$. Let $L^\perp$ be the orthogonal complement of L and let $\pi : \mathbb{R}^n \to L^\perp$ be the orthogonal projection. For $1 \leq i \leq n - 1$, define

$$\bar{F}_i = \begin{cases} \pi(F_i) & \text{if } i \leq n - a_r \\ \pi(F_{i+1}) & \text{if } i \geq n + 1 - a_r. \end{cases}$$

As $L \subset F_{n+1-a_r}$ but $L \not\subset F_{n-a_r}$, the sequence of vector spaces $\bar{F}_i$ constitutes a complete flag $\bar{F}$ in for $L^\perp$. Define $\bar{F}'$ accordingly and $\bar{F}''$ by

$$\bar{F}''_i = \begin{cases} \pi(F''_i) & \text{if } i \leq c_t \\ \pi(F''_{i+1}) & \text{if } i \geq c_t + 1. \end{cases}$$

Then, $\bar{F}$, $\bar{F}'$ and $\bar{F}''$ are transverse flags and, by linear algebra, one checks that

$$P = \pi(P) \oplus L \in \bar{C}_{a^-} \cap \bar{C}_{b^-} \cap \bar{C}_c \Longleftrightarrow \pi(P) \in \bar{C}_{(\bar{a})^-} \cap \bar{C}_{(\bar{b})^-} \cap \bar{C}_{\bar{c}}.$$

Hence,

$$\sharp(\bar{C}^F_{a^-} \cap \bar{C}^{F'}_{b^-} \cap \bar{C}^{F''}_c) = \sharp(\bar{C}^{\bar{F}}_{(\bar{a})^-} \cap \bar{C}^{\bar{F}'}_{(\bar{b})^-} \cap \bar{C}^{\bar{F}''}_{\bar{c}})$$

which, using (9.5.21), proves that $\Gamma^c_{ab} = \Gamma^{\bar{c}}_{\bar{a}\bar{b}}$.

Corollary 9.5.31 *Let a, b and c be Schubert k-symbols. Then $\Gamma^c_{ab} = 0$ unless $c_i \geq \max\{a_i + b_1 - 1, b_i + a_1 - 1\}$ for all $1 \leq i \leq k$. In particular, $\Gamma^c_{ab} = 0$ unless $c_i \geq \max\{a_i, b_i\}$ for all $1 \leq i \leq k$.*

Proof If $c_r < a_r + b_1 - 1$ for some integer r, then $\Gamma^c_{ab} = 0$ by the reduction formula for $s = 1$. As $\Gamma^c_{ab} = \Gamma^c_{ba}$, this proves the corollary. □

We now compute, for a Schubert symbol a, the expression of $w_i \smile a$ in the basis $\mathcal{S}^\sharp$. For $J \subset \{1, 2, \dots, k\}$, we define a map $a \mapsto a^J$ from $\mathbb{N}^k$ to itself by

$$a^J_i = \begin{cases} a_i + 1 & \text{if } i \in J \\ a_i & \text{if } i \notin J. \end{cases}$$

Proposition 9.5.32 *Let a be a Schubert k-symbol. The equation*

$$w_i \smile a = \sum_{\substack{J \subset \{1,2,\dots,k\} \\ \sharp J = i}} a^J \tag{9.5.22}$$

holds in $H^*(\mathrm{Gr}(k; \mathbb{R}^n))$ (with the convention of Remark 9.5.28*).*

Since in the right side of (9.5.22), we use the convention of Remark 9.5.28, Proposition 9.5.32 holds true for any n and any i ($w_i = 0$ if $i \geq k$).

Example 9.5.33

$$w_2 \smile (1,3,4,6) = \begin{cases} (2,3,5,6) + (2,3,4,7) + (1,3,5,7) & \text{in } H^6(\mathrm{Gr}(4; \mathbb{R}^n)) \text{ for } n \geq 7. \\ (2,3,5,6) & \text{in } H^6(\mathrm{Gr}(4; \mathbb{R}^6)). \end{cases}$$

Proof It suffices to prove the proposition for $n = \infty$. We identify w_i with it Schubert symbol which, by Proposition 9.5.27, is

$$w_i = (1, 2, \dots, k-i, k-i+1, \dots, k+1).$$

Then, the notation Γ^c_{a,w_i} is meaningful. Let c be a Schubert k-symbol such that $\Gamma^c_{a,w_i} \neq 0$. Then $d(c) = d(a) + i$ and, as $c_j \geq a_j$ by Corollary 9.5.31, there is

$K \subset \{1, 2, \ldots, k\}$ with $\sharp K = k - i$ such that $c_j = a_j$ for $j \in K$. By iterating the reduction formula for $s = 1$ with the indices in K, we get that $\Gamma^{c}_{a,w_i} = \Gamma^{\bar{c}}_{\bar{a},w_i}$, where $\bar{a}$ and $\bar{c}$ are Schubert i-symbols and $w_i = (2, 3, \ldots, i + 1)$. By Corollary 9.5.31, we have $c_j \geq a_j + 1$ for all $1 \leq j \leq i$. This implies that $c = a^J$ for $J = \{1, 2, \ldots, k\} - K$.

Conversely, let $J \subset \{1, 2, \ldots, k\}$ with $|J| = i$. We have to prove that $\Gamma(J) = \Gamma^{a^J}_{a,w_i} = 1$ if a^J is a Schubert symbol for $\mathrm{Gr}(k; \mathbb{R}^\infty)$. By repeating the reduction formula for $s = 1$ with all the indices not in J, we get that

$$\Gamma(J) = \Gamma^{\bar{a}^J}_{\bar{a},w_i}$$

where $\bar{a}$ is a Schubert i-symbol, $J = \{1, 2, \ldots, i\}$ and $w_i = (2, 3, \ldots, i + 1)$. By iterating again the reduction formula for $s = 1$ with the indices $i, i - 1$, etc., till 2, we get that

$$\Gamma(J) = \Gamma^{(\bar{a}_1+1)}_{(\bar{a}_1),(2)}.$$

This coefficient is equal to 1, as $(u) \smile w_1 = (u + 1)$ in $H^*(\mathrm{Gr}(1; \mathbb{R}^\infty)) \approx H^*(\mathbb{R}P^\infty)$. □

For $\lambda \in [{n \atop k}]$, let $\lambda^\perp \in [{n \atop n-k}]$ be obtained from λ by exchanging 0's and 1's and reverse the order: $100101^\perp = 010110$; in formula:

$$\lambda^\perp_j = 0 \iff \lambda_{n+1-j} = 1. \tag{9.5.23}$$

Note that $d(\lambda^\perp) = d(\lambda)$. This formal operation is related to the homeomorphism $h : \mathrm{Gr}(k; \mathbb{R}^n) \to \mathrm{Gr}(n - k; \mathbb{R}^n)$ sending k-plane P to its orthogonal complement $P^\perp$.

Lemma 9.5.34 $H^*h([\lambda]^\sharp) = [\lambda^\perp]^\sharp$.

Proof Let $F = (F_1 \subset \cdots \subset F_n)$ be a complete flag in $\mathbb{R}^n$ and let F^- be the dual flag, defined by $F^-_i = F^\perp_{n-i}$ (we add the convention that $F_0 = \{0\} = F^-_0$). To establish Lemma 9.5.34, we shall prove that $h(C^{(F)}_\lambda) = C^{(F^-)}_{\lambda^\perp}$.

Let $P \in \mathrm{Gr}(k; \mathbb{R}^n)$. Write $P = (P \cap F_i) \oplus Q_i$. Then

$$P^\perp \cap F^-_{n-i} = \{v \in F^-_{n-i} \mid \langle v, Q_i\rangle = 0\}.$$

Hence

$$\mathrm{codim}_{F^-_{n-i}}(P^\perp \cap F^-_{n-i}) = \dim Q_i = \mathrm{codim}_P(P \cap F_i)\,. \tag{9.5.24}$$

Suppose that $P \in C^{(F)}_\lambda$ for $\lambda \in [{n \atop k}]$. Then, $P^\perp \in C^{(F^-)}_\mu$ $\mu \in [{n \atop n-k}]$. We must prove that $\mu = \lambda^\perp$, that is to say ($\lambda_i = 0 \iff \mu_{n+1-i} = 1$). But, using (9.5.24)

$$\begin{aligned}\lambda_i = 0 &\iff \dim(P \cap F_i) = \dim(P \cap F_{i-1}) \\ &\iff \operatorname{codim}_{F^-_{n-i}}(P^\perp \cap F^-_{n-i}) = \operatorname{codim}_{F^-_{n+1-i}}(P^\perp \cap F^-_{n+1-i}) \\ &\iff \dim(P^\perp \cap F^-_{n+1-i}) = \dim(P^\perp \cap F^-_{n-i}) + 1 \\ &\iff \mu_{n+1-i} = 1.\end{aligned}$$

□

Let $\bar{w} = w(\zeta_k^\perp) = \mathbf{1} + \bar{w}_1 + \cdots \bar{w}_{n-k}$ be the total Stiefel-Whitney class of the tautological $(n-k)$-vector bundle over $\mathrm{Gr}(k;\mathbb{R}^n)$ (see Example 9.5.18).

Proposition 9.5.35 *Suppose that $n \geq i + k$. Then $\bar{w}_i \in H^i(\mathrm{Gr}(k;\mathbb{R}^n)$ is the class of stem $0^i 1$. Its Schubert symbol is $(1, 2, \ldots, k-1, k+i)$.*

Example: $\bar{w}_5 = [111000001]^\sharp = (1, 2, 3, 9)$ in $H^5(\mathrm{Gr}(4;\mathbb{R}^n)$ for $n \geq 9$.

Proof The homeomorphism $h : \mathrm{Gr}(k;\mathbb{R}^n) \to \mathrm{Gr}(n-k;\mathbb{R}^n)$ is covered by the tautological bundle map $\zeta_k^\perp \to \zeta_{n-k}$. Hence, $h^*\zeta_{n-k} = \zeta_k^\perp$ and thus $\bar{w}_i = H^*h(w_i(\zeta_{n-k}))$. Therefore, $\bar{w}_i = (w_i)^\perp$ by Lemma 9.5.34. As **stem** $(\lambda^\perp) =$ **stem** $(\lambda)^\perp$, Proposition 9.5.35 follows from Proposition 9.5.27. □

We now give the expression of $\bar{w}_i \smile a$ for a Schubert k-symbol a. As $\bar{w}_1 = w_1$, we can use Formula (9.5.22) for $i = 1$:

$$w_1 \smile a = \bar{w}_1 \smile a = \sum_b b \tag{9.5.25}$$

where the sum runs over all the Schubert k-symbols b such that

$$a_j \leq b_j \leq a_j + 1 \ \text{ and } \ \sum_{j=1}^{k}(b_j - a_j) = 1.$$

Example:

$$\begin{aligned}w_1^2 &= w_1 \smile (1, 2, \ldots, k-1, k+1) \\ &= (1, 2, \ldots, k-2, k, k+1) + (1, 2, \ldots, k-1, k+2) \\ &= w_2 + \bar{w}_2.\end{aligned}$$

Formula (9.5.25) admits the following generalization, called the *Pieri formula*, which is a sort of a dual of Proposition 9.5.32.

Proposition 9.5.36 (Pieri's formula) *Let a be a Schubert k-symbol. The equation*

$$\bar{w}_i \smile a = \sum_b b \tag{9.5.26}$$

holds in $H^*(\mathrm{Gr}(k;\mathbb{R}^n))$*, where the sum runs over all the Schubert k-symbols b such that*

$$a_j \le b_j < a_{j+1} \text{ and } \sum_{j=1}^{k}(b_j - a_j) = i \tag{9.5.27}$$

(with the convention of Remark 9.5.28*).*

Proof For Schubert k-symbols a, b, the proposition says that

$$\langle \bar{w}_i \smile a, b^\flat\rangle = 1 \Longleftrightarrow (a,b) \quad \text{satisfies (9.5.27)}. \tag{9.5.28}$$

(Note that the implication $\Rightarrow$ follows from Corollary 9.5.31 and from $\bar{w}_i$ being of degree i). Rewriting (9.5.28) with $\lambda, \mu \in [{n \atop k}]$ gives that $\langle \bar{w}_i \smile [\lambda]^\sharp, [\mu]\rangle = 1$ if and only if λ and μ satisfy the following pair of conditions

(i) $\lambda = A_1 10^{r_1} A_2 10^{r_2} \cdots A_s 10^{r_s} A_{s+1}$, with $\sum_{j=1}^{s} r_j = i$, and
(ii) $\mu = A_1 0^{r_1} 1 A_2 0^{r_2} 1 \cdots A_{r_s} 0^{r_s} 1 A_{s+1}$.

(Intuitively: a certain quantity of 1's are shifted by one position to the right of total amount shifting being i). The pair of conditions (i) and (ii) is equivalent to the following ones

(i) $\lambda^\perp = A_{s+1}^\perp 1^{r_s} 0\, A_s^\perp 1^{r_{s-1}} 0 \cdots A_2^\perp 1^{r_1} 0\, A_1^\perp$ with $\sum_{j=1}^{s} r_j = i$, and
(ii) $\mu^\perp = A_{s+1}^\perp 0 1^{r_s}\, A_s^\perp 0 1^{r_{s-1}} \cdots A_2^\perp 0 1^{r_1}$.

Recall from Lemma 9.5.34 and the proof of Proposition 9.5.35 that the homeomorphism $h : \mathrm{Gr}(k;\mathbb{R}^n) \to \mathrm{Gr}(n-k;\mathbb{R}^n)$ satisfies $H_*([\nu]) = [\nu^\perp]$ and $\bar{w}_i = H^*(w_i(\zeta_{n-k}))$. Therefore

$$\langle \bar{w}_i \smile [\lambda]^\sharp, [\mu]\rangle = 1 \Longleftrightarrow \langle w_i(\zeta_{n-k}) \smile [\lambda^\perp]^\sharp, [\mu^\perp]\rangle = 1. \tag{9.5.29}$$

By Proposition 9.5.32, the right hand equality in (9.5.29) is equivalent to the pair of conditions (i)$^\perp$ and (ii)$^\perp$, which proves Proposition 9.5.36. □

We finish this subsection by mentioning the *Giambelli's formula*, which express a cohomology class given by a Schubert symbol as a polynomial in the $\bar{w}_i$'s. The Giambelli and the generalized Pieri formulae together provides a procedure for computing the structure constants Γ_{ab}^c.

Proposition 9.5.37 (Giambelli's formula)

$$(a_1,\dots,a_k) = \det\left(\bar{w}_{a_i - j}\right)_{1\le i,j\le k}.$$

with the convention that $\bar{w}_u = 0$ *if* $u < 0$.

For $w_r = (1,2,\dots,k-r,k-r+1,\dots,k+1)$, Proposition 9.5.37 reproves the second formula of Lemma 9.5.19.

Proof By induction on k, starting trivially if $k = 1$. The lengthy induction step, using the Pieri formula, may be translated in our language from [75, pp. 204–205] (see also [32, p. 366]). □

9.6 Splitting Principles

Let $\alpha : \Gamma \to O(n)$ denotes the inclusion of the diagonal subgroup of $O(n)$

$$\Gamma \approx O(1) \times \cdots \times O(1) \subset O(n).$$

This induces an inclusion $B\alpha : B\Gamma \to BO(n)$ between the classifying spaces. The symmetric group Sym_n acts on $O(1) \times \cdots \times O(1)$ by permuting the factors, and then on $B\Gamma$. As in Sect. 9.5, ζ_n denotes the tautological vector bundle on $BO(n) \simeq \mathrm{Gr}(n; \mathbb{R}^\infty)$. It is the vector bundle associated to the universal $O(n)$-bundle $EO(n) \to BO(n)$.

Theorem 9.6.1 *The* **GrA**-*morphism* $H^*B\alpha : H^*(BO(n)) \to H^*(B\Gamma)$ *is injective and its image is* $H^*(B\Gamma)^{\mathrm{Sym}_n}$. *The induced vector bundle* $B\alpha^*\zeta_n$ *splits into a Whitney sum of line bundles.*

Proof We have seen in (9.5.13) that the homotopy equivalence

$$B\Gamma \simeq EO(n)\big/O(1) \times \cdots \times O(1)$$

makes $B\alpha$ homotopy equivalent to the locally trivial bundle

$$\mathrm{Fl}(1, \ldots, 1) \xrightarrow{\beta} B\Gamma \xrightarrow{B\alpha} BO(n). \tag{9.6.1}$$

We have also established in the proof of Theorem 9.5.14 that $H^*\beta$ is surjective. Hence, by Corollary 4.7.19, $H^*B\alpha$ is injective. Also, using (7.4.3) and that $O(1) \approx \{\pm 1\}$. one has a homotopy equivalence

$$B\Gamma \xrightarrow[\simeq]{\psi} BO(1) \times \cdots \times BO(1) \simeq \mathbb{R}P^\infty \times \cdots \times \mathbb{R}P^\infty$$

and thus a **GrA**-isomorphism

$$\psi^* : \mathbb{Z}_2[x_1, \cdots, x_n] \xrightarrow{\approx} H^*(B\Gamma)$$

where x_i has degree 1. By Theorem 9.5.8,

$$H^*(BO(n)) \approx \mathbb{Z}_2[w_1(\zeta_n), \ldots, w_n(\zeta_n)].$$

Note that $B\alpha$ is covered by a morphism of principal bundles

$$\begin{array}{ccc} EO(1)\times\cdots\times EO(1) & \xrightarrow{E\alpha} & EO(n) \\ \downarrow & & \downarrow \\ BO(1)\times\cdots\times BO(1) & \xrightarrow{B\alpha} & BO(n). \end{array}$$

One has a similar diagram for the associated vector bundles $\gamma = (EO(1)\times_{O(1)}\mathbb{R} \to BO(1))$ (corresponding to the tautological line bundle over $\mathbb{R}P^\infty$) and ζ_n. This implies that $B\alpha^*\zeta_n \approx \gamma\times\cdots\times\gamma$. As $w(\gamma\times\cdots\times\gamma) = \prod_{i=1}^n(1+x_i)$, one has

$$H^*B\alpha(w_i(\zeta_n)) = w_i(\gamma\times\cdots\times\gamma) = \sigma_i,$$

where σ_i is i-th elementary symmetric polynomial in the variables x_j. The second assertion of Theorem 9.6.1 follows, since the elementary symmetric polynomials **GrA**-generate $\mathbb{Z}_2[x_1,\cdots,x_n]^{\mathrm{Sym}_n} \approx H^*(B\Gamma)^{\mathrm{Sym}_n}$.

Finally, the homotopy equivalence ψ is of the form $\psi = (\psi_1,\ldots,\psi_n)$, with $\psi_i : B\Gamma \to BO(1)$. In other words, ψ coincides with the composition

$$B\Gamma \xrightarrow{\Delta} B\Gamma\times\cdots\times B\Gamma \xrightarrow{\psi_1\times\cdots\times\psi_n} BO(1)\times\cdots\times BO(1),$$

where Δ is the diagonal map. Hence,

$$B\alpha^*\zeta_n \approx \psi^*(\gamma\times\cdots\times\gamma) = \Delta^*(\psi_1^*\gamma\times\cdots\times\psi_n^*\gamma) = \psi_1^*\gamma\oplus\cdots\oplus\psi_n^*\gamma,$$

which shows that $B\alpha^*\zeta_n$ is isomorphic to a Whitney sum of line bundles. □

Theorem 9.6.1 may be generalized as follows. Consider the inclusion homomorphism

$$\alpha_{n_1,\ldots,n_r} : O(n_1)\times\cdots\times O(n_r) \to O(n)$$

sending $(A_1,\ldots,A_r)$ to the diagonal-block matrix with blocks $A_1,\ldots,A_r$. Using the homotopy equivalence $BO(n_1)\times\cdots\times BO(n_r) \simeq B(O(n_1)\times\cdots\times O(n_r))$ (see (7.4.3)), the homomorphism $\alpha_{n_1,\ldots,n_r}$ induces a continuous map

$$B\alpha_{n_1,\ldots,n_r} : BO(n_1)\times\cdots\times BO(n_r) \to BO(n).$$

Theorem 9.6.2 *The map $B\alpha_{n_1,\ldots,n_r}$ satisfies the following properties.*

(1) *The* **GrA**-*morphism*

$$H^*B\alpha_{n_1,\ldots,n_r} : H^*(BO(n)) \to H^*(BO(n_1)\times\cdots\times BO(n_r))$$

is injective.

(2) $H^*B\alpha_{n_1,\dots,n_r}(w_i) = w_i(\zeta_{n_1} \times \cdots \times \zeta_{n_r})$ *for each* $i \geq 0$. *In particular, the image of* $H^*B\alpha_{n_1,\dots,n_r}$ *is generated by* $w_i(\zeta_{n_1} \times \cdots \times \zeta_{n_r})$ $(i \geq 0)$.
(3) *The induced vector bundle* $B\alpha^*_{n_1,\dots,n_r}\zeta_n$ *splits into a Whitney sum of vector bundles of ranks* $n_1, \dots, n_r$.

Proof Using the inclusion factorization

$$O(1)^n \xrightarrow{\approx} O(1)^{n_1} \times \cdots \times O(1)^{n_r} \longrightarrow O(n_1) \times \cdots \times O(n_r) \xrightarrow{\tilde{\alpha}} O(n),$$

where $\tilde{\alpha} = \alpha_{n_1,\dots,n_r}$, the injectivity of $H^*B\tilde{\alpha}$ comes from that of $H^*B\alpha_{1,\dots,1}$, established in Theorem 9.6.1. As $B\tilde{\alpha}$ is covered by a morphism of principal bundles

$$\begin{array}{ccc} EO(n_1) \times \cdots \times EO(n_r) & \xrightarrow{\tilde{\alpha}} & EO(n) \\ \downarrow & & \downarrow \\ BO(n_1) \times \cdots \times BO(n_r) & \xrightarrow{B\tilde{\alpha}} & BO(n) \end{array},$$

one deduces (2) and (3) as in the proof of Theorem 9.6.1. □

Proposition 9.6.3 *Let* ξ *be a vector bundle over a paracompact space* X. *Then, there is a map* $f : X_\xi \to X$ *such that*

(1) H^*f *is injective.*
(2) $f^*\xi$ *splits into a Whitney sum of line bundles.*

Proposition 9.6.3 is called the *splitting principle*. For $\xi = \zeta_n$ over $BO(n)$, Theorem 9.6.1 says that on can take $BO(n)_{\zeta_n}$ and $f = B\alpha$.

Proof As X is paracompact, ξ admits an Euclidean structure and there is a classifying map $\varphi : X \to BO(n)$ for ξ, i.e. $\xi \approx \varphi^*\zeta_n$. Consider the pull-back diagram

$$\begin{array}{ccc} X_\xi & \xrightarrow{f} & X \\ \downarrow{\scriptstyle\hat{\varphi}} & & \downarrow{\scriptstyle\varphi} \\ B\Gamma & \xrightarrow{B\alpha} & BO(n) \end{array},$$

where $B\alpha$ is defined as in (9.6.1). As $B\alpha$ is a locally trivial bundle with fiber $\mathrm{Fl}(n_1, \dots, n_r)$, so is f (this is the $\mathrm{Fl}(n_1, \dots, n_r)$-bundle associated to $\mathrm{Fra}_\perp\xi$). We saw in the proof of Theorem 9.6.1 that $H^*(B\Gamma) \to H^*(\mathrm{Fl}(n_1, \dots, n_r))$ is surjective. Then, so is $H^*(X_\xi) \to H^*(\mathrm{Fl}(n_1, \dots, n_r))$. Hence, by Corollary 4.7.19, H^*f is injective. Now,

$$f^*\xi = f^*\varphi^*\zeta_n = \hat{\varphi}^*B\alpha^*\zeta_n.$$

As, by Theorem 9.6.1, $B\alpha^*\zeta_n$ is a Whitney sum of line bundles, so does $f^*\xi$. □

One consequence of the splitting principle is the uniqueness of Stiefel-Whitney classes (compare [153, Theorem 7.3] or [105, Chap. 16, Sect. 5]).

Proposition 9.6.4 *Suppose that $\tilde{w}$ is a correspondence associating, to each vector bundle ξ over a paracompact space X, a class $\tilde{w}(\xi) \in H^*(X)$, such that*

(1) *if $f : Y \to X$ be a continuous map, then $\tilde{w}(f^*\xi) = H^*f(\tilde{w}(\xi))$.*
(2) *$\tilde{w}(\xi \oplus \xi') = \tilde{w}(\xi) \smile \tilde{w}(\xi')$.*
(3) *if γ is the tautological line bundle over $\mathbb{R}P^\infty$, then $\tilde{w}(\gamma) = \mathbf{1} + a$, where $0 \neq a \in H^1(\mathbb{R}P^\infty)$.*

Then $\tilde{w} = w$, the total Stiefel-Whitney class.

Proof Condition (2) implies that

$$\text{(2.bis)} \quad \tilde{w}(\xi_1 \oplus \cdots \oplus \xi_n) = \tilde{w}(\xi_1) \smile \cdots \smile \tilde{w}(\xi_n).$$

As in the proof of Theorem 9.6.1, Conditions (1), (2.bis) and (3) imply that the map $B\alpha : B\Gamma \to BO(n)$ satisfies

$$H^*B\alpha(\tilde{w}(\zeta_n)) = (1 + x_i)^n \in H^*(B\Gamma) \approx \mathbb{Z}_2[x_1, \cdots, x_n].$$

Thus, still by the proof of Theorem 9.6.1, $H^*B\alpha(\tilde{w}(\zeta_n)) = H^*B\alpha(w(\zeta_n))$. As $H^*B\alpha$ is injective, this implies that $\tilde{w}(\zeta_n) = w(\zeta_n)$. The bundle ζ_n being universal (see 9.1.14), Condition (1) implies that $\tilde{w}(\xi) = w(\xi)$ for any vector bundle ξ over a paracompact space X. □

Another consequence of the splitting principle is the action of the Steenrod algebra on Stiefel-Whitney classes. The following proposition was proved by Wu wen-Tsün [214].

Proposition 9.6.5 *Let ξ be a vector bundle over a paracompact space X. Then*

$$\mathrm{Sq}^i w_j(\xi) = \sum_{0 \le k \le i} \binom{j-i+k-1}{k} w_{i-k}(\xi)\, w_{j+k}(\xi). \tag{9.6.2}$$

Example 9.6.6 Setting $w_i = w_i(\xi)$, we get

$$\begin{aligned}
\mathrm{Sq}^1 w_j &= w_1 w_j + (j-1) w_{j+1} \\
\mathrm{Sq}^2 w_j &= w_2 w_j + (j-2) w_1 w_{j+1} + \binom{j-1}{2} w_{j+2} \\
\mathrm{Sq}^3 w_j &= w_3 w_j + (j-3) w_2 w_{j+1} + \binom{j-2}{2} w_1 w_{j+2} + \binom{j-3}{3} w_{j+3}.
\end{aligned}$$

Proof of Proposition 9.6.5 By naturality of w and Sq, it suffices to prove (9.6.2) for $\xi = \zeta_n$, the tautological vector bundle on $BO(n)$. By Theorem 9.6.1 and its proof,

$$H^*B\alpha(w_j(\zeta_n)) = \sigma_j \in \mathbb{Z}_2[x_1, \cdots, x_n]$$

where x_j corresponds to the non-trivial element in $H^1(\mathbb{R}P^\infty)$, and σ_j is the j-th elementary symmetric polynomial in the variables x_r. As $H^*B\alpha$ is injective by Theorem 9.6.1, Formula (9.6.2) reduces to the computation of $\mathrm{Sq}^i\sigma_j$ in $\mathbb{Z}_2[x_1, \cdots, x_n]^{\mathrm{Sym}_n}$, using that $\mathrm{Sq}(x_r) = x_r + x_r^2$. This technical computation may be found in full details in [15, Theorem 7.1]. The reader may, as an exercise, prove the special cases of Example 9.6.6.

The splitting principle also gives the following result about the Stiefel-Whitney classes of a tensor product (for a more general formula, see [153, p. 87]).

Lemma 9.6.7 *Let η and ξ be vector bundles over a paracompact space X. Suppose that ξ is of rank r and that η is a line bundle. Then*

$$w(\eta \otimes \xi) = \sum_{k=0}^{r} (1 + w_1(\eta))^k w_{r-k}(\xi). \tag{9.6.3}$$

Proof Set $u = w_1(\eta)$. Suppose first that ξ splits into a Whitney sum of r line bundles ξ_j, of Stiefel-Whitney class $\mathbf{1} + v_j$. Then, letting $\sigma_k = (v_1, \ldots, v_r)$ denote the kth elementary symmetric polynomial, one has

$$\begin{aligned} w(\eta \otimes \xi) &= w(\oplus_{j=1}^r (\eta \otimes \xi_j)) \\ &= \textstyle\prod_{j=1}^r w(\eta \otimes \xi_j) && \text{by (9.4.3)} \\ &= \textstyle\prod_{j=1}^r (\mathbf{1} + u + v_j) && \text{by Proposition 9.2.4} \\ &= \textstyle\sum_{k=0}^r (\mathbf{1} + u)^k \sigma_{r-k}(v_1, \ldots, v_r). \end{aligned}$$

Since $\sigma_{r-k}(v_1, \ldots, v_r) = w_{r-k}(\xi)$, we have shown (9.6.3) when ξ splits into a Whitney sum of r line bundles. If this is not the case, Formula (9.6.3) still holds true by the splitting principle of Proposition 9.6.3. □

9.7 Complex Flag Manifolds

The plan of this section follows that of Sect. 9.5. We shall indicate the slight changes to get from the real flag manifolds to the complex ones, without repeating all the proofs.

Let $n_1, \ldots n_r$ be positive integers and let $n = n_1 + n_2 + \cdots n_r$. By the *complex flag manifold* $\mathrm{Fl}_{\mathbb{C}}(n_1, \ldots, n_r)$, we mean any smooth manifold diffeomorphic to the homogeneous space

$$\mathrm{Fl}_{\mathbb{C}}(n_1, \ldots, n_r) \approx U(n)\big/U(n_1) \times U(n_2) \times \cdots \times U(n_r). \tag{9.7.1}$$

The most usual concrete occurrence of complex flag manifolds are as below.

(1) *Nested subspaces.* $\mathrm{Fl}_{\mathbb{C}}(n_1, \ldots, n_r)$ is the set of nested complex vector subspaces $V_1 \subset \cdots \subset V_r \subset \mathbb{C}^n$ with $\dim_{\mathbb{C}} V_i = \sum_{j=1}^{i} n_j$.
(2) *Mutually orthogonal subspaces.* $\mathrm{Fl}_{\mathbb{C}}(n_1, \ldots, n_r)$ is the set of r-tuples $(W_1, \ldots, W_r)$ of complex vector subspaces $\mathbb{C}^n$ which are mutually orthogonal (for the standard Hermitian product on $\mathbb{C}^n$) and satisfy $\dim W_i = n_i$. The correspondence from this definition to Definition (1) associates to $(W_1, \ldots, W_r)$ the nested family $\{V_i\}$ where V_i is the complex vector space generated by $W_1 \cup \cdots \cup W_i$.
(3) *Isospectral Hermitian matrices.* Let $\lambda_1 > \cdots > \lambda_r$ be real numbers. Consider the manifold $HM(n)$ of all Hermitian $(n \times n)$-matrices, on which $U(n)$ acts by conjugation. Then $\mathrm{Fl}_{\mathbb{C}}(n_1, \ldots, n_r)$ occurs as the orbit of the diagonal matrix having entries λ_i with multiplicity n_i.

$$\mathrm{Fl}_{\mathbb{C}}(n_1, \ldots, n_r) = \big\{R\,\mathrm{dia}\big(\underbrace{\lambda_1, \ldots, \lambda_1}_{n_1}, \cdots, \underbrace{\lambda_r, \ldots, \lambda_r}_{n_r}\big)\, R^{-1} \mid R \in U(n)\big\}. \tag{9.7.2}$$

In other words, $\mathrm{Fl}_{\mathbb{C}}(n_1, \ldots, n_r)$ is here the space of Hermitian $(n \times n)$-matrices with characteristic polynomial equal to $\sum_{i=1}^{r} (x - \lambda_i)^{n_i}$. Indeed, two matrices in $HM(n)$ are in the same $U(n)$-orbit if and only if they have the same characteristic polynomial. The correspondence from this definition to Definition (2) associates, to a matrix M, its eigenspaces for the various eigenvalues.

Concrete definition (3) is our working definition for $\mathrm{Fl}_{\mathbb{C}}(n_1, \ldots, n_r)$ throughout this section. Special classes of flag manifolds are given by the complex Grassmannians

$$\mathrm{Gr}(k; \mathbb{C}^n) = \mathrm{Fl}_{\mathbb{C}}(k, n-k) \approx U(n)\big/U(k) \times U(n-k)$$

of complex k-planes in $\mathbb{C}^n$. This is a closed manifold of dimension

$$\dim \mathrm{Gr}(k; \mathbb{C}^n) = \dim U(n) - \dim U(k) - \dim U(n-k) = 2k(n-k).$$

For example, $\mathrm{Gr}(1; \mathbb{C}^n) \approx \mathbb{C}P^{n-1}$, of dimension $2(n-1)$.

Using Definition (3) above, our "concrete Grassmannian" will be

$$\mathrm{Gr}(k; \mathbb{C}^n) = \big\{R\,\mathrm{dia}\big(\underbrace{1, \ldots, 1}_{k}, \underbrace{0, \ldots, 0}_{n-k}\big)\, R^{-1} \mid R \in U(n)\big\}. \tag{9.7.3}$$

As, in the real case, we define the *complete complex flag manifold*

$$\mathrm{Fl}_{\mathbb{C}}(1, \ldots, 1) \approx U(n)\big/U(1) \times \cdots \times U(1)$$

with $\dim \mathrm{Fl}_{\mathbb{C}}(1, \ldots, 1) = \dim U(n) - n = n^2 - n = n(n-1)$.

As in Sect. 9.5, we define real functions on the flag manifolds by restriction of the *weighted trace* on $f : HM(n) \to \mathbb{R}$ defined by

$$f(M) = \sum_{j=1}^{n} j\, M_{jj}$$

where M_{ij} denotes the (i, j)-entry of M.

Proposition 9.7.1 *Let* $\mathrm{Fl}_{\mathbb{C}}(n_1, \dots, n_r) \subset HM(n)$ *be the complex flag manifold as presented in* (9.5.2)*. Then, the restriction* $f : \mathrm{Fl}_{\mathbb{C}}(n_1, \dots, n_r) \to \mathbb{R}$ *of the weighted trace is a perfect Morse function whose critical points are the diagonal matrices in* $\mathrm{Fl}_{\mathbb{C}}(n_1, \dots, n_r)$*. The index of the critical point* $\mathrm{dia}(x_1, \dots, x_n)$ *is twice the number of pairs* (i, j) *with* $i < j$ *and* $x_i < x_j$*. In consequence,* $\dim \mathrm{Fl}_{\mathbb{C}}(n_1, \dots, n_r) = 2 \dim \mathrm{Fl}(n_1, \dots, n_r)$ *and*

$$\mathfrak{P}_t(\mathrm{Fl}_{\mathbb{C}}(n_1, \dots, n_r)) = \mathfrak{P}_{t^2}(\mathrm{Fl}(n_1, \dots, n_r)). \tag{9.7.4}$$

Recall that $\dim \mathrm{Fl}(n_1, \dots, n_r)$ was computed in Corollary 9.5.5 and that the Poincaré polynomial $\mathfrak{P}_t(\mathrm{Fl}(n_1, \dots, n_r))$ was described in Corollary 9.5.15. Equality (9.7.4) implies the following corollary.

Corollary 9.7.2 *The cohomology groups of* $\mathrm{Fl}_{\mathbb{C}}(n_1, \dots, n_r)$ *vanish in odd degrees.*

Remark 9.7.3 The manifold $\mathrm{Fl}_{\mathbb{C}}(n_1, \dots, n_r) \subset HM(n)$ admits an $U(n)$-invariant symplectic form, induced from the non-degenerate symmetric form $(X, Y) \mapsto \mathrm{trace}\,(XY)$ on $HM(n)$ (see [12, Chap. II, Example 1.4]). The weighted trace is the moment map for the Hamiltonian circle action given by the conjugation by $\mathrm{dia}(e^{it}, e^{2it}, \dots, e^{nit})$. The involution τ given on $\mathrm{Fl}_{\mathbb{C}}(n_1, \dots, n_r)$ by the complex conjugation is anti-symplectic and anti-commutes with the circle action. Its fixed point set is $\mathrm{Fl}(n_1, \dots, n_r)$. Note that f is τ-invariant and the critical point of f or $f|\mathrm{Fl}(n_1, \dots, n_r)$ are the same. This, together with (9.7.4), is a particular case of a theorem of Duistermaat [45] (see also Remark 9.7.9).

Proof of Proposition 9.7.1 We use the injective homomorphism $r^{ij} : SU(2) \to U(n)$, introduced in the proof of Proposition 9.7.1, whose image contains the matrices

$$R_t^{ij} = r^{ij}\begin{pmatrix} \cos t & -\sin t \\ \sin t & \cos t \end{pmatrix} \quad \text{and} \quad \tilde{R}_t^{ij} = r^{ij}\begin{pmatrix} \cos t & \sqrt{-1}\sin t \\ \sqrt{-1}\sin t & \cos t \end{pmatrix} \quad (t \in \mathbb{R}).$$

Suppose that $\Delta \in \mathrm{Fl}_{\mathbb{C}}(n_1, \dots, n_r)$ is a diagonal matrix. Then, a basis of $T_\Delta \mathrm{Fl}_{\mathbb{C}}(n_1, \dots, n_r)$ is represented by the curves

$$\Delta^{ij}(t) = R_t^{ij}\, \Delta\, R_{-t}^{ij} \text{ and } \tilde{\Delta}^{ij}(t) = \tilde{R}_t^{ij}\, \Delta\, \tilde{R}_{-t}^{ij}.$$

As in the proof of Proposition 9.5.2, this shows that the critical points of f are exactly the diagonal matrices and computes the indices.

As the critical points are all of even index, the function f is a perfect Morse function by Lemma 7.6.2. One can also proceed as in the proof of Proposition 9.5.2, using that f is invariant for the action of the diagonal subgroup T of $U(n)$, which is the torus $U(1) \times \cdots \times U(1)$, and use Theorem 7.6.11. □

As in Sect. 9.5, consider the inclusion $HM(n) \subset HM(n+1)$ with image the matrices with vanishing last row and column. Seeing $\mathrm{Gr}(k; \mathbb{C}^n) \subset HM(n)$ as in (9.5.3), this gives an inclusion $\mathrm{Gr}(k; \mathbb{C}^n) \subset \mathrm{Gr}(k; \mathbb{C}^{n+1})$. The proof of the following lemma is the same as that of Lemma 9.5.7.

Lemma 9.7.4 *The homomorphism $H^j(\mathrm{Gr}(k; \mathbb{C}^{n+1})) \to H^j(\mathrm{Gr}(k; \mathbb{C}^n))$ induced by the inclusion is surjective for all j and is an isomorphism for $j \leq 2(n-k)$.*

Tautological bundles. Consider a complex flag manifold $\mathrm{Fl}_{\mathbb{C}}(n_1, \ldots, n_r)$, with $n = n_1 + \cdots + n_r$ and the following closed subgroups of $U(n)$.

$$B_i = U(n_1) \times \cdots \times \{1\} \times \cdots \times U(n_r) \subset U(n_1) \times \cdots \times U(n_r) \subset U(n).$$

Then

$$P_i = U(n)/B_i \twoheadrightarrow U(n)\big/U(n_1) \times \cdots \times U(n_r) = \mathrm{Fl}_{\mathbb{C}}(n_1, \ldots, n_r)$$

is an $U(n_i)$-principal bundle (see p. 381) over $\mathrm{Fl}_{\mathbb{C}}(n_1, \ldots, n_r)$. Its associated complex vector bundle of rank n_i, i.e. $E(\xi_i) = P_i \times_{U(n_i)} \mathbb{C}^{n_i}$, is called *$i$-th-tautological vector bundle* over $\mathrm{Fl}_{\mathbb{C}}(n_1, \ldots, n_r)$. Being associated to an $U(n_i)$-principal bundle, ξ_i is endowed with an Hermitian structure and its space of orthonormal frames $\mathrm{Fra}_{\perp}(\xi_i)$ is equal to P_i. In the mutually orthogonal subspaces presentation (2) of $\mathrm{Fl}_{\mathbb{C}}(n_1, \ldots, n_r)$, we see that

$$E(\xi_i) = \{(W_1 \ldots, W_r, v) \in \mathrm{Fl}_{\mathbb{C}}(n_1, \ldots, n_r) \times \mathbb{C}^n \mid v \in W_i\}.$$

Note that $\xi_1 \oplus \cdots \oplus \xi_r$ is trivial (see Sect. 9.5, p. 381).

The complex vector bundle ξ_1 over $\mathrm{Fl}_{\mathbb{C}}(k, n-k) = \mathrm{Gr}(k; \mathbb{C}^n)$ is called the *tautological vector bundle* over the complex Grassmannian $\mathrm{Gr}(k; \mathbb{C}^n)$; it is of (complex) rank k and is denoted by ζ or ζ_k. The space of $\mathrm{Fra}_{\perp}(\zeta_k)$ is the *complex Stiefel manifold* $\mathrm{Stief}(k, \mathbb{C}^n)$ of orthonormal k-frames in $\mathbb{C}^n$.

The inclusion $\mathbb{C}^n \approx \mathbb{C}^n \times \{0\} \hookrightarrow \mathbb{C}^{n+1}$ induces an inclusion $\mathrm{Gr}(k; \mathbb{C}^n) \hookrightarrow \mathrm{Gr}(k; \mathbb{C}^{n+1})$ and we may consider the inductive limit

$$\mathrm{Gr}(k; \mathbb{C}^\infty) = \lim_n \mathrm{Gr}(k; \mathbb{C}^n)$$

which is a CW-space. The tautological vector bundle ζ_k is also defined over $\mathrm{Gr}(k; \mathbb{C}^\infty)$ and induces that over $\mathrm{Gr}(k; \mathbb{C}^n)$ by the inclusion $\mathrm{Gr}(k; \mathbb{C}^n) \hookrightarrow \mathrm{Gr}(k; \mathbb{C}^\infty)$. It is classical that $\pi_i(\mathrm{Stief}(k, \mathbb{C}^n)) = 0$ for $i < 2(n-k)+1$ (see [181, 25.7]), thus $\mathrm{Stief}(k, \mathbb{C}^\infty) = \mathrm{Fra}(\zeta_k)$ is contractible. Hence, the $U(k)$-principal bundle $\mathrm{Stief}(k, \mathbb{C}^\infty) \to \mathrm{Gr}(k; \mathbb{C}^\infty)$ is a universal $U(k)$-principal bundle (see [181, Sect. 19.4]) and thus homotopy equivalent to the Milnor universal bundle $EU(k) \to BU(k)$. In particular, $\mathrm{Gr}(k; \mathbb{C}^\infty)$ has the homotopy type of $BU(k)$. As a consequence, any complex vector bundle of rank k over a paracompact space X is induced from ζ_k by a map $X \to \mathrm{Gr}(k; \mathbb{C}^\infty)$.

To emphasize the analogy with Sect. 9.5, we introduce the *total Chern classes* $c(\xi) \in H^{2*}(X)$ of a complex vector bundle ξ of rank k over a space X by

$$c(\xi) = \sum_{j=1}^{k} w_{2j}(\xi_{\mathbb{R}}),$$

where $\xi_{\mathbb{R}}$ is the vector bundle ξ seen as a real vector bundle of (real) rank $2k$. The component of $c(\xi)$ in $H^{2j}(X)$ is

$$c_j(\xi) = w_{2j}(\xi_{\mathbb{R}}) \in H^{2j}(X) \tag{9.7.5}$$

is called the *i-th Chern class of ξ*.

Theorem 9.7.5 *The cohomology ring of $BU(k)$ is* **GrA***-isomorphic to the polynomial ring*

$$H^*(BU(k)) = H^*(\mathrm{Gr}(k; \mathbb{C}^\infty)) \approx \mathbb{Z}_2[c_1, \dots, c_k]$$

generated by the Chern classes $c_i = c_i(\zeta_k)$ of the tautological bundle ζ_k.

Remark 9.7.6 Our Chern classes $c_j(\xi)$ are the reduction mod 2 of the *integral Chern classes* (see [153, Sect. 14] or [105, Chap. 16]). That the restriction mod 2 of $c_j(\xi)$ coincides with $w_{2j}(\xi_{\mathbb{R}})$ (whence our definition (9.7.5)) is proven in [181, Theorem 41.8]. Note that, by Theorem 9.7.5, $w_{2j+1}(\xi_{\mathbb{R}}) = 0$.

Proof of Theorem 9.7.5 It is the same as that of Theorem 9.5.8, using Proposition 9.7.1. To see that the Chern classes are algebraically independent, we use the tautological complex line bundle γ over $\mathbb{C}P^\infty$ and its n-times product γ^n over $(\mathbb{C}P^\infty)^n$.

Theorem 9.7.5 together with Lemma 9.5.7 gives the following corollary.

Corollary 9.7.7 *The cohomology ring $H^*(\mathrm{Gr}(k; \mathbb{C}^n))$ is generated, as a ring, by the Chern classes $c_1(\zeta_k), \dots, c_k(\zeta_k)$ of the tautological bundle ζ_k.*

Let $c(\xi_j) = \mathbf{1} + c_1(\xi_j) + \cdots + c_{n_j}(\xi_j) \in H^*(\mathrm{Fl}_{\mathbb{C}}(n_1, \dots, n_r))$ be the Chern class of the tautological vector bundle ξ_j. The following theorem is proven in the same way as Theorem 9.5.14. Actually, replacing $\mathbb{Z}_2[c_i(\xi_j)]$ by $\mathbb{Z}[c_i(\xi_j)]$, the statement is true for the integral cohomology (as we wrote a minus sign in the last expression).

Theorem 9.7.8 *The cohomology algebra $H^*(\mathrm{Fl}_{\mathbb{C}}(n_1, \dots, n_r))$ is* **GrA***-isomorphic to the quotient of the polynomial ring*

$$\mathbb{Z}_2[c_i(\xi_j)]\,,\ 1 \le i \le r_j\,,\ j = 1, \dots, r$$

by the ideal generated by the homogeneous components of $1 - c(\xi_1) \cdots c(\xi_r)$.

Remark 9.7.9 By Theorems 9.7.8 and 9.5.14, the correspondence $c_i(\xi_j) \mapsto w_i(\xi_j)$ provides an abstract ring isomorphism

$$H^{2*}(\mathrm{Fl}_{\mathbb{C}}(n_1, \dots, n_r)) \xrightarrow{\approx} H^*(\mathrm{Fl}(n_1, \dots, n_r)).$$

Actually, $\mathrm{Fl}_{\mathbb{C}}(n_1, \dots, n_r)$ with the complex conjugation is a *conjugation space* (see Sect. 10.2). Given Remark 9.7.3, this is established in [87, Theorem 8.3].

We have seen in Proposition 9.2.4 that the first Stiefel-Whitney class classifies the real lines bundles. The full analogue for complex line bundles requires cohomology with $\mathbb{Z}$-coefficients: the first integral Chern class provides an isomorphism $(\mathcal{L}_{\mathbb{C}}(X), \otimes) \xrightarrow{\approx} H^2(X; \mathbb{Z})$, where $\mathcal{L}_{\mathbb{C}}(X)$ be the set of isomorphism classes of complex lines bundles over a CW-complex X (see [96, pp. 62–63]). But, staying within the mod 2 cohomology, one can prove the following result.

Proposition 9.7.10 *Let ξ and ξ' be two complex line bundles over a CW-complex X. Then $c_1(\xi \otimes \xi') = c_1(\xi) + c_1(\xi')$.*

Proof The argument follows the end of the proof of Proposition 9.2.4. One has to replace $\mathbb{R}^\times$ by $\mathbb{C}^\times$ and K by $K_{\mathbb{C}} = \mathbb{C}^\times \times \mathbb{C}^\times$. The only thing to prove is that the composed map

$$B\mathbb{C}^\times \times B\mathbb{C}^\times \xrightarrow[\simeq]{P'} BK_{\mathbb{C}} \xrightarrow{B\varphi} B\mathbb{C}^\times$$

corresponding to that of Diagram (9.2.2) satisfies

$$H^*(B\varphi \circ P')(v) = v \times \mathbf{1} + \mathbf{1} \times v, \tag{9.7.6}$$

where v is the generator of $H^2(B\mathbb{C}^\times) = H^2(BU(1)) = H^2(\mathbb{C}P^\infty)$. The complex conjugation of $\mathbb{C}^\times$ induces an involution τ on $B\mathbb{C}^\times$ corresponding to the conjugation on $\mathbb{C}P^\infty$, with fixed point $\mathbb{R}P^\infty = B\mathbb{R}^\times$. The map $B\varphi \circ P'$ is τ-equivariant. Hence, Eq. (9.7.6) follows from (9.2.3), using that the inclusion $j : \mathbb{R}P^\infty \to \mathbb{C}P^\infty$ satisfies $H^*j(v) = u^2$ (see Proposition 6.1.11). □

Finally, the splitting principle results of Sect. 9.6 have their correspondents for complex bundles. One uses the inclusion of the diagonal subgroup of $U(n)$

$$\Gamma \approx U(1) \times \cdots \times U(1) \subset U(n).$$

The following result is proven in the same way as for Theorem 9.6.1.

Theorem 9.7.11 *The* **GrA**-*morphism $H^*B\alpha : H^*(BU(n)) \to H^*(B\Gamma)$ is injective and its image is $H^*(B\Gamma)^{\mathrm{Sym}_n}$. The complex vector bundle $B\alpha^*\zeta_n$ induced from the universal bundle ζ_n splits into a Whitney sum of complex line bundles.*

As for Theorems 9.6.2 and 9.7.11 generalizes in the following way for the inclusion

$$\alpha_{n_1,\dots,n_r} : U(n_1) \times \cdots \times U(n_r) \to U(n).$$

Theorem 9.7.12 *The map $B\alpha_{n_1,\dots,n_r}$ satisfies the following properties.*

(1) *The* **GrA**-*morphism*

$$H^* B\alpha_{n_1,\dots,n_r} : H^*(BU(n)) \to H^*(BU(n_1) \times \cdots \times BU(n_r))$$

is injective.

(2) $H^* B\alpha_{n_1,\dots,n_r}(c_i) = c_i(\zeta_{n_1} \times \cdots \times \zeta_{n_r})$ *for each $i \geq 0$. In particular, the image of $H^* B\alpha_{n_1,\dots,n_r}$ is generated by $c_i(\zeta_{n_1} \times \cdots \times \zeta_{n_r})$ $(i \geq 0)$.*

(3) *The induced complex vector bundle $B\alpha^*_{n_1,\dots,n_r}\zeta_n$ splits into a Whitney sum of complex vector bundles of ranks $n_1, \dots, n_r$.*

As in Sect. 9.6, we deduce from Theorem 9.7.11 the following proposition (splitting principle for complex bundles).

Proposition 9.7.13 *Let ξ be a complex vector bundle over a paracompact space X. Then, there is a map $f : X_\xi \to X$ such that*

(1) *$H^* f$ is injective.*

(2) *$f^*\xi$ splits into a Whitney sum of complex line bundles.*

As in Proposition 9.6.4, we get an axiomatic characterization of Chern classes.

Proposition 9.7.14 *Suppose that $\tilde{c}$ is a correspondence associating, to each complex vector bundle ξ over a paracompact space X, a class $\tilde{c}(\xi) \in H^{2*}(X)$, such that*

(1) *if $f : Y \to X$ be a continuous map, then $\tilde{c}(f^*\xi) = H^* f(\tilde{c}(\xi))$.*

(2) $\tilde{c}(\xi \oplus \xi') = \tilde{c}(\xi) \smile \tilde{c}(\xi')$.

(3) *if γ is the tautological complex line bundle over $\mathbb{C}P^\infty$, then $\tilde{c}(\gamma) = \mathbf{1} + a$, where $0 \neq a \in H^2(\mathbb{C}P^\infty)$.*

Then $\tilde{c} = c$, the total Chern class.

Thanks to our definition of Chern classes via Stiefel-Whitney classes, the following proposition is a direct consequence of Proposition 9.6.5.

Proposition 9.7.15 *Let ξ be a complex vector bundle over a paracompact space X. Then*

$$\mathrm{Sq}^{2i} c_j(\xi) = \sum_{0 \leq k \leq i} \binom{j-i+k-1}{k} c_{i-k}(\xi)\, c_{j+k}(\xi). \tag{9.7.7}$$

Remark 9.7.16 As in Sect. 9.5.3, the Schubert calculus may be developed for complex Grassmannians. The degrees of (co)homology classes are doubled. The Stiefel-Whitney classes w_i are replaced by the Chern classes c_i. The Stiefel-Whitney classes $\bar{w}_i$ corresponds, in the literature, to the *Segre classes*.

We finish this section with the complex analogue of Lemma 9.6.7.

Lemma 9.7.17 *Let η and ξ be complex vector bundles over a paracompact space X. Suppose that ξ is of rank r and that η is a line bundle. Then*

$$c(\eta \otimes \xi) = \sum_{k=0}^{r} (1 + c_1(\eta))^k c_{r-k}(\xi). \tag{9.7.8}$$

Proof The proof is the same as that of Lemma 9.6.7. The use of Proposition 9.2.4 has to be replaced by that of Proposition 9.7.10. □

9.8 The Wu Formula

9.8.1 Wu's Classes and Formula

Let Q be a closed manifold of dimension n. The map

$$H^{n-k}(Q) \xrightarrow{\mathrm{Sq}^k} H^n(Q) \xrightarrow{\langle -,[Q]\rangle} \mathbb{Z}_2$$

is a linear form on $H^{n-k}(Q)$. By Poincaré duality (see Theorem 5.3.12), there is a unique class $v_k(Q) \in H^k(Q)$ such that $\langle \mathrm{Sq}^k(a), [Q]\rangle = \langle v_k(Q) \smile a, [Q]\rangle$ for all $a \in H^{n-k}(Q)$. In other words,

$$\mathrm{Sq}^k(a) = v_k(Q) \smile a \tag{9.8.1}$$

for all $a \in H^{n-k}(Q)$. The left hand side of (9.8.1) vanishing if $k > n - k$, one has $v_k(Q) = 0$ if $k > n/2$. The class $v_i(Q)$ is the *i-th Wu class* of Q (for Wu classes in a more general setting, see [123, Sect. 3]). Note that $v_0(Q) = \mathbf{1}$. The *total Wu class* $v(Q)$ is defined by

$$v(Q) = \mathbf{1} + v_1(Q) + \cdots + v_{[n/2]}(Q) \in H^*(Q).$$

As an example the next lemma shows the role of $v_k(Q)$ when $n = 2k$. Let V be a $\mathbb{Z}_2$-vector space and let $B : V \times V \to V$ be a bilinear form. A *symplectic basis* of V for B is a basis $\{a_1, \dots, a_k, b_1 \dots, b_k\}$ of V such that $B(a_i, a_j) = B(b_i, b_j) = 0$ and $B(a_i, b_j) = B(b_j, a_j) = \delta_{ij}$. By convention, the empty basis for $V = \{0\}$ is also symplectic.

Lemma 9.8.1 *Let Q be a closed smooth manifold of dimension $2k$ such that its Wu class $v_k(Q)$ vanishes. Then the bilinear form $B : H^k(Q) \times H^k(Q) \to \mathbb{Z}_2$ given by $B(x, y) \mapsto \langle x \smile y, [Q]\rangle$ admits a symplectic basis.*

Note that the lemma implies that $H^k(Q)$ has even dimension and thus, by Poincaré duality, $\chi(Q)$ is even. This can be also deduced from Corollary 5.4.16 and Theorem 9.4.1 since, by the Wu formula (see below), $w_{2k}(TQ) = \mathrm{Sq}^k(v_k(Q)) = 0$.

Proof By definition of the Wu class, $v_k(Q) = 0$ is equivalent to B being alternate, i.e. $B(x, x) = 0$ for all $x \in H^k(Q)$. By Theorem 5.3.12, B is non-degenerate. We are thus reduced to prove the following classical claim: *on a $\mathbb{Z}_2$-vector space V of finite dimension, a non-degenerate bilinear form B which is alternate admits a symplectic basis.* As B is non-degenerate, there exists $a_1, b_1 \in V$ such that $B(a_1, b_1) = 1$. Hence $B(b_1, a_1) = 1$ since alternate implies symmetric in characteristic 2. One has an exact sequence

$$0 \to A \to V \xrightarrow{\phi} \mathbb{Z}_2 \oplus \mathbb{Z}_2 \to 0,$$

where ϕ is the linear map $\phi(v) = (B(a_1, v), B(v, b_1))$ and $A = \ker \phi$. As B is non-degenerate, so is its restriction to $A \times A$. This permits us to prove the claim by induction on the dim V. □

The Wu's formula below relates the Wu class of Q to the Stiefel-Whitney class $w(TQ)$ of the tangent bundle TQ of Q (often called the *Stiefel-Whitney class of Q*).

Theorem 9.8.2 (Wu's formula) *For any smooth closed manifold Q, one has*

$$w(TQ) = \mathrm{Sq}(v(Q)).$$

The Wu formula was proved by Wu wen-Tsün in 1950 [213] by direct computations in $H^*(Q \times Q)$. We follow below the proof of Milnor-Stasheff [153, Theorem 11.14] (for a proof using equivariant cohomology, see Remark 9.8.21). The computations are lightened by the use of the *slant product*

$$H^*(X \times Y) \otimes H_*(Y) \xrightarrow{/} H^*(X) \tag{9.8.2}$$

(actually: $H^k(X \times Y) \otimes H_m(Y) \xrightarrow{/} H^{k-m}(X)$) which is defined as follows. Consider the map $H^*(X) \otimes H^*(Y) \otimes H_*(Y) \to H_*(X)$ defined by the correspondence $a \otimes b \otimes \beta \mapsto \langle b, \beta\rangle a$, using the Kronecker pairing $\langle\,,\rangle$. As $H^*(X \times Y) \approx H^*(X) \otimes H^*(Y)$ by the Künneth theorem (we assume that Y is of finite cohomology type), this gives the linear map (9.8.2). The slant product is characterized by the equation

$$(a \times b)/\beta = \langle b, \beta\rangle\, a$$

for all $a \in H^*(X)$, $b \in H^*(Y)$ and $\beta \in H_*(Y)$. It is also a morphism of $H^*(X)$-modules, i.e.

$$[(u \times \mathbf{1}) \smile c]/\beta = u \smile (c/\beta) \tag{9.8.3}$$

for all $u \in H^*(X)$, $c \in H^*(X \times Y)$ and $\beta \in H_*(Y)$.

Proof of Wu's formula Consider Q as the diagonal submanifold of $M = Q \times Q$, with normal bundle $\nu = \nu(Q, M)$. By Lemma 5.4.17, TQ is isomorphic to ν.

A Riemannian metric provides a smooth bundle pair $(D(\nu), S(\nu))$ with fiber (D^r, S^{r-1}) and there is a diffeomorphism from $D(\nu)$ to a closed tubular neighbourhood W of Q in M. One has the diagram

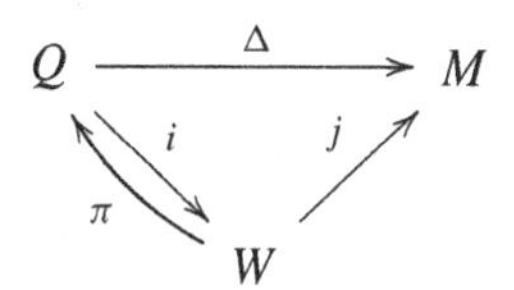

where π is the bundle projection and the other maps are inclusions. By excision,

$$H^*(M, M-Q) \xleftarrow[\approx]{j^*} H^*(W, \mathrm{Bd}\, W) \approx H^*(D(\nu), S(\nu)).$$

Hence, the Thom class of ν may be seen as an element $U \in H^q(W, \mathrm{Bd}\, W)$ satisfying $U = j^*(U')$ for a unique $U' \in H^q(M, M-Q)$. Let $U'' \in H^q(M)$ be the image of U' under the restriction homomorphisms $H^q(M, M-Q) \to H^q(M)$.

By definition of the Stiefel-Whitney class $w = w(TQ) = w(\nu)$, one has

$$\pi^*(w) \smile U = \mathrm{Sq}\, U.$$

One has $\Delta^*(\mathbf{1} \times w) = \mathbf{1} \smile w = w$, whence $j^*(\mathbf{1} \times w) = \pi^*(w)$. Hence, the equation $(\mathbf{1} \times w) \smile U' = \mathrm{Sq}\, U'$ holds true in $H^*(M, M-Q)$ which, in $H^*(M)$, implies

$$(\mathbf{1} \times w) \smile U'' = \mathrm{Sq}\, U''. \tag{9.8.4}$$

Without loss of generality, we may assume that Q is connected. Let $\mathcal{A} = \{a_1, a_2, \dots\}$ and $\mathcal{B} = \{b_1, b_2, \dots\}$ of $H^*(Q)$ be an additive bases of $H^*(Q)$ which are Poincaré dual, i.e. $\langle a_i \smile b_j, [Q] \rangle = \delta_{ij}$. We suppose that $a_0 = \mathbf{1}$. By Lemma 5.4.2 and Eq. (5.4.1), one has

$$U'' = \sum_i a_i \times b_i$$

and therefore

$$U''/[Q] = (\mathbf{1} \times b_0)/[Q] = \mathbf{1}.$$

Applying this together with Eqs. (9.8.4) and (9.8.3) gives

$$\operatorname{Sq} U''/[Q] = [(\mathbf{1} \times w) \smile U'']/[Q] = w \smile (U''/[Q]) = w\,. \tag{9.8.5}$$

We now express the Wu class $v = v(M)$ in the $\mathcal{A}$-basis: $v = \sum_i \lambda_i a_i$. Then, $\langle v \smile b_j, [Q]\rangle = \lambda_j$, which implies that

$$v = \sum_i \langle v \smile b_i, [Q]\rangle\, a_i = \sum_i \langle \operatorname{Sq} b_i, [Q]\rangle a_i, \;. \tag{9.8.6}$$

Hence, using (9.8.5), we get

$$\begin{aligned} \operatorname{Sq} v &= \sum_i \langle \operatorname{Sq} b_i, [Q]\rangle \operatorname{Sq} a_i \\ &= \sum_i \big(\operatorname{Sq} a_i \times \operatorname{Sq} b_i\big)/[Q] \\ &= \operatorname{Sq} U''/[Q] \\ &= w. \end{aligned}$$

The remainder of this subsection is devoted to general corollaries of Wu's formula. The first one says that the Stiefel-Whitney class $w(TQ)$ depends only on the homology type of Q.

Corollary 9.8.3 *Let $f : Q' \to Q$ be continuous map between smooth closed manifolds Q and Q' of the same dimension. Suppose that $H^*f : H^*(Q) \to H^*(Q')$ is surjective. Then $H^*f(w(TQ)) = w(TQ')$.*

Proof By Kronecker duality, H_*f is injective and thus $\pi_0 f$ is injective. The connected components of Q out of the image of f play no role, so one may assume that $\pi_0 f$ is a bijection. This implies that $H_*f([Q']) = [Q]$.

Let $v' = H^*f(v(Q))$. For $b \in H^*(Q)$, one has

$$\begin{aligned} \langle v(Q) \smile b, [Q]\rangle &= \langle v(Q) \smile b,\, H_*f[Q']\rangle \\ &= \langle H^*f(v(Q) \smile b), [Q']\rangle \\ &= \langle v' \smile H^*f(b), [Q']\rangle. \end{aligned}$$

On the other hand

$$\begin{aligned} \langle v(Q) \smile b, [Q]\rangle &= \langle \operatorname{Sq} b,\, H_*f[Q']\rangle \\ &= \langle \operatorname{Sq}(H^*f(b)), [Q']\rangle \\ &= \langle v(Q') \smile H^*f(b), [Q']\rangle. \end{aligned}$$

Therefore, the equality

$$\langle v' \smile H^* f(b), [Q'] \rangle = \langle v(Q') \smile H^* f(b), [Q'] \rangle$$

is valid for all $b \in H^*(Q)$. As $H^* f$ is surjective, Theorem 5.3.12, this implies that $v' = V(Q')$, so $H^* f(v(Q)) = v(Q')$. By Wu's formula,

$$w(TQ') = \mathrm{Sq}\, v(Q') = \mathrm{Sq} \circ H^* f(v(Q)) = H^* f(\mathrm{Sq}\, v(Q)) = H^* f(w(TQ)). \quad \square$$

Corollary 9.8.4 *Let $f : Q \to Q$ be continuous map of degree one. Then $H^* f(w(TQ)) = w(TQ)$.*

Proof By Proposition 5.2.8, $H_* f$ is surjective and then $H^* f$ is injective by Kronecker duality. As $H^*(Q)$ is a finite dimensional vector space, this implies that $H^* f$ (and $H_* f$) are bijective. The results then follows from Corrolary 9.8.3. □

9.8.2 Orientability and Spin Structures

A smooth manifold is *orientable* if its tangent bundle is orientable. The following corollary generalizes Proposition 4.2.3.

Corollary 9.8.5 *Let Q be a smooth closed n-dimensional manifold. Then Q is orientable if and only if $\mathrm{Sq}^1 : H^{n-1}(Q) \to H^n(Q)$ vanishes.*

Proof By Proposition 9.4.4, Q is orientable if and only if $w_1(TQ) = 0$ which, by Wu's formula, is equivalent to $v_1(Q) = 0$. By the definition of $v_1(Q)$, its vanishing is equivalent to $\mathrm{Sq}^1 : H^{n-1}(Q) \to H^n(Q)$ being zero. □

The same argument, using Proposition 9.4.7, implies the following result.

Corollary 9.8.6 *Let Q be a smooth closed n-dimensional manifold which is orientable. Then TQ admits a spin structure if and only if $\mathrm{Sq}^2 : H^{n-2}(Q) \to H^n(Q)$ vanishes.*

Example 9.8.7 A closed manifold M such that $H^*(M)$ is **GrA**-isomorphic to $H^*(\mathbb{R}P^n)$ is orientable if and only if n is odd. Indeed, let $0 \neq a \in H^1(M) = \mathbb{Z}_2$. By the Cartan formula, $\mathrm{Sq}^1(a^{n-1}) = 0$ if and only if n is odd. As $H^{n-1}(M)$ is generated by a^{n-1}, the assertion follows from Corollary 9.8.5. A similar argument, using Corollary 9.8.6, proves that a closed manifold M such that $H^*(M)$ is **GrA**-isomorphic to $H^*(\mathbb{C}P^n)$ admits a spin structure if and only if n is odd.

In the particular case $n = 4$, Corollary 9.8.6 gives the following result.

Corollary 9.8.8 *Let Q be a smooth connected closed 4-dimensional manifold which is orientable. Then,*

(1) $a \smile a = w_2(TQ) \smile a$ *for all* $a \in H^2(Q)$.
(2) TQ *admits a spin structure if and only if the cup-square map* $H^2(Q) \to H^4(Q)$ *vanishes.*
(3) $w_4(TQ) = w_2(TQ) \smile w_2(TQ)$.

Point (2) is the analogue of Proposition 4.2.3 for surfaces. In particular, $T\mathbb{C}P^2$ does not admit a spin structure.

Proof If $w_1(TQ) = 0$, then $w_2(TQ) = v_2(Q)$ by Wu's formula. Hence, $a \smile a = \mathrm{Sq}^2(a) = v_2(Q) \smile a = w_2(TQ) \smile a$ for all $a \in H^2(Q)$, which proves (a). Point (b) thus follows from Corollary 9.8.6. For Point (c), we use Wu's formula again:

$$w_4(TQ) = \mathrm{Sq}^2(v_2(Q)) = \mathrm{Sq}^2(w_2(TQ)) = w_2(TQ) \smile w_2(TQ). \qquad \square$$

Corollary 9.8.9 *Let Q be a smooth closed manifold of dimension $n \leq 7$. If TQ admits a spin structure, then $w(TQ) = \mathbf{1}$.*

Proof The proposition is obvious for $n \leq 2$. Otherwise, by Proposition 9.4.6, the existence of a spin structure implies that the restriction of TQ over the 3-skeleton of a triangulation of Q is trivial. Hence $w_3(TM)$ also vanishes which, by Wu's formula, implies that $v_i(M)$ vanishes for $i \leq 3$. As $n \leq 7$, this implies that $v(Q) = \mathbf{1}$ and thus $w(TQ) = \mathrm{Sq}\, v(Q) = \mathbf{1}$. $\square$

An interesting example is given by the projective spaces.

Proposition 9.8.10 *Let $0 \neq a \in H^1(\mathbb{R}P^n)$. The Stiefel-Whitney class of the tangent space of $\mathbb{R}P^n$ is*

$$w(T\mathbb{R}P^n) = (\mathbf{1} + a)^{n+1}$$

and the Wu class of $\mathbb{R}P^n$ is

$$v(\mathbb{R}P^n) = \sum_{i=0}^{[n/2]} \tbinom{n-i}{i} a^i.$$

Here are a few examples.

n	$v(\mathbb{R}P^n)$	$w(T\mathbb{R}P^n)$
2	$\mathbf{1}+a$	$\mathbf{1}+a+a^2$
3	$\mathbf{1}$	$\mathbf{1}$
4	$\mathbf{1}+a+a^2$	$\mathbf{1}+a+a^5$
5	$\mathbf{1}+a^2$	$\mathbf{1}+a^2+a^4$
6	$\mathbf{1}+a+a^3$	$\mathbf{1}+a+a^2+a^3+a^4+a^5+a^6$
7	$\mathbf{1}$	$\mathbf{1}$

Remark 9.8.11 The formulae of Proposition 9.8.10 imply the following.

(1) $\mathbb{R}P^n$ is orientable if and only if n is odd (this is not a surprise!). More generally, $w_{2i+1}(T\mathbb{R}P^{2k+1}) = 0$.
(2) $T\mathbb{R}P^n$ admits a spin structure if and only if $n \equiv 3 \mod 4$. In this case, there are two spin structures, since $H^1(\mathbb{R}P^n) = \mathbb{Z}_2$. For a discussion about these two structures for $\mathbb{R}P^3 \approx SO(3)$, see [130, Example 2.5, p. 87].
(3) $w(T\mathbb{R}P^n) = \mathbf{1}$ if and only if $n = 2^k - 1$. But $T\mathbb{R}P^n$ is trivial if and only if $n = 1, 3, 7$ by Adams Theorem [2, p. 21].

Proof of Proposition 9.8.10 The two formulae will be proved separately. Checking Wu's formula is left as an exercise. By (8.2.2),

$$v_i(\mathbb{R}P^n) \smile a^{n-i} = \mathrm{Sq}^i a^{n-i} = \tbinom{n-i}{i} a^n = \tbinom{n-i}{i} a^i \smile a^{n-i}.$$

which proves that $v_i(\mathbb{R}P^n) = \binom{n-i}{i} a^i$. This proves the formula for the Wu class. As for the Stiefel-Whitney class, the idea is the following. Recall that $\mathbb{R}P^n = \mathrm{Gr}(1; \mathbb{R}^{n+1}) = \mathrm{Fl}(1, n)$. Write $\gamma = \xi_1$ and $\gamma^\perp = \xi_2$ for the tautological bundles. Then,

$$T\mathbb{R}P^n \approx \hom(\gamma, \gamma^\perp) \tag{9.8.7}$$

(see [153, Lemma 4.4] for a proof). But $\gamma \oplus \gamma^\perp$ is the trivial bundle η_{n+1} of rank $n+1$. Adding to both side of (9.8.7) the bundle $\hom(\gamma, \gamma) \approx \eta_1$, we get

$$T\mathbb{R}P^n \oplus \eta_1 \approx \hom(\gamma, \eta_{n+1}).$$

The latter is the Whitney sum of $(n+1)$-copies of $\gamma^* = \hom(\gamma, \eta_1)$. But $\gamma^* \approx \gamma$, using an Euclidean metric on γ. For details (see [153, Theorem 4.5]). Hence, the formula for $w(T\mathbb{R}P^n)$ follows from (9.4.3).

Remark 9.8.12 The argument of the proof of Proposition 9.8.10 essentially works for computing the Chern class of the tangent bundle to $T\mathbb{C}P^n$ (which is a complex vector bundle). The slight difference is that $\gamma^* = \hom(\gamma, \eta_1)$ is not, as complex vector bundle, isomorphic to γ but to the conjugate bundle $\bar{\gamma}$ (the complex structure on each fiber is the conjugate of that of γ (see [153, pp. 169–170]). But this does not alter our Chern classes which are defined mod 2: $c_i(T\mathbb{C}P^n) = w_{2i}(T\mathbb{C}P^n)$. Thus

$$c(T\mathbb{C}P^n) = (\mathbf{1} + a)^{n+1} \quad \text{and} \quad v(\mathbb{C}P^n) = \sum_{i=0}^{[n/2]} \tbinom{n-i}{i} a^i$$

where $0 \neq a \in H^2(\mathbb{C}P^n)$. The first formula holds as well for the integral Chern class with a suitable choice of a generator of $H^2(\mathbb{C}P^n; \mathbb{Z})$ (see [153, Theorem 14.10]).

9.8.3 Applications to 3-Manifolds

Wu's formula has two important consequences for closed 3-dimensional manifolds. The first one is the following.

Proposition 9.8.13 *Let Q be a smooth closed 3-dimensional manifold which is orientable. Then TQ is a trivial vector bundle (in other words: Q is parallelizable).*

Proof For any smooth closed manifold, one has $w_1(TQ) = v_1(Q)$ by Wu's formula. Thus, $v_1(Q) = 0$ if Q is orientable. In dimension 3, this implies that $v(Q) = \mathbf{1}$ and, by Wu's formula again, $w(TQ) = \mathbf{1}$. The result then follows from Proposition 9.4.6. □

The second application is Postnikov's characterization of the cohomology ring of a closed connected 3-dimensional manifold [164]. Let M be such a manifold. Consider the symmetric trilinear form $\pi_M : H^1(M) \times H^1(M) \times H^1(M) \to \mathbb{Z}_2$ defined by

$$\pi_M(a, b, c) = \langle a \smile b \smile c, [M]\rangle.$$

The first observation is that π_M determines the ring structure of $H^*(M)$.

Lemma 9.8.14 *Let M and $\bar{M}$ be two closed connected 3-dimensional manifolds. Suppose that there exists an isomorphism $h^1 : H^1(M) \to H^1(\bar{M})$ such that*

$$\pi_{\bar{M}}(h^1(a), h^1(b), h^1(c)) = \pi_M(a, b, c).$$

Then, h^1 extends to a **GrA***-isomorphism $h^* : H^*(M) \to H^*(\bar{M})$.*

Proof Let $\mathcal{A} = \{a_1, \dots, a_m\}$ be a $\mathbb{Z}_2$-basis of $H^1(M)$. The set $\bar{\mathcal{A}} = \{\bar{a}_1, \dots, \bar{a}_m\}$ where $\bar{a}_i = h^1(a_i)$ is then a $\mathbb{Z}_2$-basis of $H^1(\bar{M})$. Let $\mathcal{B} = \{b_1, \dots, b_m\}$ and $\bar{\mathcal{B}} = \{\bar{b}_1, \dots, \bar{b}_m\}$ be the bases of $H^2(M)$ and $H^2(\bar{M})$ which are Poincaré dual to $\mathcal{A}$ and $\bar{\mathcal{A}}$, i.e. the equations

$$\langle a_i \smile b_j, [M]\rangle = \delta_{ij} \text{ and } \langle \bar{a}_i \smile \bar{b}_j, [\bar{M}]\rangle = \delta_{ij} \tag{9.8.8}$$

are satisfied for all i, j. Let $h^2 : H^2(M) \to H^2(\bar{M})$ be the isomorphism such that $h^2(b_i) = \bar{b}_i$ and let h^3 be the unique isomorphism from $H^3(M)$ to $H^3(\bar{M})$. This produces a **GrV**-isomorphism $h^* : H^*(M) \to H^*(\bar{M})$. To prove that h^* is a **GrA**-morphism, write $a_i \smile a_j = \sum_{l=1}^m \lambda_{ij}^l b_l$ and, using (9.8.8), note that

$$\pi_M(a_i, a_j, a_k) = \langle a_i \smile a_j \smile a_k, [M]\rangle = \sum_{l=1}^m \lambda_{ij}^l \langle b_l \smile a_k, [M]\rangle = \lambda_{ij}^k.$$

Therefore,

$$\begin{aligned}\langle h^2(a_i \smile a_j) \smile \bar{a}_k, [\bar{M}]\rangle &= \sum_{l=1}^{m} \pi_M(a_i, a_j, a_l)\langle h^2(b_l) \smile \bar{a}_k, [\bar{M}]\rangle \\ &= \sum_{l=1}^{m} \pi_M(a_i, a_j, a_l)\langle \bar{b}_l \smile \bar{a}_k, [\bar{M}]\rangle \\ &= \pi_M(a_i, a_j, a_k)\end{aligned}$$

and

$$\begin{aligned}\langle h^1(a_i) \smile h^1(a_j) \smile \bar{a}_k, [\bar{M}]\rangle &= \langle h^1(a_i) \smile h^1(a_j) \smile h^1(a_k), [\bar{M}]\rangle \\ &= \pi_{\bar{M}}(h^1(a_i), h^1(a_j), h^1(a_k)).\end{aligned}$$

Since $\pi_{\bar{M}}(h^1(a_i), h^1(a_j), h^1(a_k)) = \pi_M(a_i, a_j, a_k)$, this proves that $h^2(a_i \smile a_j) = h^1(a_i) \smile h^1(a_j)$. On the other hand, h^3 formally satisfies $\langle h^3(u), [\bar{M}]\rangle = \langle u, [M]\rangle$. Hence, the equality $h^1(a_i) \smile h^2(b_j) = h^3(a_i \smile b_j)$ follows from (9.8.8). We have thus established that h^* is a **GrA**-morphism. □

The trilinear form π_M is linked to the Wu class $v(M) \in H^1(M)$.

Lemma 9.8.15 *Let M be a closed connected 3-dimensional manifold. Then, the Wu class $v = v_1(M)$ satisfies*

$$\pi_M(v, b, c) = \pi_M(b, b, c) + \pi_M(b, c, c) \tag{9.8.9}$$

for all $b, c \in H^1(M)$.

Proof This comes from that

$$\begin{aligned}v_1(M) \smile (b \smile c) &= \mathrm{Sq}^1(b \smile c) = \mathrm{Sq}^1(b) \smile c \,+\, b \smile \mathrm{Sq}^1(c) \\ &= b \smile b \smile c + b \smile c \smile c.\end{aligned}$$

□

The following "realizability result" is due to Postnikov [164].

Proposition 9.8.16 *Let (V, π) a symmetric trilinear form, with V a finite dimensional $\mathbb{Z}_2$-vector space. Let $v \in V$ satisfying (9.8.9). Then, there exists a closed connected 3-manifold M with an isometry $(H^1(M), \pi_M) \approx (V, \pi)$, sending $v_1(M)$ onto v.*

Proof (indications) The full proof may be found in [164]. When M is orientable, the form π_M is alternate since the left hand side of (9.8.9) vanishes ($v = w_1(TM) = 0$). Hence, $\pi_M \in \bigwedge^3 H^1(M)$. An alternate form $\pi \in \bigwedge^3 V$ may be lifted to $\tilde{\pi} \in \bigwedge^3 \tilde{V}$, where $\tilde{V}$ is a free abelian group' with $\tilde{V} \otimes \mathbb{Z}_2 \approx V$. In [187], D. Sullivan constructed a closed connected orientable 3-manifold M with $(H^1(M; \mathbb{Z}), \tilde{\pi}_M) \approx (\tilde{V}, \tilde{\pi})$, which thus proves Proposition 9.8.16 in the orientable case. □

9.8.4 The Universal Class for Double Points

The material of this section is essentially a rewriting in our language of results of Haefliger [78]. Let M be a closed manifold of dimension m. Let $G = \{1, \tau\}$ acting on $M \times M$ by $\tau(x, y) = (y, x)$, with fixed point set $(M \times M)^G = \Delta_M$, the diagonal submanifold of $M \times M$. The diagonal inclusion $\delta : M \to M \times M$ induces a diffeomorphism $\bar{\delta} : M \xrightarrow{\approx} \Delta_M$. For $N > 1$, $S^N \times_G (M \times M)$ is a closed manifold containing $\mathbb{R}P^N \times \Delta_M$ as a closed submanifold of codimension m. Let $\mathrm{PD}_{G,N}(M) = \mathrm{PD}(\mathbb{R}P^N \times \Delta_M) \in H^m(S^N \times_G (M \times M))$, the Poincaré dual of $\mathbb{R}P^N \times \Delta_M$ (see Sect. 5.4.1). If N is big enough,

$$H_G^m(M \times M) \approx H^m(S^\infty \times_G (M \times M)) \to H^m(S^N \times_G (M \times M))$$

is an isomorphism. Therefore, there is a unique class $\mathrm{PD}_G(M) \in H_G^m(M \times M)$ whose image in $H^m(S^N \times_G (M \times M))$ is equal to $\mathrm{PD}_{G,N}(\Delta_M)$.

The class $\mathrm{PD}_G(M)$ is called the *universal class of double points* for continuous maps into M, a terminology justified by Lemma 9.8.17. For a space X, denote by $j : X^0 \to (X \times X)$ the inclusion of $X^0 = (X \times X) - \Delta_X$ into $(X \times X)$. As G acts freely on X^0, the quotient space $X^* = X^0/G$, called the *reduced symmetric square* of X, has the homotopy type of X_G^0 by Lemma 7.1.4.

The diffeomorphism $\bar{\delta} : M \xrightarrow{\approx} \Delta_M$ is covered by a bundle isomorphism $\tilde{\delta} : TM \xrightarrow{\approx} \nu(M \times M, \Delta_M)$ (see Lemma 5.4.17), which intertwines τ with the antipodal involution on TM. Via $\tilde{\delta}$, the sphere bundle T^1M becomes G-diffeomorphic with the boundary $\mathrm{Bd}\,W$ of a G-invariant tubular neighbourhood of Δ_M in $M \times M$. Thus,

$$W^* \approx (W - \Delta_M)_G \approx (\mathrm{Bd}W)_G \approx (T^1M)_G \approx (T^1M)/G. \qquad (9.8.10)$$

As j is G-equivariant, it induces $H_G^* j : H_G^*(X \times X) \to H_G^*(X^0)$. Let $f : Q \to M$ be a continuous map. Consider the homomorphism

$$\Psi : H_G^m(M \times M) \xrightarrow{H_G^*(f \times f)} H_G^m(Q \times Q) \xrightarrow{H_G^* j} H_G^m(Q^0) \xrightarrow{\approx} H^m(Q^*).$$

We denote by $\bar{\Psi} : H_G^m(M \times M) \to H^m(T^1 Q/G)$ the post-composition of Ψ with the homomorphism $H^m(Q^*) \to H^m(W^*) \approx H^m(T^1 N/G)$. Define

$$O_G^f = \Psi(\mathrm{PD}_G(M)) \in H_G^m(Q^*) \text{ and } \bar{O}_G^f = \bar{\Psi}(\mathrm{PD}_G(M)) \in H_G^m((T^1 Q)/G).$$

Lemma 9.8.17 *Let $f : Q \to M$ be a continuous map between closed manifolds. Then*

(1) *if f is homotopic to an embedding, then $O_G^f = 0$.*
(2) *if f is homotopic to an immersion, then $\bar{O}_G^f = 0$.*

Proof The classes O_G^f and $\bar{O}_G^f$ depend only on the homotopy class of f. If f is injective, then $(f \times f)(Q^0) \subset M^0$ and thus $(f \times f)_G(Q_G^0) \subset M_G^0 = (M \times M)_G - (\Delta_M)_G$. By Lemma 5.4.2, $\mathrm{PD}_G(M)$ has image zero in $H_G^m(M^0)$, which proves (1) (this does not use that Q is a manifold). Suppose that f is an immersion, so f is locally injective. As Q is compact, there is a G-invariant tubular neighbourhood W of Δ_Q in $Q \times Q$ such that $(f \times f)(W^0) \subset M^0$. We deduce (2) as above for (1). □

In order to get applications of Lemma 9.8.17, we now express $\mathrm{PD}_{G,N}(M)$ within the description of $H_G^m(M \times M)$ given by Proposition 8.3.3, which can be rephrased as follows. There is a **GrA**$[u]$-isomorphism from $H_G^*(M \times M)$ to $(\mathbb{Z}_2[u] \otimes \mathcal{D}) \oplus \mathcal{N}$, where $\mathcal{D}$ is the $\mathbb{Z}_2$-vector space generated by $\{x \times x \mid x \in H^k(M),\ k \geq 0\}$ so that $\rho : H_G^*(M \times M) \to H^*(M \times M)^G$ sends the elements of $\mathcal{D} \oplus \mathcal{N}$ (elements of u-degree 0) isomorphically to $H^*(M \times M)^G$. The $\mathbb{Z}_2$-vector space $\mathcal{N}$ is generated by $\{x \times y + y \times x \mid x, y \in H^*(M)\}$ and coincides with the ideal ann (u).

Proposition 9.8.18 *Using the isomorphism $H_G^m(M \times M) \approx (\mathbb{Z}_2[u] \otimes \mathcal{D}) \oplus \mathcal{N}$, we have*

(1) $\mathrm{PD}_G(M) \equiv \sum_{k=0}^{[m/2]} u^{m-2k}(v_k(M) \times v_k(M)) \mod \mathcal{N}$, *where $v_k(M)$ is the k-th Wu class of M.*
(2) $\rho(\mathrm{PD}_G(M)) = \mathrm{PD}(\Delta_M)$, *the Poincaré dual of Δ_M in $M \times M$.*

Example 9.8.19 Let $M = \mathbb{R}P^2$. One has $H^*(M) = \mathbb{Z}_2[a]/(a^3)$ and $v(M) = 1 + a$. Then, $\mathrm{PD}_G(M) \equiv u^2 + a \times a \mod \mathcal{N}$ and, according to Eq. (5.4.1), $\rho(\mathrm{PD}_G(M)) = \mathrm{PD}(\Delta_M) = \mathbf{1} \times a^2 + a \times a + a^2 \times \mathbf{1}$. Therefore,

$$\mathrm{PD}_G(\mathbb{R}P^2) = u^2 + a \times a + N(\mathbf{1} \times a^2).$$

Proof It is enough to prove (1) for $\mathrm{PD}_{G,N}(M)$ with N big enough. Let $i : Q \to P$ be the inclusion of a closed manifold Q into a compact manifold P. For $x \in H^j(P)$, one has

$$\langle x \smile \mathrm{PD}(Q), [P]\rangle = \langle x, \mathrm{PD}(Q) \frown [P]\rangle = \langle x, H_*i([Q])\rangle = \langle H^*i(x), [Q]\rangle. \tag{9.8.11}$$

We shall apply (9.8.11) to $Q = \mathbb{R}P^N \times M$ and $P = S^N \times_G (M \times M)$ and $x = u^{N-m+2i}(a \times a) \in H^{N+m}(P)$, where $a \in H^{m-i}(M)$. One has

$$\begin{aligned}\langle H^*i(x), [Q]\rangle &= \langle u^{N-m+2i} H^*i(a \times a), [Q]\rangle \\ &= \langle u^{N-m+2i} \textstyle\sum_{j=0}^{m-i} u^{m-i-j}\, \mathrm{Sq}^j(a), [Q]\rangle \quad \text{by definition of Sq(a)} \\ &= \langle u^N \mathrm{Sq}^i(a), [Q]\rangle \quad \text{only non-zero term} \\ &= \langle \mathrm{Sq}^i(a), [M]\rangle.\end{aligned} \tag{9.8.12}$$

For $y, z \in H^j(M)$, we have $(y \times y) + (z \times z) \equiv (y + z) \times (y + z) \mod \mathcal{N}$ and $u\mathcal{N} = 0$. Therefore, for $k > 0$, $u^k\,\mathrm{PD}_{G,N}(Q)$ admits an expression of the form $u^k\,\mathrm{PD}_{G,N}(Q) = u^k \sum_{j=0}^{[m/2]} u^{m-2j}(y_j \times y_j)$ with $y_j \in H^j(M)$. Hence,

$$\begin{aligned}\langle x \smile \mathrm{PD}_{G,N}(Q), [P]\rangle &= \langle u^{N-m+2i}(a \times a) \smile \sum\nolimits_{j=0}^{[m/2]} u^{m-2j}(y_j \times y_j), [P]\rangle \\ &= \langle u^N(a \times a) \smile (y_i \times y_i), [P]\rangle \\ &= \langle u^N\{(a \smile y_i) \times (a \smile y_i)\}, [P]\rangle \\ &= \langle a \smile y_i, [M]\rangle \end{aligned} \qquad (9.8.13)$$

Using (9.8.11), Formulae (9.8.12) and (9.8.13) imply that $y_i = v_i(M)$ for all $i = 0\ldots, [m/2]$. This proves (1).

For Point (2) we must prove that $\rho_N(\mathrm{PD}_{G,N}) = \mathrm{PD}(M)$ where ρ_N is induced by the fiber inclusion $M \times M \to S^N \times (M \times M) \to P$. But this map is transversal to $\mathbb{R}P^N \times \Delta_M$. Point (2) thus comes from Proposition 5.4.10. □

Proposition 9.8.18 enables us to compute the image of $\mathrm{PD}_G(M)$ under the homomorphism

$$H_G^*(M \times M) \xrightarrow{r} H_G^*((M \times M)^G) \xrightarrow{\approx} H^*(M)[u] \xrightarrow{\mathrm{ev}_1} H^*(M)$$

Corollary 9.8.20 $\mathrm{ev}_1 \circ r(\mathrm{PD}_G(M)) = w(TM)$, *the total Stiefel-Whitney class of the tangent bundle* TM.

Proof

$$\begin{aligned}\mathrm{ev}_1 \circ r(\mathrm{PD}_G(M)) &= \sum\nolimits_{k=0}^{[m/2]} \mathrm{ev}_1 \circ r\big(v_k(M) \times v_k(M)\big) && \text{by Proposition 9.8.19} \\ &= \sum\nolimits_{k=0}^{[m/2]} \mathrm{Sq}(v_k(M)) && \text{by (8.3.5)} \\ &= \mathrm{Sq}(v(M)) \\ &= w(TM) && \text{by the Wu formula.}\end{aligned}$$ □

Remark 9.8.21 The formula of Corollary 9.8.20 may be proven directly in the following way. By Lemma 5.4.4 and the considerations before (9.8.10), one has $r(\mathrm{PD}_G(M)) = e_G(TM)$, where G acts on TM by the antipodal action on each fiber. This equivariant Euler class satisfies $\mathrm{ev}_1(e_G(TM)) = w(TM)$ (see (10.4.7) in Sect. 10.4), which proves Corollary 9.8.20. Moreover, using the proof of Corollary 9.8.20, with the last line removed, gives a new proof of the Wu formula.

Lemma 9.8.22 *Let Q be a closed manifold of dimension p. There is a commutative diagram*

$$\begin{array}{ccccccccc} 0 & \longrightarrow & H_G^{*-p}(\Delta_Q) & \xrightarrow{\mathrm{Gys}_G} & H_G^*(Q \times Q) & \longrightarrow & H^*(Q^*) & \longrightarrow & 0 \\ & & \approx \downarrow H_G^*\bar{\delta} & & \downarrow r & & \downarrow & & \\ 0 & \longrightarrow & H_G^{*-p}(Q) & \xrightarrow{\Phi} & H_G^*(Q) & \longrightarrow & H^*((T^1Q)/G) & \longrightarrow & 0 \end{array}$$

where the rows are exact sequences. Here, $\Phi(a) = a \smile e_G(TQ)$, *where* G *acts on* TQ *via the antipodal involution.*

Proof The diagram comes from Proposition 5.4.10 applied, for N big, to the pair $(S^N \times_G (Q \times Q), \mathbb{R}P^N \times \Delta_Q)$. The long diagram of Proposition 5.4.10 splits into the above diagram because Φ is injective (see Proposition 7.5.14). The bottom line is the Gysin sequence for the sphere bundle $(T^1Q)_G \to Q_G$. □

The above results permit us to express an obstruction to embedding in terms of the dual Stiefel-Whitney classes. Let $f : Q \to M$ be a smooth map. Define $\bar{w}^f = \bar{w}(TQ) \smile H^*f(w(TM)) \in H^*(Q)$ where $\bar{w}(TQ)$ is the dual Stiefel-Whitney classes of TQ (see p. 389).

Proposition 9.8.23 *Let* $f : Q^q \to M^m$ *be a smooth map between closed manifolds.*

(1) *If* f *is homotopic to a smooth immersion, then* $\bar{w}^f_k = 0$ *for* $k > m - q$.
(2) *If* f *is homotopic to a smooth embedding, then*

$$(\bar{w}^f_{m-q} \times \mathbf{1}) \smile \mathrm{PD}(\Delta_Q) = H^*(f \times f)(\mathrm{PD}(\Delta_M))$$

in $H^m(Q \times Q)$.

Proof We shall argue with the help of the diagram

$$\begin{array}{ccccc}
H_G^{m-p}(\Delta_Q) & \xrightarrow[\approx]{H_G^*\bar{\delta}} & H_G^{m-p}(Q) & \xrightarrow{\mathrm{ev}_1} & H^*(Q) \\
\downarrow{\scriptstyle \mathrm{Gys}_G} & & \downarrow{\scriptstyle -\smile e_G(TQ)} & & \downarrow{\scriptstyle -\smile w(TQ)} \\
H_G^m(Q \times Q) & \xrightarrow{r} & H_G^m(Q) & \xrightarrow{\mathrm{ev}_1} & H^*(Q)
\end{array} \tag{9.8.14}$$

The left square comes from Lemma 9.8.22 and is thus commutative. So is the right square by Remark 9.8.21.

If f is homotopic to a smooth immersion, then $\bar{O}^f_G = 0$ by Lemma 9.8.17. By Lemma 9.8.22, this implies that there exists $b \in H_G^{m-q}(Q)$ such that the equations

$$\mathrm{ev}_1 \circ r \circ H_G^*(f \times f)(\mathrm{PD}_G(M)) = \mathrm{ev}_1 \circ (b \smile e_G(TQ)) = \mathrm{ev}_1(b) \smile w(TQ)$$

hold true in $H^*(Q)$. But

$$\begin{aligned}
\mathrm{ev}_1 \circ r \circ H_G^*(f \times f)(\mathrm{PD}_G(M)) &= H^*f \circ \mathrm{ev}_1 \circ r(\mathrm{PD}_G(M)) \\
&= H^*f(w(TM)) \qquad \text{by Corollary 9.8.20.}
\end{aligned}$$

Hence

$$\mathrm{ev}_1(b) \smile w(TQ) = H^*f(w(TM)). \tag{9.8.15}$$

Since $\bar{w}(TQ) \smile w(TQ) = \mathbf{1}$, multiplying both sides of (9.8.15) by $\bar{w}(TQ)$ gives

$$\mathrm{ev}_{\mathbf{1}}(b) = H^* f(w(TM)) \smile \bar{w}(TQ) = \bar{w}^f. \tag{9.8.16}$$

As b is of degree $m - q$, Eq. (9.8.16) implies that $\bar{w}_k^f = 0$ for $k > m - q$. We have thus proven (1). Also, (9.8.16) is equivalent to

$$b = \sum_{i=0}^{m-q} \bar{w}_i^f \, u^{m-q-i}. \tag{9.8.17}$$

To prove Point (2), we use the diagram

$$\begin{array}{ccccc} H_G^{m-p}(Q) & \xleftarrow[\approx]{H_G^*\bar{\delta}} & H_G^{m-p}(\Delta_Q) & \xrightarrow{\mathrm{Gys}_G} & H_G^m(Q \times Q) \\ \downarrow \rho & & \downarrow \rho & & \downarrow \rho \\ H^{m-p}(Q) & \xleftarrow[\approx]{H^*\bar{\delta}} & H^{m-p}(\Delta_Q) & \xrightarrow{\mathrm{Gys}} & H^m(Q \times Q) \end{array} \tag{9.8.18}$$

The left square is obviously commutative and the right square is so by Proposition 5.4.11, since the fiber inclusion $Q \times Q \to S^N \times_G (Q \times Q)$ is transversal to $\mathbb{R}P^N \times \Delta_Q$. If f is homotopic to a smooth embedding, then $O_G^f = 0$ by Lemma 9.8.17. Also, (1) holds and, by the above and Diagram (9.8.14), one has

$$\rho \circ H_G^*(f \times f)(\mathrm{PD}_G(M)) = \rho \circ \mathrm{Gys}_G \circ (H_G^*\bar{\delta})^{-1}(b) \tag{9.8.19}$$

By Point (2) of Proposition 9.8.18, one has

$$\rho \circ H_G^*(f \times f)(\mathrm{PD}_G(M)) = H^*(f \times f) \circ \rho(\mathrm{PD}_G(M)) = H^*(f \times f)(\mathrm{PD}(\Delta_M)). \tag{9.8.20}$$

Let $\imath : \Delta_Q \to Q \times Q$ be the inclusion and $pr_2 : Q \times Q \to Q$ be the projection onto the first factor. Then $pr_1 \circ \imath \circ \bar{\delta} = \mathrm{id}_Q$. Hence, any $a \in H^*(Q)$ satisfies

$$a = H^*\bar{\delta} \circ H^*\imath \circ H^* pr_1(a) = H^*\bar{\delta} \circ H^*\imath(a \times \mathbf{1}). \tag{9.8.21}$$

Hence,

$$\begin{aligned} \rho \circ \mathrm{Gys}_G \circ (H_G^*\bar{\delta})^{-1}(b) &= \mathrm{Gys} \circ (H^*\bar{\delta})^{-1} \circ \rho\,(b) && \text{by commutativity of (9.8.18)} \\ &= \mathrm{Gys} \circ (H^*\bar{\delta})^{-1}(\bar{w}_{m-q}^f) && \text{by (9.8.17)} \\ &= \mathrm{Gys} \circ H^*\imath(\bar{w}_{m-q}^f \times \mathbf{1}) && \text{by (9.8.21)} \\ &= (\bar{w}_{m-q}^f \times \mathbf{1}) \smile \mathrm{PD}(\Delta_Q) && \text{by Lemma 5.4.8} \end{aligned} \tag{9.8.22}$$

Combining (9.8.19), (9.8.20) and (9.8.22) provides the proof of Point (2). □

Corollary 9.8.24 *Let Q be a closed manifold of dimension q.*

(1) *If Q may be immersed in $\mathbb{R}^m$, then $\bar{w}_k(TQ) = 0$ for $k > m - q$.*
(2) *If Q may be embedded in $\mathbb{R}^m$, then $\bar{w}_{m-q}(TQ) = 0$.*

Point (1) was already proven in Proposition 9.5.22.

Proof Let $f_0 : Q \to \mathbb{R}^m$ be a smooth map. Composing with the inclusion $\mathbb{R}^m \hookrightarrow S^m = \mathbb{R}^m \cup \{\infty\}$ gives a smooth map $f : Q \to S^m$, homotopic to a constant map. Then $\bar{w}^f = w(TQ)$ and $H^*(f \times f) = 0$. Corollary 9.8.24 thus follows from Proposition 9.8.23. □

Examples 9.8.25 1. Let Q be a closed non-orientable surface. Then, $\bar{w}_1(TQ) = w_1(TQ) \neq 0$. Therefore, Q cannot be embedded in $\mathbb{R}^3$. Note that M can be embedded in $\mathbb{R}^4$ by Whitney's theorem and immersed in $\mathbb{R}^3$ using Boy's surface.
2. Let Q be a closed 4-dimensional orientable manifold which is not spin (for instance $\mathbb{C}P^2$). Then, $\bar{w}_2(TQ) = w_2(TQ) \neq 0$. Therefore, Q cannot be embedded in $\mathbb{R}^6$. Note that $\mathbb{C}P^2$ embeds in $\mathbb{R}^7$. Indeed, $\mathbb{C}P^2$ is diffeomorphic to the space $\mathrm{Fl}_{\mathbb{C}}(1,2)$ of Hermitian (3×3)-matrices with characteristic polynomial equal to $x^2(x-1)$ (see (3) on p. 405). The vector space of Hermitian (3×3)-matrices with trace 1 is isomorphic to $\mathbb{R}^8$ and each radius intersects $\mathrm{Fl}_{\mathbb{C}}(1,2)$ at most once. This gives an embedding of $\mathbb{C}P^2$ in $S^7 = \mathbb{R}^7 \cup \{\infty\}$ and thus in $\mathbb{R}^7$.
3. The quaternionic projective plane $Q = \mathbb{H}P^2$ has Wu class $v_4(Q) \neq 0$. Hence, $\bar{w}_4(TQ) = w_4(TQ) \neq 0$. Therefore, $\mathbb{H}P^2$ cannot be embedded in $\mathbb{R}^{12}$. In the same way, the octonionic projective plane $\mathbb{O}P^2$ of dimension 16 cannot be embedded in $\mathbb{R}^{24}$. Improving the technique explained in the previous example produces embeddings $\mathbb{H}P^2 \hookrightarrow \mathbb{R}^{13}$ and $\mathbb{O}P^2 \hookrightarrow \mathbb{R}^{25}$ (see [135, Sect. 3]).

9.9 Thom's Theorems

This section is a survey of some results of Thom's important paper [191], which, amongst other things, was the foundation of cobordism theory. Some proofs are almost complete and others are just sketched.

9.9.1 Representing Homology Classes by Manifolds

Theorem 9.9.1 *Let X be a topological space and $\alpha \in H_k(X)$. Then, there exists a closed smooth manifold M of dimension k and a continuous map $f : M \to X$ such that $H_* f([M]) = \alpha$.*

This theorem is due to Thom [191, Theorem III.2]. The result is wrong for integral cohomology (see [191, Theorem III.9]). This section is devoted to the proof of Theorem 9.9.1. We start with some preliminaries.

Let ξ be a vector bundle of rank r over a paracompact space Y. Let $(D(\xi), S(\xi))$ be the pair of the disk and sphere bundles associated to ξ via an Euclidean structure. The *Thom space* $T(\xi)$ of ξ is defined by

$$T(\xi) = D(\xi)/S(\xi).$$

The homeomorphism class of $T(\xi)$ does not depend on the choice of the Euclidean structure (see p. 187). Also, $S(\xi)$ has a collar neighborhood in $D(\xi)$. Hence, by Lemma 3.1.39, the pair $(D(\xi), S(\xi))$ is well cofibrant. Using Proposition 3.1.45 together with the Thom isomorphism theorem provides the following isomorphisms

$$H^k(B) \xrightarrow{\approx} H^{k+r}(D(\xi), S(\xi)) \xrightarrow{\approx} H^{k+r}(T(\xi)).$$

We now specialize to $\xi = \zeta_{r,N}$, the tautological vector bundle over the Grassmannian $\mathrm{Gr}(r; \mathbb{R}^N)$ $(N \leq \infty)$. (The Thom space $T(\zeta_{r,\infty})$ is also called $MO(r)$ in the literature; it is the r-th space of the Thom spectrum MO). We get some information on the cohomology ring $H^*(T(\zeta_{r,N}))$ using the Gysin exact sequence of the sphere bundle $q : S(\zeta_{r,N}) \to \mathrm{Gr}(r; \mathbb{R}^N)$, whose Euler class is w_r.

$$\begin{aligned} H^{i-r}(\mathrm{Gr}(r; \mathbb{R}^N)) \xrightarrow{\smile w_r} H^i(\mathrm{Gr}(r; \mathbb{R}^N)) \xrightarrow{H^*q} H^i(S(\zeta_{r,N})) \\ \to H^{i-r+1}(\mathrm{Gr}(r; \mathbb{R}^N)) \xrightarrow{\smile w_r} \cdots \end{aligned} \tag{9.9.1}$$

together with the exact Sequence of Corollary 3.1.48 for the pair $(D(\zeta_{r,N}), S(\zeta_{r,N}))$ (using that $D(\zeta_{r,N} \cong \mathrm{Gr}(r; \mathbb{R}^N))$

$$\tilde H^i(T(\zeta_{r,N})) \to \tilde H^i(\mathrm{Gr}(r; \mathbb{R}^N)) \xrightarrow{\tilde H^*q} \tilde H^i(S(\zeta_{r,N})) \to \tilde H^{i+1}(T(\zeta_{r,N})). \tag{9.9.2}$$

For $N = \infty$, $- \smile w_r$ is injective by Theorem 9.5.8. Together with Lemma 9.5.7, Sequences (9.9.1) and (9.9.2) gives the following lemma.

Lemma 9.9.2 (1) *The* **GrA**-*morphism*

$$H^*p : H^*(T(\zeta_{r,\infty})) \to H^*(\mathrm{Gr}(r; \mathbb{R}^\infty))$$

is injective and its image in positive degrees is the ideal generated by w_r. In particular, $H^*p(U) = w_r$.

(2) *The* **GrA**-*morphism $H^i(T(\zeta_{r,\infty})) \to H^i(T(\zeta_{r,N}))$ generated by the inclusion is bijective for $N \geq N - r - 1$.*

(Note that the equation $H^*p(U) = w_r$ is coherent with (4.7.21)). Consider the diagram

$$\begin{array}{ccc} \mathrm{Gr}(r;\mathbb{R}^\infty) & \xleftarrow{g} & (\mathbb{R}P^\infty)^r \\ \downarrow p & \overset{f_{w_r}}{\searrow} & \downarrow f_{a\times\cdots\times a} \\ T(\zeta_{r,\infty}) & \xrightarrow{f_U} & \mathcal{K}_r \end{array} \tag{9.9.3}$$

in which the following notations are used. If X is a CW-complex and $y \in H^r(X)$, then $f_y : X \to \mathcal{K}_r = K(\mathbb{Z}_2, r)$ denotes a map representing y. Then $w_r \in H^r(\mathrm{Gr}(r;\mathbb{R}^\infty))$ is the r-th Stiefel-Whitney class, $U \in H^r(T\zeta_{r,\infty})$ is the Thom class and $0 \neq a \in H^1(\mathbb{R}P^\infty)$. The map g classifies the r-th product of the tautological line bundle over $\mathbb{R}P^\infty$ and $p : \mathrm{Gr}(r;\mathbb{R}^\infty) \cong D(\zeta_{r,\infty}) \to T(\zeta_{r,\infty})$ is the quotient map. Diagram (9.9.3) is homotopy commutative. The commutativity of the lower triangle comes from the already mentioned equation $H^*p(U) = e(\zeta_{r,\infty}) = w_r$. For the upper triangle (see e.g. the proof of Theorem 9.5.8).

Applying the cohomology functor to Diagram (9.9.3) provides the commutative diagram

$$\begin{array}{ccc} H^*(\mathrm{Gr}(r;\mathbb{R}^\infty)) & \xrightarrow{H^*g} & H^*((\mathbb{R}P^\infty)^r) \\ \uparrow H^*p & \overset{H^*f_{w_r}}{\nwarrow} & \uparrow H^*f_{a\times\cdots\times a} \\ H^*(T(\zeta_{r,\infty})) & \xleftarrow{H^*f_U} & H^*(\mathcal{K}_r) \end{array} . \tag{9.9.4}$$

By Corollary 8.5.9, $H^*f_{a\times\cdots\times a} : H^i(\mathcal{K}_r) \to H^i((\mathbb{R}P^\infty)^r)$ is injective for $i \leq 2r$. Therefore, $H^*f_U : H^i(\mathcal{K}_r) \to H^i(T(\zeta_{r,\infty}))$ is also injective for $i \leq 2r$. As indicated in Diagram (9.9.4), H^*p and H^*g are injective. The injectivity of H^*p was proven in Lemma 9.9.2 and that of H^*g was established in the proof of Theorem 9.5.8 or in Theorem 9.6.1 and its proof.

The map H^*f_U is of course not surjective. By Lemma 9.9.2, $\dim H^{r+j}(T\zeta_{r,\infty})$ is the number of partitions of j while, for $i < r$, $\dim H^{r+j}(\mathcal{K}_r)$ is equal, by Lemma 8.5.14, to the number of partitions of j into integers of the form $2^i - 1$. Let $\mathcal{D}(j)$ be the number of partitions of j into integers with none of them of the form $2^i - 1$. For each $\omega \in \mathcal{D}(j)$ with $j \leq r$, Thom constructs a class $X_\omega \in H^{r+j}(T\zeta_{r,\infty})$ represented by a map $f_\omega : T\zeta_{r,\infty} \to \mathcal{K}_j$. Together with f_U, this gives a map

$$F : T\zeta_{r,\infty} \to Y = \mathcal{K}_r \times \prod_{j=1}^{r} \mathcal{K}_j^{\sharp\mathcal{D}(j)} = \mathcal{K}_r \times \mathcal{K}_{r+2} \times \cdots . \tag{9.9.5}$$

Thom proves that H^*F is an isomorphism up to degree $2r$. Some analogous result is proved for the cohomology with coefficients in a field of characteristic $\neq 2$ and both $T\zeta_{r,\infty}$ and Y are simply connected. This enables Thom to prove the following result (see [191, pp.35–42]).

Lemma 9.9.3 *If $N \in \mathbb{N} \cup \{\infty\}$ is big enough, there exists a map ψ from the $2r$-skeleton of Y to $T(\zeta_{r,N})$ such that the restrictions of $\psi \circ F$ and $F \circ \psi$ to the $(2r-1)$-skeleta of Y and $T\zeta_{r,N}$ respectively are homotopic to the identity.*

As a corollary, we get the following result (see [191, Corollary II.12]).

Corollary 9.9.4 *If $N \in \mathbb{N} \cup \{\infty\}$ is big enough, there exists a map ψ from the $2r$-skeleton of $\mathcal{K}_r$ to $T\zeta_{r,N}$ such that $H^*\psi(U) = \iota$, the fundamental class of $\mathcal{K}_r$.*

We are now ready to prove Theorem 9.9.1.

Proof of Theorem 9.9.1 By Theorem 3.7.4, there is a simplicial complex K_X and a map $\phi : |K_X| \to X$ such that $H_*\phi$ is an isomorphism. By Sect. 3.6, The homology of X is isomorphic to the simplicial homology of K_X. By the definition of the simplicial homology, there is a finite simplicial subcomplex K of K_X, of dimension k, such that α is in the image of $H_k(|K|) \to H_k(X)$. Now, there is a PL-embedding $\psi : |K| \to \mathbb{R}^{2k+1}$ (see e.g. [179, Theorem 3.3.9]) and the theory of smooth regular neighborhoods [94] produces a smooth compact codimension 0 submanifold W of $\mathbb{R}^{2k+1}$ which is a regular neighborhood of $\psi(|K|)$ for some $\mathcal{C}^1$-triangulation of $\mathbb{R}^{2k+1}$. In particular, W retracts by deformation on $\psi(K)$. By general position, ψ is isotopic to an embedding ψ' such that $\psi'(|K|)$ avoids some regular neighborhood W' of $\psi(|K|)$ contained in the interior of W. The closure of $W' - W$ is homeomorphic to $M \times [0, 1]$, where $M = \operatorname{Bd} W$ (see [104, Corollary 2.16.2, p. 74]). We can thus construct a map $\psi'' : |K| \to M$ such that the composite map

$$|K| \xrightarrow{\psi''} M \hookrightarrow W \to |K|$$

is isotopic to $\mathrm{id}_{|K|}$. Hence α is in the image of $H_*(M) \to H_k(X)$. Therefore, it is enough to prove Theorem 9.9.1 when X is a closed manifold of dimension $2k$.

Let $a \in H^k(X)$ be the cohomology class which is Poincaré dual to α. As a smooth manifold, X admits a $\mathcal{C}^1$-triangulation by a simplicial complex of dimension $2k$. There exists thus a continuous map $f_a : X \to \mathcal{K}_k$ representing the class a and, by cellular approximation, one may suppose that $f_a(X)$ is contained in the $2k$-skeleton of $\mathcal{K}$. Let $\tilde{f} = \psi \circ f_a : X \to T = T(\zeta_{k,N})$, where ψ is a map as provided by Corollary 9.9.4 for N large enough. Then $f_U \circ \tilde{f}$ is homotopic to f_a. Note that T is a smooth manifold except at the point $[S(\zeta_{k,N})]$. Using standard techniques of differential topology (see [95, Sects. 2.2 and 3.2]), $\tilde{f}$ is homotopic to f_1 such that f_1 is a smooth map around $f_1^{-1}(\mathrm{Gr}(r;\mathbb{R}^N))$ which is transverse to $\mathrm{Gr}(r;\mathbb{R}^N)$. Then $M = f_1^{-1}(\mathrm{Gr}(r;\mathbb{R}^N))$ is a closed submanifold of codimension k in X with normal bundle $\nu = f_1^*\zeta_{k,N}$.

Let $a' \in H^k(X)$ be the Poincaré dual of the homology class generated by $[M]$. As in Sect. 5.4.1, we consider the Thom class $U(X, M)$ of ν as an element of $H^k(X, X - M)$ and, if $j : (X, \emptyset) \to (X, X - M)$ denotes the pair inclusion, one has

$$\begin{aligned} a' &= H^*j(U(X, M)) \quad \text{by Lemma 5.4.2} \\ &= H^*f_1(U) \end{aligned}$$

$$
\begin{aligned}
&= H^*(\psi \circ f_a)(U) \\
&= H^* f_a \circ H^* \psi(U) \\
&= H^* f_a(\iota) \\
&= a.
\end{aligned}
$$

As Poincaré duality is an isomorphism, this proves that $[M] = \alpha$.

Observe that, in the proof of Theorem 9.9.1, we have established the following result, due to Thom [191, Theorem II.1, p. 29].

Proposition 9.9.5 *Let $\alpha \in H_k(X)$, where X is a closed smooth manifold of dimension $k + q > 2k$. Let $a = \mathrm{PD}(\alpha) \in H^q(X)$ be the Poincaré dual of α. Then, the following statements are equivalent.*

(1) *There exists a closed submanifold M in X such that $[M]$ represents α.*
(2) *There exists a continuous map $F : X \to T(\zeta_{q,\infty})$ such that $H^* F(U) = a$.*

9.9.2 Cobordism and Stiefel-Whitney Numbers

Let M be a (smooth, possibly disconnected) manifold of dimension n. For a polynomial $P \in \mathbb{Z}_2[X_1, \ldots, X_n]$, we set $P_M = P(w_1(TM), \ldots, w_n(TM)) \in H^*(M)$. If M is closed, the number mod 2

$$\langle P_M, [M] \rangle \in \mathbb{Z}_2$$

is called the *Stiefel-Whitney number* of M associated to P. We use the convention that $\langle a, \alpha \rangle = 0$ if $a \in H^r(X)$ and $\alpha \in H^s(X)$ with $r \neq s$.

Two closed manifolds of the same dimension are called *cobordant* if their disjoint union is the boundary of a compact manifold. One fundamental result of Thom [191, Theorema IV.3 and IV.10] is the following theorem, generalizing Corollary 5.3.10.

Theorem 9.9.6 *Two closed manifolds of the same dimension are cobordant if and only if their Stiefel-Whitney numbers coincide.*

Example 9.9.7 Let M be a closed 3-dimensional manifold. Its Wu class is $v(M) = \mathbf{1} + v_1(M) = 1 + w_1(TM)$. By Wu's formula, $w(TM) = \mathrm{Sq}(v(M)) = \mathbf{1} + w_1(TM) + w_1(TM)^2$, so $w_2(TM) = w_1(TM)^2$. The only possible non-zero Stiefel-Whitney number is then $\langle w_1(TM)^3, [M] \rangle$. But

$$
\begin{aligned}
w_1(TM)^3 &= w_1(TM)^2\, v_1(M) && \text{since } w_1(TM) = v_1(M) \\
&= \mathrm{Sq}^1(w_1(TM)^2) && \text{by definition of } v_1, \text{ since } \dim M = 3 \\
&= 0 && \text{by the Cartan formula.}
\end{aligned}
$$

Therefore, M is the boundary of a compact manifold. Note that, if M is orientable, the vanishing of its Stiefel-Whitney numbers follows from Proposition 9.8.13.

Example 9.9.8 The complex projective space $\mathbb{C}P^2$ and the manifold $\mathbb{R}P^2 \times \mathbb{R}P^2$ have the same Stiefel-Whtney numbers by Proposition 9.8.10 and Remark 9.8.11. Therefore, they are cobordant.

Example 9.9.9 Let M be a closed orientable 4-dimensional manifold. Then, $w_1(TM) = 0$ and $w_4(TM) = w_2(TM)^2$ (see Corollary 9.8.8). Its only possible non-vanishing Stiefel-Whitney number is thus

$$\langle w_4(TM), [M]\rangle = \langle e(TM), [M]\rangle = \chi(M) \mod 2$$

(using Corollary 5.4.16). Therefore, M a boundary of a (possibly non-orientable) compact 5-manifold if and only if its Euler characteristic is even.

Proof of Theorem 9.9.6 Let M_1 and M_2 be two closed manifolds of the same dimension and let $M = M_1 \dot{\cup} M_2$. For any $P \in \mathbb{Z}_2[X_1, \dots, X_n]$, one has

$$\langle P_M, [M]\rangle = \langle P_{M_1}, [M_1]\rangle + \langle P_{M_2}, [M_2]\rangle.$$

Hence, Theorem 9.9.6 is equivalent to the following statement: *a closed manifold M bounds if and only if its Stiefel-Whitney numbers vanishes.*

Suppose that $M = \mathrm{Bd}\, W$ for some compact manifold W. Then $TM \oplus \eta \approx TW_{|M}$ where η is the trivial bundle of rank 1 over M. If $j : M \to W$ denotes the inclusion, one has

$$\langle P_M, [M]\rangle = \langle H^*j(P_W), [M]\rangle = \langle P_W, H_*j([M])\rangle = 0,$$

since $H_*j([M]) = 0$ (see Eq. (5.3.6) and the end of the proof of Theorem 5.3.7).

For the converse, we shall prove that if a closed manifold M of dimension n does not bound, then at least one of its Stiefel-Whitney numbers is not zero. Let us embed M into $\mathbb{R}^{n+r}$ for r large, with normal bundle ν. Let $f : M \to \mathrm{Gr}(r; \mathbb{R}^\infty)$ be a map such that $\nu \approx f^*\zeta_{r,\infty}$. The map f induces a map $Tf : T\nu \to T\zeta_{r,\infty}$. A closed tubular neighbourhood N of M is diffeomorphic to $D(\nu)$. We consider $\mathbb{R}^{n+r} \subset S^{n+r}$. The projection $N \approx D(\nu) \to T(\nu)$ extends to a continuous map $\pi : S^{n+r} \to T\nu$ by sending the complement of N onto the point $[S(\nu)]$. This gives a map $\hat{f} = Tf \circ \pi : S^{n+r} \to T\zeta_{r,\infty}$ (called the *Pontryagin-Thom construction*). By an argument based on transversality, one can prove that, for r large enough, M bounds if and only if $\hat{f}$ is homotopic to a constant map [191, Lemma IV.7 and its proof].

Let us compose $\hat{f}$ with the map $F : T\zeta_{r,\infty} \to Y$ of (9.9.5). By Lemma 9.9.3, $\hat{f}$ is not homotopic to a constant map if and only if $F \circ \hat{f}$ is not homotopic to a constant map. As Y is a product of Eilenberg-MacLane spaces, $F \circ \hat{f}$ is not homotopic to a constant map if and only if $H^*(F \circ \hat{f}) \neq 0$. The latter implies that $H^*Tf : H^{n+r}(T\zeta_{r,\infty}) \to H^{n+r}(T\nu)$ does not vanish. Using the Thom isomorphisms, this implies that $H^*f : H^n(\mathrm{Gr}(r; \mathbb{R}^\infty)) \to H^n(M)$ does not vanish. This implies that there is a polynomial $\bar{P}$ in the Stiefel-Whitney classes of ν such that $\langle \bar{P}, [M]\rangle \neq 0$.

These classes $\bar{w}_i$ are the normal Stiefel-Whitney classes of M and, by Lemma 9.5.19, the Stiefel-Whitney classes $w_j = w_j(TM)$ have polynomial expressions in the $\bar{w}_i$. Therefore, there is a polynomial P in w_j such that $\langle P, [M]\rangle \neq 0$, producing a non-zero Stiefel-Whitney number for M.

Corollary 9.9.10 *Let M and M' be two closed smooth manifolds of the same dimension. Suppose that there exists a map $f : M' \to M$ such that $H_* f$ is an isomorphism. Then, M and M' are cobordant.*

As a consequence of Corollary 9.9.10, a $\mathbb{Z}_2$-homology sphere bounds.

Proof As $H_* f$ is an isomorphism, $\pi_0 f$ is a bijection and then $H_* f([M']) = [M]$. Let $P \in \mathbb{Z}_2[X_1, \dots, X_n]$. By Corollary 9.8.3, $H^* f(w(TM)) = w(TM')$ and then $H^* f(P_{M'}) = P_M$. Therefore,

$$\begin{aligned} \langle P_{M'}, [M']\rangle &= \langle P_{M'}, H_* f([M])\rangle \\ &= \langle H^* f(P_{M'}), [M]\rangle \\ &= \langle P_M, [M]\rangle. \end{aligned}$$

Hence, M and M' have the same Stiefel-Whitney numbers. By Theorem 9.9.6, they are cobordant. □

For closed manifolds of dimension n, "being cobordant" is an equivalence relation. The set of equivalence classes (*cobordism classes*) is denoted by $\mathfrak{N}_n$. The disjoint union endows $\mathfrak{N}_n$ with an abelian group structure, actually a $\mathbb{Z}_2$-vector space structure since $2M = M \dot{\cup} M$ is diffeomorphic to the boundary of $M \times [0, 1]$. The Cartesian product of manifolds makes $\mathfrak{N}_* = \bigoplus_n \mathfrak{N}_n$ a $\mathbb{Z}_2$-algebra, called the *cobordism ring*. A development of the results of this section and the previous one enabled Thom to compute the cobordism ring $\mathfrak{N}_*$ [191, Sect. IV]; the results are summed up in the following theorem.

Theorem 9.9.11 (1) *$\mathfrak{N}_n$ is isomorphic to* $\lim_{\to k} \pi_{n+k}(T(\zeta_{k,\infty}))$.

(2) $\dim \mathfrak{N}_n$ *is the number of partitions of n into integers with none of them of the form* $2^i - 1$.

(3) $\mathfrak{N}_*$ *is* **GrA**-*isomorphic to a polynomial algebra* $\mathbb{Z}_2[X_2, X_4, X_5, X_6, X_8, X_9, \dots]$ *with one generator X_k for each integer k not of the form* $2^i - 1$.

A representative for X_{2k} is given by the cobordism class of $\mathbb{R}P^{2k}$ [191, p. 81]. Odd dimensional generator of dimension $\neq 2^i - 1$ were first constructed by Dold [41]. For details and proofs (see [191] or [186, Chap. VI]).

For example, $\mathfrak{N}_3 = 0$, confirming Example 9.9.7. Another simple consequence of Theorem 9.9.11 is the following corollary.

Corollary 9.9.12 *Let M and N be closed manifolds which are not boundaries. Then $M \times N$ is not a boundary.*

9.10 Exercises for Chapter 9

9.1. Let ξ be a vector bundle. Prove that $\xi \oplus \xi$ is orientable. If ξ is orientable, prove that $\xi \oplus \xi$ admits a spin structure.

9.2. Let $p : \tilde{X} \to X$ be a smooth covering of a smooth manifold X, with an odd number of sheets. Show that $\tilde{X}$ is orientable if and only if X is orientable. Show that $\tilde{X}$ admits a spin structure if and only if X does.

9.3. Let $p : \tilde{M} \to M$ be a 2-fold covering of a smooth connected manifold M, with characteristic class $\omega \in H^1(M)$. Suppose that $\tilde{M}$ is orientable and that M is not orientable. Prove that $\omega = w_1(TM)$.

9.4. Let M_1 and M_2 be two closed connected manifold of the same dimension and let M be one of their connected sums (see p. 135). What is the total Stiefel-Whitney class $w(T(M))$?

9.5. Prove that $T\mathbb{R}P^4$ and $T\mathbb{C}P^2$ are indecomposable as a Whitney sum of vector bundles of smaller ranks.

9.6. Let ξ be a vector bundle over a space X, with $w(\xi) \neq 1$. Show that the smallest integer $i > 0$ such that $w_i(\xi) \neq 0$ is a power of 2.

9.7. List the critical points with their index for the weighted trace on $\mathrm{Gr}(2; \mathbb{R}^4)$ and $\mathrm{Gr}(2; \mathbb{R}^6)$. Using also Example 9.5.3, verify the statement of Lemma 9.5.7 and the second formula of Corollary 9.5.15.

9.8. Same exercise as the previous one for $\mathrm{Fl}(1, 1, 1)$.

9.9. Let $j : BO(n) \to BU(n)$ denote the inclusion. Prove that $H^* j(c_i) = w_i^2$.

9.10. Write the details for Remark 9.5.16.

9.11. Like in Example 9.5.21, find an additive basis of $H^*(\mathrm{Gr}(2; \mathbb{R}^5))$ in terms of products of Stiefel-Whitney classes. Express each of these elements in terms of Schubert symbols.

9.12. Let M be an orientable smooth closed manifold of dimension 6 or 10. Prove that $\chi(M)$ is even.

9.13. Let $f : P \to Q$ be a continuous map between n-dimensional connected closed smooth manifolds. Suppose that one of the following conditions is satisfied:

(a) P is orientable while Q is non-orientable
(b) P is spin while Q is non-spin ($w_2(TQ) \neq 0$).

Then $H^n f : H^n(Q) \to H^n(P)$ is trivial.

9.14. Let $f : S^{n+k} \to S^k$ be a smooth map. Let $x \in S^k$ be a regular value. Show that the closed manifold $f^{-1}(\{x\})$ is the boundary of a (possibly non-orientable) compact manifold.

9.15. Prove that $\mathbb{R}P^2 \times \mathbb{R}P^2$ and $\mathbb{R}P^4$ are not cobordant.

9.16. Let M be a closed n-dimensional manifold whose cohomology ring is isomorphic to that of $\mathbb{R}P^n$. Prove that M and $\mathbb{R}P^n$ are cobordant.

Chapter 10
Miscellaneous Applications and Developments

This chapter, contains various applications and developments of the techniques of mod 2 (co)homology. Most of them are somewhat original.

10.1 Actions with Scattered or Discrete Fixed Point Sets

Let X be a finite dimensional G-complex ($G = \{\mathrm{id}, \tau\}$) with $b(X) < \infty$. By Smith theory (Proposition 7.3.7), we know that $b(X^G) \le b(X)$, which implies that $\sharp(\pi_0(X^G)) \le b(X)$. Inspired by the work of Puppe [165], we study in this section the extremal case $\sharp(\pi_0(X^G)) = b(X)$ (*scattered fixed point set*). Analogous results for S^1-actions are presented at the end of this section.

Proposition 10.1.1 *Let X be a finite dimensional G-complex with $b(X) < \infty$. Suppose that $\sharp(\pi_0(X^G)) = b(X)$. Let $a \in H^k_G(X)$. Then $\mathrm{Sq}^i(a) = \binom{k}{i} u^i a$.*

Proof By Proposition 7.3.7, $H^{>0}(X^G) = 0$ and X is equivariantly formal. Therefore, X^G has the cohomology of $b(X)$ points and $(X^G)_G \approx BG \times X^G$ is homotopy equivalent to a disjoint union of $b(X)$ copies of $\mathbb{R}P^\infty$. By (8.2.2), any class $b \in H^k_G(X^G)$ satisfies $\mathrm{Sq}^i(b) = \binom{k}{i} u^i b$. As the restriction homomorphism $H^*_G(X) \to H^*_G(X^G)$ is injective by Proposition 7.3.9, this proves the assertion.

As seen in the above proof, the G-space X of Proposition 10.1.1 is equivariantly formal. Thus, ρ: $H^*_G(X) \to H^*(X)$ is surjective. As $\ker \rho = u \cdot H^*_G(X)$ by (7.1.7), Proposition 10.1.1 has the following corollary (compare [165, Corollary 1]).

Corollary 10.1.2 *Let X be a finite dimensional G-complex with $b(X) < \infty$. Suppose that $\sharp(\pi_0(X^G)) = b(X)$. Then, any $a \in H^*(X)$ satisfies $\mathrm{Sq}(a) = a$ (i.e. $\mathrm{Sq}^i = 0$ for $i > 0$). In particular, $a \smile a = 0$ if $a \in H^{>0}(X)$.*

Let us restrict the above results to the case where X is a smooth closed G-manifold. Then, X^G is a union of closed manifolds (see, e.g. [12, Corollary 2.2.2]). We have seen

J.-C. Hausmann, *Mod Two Homology and Cohomology*, Universitext,
DOI 10.1007/978-3-319-09354-3_10

in the proof of Proposition 10.1.1 that each component of X^G has the cohomology of a point. Hence X^G is a discrete set of $b(X)$ points (the smooth involution τ is called an *m-involution* in [165]). Examples include linear spheres S_0^n; if X_1 and X_2 are such G-manifolds, so is $X_1 \times X_2$ with the diagonal involution; if $\dim X_1 = \dim X_2$, the G-equivariant connected sum $X_1 \sharp X_2$ around fixed points carries an m-involution. Thus, an orientable surface carries an m-involution. Also, if X admits a G-invariant Morse function, then τ is an m-involution by Theorem 7.6.6. Corollary 10.1.2 has the following consequence for the Stiefel-Whitney class $w(TX)$ of a manifold X admitting an m-involution.

Corollary 10.1.3 *Let X be a smooth closed G-manifold such that $\sharp(\pi_0(X^G)) = b(X)$. Then $w(TX) = \mathbf{1}$.*

In consequence, a closed manifold X carrying an m-involution is orientable and admits a spin structure. Also, X is the boundary of a (possibly non-orientable) compact manifold by Thom's Theorem 9.9.6.

Proof By Corollary 10.1.2, $\mathrm{Sq}^i = 0$ for $i > 0$. Hence, the Wu class $V(X)$ is equal to $\mathbf{1}$. Therefore, using Wu's formula 9.8.2, $w(TX) = \mathrm{Sq}(V(X)) = \mathbf{1}$. □

We now generalize to smooth closed G-manifolds with X^G discrete (without asking that $\sharp X_G = b(X)$). This will lead us toward the link between closed G-manifolds with discrete fixed point set and coding theory; such a link was initiated in [165] and further developed in [121]. We start with the following lemma.

Lemma 10.1.4 *Let X^{2k+1} be a smooth closed G-manifold such that X^G is discrete. Then, $\sharp X^G$ is even.*

Proof Let $r = \sharp X^G$. Let $\tilde{W} = X - \mathrm{int}\,\mathcal{D}$ where $\mathcal{D}$ is a closed G-invariant tubular neighborhood of X^G. Then $\tilde{W}$ is a compact free G-manifold with boundary $\tilde{V}$. The orbit space $W = \tilde{W}/G$ is a compact manifold whose boundary $V = \tilde{V}/G$ is a disjoint union of r copies of $\mathbb{R}P^{2k}$. By Proposition 5.3.9, the image B of $H^k(W) \to H^k(V)$ satisfies

$$2 \dim B = \dim H^k(\mathrm{Bd}\, W) = r$$

which shows that r is even. □

Remark 10.1.5 If X is a finite dimensional G-CW-complex with $b(X) < \infty$, it is known that $b(X) \equiv b(X_G) \mod 2$ [9, Corollary 1.3.8]. If X is an odd dimensional closed manifold, then $b(X)$ is even by Poincaré duality. This provides another proof of Lemma 10.1.4

We use the notation of the proof of Lemma 10.1.4. Let $\langle\langle -, - \rangle\rangle$ be the bilinear form on $H^k(V)$ given by $\langle\langle a, b \rangle\rangle = \langle a \smile b, [V] \rangle$. By Proposition 5.3.9 and its proof, one has $\langle\langle B, B \rangle\rangle = 0$ and $r = 2 \dim B$. Labeling the r points of X^G produces an isomorphism $H^k(V) \approx \mathbb{Z}_2^r$ intertwining $\langle\langle -, - \rangle\rangle$ with the standard bilinear form on $\mathbb{Z}_2^r$. Hence, in terms of coding theory (see, e.g. [46]), B is a *binary self-dual linear code* on $\mathbb{Z}_2^r$. Choosing another labeling for the points of X^G changes B by an isometry of $\mathbb{Z}_2^r$ obtained by coordinate permutations. The class of B modulo these isometries thus provides an invariant of the G-action.

The self-dual code B has other descriptions. For instance, the diagram of inclusions

$$\begin{array}{ccc} V & \xrightarrow{i} & \mathcal{D} \\ \downarrow{\scriptstyle j} & & \downarrow \\ W & \longrightarrow & X \end{array}$$

gives rise to the commutative diagram

$$\begin{array}{ccccc} H_G^k(X) & \longrightarrow & H_G^k(\tilde{W}) \oplus H_G^k(\mathcal{D}) & \longrightarrow & H_G^k(\tilde{V}) \\ \downarrow{\scriptstyle =} & & \downarrow{\scriptstyle \approx} & & \downarrow{\scriptstyle \approx} \\ H_G^k(X) & \longrightarrow & H^k(W) \oplus H_G^k(X^G) & \longrightarrow & H^k(V) \end{array} \tag{10.1.1}$$

whose row are the Mayer-Vietoris exact sequences. Lemma 7.1.4 guarantees that the vertical maps are isomorphisms and, together with Corollary 3.8.4, implies that the map $V \simeq \tilde{V}_G \to (X^G)_G$ is, on each component, homotopy equivalent to the inclusion $\mathbb{R}P^{2k} \hookrightarrow \mathbb{R}P^\infty$. Therefore, the homomorphism $H_G^k(X^G) \to H^k(V)$ is an isomorphism. Hence, diagram-chasing in (10.1.1) shows that $B = \text{image}(H^k(W) \to H^k(V))$ coincides with

$$\text{image}(H_G^k(X) \to H_G^k(X^G))$$

(pushed into $H^k(V)$). For other descriptions of B, see [165, Sect. 2]. The following theorem is proved in [121, Theorem 3].

Theorem 10.1.6 *Every binary self-dual linear code may be obtained from a closed smooth* 3*-dimensional G-manifold X with scattered fixed point set.*

As in Sect. 7.3, the above results have analogues for S^1-actions. The proofs of Proposition 10.1.7 and Corollary 10.1.8 below are the same as for Proposition 10.1.1 and Corollary 10.1.2, replacing Propositions 7.3.7 and 7.3.9 by Propositions 7.3.12 and 7.3.14, etc. Recall that $H_{S^1}^*(pt) \approx \mathbb{Z}_2[v]$ with v of degree 2.

Proposition 10.1.7 *Let X be a finite dimensional* S^1*-complex with* $b(X) < \infty$ *and* $X^{S^1} = X^{S^0}$*. Suppose that* $\sharp(\pi_0(X^{S^1})) = b(X)$*. Then* $H_{S^1}^{odd}(X) = 0$ *and, if* $a \in H_{S^1}^{2k}(X)$*, then* $\mathrm{Sq}^{2i}(a) = \binom{k}{i} v^i a$*.*

Corollary 10.1.8 *Let X be a finite dimensional* S^1*-complex with* $b(X) < \infty$ *and* $X^{S^1} = X^{S^0}$*. Suppose that* $\sharp(\pi_0(X^{S^1})) = b(X)$*. Then,* $H^{odd}(X) = 0$ *and any* $a \in H^*(X)$ *satisfies* $\mathrm{Sq}(a) = a$*.*

Analogously to Corollary 10.1.3, one has the following result, with the same proof.

Corollary 10.1.9 *Let X be a smooth closed S^1-manifold such that $X^{S^1} = X^{S^0}$ and $\sharp(\pi_0(X^{S^1})) = b(X)$. Then, $H^{odd}(X) = 0$ and $w(TX) = \mathbf{1}$.*

In particular, the manifold X of Corollary 10.1.9 is even-dimensional. Note that this is necessary for an S^1-action admitting an isolated fixed point (the action, restricted to an invariant sphere around the fixed point has discrete stabilizers, so the sphere is odd-dimensional). The analogue of Lemma 10.1.4 is Lemma 10.1.10 below. To simplify, we restrict ourselves to semi-free actions (A Γ-action is called *semi-free* if the stabilizer of any point is either $\{\text{id}\}$ or Γ). Incidentally, the hypothesis $X^{S^1} = X^{S^0}$ is not required.

Lemma 10.1.10 *Let X be a smooth closed S^1-manifold such that X^{S^1} is discrete. Suppose that the action is semi-free. Then, $\sharp X^{S^1}$ is even.*

Proof As seen above, X is even-dimensional, say $\dim X = 2k+2$. Let $r = \sharp X^{S^1}$. Let $\tilde{W} = X - \text{int}\,\mathcal{D}$ where $\mathcal{D}$ is a closed S^1-invariant tubular neighborhood of X^{S^1}. Then $\tilde{W}$ is a compact free S^1-manifold with boundary $\tilde{V}$. The orbit space $W = \tilde{W}/G$ is then a compact manifold of dimension $2k+1$ whose boundary $V = \tilde{V}/G$ is a disjoint union of r copies of $\mathbb{C}P^k$. By Proposition 5.3.9, the image B of $H^k(W) \to H^k(V)$ satisfies

$$2\dim B = \dim H^k(\text{Bd}\, W) = r$$

which shows that r is even. □

As for the case of an involution, Lemma 10.1.10 permits us to associate, to a closed smooth semi-free S^1-manifold with X^{S^1} discrete, the self-dual linear code $B \subset H^k(V) \approx \mathbb{Z}_2^r$. One can also see B as the image of $H^k_{S^1}(X)$ in $H^k_{S^1}(X^{S^1})$.

10.2 Conjugation Spaces

Introduced in [87], conjugation spaces are equivariantly formal G-spaces ($G = \{\text{id}, \tau\}$) quite different from those with scattered fixed point sets studied in Sect. 10.1. Here, the cohomology ring of the fixed point set most resembles that of the total space. This similarity should be given by a "cohomology frame", a notion which we now define. We use the notations of Sect. 7.1 for a G-space X, for example the forgetful homomorphism $\rho : H^*_G(X) \to H^*(X)$ and the $\mathbf{GrA}[u]$-morphism r: $H^*_G(X) \to H^*_G(X^G) \approx H^*(X^G)[u]$ induced by the inclusion $X^G \subset X$.

Let (X, Y) be a G-pair. A *cohomology frame* or an H^*-frame for (X, Y) is a pair (κ, σ), where

(a) $\kappa\colon H^{2*}(X, Y) \to H^*(X^G, Y^G)$ is an additive isomorphism dividing the degrees in half; and
(b) $\sigma\colon H^{2*}(X, Y) \to H^{2*}_G(X, Y)$ is an additive section of the natural homomorphism $\rho\colon H^*_G(X, Y) \to H^*(X, Y)$,

satisfying, in $H^*(X^G) \approx H^*(X^G)[u]$, the *conjugation equation*

$$r\circ\sigma(a) = \kappa(a)u^m + lt_m \tag{10.2.1}$$

for all $a \in H^{2m}(X, Y)$ and all $m \in \mathbb{N}$; in (10.2.1), lt_m denotes any element in $H^*(X, Y)[u]$ which is of degree less than n in the variable u. An involution admitting an H^*-frame is called a *conjugation*. An even cohomology pair (i.e. $H^{odd}(X, Y) = 0$) together with a conjugation is called a *conjugation pair*. A G-space X is a *conjugation space* if the pair $(X, \emptyset)$ is a conjugation pair. The existence of the section σ implies that a conjugation space is equivariantly formal. Note that there are examples of G-spaces which admit pairs (κ, σ) satisfying (a) and (b) above but none of them satisfying the conjugation equation (see [63, Example 1]). For simplicity, we shall mostly restrict this survey to conjugation spaces; the corresponding statements for conjugation pairs may be found in [87].

Any space X such that $H^*(X) = H^*(pt) = H^*(X^G)$ is a conjugation space. For instance, a finite dimensional G-CW-complex X satisfying $H^*(X) = H^*(pt)$ and $X^G \neq \emptyset$ is a conjugation space by Corollary 7.3.8. Another easy example is the G-sphere S^{2m}_m of Example 7.1.14. Indeed, one has the following lemma.

Lemma 10.2.1 *Let X be a finite dimensional G-CW-complex. Suppose that $H^*(X) \approx H^*(S^{2n})$ and $H^n(X^G) \neq 0$. Then X is a conjugation space.*

Proof By Corollary 7.3.8, $H^*(X^G) \approx H^*(S^n)$. By Proposition 7.3.7, X is equivariantly formal and, by Proposition 7.3.9, $r\colon H^*_G(X) \to H^*_G(X^G)$ is injective. The proof of the existence of an H^*-frame then proceeds as in the proof of Corollary 7.1.17. □

An other important example is the complex projective space.

Example 10.2.2 Let $a \in H^2(\mathbb{C}P^m)$ and $b \in H^1(\mathbb{R}P^m)$ $(m \leq \infty)$. Then, the section $\sigma\colon H^*(\mathbb{C}P^m) \to H^*_G(\mathbb{C}P^m)$ of Proposition 7.1.18, together with the isomorphism $\kappa\colon H^{2*}(\mathbb{C}P^m) \to H^*(\mathbb{R}P^m)$ sending a to b makes an H^*-frame for the complex conjugation on $\mathbb{C}P^m$. By Proposition 7.1.18, the conjugation equation takes the form

$$r\circ\sigma(a^k) = (\kappa(a)u + b^2)^k = \kappa(a^k)u^k + ltk\,. \tag{10.2.2}$$

The same treatment may be done for $\mathbb{H}P^m$ or $\mathbb{O}P^2$ (see Remark 7.1.21).

These examples are actually *spherical conjugation complexes*, i.e. G-spaces obtained from the empty set by countably many successive adjunctions of collections of conjugation cells. A *conjugation cell* (of dimension $2k$) is a G-space which is G-homeomorphic to the cone over S^{2k-1}_{k-1}, i.e. to the closed disk of radius 1 in $\mathbb{R}^{2k}$, equipped with a linear involution with exactly k eigenvalues equal to -1. At each step, the collection of conjugation cells consists of cells of the same dimension but, as in [74], the adjective "spherical" is a warning that these dimensions do not need to be increasing. For less standard examples of spherical conjugation complexes, see [87, 5.3.3, p. 944].

It is proven in [87, Proposition 5.2] that *a spherical conjugation complex is a conjugation space*. For example, complex flag manifolds (with τ being the complex conjugation) are conjugation spaces because the Schubert cells (see Sect. 9.5.3) are conjugation cells. This example generalizes to co-adjoint orbits of compact Lie groups for the Chevalley involution (see [87, Sect. 8.3]) and more examples coming from Hamiltonian geometry (see [87, Sects. 8.2 and 8.4] and [86]).

Other examples may be obtained from the previous ones since the class of conjugation spaces is closed under many construction, such as

- direct products, with the diagonal G-action, when one of the factor is of finite cohomology type (see [87, Proposition 4.5]).
- inductive limits (see [87, Proposition 4.6]).
- if (X, Y, Z) is a G-triple so that (X, Y) and (Y, Z) are conjugation pairs, then (X, Z) is a conjugation pair. A direct proof using H^*-frames is given in [87, Proposition 4.1]; a shorter proof using the conjugation criterion of [158, Theorem 2.3] is provided in [157, Proposition 2.2.1].
- if $F \to E \to B$ be G-equivariant bundle (with a compact Lie group as structure group) such that F is a conjugation space and B is a spherical conjugation complex, then E is a conjugation space (see [87, Proposition 5.3]).

We now show the naturality of H^*-frames under G-equivariant maps, as proven in [87, Proposition 3.11]. Let X and Y be two conjugation spaces, with H^*-frames (κ_X, σ_X) and (κ_Y, σ_Y). Let $f\colon Y \to X$ be a G-equivariant map. We denote by $f^G\colon Y^G \to X^G$ the restriction of f to Y^G.

Proposition 10.2.3 (Naturality of H^*-frames) *The equations $H^*_G f \circ \sigma_X = \sigma_Y \circ H^* f$ and $H^* f^G \circ \kappa_X = \kappa_Y \circ H^* f$ hold true.*

Proof Let $a \in H^{2k}(X)$. As σ_X and σ_Y are sections of $\rho_X\colon H^*_G(X) \to H^*(X)$ and $\rho_Y\colon H^*_G(Y) \to H^*(Y)$ respectively, one has

$$\rho_Y \circ H^*_G f \circ \sigma_X(a) = H^* f \circ \rho_X \circ \sigma_X(a) = H^* f(a) = \rho_Y \circ \sigma_Y \circ H^* f(a)\,. \tag{10.2.3}$$

This implies that

$$H^*_G f \circ \sigma_X(a) \equiv \sigma_Y \circ H^* f(a) \mod \ker\left(\rho^{2k}_Y\colon H^{2k}_G(Y) \to H^{2k}(Y)\right).$$

The section σ produces an isomorphism $H^*_G(X) \approx H^*(X)[u]$ and ker ρ_Y is the ideal generated by u (see Remark 7.1.7). As $H^{odd}(Y) = 0$, we deduce that there exists $d_i \in H^i(Y)$ such that

$$H^*_G f \circ \sigma_X(a) = \sigma_Y \circ H^* f(a) + \sigma_Y(d_{2k-2})\, u^2 + \cdots + \sigma_Y(d_0)\, u^{2k}\,. \tag{10.2.4}$$

Let us apply r_Y to both sides of (10.2.4). For the left hand side, we get

$$\begin{aligned} r_Y \circ H^*_G f \circ \sigma_X(a) &= H^*_G f^G \circ r_X \circ \sigma_X(a) \\ &= H^*_G f^G(\kappa_X(a)\, u^k + lt_k) \quad \text{by the conjugation equation} \\ &= H^*_G f^G(\kappa_X(a))\, u^k + lt_k \quad H^*_G f^G \text{ being a } \mathbf{GrA}[u]\text{-morphism.} \end{aligned} \tag{10.2.5}$$

But, using the right hand side of (10.2.4), we get

$$r_Y \circ H^*_G f \circ \sigma_X(a) = \kappa_Y(d_0) u^{2k} + lt_{2k}. \tag{10.2.6}$$

Comparing (10.2.5) with (10.2.6) and using that κ_Y is injective implies that $d_0 = 0$. Then, (10.2.6) may be replaced by

$$r_Y \circ H^*_G f \circ \sigma_X(a) = \kappa_Y(d_2)\, u^{2k-2} + lt_{2k-2}. \tag{10.2.7}$$

Again, the comparison with (10.2.5) implies that $d_2 = 0$. This process may be continued, eventually giving that $H^*_G f \circ \sigma_X(a) = \sigma_Y \circ H^* f(a)$. Applying r_Y to the right hand member of this equation gives

$$r_Y \circ \sigma_Y \circ H^* f(a) = \kappa_Y \circ H^* f(a)\, u^k + lt_k \tag{10.2.8}$$

by the conjugation equation. Comparing the leading terms of (10.2.8) and (10.2.5) gives that $H^* f^G \circ \kappa_X(a) = \kappa_Y \circ H^* f(a)$. □

Applying Proposition 10.2.3 to $X = Y$ and $f = \mathrm{id}$, we get the following corollary.

Corollary 10.2.4 (Uniqueness of H^*-frames) *Let (κ, σ) and (κ', σ') be two H^*-* frames *for the conjugation space X. Then $(\kappa, \sigma) = (\kappa', \sigma')$.*

We can thus speak about *the* H^*-frame of a conjugation space.

Proposition 10.2.5 *Let (κ, σ) be the H^*-*frame *of a conjugation space X. Then κ and σ are multiplicative.*

Proof Let $a \in H^{2m}(X)$ and $b \in H^{2n}(X)$ One has $a \smile b = \rho \circ \sigma(a \smile b)$ and $\rho(\sigma(a) \smile \sigma(b)) = \rho(\sigma(a)) \smile \rho(\sigma(b)) = a \smile b$. Hence, $\sigma(a) \smile \sigma(b)$ is congruent to $\sigma(a \smile b)$ modulo ker ρ. The same proof as for Proposition 10.2.3 then proves the proposition (details may be found in [87, Theorem 3.3]). □

Much more difficult to prove, the following result was established in [63, Theorem 1.3].

Proposition 10.2.6 *Let* (κ, σ) *be the* H^*-frame *of a conjugation space X. Then* $\mathrm{Sq}^i \circ \kappa = \kappa \circ \mathrm{Sq}^{2i}$ *for all integer i.*

Remark 10.2.7 It is not true in general that $\sigma \circ \mathrm{Sq} = \mathrm{Sq} \circ \sigma$. For example, consider the conjugation space $\mathbb{C}P^m$ for $1 \leq m \leq \infty$, with the notations of Example 10.2.2. Of course, $\mathrm{Sq}^1(a) = 0$ and then $\sigma \circ \mathrm{Sq}^1(a) = 0$. On the other hand,

$$\begin{aligned} r \circ \mathrm{Sq}^1 \circ \sigma(a) &= \mathrm{Sq}^1 \circ r \circ \sigma(a) \\ &= \mathrm{Sq}^1\big((bu + b^2)\big) && \text{by the conjugation Eq. (10.2.2)} \\ &= \mathrm{Sq}^1(b)u + b\mathrm{Sq}^1(u) && \text{by the Cartan formula} \\ &= b^2 u + bu^2 \\ &= r(\sigma(a)\, u)\,. \end{aligned}$$

Since r is injective (see Lemma 10.2.8 below), this proves that

$$\mathrm{Sq}^1(\sigma(a)) = \sigma(a)\, u\,.$$

The following lemma is recopied with its short proof from [87, Lemma 3.8].

Lemma 10.2.8 *Let X be a conjugation space. Then the restriction homomorphism* $r\colon H^*_G(X) \to H^*_G(X^G)$ *is injective.*

Proof Suppose that r is not injective. Let $0 \neq x = \sigma(y)u^k + lt_k \in H_G^{2n+k}(X)$ be an element in $\ker r$. The conjugation equation guarantees that $k \neq 0$. We may assume that k is minimal. By the conjugation equation again, we have $0 = r(x) = \kappa(y)u^{n+k} + lt_{n+k}$. Since κ is an isomorphism, we get $y = 0$, which is a contradiction. □

The H^*-frame of a conjugation space behaves well with respect to the characteristic classes of G-conjugate-equivariant bundles. A *G-conjugate-equivariant bundle* over a G-space X (with an involution τ) is a complex vector bundle $\eta = (E \xrightarrow{p} X)$, together with an involution $\hat{\tau}$ on E such that $p \circ \hat{\tau} = \tau \circ p$ and $\hat{\tau}$ is conjugate-linear on each fiber: $\hat{\tau}(\lambda x) = \bar{\lambda}\hat{\tau}(x)$ for all $\lambda \in \mathbb{C}$ and $x \in E$. This was called a "real bundle" by Atiyah [11]. Note that $\eta^G = (E^G \xrightarrow{p} X^G)$ is a real vector bundle and $\mathrm{rank}_{\mathbb{R}}\, \eta^G = \mathrm{rank}_{\mathbb{C}}\, \eta$. The following result is proven in [87, Proposition 6.8].

Proposition 10.2.9 *Let* η *be a G-conjugate-equivariant bundle over a spherical conjugation complex X. Then* $\kappa(c(\eta)) = w(\eta^\tau)$.

A theory of (integral) equivariant Chern classes for G-conjugate-equivariant bundles over a conjugation space is developed in [160].

Another relationship between conjugation spaces and the Steenrod squares was discovered by Franz and Puppe in [63]. It is illustrated by the case of $\mathbb{C}P^m$, with the notations of Example 10.2.2, where the conjugation equation (10.2.2) for $\mathbb{C}P^m$ may be written as follows.

$$r \circ \sigma(a^k) = (\kappa(a)u + \kappa(a)^2)^k = \sum_{j=0}^{k} \tbinom{k}{j} \kappa(a)^{k+j} u^{k-j} = \sum_{j=0}^{k} \mathrm{Sq}^j(\kappa(a^k)) u^{k-j} . \qquad (10.2.9)$$

It was proven in [63, Theorem 1.1] that (10.2.9) holds true in general, leading to the following *universal conjugation equation*.

Theorem 10.2.10 *Let X be a conjugation space, with* H^*-frame (κ, σ). *Then, for* $x \in H^{2k}(X)$, *one has*

$$r \circ \sigma(x) = \sum_{j=0}^{k} \mathrm{Sq}^j(\kappa(x)) u^{k-j} . \qquad (10.2.10)$$

Note the resemblance between the right member of (10.2.10) and that of (8.3.9). For other occurrences of such an expression, see [125, Sect. 2.4].

Let $\bar{r}\colon H^*(X) \to H^*(X^G)$ be the restriction homomorphism in non-equivariant cohomology. The following corollary was observed in [63, Corollary 1.2].

Corollary 10.2.11 *For* $x \in H^*(X)$, *one has* $\bar{r}(x) = \kappa(x) \smile \kappa(x)$.

Proof Suppose that $x \in H^{2k}(X)$. Denote by $\rho^G\colon H^*_G(X^G) \to H^*(X^G)$ the forgetful homomorphism for X^G. Then

$$\begin{aligned}
\bar{r}(x) &= \bar{r} \circ \rho \circ \sigma(x) && \text{since } \rho \circ \sigma = \mathrm{id} \\
&= \rho^G \circ r \circ \sigma(X) && \text{using (7.1.5)} \\
&= \rho^G\big(\textstyle\sum_{j=0}^{k} \mathrm{Sq}^j(\kappa(x)) u^{k-j}\big) && \text{by Theorem 10.2.10} \\
&= \mathrm{Sq}^k(\kappa(x)) && \text{since } \rho^G = \mathrm{ev}_{u=0}, \text{ see (7.1.6)} \\
&= \kappa(x) \smile \kappa(x) .
\end{aligned}$$

□

Another consequence of Theorem 10.2.10 is the commutativity of the diagram.

$$\begin{array}{ccccccc}
H^*(X) & \xrightarrow{\sigma} & H^*_G(X) & \xrightarrow{r} & H^*_G(X^G) & \xleftrightarrow{\approx} & H^*(X^G)[u] \\
\downarrow{\scriptstyle\kappa} & & & & & & \downarrow{\scriptstyle \mathrm{ev}_1} \\
H^*(X) & & & \xrightarrow{\mathrm{Sq}} & & & H^*(X^G)
\end{array} \qquad (10.2.11)$$

Theorem 10.2.10 has also consequences for *conjugation manifolds*, i.e. closed manifolds X with a smooth conjugation τ. Then, X^G is a closed manifold (see, e.g. [12, Corollary 2.2.2]) whose dimension, because of the isomorphism $\kappa\colon H^*(X) \xrightarrow{\approx} H^*(X^G)$, is half of the dimension of X. By looking at the derivative of τ around a fixed point, one checks that τ preserves the orientation if and only if $\dim X \equiv 0 \bmod 4$. For various properties of conjugation manifolds, see [80, Sect. 2.7], from which we extract the following results (see also [160, Appendix A]).

Proposition 10.2.12 *Let X be a smooth conjugation manifold of dimension $2n$, with H^*-frame (κ, σ). Then κ preserves the Wu and Stiefel-Whitney classes:*

$$\kappa(v(X)) = v(X^G) \quad \textit{and} \quad \kappa(w(TX)) = w(TX^G) .$$

Proof The Wu class $v_{2i}(X)$ is characterized by the equation

$$v_{2i}(X) \smile a = \mathrm{Sq}^{2i}(a) \text{ for all } a \in H^{2n-2i}(X) . \tag{10.2.12}$$

Applying the ring isomorphism κ to (10.2.12) and using Proposition 10.2.6 gives

$$\kappa(v_{2i}(X)) \smile \kappa(a) = \mathrm{Sq}^i(\kappa(a)) \text{ for all } a \in H^{2n-2i}(X) . \tag{10.2.13}$$

As κ is bijective, (10.2.13) implies that

$$\kappa(v_{2i}(X)) \smile b = \mathrm{Sq}^i(b) \text{ for all } b \in H^{n-i}(X^G) ,$$

which implies that $\kappa(v_{2i}(X)) = v_i(X^G)$, and, as $H^{odd}(X) = 0$, that $\kappa(v(X)) = v(X^G)$. Using this and the Wu formula, one gets

$$\begin{aligned} \kappa(w(TX)) &= \kappa\circ \mathrm{Sq}(v(X)) && \text{by the Wu formula} \\ &= \mathrm{Sq}\circ\kappa(v(X)) && \text{Proposition 10.2.6} \\ &= \mathrm{Sq}(v(X^G))) && \text{as already seen} \\ &= w(TX^G) && \text{by the Wu formula.} \end{aligned}$$

□

In particular, X admits a spin structure if and only if X^G is orientable. Also, the Stiefel-Whitney numbers of X all vanish if and only if those of X^G do so. By Thom's Theorem 9.9.6, this gives the following

Corollary 10.2.13 *Let X be a conjugation manifold. Then X bounds a compact manifold if and only if X^G does so.*

Two natural problems occur for conjugation manifolds.

(i) *Given a closed connected smooth manifold M^n, does there exist a conjugation $2n$-manifold X with X^G diffeomorphic to M ?*
(ii) *Classify, up to G-diffeomorphism, conjugation manifolds with a given fixed point set.*

The circle is the fixed point set of a unique conjugation 2-manifold, namely S^2_1; the uniqueness may be proved using the Schoenflies theorem (compare [34, Theorem 4.1]).

For $n = 2$, recall that $\mathbb{R}P^2$ is the fixed point set of the conjugation manifold $\mathbb{C}P^2$ and $S^1 \times S^1$ is that of $S^2 \times S^2$. The equivariant connected sum (around a fixed point) of conjugation manifolds being again a conjugation manifold (see [87, Proposition 4.7]), any closed surface is the fixed point set of some conjugation 4-manifold (of course, $S^2 = (S^4_2)^G$). Answering Question (ii) is the main object of [80], using the following idea. For a smooth G-action on a manifold X with X^G being of codimension 2, the quotient space X/G inherits a canonical smooth structure. If X is a conjugation 4-manifold, then $H^*(X/G) \approx H^*(S^4)$. Conversely, let (Y, Σ) be a manifold pair such that Y is a 4-dimensional $\mathbb{Z}_2$-homology sphere containing Σ as a codimension 2 closed submanifold. By Alexander duality (Theorem 5.3.14), one has $H^1(Y - \Sigma) = \mathbb{Z}_2$. Thus, $Y - \Sigma$ admits a unique non-trivial 2-fold covering (see Sect. 4.3); the latter extends to a unique branched covering $X \to Y$, with branched locus Σ, and X turns out to be a conjugation 4-manifold with $X^G = M$. The final statement is thus the following ([80, Theorem A]).

Theorem 10.2.14 *The correspondence* $X \mapsto (X/G, X^G)$ *defines a bijection between*

(a) *the orientation-preserving G-diffeomorphism classes of oriented connected conjugation* 4*-manifolds, and*
(b) *the orientation-preserving G-diffeomorphism classes of smooth manifold pairs* (Y, Σ)*, where* Y *is an oriented* 4*-dimensional homology sphere and* Σ *is a closed connected surface embedded in* M.

The conjugation sphere S^4_2 corresponds to the trivial knot $S^2 \subset S^4$. Under the bijection of Theorem A, any knot $S^2 \hookrightarrow S^4$ corresponds to a conjugation 4-manifold X with $X^G \approx S^2$. In general, X is not simply connected. On the other hand, Gordon [69, 70] and Sumners [189] found infinitely many topologically distinct knots in S^4 which are the fixed point set of smooth involutions. These examples produce infinitely many topologically inequivalent smooth conjugations on S^4 (see [80, Proposition 5.12]).

If X is a simply-connected conjugation 4-manifold, it is known that X/G is at least homeomorphic to S^4 (see [80, Proposition 5.3]). In addition, X is homeomorphic (not necessarily equivariantly) to a connected sum of copies of $S^2 \times S^2$, $\mathbb{C}P^2$, and $\overline{\mathbb{C}P}^2$ (see [80, Proposition 2.17]). These are severe restrictions on a simply-connected closed smooth 4-manifold to carry a smooth conjugation.

Olbermann, in his thesis [157], was the first to address Question (i); he proved the following result (see [158, Theorem 1.2]).

Theorem 10.2.15 *Any closed smooth orientable* 3*-manifold is diffeomorphic to the fixed point-set of a conjugation* 6*-manifold.*

The case of non-orientable 3-manifolds is not known. Any 3-dimensional $\mathbb{Z}_2$-homology sphere is the fixed point of infinitely many inequivalent conjugations on S^6, [159]; this gives a partial answer to Question (ii) in this case.

Remark 10.2.16 As observed by W. Pitsch and J. Scherer, the answer to Question (i) is not always positive. For example, the octonionic projective plane $\mathbb{O}P^2$, which is a smooth closed 16-manifold (see Remark 6.1.8), is not the fixed point set of any conjugation space. Indeed, $H^*(\mathbb{O}P^2) \approx \mathbb{Z}_2[x]/(x^3)$ by Proposition 6.1.7, with degree $(x) = 8$, but, by Theorem 8.6.5, $\mathbb{Z}_2[x]/(x^3)$ is not the cohomology ring of a topological space if degree $(x) > 8$.

10.3 Chain and Polygon Spaces

Chain and polygon spaces are examples of configuration spaces, a main concept of classical mechanics. In recent decades, starting in [83, 202] (inspired by talks of Thurston on linkages [195], new interests arose for polygon spaces, in connections with Hamiltonian geometry (see e.g. [88, 91, 111, 118]), mathematical robotics [56, 142] and statistical shape theory [90]. This section contains original results on the equivariant cohomology of chain spaces, giving new proofs for known statements about their ordinary cohomology.

We use the notations of [59, 85, 90], inspired by those of statistical shape theory [112]. In order to make some formulae more readable, we may write $|J|$ for the cardinality $\sharp J$ of a finite set J.

10.3.1 Definitions and Basic Properties

Let $\ell = (\ell_1, \dots, \ell_n) \in \mathbb{R}^n_{>0}$ and let d be an integer. We define the subspace $\mathcal{C}^n_d(\ell)$ of $(S^{d-1})^{n-1}$ by

$$\mathcal{C}^n_d(\ell) = \left\{ z = (z_1, \dots, z_{n-1}) \in (S^{d-1})^{n-1} \mid \sum_{i=1}^{n-1} \ell_i z_i = \ell_n\, e_1 \right\},$$

where $e_1 = (1, 0, \dots, 0)$ is the first vector of the standard basis $e_1, \dots, e_d$ of $\mathbb{R}^d$. An element of $\mathcal{C}^n_d(\ell)$, called a *chain*, may be visualized as a configuration of $(n-1)$ successive segments in $\mathbb{R}^d$, of length $\ell_1, \dots, \ell_{n-1}$, joining the origin to $\ell_n e_1$. The vector ℓ is called the *length vector*. The *chain space* $\mathcal{C}^n_d(\ell)$ is contained in the *big chain space* $\mathcal{BC}^n_d(\ell)$ defines as follows:

$$\mathcal{BC}^n_d(\ell) = \left\{ z = (z_1, \dots, z_{n-1}) \in (S^{d-1})^{n-1} \mid \langle \sum_{i=1}^{n-1} \ell_i z_i, e_1 \rangle = \ell_n \right\},$$

(successions of $(n-1)$ segments in $\mathbb{R}^d$, of length $\ell_1, \dots, \ell_{n-1}$, joining the origin to the affine hyperplane with first coordinate ℓ_n).

The group $O(d-1)$, viewed as the subgroup of $O(d)$ stabilizing the first axis, acts naturally (on the left) upon the pair $(\mathcal{BC}^n_d(\ell), \mathcal{C}^n_d(\ell))$. The quotient $\mathcal{C}^n_d(\ell)/SO(d-1)$ is the *polygon space* $\mathcal{N}^n_d$, also defined as

$$\mathcal{N}^n_d(\ell) = \tilde{\mathcal{N}}^n_d(\ell)/SO(d),$$

where

$$\tilde{\mathcal{N}}_d^n(\ell) = \left\{ z \in (S^{d-1})^n \;\middle|\; \sum_{i=1}^{n} \ell_i z_i = 0 \right\}$$

is the *free polygon space* (called "space of polygons" in [57, 65]).

The map from $SO(d) \times \mathcal{C}_d^n(\ell)$ to $\tilde{\mathcal{N}}_d^n(\ell)$ given by

$$(A, (z_1, \ldots, z_{n-1})) \mapsto (Az_1, \ldots, Az_{n-1}, -Ae_1)$$

descends to an $SO(d)$-homeomorphism

$$SO(d) \times_{SO(d-1)} \mathcal{C}_d^n(\ell) \xrightarrow{\approx} \tilde{\mathcal{N}}_d^n(\ell)\,. \tag{10.3.1}$$

Recall that the map $SO(d) \to S^{d-1}$ given by $A \mapsto -Ae_1$ is the orthonormal oriented frame bundle for the tangent bundle to S^{d-1} (see p. 362). Thus, by (10.3.1), we get a locally trivial bundle

$$\mathcal{C}_d^n(\ell) \to \tilde{\mathcal{N}}_d^n(\ell) \to S^{d-1}\,. \tag{10.3.2}$$

When $d = 2$ the space of chains $\mathcal{C}_2^n(\ell)$ coincides with the polygon space $\mathcal{N}_2^n(\ell)$.

The axial involution τ on $\mathbb{R}^d = \mathbb{R} \times \mathbb{R}^{d-1}$ given by $\tau(t, y) = (t, -y)$ induces an involution, still called τ, on the pair $(\mathcal{BC}_d^n(\ell), \mathcal{C}_d^n(\ell))$ and on $(S^{d-1})^{n-1}$. As τ commutes with the $O(d-1)$-action on $\mathcal{C}_d^n(\ell)$, it descends to a G-action on $\mathcal{N}_d^n(\ell)$, where $G = \{\mathrm{id}, \tau\}$. A bar above a G-space denotes its orbit space:

$$\overline{\mathcal{BC}}_d^n(\ell) = \mathcal{BC}_d^n(\ell)/G\,,\; \bar{\mathcal{C}}_d^n(\ell) = \mathcal{C}_d^n(\ell)/G\,,\; \bar{\mathcal{N}}_d^n(\ell) = \mathcal{N}_d^n(\ell)/G\,.$$

We shall compute the G-equivariant cohomology of $\mathcal{BC}_d^n(\ell)$ and $\mathcal{C}_d^n(\ell)$, as algebras over $H_G^*(pt) = \mathbb{Z}_2[u]$ (u of degree 1). This uses some G-invariant Morse theory on $M = (S^{d-1})^{n-1}$. We start with the *robot arm map* $F_\ell \colon M \to \mathbb{R}^d$ defined by

$$F_\ell(z) = \sum_{i=1}^{n-1} \ell_i z_i\,, \quad z = (z_1, \ldots, z_{n-1})\,. \tag{10.3.3}$$

Consider $\mathbb{R}^d$ as the product $\mathbb{R} \times \mathbb{R}^{d-1}$, which defines the projections $p_1 \colon \mathbb{R}^d \to \mathbb{R}$ and $p_{d-1} \colon \mathbb{R}^d \to \mathbb{R}^{d-1}$. Define $f_\ell \colon M \to \mathbb{R}$ by

$$f_\ell(z) = -p_1(F_\ell(z)) = -\sum_{i=1}^{n-1} \ell_i \langle z_i, e_1 \rangle\,.$$

Note that F_ℓ is $O(d)$-equivariant and f_ℓ is $O(d-1)$-invariant. For $n = 2$, it is clear that f_ℓ is Morse function on S^{d-1}, with two critical points, namely e_1 of index 0 and $-e_1$ of index $d-1$. The following lemma follows easily.

Lemma 10.3.1 *The function* $f_\ell\colon (S^{d-1})^{n-1} \to \mathbb{R}$ *defined by*

$$f_\ell(z_1, \dots, z_{n-1}) = -\sum_{i=1}^{n-1} \ell_i \langle z_i, e_1\rangle$$

is a G-invariant Morse function, with one critical point P_J for each $J \subset \{1, \dots, n-1\}$, where $P_J = (z_1, \dots, z_{n-1})$ with z_i equal to $-e_1$ if $i \in J$ and e_1 otherwise (a collinear chain). The index of P_J is $(d-1)\,|J|$.

A length vector $\ell \in \mathbb{R}^n_{>0}$ is *generic* if $\mathcal{C}^n_1(\ell) = \emptyset$, that is to say there are no collinear chains or polygons. In this section, we shall only deal with generic length vectors.

Corollary 10.3.2 *If ℓ is a generic length vector, then $\mathcal{BC}^n_d(\ell)$, $\mathcal{C}^n_d(\ell)$ and $\tilde{\mathcal{N}}^n_d(\ell)$ are smooth closed orientable manifolds of dimension*

$$\dim \mathcal{BC}^n_d(\ell) = \dim \tilde{\mathcal{N}}^n_d(\ell) = (n-1)(d-1)-1 \text{ and } \dim \mathcal{C}^n_d(\ell) = (n-2)(d-1)-1\,.$$

Proof If ℓ is generic, then $-\ell$ is a regular value of f_ℓ. Indeed, if $p_{d-1}(F_\ell(z)) \neq 0$, this follows from the $O(d)$-equivariance of F_ℓ. If $p_{d-1}(F_\ell(z)) = 0$, then, as ℓ is generic, z is not a critical point of F_ℓ (these are the collinear configurations $z_i = \pm z_j$: see [83, Theorem 3.1]). Since $\mathcal{BC}^n_d(\ell) = f_\ell^{-1}(-\ell_n)$, this proves the assertion on $\mathcal{BC}^n_d(\ell)$ (which is orientable, having trivial normal bundle in the orientable manifold $(S^{d-1})^{n-1}$).

Define $P\colon \mathcal{BC}^n_d(\ell) \to \mathbb{R}^{d-1}$ by $P(z) = p_{d-1}(F_\ell(z))$. As seen above, as ℓ is generic, $P^{-1}(0)$ contains no critical points of F_ℓ. Therefore, P is transversal to 0 and thus $\mathcal{C}^n_d(\ell) = P^{-1}(0)$ is a closed submanifold of codimension $d-1$ of $\mathcal{BC}^n_d(\ell)$, with trivial normal bundle.

When ℓ is generic, the $O(d-1)$-action on $\mathcal{C}^n_d(\ell)$ is smooth. Hence, the bundle (10.3.2) is a smooth bundle and the assertion on $\tilde{\mathcal{N}}^n_d(\ell)$ follows form (10.3.1). For another proof that $\tilde{\mathcal{N}}^n_d(\ell)$ is a manifold, see [57, Proposition 3.1]. □

We now see how chain and polygon spaces are determined by some combinatorics of their length vector $\ell = (\ell_1, \dots, \ell_n)$. A subset J of $\{1, \dots, n\}$ is called *ℓ-short* (or just *short*) if $\sum_{i\in J} \ell_i < \sum_{i\notin J} \ell_i$. The complement of a short subset is called *long*. If ℓ is generic, subsets are either short or long. Short subsets form, with the inclusion, a poset $\mathrm{Sh}(\ell)$. Define $\mathrm{Sh}_n(\ell) = \{J \in \{1, \dots, n-1\} \mid J \cup \{n\} \in \mathrm{Sh}(\ell)\}$.

For $J \subset \{1, \dots, n\}$, let $\mathcal{H}_J$ be the hyperplane (*wall*) of $\mathbb{R}^n$ defined by

$$\mathcal{H}_J := \Big\{(\ell_1, \dots, \ell_n) \in \mathbb{R}^n \,\Big|\, \sum_{i\in J} \ell_i = \sum_{i\notin J} \ell_i\Big\}.$$

The union $\mathcal{H}(\mathbb{R}^n)$ of all these walls determines a set of open *chambers* in $(\mathbb{R}_{>0})^n$ whose union is the set of generic length vectors (a chamber is a connected component of $(\mathbb{R}_{>0})^n - \mathcal{H}(\mathbb{R}^n)$). We denote by $\mathrm{Ch}(\ell)$ the chamber of a generic length vector ℓ. Note that $\mathrm{Ch}(\ell) = \mathrm{Ch}(\ell')$ if and only if $\mathrm{Sh}(\ell) = \mathrm{Sh}(\ell')$.

Let Sym_n be the group of bijections of $\{1, \dots, n\}$; we see Sym_{n-1} as the subgroup of Sym_n formed by those bijections fixing n. If X is a set, the group Sym_n acts on the Cartesian product X^n by

$$(x_1, \dots, x_n)^\sigma = (x_{\sigma(1)}, \dots, x_{\sigma(n)}) .$$

The notation emphasizes that this action is on the right: an element $x \in X^n$ is formally a map $x : \{1, \dots, n\} \to X$ ($x_i = x(i)$) and $\sigma \in \mathrm{Sym}_n$ acts by pre-composition, i.e. $x^\sigma = x \circ \sigma$. We shall use this action on various n-tuples, in particular on length vectors.

Lemma 10.3.3 *Let $\ell = (\ell_1, \dots, \ell_n)$ and $\ell' = (\ell'_1, \dots, \ell'_n)$ be two generic length vectors. Then, the following conditions are equivalent*

(1) $\mathrm{Sh}_n(\ell)$ *and* $\mathrm{Sh}_n(\ell')$ *are poset isomorphic.*
(2) $\mathrm{Sh}(\ell)$ *and* $\mathrm{Sh}(\ell')$ *are poset isomorphic via a bijection* $\sigma \in \mathrm{Sym}_{n-1}$.
(3) $\mathrm{Ch}(\ell') = \mathrm{Ch}(\ell^\sigma)$ *for some* $\sigma \in \mathrm{Sym}_{n-1}$.

Moreover, if one of the above conditions is satisfied, there are $O(d-1)$-diffeomorphisms of manifolds pairs

$$h\colon (\mathcal{B}C^n_d(\ell), C^n_d(\ell)) \xrightarrow{\approx} (\mathcal{B}C^n_d(\ell'), C^n_d(\ell'))$$

and

$$\bar h\colon (\overline{\mathcal{B}C}^n_d(\ell), \bar C^n_d(\ell)) \xrightarrow{\approx} (\overline{\mathcal{B}C}^n_d(\ell'), \bar C^n_d(\ell')) .$$

Proof Implications (1) $\Leftarrow$ (2) $\Leftrightarrow$ (3) are obvious. Let us prove that (1) $\Rightarrow$ (2). Let $\sigma \in \mathrm{Sym}_{n-1}$ be the permutation giving the poset isomorphism $\mathrm{Sh}_n(\ell) \approx \mathrm{Sh}_n(\ell')$. Replacing ℓ by ℓ^σ, we may assume that $\mathrm{Sh}_n(\ell) = \mathrm{Sh}_n(\ell')$. We now observe that $\mathrm{Sh}_n(\ell)$ determines $\mathrm{Sh}(\ell)$. Indeed, let $J \subset \{1, \dots, n\}$. Then, either $n \in J$ or $n \in \bar J$, and thus $\mathrm{Sh}_n(\ell)$ tells us whether $J \in \mathrm{Sh}(\ell)$ (or $\bar J \in \mathrm{Sh}(\ell)$).

It remains to prove that (3) implies the existence of the $O(d-1)$-diffeomorphisms h and $\bar h$. As the G action commutes with the $O(d-1)$-action (G is naturally in the center of $O(d-1)$), it suffices to construct h, which will induce $\bar h$. If $\sigma \in \mathrm{Sym}_{n-1}$, then the correspondence $z \to z^\sigma$ defines an $O(d-1)$-diffeomorphism $(\mathcal{B}C^n_d(\ell), C^n_d(\ell)) \xrightarrow{\approx} (\mathcal{B}C^n_d(\ell^\sigma), C^n_d(\ell^\sigma))$. Replacing ℓ by ℓ^σ, we may thus assume that $\mathrm{Ch}(\ell) = \mathrm{Ch}(\ell')$ and $\sigma = \mathrm{id}$.

Consider the smooth map $L\colon (\mathbb{R}^d - \{0\})^{n-1} \to (\mathbb{R}_{>0})^n$ given, for $x = (x_1, \dots, x_{n-1})$ by

$$L(x) = \big(|x_1|, \dots, |x_{n-1}|, \langle \tilde F(x), e_1\rangle\big) ,$$

where $\tilde{F}(x) = \sum_{i=1}^{n-1} x_i$. Observe that the map $(S^{d-1})^{n-1} \to (\mathbb{R}^d - \{0\})^{n-1}$ sending $(z_1, \dots, z_{n-1})$ to $(\ell_1 z_1, \dots, \ell_{n-1} z_{n-1})$ induces a diffeomorphism $\gamma_\ell\colon \mathcal{BC}_d^n(\ell) \xrightarrow{\approx} L^{-1}(\ell)$ such that $\tilde{F} \circ \gamma = F$, the robot arm map of (10.3.3). If ℓ is generic, then ℓ is a regular value of L. Indeed, let $x = (x_1, \dots, x_{n-1}) \in L^{-1}(\ell)$. For each $i = 1, \dots, n-1$, one can construct a path $x^i(t) = (x_1^i(t), \dots, x_{n-1}^i(t)) \in (\mathbb{R}^d - \{0\})^{n-1}$ with $x(0) = x$ such that $L(x^i(t)) = (\ell_1^i(t), \dots, \ell_n^i(t))$ satisfies $\ell_j^i(t) = \ell_j$ for $j \neq i$ and $\ell_i^i(t) = \ell_i + \alpha t$ with $\alpha \neq 0$. For $i = n$, this follows from the proof of Corollary 10.3.2. Suppose that $i \leq n-1$. If x is not a lined configuration, then x_i and $\sum_{j\neq i} x_j$ are linearly independent and generate a 2-dimensional plane Π, containing $\tilde{F}(x)$. There are rotations ρ_t^i and ρ_t of Π, depending smoothly on t, such that $\rho_t^i((1+t)x_i) + \rho_t(\sum_{j\neq i} x_j) = \tilde{F}(x)$. Hence, $X^i(t)$ may be defined as

$$x_j^i(t) = \begin{cases} \rho_t^i((1+t)x_i) & \text{if } j = i \\ \rho_t(x_j) & \text{if } j \neq i\,. \end{cases}$$

Finally, if x is a lined configuration, then $\tilde{F}(x)$ and e_1 are linearly independent (since ℓ is generic). They thus generate a 2-dimensional plane Ω. Let $\bar{x}^i(t)$ defined by $\bar{x}_i^i(t) = (1+t)x_i$ and $\bar{x}_j^i(t) = x_j$ when $j \neq i$. If t is small enough, there is a unique rotation r_t of Ω such that $\langle r_t(\tilde{F}(\bar{x}^i(t)), e_1\rangle = \ell_n$. We can thus define $x_j^i(t) = \rho_t(\bar{x}_j^i(t))$.

As $\mathrm{Ch}(\ell)$ is convex, it contains the segment $[\ell, \ell']$ consisting of only generic length vectors. What has been done above shows that the map L is transversal to $[\ell, \ell']$. Therefore, $X = L^{-1}([\ell, \ell'])$ is $O(d)$-cobordism between $\mathcal{BC}_d^n(\ell)$ and $\mathcal{BC}_d^n(\ell')$. Let $p_{d-1}\colon \mathbb{R}^d \to \mathbb{R}^{d-1}$ be the projection onto the last $d-1$ coordinates. As in the proof of Corollary 10.3.2, the map $P\colon X \to \mathbb{R}^{d-1}$ defined by $P(x) = p_{d-1}(\tilde{F}(x))$ is transversal to 0. Thus, $Y = P^{-1}(0)$ is a submanifold of X of codimension $n-1$ and the pair (X, Y) is a cobordism of pairs between $(\mathcal{BC}_d^n(\ell), \mathcal{C}_d^n(\ell))$ and $(\mathcal{BC}_d^n(\ell'), \mathcal{C}_d^n(\ell'))$. The map $L\colon X \to [\ell, \ell']$ has no critical point. The standard Riemannian metric on $(\mathbb{R}^d)^n$ induces an $O(d-1)$-invariant Riemannian metric on (X, Y). Following the gradient lines of π for this metric provides the required $O(d-1)$-equivariant diffeomorphism h. □

For $n \leq 9$, a list of all chambers (modulo the action of Sym_n) was obtained in [90]. Their numbers are as follows (for $n = 10$, it was computed independently by Minfeng Wang and Dirk Schütz: see the Web complement of [90]).

n	3	4	5	6	7	8	9	10
Nb of chambers	2	3	7	21	135	2'470	175'428	52'980'624

Geometric descriptions of several chain and polygon spaces for ℓ-generic are provided in [85], as well as some general constructions. Among them, the operation of "adding a tiny edge", which we now describe. Let $\ell = (\ell_2, \dots, \ell_n)$ be a generic length vector. If $\varepsilon > 0$ is small enough, the n-tuple $\ell^+ := (\delta, \ell_2, \dots, \ell_n)$ is a generic length vector for $0 < \delta \leq \varepsilon$.

Lemma 10.3.4 *There are $O(d-1)$-equivariant diffeomorphisms*

$$\mathcal{B}\Phi\colon \mathcal{B}\mathcal{C}_d^n(\ell^+) \xrightarrow{\approx} S^{d-1} \times \mathcal{B}\mathcal{C}_d^{n-1}(\ell) \text{ and } \Phi\colon \mathcal{C}_d^n(\ell^+) \xrightarrow{\approx} S^{d-1} \times \mathcal{C}_d^{n-1}(\ell),$$

where $S^{d-1} \times \mathcal{B}\mathcal{C}_{d-1}^m(\ell)$ and $S^{d-1} \times \mathcal{C}_{d-1}^m(\ell)$ are equipped with the diagonal $O(d-1)$-action.

Proof The diffeomorphism Φ is constructed in [85, Proposition 2.1]. The construction can be easily adapted to give $\mathcal{B}\Phi$. □

10.3.2 Equivariant Cohomology

Let $M = (S^{d-1})^{n-1}$. The G-invariant Morse function $f = f_\ell\colon M \to \mathbb{R}$ of Lemma 10.3.1 satisfies the hypotheses of Proposition 7.6.13, i.e. $M^G = \operatorname{Crit} f$. Therefore, M is G-equivariantly formal and the restriction morphism

$$r\colon H_G^*(M) \to H_G^*(M^G) \approx \bigoplus_J H_G^*(P_J) \approx \bigoplus_J \mathbb{Z}_2[u_J] \tag{10.3.4}$$

is injective (this also follows from Lemma 7.3.6 and Proposition 7.3.9). The variables u_J are of degree one and the $\mathbb{Z}_2[u]$-module structure on $H_G^*(M^G)$ is given by the inclusion $u \mapsto \sum_J u_J$.

In the remainder of this section, whenever x_i ($i \in \mathbb{N}$) are formal variables in a polynomial ring and $J \subset \mathbb{N}$, we set $x_J = \prod_{j \in J} x_j$. In particular, $x_\emptyset = 1$.

Proposition 10.3.5 *For $n \geq 2$, there is a* **GrA**$[u]$*-isomorphism*

$$\mathbb{Z}_2[u, A_1 \dots, A_{n-1}, B_1, \dots, B_{n-1}]/\mathcal{I} \xrightarrow{\approx} H_G^*((S^{d-1})^{n-1}) \tag{10.3.5}$$

where the variables A_i and B_i are of degree $d-1$ and $\mathcal{I}$ is the ideal generated by the families of relators

(a) $A_i + B_i + u^{d-1}$ $\quad i = 1, \dots, n-1$
(b) $A_i^2 + A_i u^{d-1}$ $\quad i = 1, \dots, n-1$

Moreover, using (10.3.4), *one has for $J \subset \{1, 2, \dots, n-1\}$:*

$$\begin{aligned} r(A_J) &= \sum_{J \subset K} u_K^{|J|(d-1)} \\ r(B_J) &= \sum_{J \cap K = \emptyset} u_K^{|J|(d-1)} \end{aligned} \tag{10.3.6}$$

Proposition 10.3.5 generalizes Examples 7.6.9 and 7.4.5, with the slightly different notations of (10.3.4) for the equivariant cohomology of the fixed point set.

Proof The proof proceeds by induction on n. It starts with $n = 2$, using Example 7.6.9. For the induction step, set $M = (S^{d-1})^{n-1} = \hat{M} \times M_0$, where $\hat{M} = (S^{d-1})^{n-2}$ and $M_0 = S^{d-1}$. The induction hypothesis implies that

$$H_G^*(\hat{M}) \approx \mathbb{Z}_2[\hat{u}, \hat{A}_1 \ldots, \hat{A}_{n-2}, \hat{B}_1, \ldots, \hat{B}_{n-2}]/\hat{\mathcal{I}}$$

where $\hat{\mathcal{I}}$ is the ideal generated by the families $\hat{A}_i + \hat{B}_i + \hat{u}^{d-1}$ and $\hat{A}_i^2 + \hat{A}_i \hat{u}^{d-1}$ $(i = 1, \ldots, n-2)$. The $\mathbb{Z}_2[u]$-module structure is obtained by identifying u with $\hat{u}$. Also,

$$H_G^*(M_0) \approx \mathbb{Z}_2[u_0, A, B]/(A + B + u_0^{d-1}, A^2 = u_0^{d-1} A)$$

and the $\mathbb{Z}_2[u]$-module structure is obtained by identifying u with u_0. By Theorem 7.4.3, the strong equivariant cross product provides an isomorphism

$$\bar{\times}_G \colon H_G^*(\hat{M}) \otimes_{\mathbb{Z}_2[u]} H_G^*(M_0) \xrightarrow{\approx} H_G^*(M)\,.$$

Setting $A_i = \hat{A}_i \bar{\times}_G \mathbf{1}$, $B_i = \hat{B}_i \bar{\times}_G \mathbf{1}$ $(i = 1, \ldots, n-2)$, $A_{n-1} = \mathbf{1} \bar{\times}_G A$ and $B_{n-1} = \mathbf{1} \bar{\times}_G B$ gives the induction step for the isomorphism (10.3.5).

We now prove the induction step for (10.3.6). The fixed points of $\hat{M}$ are denoted by $\hat{P}_J$, indexed by $J \subset \{1, \ldots, n-2\}$. We denote the fixed point of $M_0 = S^{d-1} \subset \mathbb{R} \times \mathbb{R}^{d-1}$ by $\omega_{\min} = (1, 0)$ and $\omega_{\max} = (-1, 0)$ (corresponding to the extrema of the Morse function $(t, x) \mapsto -t$). Set $H_G^*(M_0^G) \approx \mathbb{Z}_2[u_{\min}] \oplus \mathbb{Z}_2[u_{\max}]$. For $J \subset \{1, \ldots, n-2\}$, then $P_J = \hat{P}_J \times \omega_{\min}$ and $P_{J \cup \{n-1\}} = \hat{P}_J \times \omega_{\max}$. Hence, for $i = 1, \ldots, n-2$, one has, using the obvious notations, that

$$\begin{aligned}
r(A_i) &= r(\hat{A}_i \bar{\times}_G \mathbf{1}) \\
&= \hat{r}(\hat{A}_i) \bar{\times}_G r_0(\mathbf{1}) && \text{by naturality of } \bar{\times}_G \\
&= \Big[\sum_{\substack{J \subset \{1,\ldots,n-2\} \\ i \in J}} u_J^{|J|(d-1)} \Big] \bar{\times}_G [1_{\min} + 1_{\max}] && \text{by induction hypothesis} \\
&= \sum_{\substack{J \\ i \in J}} u_J^{|J|(d-1)} + \sum_{\substack{J \cup \{n-1\} \\ i \in J}} u_J^{|J|(d-1)} && (J \subset \{1, \ldots, n-2\}) \\
&= \sum_{\substack{J \subset \{1,\ldots,n-1\} \\ i \in J}} u_J^{|J|(d-1)}\,.
\end{aligned}$$

As for $i = n-1$, one has

$$\begin{aligned} r(A_{n-1}) &= r(\mathbf{1} \bar{\times}_G A) \\ &= \hat{r}(\mathbf{1}) \bar{\times}_G r_0(A) \\ &= \Big[\sum_{J\subset\{1,\dots,n-2\}} 1_J \Big] \bar{\times}_G u_{\max}^{d-1} \\ &= \sum_{J\subset\{1,\dots,n-2\}} u_{J\cup\{n-1\}}^{d-1} \\ &= \sum_{\substack{J\subset\{1,\dots,n-1\} \\ n-1\in J}} u_J^{|J|(d-1)} . \end{aligned}$$

This proves (10.3.6) for $r(A_i)$, $i = 1,\dots,n-1$. The formula for $r(B_i)$ are deduced using relators (a). The formulae for $r(A_J)$ and $r(B_J)$ follow since r is multiplicative. □

We are now ready to compute the G-equivariant cohomology of $\mathcal{B}C_d^n(\ell)$.

Theorem 10.3.6 *Let $\ell = (\ell_1,\dots,\ell_n)$ be a generic length vector. There is a* **GrA**$[u]$*-isomorphism*

$$\mathbb{Z}_2[u, A_1 \dots, A_{n-1}, B_1, \dots, B_{n-1}]\big/\mathcal{I}_\ell \xrightarrow{\approx} H_G^*(\mathcal{B}C_d^n(\ell)) \approx H^*(\overline{\mathcal{B}C}_d^n(\ell))$$

where the variables A_i and B_i are of degree $d-1$ and $\mathcal{I}_\ell$ is the ideal generated by the families of relators

(a) $A_i + B_i + u^{d-1}$	$i = 1,\dots,n-1$	
(b) $A_i^2 + A_i u^{d-1}$	$i = 1,\dots,n-1$	
(c) A_J	$J \subset \{1,\dots,n-1\}$	*and $J \cup \{n\}$ is long*
(d) B_J	$J \subset \{1,\dots,n-1\}$	*and J is long.*

Proof Let $M = (S^{d-1})^{n-1}$, $M_- = f^{-1}((-\infty, -\ell_n])$ and $M_+ = f^{-1}([-\ell_n, \infty])$, with the inclusions $j_\pm \colon M_\pm \to M$ ($f = f_\ell$, the Morse function of Lemma 10.3.1). One has $M_- \cap M_+ = \mathcal{B} = \mathcal{B}C_d^n(\ell) = f^{-1}(-\ell_n)$. The G-invariant Morse function $f\colon M \to \mathbb{R}$ satisfies the hypotheses of Proposition 7.6.13. The latter implies that the morphism $H_G^*(M) \to H_G^*(\mathcal{B})$ induced by the inclusion is surjective with kernel equal to $\ker H_G^* j_- + \ker H_G^* j_+$. By Proposition 10.3.5, $H_G^*(M)$ is **GrA**$[u]$-isomorphic to $\mathbb{Z}_2[u, A_1 \dots, A_{n-1}, B_1, \dots, B_{n-1}]\big/\mathcal{I}$ where $\mathcal{I}$ is the ideal generated by families (a) and (b). We shall prove that $\ker H_G^* j_-$ is the ideal generated by relators (c) and that $\ker H_G^* j_+$ is the ideal generated by relators (d).

The critical point P_J satisfies

$$f(P_J) = \sum_{i\in J} \ell_i - \sum_{i\notin J} \ell_i .$$

Therefore,

$$P_J \in M_- \iff f(P_J) < -\ell_n \iff J \cup \{n\} \text{is short}\,. \tag{10.3.7}$$

Therefore, one has a commutative diagram

$$\begin{array}{ccccc} H^*_G(M) & \xrightarrow{r} & H^*_G(M^G) & \xrightarrow{\approx} & \bigoplus_{J\subset\{1,\dots,n-1\}} \mathbb{Z}_2[u_J] \\ \downarrow{\scriptstyle H^*_G j_-} & & \downarrow{\scriptstyle H^*_G j_-^G} & & \downarrow{\scriptstyle \mathrm{pr}_-} \\ H^*_G(M_-) & \xrightarrow{r_-} & H^*_G(M^G_-) & \xrightarrow{\approx} & \bigoplus_{\substack{J\subset\{1,\dots,n-1\}\\ J\cup\{n\}\text{ short}}} \mathbb{Z}_2[u_J] \end{array} \tag{10.3.8}$$

That r and r_- are injective follows from Theorem 7.6.6. Hence, for $x \in H^*_G(M)$, $H^*j_-(A_J) = 0$ if and only if $\mathrm{pr}_- \circ r(x) = 0$. Since M is equivariantly formal (by Theorem 7.6.6 again), Theorem 10.3.6 implies that

$$H^*(M) \approx H^*_G(M)/(u) \approx \mathbb{Z}_2[A_1, \dots, A_{n-1}]/(A_i^2)\,.$$

By the Leray-Hirsch theorem, $H^*_G(M)/(u)$ is then isomorphic to the free $\mathbb{Z}_2[u]$-module with basis $\{A_j \mid J \subset \{1, \dots, n-1\}\}$ (or $\{B_j \mid J \subset \{1, \dots, n-1\}\}$). Thus, $x \in H^*_G(M)$ may be uniquely written as $x = \sum_{J\subset\{1,\dots,n-1\}} \lambda_J A_J$, with $\lambda_J \in \mathbb{Z}_2[u]$. Let $J_0 \subset \{1, \dots, n-1\}$ minimal (for the inclusion) such that $\lambda_{J_0} \neq 0$. By (10.3.7), one has

$$r(x) = \lambda_{J_0} u_{J_0} \mod \bigoplus_{\substack{J\subset\{1,\dots,n-1\}\\ J\neq J_0}} \mathbb{Z}_2[u_J]\,.$$

Hence, if $x \in \ker H^*_G j_-$, we deduce using Diagram (10.3.8) that $J_0 \cup \{n\}$ is long. Therefore, $\lambda_{J_0} A_{J_0} \in \ker H^*_G j_-$ and $x + \lambda_{J_0} A_{J_0} \in \ker H^*_G j_-$. Repeating the above argument with $x + \lambda_{J_0}$ and so on proves that

$$x = \sum_{J\subset\{1,\dots,n-1\}} \lambda_J A_J \in \ker H^*_G j_- \iff \lambda_J = 0 \text{ whenever } J \cup \{n\} \text{ is short}\,.$$

This proves that $\ker H^*_G j_-$ is the $\mathbb{Z}_2[u]$-module generated by relators (c) (since $A_J A_K = A_{J\cup K}$, this is an ideal).

In the same way, we prove that $\ker H^*_G j_+$ is the $\mathbb{Z}_2[u]$-module generated by relators (d). Details are left to the reader. □

Corollary 10.3.7 *For a generic length vector $\ell = (\ell_1, \dots, \ell_n)$, there is a* **GrA**$[u]$-*isomorphism*

$$\mathbb{Z}_2[u, A_1 \dots, A_{n-1}]/\hat{\mathcal{I}}_\ell \xrightarrow{\approx} H^*_G(\mathcal{B}C^n_d(\ell))$$

where the variables A_i are of degree $d-1$ and $\hat{\mathcal{I}}_\ell$ is the ideal generated by the families of relators

(1) $A_i^2 + A_i u^{d-1}$	$i = 1, \dots, n-1$	
(2) A_J	$J \subset \{1, \dots, n-1\}$	*and $J \cup \{n\}$ is long*
(3) $u^{d-1} \sum_{K \subset J} A_K \, u^{(\lvert J-K \rvert -1)(d-1)}$	$J \subset \{1, \dots, n-1\}$	*and J is long*

Note that, by (2), only the sets $K \subset J$ with $K \cup \{n\}$ being short occur in the sum of Relators (3).

Proof This presentation of $H^*_G(\mathcal{B}C^n_d(\ell))$ is algebraically deduced from that of Theorem 10.3.6. The generators B_i are eliminated using relators (a). Realtors (b) and (c) become respectively (1) and (2). Relators (d) become relators (3). Indeed,

$$\begin{aligned} B_J &= \textstyle\prod_{i \in J} B_i \\ &= \textstyle\prod_{i \in J} (A_i + u^{d-1}) && \text{using (a)} \\ &= \textstyle\sum_{K \subset J} A_K \, u^{\lvert J-K \rvert (d-1)} && \text{plain extension} \\ &= u^{d-1} \sum_{K \subset J} A_K \, u^{(\lvert J-K \rvert -1)(d-1)} && \text{as } A_J = 0 \text{ since } J \text{ is long.} \end{aligned}$$

□

Example 10.3.8 Elementary geometry easily shows that $\mathcal{B}C^n_d(\ell) = \emptyset$ if and only if $\{n\}$ is long (see also Example 10.3.23 below). In Corollary 10.3.7, we see that if $\{n\}$ is long, then relator (2) for $J = \emptyset$ implies that $1 \in \hat{\mathcal{I}}_\ell$ and thus $H^*_G(\mathcal{B}C^n_d(\ell)) = 0$. Compare Example 10.3.21.

Example 10.3.9 Suppose that ℓ is generic and that $\ell_n = -\alpha + \sum_{i=1}^{n-1} \ell_i$, with $\alpha > 0$ small enough such that $J \cup \{n\}$ is short only for $J = \emptyset$. Hence, relators (2) imply that $A_i = 0$ for $i = 1, \dots, n-1$. The only subset J of $\{1, \dots, n-1\}$ which is long is $\{1, \dots, n-1\}$ itself. Thus, the family of relators (3) contains one element, in which the only non-zero term in the sum occurs for $K = \emptyset$. This relator has thus the form $u^{(n-1)(d-1)}$ and we get

$$H^*_G(\mathcal{B}C^n_d(\ell)) \approx \mathbb{Z}_2[u]/(u^{(n-1)(d-1)}) \,. \tag{10.3.9}$$

Notice that, with our hypothesis, $-\ell_n$ is a regular value of f which is just above a minimum. Thus, $\mathcal{B}C^n_d(\ell) = f^{-1}(-\ell)$ is G-diffeomorphic to the sphere $S^{(n-1)(d-1)-1}$ endowed with the antipodal involution (the isotropy representation of G on the tangent space to M at the minimum $P_\emptyset$ of f). As the G-action on $\mathcal{B}C^n_d(\ell)$ is free, one has

$$H_G^*(\mathcal{B}C_d^n(\ell)) \approx H^*(\overline{\mathcal{B}C}_d^n(\ell)/G) \approx H^*(\mathbb{R}P^{(n-1)(d-1)-1})$$

which is coherent with (10.3.9).

Example 10.3.9 will help us to compute $H_G^*(C_d^n(\ell))$, after introducing some preliminary material. We use the robot arm map $F_\ell\colon M = (S^{d-1})^{n-1} \to \mathbb{R}^d = \mathbb{R}\times\mathbb{R}^{d-1}$ defined in (10.3.3). Let $N = F_\ell^{-1}(\mathbb{R} \times 0)$. If $\ell' = (\ell_1, \dots, \ell_{n-1})$ is itself generic, then N is a closed submanifold of codimension $d-1$ in M. Indeed, except at $F_\ell^{-1}(0)$, the robot arm map is clearly transverse to $\mathbb{R} \times 0$ (use that F is $SO(d)$-equivariant). If ℓ' is generic, then 0 is a regular value of F_ℓ (see the proof of Corollary 10.3.2). Hence, F_ℓ is everywhere transversal to $\mathbb{R} \times 0$.

A slight change of e.g. ℓ_1 (which does not change the G-diffeomorphism type of the pair $(\mathcal{B}C_d^n(\ell), C_d^n(\ell))$ by Lemma 10.3.3) will make ℓ' is generic. Hence, without loss of generality, one may assume that N is a closed G-invariant submanifold of M. One has $C_d^n(\ell) = N_- \cap N_+$ where $N_\pm = N \cap M_\pm$ (notation of the proof of Corollary 10.3.7).

There is a G-equivariant map $\phi_t\colon M_- \to M_-$ such that $\phi_0 = \mathrm{id}$ and $\phi_1(M_-) = N_-$. Indeed, for $z \in M_-$, denote by Π_z the 2-plane in $\mathbb{R}^d$ generated by e_1 and $F_\ell(z)$. Define $\rho_t(z) \in SO(d)$ to be the rotation of angle $\cos^{-1}(t|F_\ell(z)|/|f_\ell(z)|)$ on Π_z, the identity on $\Pi_z^\perp$ and such that $\langle \rho_t \circ F_\ell(z), e_1\rangle \geq f_\ell(z)$. The retraction by deformation ϕ_t is defined by $\phi_t(z_1, \dots, z_{n-1}) = (\rho_t(z_1), \dots, \rho_t(z_{n-1}))$. The existence of ϕ_t implies that

$$H_G^*(M_-) \approx H_G^*(N_-)\,. \tag{10.3.10}$$

The restriction of $f_{\ell,N}\colon N \to \mathbb{R}$ of f_ℓ is also a G-invariant Morse function on M, with $\mathrm{Crit}\, f_{\ell,N} = \mathrm{Crit}\, f_\ell$ (see [83, Sect. 3]; the index of a critical point P is different for $f_{\ell,N}$ and f when $f_\ell(P) < 0$).

Proposition 10.3.10 *Let $\ell = (\ell_1, \dots, \ell_n)$ be a generic length vector and let $i\colon C_d^n(\ell) \to \mathcal{B}C_d^n(\ell)$ be the inclusion. Then, $H_G^* i\colon H_G^*(\mathcal{B}C_d^n(\ell)) \to H_G^*(C_d^n(\ell))$ is surjective, with kernel equal to* Ann (u^{d-1}), *the annihilator of u^{d-1}.*

Proof Let $(\mathcal{B}, \mathcal{C}) = (\mathcal{B}C_d^n(\ell), C_d^n(\ell))$. Consider the commutative diagram

$$\begin{array}{ccc} H_G^*(M_-) & \twoheadrightarrow & H_G^*(\mathcal{B}) \\ \approx \downarrow H_G^* j & & \downarrow H_G^* i \\ H_G^*(N_-) & \twoheadrightarrow & H_G^*(\mathcal{C}) \end{array}$$

where all the arrows are induced by the inclusions. The horizontal maps are indicated to be surjective: this follows from Proposition 7.6.13 since $\mathrm{Crit} f = M^G = N^G$. That $H_G^* j$ is an isomorphism was noticed in (10.3.10). Hence, $H_G^* i$ is surjective.

Let $\bar{\mathcal{B}} = \mathcal{B}/G$ and $\bar{\mathcal{C}} = \mathcal{C}/G$. As the G-action on $(\mathcal{B}, \mathcal{C})$ is free, the vertical maps in the diagram

$$\begin{array}{ccc} \mathcal{C}_G & \xrightarrow{i_G} & \mathcal{B}_G \\ \downarrow & & \downarrow \\ \bar{\mathcal{C}} & \xrightarrow{\bar{i}} & \bar{\mathcal{B}} \end{array}$$

are homotopy equivalences (see Lemma 7.1.4). As $\bar{\mathcal{B}}$ and $\bar{\mathcal{C}}$ are smooth closed manifolds, Proposition 5.4.5 implies that $\ker H^*_G i$ is the annihilator of the Poincaré dual $\mathrm{PD}(\bar{\mathcal{C}}) \in H^{d-1}(\bar{\mathcal{B}}) \approx H^{d-1}_G(\mathcal{B})$. It thus remains to show that $\mathrm{PD}(\bar{\mathcal{C}}) = u^{d-1}$.

Let $p_{d-1} : \mathbb{R}^d = \mathbb{R} \times \mathbb{R}^{d-1} \to \mathbb{R}^{d-1}$ be the projection onto the second factor. The map $\varphi \colon \mathcal{B} \to (\mathbb{R}^{d-1})^{n-1} - \{0\}$ defined by

$$\varphi(z_1, \dots, z_{n-1}) = (p_{d-1}(z_1), \dots, p_{d-1}(z_{n-1}))$$

is smooth, G-equivariant (for the involution $x \mapsto -x$ on $(\mathbb{R}^{d-1})^{n-1}$) and satisfies $\mathcal{C} = \varphi^{-1}((\mathbb{R}^{d-1})^{n-2} - \{0\})$. It thus descends to a smooth map $\bar{\varphi} \colon \bar{\mathcal{B}} \to \mathbb{R}P^{(n-1)(d-1)-1}$, such that $\bar{\mathcal{C}} = \bar{\varphi}^{-1}(\mathbb{R}P^{(n-2)(d-1)-1})$. As in the proof of Corollary 10.3.2, one shows that $\bar{\varphi}$ is transversal to $\mathbb{R}P^{(n-2)(d-1)-1}$. By Proposition 5.4.5, one has

$$\mathrm{PD}(\bar{\mathcal{C}}) = H^*\bar{\varphi}(\mathrm{PD}(\mathbb{R}P^{(n-2)(d-1)-1})) = H^*\bar{\varphi}(u^{d-1}) = u^{d-1}\,.$$

The two occurrences of the letter u in the above formulae is a slight abuse of language, permitted by the considerations of Lemma 7.1.4: the G-action under consideration are all free and in the commutative diagram

$$\begin{array}{ccc} H^*(\mathbb{R}P^{(n-1)(d-1)-1}) & \xrightarrow{\approx} & H^*_G((\mathbb{R}^{d-1})^{n-1} - \{0\}) \\ \Big\downarrow H^*\bar{\varphi} & & \Big\downarrow H^*_G\varphi \\ H^*(\bar{\mathcal{B}}) & \xrightarrow{\approx} & H^*_G(\mathcal{B}) \end{array}$$

the generator of $H^1(\mathbb{R}P^{(n-1)(d-1)-1})$ is sent to u. □

We are now ready to compute $H^*_G(\mathcal{C}^n_d(\ell))$.

Theorem 10.3.11 *For a generic length vector $\ell = (\ell_1 \dots, \ell_n)$, there is a* **GrA**$[u]$-*isomorphism*

$$\mathbb{Z}_2[u, A_1 \dots, A_{n-1}]/\bar{\mathcal{I}}_\ell \xrightarrow{\approx} H^*_G(\mathcal{C}^n_d(\ell)) \approx H^*(\bar{\mathcal{C}}^n_d(\ell))$$

where the variables A_i are of degree $d-1$ and $\bar{\mathcal{I}}_\ell$ is the ideal generated by the families of relators

(1) $A_i^2 + A_i u^{d-1}$ $\quad i = 1, \ldots, n-1$

(2) A_J $\quad J \subset \{1, \ldots, n-1\}$ *and* $J \cup \{n\}$ *is long*

(3') $\sum_{K \subset J} A_K\, u^{(|J-K|-1)(d-1)}$ $\quad J \subset \{1, \ldots, n-1\}$ *and* J *is long*

Proof We use the notations of the proof Theorem 10.3.6, with $M = (S^{d-1})^{n-1}$, etc. Recall from Proposition 10.3.5 that

$$H_G^*(M) \approx \mathbb{Z}_2[u, A_1 \ldots, A_{n-1}]/\mathcal{I}_1$$

where $\mathcal{I}_1$ is the ideal generated by relators (1). Denote by $\mathcal{I}_2$, $\mathcal{I}_{3'}$ and $\mathcal{I}_3$ the ideals of $H^*(M)$ generated by, respectively, relators (2), (3') and relators (3) of Corollary 10.3.7. It was shown in Theorem 10.3.6 and Corollary 10.3.7 that

$$\mathcal{J} = \ker(H_G^*(M) \to H_G^*(\mathcal{BC}_d^n(\ell)) = \mathcal{I}_2 + \mathcal{I}_3\,. \tag{10.3.11}$$

In view of Proposition 10.3.10, we have to prove that the "quotient ideal"

$$\bar{\mathcal{J}} = \{x \in H_G^*(M) \mid u^{d-1}x \in \mathcal{J}\}$$

is equal to $\mathcal{I}_2 + \mathcal{I}_{3'}$.

That $\mathcal{I}_2 + \mathcal{I}_{3'} \subset \bar{\mathcal{J}}$ is obvious. For the reverse inclusion, let $x \in \bar{\mathcal{J}}$. By (10.3.11), one has $u^{d-1}x = y_2 + y_3$ for some $y_2 \in \mathcal{I}_2$ and $y_3 \in \mathcal{I}_3$. As $\mathcal{I}_3 = u^{d-1}\mathcal{I}_{3'}$, we can write $y_3 = u^{d-1}y_{3'}$ with $y_{3'} \in \mathcal{I}_{3'}$. Let $z = y + y_{3'}$. Then $u^{d-1}z \in \mathcal{I}_2$. We shall prove that $z \in \mathcal{I}_2$.

As noticed in the proof of Theorem 10.3.6, $H_G^*(M)$ is the free $Z[u]$-module generated by A_J ($J \subset \{1, \ldots, n-1\}$). Thus, z admits a unique expression

$$z = \sum_{J \subset \{1,\ldots,n-1\}} \lambda_J A_J\,,$$

with $\lambda_J \in \mathbb{Z}_2[u]$. Hence,

$$u^{d-1}z = \sum_{J \subset \{1,\ldots,n-1\}} (u^{d-1}\lambda_J)A_J\,. \tag{10.3.12}$$

But, as $u^{d-1}z \in \mathcal{I}_2$, one has

$$u^{d-1}z = \sum_{\substack{J \subset \{1,\ldots,n-1\} \\ J \cup \{n\} long}} \mu_J A_J\,. \tag{10.3.13}$$

As, $H_G^*(M)$ is the free $Z[u]$-module generated by the classes A_J, one deduces from (10.3.12) and (10.3.13) that $\lambda_J = 0$ if $J \cup \{n\}$ is short. Thus, $z \in \mathcal{I}_2$.

The equality $\bar{\mathcal{J}} = \mathcal{I}_2 + \mathcal{I}_{3'}$ may be also obtained using a partial Groebner calculus with respect to the variable u, as presented in [89, Sect. 6]. □

Remark 10.3.12 In the case $d = 2$, where $\bar{\mathcal{C}}_2^n(\ell) = \bar{\mathcal{N}}_2^n(\ell)$, the presentation of $H^*(\bar{\mathcal{C}}_2^n(\ell))$ of Theorem 10.3.11 was obtained in [89, Corollary 9.2], using techniques of toric manifolds.

Example 10.3.13 It is easy to see that $\mathcal{C}_d^n(\ell) = \emptyset$ if and only $\{k\}$ is long for some $k \in \{1, \ldots, n\}$. If $k = n$ then relator (2) for $J = \emptyset$ implies that $1 \in \bar{\mathcal{I}}_\ell$ and thus $H_G^*(\mathcal{C}_d^n(\ell)) = 0$. If $k < n$, it is relator (3') for $J = \{k\}$ which implies that $1 \in \bar{\mathcal{I}}_\ell$.

Example 10.3.14 Let $\ell = (1, 1, 1, \varepsilon)$, with $\varepsilon < 1$. The presentation of $H_G^*(\mathcal{BC}_d^4(\ell))$ given by Corollary 10.3.7 takes the form

$$H_G^*(\mathcal{BC}_d^n(\ell)) \approx \mathbb{Z}_2[u, A_1, A_2, A_3]/\hat{\mathcal{I}}_\ell$$

with $\hat{\mathcal{I}}_\ell$ being the ideal generated by $A_i^2 + A_i u^{d-1}$ $(i = 1, 2, 3)$, A_J for $|J| = 2$, and relators (3) for $J = \{1, 2\}, \{1, 3\}$ and $\{2, 3\}$, which are

$$\begin{aligned} &u^{d-1}(u^{d-1} + A_1 + A_2) \\ &u^{d-1}(u^{d-1} + A_2 + A_3) \\ &u^{d-1}(u^{d-1} + A_1 + A_3)\,. \end{aligned}$$

The sum of these relators equals $u^{2(d-1)}$ which thus belongs to $\hat{\mathcal{I}}_\ell$. Relator (3) for $J = \{1, 2, 3\}$ does not bring new generators for $\hat{\mathcal{I}}_\ell$.

The presentation of $H_G^*(\mathcal{C}_d^4(\ell))$ given by Theorem 10.3.11 is similar, with relators (3) replaced by relators (3'):

$$\begin{aligned} &u^{d-1} + A_1 + A_2 \\ &u^{d-1} + A_2 + A_3 \\ &u^{d-1} + A_1 + A_3\,. \end{aligned}$$

The sum of these relators being equal to u^{d-1}, we get that the three classes $A_i \in H_G^{d-1}(\mathcal{BC}_d^4(\ell))$ are mapped to the same class $A \in H_G^{d-1}(\mathcal{C}_d^4(\ell))$. Therefore,

$$H_G^*(\mathcal{C}_d^4(\ell)) \approx \mathbb{Z}_2[u, A]/(u^{d-1}, A^2)\,. \tag{10.3.14}$$

Note that $\mathcal{C}_d^4(\ell)$ is G-diffeomorphic to the unit tangent space $T^1 S^{d-1}$ (by orthonormalizing (z_1, z_2)). Thus, (10.3.14) is a presentation of $H^*(\bar{\mathcal{C}}_d^4(\ell)) \approx H^*((T^1 S^{d-1})/G)$.

In the presentations of $H_G^*(\mathcal{BC}_d^n(\ell))$ and $H_G^*(\mathcal{C}_d^n(\ell))$ given in Corollary 10.3.7 and Theorem 10.3.11, the integer d is only used to fix the degree of the variables A_i. Here is an application of that.

Lemma 10.3.15 *Let $\ell = (\ell_1, \dots, \ell_n)$ be a generic length vector. For $d \geq 2$, there is an isomorphism of graded rings*

$$\Psi_d^{\mathcal{BC}}: H_G^*(\mathcal{BC}_2^n(\ell)) \xrightarrow{\approx} H_G^{*(d-1)}(\mathcal{BC}_d^n(\ell)) \text{ and } \Psi_d^{\mathcal{C}}: H_G^*(\mathcal{C}_2^n(\ell)) \xrightarrow{\approx} H_G^{*(d-1)}(\mathcal{C}_d^n(\ell))$$

which multiply the degrees by $d - 1$.

Proof For an integer $a \geq 2$, set $M_a = (S^{a-1})^{n-1}$ and $\mathcal{BC}_a = \mathcal{BC}_a^n(\ell)$. As M_a is equivariantly formal, Corollary 4.7.20 applied to the bundle $M_a \to (M_a)_G \to \mathbb{R}P^\infty$ implies that

$$\mathfrak{P}_t(H_G^*(M_a)) = \mathfrak{P}_t(M_a) \cdot \mathfrak{P}_t(\mathbb{R}P^\infty) = \frac{(1 + t^{a-1})^{n-1}}{1 - t}.$$

Hence

$$\mathfrak{P}_t(H_G^{*(d-1)}(M_d)) = \frac{(1 + t^{d-1})^{n-1}}{1 - t^{d-1}} = \mathfrak{P}_{t^{d-1}}(H_G^*(M_2))$$

which implies that

$$\dim H_G^p(M_2) = \dim H_G^{p(d-1)}(M_d) \tag{10.3.15}$$

for all $p \in \mathbb{N}$.

By eliminating the variables B_i in the presentation of $H_G^*(M_a)$ given in Proposition 10.3.5, we get the presentation

$$H_G^*(M_a) \approx \mathbb{Z}_2[u_a, A_1^a \dots, A_{n-1}^a] / \left((A_i^a)^2 = u_a^{a-1} A_i^a\right),$$

where A_i^a is of degree $a - 1$ and u_a is of degree 1. Therefore, the correspondences $u_2 \mapsto u_d^{d-1}$ and $A_i^2 \mapsto A_i^a$ define a homomorphism of graded rings $\tilde{\Psi}_d: H_G^*(M_2) \to H_G^{*(d-1)}(M_d)$, multiplying the degrees by $d - 1$, which is clearly surjective. By (10.3.15), $\tilde{\Psi}_d$ is an isomorphism.

By Corollary 10.3.7, $H_G^*(\mathcal{BC}_a)$ is the quotient of $H_G^*(M_a)$ by $\mathcal{I}_2^a + \mathcal{I}_3^a$, where $\mathcal{I}_j^a$ is the ideal generated by relators (j) of Corollary 10.3.7. As $\tilde{\Psi}_d(\mathcal{I}_j^2) = \mathcal{I}_j^d \cap H_G^{*(d-1)}(M_d)$, the isomorphism $\tilde{\Psi}_d$ descends to the required isomorphism $\Psi_d^{\mathcal{BC}}$. In the same way, we construct $\Psi_d^{\mathcal{C}}$ using Theorem 10.3.11. □

10.3.3 Non-equivariant Cohomology

The G-cohomology computations of Corollary 10.3.7 and Theorem 10.3.11 give some information on the non-equivariant cohomology of $\mathcal{BC}_d^n(\ell)$ and $\mathcal{C}_d^n(\ell)$. We start with the big chain space.

Theorem 10.3.16 *Let $\ell = (\ell_1, \ldots, \ell_n)$ be a generic length vector. The Poincaré polynomial of $\mathcal{BC}_d^n(\ell)$ is*

$$\mathfrak{P}_t(\mathcal{BC}_d^n(\ell)) = \sum_{\substack{J\subset\{1,\ldots,n-1\}\\ J\cup\{n\}\text{short}}} t^{|J|(d-1)} + \sum_{\substack{K\subset\{1,\ldots,n-1\}\\ K\text{ long}}} t^{|K|(d-1)-1}. \tag{10.3.16}$$

The proof of Theorem 10.3.16 makes use the simplicial complex $\mathrm{Sh}_n^\times(\ell)$ whose simplexes are the non-empty subsets of the poset $\mathrm{Sh}_n(\ell)$.

Proof Let $\mathcal{BC}_d = \mathcal{BC}_d^n(\ell)$. By (7.1.8), one has a short exact sequence

$$0 \to H_G^*(\mathcal{BC}_d)/(u) \xrightarrow{\rho} H^*(\mathcal{BC}_d) \xrightarrow{\mathrm{tr}^*} \mathrm{Ann}\,(u) \to 0\,, \tag{10.3.17}$$

whence

$$\mathfrak{P}_t(\mathcal{BC}_d) = \mathfrak{P}_t(H_G^*(\mathcal{BC}_d)/(u)) + \mathfrak{P}_t(\mathrm{Ann}\,(u))\,. \tag{10.3.18}$$

The presentation of $H_G^*(\mathcal{BC}_d)$ given in Corollary 10.3.7 implies that

$$H_G^*(\mathcal{BC}_d)/(u) \approx \mathbb{Z}_2[A_1, \ldots, A_{n-1}]/\mathcal{J} \tag{10.3.19}$$

where $\mathcal{J}$ is the ideal generated by the squares A_i^2 of the variables and the monomials A_J when $J \cup \{n\}$ is long. Therefore,

$$H_G^*(\mathcal{BC}_d)/(u) \approx \Lambda_{d-1}(\mathrm{Sh}_n^\times(\ell))\,, \tag{10.3.20}$$

the face exterior algebra of the simplicial complex $\mathrm{Sh}_n^\times(\ell)$ (see Sect. 4.7.8). Then, by Corollary 4.7.52,

$$\mathfrak{P}_t(H_G^*(\mathcal{BC}_d)/(u)) = \mathfrak{P}_t(\Lambda_{d-1}(\mathrm{Sh}_n^\times(\ell))) = \sum_{\sigma\in\mathcal{S}(\mathrm{Sh}_n^\times(\ell))} t^{(\dim\sigma+1)(d-1)} = \sum_{J\in\mathrm{Sh}_n(\ell)} t^{|J|(d-1)}\,. \tag{10.3.21}$$

Let us assume that $d \geq 3$. The graded algebra $H_G^*(\mathcal{BC}_d)/(u)$ is concentrated in degrees $*(d-1)$. We claim that $\mathrm{Ann}\,(u)$ is concentrated in degrees $*(d-1)-1$. Indeed, let us write $H_G^*(\mathcal{BC}_d)$ as the quotient $H^*(M)/\mathcal{J}$ as in the proof of Theorem 10.3.11. A class $0 \neq z \in H_G^p(\mathcal{BC}_d)$ is the image of $\tilde{z} \in H_G^*(M)$. As M is equivariantly formal, one has $u\tilde{z} \neq 0$. Hence, if $z \in \mathrm{Ann}\,(u)$, one has $0 \neq u\tilde{z} \in \mathcal{J}$. As the ideal $\mathcal{J}$ is concentrated in degrees $*(d-1)$, we deduce that $p = q(d-1)-1$. Together with (10.3.17), this implies that

$$H^{*(d-1)}(\mathcal{BC}_d) \approx H_G^*(\mathcal{BC})/(u)\,,\; H^{*(d-1)-1}(\mathcal{BC}_d) \approx \mathrm{Ann}\,(u) \tag{10.3.22}$$

and $H^*(\mathcal{BC}_d)$ vanishes in other degrees. Since $\dim \mathcal{BC}_d = (n-1)(d-1)-1$, Poincaré duality gives the formula

$$\mathfrak{P}_t(\mathcal{BC}_d) = \sum_{J\in \mathrm{Sh}_n(\ell)} t^{|J|(d-1)} + \sum_{J\in \mathrm{Sh}_n(\ell)} t^{(n-1-|J|)(d-1)-1}. \tag{10.3.23}$$

Using (10.3.17), we thus get, for $d \geq 3$, that

$$\mathfrak{P}_t(\mathrm{Ann}\,(u)) = \sum_{J\in \mathrm{Sh}_n(\ell)} t^{((n-1-|J|)(d-1)-1)(d-1)} = \sum_{\substack{K\subset\{1,\dots,n-1\}\\ K \text{ long}}} t^{|K|(d-1)-1}, \tag{10.3.24}$$

where the last equality is obtained by re-indexing the sum with $K = \{1,\dots,n-1\}-J$. It remains to prove that (10.3.24) is also valid when $d = 2$.

Let us fix some integer $d \geq 3$. By Lemma 10.3.15 and its proof, there is an isomorphism of graded rings $\Psi_d^{\mathcal{BC}}\colon H^*(\mathcal{BC}_2) \xrightarrow{\approx} H^{*(d-1)}(\mathcal{BC}_d)$ such that

$$\Psi_d^{\mathcal{BC}}\big(\mathrm{Ann}\,(u; H^*(\mathcal{BC}_2)\big) = \mathrm{Ann}\,(u^{d-1}; H^{*(d-1)}(\mathcal{BC}_d)), \tag{10.3.25}$$

where the second argument in Ann( ) specifies the ring in which the first argument is considered. As the relators of the presentation of $H^*(\mathcal{BC}_d)$ given in Corollary 10.3.7 are in degree $*(d-1)$, the correspondence $x \mapsto u^{d-2}x$ provides, for every $p \geq 0$, an isomorphism of $\mathbb{Z}_2$-vector spaces

$$\Phi_d\colon H^{p(d-1)}(\mathcal{BC}_d) \xrightarrow{\approx} H^{(p+1)(d-1)-1}(\mathcal{BC}_d).$$

We thus get an isomorphism of $\mathbb{Z}_2$-vector spaces

$$\Phi_d\colon H^{*(d-1)}(\mathcal{BC}_d) \xrightarrow{\approx} H^{*(d-1)-1}(\mathcal{BC}_d)$$

multiplying the degrees by $d-2$ and satisfying

$$\Psi_d\big(\mathrm{Ann}\,(u^{d-1}; H^{*(d-1)}(\mathcal{BC}_d))\big) = \mathrm{Ann}\,(u; H^*(\mathcal{BC}_d)). \tag{10.3.26}$$

From (10.3.25) and (10.3.26), we get

$$t^{d-2}\, \mathfrak{P}_{t^{d-1}}\big(\mathrm{Ann}\,(u; H^*(\mathcal{BC}_2))\big) = \mathfrak{P}_t(\mathrm{Ann}\,(u; H^*(\mathcal{BC}_d))). \tag{10.3.27}$$

The right hand of (10.3.27) being given by (10.3.24), we checks that

$$\mathfrak{P}_t(\mathrm{Ann}\,(u; H^*(\mathcal{BC}_2))) = \sum_{\substack{K\subset\{1,\dots,n-1\}\\ K \text{ long}}} t^{|K|-1}$$

is the unique solution of Equation (10.3.27).

We have thus proven that (10.3.24) is valid for all $d \geq 2$. Together with (10.3.21) and (10.3.18), this establishes the proposition. □

Below is the counterpart of Theorem 10.3.16 for chain spaces. It requires the length vector ℓ be *dominated*, i.e. satisfying $\ell_n \geq \ell_i$ for $i \leq n$.

Theorem 10.3.17 *Let ℓ be a generic length vector which is dominated. Then, the Poincaré polynomial of $\mathcal{C}^n_d(\ell)$ is*

$$\mathfrak{P}_t(\mathcal{C}^n_d(\ell)) = \sum_{\substack{J\subset\{1,\dots,n-1\}\\ J\cup\{n\}\text{short}}} t^{|J|(d-1)} + \sum_{\substack{K\subset\{1,\dots,n-1\}\\ K\text{ long}}} t^{(|K|-1)(d-1)-1}. \tag{10.3.28}$$

Theorem 10.3.17 above reproves the computations of the Betti numbers of $\mathcal{C}^n_d(\ell)$ obtained by other methods in [60, Theorem 1] and [58, Theorem 2.1].

Proof The proof is the same as that of Theorem 10.3.16, using the isomorphism $\Psi^{\mathcal{C}}_d$ of Lemma 10.3.15, instead of $\Psi^{\mathcal{BC}}_d$. The hypothesis that ℓ is dominated is used to obtain the analogue of Equation (10.3.20), namely

$$H^*_G(\mathcal{C}^n_d(\ell))/(u) \approx \Lambda_{d-1}(\mathrm{Sh}^{\times}_n(\ell)). \tag{10.3.29}$$

Indeed, let $J \subset \{1,\dots,n-1\}$ be a long subset and $k \in J$. As ℓ is dominated, the set $(J - \{k\}) \cup \{n\}$ is long. Therefore, the constant terms in relators (3') of Theorem 10.3.11 vanish and these relators are all multiples of u^{d-1}. Equation (10.3.29) thus follows from Theorem 10.3.11. □

Example 10.3.18 The length vector $\ell = (1, 1, \dots, 1)$ is dominated and is generic if $n = 2r+1$. A subset J of $\{1,\dots,n\}$ is short if and only if $|J| \leq r$. Hence, for $d = 2$, Eq. (10.3.28) gives

$$\mathfrak{P}_t(\mathcal{C}^{2r+1}_2(1,\dots,1)) = \mathfrak{P}_t(\mathcal{N}^{2r+1}_2(1,\dots,1)) = \sum_{k\leq r-1} \binom{n-1}{k} t^k + \sum_{k\geq r-1} \binom{n-1}{k+2} t^k.$$

This formula was first proven in [110, Theorem C].

Remark 10.3.19 The hypothesis that ℓ is dominated is necessary (for any d) in Theorem 10.3.17, as shown by the example $\mathcal{C} = \mathcal{C}^4_d(\ell)$ for $\ell = (1, 1, 1, \varepsilon)$ (see Example 10.3.14). As $\mathcal{C}$ is diffeomorphic to the unit tangent space T^1S^{d-1}, one has $\mathfrak{P}_t(\mathcal{C}) = 1 + t^{d-2} + t^{d-1} + t^{2(d-1)-1}$, as seen in Example 5.4.4, while Theorem 10.3.17 would give $1 + 3t^{d-2} + 3t^{d-1} + t^{2(d-1)-1}$. What goes wrong is Formula (10.3.29). Using the presentation of $H^*_G(\mathcal{C})$ given in Theorem 10.3.11, one gets that $H^{*(d-1)}(\mathcal{C})$ is the quotient of $\Lambda_{d-1}(\mathrm{Sh}^{\times}_n(\ell))$ by the constant terms of relators (3') in Theorem 10.3.11, namely $\sum_{j\in J} A_{J-\{j\}}$ for all $J \subset \{1,\dots,n-1\}$ which are long.

Remark 10.3.20 Let $\mathcal{C} = \mathcal{C}_d^n(\ell)$ with ℓ generic and dominated. As observed in [65, Proposition A.2.4], $H^*(\mathcal{C})$ is determined by $H^{*(d-1)}(\mathcal{C})$ when $d > 3$, using Poincaré duality. Indeed, by Theorem 10.3.11 and Eq. (10.3.29), $\mathcal{Z} = \{Z_J = \rho(A_J) \mid J \in \mathrm{Sh}_n(\ell)\}$ is a $\mathbb{Z}_2$-basis of $H^{*(d-1)}(\mathcal{C})$ ($Z_\emptyset = \mathbf{1}$). The bilinear map $H^{*(d-1)}(\mathcal{C}) \times H^{*(d-1)-1}(\mathcal{C}) \to \mathbb{Z}_2$ given by

$$(x, y) \mapsto \langle x \smile y, [\mathcal{C}] \rangle$$

is non degenerate (see Theorem 5.3.12) and thus identifies $H^{*(d-1)-1}(\mathcal{C})$ with $H^{*(d-1)}(\mathcal{C})^\sharp$. Let $\mathcal{Y} = \{Y_J \mid J \in \mathrm{Sh}_n(\ell)\}$ be the $\mathbb{Z}_2$-basis of $H^{*(d-1)-1}(\mathcal{C})$ which is dual to $\mathcal{Z}$ under this identification. In particular, $Y_\emptyset = [\mathcal{C}]$, the generator of $H^{(n-2)(d-1)-1}(\mathcal{C}) = \mathbb{Z}_2$ (we say that $\mathcal{Y}$ is the *Poincaré dual basis* to the basis $\mathcal{Z}$). One has then the multiplication table.

$$Z_J \smile Z_K = \begin{cases} Z_{J\cup K} & \text{if } J \cap K = \emptyset \text{ and } J \cup K \in \mathrm{Sh}_n(\ell) \\ 0 & \text{otherwise,} \end{cases} \tag{10.3.30}$$

$$Z_J \smile Y_K = \begin{cases} Y_{J-K} & \text{if } K \subset J \\ 0 & \text{otherwise} \end{cases} \tag{10.3.31}$$

and

$$Y_J \smile Y_K = 0\,. \tag{10.3.32}$$

Indeed, (10.3.30) comes from the corresponding relation amongst the classes A_J. Formula (10.3.32) is true for dimensional reasons, since $d > 3$. For (10.3.31), note that, $Z_J \smile Y_K \in H^{(n-2-(|K|-|J|)(d-1)-1}(\mathcal{C})$ and hence may be uniquely written as a linear combination

$$Z_J \smile Y_K = \sum_{L\in\mathcal{L}} \lambda_L Y_L\,,$$

where $\mathcal{L}$ is the set $L \in \mathrm{Sh}_n(\ell)$ with $|L| = |K| - |J|$. If $I \in \mathcal{L}$, one has on one hand

$$\langle Z_I \smile \sum_{L\in\mathcal{L}} \lambda_L Y_L, [\mathcal{C}] \rangle = \lambda_I$$

and on the other hand

$$\langle Z_I \smile (Z_J \smile Y_K), [\mathcal{C}] \rangle = \langle Z_{I\cup J} \smile Y_K, [\mathcal{C}] \rangle = \begin{cases} 1 & \text{if } J \cap K = \emptyset \text{ and } I \cup J = k \\ 0 & \text{otherwise.} \end{cases}$$

This shows that $\lambda_L = 1$ if and only if $L = K - J$.

We finish this subsection with some illustrations and applications of Theorems 10.3.16 and 10.3.17. The *lopsidedness* lops (ℓ) of a length vector $\ell = (\ell_1, \dots, \ell_n)$ is defined by

$$\text{lops}\,(\ell) = \inf\{k \mid \exists\, J \subset \{1, \dots, n-1\} \text{ with } J \text{ long and } |J| = k\}\,,$$

with the convention that $\inf \emptyset = 0$. The terminology is inspired by that of [91]. If lops $(\ell) > 0$, one has

$$\dim \text{Sh}_n^{\times}(\ell) = n - 2 - \text{lops}\,(\ell)\,. \tag{10.3.33}$$

Example 10.3.21 For a generic length vector $\ell = (\ell_1, \dots, \ell_n)$, the condition lops $(\ell) = 0$ is equivalent to $\{n\}$ being long. By Theorem 10.3.16 this is equivalent to $\mathcal{BC}_d^n(\ell) = \emptyset$: otherwise $J = \emptyset$ produces a non-zero summand in the first sum of (10.3.20) (Compare Example 10.3.8). The chamber of ℓ is unique, represented by e.g. $\ell^0 = (\varepsilon, \dots, \varepsilon, 1)$ with $\varepsilon < 1/(n-1)$.

Example 10.3.22 Let $\ell = (\ell_1, \dots, \ell_n)$ be a generic length vector with lops $(\ell) = 1$. From the second sum of (10.3.20), we see that this is equivalent to $\tilde{H}^{d-2}(\mathcal{BC}_d^n(\ell)) \neq 0$ (the reduced cohomology is relevant for the cases $d = 2$, where it says that $\mathcal{BC}_2^n(\ell)$ is not connected). We check that $\text{Sh}_n(\ell)$ is poset isomorphic to $\text{Sh}_n(\ell^0)$ where $\ell^0 = (\varepsilon, \dots, \varepsilon, 2, 1)$, with $\varepsilon < 1/(n-2)$. By Lemma 10.3.3, the chamber of ℓ is well determined modulo the action of Sym_{n-1}. The $O(d-1)$-diffeomorphism type of $\mathcal{BC}_d^n(\ell_0)$ may be easily described. It is clear that $\mathcal{BC}_d^2(2, 1) \approx S^{d-2}$. Therefore, by Lemma 10.3.4, $\mathcal{BC}_d^n(\ell_0) \approx (S^{d-1})^{n-2} \times S^{d-2}$. When $d = 2$, this is the only case where $\mathcal{BC}_2^n(\ell)$ is not connected, as shown by Theorem 10.3.16. Note that $\mathcal{C}_d^n(\ell_0)$ is empty by Theorem 10.3.17, which is coherent with Proposition 10.3.10.

Example 10.3.23 Let $\ell = (\ell_1, \dots, \ell_n)$ be a dominated generic length vector with lops $(\ell) = 2$. We check that $\text{Sh}_n(\ell)$ is poset isomorphic to $\text{Sh}_n(\ell^0)$ where $\ell^0 = (\varepsilon, \dots, \varepsilon, 1, 1, 1)$, with $\varepsilon < 1/(n-3)$ ([85, Remark 2.4]). By Lemma 10.3.3, the chamber of ℓ is well determined modulo the action of Sym_{n-1}. As in the previous example, we can describe the $O(d-1)$-diffeomorphism type of $\mathcal{BC}_d^n(\ell_0)$. Suppose first that $n = 3$. The Morse function $f\colon (S^{(d-1)})^2 \to [-3, 3]$ of Lemma 10.3.1 has no critical point between its minimum and the level set $f^{-1}(-1) = \mathcal{BC}_d^3(1, 1, 1)$. By the Morse Lemma, $\mathcal{BC}_d^3(1, 1, 1)$ is diffeomorphic to $S^{2(d-1)-1}$. Using Lemma 10.3.4, we deduce that $\mathcal{BC}_d^n(\ell_0) \approx (S^{d-1})^{n-3} \times S^{2(d-1)-1}$. In the same way, one proves that $\mathcal{C}_d^n(\ell_0) \approx (S^{d-1})^{n-3} \times S^{d-2}$. Using Formula (10.3.28), we see that lops $(\ell) = 2$ if and only if $\mathcal{C}_2^n(\ell)$ is not connected.

The following lemma uses the nilpotency class **nil** introduced in Sect. 4.4.

Lemma 10.3.24 *Let $\ell = (\ell_1, \dots, \ell_n)$ be a generic length vector. Then*

(a) *If* lops $(\ell) > 1$, *then* lops $(\ell) = n - \textbf{nil}\, H^{>0}(\mathcal{BC}_d^n(\ell)) + 1$.

(b) *Suppose that ℓ is dominated. If $d > 2$ or* $\text{lops}(\ell) > 2$, *then* $\text{lops}(\ell) = n - \mathbf{nil}\, H^{>0}(C_d^n(\ell)) + 1$.

Proof Let $\mathcal{BC} = \mathcal{BC}_d^n(\ell)$. Suppose that $\text{lops}(\ell) = k \geq 2$. By (10.3.17) and (10.3.20), the algebra $H^*(\mathcal{BC})$ contains a copy of $H_G^*(\mathcal{BC})/(u) \approx \Lambda_{d-1}(\text{Sh}_n^\times(\ell))$. By (10.3.33), $\dim \text{Sh}_n^\times(\ell) = n - 2 - k$. Therefore, there exists $x_1, \dots, x_{n-k-1} \in H^{d-1}(\mathcal{BC})$ whose cup product v does not vanish in $H^{(n-k-1)(d-1)}(\mathcal{BC})$. By Poincaré duality Theorem 5.3.12, there is $w \in H^{k(d-1)-1}(\mathcal{BC})$ such that $v \smile w \neq 0$ in $H^{(n-1)(d-1)-1}(\mathcal{BC})$. As $k \geq 2$ the number $k(d-1)-1$ is strictly positive since $d \geq 2$. Thus, $v \smile w$ is a non-vanishing cup product of length $n-k$. Such a length is the maximal possible, as seen using Sequence (10.3.17). Hence, $\mathbf{nil}\, H^{>0}(\mathcal{BC}_d^n(\ell)) = n-k+1$. This proves (a).

The proof of (b) is similar, using Theorem 10.3.17 and its proof instead of Theorem 10.3.16. As $\dim C = (n-2)(d-1) - 1$, the class v is of degree $(k-1)(d-1) - 1$. The latter is strictly positive if $d > 2$ or $k > 2$. □

Corollary 10.3.25 *Let $\ell = (\ell_1, \dots, \ell_n)$ and $\ell' = (\ell'_1, \dots, \ell'_n)$ be two generic length vectors. Suppose that, for some $d \geq 2$, there exists a* **GrA***-isomorphism $H^*(\mathcal{BC}_d^n(\ell)) \approx H^*(\mathcal{BC}_d^n(\ell'))$. Then*

$$\text{lops}(\ell) = \text{lops}(\ell')\,.$$

If ℓ and ℓ' are both dominated, then the above equality holds true if there exists a **GrA***-isomorphism $H^*(C_d^n(\ell)) \approx H^*(C_d^n(\ell'))$.*

Proof For the big chain space $\mathcal{BC}_d^n()$, this is follows from Lemma 10.3.24, except when $d = 2$ and $\text{lops}(\ell) \leq 1$, cases which are covered by Examples 10.3.21 and 10.3.22. The argument for $C_d^n()$ is quite similar. The case $\text{lops}(\ell) = 2$ is covered by Example 10.3.23. The case $\text{lops}(\ell) = 1$ is not possible if ℓ is dominated, so $\text{lops}(\ell) = 0$ is equivalent to $C_d^n(\ell) = \emptyset$. □

10.3.4 The Inverse Problem

By Lemma 10.3.3, the diffeomorphism type of $\mathcal{BC}_d^n(\ell)$ or $C_d^n(\ell)$ is determined by the chamber $\text{Ch}(\ell)$ (up to the action of Sym_{n-1}). The *inverse problem* consists of recovering $\text{Ch}(\ell)$ by algebraic topology invariants of $\mathcal{BC}_d^n(\ell)$ (or $C_d^n(\ell)$). We start by the big chain space.

Proposition 10.3.26 *Let $\ell = (\ell_1, \dots, \ell_n)$ and $\ell' = (\ell'_1, \dots, \ell'_n)$ be two generic length vectors. Then, the following conditions are equivalent.*

(1) $\text{Ch}(\ell') = \text{Ch}(\ell^\sigma)$ *for some* $\sigma \in \text{Sym}_{n-1}$.
(2) $\mathcal{BC}_d^n(\ell)$ *and* $\mathcal{BC}_d^n(\ell')$ *are $O(d-1)$-diffeomorphic.*
(3) $H_G^*(\mathcal{BC}_d^n(\ell))$ *and* $H_G^*(\mathcal{BC}_d^n(\ell'))$ *are* **GrA**$[u]$*-isomorphic.*

Moreover, if $d > 2$ or $n > 3$, any condition (1)–(3) above is equivalent to

(4) *$H^*(\overline{\mathcal{BC}}^n_d(\ell))$ and $H^*(\overline{\mathcal{BC}}^n_d(\ell'))$ are* **GrA**-*isomorphic.*

Finally, if $d > 2$ or if $\mathrm{lops}(\ell) \neq 2$, *then any condition (1)–(3) above is equivalent to*

(5) *$H^*(\mathcal{BC}^n_d(\ell))$ and $H^*(\mathcal{BC}^n_d(\ell'))$ are* **GrA**-*isomorphic.*

That (5) implies (1) is not known in general if $d = 2$. That (4) implies (3) is wrong if $n = 3$ and $d = 2$. Indeed, $\overline{\mathcal{BC}}^3_2(1, 3, 1)$ and $\overline{\mathcal{BC}}^3_2(1, 1, 1)$ are connected closed 1-dimensional manifolds, thus both diffeomorphic to S^1 but, by Corollary 10.3.7, one has

$$H^*_G(\mathcal{BC}^3_2(1, 3, 1)) \approx \mathbb{Z}_2[u, A_1]/(A_1^2, u) \text{ while } H^*_G(\mathcal{BC}^3_2(1, 1, 1)) \approx \mathbb{Z}_2[u]/(u^2)\,.$$

Implications like (4) $\Rightarrow$ (2) or (5) $\Rightarrow$ (2) are in the spirit of Proposition 4.2.5: characterizing a closed manifold (within some class) by algebraic topology tools. This was the historical goal of algebraic topology (see p. 201).

Proof As ℓ and ℓ' are generic, one has $H^*(\overline{\mathcal{BC}}^n_d(\ell)) \approx H^*_G(\mathcal{BC}^n_d(\ell))$ and the same for ℓ'. The following implications are then obvious, except (a) which was established in Lemma 10.3.3.

$$\begin{array}{ccccccc} (1) & \overset{(a)}{\Longrightarrow} & (2) & \Longrightarrow & (3) & \Longrightarrow & (4) \\ & & & \searrow & & & \\ & & & & (5) & & \end{array}$$

We shall now prove that (3) $\Rightarrow$ (1), (4) $\Rightarrow$ (3) and finally (5) $\Rightarrow$ (1).

(3) $\Rightarrow$ (1). A **GrA**$[u]$-isomorphism $H^*_G(\mathcal{BC}^n_d(\ell)) \xrightarrow{\approx} H^*_G(\mathcal{BC}^n_d(\ell'))$ descends to a **GrA**-isomorphism : $H^*_G(\mathcal{BC}^n_d(\ell))/(u) \xrightarrow{\approx} H^*_G(\mathcal{BC}^n_d(\ell'))/(u)$. By (10.3.20), this implies that $\Lambda_{d-1}(\mathrm{Sh}^{\times}_n(\ell))$ and $\Lambda_{d-1}(\mathrm{Sh}^{\times}_n(\ell'))$ are **GrA**-isomorphic. Using Lemma 4.7.51 and Proposition 4.7.50, we deduce that the simplicial complexes $\mathrm{Sh}^{\times}_n(\ell)$ and $\mathrm{Sh}^{\times}_n(\ell')$ are isomorphic. It follows that $\mathrm{Sh}_n(\ell)$ and $\mathrm{Sh}_n(\ell')$ are poset isomorphic. By Lemma 10.3.3, this implies (1).

(4) $\Rightarrow$ (3). Let β: $H^*(\overline{\mathcal{BC}}^n_d(\ell)) \xrightarrow{\approx} H^*(\overline{\mathcal{BC}}^n_d(\ell'))$ is a **GrA**-isomorphism and let $\beta(u) = v$. We must prove that $v = u$. This is obvious for $d > 2$ since $H^1(\overline{\mathcal{BC}}^n_d(\ell')) = \mathbb{Z}_2$. If $d = 2$, Corollary 10.3.7 implies that

$$xu = xv = x^2 \tag{10.3.34}$$

for all $x \in H^1(\overline{\mathcal{BC}}^n_2(\ell'))$. By Corollary 10.3.7 again, $H^*(\overline{\mathcal{BC}}^n_2(\ell'))$ is generated in degree 1, so (10.3.34) implies that $(u + v)x = 0$ for all $x \in H^*(\overline{\mathcal{BC}}^n_2(\ell'))$. By Corollary 10.3.2, $\overline{\mathcal{BC}}^n_2(\ell')$ is a closed manifold of dimension > 1 (since $n \geq 4$). We conclude that $v = u$ by Poincaré duality, using Theorem 5.3.12.

(5)⇒(1). Suppose first that $d > 2$. By (10.3.22) and (10.3.20), Condition (5) implies that $\Lambda_{d-1}(\mathrm{Sh}_n^\times(\ell))$ and $\Lambda_{d-1}(\mathrm{Sh}_n^\times(\ell'))$ are **GrA**-isomorphic. The argument is then the same as that for (3) ⇒(1).

We now assume that $d = 2$. By Corollary 10.3.25, Condition (5) implies that $\mathrm{lops}\,(\ell) = \mathrm{lops}\,(\ell')$. Let $L = \mathrm{lops}\,(\ell) = \mathrm{lops}\,(\ell')$. The cases $L = 0, 1$ were treated in Examples 10.3.21 and 10.3.22. Let us assume that $L > 2$. By Theorem 10.3.16 and its proof, the subalgebra of $H^*(\mathcal{B}C_2^n(\ell))$ (respectively: $H^*(\mathcal{B}C_2^n(\ell'))$) generated by the elements of degree one is isomorphic to $\Lambda_1(\mathrm{Sh}_n^\times(\ell))$ (respectively: $\Lambda_1(\mathrm{Sh}_n^\times(\ell'))$). By Condition (5), this implies that $\Lambda_1(\mathrm{Sh}_n^\times(\ell))$ and $\Lambda_1(\mathrm{Sh}_n^\times(\ell'))$ are **GrA**-isomorphic and the proof that $\mathrm{Ch}(\ell') = \mathrm{Ch}(\ell^\sigma)$ proceeds as that for (3) ⇒(1). □

Here is the analogue of Proposition 10.3.26 for the chain spaces.

Proposition 10.3.27 *Let $\ell = (\ell_1, \ldots, \ell_n)$ and $\ell' = (\ell'_1, \ldots, \ell'_n)$ be two generic length vectors. Suppose that ℓ and ℓ' are dominated. Then, the following conditions are equivalent.*

(1) $\mathrm{Ch}(\ell') = \mathrm{Ch}(\ell^\sigma)$ *for some* $\sigma \in \mathrm{Sym}_{n-1}$.
(2) $C_d^n(\ell)$ *and* $C_d^n(\ell')$ *are* $O(d-1)$*-diffeomorphic.*
(3) $H_G^*(C_d^n(\ell))$ *and* $H_G^*(C_d^n(\ell'))$ *are* **GrA**$[u]$*-isomorphic.*

Moreover, if $d > 2$ or $n > 4$, then any condition (1)–(3) above is equivalent to

(4) $H^*(\bar{C}_d^n(\ell))$ *and* $H^*(\bar{C}_d^n(\ell'))$ *are* **GrA**-*isomorphic.*

Finally, if $d > 2$ or if $\mathrm{lops}\,(\ell) \neq 3$, *then any condition (1)–(3) above is equivalent to*

(5) $H^*(C_d^n(\ell))$ *and* $H^*(C_d^n(\ell'))$ *are* **GrA**-*isomorphic.*

Proof The proof is the same as that of Proposition 10.3.26, except for the following small differences. For (3) ⇒(1), instead of (10.3.16), one uses Equation (10.3.29), using that ℓ and ℓ' are dominated. For (4) ⇒(3), the hypothesis that $n > 4$ guarantees that $\dim C_2^n() > 1$. For (5) ⇒ (1), one uses Theorem 10.3.17 instead of Theorem 10.3.16. □

Remark 10.3.28 In Proposition 10.3.27, implication (4) ⇒ (1) is wrong for $d = 2$ and $n = 4$: $\mathcal{B}C_2^4(1, 1, 1, 2)$ and $\mathcal{B}C_2^4(1, 2, 2, 2)$ are connected closed 1-dimensional manifolds, thus both diffeomorphic to S^1. Implication (5) ⇒ (1) is not known in general if $d = 2$. It is however true if one uses the integral cohomology: this difficult result, conjectured by Walker in 1985 [202] was proved by Schütz in 2010 [170], after being established when $\mathrm{lops}\,(\ell) \neq 3$ in [58, Theorem 4] (length vectors with lopsidedness > 3 are called *normal* in [58, 170]).

The hypothesis that ℓ is dominated in Proposition 10.3.27 is essential, as seen by Proposition 10.3.29 and Lemma 10.3.30 below.

Proposition 10.3.29 *Let ℓ be a generic length vector and let $\sigma \in \mathrm{Sym}_n$. If $d \neq 3$, then $H^*(C_d^n(\ell))$ and $H^*(C_d^n(\ell^\sigma))$ are* **GrA**–*isomorphic.*

Proposition 10.3.29 was first proved by V. Fromm in his thesis [65, Cor. 1.2.5]. It is wrong if $d = 3$: for ε small, $\mathcal{C}_3^4(\varepsilon, 1, 1, 1)$ is diffeomorphic to $S^2 \times S^1$ (see Example 10.3.23) while $\mathcal{C}_3^4(1, 1, 1, \varepsilon)$ is diffeomorphic to $T^1 S^2 \approx \mathbb{R}P^3$ (see Example 10.3.14).

We give below a proof of Proposition 10.3.29 based on an idea of D. Schütz, using the following lemma.

Lemma 10.3.30 *If $d = 2, 4, 8$, then $\mathcal{C}_d^n(\ell)$ is diffeomorphic to $\mathcal{C}_d^n(\ell^\sigma)$ for any $\sigma \in$ Sym_n.*

The hypothesis $d = 2, 4, 8$ is essential in the above lemma. Indeed, for ε small, $\mathcal{C}_d^4(\varepsilon, 1, 1, 1)$ is diffeomorphic to $S^{d-1} \times S^{d-2}$ (see Example 10.3.23) while $\mathcal{C}_d^4(1, 1, 1, \varepsilon)$ is diffeomorphic to $T^1 S^{d-1}$ (see Example 10.3.14). As $d \geq 2$, these two spaces have the same homotopy type only when $d = 2, 4, 8$ (see Example 5.4.18).

Proof Identifying $\mathbb{R}^d$ with $\mathbb{C}$, $\mathbb{H}$ or $\mathbb{O}$, we get a smooth multiplication on S^{d-1} with e_1 as unit element. Consider the smooth map $\pi\colon \tilde{\mathcal{N}}_d^n(\ell) \to \mathcal{C}_d^n(\ell)$ given by

$$\pi(z_1 \ldots, z_n) = -z_n^{-1}(z_1, \ldots, z_{n-1}) .$$

The embedding $j\colon \mathcal{C}_d^n(\ell) \to \tilde{\mathcal{N}}_d^n(\ell)$ given by

$$j(z_1, \ldots, z_{n-1}) = (z_1, \ldots, z_{n-1}, -e_1)$$

is a section of π. Consider the composed map

$$\mathcal{C}_d^n(\ell) \xrightarrow{\ j\ } \tilde{\mathcal{N}}_d^n(\ell) \xrightarrow[\approx]{h^\sigma} \tilde{\mathcal{N}}_d^n(\ell^\sigma) \xrightarrow{\ \pi\ } \mathcal{C}_d^n(\ell^\sigma)$$

where $h^\sigma(z) = z^\sigma$. Then, $\pi \circ h^\sigma \circ j$ is a diffeomorphism: a direct computation shows that its inverse is $\pi \circ (h^\sigma)^{-1} \circ j$. □

Proof of Proposition 10.3.29 The case $d = 2$ is covered by Lemma 10.3.30, so we assume that $d \geq 4$. As observe in Remark 10.3.19, one has a **GrA**-isomorphism

$$H^{*(d-1)}(\mathcal{C}_d^n(\ell) \approx \Omega_d(\ell) \tag{10.3.35}$$

where $\Omega_d(\ell)$ is the quotient of $\mathbb{Z}_2[A_1, \ldots, A_{n-1}]$ (A_i of degree $d-1$) by the ideal generated by A_i^2, A_J when $J \cup \{n\}$ is ℓ-long and $\sum_{j\in J} A_{J-\{j\}}$ when J is ℓ-long. By Lemma 10.3.30, there exists a ring isomorphism $q_4\colon \Omega_4(\ell) \xrightarrow{\approx} \Omega_4(\ell^\sigma)$. But, in the definition of $\Omega_d()$, the integer d is only used to fix the degree of the variables A_i, and thus the grading of $\Omega_d()$ (as non-graded rings, the rings $\Omega_d(\ell)$ are isomorphic for all d). Therefore, the isomorphism q_4 defines a **GrA**-isomorphism $q_d\colon \Omega_d(\ell) \xrightarrow{\approx} \Omega_d(\ell^\sigma)$ for all $d \geq 2$. Together with (10.3.35), this gives a **GrA**-isomorphism $q_d\colon H^{*(d-1)}(\mathcal{C}_d^n(\ell)) \xrightarrow{\approx} H^{*(d-1)}(\mathcal{C}_d^n(\ell^\sigma))$ when $d \geq 3$.

Without loss of generality, we may assume that ℓ is dominated. Remark 10.3.20 thus provides an additive basis $\mathcal{Z} \cup \mathcal{Y}$ of $H^*(\mathcal{C}_d^n(\ell))$. The set $\mathcal{Z}' = \{Z_J = q_d(Z_J) \mid J \in \mathrm{Sh}(\ell)\}$ is then a $\mathbb{Z}_2$-basis of $H^{*(d-1)}(\mathcal{C}_d^n(\ell^\sigma)$. Let $\mathcal{Y}' = \{Y_J' \mid J \in \mathrm{Sh}(\ell)\}$ be the basis of $H^{*(d-1)-1}(\mathcal{C}_d^n(\ell^\sigma)$ which is Poincaré dual to $\mathcal{Z}'$, as in Remark 10.3.20. Relations (10.3.30)–(10.3.32) hold true both for $\mathcal{Z} \cup \mathcal{Y}$ in $H^*(\mathcal{C}_d^n(\ell))$ and for $\mathcal{Z}' \cup \mathcal{Y}'$ in $H^*(\mathcal{C}_d^n(\ell^\sigma))$. Therefore, q_d extends to a **GrA**-isomorphism $\bar{q}_d\colon H^*(\mathcal{C}_d^n(\ell)) \xrightarrow{\approx} H^*(\mathcal{C}_d^n(\ell^\sigma))$ when $d > 3$.

10.3.5 Spatial Polygon Spaces and Conjugation Spaces

The integral cohomology ring of the *spatial polygon space* $\mathcal{N}_3^n(\ell)$ has been computed in [89, Theorem 6.4]. The result is as follows.

Theorem 10.3.31 *Let $\ell = (\ell_1 \dots, \ell_n)$ be a generic length vector. Then, there is a graded ring isomorphism*

$$\mathbb{Z}[v, A_1 \dots, A_{n-1}]/\bar{\mathcal{I}}_\ell \xrightarrow{\approx} H^*(\mathcal{N}_3^n(\ell); \mathbb{Z})$$

where the variables v and A_i are of degree 2 and $\bar{\mathcal{I}}_\ell$ is the ideal generated by the families of relators

$$\begin{array}{lll} (1)\ A_i^2 + A_i v & i = 1, \dots, n-1 & \\ (2)\ A_J & J \subset \{1, \dots, n-1\} & \textit{and } J \cup \{n\} \textit{ is long} \\ (3)\ \sum\limits_{K \subset J} A_K\, v^{|J-K|-1} & J \subset \{1, \dots, n-1\} & \textit{and } J \textit{ is long} \end{array}$$

Theorem 10.3.31 says in particular that $H^{odd}(\mathcal{N}_3^n(\ell); \mathbb{Z}) = 0$. The Bockstein exact sequence for $0 \to \mathbb{Z} \to \mathbb{Z} \to \mathbb{Z}^2 \to 0$ (see [179, Chap. 5, Sect. 2, Theorem 11]) thus implies that $H^*(\mathcal{N}_3^n(\ell)) \approx H^*(\mathcal{N}_3^n(\ell); \mathbb{Z}) \otimes \mathbb{Z}_2$. Using Theorem 10.3.11, the correspondence $A_i \mapsto A_i$ and $v \mapsto u$ thus provides a graded ring isomorphism

$$H^{2*}(\mathcal{N}_3^n(\ell)) \xrightarrow{\approx} H^*(\bar{\mathcal{C}}_2^n(\ell)) \approx H^*(\bar{\mathcal{N}}_2^n(\ell))$$

which divides the degrees by half. This suggests that $\mathcal{N}_3^n(\ell)$ is a conjugation space, which we shall prove below. The involution $\hat{\tau}$ on $\mathbb{R}^3$ given by the reflection through the horizontal plane induces an involution, still called $\hat{\tau}$, on $\mathcal{N}_3^n(\ell)$, with fixed point set equal to $\bar{\mathcal{N}}_2^n(\ell)$.

Proposition 10.3.32 *Let $\ell = (\ell_1, \dots, \ell_n)$ be a generic length vector. Then, the space $\mathcal{N}_3^n(\ell)$ endowed with the involution $\hat{\tau}$ is a conjugation manifold.*

The proofs of this proposition use some Hamiltonian geometry, directly or indirectly, so their understanding requires some knowledge in the subject, as presented in e.g. [12, Chapters 2 and 3]. Recall that, if ℓ is generic, $\mathcal{N}_3^n(\ell)$ is a symplectic manifold [88, 89, 111, 118]. The proof of Proposition 10.3.32 uses the following lemma.

Lemma 10.3.33 *Let $\ell = (\ell_1, \dots, \ell_n)$ be a generic length vector. Then,*

(a) *$\mathcal{N}_3^n(\ell)$ is simply connected.*
(b) *$\hat{\tau}$ induces the multiplication by (-1) on $H^2(\mathcal{N}_3^n(\ell); \mathbb{Z})$.*
(c) *$\hat{\tau}$ is anti-symplectic.*

Proof The proof of (a) and (b) proceeds by induction on n. When $n = 3$, there are two chambers up to permutation, represented by $\ell^0 = (1, 1, 2)$ and $\ell^1 = (1, 1, 1)$. As $\mathcal{N}_3^3(\ell_0) = \emptyset$ and $\mathcal{N}_3^3(\ell_1) = \bar{\mathcal{N}}_2^3(\ell_1) = pt$, (a) and (b) are true. For the induction step, we consider the diagonal-length function $\delta\colon \mathcal{N}_3^n(\ell) \to \mathbb{R}$ given by $\delta(z) = |\ell_n z_n - \ell_{n-1} z_{n-1}|$. By Lemma 10.3.3, we can slightly change ℓ_{n-1} without modifying the $O(2)$-diffeomorphism type of $\mathcal{C}_3^n(\ell)$ and thus the $\hat{\tau}$-equivariant diffeomorphism type of $\mathcal{N}_3^n(\ell)$. Therefore, we can assume that $\ell_{n-1} \neq \ell_n$, in which case δ is a smooth map. Using [83, Theorem 3.2], we deduce that δ is Morse-Bott function. The critical points are of even index and are isolated, except possibly for the two extrema. The preimage of the maximum is either a point or $\mathcal{N}_3^{n-1}(\ell_1, \dots, \ell_{n-2}, \ell_{n-1} + \ell_n)$ and the preimage of the minimum is either a point or $\mathcal{N}_3^{n-1}(\ell_1, \dots, \ell_{n-2}, |\ell_{n-1} - \ell_n|)$. This proves (a) by induction.

The restriction δ' of δ to $\bar{\mathcal{N}}_2^n(\ell) \subset \mathcal{N}_3^{n-1}(\ell)$ is also Morse-Bott with $\mathrm{Crit}\,\delta' = \mathrm{Crit}\,\delta$ but the index of each critical point is divided in half. Thus, passing a critical point of δ index 2 corresponds to add a conjugation 2-cell. This proves (b) by induction on n.

To prove (c), we use that $\mathcal{N}_3^{n-1}(\ell)$ is the $SO(3)$-symplectic reduction at 0 of $\prod_{i=0}^n S^2_{\ell_i}$ where $S^2_{\ell_i}$ is the standard 2 sphere equipped with the $SO(3)$-homogeneous symplectic form with symplectic volume equal to $2\ell_i$ (see [89]). The involution $\hat{\tau}$ is clearly anti-symplectic on $S^2_{\ell_i}$ and this property descends to the symplectic reduction. □

Proof of Proposition 10.3.32 We give below three proofs. The first one is the one indicated in [87, Example 8.7].

1st proof. By induction on n, using the function $\delta\colon \mathcal{N}_3^n(\ell) \to \mathbb{R}$ of the proof of Lemma 10.3.33. For $n = 3$, $\mathcal{N}_3^3(\ell)$ is either empty or a point, which starts the induction. The induction step uses that δ is the moment map of an S^1-Hamiltonian action on $\mathcal{N}_3^n(\ell)$ [118] and called the *bending flow* [111]. It may be visualized as a rotation of z_{n-1} and z_n at constant speed around of axis $\ell_n z_n + \ell_{n-1} z_{n-1}$, leaving the other z_i's fixed. As a moment map for a circle action, it satisfies $\mathrm{Crit}\,\delta = (\mathcal{N}_3^n(\ell))^{S^1}$. We have seen that the critical points of δ are isolated or polygon spaces with fewer edges. Hence, by induction hypothesis, $(\mathcal{N}_3^n(\ell))^{S^1}$ is a conjugation space. The involution $\hat{\tau}$ is anti-symplectic by Lemma 10.3.33 and satisfies $\hat{\tau}(\gamma z) = \gamma^{-1}\hat{\tau}(z)$ for all $z \in \mathcal{N}_3^3(\ell)$ and $\gamma \in S^1$. That $\mathcal{N}_3^n(\ell)$ is a conjugation manifold thus follows from [87, Theorem 8.3].

2nd proof. We use that $\mathcal{N}_3^n(\ell)$ is a symplectic reduction for the Hamiltonian action of the maximal torus of $U(n)$ on $\mathrm{Gr}(2;\mathbb{C}^n)$ (see [88, Theorem 4.4]). The complex conjugation on $\mathrm{Gr}(2;\mathbb{C}^n)$ is anti-symplectic, with fixed point set equal to $\mathrm{Gr}(2;\mathbb{R}^n)$, descends to the involution $\hat{\tau}$ on $\mathcal{N}_3^n(\ell)$. The manifold $\mathrm{Gr}(2;\mathbb{C}^n)$ with the complex conjugation is a conjugation space (see p. 438 or Remark 9.7.9). That $\mathcal{N}_3^n(\ell)$ is a conjugation space thus follows from [87, Theorem 8.12].

3rd proof. By Theorem 10.3.31, $H^{odd}(\mathcal{N}_3^n(\ell)) = 0$ and $H^*(\mathcal{N}_3^n(\ell))$ is generated by $H^2(\mathcal{N}_3^n(\ell))$. By Lemma 10.3.33, $\mathcal{N}_3^n(\ell)$ is simply connected and $H^2\hat{\tau}$ is multiplication by (-1). Therefore, $\mathcal{N}_3^n(\ell)$ is a conjugation manifold by results of V. Puppe (see [166, Theorem 5 and Remark 2]).

Remark 10.3.34 The quotient $\mathcal{BN}_3^n(\ell) = \mathcal{BC}_3^n(\ell)/SO(2)$ is called in [89] the *abelian polygon space* (being an S^1-symplectic reduction while $\mathcal{N}_3^n(\ell)$ is an $SO(3)$-symplectic reduction). In [89, Theorem 6.4], the integral cohomology ring of $\mathcal{BN}_3^n(\ell)$ is computed: the statement is as that of Theorem 10.3.31 but with the relators of Corollary 10.3.7. The involution $\hat{\tau}$ is defined on $\mathcal{BN}_3^n(\ell)$ with fixed point set $\mathcal{BC}_2^n(\ell)$. The 3rd proof of Proposition 10.3.32 may be easily adapted to show that $\hat{\tau}$ is a conjugation on $\mathcal{BN}_3^n(\ell)$.

10.4 Equivariant Characteristic Classes

Let Γ be a topological group and let X be a Γ-space. Let $\xi = (p : E \to X)$ be a Γ-equivariant vector bundle over X. Recall from Lemma 7.5.6 that ξ is then induced by $i: X \to X_\Gamma$ from the vector bundle ξ_Γ over X_Γ. The Stiefel-Whitney classes of ξ_Γ are called the *equivariant Stiefel-Whitney classes* of ξ. Hence, $w(\xi) = \rho(w(\xi_\Gamma))$ where $\rho: H^*_\Gamma(X) \to H^*(X)$ is the forgetful homomorphism.

An important example is given by the tautological bundle ξ_j $(j = 1, \dots, r)$ over the flag manifold $\mathrm{Fl}(n_1, \dots, n_r)$ (see Sect. 9.5.2), which is an $O(n)$-equivariant vector bundle of rank n_j.

Proposition 10.4.1 *Let R be a closed subgroup of $O(n)$. Then, as an $H^*(BR)$-algebra, $H^*_R(\mathrm{Fl}(n_1, \dots, n_r))$ is generated by the equivariant Stiefel-Whitney classes $w_i((\xi_j)_R)$ $(j = 1 \dots, r,\ i = 1, \dots, n_j)$ of the tautological bundles. In particular, $\mathrm{Fl}(n_1, \dots, n_r)$ is R-equivariantly formal.*

Proof We have noticed above that $w_i(\xi_j) = \rho(w_i((\xi_j)_R))$. By Theorem 9.5.15, $H^*(\mathrm{Fl}(n_1, \dots, n_r))$ is additively generated by the monomials in the $w_i(\xi_j)$'s. Such a monomial is the image by ρ of the corresponding monomial in the $w_i((\xi_j)_R)$'s and $\mathrm{Fl}(n_1, \dots, n_r)$ is thus R-equivariantly formal. By the Leray-Hirsch theorem 4.7.17, the monomials in the $w_i((\xi_j)R)$'s generate $H^*_R(\mathrm{Fl}(n_1, \dots, n_r))$ as an $H^*(BR)$-module, whence the proposition. □

We specialize to the Grassmannian $\mathrm{Gr}(k;\mathbb{R}^n) = \mathrm{Fl}(k, n-k)$ with $R = T_2$, the maximal 2-torus of diagonal matrices in $O(n)$. By (7.5.14), $H^*(BT_2) \approx \mathbb{Z}_2$

$[u_1, \ldots, u_n]$, with $\deg(u_i) = 1$. The tautological bundles are $\zeta = \xi_1$ and $\zeta^\perp = \xi_2$ (see Example 9.5.20). As $\mathrm{Gr}(k; \mathbb{R}^n)$ is T_2-equivariantly formal, the restriction to the fixed points $r\colon H^*_{T_2}(\mathrm{Gr}(k; \mathbb{R}^n)) \to H^*_{T_2}(\mathrm{Gr}(k; \mathbb{R}^n)^{T_2})$ is injective by Theorem 7.6.6. The fixed point set $\mathrm{Gr}(k; \mathbb{R}^n)^{T_2}$ is discrete and in bijection with $[{n \atop k}]$, the set of binary words $\lambda = \lambda_1 \cdots \lambda_n$ such that $\sum \lambda_i = k$. We identify $\mathrm{Gr}(k; \mathbb{R}^n)^{T_2}$ with $[{n \atop k}]$ via this bijection, which associates to λ the coordinate k-plane

$$\Pi_\lambda = \{(t_1, \ldots, t_n) \in \mathbb{R}^n \mid t_i = 0 \text{ if } \lambda_i = 0\}.$$

Note that Π_λ is the "center" of the Schubert cell C^F_λ for the standard complete flag F in $\mathbb{R}^n$ (see Sect. 9.5.3).

Proposition 10.4.2 *For $\lambda = \lambda_1 \cdots \lambda_n \in [{n \atop k}]$, the restriction homomorphism*

$$r_\lambda\colon H^*_{T_2}(\mathrm{Gr}(k; \mathbb{R}^n)) \to H^*_{T_2}(\{\lambda\}) \approx \mathbb{Z}_2[u_1, \ldots, u_n]$$

satisfies

$$r_\lambda(w(\zeta_{T_2})) = \prod_{\lambda_i = 1} (1 + u_i) \text{ and } r_\lambda(w((\zeta^\perp)_{T_2}) = \prod_{\lambda_i = 0} (1 + u_i).$$

We may note that the classes $r_\lambda(w(\zeta_{T_2}))$ and $r_\lambda(w(\zeta^\perp_{T_2}))$ satisfy the GKM-conditions (see p. 483)

Proof Let $E(\zeta)_\lambda$ be the fiber of ζ over π_λ, seen as a T_2-equivariant vector bundle over λ. One has a T_2-equivariant isomorphism

$$E(\zeta)_\lambda \approx \bigoplus_{\lambda_i = 1} L(u_i)$$

where $L(u_i)$ is the T_2-equivariant line bundle over λ, on which T_2 acts via the homomorphism $\mathrm{dia}(\delta_1, \ldots, \delta_n) \mapsto \delta_i$. This homomorphism is associated to u_i under the bijection of (7.5.8). By Lemma 7.5.11, $w(L(u_i)_{T_2}) = 1 + u_i$ and, using Lemma 7.5.7 and (9.4.3), we get

$$r_\lambda(w(\zeta_{T_2})) = w((E(\zeta)_\lambda)_{T_2}) = \prod_{\lambda_i = 1} (1 + u_i).$$

The proof of the assertion for $\zeta^\perp$ is similar, since

$$E(\zeta^\perp)_\lambda = \{(t_1, \ldots, t_n) \in \mathbb{R}^n \mid t_i = 0 \quad \text{if } \lambda_i = 1\}.$$ □

Remark 10.4.3 Proposition 10.4.2 implies that the relation

$$w(\zeta_{T_2})\, w((\zeta^{\perp})_{T_2}) = \prod_{i=1}^{n}(1+u_i)$$

holds true in $H^*_{T_2}(\mathrm{Gr}(k;\mathbb{R}^n)$. In fact, this provides a presentation of $H^*_{T_2}(\mathrm{Gr}(k;\mathbb{R}^n)$ (see Corollary 10.5.7 and Remark 10.5.8).

Example 10.4.4 Consider $\mathrm{Gr}(1;\mathbb{R}^3) \approx \mathbb{R}P^2$. The T^2-fixed points are in bijection with $\{100, 010, 001\}$. As a $\mathbb{Z}_2[u_1,u_2,u_3]$-algebra, $H^*_{T_2}(\mathbb{R}P^2)$ is generated by $w_1 = w_1(\zeta_{T_2})$ and $\bar{w}_i = w_i(\zeta^{\perp}_{T_2})$ $(i = 1, 2)$. The value of r_λ for these classes are given in Table 10.1.

One checks that the relations

$$w_1 + \bar{w}_1 = \sigma_1\ ,\ \ \bar{w}_2 + w_1\bar{w}_1 = \sigma_2\ ,\ \ w_1\bar{w}_2 = \sigma_3\,. \tag{10.4.1}$$

are satisfied. For a generalization to $\mathrm{Gr}(1;\mathbb{R}^n)$, see Example 10.5.9.

Example 10.4.5 The $\mathbb{Z}_2[u_1,\dots,u_4]$-algebra $H^*_{T_2}(\mathrm{Gr}(2;\mathbb{R}^4))$ is generated by $w_i = w_i(\zeta_{T_2})$ and $\bar{w}_i = w_i(\zeta^{\perp}_{T_2})$ $(i = 1, 2)$. The value of r_λ for these classes are given in Table 10.2.

The following relations are thus satisfied (compare Example 10.5.10).

$$\begin{aligned} w_1 + \bar{w}_1 &= \sigma_1 \\ w_2 + w_1\bar{w}_1 + \bar{w}_2 &= \sigma_2 \\ w_2\bar{w}_1 + w_1\bar{w}_2 &= \sigma_3 \\ w_2\bar{w}_2 &= \sigma_4\,. \end{aligned} \tag{10.4.2}$$

Table 10.1 The map r_λ for $\mathrm{Gr}(1;\mathbb{R}^3)$ (Example 10.4.4)

	w_1	$\bar{w}_1$	$\bar{w}_2$
100	u_1	u_2+u_3	u_2u_3
010	u_2	u_1+u_3	u_1u_3
001	u_3	u_1+u_2	u_1u_2

Table 10.2 The map r_λ for $\mathrm{Gr}(2;\mathbb{R}^4)$ (Example 10.4.5)

	w_1	$\bar{w}_1$	w_2	$\bar{w}_2$
1100	u_1+u_2	u_3+u_4	u_1u_2	u_3u_4
1010	u_1+u_3	u_2+u_4	u_1u_3	u_2u_4
1001	u_1+u_4	u_2+u_3	u_1u_4	u_2u_3
0110	u_2+u_3	u_1+u_4	u_2u_3	u_1u_4
0101	u_2+u_4	u_1+u_3	u_2u_4	u_1u_3
0011	u_3+u_4	u_1+u_2	u_3u_4	u_1u_2

The above results have their analogues in the complex case. The same proof as for Proposition 10.4.1 gives the following proposition.

Proposition 10.4.6 *Let R be a closed subgroup of $U(n)$. Then, as an $H^*(BR)$-algebra, $H^*_R(\mathrm{Fl}_{\mathbb{C}}(n_1,\dots,n_r))$ is generated by the equivariant Chern classes $c_i((\xi_j)_R)$ ($j = 1\dots,r$, $i = 1,\dots,n_j$) of the tautological bundles. In particular, $\mathrm{Fl}_{\mathbb{C}}(n_1,\dots,n_r)$ is R-equivariantly formal.*

As in the real case, we specialize to the Grassmannian $\mathrm{Gr}(k;\mathbb{C}^n) = \mathrm{Fl}_{\mathbb{C}}(k,n-k)$ and their tautological bundles $\zeta = \xi_1$ and $\zeta^\perp = \xi_2$, with $R = T$, the maximal torus of diagonal matrices in $U(n)$. As seen in (7.5.14), $H^*(BT) \approx \mathbb{Z}_2[v_1,\dots,v_n]$, with $\deg(v_i) = 2$. The associated 2-torus is the maximal 2-torus T_2 of diagonal matrices in $O(n)$, and the action of T on $\mathrm{Gr}(k;\mathbb{C}^n)$ shows that $\mathrm{Gr}(k;\mathbb{C}^n)^T = \mathrm{Gr}(k;\mathbb{C}^n)^{T_2}$. As $\mathrm{Gr}(k;\mathbb{C}^n)$ is T-equivariantly formal, the restriction to the fixed points $r\colon H^*_T(\mathrm{Gr}(k;\mathbb{C}^n)) \to H^*_T(\mathrm{Gr}(k;\mathbb{C}^n)^T)$ is injective by Theorem 7.6.11. The same proof of Proposition 10.4.2 gives the analogous formulae

$$r_\lambda(c(\zeta_T)) = \prod_{\lambda_i=1}(1+v_i) \text{ and } r_\lambda(c(\zeta^\perp)_T) = \prod_{\lambda_i=0}(1+v_i)\,.$$

We now study the equivariant characteristic classes of a rigid Γ-bundle (conversations with T. Holm were useful for this part). Recall that a Γ-equivariant vector bundle $\xi = (E \xrightarrow{p} X)$ is called *rigid* if the Γ-action on X is trivial (see p. 303). Since then $X_\Gamma \approx B\Gamma \times X$, the equivariant Stiefel-Whitney class $w(\xi_\Gamma)$ belongs to $H^*(X_\Gamma) \approx H^*(B\Gamma)\otimes H^*(X)$. If η is a Γ-equivariant vector bundle over a space Y which is Γ-equivariantly formal, and if Γ is a 2-torus for instance, then $r(w(\eta_\Gamma)) = w(\eta_{|Y^\Gamma})$, $r\colon H^*_\Gamma(Y) \to H^*(Y^\Gamma)$ is injective (see Theorem 7.6.6) and $\eta_{|Y^\Gamma}$ is rigid. Hence, in such cases, rigid equivariant vector bundles play an important role.

Let $\xi = (E \xrightarrow{p} X)$ be a rigid Γ-equivariant vector bundle, where Γ is a 2-torus. Let $\chi\colon \Gamma \to O(1) \approx \{\pm 1\}$ be a homomorphism. We call ξ a *weight* Γ *-bundle with respect to* χ (or just a χ *-weight* Γ *-bundle*) if $\gamma\cdot v = \chi(\gamma)v$ for all $\gamma\in\Gamma$ and $v \in E$.

Lemma 10.4.7 *If Γ as 2-torus, then any rigid Γ-equivariant vector bundle over a locally contractible space decomposes into a Whitney sum $\xi = \bigoplus_{\chi\in\hom(\Gamma,O(1))}\xi^\chi$ of weight subbundles.*

Proof Let $\xi = (E \xrightarrow{p} X)$ and let $x \in X$. Let $\varphi\colon p^{-1}(U) \xrightarrow{\approx} U\times\mathbb{R}^r$ be a trivialization of ξ over an open set $U \subset X$. Such trivialization is of the form $\varphi(v) = (p(v),\varphi_2(v))$, where $\varphi_2\colon p^{-1}(U) \to \mathbb{R}^r$ is a continuous map which is a linear isomorphism on each fiber, and there is a bijection from the set of trivializations of ξ over U and such maps. A continuous map $y \mapsto A^\varphi_y$ from U to $O(r)$ is thus defined by the equation $\varphi_2(\gamma v) = A^\varphi_{p(v)}(\gamma)\varphi_2(v)$, required to be valid for all $\gamma\in\Gamma$ and $v\in p^{-1}(U)$. If U retracts by deformation onto $b\in U$, a classical folklore fact about representation theory says that there is a continuous map $y\mapsto g_y$ from U to $O(r)$, with $g_b = \mathrm{id}$, such that $A^\varphi_y(\gamma) = g_y A^\varphi_b(\gamma) g_y^{-1}$ (see e.g. [79, Lemma 1.2]). We can thus construct a new trivialization $\tilde{\varphi}(v) = (p(v), \tilde{\varphi}_2(v))$ of ξ over U by setting $\tilde{\varphi}_2(v) = g_{p(v)}^{-1}\varphi_2(v)$. One checks that

$$\tilde{\varphi}_2(\gamma v) = A_b^{\varphi}(\gamma)v\,, \tag{10.4.3}$$

in other words, the map $y \mapsto A_y^{\tilde{\varphi}}$ is constant over U. As X is locally contractible, there are such trivializations $(U_x, \tilde{\varphi}_x)$ as above around each $x \in X$. Define

$$E^\chi = \{v \in E \mid \gamma \cdot v = \chi(\gamma)v, \ \forall \gamma \in \Gamma\}\,.$$

As Γ as 2-torus, the vector space E_x decomposes into a direct sum of weight subspaces

$$E_x = \bigoplus_{\chi \in \hom(\Gamma, O(1))} E_x^\chi \tag{10.4.4}$$

(see the proof of Lemma 7.5.13). Clearly, $E^\chi \cap p^{-1}(x) = E_x^\chi$ and, using (10.4.3), E^χ is the total space of a χ-weight subbundle of ξ. By (10.4.4), this proves the proposition. □

By Lemma 7.5.7, ξ_Γ is the Whitney sum of the bundles ξ_Γ^χ. Hence, to compute the Stiefel-Whitney classes of rigid Γ-equivariant vector bundles, it suffices to know those of χ-weight bundles. The following proposition provides the answer. We identify $\hom(\Gamma, O(1))$ with $H^1(B\Gamma)$ via the bijection of (7.5.8).

Proposition 10.4.8 *Let Γ be a 2-torus and let $\chi \in H^1(B\Gamma)$. Let ξ be a χ-weight Γ-bundle of rank r over X. Then, in $H_\Gamma^*(X) \approx H^*(B\Gamma) \otimes H^*(X)$, one has*

$$w(\xi_\Gamma) = \sum_{k=0}^{r} \left[(1+\chi)^k \times w_{r-k}(\xi)\right]. \tag{10.4.5}$$

Proof Let $\check{\chi} = (\mathbb{R} \to pt)$ be the χ-weight line bundle over a point and let $\check{\chi}_X = q^*\check{\chi}$, where $q\colon X \to pt$ is the constant map. Hence, $\check{\chi}_X$ is a χ-weight bundle over X whose underlying bundle is a trivial line bundle. Note that any bundle η over X may be endowed with a structure of a χ-weight bundle using the isomorphism $\eta \approx \eta \otimes \check{\chi}_X$. In fact, using the definition of the tensor product bundle of (7.5.5), we see that any χ-weight bundle ξ is obtained this way: $\xi \approx_\Gamma \hat{\xi} \otimes \check{\chi}_X$, where $\hat{\xi}$ is the bundle ξ endowed with the trivial Γ-action. By Lemma 7.5.7, one has

$$\xi_\Gamma \approx \hat{\xi}_\Gamma \otimes (\check{\chi}_X)_\Gamma\,. \tag{10.4.6}$$

Consider the two projections $\pi_{B\Gamma}\colon X_\Gamma \to B_\Gamma$ and $\pi_X\colon X_\Gamma \to X$ (using that $X_\Gamma \approx B\Gamma \times X$). As Γ-acts trivially on $E(\hat{\xi})$, one checks easily that $\hat{\xi}_\Gamma = \pi_X^*\xi$ and thus $w(\hat{\xi}_\Gamma) = \mathbf{1} \times w(\xi)$. By (7.5.4), $(\check{\chi}_X)_\Gamma = \pi_{B\Gamma}^*\check{\chi}_\Gamma$ and, by Lemma 7.5.11, $w(\check{\chi}_\Gamma) = \mathbf{1}+\chi$. Hence, $w((\check{\chi}_X)_\Gamma) = (\mathbf{1}+\chi) \times \mathbf{1}$. As $(\check{\chi}_X)_\Gamma$ is a line bundle, the proposition follows from (10.4.6) together with Lemma 9.6.7. □

Example 10.4.9 Let ξ be a vector bundle of rank r over X. We let $G = \{1, \tau\}$ act on ξ by $\tau(v) = -v$. Hence, ξ is an u-weight for u the generator of $H^1(BG)$. Using the identification $H^*_G(X) \approx H^*(X)[u]$, Formula (10.4.5) becomes

$$w(\xi_G) = \sum_{k=0}^{r} \big[w_{r-k}(\xi)(1+u)^k \big].$$

Thus,

$$w_j(\xi_G) = w_j(\xi) + w_{j-1}(\xi)u + \cdots + w_1(\xi)u^{j-1} + u^j .$$

In particular, the evaluation ev_1 at $u = \mathbf{1}$ of the equivariant Euler class $e_G(\xi) = e(\xi_G)$ satisfies

$$\mathrm{ev}_{\mathbf{1}}(e_G(\xi)) = \mathrm{ev}_{\mathbf{1}}(e(\xi_G)) = w(\xi) . \tag{10.4.7}$$

Proposition 10.4.8 has its analogue for T-equivariant complex vector bundles, where T is a torus. Let $\chi \in \hom(T, U(1))$, giving rise to $\kappa(\chi) \in H^2(BT)$ (see (7.5.11)). Then, if ξ be a χ-weight complex vector T-bundle of rank r over X, its equivariant Chern class $c(\xi_T)$ satisfies

$$c(\xi_T) = \sum_{k=0}^{r} \big[(1+\kappa(\chi))^k \times c_{r-k}(\xi) \big] . \tag{10.4.8}$$

in $H^*_T(X) \approx H^*(BT) \otimes H^*(X)$. The proof is the same as for Proposition 10.4.8, using at the end Lemma 9.7.17 instead of Lemma 9.6.7.

10.5 The Equivariant Cohomology of Certain Homogeneous Spaces

The title of the section paraphrases that of Borel's famous paper [15]. The main goal is to prove and exemplify Theorem 10.5.1 below, due to A. Knutson (unpublished). In addition, at the end of the section, we study the so-called GKM-conditions for the flag manifolds.

Theorem 10.5.1 (A. Knutson)

(1) *Let Γ_1, Γ_2 be two closed subgroups of a compact Lie group Γ. Suppose that Γ/Γ_1 or Γ/Γ_2 is Γ-equivariantly formal. Then, there is an* **GrA**-*isomorphism*

$$\Xi_{\Gamma_1,\Gamma,\Gamma_2} \colon H^*_{\Gamma_1}(\Gamma/\Gamma_2) \xrightarrow{\approx} H^*(B\Gamma_1) \otimes_{H^*(B\Gamma)} H^*(B\Gamma_2) . \tag{10.5.1}$$

(2) *Let* $(\Gamma, \Gamma_1, \Gamma_2)$ *and* $(\Gamma', \Gamma'_1, \Gamma'_2)$ *be two data as in (1). Let* $\Phi\colon \Gamma \to \Gamma'$ *be a continuous homomorphism such that* $\Phi(\Gamma_i) \subset \Gamma'_i$. *Denote by* $\Phi_i\colon \Gamma_i \to \Gamma'_i$ *the restriction of* Φ. *Set* $\Xi = \Xi_{\Gamma_1,\Gamma,\Gamma_2}$ *and* $\Xi' = \Xi_{\Gamma'_1,\Gamma',\Gamma'_2}$. *Then the diagram*

$$\begin{array}{ccc} H^*_{\Gamma'_1}(\Gamma'/\Gamma'_2) & \xrightarrow[\approx]{\Xi'} & H^*(B\Gamma'_1) \otimes_{H^*(B\Gamma')} H^*(B\Gamma'_2) \\ \downarrow{\scriptstyle \Phi^*} & & \downarrow{\scriptstyle \Phi_1^* \otimes \Phi_2^*} \\ H^*_{\Gamma_1}(\Gamma/\Gamma_2) & \xrightarrow[\approx]{\Xi} & H^*(B\Gamma_1) \otimes_{H^*(B\Gamma)} H^*(B\Gamma_2) \end{array} \tag{10.5.2}$$

is commutative (*the vertical arrows are induced by* Φ *and* Φ_i, *using the functorialities of the Borel construction*).

Remark 10.5.2 Point (3) says that $\Xi_{\Gamma_1,\Gamma,\Gamma_2}$ is an isomorphism of $H^*(B\Gamma_1)$-modules with respect to the isomorphism $\Xi_{\Gamma_1,\Gamma,\Gamma}$. With the obvious vertical identification in the commutative diagram

$$\begin{array}{ccc} H^*_{\Gamma_1}(\Gamma/\Gamma) & \xrightarrow[\approx]{\Xi_{\Gamma_1,\Gamma,\Gamma}} & H^*(B\Gamma_1) \otimes_{H^*(B\Gamma)} H^*(B\Gamma) \\ \uparrow{\scriptstyle \approx} & & \downarrow{\scriptstyle \approx} \\ H^*(B\Gamma_1) & \overset{\Xi_{\Gamma_1}}{\underset{\approx}{\dashrightarrow}} & H^*(B\Gamma_1) \end{array},$$

the isomorphism $\Xi_{\Gamma_1,\Gamma,\Gamma}$ is identified with a **GrA**-automorphism Ξ_{Γ_1} of $H^*(B\Gamma_1)$. We do not know in general whether Ξ_{Γ_1} coincides with the identity. This is however the case in the following cases:

(1) $\Gamma_1 = \{\pm 1\}$, since $H^*(B\{\pm 1\}) \approx \mathbb{Z}_2[u]$.
(2) Γ_1 a 2-torus. One uses (1), the naturality of Ξ_{Γ_1} and that $H^1(B\Gamma_1) \approx \hom(\Gamma_1, \{\pm 1\}$ (see (7.5.8)).
(3) $\Gamma_1 = O(n)$. One uses (2), the naturality of Ξ_{Γ_1} and that $H^*(BO(n)) \to H^*(BT_2)$ is injective, where T_2 is a maximal 2-torus of $O(n)$ (see Theorem 9.6.1).
(4) Γ_1 a torus or $U(n)$. The argument is analogous to (1)–(3) above.

Proof of Theorem 10.5.1 Let $\Gamma_1 \times \Gamma_2$ acts on Γ by $(\gamma_1, \gamma_2) \cdot \gamma = \gamma_1 \gamma \gamma_2^{-1}$. The kernel of the projection $\Gamma_1 \times \Gamma_2 \to \Gamma_1$ acts freely on Γ. By Lemma 7.2.7, one has a **GrA**-isomorphism

$$H^*_{\Gamma_1}(\Gamma/\Gamma_2) \xrightarrow{\approx} H^*_{\Gamma_1 \times \Gamma_2}(\Gamma)\,. \tag{10.5.3}$$

Let $\Gamma_1 \times \Gamma \times \Gamma_2$ acts on $\Gamma \times \Gamma$ by $(\gamma_1, \beta, \gamma_2) \cdot (\gamma', \gamma'') = (\gamma_1 \gamma' \beta^{-1}, \beta \gamma'' \gamma_2^{-1})$. Then $\Gamma = 1 \times \Gamma \times 1$ acts freely on $\Gamma \times \Gamma$ and the multiplication $\mu\colon \Gamma \times \Gamma \to \Gamma$ coincides with the quotient map $\Gamma \times \Gamma \to \Gamma \times \Gamma)/\Gamma$. Hence, as above, one has a **GrA**-isomorphism

$$\mu^*\colon H^*_{\Gamma_1\times\Gamma_2}(\Gamma) \xrightarrow{\approx} H^*_{\Gamma_1\times\Gamma\times\Gamma_2}(\Gamma\times\Gamma)\,. \tag{10.5.4}$$

The kernel of the projection $\Gamma_1\times\Gamma\times\Gamma_2\to\Gamma$ acts freely on $\Gamma\times\Gamma$. By Lemma 7.2.7 again, one has a **GrA**-isomorphism

$$H^*_\Gamma(\Gamma_1\backslash\Gamma\times\Gamma/\Gamma_2) \xrightarrow{\approx} H^*_{\Gamma_1\times\Gamma\times\Gamma_2}(\Gamma\times\Gamma)\,. \tag{10.5.5}$$

Now, $\Gamma_1\backslash\Gamma$ and Γ/Γ_2 being closed smooth manifolds, they are equivalent to finite Γ-CW-complexes (see [107]). If $\Gamma_1\backslash\Gamma$ or Γ/Γ_2 is Γ-equivariantly formal, then the equivariant Künneth theorem 7.4.3 holds true, telling us that the strong equivariant cross product gives a **GrA**-isomorphism

$$H^*_\Gamma(\Gamma_1\backslash\Gamma)\otimes_{H^*_\Gamma(pt)} H^*_\Gamma(\Gamma/\Gamma_2) \xrightarrow{\approx} H^*_\Gamma(\Gamma_1\backslash\Gamma\times\Gamma/\Gamma_2)\,. \tag{10.5.6}$$

Using Example 7.2.4, we get a final **GrA**-isomorphism

$$H^*_\Gamma(\Gamma_1\backslash\Gamma)\otimes_{H^*_\Gamma(pt)} H^*_\Gamma(\Gamma/\Gamma_2) \approx H^*(B\Gamma_1)\otimes_{H^*(B\Gamma)} H^*(B\Gamma_2)\,. \tag{10.5.7}$$

Combining the isomorphisms (10.5.3)–(10.5.7) provides the **GrA**-isomorphism $\Xi_{\Gamma_1,\Gamma,\Gamma_2}$. Point (2) comes from the functoriality of the above constructions and Point (3) is just an observation about Diagram (10.5.2).

Remark 10.5.3 As the right member of (10.5.1) is symmetric in Γ_1 and Γ_2, one has a **GrA**-isomorphism

$$H^*_{\Gamma_1}(\Gamma/\Gamma_2) \approx H^*_{\Gamma_2}(\Gamma/\Gamma_1)\,.$$

This can be more easily deduced from (10.5.3) and thus, does not require Γ/Γ_1 or Γ/Γ_2 being Γ-equivariantly formal.

The first two applications of Theorem 10.5.1 concern the extreme cases $\Gamma_1=\Gamma$ and $\Gamma_1=1$. The space $\Gamma/\Gamma=pt$ is Γ-equivariantly formal. Therefore, for any closed subgroup Γ_2 of Γ, one has the **GrA**-isomorphism

$$H^*_\Gamma(\Gamma/\Gamma_2)\approx H^*(B\Gamma)\otimes_{H^*(B\Gamma)} H^*(B\Gamma_2)\approx H^*(B\Gamma_2)\,. \tag{10.5.8}$$

Note that one already has an identification $H^*_\Gamma(\Gamma/\Gamma_2)\approx H^*(B\Gamma_2)$ established in Example 7.2.4. We do not know whether these two identifications are the same (it can be proved for special cases, as in Remark 10.5.2). In the case $\Gamma_1=1$, we must assume that Γ/Γ_2 is Γ-equivariantly formal. We then get **GrA**-isomorphism

$$H^*(\Gamma/\Gamma_2) \approx \mathbb{Z}_2 \otimes_{H^*(B\Gamma)} H^*(B\Gamma_2) \approx H^*(B\Gamma_2)\big/\text{image}\left(H^*(B\Gamma) \to H^*(B\Gamma_2)\right). \tag{10.5.9}$$

We now concentrate on flag manifolds. The real flag manifold is $O(n)$-equivariantly formal by Proposition 10.4.1. The complex flag manifold is $U(n)$-equivariantly formal by Proposition 10.4.6. Below is a choice of examples.

Example 10.5.4 Let $\beta_1 \colon \Gamma_1 \to O(n)$ denote the inclusion of the closed subgroup Γ_1 in $O(n)$. Theorem 10.5.1 together with Proposition 10.4.1 implies that

$$H^*_{\Gamma_1}(\mathrm{Fl}(n_1, \dots, n_r)) \approx H^*(B\Gamma_1) \otimes_{H^*(BO(n))} H^*(BO(n_1) \times \cdots \times BO(n_r)).$$

Recall from Theorem 9.5.8 that $H^*(BO(n)) \approx \mathbb{Z}_2[w_1, \dots, w_n]$, where $w_i = w_i(\zeta_n)$, the Stiefel-Whitney classes of the tautological bundle ζ_n. Also, $H^*(BO(n_j)) \approx \mathbb{Z}_2[w_1(\zeta_{n_j}), \dots, w_{n_j}(\zeta_{n_j})]$. Using the Künneth formula, one has

$$H^*(BO(n_1) \times \cdots \times BO(n_r)) \approx \mathbb{Z}_2[w_i(\zeta_{n_j})] \quad (j = 1, \dots, r\,,\; i = 1, \dots, n_j).$$

By Theorem 9.6.2, the homomorphism $H^*(BO(n)) \to H^*(BO(n_1) \times \cdots \times BO(n_r))$ sends w_i to $w_i(\zeta_{n_1} \times \cdots \times \zeta_{n_r})$. Hence

$$H^*_{\Gamma_1}(\mathrm{Fl}(n_1, \dots, n_r)) \approx H^*(B\Gamma_1)[w_i(\zeta_{n_j}) \mid j = 1, \dots, r\,,\; i = 1, \dots, n_j]\big/\mathcal{I}\,, \tag{10.5.10}$$

where $\mathcal{I}$ is the ideal generated by

$$w_*(\zeta_{n_1} \times \cdots \times \zeta_{n_r}) + H^*\beta_1(w_*) \quad (w_* = 1 + w_1 + w_2 + \cdots).$$

In the particular case of the complete flag manifold $\mathrm{Fl}(1, \dots, 1)$, Isomorphism (10.5.10) takes the form

$$H^*_{\Gamma_1}(\mathrm{Fl}(1, \dots, 1)) \approx H^*(B\Gamma_1)[x_1, \dots, x_n]\big/(\sigma_i(x_1, \dots, x_n) = H^*\beta_1(w_i))\,, \tag{10.5.11}$$

where σ_i is the i-th elementary symmetric polynomial in the variables x_j (see Example 9.5.17).

Example 10.5.5 Let $\Gamma_1 = T_2 \approx O(1) \times \cdots \times O(1)$ be the maximal 2-torus of the diagonal matrices in $O(n)$. By Theorem 9.6.1 and its proof, $H^*(BT_2) \approx \mathbb{Z}_2[u_1, \dots, u_n]$, with $\deg(u_i) = 1$, and $H^*\beta_1(w_i) = \sigma_i(u_1, \dots, u_n)$, where σ_i denotes the i-th elementary symmetric polynomial. Also, $O(n)/T_2 \approx \mathrm{Fl}(1, \dots, 1)$ is $O(n)$-equivariantly formal by Proposition 10.4.1. Therefore, for *any closed subgroup* Γ_2

in $O(n)$, Theorem 10.5.1 provides the isomorphism

$$\Xi = \Xi_{T_2,O(n),\Gamma_2}\colon H^*_{T_2}(O(n)/\Gamma_2) \xrightarrow{\approx} \mathbb{Z}_2[u_1,\dots,u_n] \otimes_{H^*(BO(n))} H^*(B\Gamma_2)\,. \tag{10.5.12}$$

For instance, (10.5.10) implies that

$$H^*_{T_2}(\mathrm{Fl}(n_1,\dots,n_r)) \approx \mathbb{Z}_2[u_1,\dots,u_n][w_i(\zeta_{n_j}) \mid j=1,\dots,r,\ i=1,\dots,n_j]\big/\mathcal{I}\,, \tag{10.5.13}$$

where $\mathcal{I}$ is the ideal generated by $w_i(\zeta_{n_1}\oplus\cdots\oplus\zeta_{n_r}) - \sigma_i(u_1,\dots,u_n)$ $(i=1,\dots,n)$. In the particular case of the full flag manifold $\mathrm{Fl}(1,\dots,1)$, we thus get that

$$H^*_{T_2}(\mathrm{Fl}(1,\dots,1)) \approx \mathbb{Z}_2[u_1,\dots,u_n,x_1,\dots,x_n]\big/\mathcal{I}\,, \tag{10.5.14}$$

where $\mathcal{I}$ is the ideal generated by

$$\sigma_i(u_1,\dots,u_n,) = \sigma_i(x_1,\dots,x_n,) \quad (i=1,\dots,n)\,.$$

Another example is given by the Stiefel manifold $\mathrm{Stief}(k,\mathbb{R}^n) = O(n)/O(n-k)$ of orthonormal k-frames in $\mathbb{R}^n$. Here, $H^*(BO(n)) \to H^*(BO(n-k))$ is just the obvious epimorphism $\mathbb{Z}_2[w_1\dots,w_n] \to \mathbb{Z}_2[w_1\dots,w_{n-k}]$. Hence, (10.5.12) implies that

$$H^*_{T_2}(\mathrm{Stief}(k,\mathbb{R}^n)) \approx \mathbb{Z}_2[u_1,\dots,u_n]/(\sigma_{n-k+1},\dots,\sigma_n)\,. \tag{10.5.15}$$

Note that $\mathrm{Stief}(k,\mathbb{R}^n))$ is not $O(n)$-equivariantly formal. At the contrary,

$$\rho\colon H^*_{T_2}(\mathrm{Stief}(k,\mathbb{R}^n)) \to H^*(\mathrm{Stief}(k,\mathbb{R}^n))$$

is the zero homomorphism in positive degrees.

It is reasonable to conjecture that the isomorphism $\Xi_{\Gamma_1,O(n),O(n_1)\times\cdots\times O(n_r)}$ of Theorem 10.5.1 identifies the equivariant Stiefel-Whitney class $w_i((\xi_j)_{\Gamma_1})$ (see Sect. 10.4) with $\mathbf{1}\otimes w_i(\zeta_j)$. The following proposition proves it for the maximal 2-torus T_2:

Proposition 10.5.6 *Under the isomorphism* $\Xi = \Xi_{T_2,O(n),O(n_1)\times\cdots\times O(n_r)}$, *one has*

$$\Xi(w_i((\xi_j)_{T_2})) = \mathbf{1}\otimes w_i(\zeta_j)\,.$$

Proof We start with the Grassmannian $\mathrm{Gr}(k;\mathbb{R}^n) = \mathrm{Fl}(k,n-k)$. For the inclusion $\Phi\colon (T_2,O(n),1\times O(n-k)) \to (T_2,O(n),O(k)\times O(n-k))$, Diagram (10.5.2) has the form

$$\begin{array}{ccc} H^*_{T_2}(\mathrm{Gr}(k;\mathbb{R}^n)) & \xrightarrow[\approx]{\Xi} & H^*(BT_2) \otimes_{H^*(BO(n))} H^*(BO(k) \times BO(n-k)) \\ \Big\downarrow \Phi^* & & \Big\downarrow \mathrm{id} \otimes \Phi_2^* \\ H^*_{T_2}(\mathrm{Stief}(k,\mathbb{R}^n)) & \xrightarrow[\approx]{\Xi} & H^*(BT_2) \otimes_{H^*(BO(n))} H^*(BO(n-k)) \end{array}$$

(we do not write the indices to the isomorphisms Ξ). The $O(n-k)$-principal bundle $\mathrm{Stief}(k,\mathbb{R}^n) \to \mathrm{Gr}(k;\mathbb{R}^n)$ is ξ_1, so $\mathrm{Stief}(k,\mathbb{R}^n)_{T_2} \to \mathrm{Gr}(k;\mathbb{R}^n)_{T_2}$ is $(\xi_1)_{T_2}$. Hence, $w_1((\xi_1)_{T_2}) \in \ker \Phi^*$. Using (10.5.15), one sees that $\ker(\mathrm{id} \otimes \Phi_2^*)$ is the $\mathbb{Z}_2$-vector space generated by $\mathbf{1} \otimes w_1(\zeta_1)$. Therefore, $\Xi(w_1((\xi_1)_{T_2})) = \mathbf{1} \otimes w_1(\zeta_1)$. A symmetric argument shows that $\Xi(w_1((\xi_2)_{T_2})) = \mathbf{1} \otimes w_1(\zeta_2)$.

This starts an induction argument on k to prove the proposition for $\mathrm{Gr}(k;\mathbb{R}^{k+1}) = \mathrm{Fl}(k,1)$. The induction step uses Diagram (10.5.2) for the inclusion

$$\Phi\colon (T_2, O(k+1), O(k) \times O(1)) \to (T_2, O(k+2), O(k+1) \times O(1))$$

which looks like

$$\begin{array}{ccc} H^*_{T_2}(\mathrm{Gr}(k+1;\mathbb{R}^{k+2})) & \xrightarrow[\approx]{\Xi} & H^*(BT_2) \otimes_{H^*(BO(k+2))} H^*(BO(k+1) \times BO(1)) \\ \Big\downarrow \Phi^* & & \Big\downarrow \mathrm{id} \otimes \Phi_2^* \\ H^*_{T_2}(\mathrm{Gr}(k;\mathbb{R}^{k+1})) & \xrightarrow[\approx]{\Xi} & H^*(BT_2) \otimes_{H^*(BO(k+1))} H^*(BO(k) \times BO(1)) \end{array}$$

By induction hypothesis, $\Xi(w_i((\xi_1)_{T_2})) = \mathbf{1} \otimes w_i(\zeta_1)$ for $i \leq k$ and $\Xi(w_1((\xi_2)_{T_2})) = \mathbf{1} \otimes w_1(\zeta_2)$. By Proposition 10.4.1, $w_{k+1}((\xi_1)_{T_2})$ is in $\ker \Phi^*$. On the other hand, $\ker(\mathrm{id} \otimes \Phi_2^*)$ is, in degree $k+1$, the $\mathbb{Z}_2$-vector space generated by $\mathbf{1} \otimes w_{k+1}(\zeta_1)$. This proves that $\Xi(w_{k+1}((\xi_1)_{T_2})) = \mathbf{1} \otimes w_{k+1}(\zeta_1)$.

We now prove the proposition for $\mathrm{Gr}(k;\mathbb{R}^n) = \mathrm{Fl}(k, n-k)$, by induction on n. The previous argument starts the induction for $n = k+1$. The induction step proceeds as above, using Diagram (10.5.2) for the inclusion

$$\Phi\colon (T_2, O(n), O(k) \times O(n-k)) \to (T_2, O(n+1), O(k) \times O(n+1-k))$$

and checking $\ker \Phi^*$ and $\ker(\mathrm{id} \otimes \Phi_2^*)$ in degree $n-k+1$.

Finally, consider the map $\pi_j\colon \mathrm{Fl}(n_1,\dots,n_r) \to \mathrm{Gr}(n_j;\mathbb{R}^n)$ defined, in the mutually orthogonal subspaces presentation of $\mathrm{Fl}(n_1,\dots,n_r)$ given in (2) p. 376, by $\pi_j(W_1,\dots,W_r) = W_j$. Take the simplest permutation matrix $\sigma \in O(n)$ so that $\sigma(0 \times \mathbb{R}^{n_j} \times 0) = \mathbb{R}^{n_j} \times 0$. The conjugation with σ gives an homomorphism

$$\Phi\colon (T_2, O(n), O(n_1) \times \cdots \times O(n_r)) \to (T_2, O(n), O(n_1), O(n_j) \times O(n-n_j))$$

such that $\Phi_2(1 \times O(n_j) \times 1) = O(n_j) \times 1$ and, for $k \neq j$, $\Phi_2(1 \times O(n_k) \times 1) \subset 1 \times O(n - n_j)$. Diagram (10.5.2) for Φ has the form

$$\begin{array}{ccc} H^*_{T_2}(\mathrm{Gr}(n_j;\mathbb{R}^n)) & \xrightarrow[\approx]{\Xi} & H^*(BT_2) \otimes_{H^*(BO(n_j))} H^*(BO(n_j) \times BO(n-n_j)) \\ \big\downarrow \pi_j^* & & \big\downarrow \mathrm{id} \otimes \Phi_2^* \\ H^*_{T_2}(\mathrm{Fl}(n_1,\dots,n_r)) & \xrightarrow[\approx]{\Xi} & H^*(BT_2) \otimes_{H^*(BO(n))} H^*(BO(n_1) \times \cdots \times BO(n_r)) \end{array}$$

Therefore,

$$\begin{aligned} \Xi(w_i((\xi_j)_{T_2})) &= \Xi \circ \pi_j^*(w_i((\xi_1)_{T_2})) && \text{since } \xi_j = \pi_j^* \xi_1 \\ &= (\mathrm{id} \otimes \Phi_2^*) \circ \Xi(w_i((\xi_1)_{T_2})) \\ &= (\mathrm{id} \otimes \Phi_2^*)(\mathbf{1} \otimes w_i(\zeta_1)) && \text{(case } \mathrm{Gr}(n_j;\mathbb{R}^n) \text{ done above)} \\ &= \mathbf{1} \otimes w_i(\zeta_j) && \text{since } \xi_j = \pi_j^* \xi_1 \end{aligned}$$

□

Using (10.5.13), Proposition 10.5.6 has the following corollary.

Corollary 10.5.7 *One has an isomorphism of $\mathbb{Z}_2[u_1,\dots,u_n]$-algebra*

$$H^*_{T_2}(\mathrm{Fl}(n_1,\dots,n_r)) \approx \mathbb{Z}_2[u_1,\dots,u_n][w_i((\xi_j)_{T_2}) \mid j = 1,\dots,r\,,\ i = 1,\dots,n_j]\big/\mathcal{I}\,,$$

where $\mathcal{I}$ is the ideal generated by

$$w_i((\xi_1)_{T_2} \oplus \cdots \oplus (\xi_r)_{T_2}) - \sigma_i(u_1,\dots,u_n) \quad (i = 1,\dots,n)\,. \tag{10.5.16}$$

Remark 10.5.8 The vanishing of the generators of $\mathcal{I}$ is equivalent to the relation

$$w((\xi_1)_{T_2}) \cdots w((\xi_r)_{T_2}) = \prod_{i=1}^{n} (1 + u_i)$$

holding in $H^*_{T_2}(\mathrm{Fl}(n_1,\dots,n_r))$. This relation may be obtained in the following way. The trivializing map of (9.5.8) provides a morphism of T_2-equivariant bundles

$$\begin{array}{ccc} E(\xi_1 \oplus \cdots \oplus \xi_r) & \longrightarrow & \mathbb{R}^n \\ \big\downarrow & & \big\downarrow \\ \mathrm{Fl}(n_1,\dots,n_r) & \xrightarrow{f} & pt \end{array}$$

where T_2 acts on $\mathbb{R}^n$ via the standard action of $O(n)$. Using Example 7.5.12, one has

$$w((\xi_1)_{T_2} \oplus \cdots \oplus (\xi_r)_{T_2}) = f^*(w((\mathbb{R}^n)_{T_2}) = \prod_{i=1}^{n} (1 + u_i)\,.$$

Example 10.5.9 Let $\Gamma = O(n)$ and $\Gamma_2 = O(1) \times O(n-1)$, so $\Gamma/\Gamma_2 \approx \mathbb{R}P^{n-1}$. Then $H^*(BO(1)) \approx \mathbb{Z}_2[w_1]$ and $H^*(BO(n-1)) \approx \mathbb{Z}_2[\bar{w}_1, \dots, \bar{w}_{n-1}]$, where w_i and $\bar{w}_i$ are the Stiefel-Whitney classes of the tautological bundles ζ_1 and ζ_{n-1}. If $\Gamma_1 = T_2$, we get, as in (10.5.14) that $H^*_{T_2}(\mathbb{R}P^{n-1})$ is the quotient of

$$\mathbb{Z}_2[u_1, \dots, u_n][w_1, \bar{w}_1, \dots, \bar{w}_{n-1}]$$

by the relations

$$w_1 + \bar{w}_1 = \sigma_1\,,\ \bar{w}_k + w_1\bar{w}_{k-1} = \sigma_k\ (k = 2, \dots, n-1)\,,\ w_1\bar{w}_{n-1} = \sigma_n\,. \tag{10.5.17}$$

These relations are the same as those of (10.4.1), no wonder given Corollary 10.5.7. As $\mathbb{R}P^{n-1}$ is T_2-equivariantly formal, $H^*(\mathbb{R}P^{n-1})$ is the quotient of $H^*_{T_2}(\mathbb{R}P^{n-1})$ by the relations $u_i = 0$. Relations (10.5.17) becomes $\bar{w}_k = w_1^k$ and $w_1^n = 0$. Thus, $H^*(\mathbb{R}P^{n-1}) \approx \mathbb{Z}_2[w_1]/(w_1^n)$ as expected.

Example 10.5.10 Let $\Gamma = O(4)$ and $\Gamma_2 = O(2) \times O(2)$, so $\Gamma/\Gamma_2 \approx \mathrm{Gr}(2; \mathbb{R}^4)$. As in Example 10.5.9, we see that $H^*_{T_2}(\mathrm{Gr}(2; \mathbb{R}^4))$ is isomorphic to the quotient of $\mathbb{Z}_2[u_1, \dots, u_4][w_1, w_2, \bar{w}_1, \bar{w}_2]$ by the relations

$$\begin{aligned} w_1 + \bar{w}_1 &= \sigma_1 \\ w_2 + w_1\bar{w}_1 + \bar{w}_2 &= \sigma_2 \\ w_2\bar{w}_1 + w_1\bar{w}_2 &= \sigma_3 \\ w_2\bar{w}_2 &= \sigma_4\,. \end{aligned} \tag{10.5.18}$$

These relations are the same as those of (10.4.2) which is coherent with Corollary 10.5.7. As in Example 10.5.9, we get a presentation of $H^*(\mathrm{Gr}(2; \mathbb{R}^4))$ by setting $u_i = 0$. This presentation is equivalent to that of Example 9.5.23.

Example 10.5.11 In the case $\Gamma_1 = \{1\}$ (the trivial subgroup), one has $H^*_{\{1\}}(\mathrm{Fl}(n_1, \dots, n_r)) \approx H^*(\mathrm{Fl}(n_1, \dots, n_r))$ and the above coincides with some theorems of Sect. 9.5.2 (e.g. Theorem 9.5.14 and Example 9.5.17).

Example 10.5.12 The analogues of the above examples works for the complex flag manifolds $\mathrm{Fl}_{\mathbb{C}}(n_1, \dots, n_r)$, where $O(n)$ is replaced by $U(n)$, T_2 is replaced by the maximal torus $T \approx (S^1)^n$ of the diagonal matrices in $U(n)$ and the Stiefel-Whitney classes are replaced by the Chern classes. The variables x_i and y_i have degree 2 instead of 1. The analogue of Proposition 10.5.6 says that $\Xi(c_i((\xi_j)_{T_2})) = \mathbf{1} \otimes c_i(\zeta_j)$. One can show that these results are valid for the cohomology with coefficients in any field.

Example 10.5.13 Consider the case where $\Gamma = S^3 \subset \mathbb{C}^2$ and $\Gamma_1 = \Gamma_2 = S^1 \subset \mathbb{R}^2$. Thus, $\Gamma/\Gamma_2 \approx S^2$. By (7) p. 275, the inclusion $i\colon S^2 \to (S^2)_\Gamma$ is homotopy equivalent to a principal bundle with structure group S^3. By Proposition 4.7.23, $H^2 i$ is an isomorphism and therefore S^2 is S^3-equivariantly formal. Now, $BS^1 \approx \mathbb{C}P^\infty$ and $BS^3 \approx \mathbb{H}P^\infty$. Hence, $H^*(B\Gamma_1) \approx \mathbb{Z}_2[x]$, $H^*(B\Gamma_2) \approx \mathbb{Z}_2[y]$ and $H^*(B\Gamma) \approx \mathbb{Z}_2[p]$, where x and y are of degree 2 and p of degree 4. The inclusion $\alpha_i\colon \Gamma_i \to \Gamma$ satisfies $H^*B\alpha_1(p) = x^2$ and $H^*B\alpha_2(p) = y^2$ (see Proposition 6.1.11). Therefore, using Theorem 10.5.1, one gets

$$H^*_{S^1}(S^2) \approx \mathbb{Z}_2[x,y]/(x^2+y^2)\,.$$

We finish this section by studying the GKM-conditions for the flag manifolds. Viewing $\mathrm{Fl}(n_1,\dots,n_r) \subset SM(n)$ using (9.5.2), the fixed point set $\mathrm{Fl}(n_1,\dots,n_r)^{T_2}$ is clearly formed by the diagonal matrices. If $\Delta = \mathrm{dia}(x_1\ \dots, x_n)$ is a diagonal matrix and $\sigma \in \mathrm{Sym}_n$, we set $\Delta^\sigma = \mathrm{dia}(x_{\sigma(1)}\ \dots, x_{\sigma(n)})$. We say that a class $a \in H^*_{T_2}(\mathrm{Fl}(n_1,\dots,n_r)^{T_2})$ *satisfies the GKM-conditions* if, for all transposition $\tau = (i,j)$, the class $a_\Delta - a_{\Delta^\tau}$ is a multiple $u_i - u_j$. This is an *ad hoc* formulation for $\mathrm{Fl}(n_1,\dots,n_r)$ (in the spirit of [119]) of the conditions introduced in [71] by M. Goresky, R. Kottwitz and R. MacPherson (whence the initials *GKM*). The importance of the GKM-conditions is illustrated in the following result.

Proposition 10.5.14 *The image of*

$$r\colon H^*_{T_2}(\mathrm{Fl}(n_1,\dots,n_r)) \to H^*_{T_2}(\mathrm{Fl}(n_1,\dots,n_r)^{T_2})$$

is the set of classes satisfying the GKM-conditions.

Proof We first prove that the classes in the image of r satisfy the GKM-conditions. Let $\tau = (i,j)$ with $i<j$. The GKM-condition for τ is trivial if $\Delta^\tau = \Delta$ (i.e., when $x_i = x_j$). We may thus assume that $x_i \neq x_j$. In the proof of Proposition 9.7.1, we have introduced an embedding $r^{ij}\colon SO(2) \to SO(n)$. The orbit of Δ under the action of $r^{ij}(SO(2))$ on $\mathrm{Fl}(n_1,\dots,n_r)$ by conjugation is a circle C_{ij} joining Δ to Δ^τ. This circle is T_2-invariant and is actually an (u_i-u_j)-circle in the sense of Example 7.6.10, with fixed points $\{\Delta, \Delta^\tau\}$. One has a commutative diagram

$$\begin{array}{ccccc}
H^*_{T_2}(\mathrm{Fl}(n_1,\dots,n_r)) & \xrightarrow{r} & H^*_{T_2}(\mathrm{Fl}(n_1,\dots,n_r)^{T_2}) & \xrightarrow{\approx} & \bigoplus_{\mathrm{Fl}(n_1,\dots,n_r)^{T_2}} \mathbb{Z}_2[u_1,\dots,u_n] \\
\downarrow & & \downarrow & & \\
H^*_{T_2}(C_{ij}) & \xrightarrow{r} & H^*_{T_2}(C_{ij}^{T_2}) & \xrightarrow{\approx} & \bigoplus_{\Delta,\Delta^\tau} \mathbb{Z}_2[u_1,\dots,u_n]
\end{array}$$

where all the vertical arrows are induced by inclusions. The right vertical arrow is just the projection. Therefore, the GKM-condition for τ comes from Example 7.6.10.

For the converse, we use the weighted trace $f(M) = \sum_{j=1}^n j\,M_{jj}$ which is, by Proposition 9.5.2, a Morse function with $\mathrm{Crit} f = \mathrm{Fl}(n_1,\dots,n_r)^{T_2}$. Let $W_x = f^{-1}(-\infty, x]$ and let $\mathcal{T}_x$ be the set of those transpositions τ such that Δ and Δ^τ are in W_x. We claim that a class $a \in H^*_{T_2}(W_x^{T_2})$ is in the image of $r\colon H^*_{T_2}(W_x) \to H^*_{T_2}(W_x^{T_2})$ if and only if it satisfies the GKM-conditions for $\mathcal{T}_x$. The "only if" part is proven as above since, if $(i,j) \in \mathcal{T}_x$, then $C_{ij} \subset W_x$. The proof of the "if" part proceeds by induction on the number n_x of critical values of f in W_x, starting trivially if $n_x = 0$ or 1. For the induction step, choose $z < y$ such that $n_z = n_y - 1$. Let $M_{z,y} = W_y - W_z$. As in (7.6.14), one has the commutative diagram

$$\begin{array}{ccccccccc} 0 & \longrightarrow & H^*_{T_2}(W_y, W_z) & \xrightarrow{\tilde{\alpha}} & H^*_{T_2}(W_y) & \xrightarrow{\tilde{\beta}} & H^*_{T_2}(W_z) & \longrightarrow & 0 \\ & & \downarrow{\scriptstyle r_{z,y}} & & \downarrow{\scriptstyle r_y} & & \downarrow{\scriptstyle r_z} & & \\ 0 & \longrightarrow & H^*_{T_2}(M_{z,y}) & \xrightarrow{\alpha} & H^*_{T_2}(W_y^{T_2}) & \xrightarrow{\beta} & H^*_{T_2}(W_z^{T_2}) & \longrightarrow & 0 \end{array} \tag{10.5.19}$$

where all arrows are induced by the inclusions and where the horizontal lines are exact.

If $a \in H^*_{T_2}(W_y^{T_2})$ satisfies the GKM-conditions for $\mathcal{T}_y$, so does $\bar{a} = \beta(a)$ for $\mathcal{T}_z$. By induction hypothesis, $\bar{a} = r_z(\bar{b})$ for some $\bar{b} \in H^*_{T_2}(W_z)$. As $\tilde{\beta}$ is surjective, there exists $b \in H^*_{T_2}(W_y)$ such that $a - r_y(b) = \alpha(c)$ for some $c \in H^*_{T_2}(W_{z,y}^{T_2})$. By the "only if" part the class $r_y(b)$ satisfies the GKM-conditions for $\mathcal{T}_y$, and then so does $a - r_y(b)$. let $D \in M_{z,y}^{T_2}$. Let $\mathcal{T}_D$ be the set of transpositions in $\mathcal{T}_y$ such that $D^\tau \neq D$. For each $(i,j) \in \mathcal{T}_D$, the class $(a - r_y(b))_D$ is a multiple of $u_i - u_j$ (since $(a - r_y(b))_\Delta = 0$ when $\Delta \neq D$). Since α is injective, the class c_D is a multiple of $u_i - u_j$ for each $(i,j) \in \mathcal{T}_D$. By the proof of Proposition 9.5.2, the negative normal bundle $\nu^-(D)$ for f at D is the Whitney sum

$$\nu^-(D) = \bigoplus_{(i,j)\in\mathcal{T}_D} T_D C_{ij} \,.$$

As $\mathbb{Z}_2[u_1, \dots, u_n]$ is a unique factorization domain, the class c_D is a multiple of

$$\prod_{(i,j)\in\mathcal{T}_D} (u_i - u_j) = \prod_{(i,j)\in\mathcal{T}_D} e(T_D C_{ij}) = e(\nu^-(D)) \,.$$

This can be done for any $D \in M_{z,y}^{T_2}$. Using Diagram (7.6.13), we deduced that there exists $\tilde{c} \in H^*_{T_2}(W_y, W_z)$ such that $r_{z,y}(\tilde{c}) = c$. Hence, $a = r_y(\tilde{\alpha}(\tilde{c}) + \tilde{b})$. □

Analogous GKM-relations hold true for the T-equivariant cohomology of the complex flag manifold $\mathrm{Fl}_{\mathbb{C}}(n_1, \dots, n_r)$. Here, T is the maximal torus of diagonal matrix in $U(n)$. It is naturally isomorphic to $U(1)^n$ and thus $H^*(BT)$ is isomorphic to $\mathbb{Z}_2[v_1, \dots, v_n]$ where $v_i \in H^2(BT)$ is the class associated, under the map κ of (7.5.11), to the projection of $U(1)^n$ onto its i-th factor. As in the real case, we see $\mathrm{Fl}_{\mathbb{C}}(n_1, \dots, n_r) \subset HM(n)$ (using (9.7.2)), then $\mathrm{Fl}_{\mathbb{C}}(n_1, \dots, n_r)^T$ is formed by the diagonal matrices. We say that a class $a \in H^*_T(\mathrm{Fl}_{\mathbb{C}}(n_1, \dots, n_r)^T)$ *satisfies the GKM-conditions* if, for all transposition $\tau = (i,j)$, the class $a_\Delta - a_{\Delta^\tau}$ is a multiple $v_i - v_j$.

Proposition 10.5.15 *The image of injective homomorphism of* $\mathbb{Z}_2[v_1, \dots, v_n]$ *-algebras*

$$r\colon H_T^*(\mathrm{Fl}_{\mathbb{C}}(n_1, \dots, n_r)) \to H_T^*(\mathrm{Fl}_{\mathbb{C}}(n_1, \dots, n_r)^T)$$

is the set of classes satisfying the GKM-conditions.

Proof We use the injective homomorphism $r^{ij}\colon SU(2) \to U(n)$, introduced in the proof of Proposition 9.7.1. The orbit of Δ under the action of $r^{ij}(SU(2))$ on $\mathrm{Fl}_{\mathbb{C}}(n_1, \dots, n_r)$ by conjugation is a 2-sphere S_{ij} (diffeomorphic to $SU(2)/U(1) \approx S^2$), whose intersection with $\mathrm{Fl}_{\mathbb{C}}(n_1, \dots, n_r)^T$ is $\{\Delta, \Delta^\tau\}$. This sphere is T-invariant and is actually a χ-sphere in the sense of Example 7.6.12, with $\kappa(\chi) = v_i - v_j$. Note that $\mathrm{Fl}_{\mathbb{C}}(n_1, \dots, n_r)^T = \mathrm{Fl}_{\mathbb{C}}(n_1, \dots, n_r)^{T_2}$. The proof of Proposition 10.5.15 then goes as that of Proposition 10.5.14, replacing C_{ij} by S_{ij} and the material of Proposition 9.5.2 and Example 7.6.10 by that of Proposition 9.7.1 and Example 7.6.12. □

10.6 The Kervaire Invariant

In 1960, Michel Kervaire introduced an invariant for framed manifolds which enabled him to construct the first topological manifold admitting no smooth structure (see Theorem 10.6.12). The computation of the Kervaire invariant then led to one of the most important problems in homotopy theory (see Theorem 10.6.11). In this section, we give a survey of the geometric side of the Kervaire invariant, using the surgery point of view of Wall (though well known by specialists, such a presentation is new in the literature). The stable homotopy aspect of the invariant is only briefly mentioned, being beyond the scope of this book. The notes of Weber [205] were helpful for preparing this section. We start with the invariant introduced by Arf [10] in order to classify quadratic forms in characteristic 2.

Let V be a finite dimensional $\mathbb{Z}_2$-vector space. A *quadratic form* on V is a map $q\colon V \to \mathbb{Z}_2$ such that the expression

$$B(x, y) = q(x) + q(y) + q(x + y) \tag{10.6.1}$$

defines a bilinear form B on V, the *bilinear form associated to* q. It is obviously symmetric and alternate, i.e. $B(x, x) = 0$.

For $i \in \mathbb{Z}_2$, let $\alpha_i(q) = \sharp\, q^{-1}(i)$. The *majority* (or *democratic*) *invariant* of q is the element $\mathrm{maj}(q) \in \mathbb{Z}_2$ defined by

$$\mathrm{maj}(q) = \begin{cases} 1 & \text{if } \alpha_1(q) > \alpha_0(q) \\ 0 & \text{otherwise.} \end{cases}$$

In fact, $\alpha_0(q) \neq \alpha_1(q)$ when B is non-degenerate (see Proposition 10.6.1).

Suppose that the associated bilinear form B is non-degenerate. Since it is alternate, there exists a symplectic basis $\{a_1, \dots, a_k, b_1 \dots, b_k\}$ of V for B (see the proof of Lemma 9.8.1). The *Arf invariant* $\mathrm{Arf}(q) \in \mathbb{Z}_2$ *of* q is defined by

$$\mathrm{Arf}(q) = \sum_{i=1}^{k} q(a_i)q(b_i)\,.$$

That $\mathrm{Arf}(q)$ is independent of the choice of the symplectic basis follows from Point (2) of the following proposition. Two quadratic forms q and q' on V are *equivalent* if there is an automorphism h of V such that $q' = q \circ h$.

Proposition 10.6.1 *Let V be a finite dimensional $\mathbb{Z}_2$-vector space. Let q and q' be two quadratic forms on V with non-degenerate associated bilinear forms. Then*

(1) $\alpha_0(q) \neq \alpha_1(q)$.
(2) $\mathrm{Arf}(q) = \mathrm{maj}(q)$.
(3) *q and q' are equivalent if and only if* $\mathrm{Arf}(q) = \mathrm{Arf}(q')$.

Proof (Compare [24, Sect. III.1].) Suppose first that $\dim V = 2$. Let $\mathcal{A} = \{a, b\}$ be a symplectic basis for B on V with which we compute $\mathrm{Arf}(q) = q(a)q(b)$. Suppose that $\mathrm{Arf}(q) = 0$. There are two cases:

(i) $q(a) + q(b) = 0$, hence $\mathrm{maj}(q) = 0$.
(ii) $q(a) + q(b) = 1$. By (10.6.1), one has $q(a + b) = 0$ and $\mathrm{maj}(q) = 0$. By symmetry, one may assume that $q(a) = 0$ and $q(b) = 1$. Then, the changing of basis $a' = a$ and $b' = a + b$ makes this form equivalent to that of (i).

If $\mathrm{Arf}(q) = 1$, then $q(a) = q(b) = 1$, thus $q(a+b) = 1$ and $\mathrm{maj}(q) = 1$. Points (1), (2) and (3) are thus proven when $\dim V = 2$. We denote by q_i the quadratic form for which $\mathrm{Arf}(q_i) = i$.

We now prove (1) and (2) by induction on $\dim V$. Let $\mathcal{B} = \{a_1, \dots, a_k, b_1 \dots, b_k\}$ be a basis of V which is symplectic for B. One has $V = \bar{V} \oplus \hat{V}$ where $\bar{V}$ is generated by $\{a_j, b_j \mid j \leq k-1\}$ and $\hat{V}$ is generated by $\{a_k, b_k\}$. The restriction of q to $\bar{V}$ (respectively: $\hat{V}$) is denoted by $\bar{q}$ (respectively: $\hat{q}$). We have

$$\begin{aligned} \mathrm{Arf}(q) &= \mathrm{Arf}(\bar{q}) + \mathrm{Arf}(\hat{q}) \text{ using the basis } \mathcal{B} \\ &= \mathrm{maj}(\bar{q}) + \mathrm{maj}(\hat{q}) \text{ by induction hypothesis.} \end{aligned}$$

It thus suffices to prove that $\mathrm{maj}(\bar{q}) + \mathrm{maj}(\hat{q}) = \mathrm{maj}(q)$. Suppose that $\hat{q} = q_1$. One has $\alpha_1(q_1) = 3$ and $\alpha_0(q_1) = 1$. Therefore,

$$\begin{cases} \alpha_1(q) = 3\alpha_0(\bar{q}) + \alpha_1(\bar{q}) \\ \alpha_0(q) = 3\alpha_1(\bar{q}) + \alpha_0(\bar{q})\,. \end{cases}$$

By induction hypothesis, $\alpha_0(\bar{q}) \neq \alpha_1(\bar{q})$. Therefore, $\mathrm{maj}(q) \neq \mathrm{maj}(\bar{q})$, which implies that

$$\mathrm{maj}(q) = \mathrm{maj}(\bar{q}) + 1 = \mathrm{maj}(\bar{q}) + \mathrm{maj}(q_1)\,.$$

If $\hat{q} = q_0$, a similar argument shows that $\mathrm{maj}(q) = \mathrm{maj}(\bar{q}) = \mathrm{maj}(\bar{q}) + \mathrm{maj}(q_0)$. Point (1) is thus established.

It remains to prove Point (3). If q and q' are equivalent, it is obvious that $\mathrm{maj}(q) = \mathrm{maj}(q')$. Conversely, we easily deduce from above that q is an orthogonal sum of q_0's and q_1's and that $\mathrm{Arf}(q)$ is the numbers of q_1's mod 2. That q and q' are equivalent when $\mathrm{Arf}(q) = \mathrm{Arf}(q')$ then comes from the equivalence

$$q_1 \boxplus q_1 \simeq q_0 \boxplus q_0$$

which is achieved by the automorphism

$$h(a_1) = a_1 + a_2\ ,\ h(a_2) = a_2\ ,\ h(b_1) = b_1\ ,\ h(b_2) = b_1 + b_2\,. \tag{10.6.2}$$

□

We now make some preparations for the Kervaire invariant. Let ξ and ξ' be two vector bundles over the same space X and let $\eta_r = (pr_X\colon X \times \mathbb{R}^r \to X)$ denote the product vector bundle of rank r over X. A *stable isomorphism* from ξ to ξ' is a family of isomorphisms $h_{r,r'}\colon \xi \oplus \eta^r \to \xi' \oplus \eta^{r'}$ for each r, r' sufficiently large, such that if $s \geq r$ and $s' \geq r'$, the diagram

$$\begin{array}{ccc} \xi \oplus \eta^r & \xrightarrow[\approx]{h_{r,r'}} & \xi' \oplus \eta^{r'} \\ \downarrow & & \downarrow \\ \xi \oplus \eta^s & \xrightarrow[\approx]{h_{s,s'}} & \xi' \oplus \eta^{s'} \end{array}$$

is commutative, where the vertical arrows are the inclusion morphisms. A *stable trivialization* of a vector bundle ξ over X is a stable isomorphism of ξ with a product bundle. A vector bundle admitting a stable trivialization is called *stably trivial.*

A *framed manifold* is a smooth manifold M together with a smooth stable trivialization of its tangent bundle TM, called a *stable framing of* M. Two framed manifolds M_1 and M_2 of the same dimension m and with $\mathrm{Bd}\, M_1 = \mathrm{Bd}\, M_2$ are *framed cobordant* if there is exists a framed manifold W^{m+1} such that

$$\mathrm{Bd}\, W = M_1 \cup M_2 \text{ and } M_1 \cap M_2 = \mathrm{Bd}\, M_1 = \mathrm{Bd}\, M_2$$

and whose stable framing extends those of M_1 and M_2. The set of framed cobordism classes of framed *closed* manifolds of dimension m is denoted by Ω_m^{fr}. It is an abelian group for the disjoint union.

Let M^m be a closed manifold. Let ν_M^k be the normal bundle of an embedding of M into $\mathbb{R}^{n+k}$. If $k > m$, such an embedding is unique up to isotopy, so the stable isomorphism class of ν_M^k is well defined. There is a canonical stable trivialization h_M of $TM \oplus \nu_M^k$ since the latter is the restriction of $T\mathbb{R}^n$ to M. If h: $TM \oplus \eta^r \xrightarrow{\approx} \eta^{m+r}$ is represents a stable isomorphism, the stable isomorphism represented by

$$\eta^{m+k+r} \xleftarrow[\approx]{h_M} TM \oplus \nu_M^k \oplus \eta^r \xrightarrow[\approx]{} TM \oplus \eta^r \oplus \nu_M^k \xrightarrow[\approx]{h} \eta^{m+r} \oplus \nu_M^k$$

represents a stable trivialization of ν_M^k. For k large enough, such a stable trivialization gives a vector bundle morphism

$$\begin{array}{ccc} E(\nu_M^k) & \longrightarrow & \mathbb{R}^k \\ \downarrow & & \downarrow \\ M & \longrightarrow & pt \end{array}$$

Applying the Pontryagin-Thom construction (see p. 430) to this morphism gives an element of $\pi_{m+k}(S^k)$. This produces an isomorphism $\Omega_m^{fr} \approx \pi_m^S$ from the framed cobordism onto the *m-stem*

$$\pi_m^S = \varinjlim_k \pi_{m+k}(S^k) \qquad (10.6.3)$$

(see [152, Sect. 7]).

Example 10.6.2 Let (S^n, F) be the standard sphere equipped with a stable framing F. The comparison between F and the standard stable framing F_0 (extending to D^{n+1}) takes the form $F = \lambda_F \cdot F_0$, where λ_F: $S^n \to SO$ is a smooth map, whose class $[\lambda_F] \in \pi_n(SO)$ is unique (compare Lemma 9.1.2). The correspondence $\lambda_F \mapsto [S^n, F] \in \Omega_{fr}^n \approx \pi_n^S$ gives a map J_n: $\pi_n(SO) \to \pi_n^S$ which coincides with the J-homomorphism of Whitehead [207].

We now describe framed surgery, following the point of view of Wall [204] (for another approach, see p. 495). Let β: $S^j \times D^{m-j} \to M^m$ $(0 \le j \le m)$ be a smooth embedding. We consider the $(m+1)$-dimensional manifold

$$W_\beta = M \times [0,1] \cup_\beta D^{j+1} \times D^{m-j}$$

where β is seen having image in $M \times \{1\}$. The corners of W_β may be smoothed in a canonical way (see [17, Appendix, Theorem 6.2]) and thus W is a smooth cobordism between M and M_β where

$$M_\beta = M - \mathrm{int}(\mathrm{im}\beta) \cup_{\beta|S^j \times S^{m-j-1}} D^{j+1} \times S^{m-j-1}.$$

The manifold M_β is said being obtained from M by a *surgery* using β. If M is endowed with a stable framing F which extends to a stable framing of W_β, we say that the surgery on β is a *stably framed surgery* (for the framing F). Let M^m be a manifold and let $\alpha\colon S^j \to M$ be a continuous map, with $j < m-1$. We wish to perform a surgery on M using a smooth embedding $\beta\colon S^j \times D^{m-j} \to M$ so that the restriction of β to $S^j \times \{0\}$ is homotopic to α. Let $\mathrm{Imm}(S^j \times D^{m-j}, M)$ be the set of regular homotopy classes of smooth immersions from $S^j \times D^{m-j}$ into M. The restriction to $S^j \times \{0\}$ provides a map

$$\rho\colon \mathrm{Imm}(S^j \times D^{m-j}, M) \to [S^j, M].$$

Proposition 10.6.3 *Let M^m be a manifold and let $j < m-1$. Then*

(a) *a stable framing F of M provides a map*

$$\phi_F\colon [S^j, M] \to \mathrm{Imm}(S^j \times D^{m-j}, M)$$

which is a section of ρ, i.e. $\rho\circ\phi_F(a) = a$ for all $a \in [S^j, M]$.

(b) *Suppose that $\phi_F(a)$ contains an embedding β. Then, the surgery on β is a stably framed surgery for the framing F.*

Proof Let $\alpha\colon S^j \to M$ be a continuous map, which we extend to $\alpha_1\colon S^j \times D^{m-j} \to M$ by $\alpha_1(x,z) = \alpha(x)$. The framing F of M gives rise to a stable trivialization of α_1^*TM. On the other hand, $T(S^j \times D^{m-j})$ has a canonical stable trivialization. Comparing these two trivializations gives a stable isomorphism α_1^S from $T(S^j \times D^{m-j})$ to α_1^*TM. Since $j < m-1$, $\pi_j(GL(m;\mathbb{R})) \to \pi_j(GL(m+N;\mathbb{R}))$ is an isomorphism and thus α_1^S is induced by a unique isomorphism from $T(S^j \times D^{m-j})$ to α_1^*TM, giving rise to an injective bundle map $\alpha_1'\colon T(S^j \times D^{m-j}) \to TM$. Assertion (a) then follows from the classification of immersions [93, Sect. 5], saying that $\mathrm{Imm}(S^j \times D^{m-j}, M)$ is (by the tangent map) in bijection with the set of injective bundle maps from $T(S^j \times D^{m-j})$ into TM (this also uses that $j < m-1$).

If β is as in (b), the equation $T\beta = \alpha_1'$ is satisfied. This is exactly what is needed to extend the stable framing F over W_β. For more details, see [204, Theorem 1.1]. □

Proposition 10.6.4 (Surgery below the middle dimension) *A stably framed compact m-dimensional manifold M^m is stably framed cobordant to a manifold M' which is $([m/2]-1)$-connected (i.e. $\pi_i(M') = 0$ for $i \le m/2-1$).*

Proof Let F be the stable framing of M and let $a \in [S^k, M]$. If $k < m/2$, general position implies that $\phi_F(a)$ contains an embedding $\alpha\colon S^k \times D^{m-k} \to M$ using which, by Proposition 10.6.3, a stably framed surgery may be performed on M. Up to homotopy equivalence, W_α is obtained from M by attaching a $(k+1)$-cell and from M_α by attaching a $(m-k)$-cell. Hence, the inclusions $i\colon M \to W_\alpha$ and $j\colon M_\alpha \to W_\alpha$ satisfy

- $\pi_* i\colon \pi_p(M) \to \pi_p(W_\alpha)$ is an isomorphism for $p \le k-1$. For $p = k$, it is surjective and kills $[\alpha]$.

- $\pi_* j$: $\pi_p(M_\alpha) \to \pi_p(W_\alpha)$ is an isomorphism for $p \leq m - k - 1$.

Therefore, if $k \leq m/2 - 1$, then $\pi_p(M_\alpha)$ is isomorphic to $\pi_p(M)$ for $p \leq k-1$ and, if $[\alpha] \neq 0$, $\pi_k(M_\alpha)$ is isomorphic to a strict quotient of $\pi_k(M)$. In particular, if M is $(k-1)$-connected, so is M_α and, in addition, the class $[\alpha]$ has been "killed". Therefore, by a finite sequence of framed surgeries, one may obtain a stably framed manifold M' which is $[m/2]-1$-connected. For more details, see e.g. [204, Theorem 1.2]. □

We now treat the middle dimensional surgery for an $(k-1)$-connected stably framed closed manifold M^m with $m = 2k$ (k odd). Consider the Hurewicz homomorphisms $\mathbf{h}$: $\pi_k(M) \to H_k(M;\mathbb{Z})$ and $\mathbf{h}_2$: $\pi_k(M) \to H_k(M)$ (sending $[\gamma\colon S^k \to M]$ to $H_*\gamma([S^k])$); since $k \geq 3$, we do not worry about base points). By the Hurewicz theorem [82, Theorem 4.32], $\mathbf{h}$ is an isomorphism and, the universal coefficient theorem [82, Theorem 3B.5], $H_k(M) \approx H_k(M;\mathbb{Z}) \otimes \mathbb{Z}_2$. Hence, $\mathbf{h}_2$ descends to an isomorphism

$$\pi_k(M)/2\,\pi_k(M) \xrightarrow{\approx} H_k(M)\,. \tag{10.6.4}$$

Let β: $V^v \to N^n$ be an immersion of smooth manifolds. The *self-intersection* $SI(\beta)$ of β is defined by

$$SI(\beta) = \{x \in \beta(V) \mid \sharp\beta^{-1}(x) > 1\}\,.$$

When β is in general position and $3v < 2n$, then $\sharp\beta^{-1}(x) \leq 2$ and $SI(\beta)$ is a $(2v-n)$-dimensional submanifold of N (see [77, Theorem 2.5])

Let $a \in \pi_k(M)$. Choose an immersion α: $S^k \times D^k \to M$ in general position representing $\phi_F(a)$ (where F is the stable framing of M). Let α_0 be the restrictions of α to $S^k \times \{0\}$. The self-intersection $SI(\alpha_0)$ is thus a finite number of points. Define

$$\tilde{q}(\alpha) = \sharp SI(\alpha_0) \mod 2\,.$$

Proposition 10.6.5 *Let $m = 2k \geq 6$ with k odd. Let M^m be $(k-1)$-connected manifold endowed with a stable framing F. Then*

(a) *the above correspondence $a \mapsto \tilde{q}(\alpha)$ induces, via* (10.6.4), *a well defined map*

$$q = q_M : H_k(M) \to \mathbb{Z}_2\,.$$

(b) *for all $a, b \in H_k(M)$ one has*

$$q(a+b) = q(a) + q(b) + a \cdot b$$

where $a \cdot b$ denotes the (absolute) intersection form (see Sect. 5.3.3).

(c) *$q(a) = 0$ if and only if $\phi_F(a)$ contains an embedding.*

Proof Let α and $\bar{\alpha}$ be two immersions representing $\phi_F(a)$ which are in general position. There are thus joined by an immersion A: $S^k \times D^k \times I \to M \times I$ which we also assumed to be in general position. Then, $SI(A_0)$ is a compact 1-manifold with boundary $SI(\alpha_0) \dot\cup SI(\bar\alpha_0)$. As an arc has two ends, one has $\tilde q(\alpha) = \tilde q(\bar\alpha)$. Thus, $\tilde q(\alpha)$ depends only on $a \in \pi_k(M)$, so we can write $\tilde q(a)$. In order to prove Point (a), it remains to establish that $\tilde q(2a) = 0$, which will be done together with the proof of (b).

Let $a, b \in \pi_k(M)$. Let $\alpha \in \phi_F(a)$ and $\beta \in \phi_F(b)$ be two immersions in general position. Then, $\phi_f(a+b)$ may be represented by an immersion γ obtained by connected sum of α and β along a tube $S^{k-1} \times D^k \times I$, disjoint from the images of α and β except at its ends. Let $B_0(\alpha, \beta) \in \mathbb{Z}_2$ defined by

$$B_0(\alpha, \beta) = \sharp\,[(\alpha_0(S^k) \cap \beta_0(S^k)] \mod 2\,.$$

Obviously, one has

$$\tilde q(a+b) = \tilde q(a) + \tilde q(b) + B_0(\alpha, \beta)\,. \tag{10.6.5}$$

We claim that

$$B_0(\alpha, \beta) = a \cdot b\,. \tag{10.6.6}$$

Indeed, if α_0 and β_0 are embeddings, this is just Corollary 5.4.13. We claim that a and b may be represented by embeddings $\tilde\alpha_0$ and $\tilde\beta_0$ such that $\tilde\alpha_0(S^k) \cap \tilde\beta_0(S^k) = \alpha_0(S^k) \cap \beta_0(S^k)$. Suppose first that $\sharp SI(\alpha_0)$ is even. The points of $SI(\alpha_0)$ may be be pairwise eliminated by Whitney procedure (since M is simply connected: see [210, Theorem 4 and its proof]). This produces a regular homotopy α_0^t: $S^k \to M$ with $\alpha_0^0 = \alpha_0$, so that $\tilde\alpha_0 = \alpha_0^1$ is an embedding. The control on the Whitney process guarantees that the intersection of $\alpha_0^t(S^k)$ with $\beta(S^k)$ is constant in t. If $\sharp SI(\alpha_0)$ is odd, we use that there exists an immersion μ: $S^k \to \mathbb{R}^m$ with $\sharp SI(\mu) = 1$ (see [210, Sect. I.2]). We can compose μ with a chart $\mathbb{R}^m \hookrightarrow M$ whose range is away from $\alpha_0(S^k) \cup \beta_0(S^k)$, obtaining an immersion μ': $S^k \to M$ which, as a map, is null-homotopic. Let $\bar\alpha$: $S^k \to M$ be the immersion obtained by connected sum of α_0 and μ'. Then, $\bar\alpha_0$ represents a and $\sharp SI(\bar\alpha_0)$ is even, so we can proceed as in the previous case. The whole process may be independently applied to β_0. This proves the claim and then Eq. (10.6.6).

Equations (10.6.5) and (10.6.6) imply that $\tilde q(2a) = 0$, because

$$a \cdot a = \langle \mathrm{PD}(a) \smile \mathrm{PD}(a), [M]\rangle = 0\,.$$

Indeed, each $x \in H_k(M)$ satisfies $x \smile x = \mathrm{Sq}^k(x) = x \smile v_k(M)$, where $v_k(M)$ is the Wu class. But, as TM is stably trivial, its Stiefel-Whitney class satisfies $w(TM) = \mathbf{1}$, which implies that $v(M) = \mathbf{1}$ by the Wu formula. This proves (a) and, thus, Eqs. (10.6.5) and (10.6.6) imply (b).

For Point (c), suppose that $q(a) = 0$ and let $\alpha \in \phi_F(a)$ be an immersion in general position. The self-intersection $SI(\alpha_0)$ then consists of an even number of double points. These points can then be pairwise eliminated by the Whitney procedure as explained above. □

By Point (b) of Proposition 10.6.5, $q_M: H_k(M) \to \mathbb{Z}_2$ is a quadratic form associated to the absolute intersection form of M. If $\mathrm{Bd}\, M$ is either empty or a $\mathbb{Z}_2$-homology sphere, the intersection form is non-degenerate (see Proposition 5.3.11). Its Arf invariant $\mathrm{Arf}(q_M)$ is thus defined and is called the *Kervaire invariant* $c(M)$ of M. The case where $\mathrm{Bd}\, M$ is a $\mathbb{Z}_2$-homology sphere is too general for our purpose so we introduce the following definition: a compact manifold M is *almost closed* if $\mathrm{Bd}\, M$ is either empty or a homotopy sphere.

Proposition 10.6.6 *Let $m = 2k \geq 6$ (k odd). Let M_0 and M_1 be two $(k-1)$-connected stably framed almost closed manifolds of dimension m. If M_0 and M_1 are stably framed cobordant, then $c(M_0) = c(M_1)$.*

Proof Let W_0^{m+1} be a stably framed cobordism between M_0 and M_1. If $\mathrm{Bd}\, M_0$ (and thus $\mathrm{Bd}\, M_1$) is empty, we remove a tube $D^m \times I$ out of W_0, getting a stably framed manifold W^{m+1} whose boundary is the connected sum $M = M_0 \sharp M_1$. If $\mathrm{Bd}\, M_0$ (and thus $\mathrm{Bd}\, M_1$) is a homotopy sphere, we set $W = W_0$ and $M = \mathrm{Bd}\, W$. The manifold M is a $(k-1)$-connected stably framed closed manifold and, clearly, $c(M) = c(M_0) + c(M_1)$. It thus suffices to prove that $c(M) = 0$. By surgery below the middle dimension (see Proposition 10.6.4), we may assume that W is $(k-1)$-connected. To prove that $c(M) = 0$, it is enough to show that $q(\mathcal{B}) = 0$ where $\mathcal{B} = \ker(H_k(M) \to H_k(W))$. Indeed, Proposition 5.3.9 and Kronecker duality imply that $2 \dim \mathcal{B} = \dim H_k(M)$. The vanishing of $q(\mathcal{B})$ thus implies, using Proposition 10.6.1, that $c(M) = \mathrm{Arf}(q) = \mathrm{maj}(q) = 0$.

Let h_* and h^* denote the integral (co)homology. As M is $(k-1)$-connected and almost closed, the Hurewicz theorem, the integral Poincaré duality and the universal coefficient theorem imply that

$$\pi_k(M) \approx h_k(M) \approx h^k(M, \mathrm{Bd}\, M) \approx h^k(M) \approx \mathrm{hom}(h_k(M); \mathbb{Z})\,. \tag{10.6.7}$$

Therefore, all the groups in (10.6.7) are free abelian and the isomorphism $h_k(M) \approx h^k(M)$ sends $\mathcal{B}_{\mathbb{Z}} = \ker(h_k(M) \to h_k(W))$ onto $\mathcal{B}_{\mathbb{Z}} = \mathrm{Image}\big(h^k(W) \to h^k(M)\big)$. As $h^k(M)$ is free abelian, the same proof as for Proposition 5.3.9 shows that $\mathcal{B}_{\mathbb{Z}}$ is a direct summand of $h^k(M)$ and $\mathrm{rank}\, h^k(M) = 2\, \mathrm{rank}\, \mathcal{B}_{\mathbb{Z}}$ (all groups in the analogue of Diagram (5.3.7) are free abelian). Hence, $\mathcal{B}_{\mathbb{Z}}$ is a direct summand of $h_k(M)$. The homomorphism $\mathcal{B}_{\mathbb{Z}}/2\,\mathcal{B}_{\mathbb{Z}} \to B$ is then injective, where $B = \mathrm{Image}\big(H^k(W) \to H^k(M)\big)$. By Proposition 5.3.9, $\dim H_k(M) = 2 \dim \mathcal{B}$, so $\mathcal{B}$ and $\mathcal{B}_{\mathbb{Z}}/2\,\mathcal{B}_{\mathbb{Z}}$ have the same dimension. This shows that the homomorphism $\mathcal{B}_{\mathbb{Z}} \to \mathcal{B}$ is surjective. Consider the commutative diagram

$$\begin{array}{ccccccc}
\pi_{k+1}(W) & \longrightarrow & \pi_{k+1}(W, M) & \longrightarrow & \pi_k(M) & \longrightarrow & \pi_k(W) \\
\downarrow & & \downarrow & & \downarrow \approx & & \downarrow \approx \\
h_{k+1}(W) & \longrightarrow & h_{k+1}(W, M) & \longrightarrow & h_k(M) & \longrightarrow & h_k(W)
\end{array}$$

whose rows are exact. The bijectivities and surjectivity seen on vertical arrows come from the Hurewicz-Whitehead theorem [179, Chap. 7, Sect. 5, Theorem 9], since M and W are $(k-1)$-connected. By five-lemma's arguments, we deduce that $\pi_{k+1}(W, M) \to \mathcal{B}_{\mathbb{Z}} \to \mathcal{B}$ is surjective.

Let $b \in B$. By the above, there is a map $\beta\colon S^k \to M$, representing b, which extends to $\gamma\colon D^{k+1} \to W$. Using the stable framing of W, the pair if map (γ, β) determines, in its homotopy class, a pair of immersion $(\hat{\gamma}, \hat{\beta})\colon (D^{k+1}, S^k) \to (W, M)$ which we may assume to be in general position. The self-intersection $SI(\hat{\gamma}_0)$ is a compact 1-dimensional manifold whose boundary is $SI(\hat{\beta}_0)$. As an arc has two ends, one has $q(b) = \tilde{q}(\hat{\beta}_0) = 0$. □

Example 10.6.7 The sphere S^k has a standard stable framing. Let $M = S^k \times S^k$ (k odd), with the product stable framing. The manifold M is $(k-1)$-connected and, by the Künneth formula, $H^k(M) \approx H^k(S^k) \otimes H^0(S^k) \oplus H^0(S^k) \otimes H^k(S^k)$, with generators $a = [S^k] \otimes \mathbf{1}$ and $b = \mathbf{1} \otimes [S^k]$. As M is the boundary of $S^k \times D^{k+1}$ and of $D^{k+1} \times S^k$, one has $c(M) = 0$ and $q(a) = q(b) = 0$ by Proposition 10.6.6 and its proof. As $a \cdot b = 1$, one has $q(a+b) = 1$. Note that $a + b$ is represented by the diagonal manifold of $S^k \times S^k$.

Proposition 10.6.5 permits us to define the Kervaire invariant for any stably framed almost closed manifold M^{2k} (k odd), as $c(M) = c(M')$ where M' is a $(k-1)$-connected manifold stably framed cobordant to M. For closed manifolds, this gives a map

$$c\colon \Omega_{2k}^{fr} \to \mathbb{Z}_2 \quad (k \text{ odd})\,. \tag{10.6.8}$$

which is a homomorphism. Indeed, the sum in Ω_{2k}^{fr} may be represented by the connected sum. If M_1 and M_2 are $(k-1)$ connected stably framed closed manifolds, then $M = M_1 \sharp M_2$ is $(k-1)$-connected and $c(M) = c(M_1) + c(M_2)$.

Proposition 10.6.8 *Let $m = 2k \geq 6$ (k odd). Let M^m be a stably framed almost closed manifold. Then, $c(M) = 0$ if and only if M is stably framed cobordant to a contractible manifold* (*if* Bd M *is not empty*) *or to a homotopy sphere* (*if* Bd M *is empty*).

Proof The "if" part is obvious since a contractible manifold of a homotopy sphere is $(k-1)$-connected and its middle dimensional homology vanish.

The proof of the converse uses the integral (co)homology, denoted by h_* and h^*. By surgery below the middle dimension, we may suppose that M is $(k-1)$-connected. As seen in (10.6.7), $h_k(M)$ is free abelian. Consider the integral intersection form on $h_k(M)$ given by $a \mathbin{\tilde{\cdot}} b = \langle \mathrm{PD}(a) \smile \mathrm{PD}(b), [M]_{\mathbb{Z}}\rangle$ (for some choice of a generator $[M]_{\mathbb{Z}} \in H_m(M, \mathrm{Bd}\, M)$). This form is unimodular since M is almost closed (same proof as for Proposition 5.3.11, or see [97, p. 58]). As k is odd, the integral intersection form is alternate. Hence, $h_k(M)$ admits a skew-symplectic basis, i.e. a basis $\tilde{a}_1, \tilde{b}_1, \dots, \tilde{a}_p, \tilde{b}_b$ such that $\tilde{a}_i \mathbin{\tilde{\cdot}} \tilde{a}_j = \tilde{b}_i \mathbin{\tilde{\cdot}} \tilde{b}_j = 0$ and $\tilde{a}_i \mathbin{\tilde{\cdot}} \tilde{b}_j = \pm\delta_{ij}$

(same proof as that of Lemma 9.8.1, or see [156, IV.1]). Under the isomorphism $h_k(M)/2h_k(M) \approx H_k(M)$ the basis $\{\tilde{a}_i, \tilde{b}_i\}$ gives a symplectic basis $\{a_i, b_i\}$ for the $\mathbb{Z}_2$-intersection form.

By changing the basis $\{\tilde{a}_i, \tilde{b}_i\}$, we may assume that $q(a_1) = 0$. Indeed, this can be achieved if $q(a_1)q(b_1) = 0$ (by exchanging $\tilde{a}_1$ with $\tilde{b}_1$ if necessary). Otherwise, as $\mathrm{Arf}(q) = c(M) = 0$, there exists $j \neq 1$ such that $q(a_j)q(b_j) = 1$ (say $j = 2$). The basis change of (10.6.2) then does the job.

We are thus in position to perform a stably framed surgery on an embedding $\hat{\alpha}\colon S^k \times D^k \to M$ representing $\tilde{a}_1$, giving a stably framed manifold M'. Let $M_0 = M - \mathrm{int}(\hat{\alpha}(S^k \times D^k))$, contained in both M and M'.

By excision, $h_j(M, M_0) \approx h_j(S^k \times D^k, S^k \times S^{k-1})$ vanishes except for $j = k$ where it is infinite cyclic. As M is $(k-1)$-connected, the integral homology sequence of the pair (M, M_0) yields to the exact sequence

$$0 \to h_k(M_0) \to h_k(M) \xrightarrow{h_*j} h_k(M, M_0) \to h_{k-1}(M_0) \to 0 \tag{10.6.9}$$

where $j\colon (M, \emptyset) \to (M.M_0)$ denotes the pair inclusion. Since M is almost closed, Poincaré and Kronecker dualities provide the commutative diagram

$$\begin{array}{ccccc}
h_k(M) & \xrightarrow{h_*j} & h_k(M, M_0) & \xleftarrow{\approx} & h_k(S^k \times D^k, S^k \times S^{k-1}) \\
\approx \uparrow \frown[M] & & & & \approx \uparrow \frown[S^k \times D^k] \\
h^k(M, \mathrm{Bd}\, M) & \xrightarrow{\approx} & h^k(M) & \xrightarrow{h^*\hat{\alpha}} & h^k(S^k \times D^k) \\
 & & \approx \downarrow \mathbf{k} & & \approx \downarrow \mathbf{k} \\
 & & \hom(h_k(M); \mathbb{Z}) & \xrightarrow{(h_*\hat{\alpha})^\sharp} & \hom(h_k(S^k \times D^k); \mathbb{Z})
\end{array}$$

As $h_k(M)$ is free abelian, $h_*\hat{\alpha}\colon h_k(S^k \times D^k) \to h_k(M)$ is injective, so $h^*\hat{\alpha}$ is surjective. Hence, h_*j is surjective and, by (10.6.9), $h_{k-1}(M_0) = 0$. Note that $h_i(M_0) \approx h_i(M) = 0$ for $i < k$ and, since $k \geq 2$, van Kampen's theorem applied to

$$M \approx M_0 \cup (S^k \times D^k) \text{ and } M_0 \cap (S^k \times D^k) \approx S^k \times S^{k-1}$$

implies that M_0 is simply connected. Therefore, M_0 is $(k-1)$-connected. Also, from (10.6.9), we deduce that $h_k(M_0)$ is free with rank $h_k(M_0) = \mathrm{rank}\, h_k(M) - 1$.

From similar considerations with the pair (M', M_0), we deduce that M' is $(k-1)$-connected and that $h_k(M')$ is free with rank $h_k(M') = \mathrm{rank}\, h_k(M) - 2$. The above process may be repeated with M' showing that, thanks to a finite number of stably framed surgeries, M is stably framed cobordant to $\bar{M}$ which is k-connected. By Poincaré duality, $h_*(\bar{M}) = 0$ if $* \leq m-1$. If $\mathrm{Bd}\, M$ is not empty, then $h_*(\bar{M}) \approx h_*(pt)$ and, as $\bar{M}$ is simply connected, $\bar{M}$ is contractible by the Hurewicz-Whitehead theorem [82, Proposition 4.74]. If $\bar{M}$ is closed, then $h_*(\bar{M}) \approx h_*(S^m)$ and by the Hurewicz theorem, $\pi_m(\bar{M}) \approx h_m(\bar{M}) \approx \mathbb{Z}$. If $\gamma: S^m \to \bar{M}$ represents a generator of $\pi_m(\bar{M})$, then γ is a homotopy equivalence by the Hurewicz-Whitehead theorem. Hence $\bar{M}$ is a homotopy sphere. □

Corollary 10.6.9 *Let $m = 2k \geq 6$ with k odd. Let M^m be a compact stably framed manifold whose boundary is a homotopy sphere. Suppose that $c(M) = 0$. Then $\Sigma = \mathrm{Bd}\, M$ is diffeomorphic to the standard sphere S^{m-1}.*

Proof By Proposition 10.6.8, the homotopy sphere Σ is also equal to $\mathrm{Bd}\, \bar{M}$ where $\bar{M}$ is contractible. Since Σ is simply connected, it is a consequence of the h-cobordism that $\bar{M}$ is diffeomorphic to D^m (see [151, Proposition A, p. 108]). Hence, Σ is diffeomorphic to S^{m-1}. □

We have so far presented the stably framed surgery under the point of view of Wall [204], using the theory of immersions. An earlier approach was introduced by Milnor in [148] and developed in [24, 113] (for a presentation of the Kervaire invariant in this framework, see [120, 133]). With this method, in order to perform a stably framed surgery on a class $a \in \pi_j(M^m)$, we first represent a by an embedding $\alpha: S^j \to M$ (this is possible when $m > 2j$ and when $m = 2j$ if M is simply connected [77]). The stable framing of M gives a trivialization of $\eta_N \oplus TM$ where η_N is the trivial bundle of rank N (N large). We thus get a trivialization F of $\eta_N \oplus \alpha^*TM \approx \eta_N \oplus TS^j \oplus \nu_\alpha$. The vector bundle $\eta_1 \oplus TS^j$ is the restriction of the tangent bundle to $\mathbb{R}^{j+1}$ and therefore has a canonical field of orthonormal frames. Thus, $\eta_N \oplus TS^j = \eta_1 \oplus \eta_{N-1} \oplus TS^j$ has a canonical field of orthonormal frames. Together with the above trivialization F, this gives an element $[\hat{\alpha}]$ of the homotopy group $\pi_j(\mathrm{Stief(j+N, \mathbb{R}^{m+N})})$ of the Stiefel manifold $\mathrm{Stief}(j+N, \mathbb{R}^{m+N})$ which is shown to be the obstruction to perform a stably framed surgery on the class a. Homotopy groups of Stiefel manifold are known (see [181, Theorem 25.6]). For $2j < m$, $\pi_j(\mathrm{Stief(j+N, \mathbb{R}^{m+N})}) = 0$, whence the surgery below the middle dimension. For $m = 2k$ with k odd, one has $\pi_k(\mathrm{Stief(k+N, \mathbb{R}^{2k+N})}) = \mathbb{Z}_2$. This gives the quadratic form $q: H_k(M) \to \mathbb{Z}_2$, by $q(a) = [\hat{\alpha}]$. Consider the principal bundle

$$\mathcal{P} = \big(SO(k) \to SO(2k+N) \to \mathrm{Stief}(k+N, \mathbb{R}^{2k+N})\big)$$

and its homotopy exact sequence

$$\pi_k(SO(2k+N)) \xrightarrow{j_*} \pi_k(\mathrm{Stief(k+N, \mathbb{R}^{2k+N})}) \xrightarrow{\partial} \pi_{k-1}(SO(k)) \xrightarrow{i_*} \pi_{k-1}(SO(2k+N))\,.$$

Changing the stable framing of M adds to $q(a)$ an element in the image of j_*. The $SO(k)$-principal bundle $\hat{\alpha}^*\mathcal{P}$ is the bundle of orthonormal frames in ν_k. Hence, $\partial(q(a)) \in \pi_{k-1}(SO(k)) \approx \pi_k(BSO(k))$ classifies ν_α. If k is odd and $k \neq 1, 3, 7$, then $\ker i_* \approx \mathbb{Z}_2$ (see [24, Corollary IV.1.11]) and thus ∂ is injective. This proves the following result.

Lemma 10.6.10 *Let M be a $(k-1)$-connected $2k$-dimensional stably framed manifold, with k odd and $k \neq 1, 3, 7$. Then, a class $a \in H_k(M)$ satisfies $q(a) = 0$ if and only if it is representable by a k-sphere in M with trivial normal bundle.*

The Kervaire invariant has several other definitions (see, e.g. [23, 123] and also the proof of Theorem 10.6.12). These more homotopic descriptions were much used for computing the image of the Kervaire invariant $c\colon \Omega^m_{fr} \approx \pi^S_m \to \mathbb{Z}_2$, an outstanding problems in stable homotopy theory. One of the main advance was due to Browder [23]. An almost complete solution (except for $m = 126$) was provided by Hill, Hopkins and Ravenel in 2009, who proved the following theorem (see [92]).

Theorem 10.6.11 *Let $m = 2k$ with k odd. Then, the Kervaire invariant $c\colon \Omega^m_{fr} \to \mathbb{Z}_2$ is surjective if $m = 2, 6, 14, 30, 62$ and possibly 126. In all other dimensions, the Kervaire invariant vanishes.*

The vanishing of the Kervaire invariant in dimension m implies the existence of a closed PL-manifold K^m which does not have the homotopy type of a closed smooth manifolds. Below is a description of the construction of K^m, originally due to Kervaire [114].

Let $p_i\colon \mathcal{D}_i \to S^k$ $(i = 1, 2)$ be two copies of the unit disk bundle associated to the tangent disk bundle to S^k $(k \geq 3)$. Let A be a closed k-disk in S^k. Choose trivializations $\mu_i\colon p_i^{-1}(A) \xrightarrow{\approx} A \times D^k \approx D^k \times D^k$ over A (they are unique up to isotopy: see Lemma 9.1.2). Let K_0 be the quotient space of $\mathcal{D}_1 \dot\cup \mathcal{D}_2$ under the identification $\mu_1(x, y) = \mu_2(y, x)$. After rounding the corners (see [17, Appendix, Theorem 6.2]), K_0 is a smooth compact manifold of dimension $m = 2k$, with boundary $\Sigma_0 = \Sigma_0^{m-1}$. This is an example of the so called *plumbing technique* (see [97, Sect. 8], [120]).

The boundary manifold Σ_0 is a homotopy sphere. Indeed, K_0 is $(k-1)$-connected and $h_k(K_0)$ is free abelian. The integral intersection form clearly induces an isomorphism $h_k(K_0) \xrightarrow{\approx} \hom(h_k(K_0), \mathbb{Z})$. By the analogue for the integral homology of Proposition 5.3.11, we deduce that $h_*(\Sigma_0) \approx h_*(S^{m-1})$ (compare [97, p. 58]). Moreover, K_0 is a thickening of $S^k \vee S^k$. By general position (since $k \geq 3$), one has $\pi_1(\Sigma_0) \approx \pi_1(K_0) = 1$. Hence, Σ_0 is a homotopy sphere (see the proof of Proposition 10.6.8).

We claim that $c(K_0) = 1$. Indeed, $H_k(K_0)$ has a symplectic basis a, b represented by the two copies of S^k. Recall that TS^k is isomorphic to the normal bundle of the diagonal sphere in $S^k \times S^k$ (see Lemma 5.4.14). It then follows from Example 10.6.7 (or from Lemma 10.6.10 if $k \neq 1, 3, 7$) that $q_{K_0}(a) = q_{K_0}(b) = 1$. Therefore $c(K_0) = 1$.

The smooth structure on K_0 determines a unique PL-structure (see (3) on p. 203). The homotopy sphere Σ_0 is PL-isomorphic to the standard sphere by Smale's theorem [177]. Let K be the PL-manifold obtained by gluing to K_0 the cone over Σ_0. The homotopy sphere $\Sigma_0 = \Sigma_0^{m-1}$ is called the *Kervaire sphere* while the PL-manifold $K = K^m$ is called the *Kervaire manifold*.

Theorem 10.6.12 *Let $m = 2k \geq 10$ with k odd. Then the following assertions are equivalent.*

(a) *The Kervaire invariant $c\colon \Omega^m_{fr} \to \mathbb{Z}_2$ vanishes.*
(b) *The Kervaire sphere Σ_0^{m-1} is not diffeomorphic to the standard sphere.*

(c) *The Kervaire manifold K^m does not have the homotopy type of a smooth closed manifold.*

Thus, according to Theorem 10.6.11, (b) and (c) are true for $10 \leq m \neq 14, 30, 62$ and possibly 126. Theorem 10.6.12 goes back to Kervaire [114] who proved it in 1960 for $m = 10$, constructing the first example in history of a topological closed manifold not admitting any smooth structure (even up to homotopy type). Together with the discovery by J. Milnor in 1956 of several smooth structures on the 7-sphere, Kervaire's result, was quite influential in the history of differential topology (see [53]).

Proof (c) $\Rightarrow$ (b). Suppose that (b) is not true, so there is a diffeomorphism $\varphi\colon S^{m-1} \to \Sigma_0$. Then $K_0 \cup_\varphi D^m$ is a smooth closed manifold which is homeomorphic to K. This contradicts (c).

(b) $\Rightarrow$ (a). If (a) is not true, there is a closed smooth stably framed manifold N^m with $c(N) = 1$. Let N_0 be N with an open m-disk removed. Thus, N_0 is a smooth almost closed stably framed manifold with $c(N_0) = 1$. Let P be the boundary connected sum of N_0 and K_0. The boundary of P is $\Sigma_0 \sharp S^{m-1}$, diffeomorphic to Σ_0. But P is a smooth almost closed stably framed manifold with $c(P) = c(N_0) + c(K_0) = 0$. By Corollary 10.6.9, Σ_0 is diffeomorphic to S^{m-1}.

(a) $\Rightarrow$ (c). This is more complicated and we just sketch the idea of the proof. Suppose that (c) is not true, so there is a smooth closed manifold M and a homotopy equivalence $f\colon K^m \to M$. Hence, M is of dimension m and is $(k-1)$-connected. Let x be the cone point of K and $y = f(x)$. Let D be an open m-disk in M around y and let $M_0 = M - D$. Using boundary collars in K_0 and M_0, one can construct continuous maps of pairs $f_K\colon (K_0, \Sigma_0) \to (K, x)$ and $f_M\colon (M_0, S^{m-1}) \to (M, y)$. These maps induce isomorphism

$$\begin{array}{ccc} H_k(K) & \xrightarrow[\approx]{H_*f} & H_k(M) \\ {\scriptstyle\approx}\Big\downarrow {\scriptstyle H_*f_K} & & {\scriptstyle\approx}\Big\downarrow {\scriptstyle H_*f_M} \\ H_k(K_0) & \xleftarrow[\approx]{\Psi} & H_k(M_0) \end{array} \tag{10.6.10}$$

where Ψ is defined to make the diagram commutative. For the sake of this proof, an m-dimensional relative smooth manifold (X, Y) is called *acceptable* if X is $(k-1)$-connected and $H^i(X, Y; G) = 0$ for $k < i < n$ for all coefficient groups G. The above pairs (K, x), (M, y), (K_0, Σ_0) and (M_0, S^{m-1}) are acceptable. If (X, Y) is an acceptable pair, Kervaire and Milnor in [113, pp. 531–534] defined a map $\psi_{X,Y}\colon H_k(X) \to \mathbb{Z}_2$ (this is the map $\psi_0\colon H_k(X; \mathbb{Z}) \to \mathbb{Z}_2$ of [113, p. 534] which descents to $H_k(X)$). The following two properties hold true.

(i) Let $g\colon (X, Y) \to (X', Y')$ be a continuous map between acceptable pairs. If g induces an isomorphism $H^*(X', Y'; \mathbb{Z}) \to H^*(X, Y; \mathbb{Z})$, then

$$\psi_{X,Y} \circ H_*g = \psi_{X',Y'}\,.$$

(ii) Let (X, Y) be an acceptable pair with X stably parallelizable. Then $\psi_{X,Y}(a) = 0$ if and only if a is representable by an embedded k-sphere with trivial normal bundle [113, Lemma 8.3 and p. 534].

Another fact is that, as M is homotopy equivalent to K, it is stably parallelizable (see [25, Theorem A2.1]). Therefore, the quadratic form q_M and q_{M_0} are defined. Since (a) is true and $m \geq 10$, one has $m \neq 14$ by Theorem 10.6.11 and thus Lemma 10.6.10 applies. Therefore, for $a \in H_k(M_0) \approx H_k(M)$, one has

$$\begin{aligned} q_M(a) &= q_{M_0}(a) \\ &= \psi_{M,M_0}(a) && \text{by (ii) and Lemma 10.6.10} \\ &= \psi_{K,K_0} \circ \Psi(a) && \text{by (10.6.10) and (i)} \\ &= q_{K_0} \circ \Psi(a) && \text{by (ii) and Lemma 10.6.10.} \end{aligned}$$

Hence $c(M) = \text{maj}(q_M) = \text{maj}(q_{K_0}) = 1$, which contradicts Assertion (a). □

Besides giving the Kervaire invariant for a stably framed manifold, the Arf invariant of a quadratic form determines the surgery obstruction group $L_2(\pi) \approx \mathbb{Z}_2$ for π of order ≤ 2 [24, 203]. It has also applications in classical and high dimensional knot theory (see [205] for a survey on these works).

10.7 Exercises for Chapter 10

10.1. Which closed surfaces admit a involution with scattered fixed point set?

10.2. Let X be a conjugation space. Prove that κ: $H^0(X) \to H^0(X^G)$ coincides with the homomorphism induced by the inclusion $X^G \hookrightarrow X$. Deduce that a G-space Y is a conjugation space if and only if each path-connected component of Y is a conjugation space.

10.3. Prove that the definition of a conjugation space may be expressed with the reduced cohomology.

10.4. Let X and Y be two conjugation spaces which are G-equivariantly well pointed. Prove that $X \vee Y$ is a conjugation space.

10.5. Let X_i $(i = 1, 2)$ be two conjugation spaces which are G-equivariantly well pointed. Suppose that X_2 has finite cohomology type. Prove that $X_1 \wedge X_2$ is a conjugation space.

10.6. Let X be a conjugation space which is equivariantly well pointed. Construct a conjugation space Z such that Z^G is homeomorphic to $X^G \wedge S^1$.

10.7. Prove that there is no conjugation space X such that X^G is homotopy equivalent to $\Sigma^k \mathbb{O}P^2$. [Hint: see Remark 10.2.16.]

10.8. Consider the generic length vector $\ell = (1, \dots, 1, n-2) \in \mathbb{R}^n_{>0}$. Let $\mathcal{C} = \mathcal{C}^n_d(\ell)$ and $\bar{\mathcal{C}} = \bar{\mathcal{C}}^n_d(\ell)$. Compute $\mathfrak{P}_t(\mathcal{C})$ using Theorem 10.3.17. Compute $H^*(\bar{\mathcal{C}})$ using the transfer exact sequence or Theorem 10.3.11.

10.9. Let $\ell = (1, 1, 2, 2, 3)$. Compute $\mathfrak{P}_t(\mathcal{C}^5_d(\ell))$. What are $\mathcal{C}^5_2(\ell)$ and $\bar{\mathcal{C}}^5_2(\ell)$?

10.10. For which generic length vector ℓ is the chain space $\bar{C}_2^n(\ell)$ an orientable manifold? [Hint: this depends on n and on the lopsidedness lops (ℓ).]

10.11. For which generic length vector ℓ does the spatial polygon space $\mathcal{N}_3^n(\ell)$ admit a spin structure?

10.12. Let $\ell = (\ell_1, \ldots, \ell_n)$ be a generic length vector with lops $(\ell) \geq 2$. Prove that $u^{\text{lops}\,(\ell)-2} \neq 0$ in $H^*(\bar{C}_2^n(\ell)) \approx H_G^*(C_2^n(\ell))$.

10.13. Do Example 10.4.5 for $\mathrm{Gr}(2; \mathbb{R}^5)$.

10.14. Use Theorem 10.5.1 to compute $H^*_{\Gamma_1}(\Gamma/\Gamma_2)$ for $\Gamma_1 = \Gamma_2 = \{\pm 1\}$, the center of $\Gamma = SU(2)$. Prove that $H^*_{\Gamma_1}(\Gamma/\Gamma_2) \approx H^*(\mathbb{R}P^3)[u]$. Is that surprising?

Chapter 11
Hints and Answers for Some Exercises

11.1 Exercises for Chapter 2

2.1. $\mathcal{F}'_n$ contains $(n+1)!$ n-simplexes (by induction on n).

2.2. $\mathfrak{P}_t(\mathcal{F}_n^k) = 1 + \binom{n}{k+1} t^k$. By induction on k, starting with $\mathfrak{P}_t(\mathcal{F}_n^0) = 1 + n$ and with

$$b_{k+1}(\mathcal{F}_n^{k+1}) = \dim C_{k+1}(\mathcal{F}_n) - b_k(\mathcal{F}_n^k) = \binom{n+1}{k+2} - \binom{n}{k+1} = \binom{n}{k+2}.$$

as induction step. For the Euler characteristic, evaluate $\mathfrak{P}_t(\mathcal{F}_n^k)$ at $t = -1$.

2.3. (a) X is isometric to $A\dot{\cup}B$ where $A = \{(0,0,0),(0,1,0),(1,1,1),(1,0,1)\}$ and $B = \{(1,0,0),(1,1,0),(0,1,1),(0,0,1)\}$. The complex A_ε and B_ε are both quadrilaterals, and $X_\varepsilon = A_\varepsilon * B_\varepsilon$. Thus X_ε is isomorphic to the double suspension of B_ε, so $|X_\varepsilon| \approx S^3$.

2.4. $\ell = (1,1,1,1,3)$: $\mathrm{Sh}(\ell) = \{5\}\dot{\cup}\dot{\mathcal{F}}\{1,2,3,4\}$. $\chi(\mathrm{Sh}(\ell)) = 3$. $\mathfrak{P}_t(\mathrm{Sh}(\ell)) = 2 + t^2$.
$\ell = (1,1,3,3,3)$: the maximal simplexes of $\mathrm{Sh}(\ell)$ are $\{1,2,3\}$, $\{1,2,4\}$ and $\{1,2,5\}$. $\chi(\mathrm{Sh}(\ell)) = 1$. $\mathfrak{P}_t(\mathrm{Sh}(\ell)) = 1$.
$\ell = (1,1,1,1,1)$: $\mathrm{Sh}(\ell)$ is the 1-skeleton of $\mathcal{F}\{1,2,3,4,5\}$. $\chi(\mathrm{Sh}(\ell)) = -5$. $\mathfrak{P}_t(\mathrm{Sh}(\ell)) = 1 + 6t$.

2.5. $a = \{\{2z\} \mid z \in \mathbb{Z}\}$ and $a = \{\{2z+1\} \mid z \in \mathbb{Z}\}$.

2.10. By definition of a pseudomanifold, there exists a sequence $\sigma = \sigma_0, \dots, \sigma_m = \sigma'$ of n-simplexes such that, for $i \leq 1 < m$, σ_i and σ_{i+1} have an $(n-1)$-face in common, called μ_i. Then, $a = \{\mu_0, \dots, \mu_{m-1}\}$ satisfies $\delta(a) = \{\sigma, \sigma'\}$.

2.11. If $\gamma = \partial(\alpha)$, then $\gamma = \partial(\alpha + [M])$. If also $\gamma = \partial(\alpha')$, then $\alpha + \alpha' \in Z_m(M)$, therefore $\alpha + \alpha' = 0$ or $[M]$ by Proposition 2.4.4.

2.12. This is the content of Proposition 2.5.7.

2.15. $H_n(M) = 0$. Indeed, if $0 \neq a \in C_n(M)$, there are $\sigma \in a$ and $\sigma' \in \mathcal{S}(M) - a$ (since M is infinite). Using Point (c) of the definition of a pseudomanifold, one can suppose that $\sigma \cap \sigma' = \mu \in \mathcal{S}_{n-1}(M)$. Then $\mu \in \partial a$. Thus, $Z_n(M) = 0$.

J.-C. Hausmann, *Mod Two Homology and Cohomology*, Universitext,
DOI 10.1007/978-3-319-09354-3_11

2.17. One has $\mathcal{S}(K)-\mathcal{S}(K_2)=\mathcal{S}(K_1)-\mathcal{S}(K_0)$, whence $H_*(K_1,K_0)\approx H_*(K,K_2)$. The diagram for (b) is

$$\begin{array}{ccccccc} H_r(K_0) & \longrightarrow & H_r(K_1) & \longrightarrow & H_r(K_1,K_0) & \xrightarrow{\partial_*^{01}} & H_{r-1}(K_0) \\ \downarrow & & \downarrow & & \approx\downarrow i_* & & \downarrow \\ H_r(K_2) & \longrightarrow & H_r(K) & \xrightarrow{j_*} & H_r(K,K_2) & \xrightarrow{\partial_*} & H_{r-1}(K_2) \end{array} .$$

One checks that $\partial_{MV}=\partial_*^{01}\circ i_*^{-1}\circ j_*$ works as a Mayer-Vietoris connecting homomorphism.

2.19. By Mayer-Vietoris, $\mathfrak{P}_t(M)=\mathfrak{P}_t(M_1)+\mathfrak{P}_t(M_2)-t^n-1$.

11.2 Exercises for Chapter 3

3.1. In the triangulation of $\Delta^m\times I$ used in the proof of Proposition 3.1.30, the number of $(m+1)$-simplexes is equal to $m+1$. In that used in the proof of Lemma 3.1.35, this number is $\sum_{k=1}^{m+1}k!$.

3.2. Point (b) is proven by induction on the skeleta of (A,B), using (a) and that $D^n=CS^{n-1}$. For (c), let X be a contractible space. Thus, there exists $x_0\in X$ and a continuous map $F\colon X\times I\to X$ with $F(x,0)=x$ and $F(x,1)=x_0$. Note that the standard simplex Δ^{n+1} is homeomorphic to the cone on Δ^n. Therefore, a point of Δ^{n+1} has coordinates $[z,t]$, where $(z,t)\in\Delta^n\times I$, with $[z,1]=[z',1]$ for all $z,z'\in\Delta^n$. A linear map $D\colon C_m(X)\to C_{m+1}(X)$ is then defined, for $\sigma\in\mathcal{S}_m(X)$, by $D(\sigma)([z,t])=F(\sigma(z),t)$. It satisfies $\partial D(\alpha)=D(\partial\alpha)+\alpha$ for all $\alpha\in C_m(X)$ and all $m\geq 1$. This implies that $Z_m(X)=B_m(X)$ for $m\geq 1$.

3.3. Using the homology sequence of $(X,X-A)$: $\mathfrak{P}_t(S^2-A)=1+(n-1)t$ and $\mathfrak{P}_t(T^2-A)=1+(n+1)t$.

3.4. $(\mathbb{R}P^2,\mathbb{R}P^1)$ and $(\mathbb{R}P^2,S)$ where S is the boundary of a 2-disk in $\mathbb{R}P^2$.

3.5. Define $h(x,t)=x$, $u(A)=0$ and $u(X-A)=1$. Then (u,h) is a presentation of (X,A) as a well cofibrant pair.

3.6. Let M be the Möbius band. Then $H_1(\mathrm{Bd}M)\to H_1(M)$ is the zero homomorphism, contradicting Lemma 3.3.1 if there were a continuous retraction of M onto $\mathrm{Bd}M$).

3.8. There is a unique smaller arc of great circle joining $f(x)$ to $f(-x)$, whence a homotopy to a map $\bar f$ satisfying $\bar f(x)=\bar f(-x)$. This means that $\bar f$ factors through $S^n\to\mathbb{R}P^n$ which is of degree 0.

3.9. It is a braid diagram looking locally like

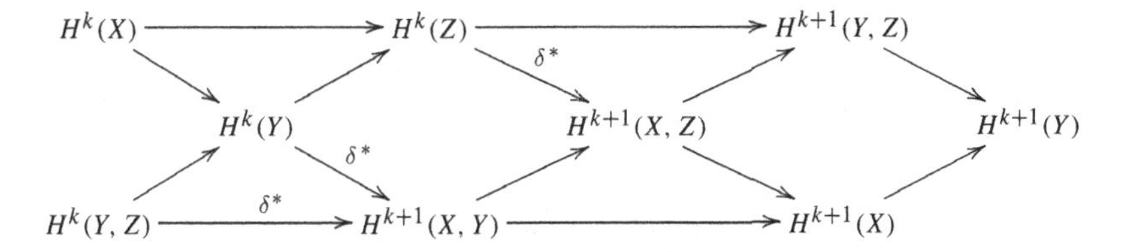

3.11. Like for Exercise 2.17.

3.13. If X is countable, so is its cellular chain complex and then $H_*(X)$ is countable. But if, for example, X has infinitely many connected components, then $H^0(X)$ is not countable by Corollary 3.1.12.

3.14. There is a retraction $r_n\colon B \to C_n$ sending $B - C_n$ to 0. This gives a homomorphism $r_* = \prod H_* r_n\colon H_1(B) \to \prod_{n=1}^{\infty} H_1(C_n) = \prod_{n=1}^{\infty} \mathbb{Z}_2$. Dividing $[0, 1]$ using the intervals $I_n = [1/(n+1), 1/n]$, we can construct various paths in B so that the generator of $H_1(I_n, \mathrm{Bd}\, I_n)$ is sent to that of $H_1(C_n)$ or to 0. Any element of $\prod_{n=1}^{\infty} H_1(C_n)$ may thus be realized, proving that r_* is surjective.
For (b), let R_k be the set of homotopy classes of maps from B to $\mathbb{R}P^\infty$ sending C_n onto a point for $n > k$. This set is finite. As $\mathbb{R}P^\infty$ is locally contractible, the map $\bigcup_{n=1}^{\infty} R_n \to [B, \mathbb{R}P^\infty]$ is surjective, which proves (b).
If B has the homotopy type of a CW-complex, then $H^1(B) \approx [B, \mathbb{R}P^\infty]$ by Proposition 3.8.3. As the latter is countable by (b), this contradicts (a) by Kronecker duality.

3.15. Let $\mathcal{B}_q$ be a $\mathbb{Z}_2$ basis of R_q and let X_q be a bouquet of q-spheres indexed by $\mathcal{B}_q$. The bouquet $X = \bigvee_{q \geq 1} X_q$ satisfies $H_q(X) \approx R_q$.

3.16. Since $1 - m + n = \chi(X) = 1 - b_1(X) + b_2(X)$.

3.18. Note that $H_1(X_p) \approx \big(\pi_1(X_P)/[\pi_1(X_P), \pi_1(X_P)]\big) \otimes \mathbb{Z}_2$. Thus, $b_1(X_{P_1}) = b_1(X_{P_2}) = 1$ and $b_1(X_{P_3}) = 0$. As $\chi(X_{P_1}) = \chi(X_{P_2}) = 0$ and $\chi(X_{P_3}) = 1$, we get $\mathfrak{P}_t(X_{P_1}) = \mathfrak{P}_t(X_{P_2}) = 1 + t$ and $\mathfrak{P}_t(X_{P_3}) = 1$.

3.20. Since X is not contractible, $r \circ j$ is not homotopic to a constant map, and nor is j. One deduces that $H^n(X) \approx \mathbb{Z}_2$ and then $j \circ r$ is homotopic to the identity by Exercise 3.19.

11.3 Exercises for Chapter 4

4.5. As in Sect. 4.3.3, one defines a *transfer homomorphism* $\mathrm{tr}_*\ H_*(X) \to H_*(\tilde{X})$ by sending each singular simplex $\sigma\colon \Delta^m \to X$ to the set of its liftings into $\tilde{X}$. As the number of these liftings is odd, one has $H_* p \circ \mathrm{tr}_* = \mathrm{id}$. Hence $H_* p$ is surjective and $H^* p$ is injective by Corollary 2.3.11.

4.6. Use Proposition 4.2.3.

4.7. From T to K, there is no map of degree one by Exercise 4.6. The same kind of argument shows that there is no map of degree one from $S^1 \times S^2$ to $\mathbb{R}P^3$ (the map $a \mapsto a^3$ is non trivial in $H^*(\mathbb{R}P^3)$ while it is trivial in $H^*(S^1 \times S^2)$).

For the other directions, let $f: K \to T$ be a map. By Proposition 4.3.10, $H^1 f$ cannot be injective which, given the ring structure of $H^*(T)$, implies that f is not of degree one. For a map $g: S^1 \times S^2 \to \mathbb{R}P^3$, we see above that $H^1 f = 0$. Thus, the composition of g with the inclusion $j: \mathbb{R}P^3 \to \mathbb{R}P^\infty$ is homotopic to a constant map by Proposition 3.8.3. But $H^3 j$ is an isomorphism by Proposition 4.3.10, so g is of degree 0.

4.8. $\hat{M}$ is the a connected sum of M and $\mathbb{R}P^n$. By Propositions 4.2.1 and 4.3.10, one has $H^*(\hat{M}) \approx H^*(M)[a]/(a^m + [M]^\sharp)$.

4.9. By Sect. 4.2.4 and Proposition 4.2.1, one has

$$H^*((S^1 \times S^1) \sharp \mathbb{R}P^2) \approx \mathbb{Z}_2[a,b,c]/(a^2, b^2, c^3, ab + c^2, ac, bc),$$

with a, b and c of degree 1, and

$$H^*(\mathbb{R}P^2 \sharp \mathbb{R}P^2 \sharp \mathbb{R}P^2) \approx \mathbb{Z}_2[x,y,z]/(x^3, y^3, z^3, x^2{+}y^2, y^2{+}z^2, xy, xz, yz)$$

with x, y and z of degree 1. An isomorphism is given by $a \mapsto x+y$, $b \mapsto x+z$ and $c \mapsto x+y+z$.

4.11. Use Proposition 4.7.11 and (4.7.8).

4.12. (a) $H^*(X) \approx \mathbb{Z}_2[x_1, \dots, x_n]$, with x_i of degree 1 (Künneth theorem). For (b) and (c), set $H^*(\mathbb{C}P^2) \approx \mathbb{Z}_2[a]/(a^3)$ and $H^*(\mathbb{C}P^3) \approx \mathbb{Z}_2[b]/(b^4)$, with a and b of degree 2. Then

(b) As a $\mathbb{Z}_2$-vector space, $\tilde{H}^*(Y)$ admits the basis $ab, a^2b, ab^2, ab^3, a^2b^2, a^2b^3$ (see Proposition 4.7.11; we write xy for the reduced cross product of x and y). The only non trivial cup products are $(ab)^2 = a^2b^2$ and $(ab)(ab^2)) = a^2b^3$.

(c) $\tilde{H}^*(Z) \approx \tilde{H}^*(\Sigma Y)$ (see p. 172). All cup products vanish.

4.15. The space $X = \mathbb{R}P^m \times \mathbb{R}P^n$ is a connected CW-complex of dimension $m+1$, thus **cat** $(X) \leq m + n + 1$ by Proposition 4.4.1. By Proposition 4.3.10 and the Künneth theorem, $H^*(X) \approx \mathbb{Z}_2[a,b]/(a^{m+1}, b^{n+1})$. Thus, $a^m b^n \neq 0$, so **nil** $H^{>0}(X) \geq m + n + 1$. Proposition 25.2 thus implies that **cat** $(X) = m + n + 1$.

4.16. **cat** $(T^n) = n + 1$ (same arguments as for Exercise 4.15).

4.17. For continuous maps $f: X \to X'$ and $g: Y \to Y'$, the diagram

$$\begin{array}{ccc} H_*(X \times Y) & \xrightarrow{\times} & H_*(X) \otimes H_*(Y) \\ \downarrow{\scriptstyle H_*(f\times g)} & & \downarrow{\scriptstyle H_*f\otimes H_*g} \\ H_*(X' \times Y') & \xrightarrow{\times} & H_*(X') \otimes H_*(Y') \end{array}$$

is commutative. This is proven using Formula (4.6.8).

4.18. Using the Künneth formula, the condition $\mathfrak{P}_t(X) = \mathfrak{P}_t(Y)$ amounts to the polynomial equality $\prod_{i=1}^{r}(1 + a_i\, t) = \prod_{j=1}^{s}(1 + b_j\, t)$. Therefore $r = s$ and $\sigma_k(a_1, \dots, a_r) = \sigma_k(b_1, \dots, b_r)$ for $k = 1 \dots, r$, where σ_k is the kth elementary symmetric function. This implies the polynomial equality

$$\prod_{i=1}^{r}(x - a_i) = \prod_{i=1}^{r}(x - b_i)\,,$$

which implies that $b_i = a_{\alpha(i)}$.

4.19. For $u \in H^*(X)$ and $v \in H^*(Y)$, one has

$$\begin{aligned}
\langle u \otimes v, (a \frown \alpha) \otimes (b \frown \beta)\rangle &= \langle u, a \frown \alpha\rangle\langle v, b \frown \beta\rangle && \text{by (4.6.7)}\\
&= \langle u \smile a, \alpha\rangle\langle v \smile b, \beta\rangle \\
&= \langle (u \smile a) \otimes (v \smile b), \alpha \otimes \beta\rangle && \text{by (4.6.7)}\\
&= \langle (u \smile a) \times (v \smile b), \underline{\times}^{-1}(\alpha \otimes \beta)\rangle && \text{by (4.6.8)}\\
&= \langle (u \times v) \smile (a \times b), \underline{\times}^{-1}(\alpha \otimes \beta)\rangle && \text{by Remark 4.6.4}\\
&= \langle u \times v, (a \times b) \frown \underline{\times}^{-1}(\alpha \otimes \beta)\rangle \\
&= \langle u \otimes v, \underline{\times}\big((a \times b) \frown \underline{\times}^{-1}(\alpha \otimes \beta)\big)\rangle && \text{by (4.6.8).}
\end{aligned}$$

As the elements $u \otimes v$ generate $H^*(X) \otimes H^*(Y)$, this proves the formula.

4.20. Using Lemma 4.7.2 and Eq. (4.6.8).

4.21. Let $i\colon E_0 \to E$ and $j\colon (E, \emptyset) \to (E, E_0)$ denote the inclusions. The map H^*i and H^*j are morphisms of $H^*(B)$-algebras. To see that $\delta^*\colon H^*(E_0) \to H^{*+1}(E, E_0)$ is a morphism of $H^*(B)$-modules, let $b \in H^r(B)$ and $v \in H^s(E, E_0)$. Then,

$$\delta^*(H^*p_0(b) \smile v) = \delta^*(H^*i \circ H^*p(b) \smile v) = H^*p(b) \smile \delta^*(v)\,,$$

holds true in $H^{r+s}(E, E_0)$, thanks to the singular cohomology analogue of Lemma 4.1.9.

4.22. If d is odd, then $\chi(\mathbf{F}_d(K)) = 1 - \chi(K)$. If d is even, $\chi(\mathbf{F}_d(K)) = 1 + b(K)$, where (K) is the total Betti number of K. This may be deduced from Corollary 4.7.52.

4.23. The Gysin sequence implies (a) and (b). Actually, $e = 0$ implies that the restriction homomorphism $H^*(E) \to H(\Sigma)$ is surjective (see Proposition 4.7.35), so the isomorphism ϕ of (b) is an isomorphism of $H^*(S^m)$-modules by the Leray-Hirsch theorem. If $n > 2$, ϕ is a **GrA**-isomorphism for degree's reasons. The hypothesis $n > 2$ in (c) is necessary (see Remark 4.7.34).

4.24. Via a Riemannian metric, we identify $D(\nu)$ with a closed tubular neighbourhood of Q in M. For $x \in Q$, let $(D_x, S_x) \approx (D^r, S^{r-1})$ be the fiber of $(D(\nu), S(\nu))$ over x. The pair inclusions $(D_x, S_x) \hookrightarrow (D(\nu), S(\nu)) \hookrightarrow (M, M - Q)$ gives rise to the commutative diagram

$$\begin{array}{ccccc} H^{r-1}(M-Q) & \longrightarrow & H^{r-1}(S(\nu)) & \longrightarrow & H^{r-1}(S_x) \\ \downarrow{\scriptstyle \delta^*} & & \downarrow{\scriptstyle \delta^*} & & \approx\downarrow{\scriptstyle \delta^*} \\ H^r(M, M-Q) & \xrightarrow[\approx]{} & H^r(D(\nu), S(\nu)) & \twoheadrightarrow & H^r(D_x, S_x) \end{array} .$$

The left vertical arrow is surjective since $H^r(M) = 0$. Thus, $H^{r-1}(S(\nu)) \to H^{r-1}(S_x)$ is onto, which implies that $e(\nu) = 0$ by Proposition 4.7.35. Compare Exercise 5.11.

11.4 Exercises for Chapter 5

5.1. Let X and Y be two homology manifolds and let $(x, y) \in X \times Y$. As $(X \times (Y - y)) \cup ((X - x) \times Y) = X \times Y - (x, y)$, one can use a relative Künneth theorem like Theorem 4.6.10 (see the comments after this theorem for the hypotheses that we use, i.e. that $(X \times (Y - y), (X - x) \times Y)$ being excisive in $X \times Y$, which is true since they are both open).

5.4. It is enough to prove it for a manifold X connected with $Y = \operatorname{Bd} X \neq \emptyset$. If $n = \dim X$, Theorem 5.3.7 implies that $H_n(X) \approx H^0(X, Y) = 0$ and $H_n(X, Y) \approx H^0(X) = \mathbb{Z}_2$. Therefore, $H_{n-1}(Y) \to H_{n-1}(X)$ is not injective by the homology sequence of (X, Y), which would be the cases if there were a retraction from X onto Y (see Lemma 3.3.1).

5.5. Only the connected component of the point plays a role, so we can suppose that M is connected. Instead of removing a point, we remove a closed n-disk D around it. Thus $M = M_0 \cup D$, where $M_0 = M - \operatorname{int} D$. As $H_n(M) \approx \mathbb{Z}_2 \approx H_n(M, M_0)$, the result follows from the homology sequence of (M, M_0).

5.6. Since $[N] = [N_1 + [N_2] = \partial[M]$ in $H_{n-1}(M)$. If $N_2 = \emptyset$, then $H_* f([N_1]) = 0$.

5.8. There exists no degree one map from Σ_m to $\bar{\Sigma}_n$ by Propositions 5.2.8 and 4.2.3. The same arguments show that if there is a degree one map from $\bar{\Sigma}_n$ to Σ_m, then $m \leq [(n-1)/]$. The converse is true, using that $\mathbb{R}P^2 \sharp \mathbb{R}P^2 \sharp \mathbb{R}P^2$ and $(S^1 \times S^1) \sharp \mathbb{R}P^2$ are homeomorphic (see [136, Lemma 7.1]).

5.9. No, because $H^* f$ would be injective by Proposition 5.2.8, and, by the Künneth theorem, there is no element $a \in H^1(M)$ with $a^m \neq 0$.

5.10. Follows from Corollary 5.4.13.

5.11. Let ν be the normal bundle to Q. That $H_* i([Q]) = 0$ implies that the Poincaré dual PD(Q) of Q vanishes, whence $e(\nu) = 0$ by Lemma 5.4.4. Compare Exercise 4.24.

5.13. Let $A = \mathrm{PD}(Q_1) \times \mathrm{PD}(Q_2)$. For dimensional reasons, $[M_1 \times M_2] = \underline{\times}^{-1}([M_1] \otimes [M_2])$ and $[Q_1 \times Q_2] = \underline{\times}^{-1}([Q_1] \otimes [Q_2])$. Therefore,

$$\begin{aligned}
\underline{\times}(A \frown [M_1 \times M_2]) &= \underline{\times}(\mathrm{PD}(Q_1) \times \mathrm{PD}(Q_2) \frown \underline{\times}^{-1}([M_1] \otimes [M_2])) \\
&= (\mathrm{PD}(Q_1) \frown [M_1]) \otimes (\mathrm{PD}(Q_2) \frown [M_2]) \qquad \text{by (4.8.1)} \\
&= H_* i_1([Q_1]) \otimes H_* i_2([Q_2]) \\
&= (H_* i_1 \otimes H_* i_2)([Q_1] \otimes [Q_2]) \\
&= \underline{\times}(H_* i([Q_1 \times Q_2])) \qquad \text{by Exercise 4.17}
\end{aligned}$$

where $i \colon Q_1 \times Q_2 \to M_1 \times M_2$ denotes the inclusion. As $\underline{\times}$ is an isomorphism, this proves (5.5.1).

5.14. By Exercise 5.13, $\mathrm{PD}(\{x\} \times M') = \mathrm{PD}(\{x\}) \times \mathrm{PD}(M') = [M_x]^\sharp \times \mathbf{1}$, where M_x is the connected component of x in M.

5.18. Let $a \in Q$. The map associating to $(t, x) \in I \times Q'$ the point $(\sqrt{1-t^2}\, a, t^2 x)$ parameterizes a smooth $(q'+1)$-disk D'_a in Σ, with boundary Q', which intersects Q transversally at a. By Proposition 5.4.22, $l(Q, Q') = 1$.

11.5 Exercises for Chapter 6

6.1. $\mathbb{K}P^n$ may be covered by $n+1$ chart domains which are contractible (see (6.1.1) for $\mathbb{K} = \mathbb{C}$; the same formalism works for $\mathbb{K} = \mathbb{H}$). On the other hand, $\mathbf{nil}\, H^{>0}(\mathbb{K}P^n) = n+1$, thus $\mathbf{cat}\,(\mathbb{K}P^n) = n+1$ by Proposition 4.4.2, like for $\mathbb{K} = \mathbb{R}$: see Corollary 4.4.3).

6.3. Follows from Proposition 6.1.11.

6.4. The two parts of CC_f give rise to the Mayer-Vietoris sequence

$$\longrightarrow H^k(CC_f) \longrightarrow H^k(Y) \oplus H^k(Y) \xrightarrow{\varphi} H^k(X) \longrightarrow H^{k+1}(CC_f) \longrightarrow$$

where $\varphi = H^* f + H^* f$. For $f = j$, this sequence together Proposition 6.1.11 gives the Poincaré series of CC_j:

$$\mathfrak{P}_t(CC_j) = \sum_{k=0}^{\infty} t^{4k} + \sum_{k=1}^{\infty} t^{4k+3} = \frac{1+t^3}{1-t^4}\,.$$

Actually, one has a **GrA**-isomorphism $H^*(CC_j) \approx \mathbb{Z}_2[a,b]/(b^2)$ where a is of degree 4 and b of degree 3 (see Exercise 7.20).

6.5. By the Künneth theorem and Proposition 4.2.1, $H^*(X) \approx \mathbb{Z}_2[a,b]\big/(a^2, b^2)$ and $H^*(X) \approx \mathbb{Z}_2[c,d]\big/(c^3, d^3, cd, c^2+d^2)$ (generators in degree 4). Thus, $\mathfrak{P}_t(X) = \mathfrak{P}_t(Y) = 1 + 2t^4 + t^8$. But the cup-square map vanishes in $H^*(X)$ while not in $H^*(Y)$. (If the degree of the generators are set to 1, we get the cohomology of the torus for X and of the Klein bottle for Y).

6.6. Use the Hopf vector bundle $\gamma_{\mathbb{K}}$ over $\mathbb{K}P^1$. For $n = 2k$, one can take $\xi = \gamma_{\mathbb{C}} \times \cdots \times \gamma_{\mathbb{C}}$ (k times) over $(\mathbb{C}P^1)^k$. For $n = 2k+1$, take $\xi = \gamma_{\mathbb{C}} \times \cdots \times \gamma_{\mathbb{C}} \times \gamma_{\mathbb{R}}$ over $(\mathbb{C}P^1)^k \times \mathbb{R}P^1$.

6.7. There exits a map $f\colon X \to S^n$ such that $a = H^*f(\iota)$ (see the proof of Proposition 3.8.1). One can then take $\xi = f^*\gamma_{\mathbb{K}}$ where $\gamma_{\mathbb{K}}$ is the vector Hopf bundle over $\mathbb{K}P^1$ for $\mathbb{K} = \mathbb{R}, \mathbb{C}, \mathbb{H}$ and $\mathbb{O}$.

6.10. Follows from Exercise 4.24.

6.13. The maps h and f extend to maps on mapping cones $C_{g\circ h} \xrightarrow{\hat{h}} C_g$ and $C_g \xrightarrow{\hat{f}} C_{f\circ g}$. The formula follows from the functionality of the cup product.

6.14. Let $y' \in S^m$ be another regular value for f. We may suppose that j is transversal to $Q' = f^{-1}(\{y'\})$. Then,

$$\begin{aligned} \text{Hopf}\,(f) &= l(Q, Q') && \text{by Proposition 6.3.7} \\ &= \sharp\, j^{-1}(Q') \quad \bmod 2 && \text{by Proposition 5.4.22} \\ &= \sharp\,(f\circ j)^{-1}(\{y'\}) \bmod 2 && \\ &= \deg(f\circ j) && \text{as in Proposition 3.2.6.} \end{aligned}$$

11.6 Exercises for Chapter 7

7.2. Follows from Corollary 7.1.17. Note that this is wrong for $p = 0$ (see Example 7.1.16).

7.3. As X is equivariantly formal, there is a $\mathbf{GrV}[u]$-isomorphism between $H^*(X_G)$ and $\mathbb{Z}_2[u][\mathcal{B}]$, where $\mathcal{B}$ is a $\mathbf{GrV}$ basis of $H^*(X)$. Therefore,

$$\mathfrak{P}_t(X_G) = \frac{\mathfrak{P}_t(X)}{1-t} \text{ and } \mathfrak{P}_t(u\cdot H^*(X_G)) = \frac{t\,\mathfrak{P}_t(X)}{1-t}.$$

7.6. By Proposition 7.1.12, $\tilde{\rho}\colon \tilde{H}^d_G(X) \to \tilde{H}^d(X)^G$ is an isomorphism and, as $d \geq 1$, $\tilde{H}^d(X)^G$ is the diagonal subgroup of $H^d(X) \approx \mathbb{Z}_2 \oplus \mathbb{Z}_2$. By Sequence (7.1.8), there exists $0 \neq a \in H^d_G(X)$ and $\text{Ann}\,(u) = \mathbb{Z}_2 a$. Sequence (7.1.8) also implies that $H^*_G(X) \approx \mathbb{Z}_2[u,a]\big/(ua)$. If $a^2 \neq 0$, then $a^2 = u^{2d}$ which would contradict the equation $ua = 0$; indeed $ua^2 = u^{2d+1} \neq 0$ by Corollary 7.1.5.

7.7. The pair $(SU(2), \Delta)$ is isomorphic to (S^3, S^1), the groups of units quaternions and complex numbers. For Γ a topological group, let $E_n\Gamma$ be the join of n copies of Γ, on which Γ acts diagonally. Then, for $n \geq 1$, $E_n(S^3)/S^3 \approx S^{4(n-1)+3}/S^3 \approx \mathbb{H}P^{n-1}$ and $E_n(S^1)/S^1 \approx S^{4(n-1)+1}/S^1 \approx \mathbb{C}P^{n-1}$. Thus, there are homotopy equivalences $BS^3 \simeq E_\infty S^3/S^3 \simeq \mathbb{H}P^\infty$ and $BS^1 \simeq E_\infty S^1/S^1 \simeq \mathbb{C}P^\infty$.

7.8. One check that $(X \times Y)_\Gamma \approx X_\Gamma \times Y$. The result then follows from the Künneth theorem 4.6.7. One can also use the equivariant Künneth theorem 7.4.3 with an additional hypothesis on Γ.

7.9. The hypotheses of the Künneth theorems 4.6.7 and 7.4.3 are satisfied (see the end of Remark 7.4.4). Therefore, there is a commutative diagram

$$\begin{array}{ccc} H^*_\Gamma(X) \otimes_{H^*_\Gamma(pt)} H^*_\Gamma(Y) & \xrightarrow[\approx]{\bar{\times}_\Gamma} & H^*_\Gamma(X \times Y) \\ \Big\downarrow \rho_X \otimes \rho_Y & & \Big\downarrow \rho_{X \times Y} \\ H^*(X) \otimes H^*(Y) & \xrightarrow[\approx]{\times} & H^*(X \times Y) \end{array}$$

As X and Y are equivariantly formal, $\rho_X \otimes \rho_Y$ is surjective and thus, $\rho_{X \times Y}$ is surjective.

7.10. By Exercise 7.9 and (7.1.16), one has a **GrA**$[u]$-isomorphism

$$H^*_G(X) \approx H^*_G(S^1_0) \otimes H^*(Y) \approx \mathbb{Z}_2[u,a]/(a^2+ua) \otimes H^*(Y) \approx H^*(Y)[a]/(a^2+ua)$$

with a of degree 1.The Mayer-Vietoris sequence in question is equivalent to

$$0 \longrightarrow H^*_G(X) \xrightarrow{r} H^*_G(X^G) \xrightarrow{J} H^*(Y) \longrightarrow 0$$

with the **GrA**$[u]$-isomorphism $H^*_G(X^G) \approx H^*(Y)[u_1] \oplus H^*(Y)[u_2]$. One has $J(b,c) = b + c$ for $b, c \in H^*(Y)$ (proving that J is onto) and $J(u_1, 0) = J(0, u_2) = 0$. One has $r(u) = (u_1, u_2)$, $r(b) = (b, b)$ for $b \in H^*(Y)$ and $r(a) = (0, u_2)$.

7.11. By the equivariant Künneth theorem 7.4.3, $H^*_G(X) \approx \mathbb{Z}_2[u,a,b]/(a^2 + ua, b^2 + ub)$ with a and b of degree 1. The critical points of f, labeled by there value, are $x_{-3} = (-1,-1)$ (index 0), $x_{-1} = (1,-1)$ and $x_1 = (-1,1)$ (index 1) and $x_3 = (1,1)$ (index 2). Therefore

$$H^*_G(X^G) \approx \mathbb{Z}_2[u_{-3}] \oplus \mathbb{Z}_2[u_{-1}] \oplus \mathbb{Z}_2[u_1] \oplus \mathbb{Z}_2[u_3]\,.$$

One has $r(u) = (u_{-3}, u_{-1}, u_1, u_3)$, $r(a) = (0, u_{-1}, 0, u_3)$ and $r(b) = (0, 0, u_1, u_3)$, whence $r(ab) = (0,0,0,u_3^2)$.

7.12. $X = \mathbb{K}P^2$ ($\mathbb{K} = \mathbb{R}, \mathbb{C}, \mathbb{H}$ or $\mathbb{O}$) with the involution $\tau(x_0 : x_1 : x_2) = (-x_0 : x_1 : x_2)$. Related to this exercise is Exercise 8.9.

7.13. One has $(P^n_p)^G = P^p_p \dot\cup P^{n-p-1}_{n-p-1}$. The total Betti number thus satisfy $b(P^n_p) = b((P^n_p)^G)$. This proves (a) by Proposition 7.3.7.

For (b), set $H^*(P^p_p) = \mathbb{Z}_2[u, b_1]/(b_1^{p+1})$, $H^*(P^{n-p-1}_{n-p-1}) = \mathbb{Z}_2[u, b_2]/(b_2^{n-p})$ and $H^*(P^n_p) = \mathbb{Z}_2[a]/(a^{n+1})$. Thus, as a $\mathbb{Z}_2$-module, $H^*_G(P^n_p)$ is generated by the powers of an element $A \in H^1_g(P^n_p)$ such that $\rho(A) = a$. There are two such elements, A and B, with the relation $B = A + u$. One has $r(A) =$

$(b_1 + u, b_2)$ or $r(A) = (b_1, b_2 + u)$. (use that $\bar{r}: H^*(P^n_p) \to H^*((P^n_p)^G)$ satisfy $\bar{r}(a) = (b_1, b_2)$; but $(b_1, b_2)^n = 0$ in $H^n_G((P^n_p)^G)$ and $(b_1+u, b_2+u)^n \in (u)$, contradicting $\rho(A^n) = a^n \neq 0$). By exchanging A and B if necessary, one may suppose that $r(A) = (b_1 + u, b_2)$.
For (c), one uses that $r(A) = (u, b_2)$ (see (b)). Hence, $r(A^k) = (u^k, 0)$ for $k \geq n-1$. Therefore, the relation $r(A^{n+1}) = r(uA^n)$ holds true in $H^*_G((P^n_p)^G)$. As r is injective, the relation $A^{n+1} = uA^n$ holds true in $H^*_G(P^n_p)$. Therefore, there is a surjective **GrA**$[u]$-morphism $h: \mathbb{Z}_2[u][A]/(A^{n+1} + uA^n) \to H^*_G(P^n_p)$. But $\mathbb{Z}_2[u][A]/(A^{n+1} + uA^n)$ and $H^*_G(P^n_p)$ (which is a free $\mathbb{Z}_2$-module over $\{A, \dots, A^n\}$) have the same Poincaré series. Hence, h is an isomorphism. The second presentation follows by substituting $B + u$ to A. Point (d) is proved in the same way.

7.14. The isomorphism is given by $A \mapsto B + u$, with inverse given by $B \mapsto A + u$. Note that R and S are the two presentations for $H^*_G(P^2_0)$ given in Exercise 7.14 (c). Take $X = \mathbb{H}P^2$ with $\tau(x_0 : x_1 : x_2) = (-x_0 : x_1 : x_2)$.

7.16. If we consider S^1 acting on $\mathbb{C}P^n$ by

$$g \cdot (x_0 : \cdots : x_p :_{p+1} : \cdots : x_n) = (gx_0 : \cdots : gx_p : x_{p+1} : \cdots : x_n),$$

the statements the are all the same, the degree of each cohomology classes being doubled.

7.18. Write ΣX for $\Sigma_\Gamma X$. We see X as the union of two cones C_-X and C_+X glued over their common base X. The homomorphism Σ^*_Γ comes from the composition

$$H^*_\Gamma(X) \xrightarrow{\delta^*} H^{*+1}_\Gamma(C_-X, X) \xleftarrow{\approx} H^{*+1}_\Gamma(\Sigma X, C_+X) \longrightarrow H^{*+1}_\Gamma(\Sigma X)$$

The homomorphism δ^* comes from the cohomology sequence of the pair $((C_-X)_\Gamma, X_\Gamma)$. As C_X is Γ-equivariantly contractible, the homomorphism δ^* descends to an injection $\tilde{\delta}^*: \tilde{H}^*_\Gamma(X) \to H^{*+1}_\Gamma(C_-X, X)$. The middle arrow is induced by the pair inclusion $(C_-X, X) \to (\Sigma X, C_+X)$ and is an isomorphism by excision on the Borel constructions. By (7.2.10), the last arrow becomes an isomorphism when composed by $H^{*+1}_\Gamma(\Sigma X) \to \tilde{H}^{*+1}_\Gamma(\Sigma X)$. Whence the homomorphism Σ^*_Γ which is injective. That Σ^*_Γ is a morphism of $H^*_\Gamma(pt)$-module comes from Exercise 4.21 applied to the bundle pair $((C_-X)_\Gamma, X_\Gamma)$ over $B\Gamma$. The morphism Σ^*_Γ is functorial in both X and in Γ. In particular, the trivial homomorphism $\Gamma \to \{1\}$ gives the commutative diagram

$$\begin{array}{ccc} \tilde{H}^*_\Gamma(X) & \xrightarrow{\Sigma^*_\Gamma} & \tilde{H}^{*+1}_\Gamma(\Sigma_\Gamma X) \\ \downarrow{\scriptstyle\rho} & & \downarrow{\scriptstyle\rho_\Sigma} \\ \tilde{H}^*(X) & \xrightarrow[\approx]{\Sigma^*} & \tilde{H}^{*+1}(\Sigma X) \end{array} \quad . \tag{11.6.1}$$

When $\Gamma = G = X$, one has $\tilde{H}^*_G(G) = 0$ by Lemma 7.1.14 while $\tilde{H}^1_G(\Sigma_G G) = \mathbb{Z}_2$, since $\Sigma_G G$ is G-homeomorphic to the sphere S^1_0 of Example 7.1.4.
For (c), let $\sigma\colon H^*(X) \to H^*_\Gamma(X)$ be a section of ρ. Using the commutativity of Diagram (11.1), one defines a section σ_Σ of ρ_Σ by $\sigma_\Sigma = \Sigma^*_\Gamma \circ \sigma \circ (\Sigma^*)^{-1}$. Therefore, ΣX is Γ-equivariantly formal and, as $H^*_\Gamma(pt)$-module, $\tilde{H}^{*+1}_\Gamma(\Sigma X)$ is generated by Image(σ_Σ). As Image$(\sigma_\Sigma) \subset$ Image$(\Sigma^*\Gamma)$, the morphism Σ^*_Γ is an isomorphism.

7.19. The G-space $\check{\Sigma} X$ is a finite dimensional G-complex satisfying $b(\check{\Sigma} X) = b((\check{\Sigma} X)^G) < \infty$ (since $(\check{\Sigma} X)^G = X^G$). Therefore, $\check{\Sigma} X$ is equivariantly formal by Proposition 7.3.7 and thus $r\colon \tilde{H}^*_G(\check{\Sigma} X) \to \tilde{H}^*_G((\check{\Sigma} X)^G)$ is injective by Proposition 7.3.9. As r factor through $\tilde{H}^*_G i$, we know that $\tilde{H}^*_G i$ is injective. By Proposition 7.3.7 again, X is itself equivariantly formal and thus ρ is surjective. As $\tilde{H}^* i = 0$, the image of $\tilde{H}^*_G i$ is contained in ker ρ. By (7.1.8), the later is equal to the ideal (u) generated by u. By comparing the Poincaré polynomials of $\tilde{H}^*_G(\check{\Sigma} X)$ with that of (u) (see Exercise 7.3) we deduce that the image of $H^*_G i$ is actually equal to (u).

7.20. Let Δ be the subgroup of $SU(2)$ formed by the diagonal matrices and let $\Delta^+ = \{\mathrm{dia}(e^{i\alpha}, e^{-i\alpha}) \mid \alpha \in [0, \pi]\}$. Each conjugation class in Γ has a unique representative in Δ^+, so X/Γ is homeomorphic to $[0, \pi]$. Hence, each class in $E\Gamma \times_\Gamma X$ has representatives of the form (z, b) with $b \in \Delta^+$. We deduce that X_Γ is homeomorphic to $(E\Gamma/\Delta \times [0, \pi]) \cup (E\Gamma/\Gamma \times \{0, \pi\})$, that is $X_\Gamma \approx CC_{j_0}$ where j_0 is the quotient map $E\Gamma/\Delta \to E\Gamma/\Gamma$. The latter is homotopy equivalent to the map $j_1\colon B\Delta \to B\Gamma$ induced by the inclusion $\Delta \to \Gamma$ (see Example 7.2.4). By Exercise 7.7, j_1 is equivalent to the inclusion $j\colon \mathbb{C}P^\infty \to \mathbb{H}P^\infty$. This proves (a). From the correction of Exercise 6.4 (see p. 507), we have

$$\mathfrak{P}_t(H^*_\Gamma(X)) = \frac{(1+t^3)}{1-t^4} = \mathfrak{P}_t(H^*_\Gamma(pt)) \cdot \mathfrak{P}_t(X)\,.$$

This equality implies that X is Γ-equivariantly formal (see the comment following Corollary 4.7.20). The unique section $H^*(X) \to H^*_\Gamma(X)$ of ρ is multiplicative for degree's reasons. Hence, the Leray-Hirsch theorem gives the $H^*_\Gamma(pt)$-algebra isomorphisms of (c).

7.21. $(X_\Gamma, (X_1)_\Gamma, (X_2)_\Gamma, (X_0)_\Gamma)$ is a Mayer-Vietoris data and $X_\Gamma = (X_1)_\Gamma \cup (X_2)_\Gamma$. By Lemma 7.2.11, $((X_i)_\Gamma, (X_0)_\Gamma)$ is a well cofibrant for $i = 1, 2$. The Mayer-Vietoris is then given by Lemma 3.1.38 and Proposition 3.1.51.

7.22. Let $z \in X \vee Y$ be the image of x and y. Using the Mayer-Vietoris sequence of Exercise 7.21, one has a commutative diagram

$$\begin{array}{ccccccccc} 0 & \longrightarrow & H^*_\Gamma(z) & \xrightarrow{\Delta} & H^*_\Gamma(z) \oplus H^*_\Gamma(z) & \xrightarrow{+} & H^*_\Gamma(z) & \longrightarrow & 0 \\ & & \downarrow & & \downarrow & & \downarrow \approx & & \\ 0 & \longrightarrow & H^*_\Gamma(X \vee Y) & \longrightarrow & H^*_\Gamma(X) \oplus H^*_\Gamma(Y) & \longrightarrow & H^*_\Gamma(z) & \longrightarrow & 0 \end{array}$$

where the vertical arrows are induced by the constant maps. Since the right vertical map is an isomorphism, we check that the quotient of the second row by the first one is an exact sequence (a particular case of the *snake lemma*: see [28, Lemma 3.3]). Thus, the H^*_Γ-algebra homomorphism $(\tilde{H}^*_\Gamma i_X, \tilde{H}^*_\Gamma i_Y)\colon \tilde{H}^*_\Gamma(X \vee Y) \to \tilde{H}^*_\Gamma(X) \oplus \tilde{H}^*_\Gamma(Y)$ induced by the inclusions is an isomorphism.

11.7 Exercises for Chapter 8

8.1. No, since Sq^1 vanishes in $H^*(\Sigma T)$ while not in $H^*(\Sigma K)$.

8.2. $\mathrm{Sq}^6(a^5b^7) = a^{10}b^8 + a^9b^9 + a^6b^{12} + a^5b^{13}$ (use that $\mathrm{Sq}(a^5b^7) = \mathrm{Sq}(a)^5\mathrm{Sq}(b)^7 = (a+a^2)^5(b+b^2)^7$).

8.3. Comes form Lemma 8.5.1.

8.5. Follows from the Adem relations. Remark: by [1, Sect. 2.5], any finite set in $\mathcal{A}$ generates a finite subalgebra; hence, all elements of $\mathcal{A}$ are nilpotent. For estimates of nilpotency heights, see [212, Sect. 2.6].

8.6. If $a \in H^1(\mathbb{R}P^\infty)$ is the generator then, as $\mathcal{A}$-module, $H^*(\mathbb{R}P^\infty)$ admits $\{a^r \mid r = 2^{n-1}\}$ as a minimal set of generators. This follows from (8.2.2.). Note: if $b(X)$ is infinite, then $H^*(X)$ is not finitely generated as an $\mathcal{A}$-module [172].

8.7. Follows from Proposition 8.6.2 and Theorem 8.6.5.

8.8. We may assume that $r > 1$, otherwise there is nothing to prove. If U is an open set of M over which p is trivial, then $p^{-1}(U)$ is homeomorphic to $U \times S^{r-1}$, which implies that $\dim M = n - r$ (using Theorem 3.3.4). For $r > 1$, the Gysin exact sequence implies that $H^*(M)$ is a **GrA**-quotient of $\mathbb{Z}_2[e]$, where $e \in H^r(M)$ is the Euler class of p. Hence, $n = kr$ and, using Exercise 8.7, $r = 1, 2, 4, 8$.

8.9. One has $n = \dim M = 2r$ by Corollary 5.2.5. By Poincaré duality, $\mathfrak{P}_t(M) = 1 + t^r + t^{2r}$ with non trivial cup-square. By Exercise 8.7, $n = 2^s$ with $s = 1, 2, 3, 4$.

8.10. The map f must have at least two critical points on each connected component of M. Since $M_{-1} = pt$, M is connected. We use the notation of Sect. 7.6: $M_{x,y} = f^{-1}([x, y])$, and $M_x = f^{-1}(\{x\})$. There are G-equivariant diffeomorphisms $\phi_-\colon (D(\nu_{-1}), S(\nu_{-1})) \to (M_{-1,0}, M_0)$ and $\phi_+\colon (D(\nu_1), S(\nu_1)) \to (M_{0,1}, M_0)$, where ν_x is the normal bundle to M_x in M. The pair $(D(\nu_{-1}), S(\nu_{-1}))$ is G-equivariantly diffeomorphic to (D^n, S^{n-1}) endowed with the antipodal involution. We thus have a smooth locally trivial bundle $S^{r-1} \xrightarrow{p} S^{n-1} \to M_1$, where r is the codimension of M_1 in M.

Since n is odd, Exercise 8.8 implies that $r = 1$, so p is a 2-fold covering. But p is G-invariant, so it can be identified with the quotient map $S^{n-1} \to M_1$. Therefore, M_1 is diffeomorphic to $\mathbb{R}P^{n-1}$ and ν_1 is the Hopf bundle η. Thus, M is diffeomorphic to $D^n \cup_{\phi_+^{-1}\circ\phi_-} D(\eta)$ which is homeomorphic to $\mathbb{R}P^n$. Note that M may not be diffeomorphic to $\mathbb{R}P^n$ if the diffeomorphism $\phi_+^{-1}\circ\phi_-\colon S^{n-1} \to S(\eta) = S^{n-1}$ does not extend to a diffeomorphism of D^n.

11.8 Exercises for Chapter 9

9.1. Use that $w_1(\xi\oplus\xi) = w_1(\xi)+w_1(\xi) = 0$ and, if $w_1(\xi) = 0$, that $w_2(\xi\oplus\xi) = w_2(\xi) + w_2(\xi) = 0$.

9.2. As p is a local diffeomorphism, one has $T\tilde{X} = p^*TX$. Hence $w_i(T\tilde{X}) = H^*p(w_i(TX))$. But H^*p is injective by Exercise 4.5. The result follows from Propositions 9.4.4 and 9.4.7.

9.3. Since p is a local diffeomorphism, one has $T\tilde{M} \approx p^*TM$. Hence, $H^*p(w_1(TM)) = w_1(T\tilde{M}) = 0$. Thus, $0 \neq w_1(TM) \in \ker(H^*p\colon H^1(M) \to H^1(\tilde{M}))$ which, by the transfer exact sequence, is generated by w.

9.4. Let $m = \dim M$. Proposition 4.2.1 and its proof provide an epimorphism $\hat{\alpha}\colon H^{>0}(M_1) \oplus H^{>0}(M_2) \to H^{>0}(M)$. For $0 < i < m$, $\hat{\alpha}$ is induced by the inclusions of M_i minus a disk into M. This implies that, for $0 < i < \dim M_i$, $w_i(M)$ is the image of $(w_i(M_1), w_i(M_2)$. The same holds true for $i = m$, using that $\chi(M) \equiv \chi(M_1) + \chi(M_2) \mod 2$ together with Corollary 5.4.16.

9.5. The Stiefel-Whitney classes $w(T\mathbb{R}P^4)$ (see Proposition 9.8.10) and $w(T\mathbb{C}P^2)$ (see Remark 9.8.12) are incompatible with such a decomposition, given Formula (9.4.3).

9.6. Given Formula (9.4.1), this follows from Exercise 8.3.

9.9. Using the tautological bundles, as for Proposition 6.1.11. Using that $BU(n)$ is a conjugation space (see p. 438), this also follows from Corollary 10.2.11.

9.11. $H^*(\mathrm{Gr}(2;\mathbb{R}^5))$ has 10 basis elements: $\mathbf{1} = (1,2)$, $w_1 = (1,3)$, $w_1^2 = (2,3) + (1,4)$, $w_2 = (2,3)$, $w_1^3 = (1,5)$, $w_1w_2 = (2,4)$, $w_2^2 = (3,4)$, $w_1^2w_2 = (3,4) + (2,5)$ ($w_1^4 = w_2^2 + w_1^2w_2$), $w_1^5 = w_1w_2^2 = (3,5)$ ($w_1^3w_2 = 0$), $w_2^2w_1^2 = (4,5)$. The expression in the Schubert symbols were found using Propositions 9.5.27 and 9.5.32.

9.12. Since M is orientable, $\mathrm{Sq}^1 : H^5(M) \to H^6(M)$ vanishes (see Corollary 9.8.5). By Wu's formula and the Adem relations, $w_6(TM) = \mathrm{Sq}^3(v_3(M)) = \mathrm{Sq}^1\mathrm{Sq}^2(v_3(M)) = 0$. The result follows from Corollary 5.4.16. The same argument works when $\dim M = 10$, using the Adem relation $\mathrm{Sq}^5 = \mathrm{Sq}^1\mathrm{Sq}^4$.

9.13. By Corollary 9.8.5, $\mathrm{Sq}^1\colon H^{n-1}(Q) \to H^n(Q)$ is surjective while $\mathrm{Sq}^1\colon H^{n-1}(P) \to H^n(P)$ vanishes. As $\mathrm{Sq}^1\circ H^{n-1}f = H^n f\circ\mathrm{Sq}^1$, this proves the statement under Hypothesis (a). The argument for Hypothesis (b) is similar, using Corollary 9.8.6.

9.14. The normal bundle ν_M of $M = f^{-1}(\{x\})$ in S^{n+k} is trivial, since $\nu_M = f^* T_x S^k$. Thus, M admits an embedding into $\mathbb{R}^{n+k+1}$ with trivial normal bundle. Therefore, $\bar{w}(TM) = \mathbf{1}$ and thus $w(TM) = \mathbf{1}$. The results follows from Thom's theorem 9.9.7.

9.15. By Thom's theorem 9.9.7, since they do not have the same Stiefel-Whitney numbers.

9.16. The manifold M is of dimension n. Let $a \in H^1(M)$ be the non-zero class. By Proposition 3.8.3, there exits a map $f: M \to \mathbb{R}P^\infty$ such that $a = H^* f(\iota)$. By cellular approximation (or by the proof of Proposition 3.8.3), the image of f is contained in $\mathbb{R}P^n$ and thus $H^* f: H^*(\mathbb{R}P^n) \to H^*(M)$ is an isomorphism. The result then follows from Corollary 9.9.10.

11.9 Exercises for Chapter 10

10.1. As usually drawn in $\mathbb{R}^3$, the orientable surface Σ_g of genus g admits an axial symmetry, with a scattered fixed point set, i.e. $2g + 2 = b(\Sigma_g)$ fixed points (this corresponds to a connected sum of g copies of $S^1 \times S^1$ with the involution $\tau(z_1, z_2) = (\bar{z}_1, \bar{z}_2)$). A non-orientable surface does not admit any involution with scattered fixed point set by Proposition 4.2.3 and Corollary 10.1.2 (or by Corollary 10.1.3 when the involution is smooth).

10.2. The first assertion comes from Diagram (10.2.11) (for a more elementary argument, see [87, Remark 3.1]). It implies that the H^*-frame respects the path-connected component of X, proving the "only if" part of last assertion. The "if" part is obvious.

10.3. Let (κ, σ) be the H^*-frame of a conjugation space X. The isomorphism κ satisfies $\kappa(\mathbf{1}) = \mathbf{1}$ by Exercise 10.2 (or by Proposition 10.2.5). Therefore, κ descends to a **GrV**-isomorphism $\tilde{\kappa}: \tilde{H}^{2*}(X) \to \tilde{H}^*(X^G)$. Also, $\sigma(\mathbf{1}) = \mathbf{1}$, so it descends to $\tilde{\sigma}: \tilde{H}^*(X) \to \tilde{H}^*_G(X)$ which is a section of $\tilde{\rho}: \tilde{H}^*_G(X) \to \tilde{H}^*(X)$. The conjugation equation holds true for $(\tilde{\kappa}, \tilde{\sigma})$ (call $(\tilde{\kappa}, \tilde{\sigma})$ a $\tilde{H}^*$-frame for X). Conversely, such a $\tilde{H}^*$-frame determines an H^*-frame (it helps to assume X path-connected, which we can do by Exercise 10.2).

10.4. Use Exercises 7.22 and 10.3.

10.5. By Exercise 10.3, one can use the reduced cohomology. Let $(\tilde{\kappa}_i, \tilde{\sigma}_i)$ be the $\tilde{H}^*$-frame for X_i. Obviously, $(X_1 \wedge X_2)^G = X_1^G \wedge X_2^G$ and X_2^G is of finite cohomology type (using $\tilde{\kappa}_2$). Using Proposition 4.7.11 and Lemma 7.4.9, one has commutative diagrams

$$\begin{array}{ccc}
\tilde{H}^{2*}(X_1) \otimes \tilde{H}^{2*}(X_2) & \xrightarrow[\approx]{\tilde{\times}} & \tilde{H}^{2*}(X_1 \wedge X_2) \\
\approx \big\downarrow \tilde{\kappa}_1 \otimes \tilde{\kappa}_2 & & \approx \big\downarrow \tilde{\kappa} \\
\tilde{H}^*(X_1^G) \otimes \tilde{H}^*(X_2^G) & \xrightarrow[\approx]{\tilde{\times}} & \tilde{H}^*(X_1^G \wedge X_2^G)
\end{array}$$

and

$$\begin{array}{ccc}
\tilde{H}^{2*}(X_1) \otimes \tilde{H}^{2*}(X_2) & \xrightarrow{\tilde{\sigma}_1 \otimes \tilde{\sigma}_2} & \tilde{H}_G^{2*}(X_1) \otimes \tilde{H}_G^{2*}(X_2) \\
\approx \downarrow \tilde{\times} & & \downarrow \tilde{\times}_G \\
\tilde{H}^{2*}(X_1 \wedge X_2) & \xrightarrow{\tilde{\sigma}} & \tilde{H}_G^{2*}(X_1 \wedge X_2)
\end{array}$$

which define $(\tilde{\kappa}, \tilde{\sigma})$. Checking the conjugation equation is straightforward.

10.6. $Z = X \wedge S^2_1$, where S^2_1 is the 2-sphere with the linear involution of Example 7.1.14. This uses Exercise 10.5 .

10.7. In $H^*(X)$, Sq^{16} should not vanish by Proposition 10.2.6. This contradicts Theorem 8.6.5.

10.8. The two sums in Proposition 10.3.7 have only one term and $\mathfrak{P}_t(\mathcal{C}) = 1 + t^{(n-2)(d-1)-1}$. Hence, $H^*(\mathcal{C}) \approx H^*(S^{(n-2)(d-1)-1})$. By the transfer exact sequence, $H^*(\bar{\mathcal{C}})$ is **GrA**-isomorphic to $H^*(\mathbb{R}P^{(n-2)(d-1)-1})$. Theorem 10.3.11 gives the **GrA**$[u]$-presentation $H^*_G(\mathcal{C}) \approx \mathbb{Z}_2[u]/(u^{(n-2)(d-1)})$. Actually, arguments like in Example 10.3.9 enable us to prove that $\mathcal{C} \approx S^{(n-2)(d-1)-1}$ and $\bar{\mathcal{C}} \approx \mathbb{R}P^{(n-2)(d-1)-1}$ (see e.g. [85, Example 2.6]).

10.9. By Proposition 10.3.7, $\mathfrak{P}_t(\mathcal{C}^5_d(\ell)) = 1+2t^{d-1}+2t^{2(d-1)-1}+t^{3(d-1)-1}$. When $d = 2$, $\mathcal{C}^5_2(\ell)$ is an orientable surface by Corollary 10.3.2. As $\mathfrak{P}_t(\mathcal{C}^5_2(\ell) = 1 + 4t + t^2$, $\mathcal{C}^5_2(\ell)$ is an orientable surface of genus 2 and thus $\bar{\mathcal{C}}^5_2(\ell)$ is a non-orientable surface of genus 3.

10.10. Let $\bar{\mathcal{C}} = \bar{\mathcal{C}}^n_2(\ell)$ and $\mathcal{C} = \mathcal{C}^n_2(\ell)$. Since ℓ is generic, $H^*(\bar{\mathcal{C}}) \approx H^*_G(\mathcal{C})$ and $\bar{\mathcal{C}}$ is a closed manifold of dimension $n - 3$ (see Corollary 10.3.2). By Lemma 10.3.3, we may suppose that ℓ is dominated. If lops $(\ell) \le 1$, $\bar{\mathcal{C}}$ is empty (see Examples 10.3.21 and 10.3.22). If lops $(\ell) = 2$, then $\bar{\mathcal{C}}^n_d(\ell)$ is orientable, being diffeomorphic to $(S^1)^{n-3}$ (see Example 10.3.23). Suppose that lops $(\ell) \ge 3$. By (10.3.28), $\mathcal{C}$ is connected and, since $p\colon \mathcal{C} \to \bar{\mathcal{C}}$ is of degree 0, one has $H^{n-3}(\bar{\mathcal{C}}) = u \cdot H^{n-4}(\bar{\mathcal{C}})$ by the transfer exact sequence. As $H^*_G(\mathcal{C})$ is **GrA**-generated in degree 1 (see Theorem 10.3.11), there exists $x_1, \dots, x_{n-4} \in H^1(\bar{\mathcal{C}})$ such that $x_1 \cdots x_{n-4}u \neq 0$ in $H^{n-3}_G(\bar{\mathcal{C}}) \approx \mathbb{Z}_2$. Then $\mathrm{Sq}^1(x_1 \cdots x_{n-4}) = (n-4)x_1 \cdots x_{n-4}u$ by Theorem 10.3.7. Note that the monomials of the form $x_1, \dots, x_{n-4}$ generate $H^{n-4}(\bar{\mathcal{C}})$. Using Corollary 9.8.5, we see that, for ℓ generic and dominated, $\bar{\mathcal{C}}^n_2(\ell)$ is non-orientable if and only if lops $(\ell) \ge 3$ and n is odd.

10.11. By Proposition 10.3.32, $\mathcal{N}^n_3(\ell)$ is a conjugation manifold with $\mathcal{N}^n_3(\ell)^G = \bar{\mathcal{C}}^n_2(\ell)$. Using Proposition 10.2.12, the answer is analogous to that of Exercise 10.10: for ℓ generic and dominated, $\mathcal{N}^n_3(\ell)$ has no spin structure if and only if lops $(\ell) \ge 3$ and n is odd.

10.12. Let $\bar{\mathcal{C}} = \bar{\mathcal{C}}^n_2(\ell)$. By (10.3.33), the simplicial complex $\mathrm{Sh}^\times_n(\ell)$ (definition p. 459) satisfies $\dim \mathrm{Sh}^\times_n(\ell) = n-2-\text{lops}(\ell)$. Therefore, in the face exterior algebra $\Lambda_1(\mathrm{Sh}^\times_n(\ell))$ (see Sect. 4.7.8), there exists at most $n - 2 - \text{lops}(\ell)$ elements whose product is not 0. By (10.3.20), $H_G^*(\bar{C})/(u) \approx \Lambda$. As $H^{n-3}(\bar{C}) \neq 0$, this implies that $u^{\mathrm{lops}\,(\ell)-2} \neq 0$.

10.14. Like in Example 10.5.13, Theorem 10.5.1 gives the presentation $H_{\Gamma_1}^*(\Gamma/\Gamma_2) \approx \mathbb{Z}_2[u, a]/(u^4 + a^4)$, where a and u are of degree one. Setting $b = a + u$, we get $H_{\Gamma_1}^*(\Gamma/\Gamma_2) \approx \mathbb{Z}_2[u, b]/(b^4)$ which is isomorphic to $H^*(\mathbb{R}P^3)[u]$. This is not surprising since $SU(2)/\{\pm 1\} \approx SO(3) \approx \mathbb{R}P^3$ and that $\{\pm 1\}$ acts trivially on $SU(2)/\{\pm 1\}$.

References

1. Adams, J.F.: On the structure and applications of the Steenrod algebra. Comment. Math. Helv. **32**, 180–214 (1958)
2. Adams, J.F.: On the non-existence of elements of Hopf invariant one. Ann. of Math. **2**(72), 20–104 (1960)
3. Adem, A., Milgram, R.J.: Cohomology of finite groups. Grundlehren der Mathematischen Wissenschaften [Fundamental Principles of Mathematical Sciences], vol. 309. Springer, Berlin (1994)
4. Adem, J.: The iteration of the Steenrod squares in algebraic topology. Proc. Nat. Acad. Sci. U. S. A. **38**, 720–726 (1952)
5. Adem, J.: The relations on Steenrod powers of cohomology classes. Algebraic geometry and topology. A symposium in honor of S. Lefschetz, pp. 191–238. Princeton University Press, Princeton, NJ (1957)
6. Adem, J., Gitler, S., James, I.M.: On axial maps of a certain type. Bol. Soc. Mat. Mexicana **2**(17), 59–62 (1972)
7. Alexander, J.W.: A proof and extension of the Jordan-Brouwer separation theorem. Trans. Amer. Math. Soc. **23**(4), 333–349 (1922)
8. Alexander, J.W.: Some problems in topology. Verh. Internat. Math. Kongr. Zürich, **1**, 249–257 (1932)
9. Allday, C., Puppe, V.: Cohomological methods in transformation groups. Cambridge Studies in Advanced Mathematics. 32, vol. 11, pp. 470. Cambridge University Press, Cambridge (1993)
10. Arf, C.: Untersuchungen über quadratische Formen in Körpern der Charakteristik 2. I. J. Reine Angew. Math. **183**, 148–167 (1941)
11. Atiyah, M.F.: K-theory and reality. Quart. J. Math. Oxford Ser. **2**(17), 367–386 (1966)
12. Audin, M.: The topology of torus actions on symplectic manifolds. Progress in Mathematics, vol. 93. Birkhäuser Verlag, Basel (1991)
13. Banyaga, A., Hurtubise, D.: Lectures on morse homology. Kluwer Texts in the Mathematical Sciences, vol. 29. Kluwer Academic Publishers Group, Dordrecht (2004)
14. Beyer, W.A., Zardecki, A.: The early history of the ham sandwich theorem. Amer. Math. Monthly **111**(1), 58–61 (2004)
15. Borel, A.: La cohomologie mod 2 de certains espaces homogènes. Comment. Math. Helv. **27**, 165–197 (1953)
16. Borel, A., Haefliger, A.: La classe d'homologie fondamentale d'un espace analytique. Bull. Soc. Math. France **89**, 461–513 (1961)

J.-C. Hausmann, *Mod Two Homology and Cohomology*, Universitext,
DOI 10.1007/978-3-319-09354-3

17. Borel, A., Serre, J.-P.: Corners and arithmetic groups. Comment. Math. Helv. **48**, 436– 491. Avec un appendice: Arrondissement des variétés à coins, par A. Douady et L, Hérault (1973)
18. Bott, R.: Lectures on Morse theory, old and new. Bull. Amer. Math. Soc. (N.S.) **7**(2), 331–358 (1982)
19. Bott, R., Tu, L.W.: Differential Forms in Algebraic Topology. Springer, New York (1982) (vol. 82 of Graduate Texts in Mathematics)
20. Bredon, G.E.: Introduction to Compact Transformation Groups. Pure and Applied Mathematics, vol. 46. Academic Press, New York (1972)
21. Bröcker, T., Tom Dieck, T.: Representations of compact Lie groups. Graduate Texts in Mathematics, vol. 98. Springer, New York (1995)
22. Brouwer, L.E.J.: Collected works. In: Freudenthal, H. (ed.) Geometry, Analysis, Topology and Mechanics, vol. 2. North-Holland Publishing Co., Amsterdam (1976)
23. Browder, W.: The Kervaire invariant of framed manifolds and its generalization. Ann. of Math. **2**(90), 157–186 (1969)
24. Browder, W.: *Surgery on Simply-Connected Manifolds* (Springer, NewYork, 1972). Ergebnisse der Mathematik und ihrer Grenzgebiete, Band 65
25. Brown, E.H. Jr., Peterson, F.P.: The Kervaire invariant of $(8k+2)$-manifolds. Amer. J. Math. **88**, 815–826 (1966)
26. Brown, K.S:. Cohomology of groups. Graduate Texts in Mathematics, vol. 87. Springer, New York (1982)
27. Bullett, S.R., Macdonald, I.G.: On the Adem relations. Topology **21**(3), 329–332 (1982)
28. Cartan, H., Eilenberg, S.: Homological algebra. Princeton University Press, Princeton, N. J. (1956)
29. Čech, E.: Multiplications on a complex. Ann. of Math. (2) **37**(3), 681–697 (1936)
30. Charney, R.: An introduction to right-angled Artin groups. Geom. Dedicata **125**, 141–158 (2007)
31. Chern, S.: Characteristic classes of Hermitian manifolds. Ann. of Math. **2**(47), 85–121 (1946)
32. Chern, S.: On the multiplication in the characteristic ring of a sphere bundle. Ann. of Math. **2**(49), 362–372 (1948)
33. Cohen, D.E.: On the Adem relations. Proc. Cambridge Philos. Soc. **57**, 265–267 (1961)
34. Constantin, A., Kolev, B.: The theorem of Kerékjártó on periodic homeomorphisms of the disc and the sphere. Enseign. Math. (2), **40**(3–4), 193–200 (1994)
35. Cornea, O., Lupton, G., Oprea, J., Tanré, D.: Lusternik-Schnirelmann category, Mathematical Surveys and Monographs, vol. 103. American Mathematical Society, Providence (2003)
36. Davis, D.M.: Table of immersions and embeddings of real projective spaces. Available on Donald Davis's home page at Lehigh University. http://www.lehigh.edu/~dmd1/immtable
37. tom Dieck, T.: Transformation Groups. Walter de Gruyter and Co., Berlin (1987)
38. tom Dieck, T.: Algebraic topology. In: EMS Textbooks in Mathematics. European Mathematical Society (EMS), Zürich (2008)
39. tom Dieck, T., Kamps, K.H., Puppe, D.: Homotopietheorie. In: Lecture Notes in Mathematics, vol. 157. Springer-Verlag, Berlin (1970)
40. Dieudonné, J.: A History of Algebraic and Differential Topology. 1900–1960. Birkhäuser-Boston Inc., Boston (1989)
41. Dold, A.: Erzeugende der Thomschen Algebra $\mathfrak{N}$. Math. Z. **65**, 25–35 (1956)
42. Dold, A.: Partitions of unity in the theory of fibrations. Ann. of Math. **2**(78), 223–255 (1963)
43. Dold, A.: Lectures on algebraic topology. In: Classics in Mathematics. Springer, Berlin (1995) (Reprint of the 1972 edition)
44. Dugundji, J.: Topology. Allyn and Bacon Inc., Boston (1966)
45. Duistermaat, J.J.: Convexity and tightness for restrictions of Hamiltonian functions to fixed point sets of an antisymplectic involution. Trans. Amer. Math. Soc. **275**(1), 417–429 (1983)
46. Ebeling, W.: Lattices and codes. Advanced Lectures in Mathematics. Friedr. Vieweg & Sohn, Braunschweig. A course partially based on lectures by F. Hirzebruch (1994)
47. Ehresmann, C.: Sur la topologie de certains espaces homogènes. Ann. of Math. (2) **35**(2), 396–443 (1934)

48. Ehresmann, C.: Sur la topologie de certaines variétés algébriques réelles. J. Math. Pures Appl. **IX**, 69–100 (1937)
49. Eilenberg, S.: Singular homology theory. Ann. of Math. **2**(45), 407–447 (1944)
50. Eilenberg, S., MacLane, S.: Acyclic models. Amer. J. Math. **75**, 189–199 (1953)
51. Eilenberg, S., Steenrod, N.: Foundations of Algebraic Topology. Princeton University Press, Princeton (1952)
52. Eilenberg, S., Steenrod, N.E.: Axiomatic approach to homology theory. Proc. Nat. Acad. Sci. U. S. A. **31**, 117–120 (1945)
53. Eliahou, S., de la Harpe, P., Hausmann, J.-C., Weber, C.: Michel Kervaire 1927–2007. Notices Amer. Math. Soc. **55**(8), 960–961 (2008)
54. Farber, M.: Topological complexity of motion planning. Discrete Comput. Geom. **29**(2), 211–221 (2003)
55. Farber, M.: Topology of robot motion planning. In: Morse Theoretic Methods in Nonlinear Analysis and in Symplectic Topology, NATO Science Series II: Mathematics, Physics and Chemistry, vol. 217, pp. 185–230. Springer, Dordrecht (2006)
56. Farber, M.: Invitation to topological robotics. European Mathematical Society (EMS), Zürich, Zurich Lectures in Advanced Mathematics (2008)
57. Farber, M., Fromm, V.: The topology of spaces of polygons. Trans. Amer. Math. Soc. **365**(6), 3097–3114 (2013)
58. Farber, M., Hausmann, J.-C., Schütz, D.: On the conjecture of Kevin Walker. J. Topol. Anal. **1**(1), 65–86 (2009)
59. Farber, M., Hausmann, J.-C., Schütz, D.: The Walker conjecture for chains in $\mathbb{R}^d$. Math. Proc. Camb. Philos. Soc. **151**(2), 283–292 (2011)
60. Farber, M., Schütz, D.: Homology of planar polygon spaces. Geom. Dedicata **125**, 75–92 (2007)
61. Farber, M., Tabachnikov, S., Yuzvinsky, S.: Topological robotics: motion planning in projective spaces. Int. Math. Res. Not. **34**, 1853–1870 (2003)
62. Ferry, S.C.: Constructing UV^k-maps between spheres. Proc. Amer. Math. Soc. **120**(1), 329–332 (1994)
63. Franz, M., Puppe, V.: Steenrod squares on conjugation spaces. C. R. Math. Acad. Sci. Paris **342**(3), 187–190 (2006)
64. Fritsch, R., Piccinini, R.A.: Cellular Structures in Topology, Volume 19 of Cambridge Studiesin Advanced Mathematics. Cambridge University Press, Cambridge (1990)
65. Fromm, V.: The topology of spaces of polygons. Ph.D. thesis, Durham University. http://etheses.dur.ac.uk/3208/ (2011)
66. Fulton, W.: Young Tableaux. With Applications to Representation Theory and Geometry. London Mathematical Society Student Texts, vol. 35. Cambridge University Press, Cambridge (1997)
67. Giever, J.B.: On the equivalence of two singular homology theories. Ann. of Math. **2**(51), 178–191 (1950)
68. Gonzalez, J., Landweber, P.: Symmetric topological complexity of projective and lens spaces. Algebr. Geom. Topol. 9, 473–494 (2009) (electronic)
69. Gordon, McA.: On the higher-dimensional Smith conjecture. Proc. London Math. Soc. **3**(29), 98–110 (1974)
70. Gordon, McA.: Knots in the 4-sphere. Comment. Math. Helv. **51**(4), 585–596 (1976)
71. Goresky, M., Kottwitz, R., MacPherson, R.: Equivariant cohomology, Koszul duality, and the localization theorem. Invent. Math. **131**(1), 25–83 (1998)
72. Gottlieb, D.H.: Fiber bundles with cross-sections and noncollapsing spectral sequences. Illinois J. Math. **21**(1), 176–177 (1977)
73. Gray, B.: Homotopy Theory. Academic Press (Harcourt Brace Jovanovich Publishers), New York (1975)
74. Greenberg, M.J., Harper, J.R.: *Algebraic topology, volume 58 of Mathematics Lecture Note Series. Advanced Book Program* (Benjamin/Cummings Publishing Co. Inc., Reading, 1981) (A first course)

75. Griffiths, P., Harris, J.: Principles of Algebraic Geometry. Pure and Applied Mathematics.Wiley-Interscience, New York (1978)
76. Gubeladze, J.: The isomorphism problem for commutative monoid rings. J. Pure Appl. Algebra **129**(1), 35–65 (1998)
77. Haefliger, A.: Plongements différentiables de variétés dans variétés. Comment. Math. Helv. **36**, 47–82 (1961)
78. Haefliger, A.: Points multiples d'une application et produit cyclique réduit. Amer. J. Math. **83**, 57–70 (1961)
79. Hambleton, I., Hausmann, J.-C.: Equivariant bundles and isotropy representations. Groups Geom. Dyn. **4**(1), 127–162 (2010)
80. Hambleton, I., Hausmann, J.-C.: Conjugation spaces and 4-manifolds. Math. Z. **269**(1–2), 521–541 (2011)
81. Hatcher, A.: Vector Bundles and K-Theory. Available on Allen Hatcher's home page at Cornell University
82. Hatcher, A.: Algebraic topology. Cambridge University Press, Cambridge (2002)
83. Hausmann, J.-C.: Sur la topologie des bras articulés. In Algebraic topology Poznań 1989, volume 1474 of Lecture Notes in Math (Springer, Berlin, 1991), pp. 146–159
84. Hausmann, J.-C.: On the Vietoris-Rips complexes and a cohomology theory for metric spaces. In: Prospects in Topology (Princeton, NJ, 1994), volume 138 of Ann. of Math. Stud., pp. 175–188. Princeton University Press, Princeton (1995)
85. Hausmann, J.-C.: *Geometric descriptions of polygon and chain spaces, in Topology and robotics, volume 438 of Contemp. Math.* (American Mathametical Society, Providence, 2007), pp. 47–57
86. Hausmann, J.-C., Holm, T.: *Conjugation spaces and edges of compatible torus actions, in Geometric aspects of analysis and mechanics, volume 292 of Progr* (Math. Birkhäuser/Springer, New York, 2011), pp. 179–198
87. Hausmann, J.-C., Holm, T., Puppe, V.: Conjugation spaces. Algebr. Geom. Topol. **5**:923–964 (2005) (electronic)
88. Hausmann, J.-C., Knutson, A.: Polygon spaces and Grassmannians. Enseign. Math. (2), **43**(1–2), 173–198 (1997)
89. Hausmann, J.-C., Knutson, A.: The cohomology ring of polygon spaces. Ann. Inst. Fourier (Grenoble) **48**(1), 281–321 (1998)
90. Hausmann, J.-C., Rodriguez, E.: The space of clouds in Euclidean space. Exp. Math. **13**(1):31–47, 2004. http://www.unige.ch/math/folks/hausmann/polygones
91. Hausmann, J.-C., Tolman, S.: Maximal Hamiltonian tori for polygon spaces. Ann. Inst. Fourier (Grenoble) **53**(6), 1925–1939 (2003)
92. Hill, M.A., Hopkins, M.J., Ravenel, D.C.: The Arf-Kervaire invariant problem in algebraic topology: introduction, in *Current Developments in Mathematics, 2009* (International Press, Somerville, 2010), pp. 23–57
93. Hirsch, M.W.: Immersions of manifolds. Trans. Amer. Math. Soc. **93**, 242–276 (1959)
94. Hirsch, M.W.: Smooth regular neighborhoods. Ann. of Math. **2**(76), 524–530 (1962)
95. Hirsch, M.W.: Differential topology. Graduate texts in mathematics, No. 33 (Springer-verlag, New York, 1976)
96. Hirzebruch, F.: Topological Methods in Algebraic Geometry. Third enlarged edition. New appendix and translation from the second German edition by R.L.E. Schwarzenberger, with an additional section by A. Borel. Die Grundlehren der Mathematischen Wissenschaften, Band 131. Springer, New York (1966)
97. Hirzebruch, F., Mayer, K.H.: O(n)-Mannigfaltigkeiten, exotische Sphären und Singularitäten. Lecture Notes in Mathematics, No. 57 (Springer, Berlin, 1968)
98. Hopf, H.: Zur Algebra der Abbildungen von Mannigfaltigkeiten. J. Reine Angew. Math. **163**, 71–88 (1930)
99. Hopf, H.: Über die Abbildungen der dreidimensionalen Sphären auf die Kugelfläche. Math. Ann. **104**(1), 637–665 (1931)

100. Hopf, H.: Über die Abbildungen von Sphären auf Sphären niedrigerer Dimension. Fundam. Math. **25**, 427–440 (1935)
101. Hopf, H.: Ein topologischer Beitrag zur reellen Algebra. Comment. Math. Helv. **13**, 219–239 (1941)
102. Hopf, H.: Einige persönliche Erinnerungen aus der Vorgeschichte der heutigen Topologie. In: Colloque de Topologie (Brussels, 1964), pp. 8–16. Librairie Universitaire, Louvain (1966)
103. Hsiang, Wu-chung: Hsiang, Wu-yi: Differentiable actions of compact connected classical groups. I. Amer. J. Math. **89**, 705–786 (1967)
104. Hudson, J.F.P.: Piecewise Linear Topology. University of Chicago Lecture Notes prepared with the assistance of J. L. Shaneson and J. Lees. W. A. Benjamin Inc, New York (1969)
105. Husemoller, D.: Fibre bundles. In: Graduate Texts in Mathematics, vol. 20, 3rd edn. Springer, New York (1994)
106. Illman, S.: Smooth equivariant triangulations of G-manifolds for G a finite group. Math. Ann. **233**(3), 199–220 (1978)
107. Illman, S.: The equivariant triangulation theorem for actions of compact Lie groups. Math. Ann. **262**(4), 487–501 (1983)
108. James, I.M.: On category, in the sense of Lusternik-Schnirelmann. Topology **17**(4), 331–348 (1978)
109. James, I.M., Whitehead, J.H.C.: The homotopy theory of sphere bundles over spheres. II. Proc. London Math. Soc. **3**(5), 148–166 (1955)
110. Kamiyama, Y., Tezuka, M., Toma, T.: Homology of the configuration spaces of quasi-equilateral polygon linkages. Trans. Amer. Math. Soc. **350**(12), 4869–4896 (1998)
111. Kapovich, M., Millson, J.J.: The symplectic geometry of polygons in Euclidean space. J. Differential Geom. **44**(3), 479–513 (1996)
112. Kendall, D.G., Barden, D., Carne, T.K., Le, H.: Shape and shape theory. Wiley Series in Probability and Statistics. John Wiley and Sons Ltd., Chichester (1999)
113. Kervaire, M., Milnor, J.W.: Groups of homotopy spheres. I. Ann. of Math. **2**(77), 504–537 (1963)
114. Kervaire, M.A.: A manifold which does not admit any differentiable structure. Comment. Math. Helv. **34**, 257–270 (1960)
115. Kervaire, M.A.: Smooth homology spheres and their fundamental groups. Trans. Amer. Math. Soc. **144**, 67–72 (1969)
116. Kirby, R.C., Siebenmann, L.C.: Foundational Essays on Topological Manifolds, Smoothings, and Triangulations. Princeton University Press, Princeton, NewJersey (1977) (Annals of Mathematics Studies, No. 88)
117. Kirwan, F.C.: Cohomology of quotients in symplectic and algebraic geometry. Mathematical Notes. vol. 31. Princeton University Press, Princeton, NJ (1984)
118. Klyachko, A.A.: Spatial polygons and stable configurations of points in the projective line, in *Algebraic geometry and its applications (Yaroslavl 1992)*, Aspects Math., E25 (Vieweg, Braunschweig, 1994), pp. 67–84
119. Knutson, A., Tao, T.: Puzzles and (equivariant) cohomology of Grassmannians. Duke Math. J. **119**, 221–260 (2003)
120. Kosinski, A.A.: Differential Manifolds. Academic, Boston, MA (1993) (vol. 138 of Pure and Applied Mathematics)
121. Kreck, M., Puppe, V.: Involutions on 3-manifolds and self-dual, binary codes. Homology, Homotopy Appl. **10**(2), 139–148 (2008)
122. Lang, S.: Algebra, 2nd edn. Addison-Wesley Publishing Company Advanced Book Program, Reading, MA (1984)
123. Lannes, J.: Sur l'invariant de Kervaire des variétés fermées stablement parallélisées. Ann. Sci. École Norm. Sup. (4), **14**(2), 183–197 (1981)
124. Lannes, J.: Sur les espaces fonctionnels dont la source est le classifiant d'un p-groupe abélien élémentaire. Inst. Hautes Études Sci. Publ. Math. (75), 135–244, (1992). With an appendix by Michel Zisman

125. Lannes, J., Zarati, S.: Sur les foncteurs dérivés de la déstabilisation. Math. Z. **194**(1), 25–59 (1987)
126. Lashof, R.K.: Equivariant bundles. Illinois J. Math. **26**(2), 257–271 (1982)
127. Lashof, R.K., May, J.P.: Generalized equivariant bundles. Bull. Soc. Math. Belg. Sér. A **38**(265–271), 1986 (1987)
128. Latour, F.: Double suspension d'une sphère d'homologie [d'après R. Edwards]. In: Séminaire Bourbaki, 30e année (1977/78), vol. 710 of Lecture Notes in Math., pages Exp. No. 515, pp. 169–186. Springer, Berlin (1979)
129. Latschev, J.: Vietoris-Rips complexes of metric spaces near a closed Riemannian manifold. Arch. Math. (Basel) **77**(6), 522–528 (2001)
130. Lawson, H.B. Jr., Michelsohn, M.-L.: Spin Geometry. Princeton Mathematical Series, vol. 38. Princeton University Press, Princeton (1989)
131. Lebesgue, H.: Oeuvres scientifiques (en cinq volumes), vol. IV. L'Enseignement mathématique, Geneva (1973) (Sous la rédaction de François Châtelet et Gustave Choquet)
132. Lefschetz, S.: Introduction to Topology (Princeton Mathematical Series, vol. 11). Princeton University Press, Princeton, New Jersey (1949)
133. Levine, J.P.: Lectures on groups of homotopy spheres. In: Algebraic and Geometric Topology, New Brunswick, New Jersey, 1983, vol. 1126 of Lecture Notes in Math., pp. 62–95. Springer, Berlin (1985)
134. MacLane, S.: Categories for the Working Mathematician, Volume 5 of Graduate Texts in Mathematics, 2nd edn. Springer, New York (1998)
135. Massey, W.S.: Imbeddings of projective planes and related manifolds in spheres. Indiana Univ. Math. J. 23, 791–812 (1973/74)
136. Massey, W.S.: A Basic Course in Algebraic Topology, Volume 127 of Graduate Texts in Mathematics. Springer-Verlag, New York (1991)
137. Massey, W.S.: A history of cohomology theory. In: History of Topology, pp. 579–603. North-Holland, Amsterdam (1999)
138. Matoušek, J.: Lectures on Discrete Geometry. Graduate Texts in Mathematics, vol. 212. Springer, New York (2002)
139. May, J.P.: Equivariant homotopy and cohomology theory. CBMS Regional Conference Series in Mathematics., vol. 91. Published for the Conference Board of the Mathematical Sciences, Washington, DC (1996). With contributions by Cole, M., Comezaña, G., Costenoble, S., Elmendorf, A.D., Greenlees, J.P.C., Lewis, L.G. Jr., Piacenza, J.R.J., Triantafillou, G., Waner, S.
140. May, J.P.: A concise course in algebraic topology. Chicago Lectures in Mathematics. University of Chicago Press, Chicago, IL (1999)
141. McCleary, J.: A user's guide to spectral sequences. In: Cambridge Studies in Advanced Mathematics, vol. 58, 2nd edn. Cambridge University Press, Cambridge (2001)
142. Milgram, R.J., Trinkle, J.C.: The geometry of configuration spaces for closed chains in two and three dimensions. Homology Homotopy Appl. **6**(1), 237–267 (2004)
143. Miller, H.: The Sullivan conjecture on maps from classifying spaces. Ann. Math. (2) **120**(1), 39–87 (1984)
144. Milnor, J.W.: Construction of universal bundles. II. Ann. of Math. **2**(63), 430–436 (1956)
145. Milnor, J.W.: On manifolds homeomorphic to the 7-sphere. Ann. of Math. **2**(64), 399–405 (1956)
146. Milnor, J.W.: The geometric realization of a semi-simplicial complex. Ann. of Math. **2**(65), 357–362 (1957)
147. Milnor, J.W.: On spaces having the homotopy type of CW-complex. Trans. Amer. Math. Soc. **90**, 272–280 (1959)
148. Milnor, J.W.: A procedure for killing homotopy groups of differentiable manifolds, in *Proceedings of the Symposium Pure Mathematics, vol. III* (American Mathematical Society, Providence, 1961), pp. 39–55
149. Milnor, J.W.: Morse theory. Based on lecture notes by M. Spivak and R. Wells. Annals of Mathematics Studies, No. 51. Princeton University Press, Princeton, N.J. (1963)

150. Milnor, J.W.: Spin structures on manifolds. Enseignement Math. **2**(9), 198–203 (1963)
151. Milnor, J.W.: Lectures on the h-cobordism theorem. Notes by L. Siebenmann and J. Sondow. Princeton University Press, Princeton, N.J. (1965)
152. Milnor, J.W.: Topology from the differentiable viewpoint. In: David, W.W. (ed.) Based on Notes. The University Press of Virginia, Charlottesville, VA (1965)
153. Milnor, J.W., Stasheff, J.D.: Characteristic Classes. Princeton University Press, Princeton (1974) Annals of Mathematics Studies, No. 76
154. Mosher, R.E., Tangora, M.C.: Cohomology operations and applications in homotopy theory. Harper and Row Publishers, New York (1968)
155. Munkres, J.R.: Elements of algebraic topology. Addison-Wesley Publishing Company, Menlo Park, CA (1984)
156. Newman, M.: Integral Matrices, in *Pure and Applied Mathematics, vol. 45* (Academic Press, New York, 1972)
157. Olbermann, M.: Conjugations on 6-manifolds, Ph.D. thesis, University of Heidelberg. http://www.ub.uni-heidelberg.de/archiv/7450 (2007)
158. Olbermann, M.: Conjugations on 6-manifolds. Math. Ann. **342**(2), 255–271 (2008)
159. Olbermann, M.: Involutions on S^6 with 3-dimensional fixed point set. Algebr. Geom. Topol. **10**(4), 1905–1932 (2010)
160. Pitsch, W., Scherer, J.: Conjugation spaces and equivariant Chern classes. Bull. Belg. Math. Soc. Simon Stevin **20**, 77–90 (2013)
161. Poincaré, H.: Analysis situs. J. Ec. Polytech. **21**, 1–123 (1895)
162. Poincaré, H.: Complément à l'Analysis situs. Palermo Rend. **13**, 285–343 (1899)
163. Pont, J.-C.: La topologie algébrique des origines à Poincaré. Presses Universitaires de France, Paris (1974)
164. Postnikov, M.M.: The structure of the ring of intersections of three-dimensional manifolds. Doklady Akad. Nauk. SSSR (N.S.) **61**, 795–797 (1948)
165. Puppe, V.: Group actions and codes. Canad. J. Math. **53**(1), 212–224 (2001)
166. Puppe, V.: Do manifolds have little symmetry? J. Fixed Point Theory Appl. **2**(1), 85–96 (2007)
167. Quillen, D.: The mod 2 cohomology rings of extra-special 2-groups and the spinor groups. Math. Ann. **194**, 197–212 (1971)
168. Ranicki, A.A.: On the Hauptvermutung. In: The Hauptvermutung Book, pp. 3–31. Kluwer Academic, Dordrecht (1996) (vol. 1 of K-Monogr. Math.)
169. Ronga, F.: Schubert calculus according to Schubert. arXiv:math/0608784v1[math.AG]
170. Schütz, D.: The isomorphism problem for planar polygon spaces. J. Topol. **3**(3), 713–742 (2010)
171. Schwartz, L.: Unstable modules over the Steenrod algebra and Sullivan's fixed point set conjecture. Chicago Lectures in Mathematics. University of Chicago Press, Chicago, IL (1994)
172. Schwartz, L.: À propos de la conjecture de non-réalisation due à N. Kuhn. Invent. Math. **134**(1), 211–227 (1998)
173. Scull, L.: Equivariant formality for actions of torus groups. Canad. J. Math. **56**(6), 1290–1307 (2004)
174. Seifert, H., Threlfall, W.: Lehrbuch der Topologie. B. G. Teubner, Leipzig und Berlin (1934)
175. Serre, J.-P.: Cohomologie modulo 2 des complexes d'Eilenberg-MacLane. Comment. Math. Helv. **27**, 198–232 (1953)
176. Seymour, R.M.: On G-cohomology theories and Künneth formulae, in *Current trends in algebraic topology, Part 1 (London, Ont., 1981), vol. 2. CMS Conferences Proceedings* (American Mathematical Society, Providence, RI, 1982), pp. 257–271
177. Smale, S.: Generalized Poincaré's conjecture in dimensions greater than four. Ann. of Math. **2**(74), 391–406 (1961)
178. Smith, P.A.: Transformations of finite period. Ann. of Math. **2**(39), 127–164 (1938)
179. Spanier, E.H.: Algebraic topology. Springer-Verlag, New York (1981)
180. Spivak, M.: A Comprehensive Introduction to Differential Geometry, vol. V, 2nd edn. Publish or Perish Inc., Wilmington, USA (1979)
181. Steenrod, N.: The Topology of Fibre Bundles. Princeton University Press, New Jersey (1951)

182. Steenrod, N.E.: Cohomology invariants of mappings. Ann. of Math. **2**(50), 954–988 (1949)
183. Steenrod, N.E.: The work and influence of Professor S. Lefschetz in algebraic topology, in *Algebraic Geometry and Topology. A Symposium in Honor of S. Lefschetz* (Princeton University Press, New Jersey, 1957), pp. 24–43
184. Steenrod, N.E.: Cohomology operations. Lectures by N. E. Steenrod written and revised by D. B. A. Epstein. Annals of Mathematics Studies, vol. 50. Princeton University Press, Princeton (1962)
185. Steenrod, N.E.: A convenient category of topological spaces. Michigan Math. J. **14**, 133–152 (1967)
186. Stong, R.E.: Notes on cobordism theory. Mathematical notes. Princeton University Press, Princeton, N.J. (1968)
187. Sullivan, D.P.: On the intersection ring of compact three manifolds. Topology **14**(3), 275–277 (1975)
188. Sullivan, D.P.: Geometric Topology: Localization, Periodicity and Galois Symmetry, Volume 8 of K-Monographs in Mathematics. Springer, Dordrecht (2005) The 1970 MIT notes, Edited and with a preface by Andrew Ranicki
189. Sumners, D.W.: Smooth Z_p-actions on spheres which leave knots pointwise fixed. Trans. Amer. Math. Soc. **205**, 193–203 (1975)
190. Thom, R.: Espaces fibrés en sphères et carrés de Steenrod. Ann. Sci. Ecole Norm. Sup. **3**(69), 109–182 (1952)
191. Thom, R.: Quelques propriétés globales des variétés différentiables. Comment. Math. Helv. **28**, 17–86 (1954)
192. Thom, R.: Un lemme sur les applications différentiables. Bol. Soc. Mat. Mexicana **2**(1), 59–71 (1956)
193. Thomas, E.: On the cohomology groups of the classifying space for the stable spinor groups. Bol. Soc. Mat. Mexicana **2**(7), 57–69 (1962)
194. Thomas, E.: On the mod 2 cohomology of certain H-spaces. Comment. Math. Helv. **37**, 132–140 (1962)
195. Thurston, W., Meeks, J.: The mathematics of three-dimensional manifolds. Scientific American **251**(1), 94–106 (July 1984)
196. Tietze, H.: Über die topologischen Invarianten mehrdimensionaler Mannigfaltigkeiten. Monatsh. Math. Phys. **19**(1), 1–118 (1908)
197. Toda, H.: Composition methods in homotopy groups of spheres. Annals of Mathematics Studies, vol. 49. Princeton University Press, Princeton (1962)
198. Tolman, S., Weitsman, J.: On the cohomology rings of hamiltonian T-spaces, in *Northern California Symplectic Geometry Seminar, vol. 196. Amer. Math. Soc. Transl. Ser. 2* (American Mathematical Society, Providence, RI, 1999), pp. 251–258
199. Tolman, S., Weitsman, J.: The cohomology rings of symplectic quotients. Comm. Anal. Geom. **11**(4), 751–773 (2003)
200. Veblen, O., Alexander, J.W. II.: Manifolds of N dimensions. Ann. Math. (2), **14**(1–4), 163–178 (1912/1913)
201. Vietoris, L.: Über den höheren Zusammenhang kompakter Räume und eine Klasse von zusammenhangstreuen Abbildungen. Math. Ann. **97**(1), 454–472 (1927)
202. Walker, K.: Configuration spaces of linkages. Bachelor's thesis, Princeton. http://canyon23.net/math/1985thesis.pdf (1985)
203. Wall, C.T.C.: Surgery of non-simply-connected manifolds. Ann. of Math. **2**(84), 217–276 (1966)
204. Wall, C.T.C.: *Surgery on Compact Manifolds* (Academic Press, London, 1970) (London Mathematical Society Monographs, No. 1)
205. Weber, C.: *The Robertello-Arf invariant of knots Preprint* (University of Geneva, Geneva)
206. Weinberger, S.: Homology manifolds. In: Sher, R.B., Daverman, R.J. (eds.) Handbook of Geometric Topology, pp. 1085–1102. North-Holland, Amsterdam (2002)
207. Whitehead, G.W.: *Elements of Homotopy Theory, Volume 61 of Graduate Texts in Mathematics* (Springer, New York, 1978)

208. Whitehead, J.H.C.: On C^1-complexes. Ann. of Math. **2**(41), 809–824 (1940)
209. Whitney, H.: On products in a complex. Ann. of Math. (2) **39**(2), 7–432 (1938)
210. Whitney, H.: The singularities of a smooth n-manifold in (2n–1)-space. Ann. of Math. **2**(45), 247–293 (1944)
211. Whitney, H.: Moscow 1935: Topology moving toward America, in *A Century of Mathematics in America, Part I, History of Mathematics, vol. 1* (American Mathematical Society, Providence, 1988), pp. 97–117
212. Wood, R.M.W.: Problems in the Steenrod algebra. Bull. London Math. Soc. **30**(5), 449–517 (1998)
213. Wen-tsün, Wu: Classes caractéristiques et i-carrés d'une variété. C. R. Acad. Sci. Paris **230**, 508–511 (1950)
214. Wen-tsün, Wu: Les i-carrés dans une variété grassmannienne. C. R. Acad. Sci. Paris **230**, 918–920 (1950)

Index

J.-C. Hausmann, *Mod Two Homology and Cohomology*, Universitext,
DOI 10.1007/978-3-319-09354-3